T0401825

SpringerWienNewYork

Marina V. Rodnina
Wolfgang Wintermeyer
Rachel Green

Editors

Ribosomes

Structure, Function, and Dynamics

SpringerWienNewYork

Prof. Dr. Marina Rodnina
Prof. Dr. Wolfgang Wintermeyer
Dept. of Physical Biochemistry, MPI for Biophysical Chemistry,
Goettingen, Germany

Prof. Dr. Rachel Green
Dept. of Molecular Biology and Genetics
Howard Hughes Medical Institute,
Johns Hopkins University School of Medicine, Baltimore, MD, USA

© 2011 Springer-Verlag/Wien
Printed in Austria

SpringerWienNewYork is part of Springer Science+Business Media
springer.at

Coverdesign: WMX Design GmbH, 69126 Heidelberg, Germany
Layout and Typesetting: JungCrossmedia Publishing GmbH, 35633 Lahnau, Germany
Printing: Holzhausen Druck GmbH, 1140 Wien, Austria

Printed on acid-free and clorine-free bleached paper

SPIN: 80011232

Library of Congress Control Number: 2011928714

ISBN 978-3-7091-0214-5 SpringerWienNewYork

Preface

The breakthrough discovery through which ribosome research entered a new era was the determination, in 2000, of high-resolution crystal structures of subunits of the ribosome, an achievement honored with the Nobel Prize in Chemistry 2009 for Ada Yonath, Tom Steitz, and Venki Ramakrishnan. The articles of this book present research covering about one decade, since the last such book, "The Ribosome", which was published in 2001 associated with a Cold Spring Harbor Symposium series of lectures. The articles are based on presentations given at the meeting "Ribosomes 2010" which was held in Orvieto, Italy, May 3 to 7, and brought together 320 scientists from all areas of ribosome research. Not included in the book are articles about recoding, as the topic was covered very recently in the 2010 book "Recoding" from this same series.

In the first section of this book, new structures of functional complexes of bacterial ribosomes are presented, with an early look at the first crystal structure of a eukaryotic (yeast) ribosome. Next follow chapters focused on structural insights into translation initiation, ribosome recycling, and polysome structure. Subsequent sections deal with mechanisms of decoding, including fidelity, and the chemistry of catalysis in the peptidyl transferase center. In the next section are a number of chapters where the recurrent theme is the structural dynamics of the ribosome. Single-molecule fluorescence methods on the one hand and single-particle cryo-electron microscopic image reconstruction on the other reveal a highly dynamic picture of the ribosome, exhibiting the energy landscape of a processive Brownian machine. The fate of the nascent peptide emerging through the exit tunnel of the ribosome and its interactions inside and outside the tunnel is described in a number of articles. A final chapter on the evolution of ribosomes presents a molecular paleontology approach suggesting how a small, relatively simple primordial RNA molecule may have evolved to become the large and complex structure that we know as the modern day ribosome.

The editors would like to express their gratitude to the authors for their cooperation in preparing their chapters and for providing the basis for a comprehensive overview of ribosome research as of 2010. We hope that the book, as previous "ribosome books", will for a long time serve as a source of information for researchers in the field and for those who are interested in entering this field which becomes ever more fascinating.

Göttingen and Baltimore, June 2011

Marina V. Rodnina
Wolfgang Wintermeyer
Rachel Green

Contents

Section I Ribosome structure

Anat Bashan and Ada Yonath

1. Introduction

Ribosome research took off as soon as ribosomes were identified. At the end of the seventies the extensive biochemical studies yielded illuminating findings about the overall nature of ribosome function. These studies showed that all ribosomes are composed of two unequal subunits. The small subunit in bacteria, denoted as 30S, contains an RNA chain (16S) of about 1500 nucleotides and 20–21 different proteins, whereas the large subunit, denoted as 50S, contains two RNA chains (23S and 5S RNA) of about 3000 nucleotides in total, and 31–35 different proteins. In all organisms the two subunits exist independently and associate to form functionally active ribosomes.

The substrates engaged in protein formation are aminoacylated, peptidylated and deacylated (exiting) tRNA molecules. The three-dimensional structures of all tRNA molecules from all living cells across evolution are alike, although each of them has features specific to its cognate amino acid. The tRNAs are double helical L-shaped RNA molecules in a stem-elbow-stem organization, and contain a loop that comprises the anticodon complementing the three-nucleotide codon on the mRNA. About 70 Å away, at their 3' end, tRNAs contain a single strand with the universal sequence CCA, to which the cognate amino acid is attached by an ester bond. The tRNA molecules are non-ribosomal entities that bring together the two subunits, as all three of their binding sites, A (aminoacyl), P (peptidyl), and E (exit), reside on both subunits. The small subunit provides the path along which the mRNA progresses, the decoding center and the mechanism controlling translation fidelity, and the large subunit contains the site for the main ribosomal catalytic function, namely the polymerization of amino acids.

Although an overall description of protein biosynthesis was available by the end of the seventies, detailed functional information was not available because of the lack of three-dimensional molecular structures. Indeed, the common hypotheses about the mode of ribosome function underwent significant alterations once three-dimensional structures became available. Striking examples for conceptual revolutions in the understanding of ribosomal function (Wekselman et al., 2008) relate to the functional contribution of the different ribosomal components and the path taken by nascent chains.

In the middle of the last century, RNA-rich particles, called "Palade particles", were identified in the vicinity of the endoplasmic reticulum (Palade, 1955; Watson, 1963). These were proposed to be involved in gene expression, suggesting that proteins are made by an RNA machine (Crick, 1968). Nevertheless, it was commonly assumed for over four decades that decoding of the genetic code and peptide bond formation are performed by r-proteins, while r-RNA provides the ribosomal scaffold (Garrett and Wittmann, 1973). The proposition that RNA provides the ribosome catalytic activities (Noller et al., 1992) was met first with skepticism. Modest acceptance of this idea was achieved as several functional roles played by RNA molecules in various life processes were identified around the last decade of the 20th century, including peptide bond formation *in vitro* by selected ribozymes (Zhang and Cech, 1997) and spontaneous conjugation of amino acids with oligonucleotides (Illangasekare et al., 1995). Finally, at the turn of the third millennium, the emerging high-resolution structures verified the notion that both the decoding center and the site of peptide bond formation (called peptidyl transferase center or PTC) reside in regions where rRNA predominates.

The three-dimensional crystal structures of ribosomes and their various complexes illuminated the molecular basis for most of the mechanisms involved in ribosome function. These showed how the assembly of the initiation complex occurs (Simonetti et al., 2008; Simonetti et al., 2009), revealed the decoding mechanism (reviewed in (Ramakrishnan, 2008; Demeshkina et al., 2010)), the mRNA progression mode, including the narrowing of the downstream mRNA tunnel that occurs upon the transition from initiation to elongation (Yusupova et al., 2006; Jenner et al., 2010), identified the relative positions of A-, P- and E-site tRNAs (Yusupov et al., 2001) and shed light on the way the initiation, elongation, termination, and recycling factors modulate ribosome function (Carter et al., 2001; Pioletti et al., 2001; Wilson et al., 2005; Borovinskaya et al., 2007; Laurberg et al., 2008; Weixlbaumer et al., 2008; Schmeing et al., 2009). In addition, the positions of the tRNA molecules within the PTC (Woolhead et al., 2004; Blaha et al., 2009; Voorhees et al., 2009), the conformational rearrangements that E-site tRNA undergoes while exiting the ribosome (Jenner et al., 2007), and the architectural and dynamic elements required for amino acid polymerization were determined (Bashan et al., 2003a; Bashan and Yonath, 2008b). Thus, it appears that the main catalytic activities of the ribosome provide the framework for proper positioning of all participants in the protein biosynthetic process, thus enabling decoding, successive peptide bond formation and the protection of the nascent protein chains.

This article focuses on ribosome crystallography and on the functional implications evolving from these studies. It describes snapshots from the chronological progress of ribosomal crystallography as a semi historical report. It highlights selected events occurring during the long way from the initial ribosome crystallization including the introduction of innovations in the procedures required for the determination of the ribosomal structures, such as cryo bio-crystallography and the use of heavy atom clusters [reviewed in (Gluehmann et al., 2001)]. The article also focuses on selected structural and dynamic properties of the ribosome that enable its function as an efficient machine and illuminates several key ribosomal strategies for efficient usage of resources and for minimizing protein production under non-optimal conditions. Additionally, this article discusses modes of action of antibiotics that hamper ribosome function and suggests mechanisms for the acquisition of antibiotic resistance. It also addresses issues concerning the origin of translation, as

can be deduced from the universal structural element that embraces the ribosomal active site and possesses internal symmetry within the otherwise asymmetric contemporary ribosome.

2. Hibernating bears stimulated ribosome crystallization

Once it was found that ribosomes are the molecular machines translating the genetic code and initial knowledge of the chemical composition of the *E. coli* ribosome became partially available, attempts at ribosome crystallization were made worldwide. For over two decades these attempts were unproductive. Owing to repeated failures the crystallization of ribosomes was considered a formidable task. The extreme difficulties in ribosome crystallization stemmed from their marked tendency to deteriorate, their high degree of internal mobility, their flexibility, their functional heterogeneity, their chemical complexity, their large size and their asymmetric nature.

Nevertheless, the finding that large amounts of ribosomes in hibernating bears are packed in an orderly fashion on the inner side of their cell membranes indicated that ribosomes can assemble in periodical arrangements *in vivo*. Similar observations were made in shock-cooled fertilized eggs (Unwin and Taddei, 1977; Milligan and Unwin, 1986). These phenomena were associated with cold or similar shocks, and were rationalized as a strategy taken by organisms under stress for storing pools of functionally active ribosomes that can be utilized when the stressful conditions are removed. Indeed, structural studies, performed on samples obtained from shock-cooled fertilized eggs led later to the visualization of ribosomal internal features (Milligan and Unwin, 1986) (see below).

Extending the level of order from membrane-supported two-dimensional monolayers produced *in vivo* to three-dimensional crystals grown *in vitro* was not trivial, but became successful with the introduction of uncommon crystallization strategies. These were based on the interpretation of the life cycle of the winter sleeping bears, which regularly pack and unpack their ribosomes each year, as part of their normal life cycle. The fact that these processes are associated with living organisms that require functionally active ribosomes immediately when awaking from a state of winter sleep suggested (i) that the integrity of highly active ribosomes can be maintained for relatively long periods without undergoing significant deterioration and (ii) that the ribosomes can be periodically ordered in three-dimensions. In other words, ribosomes seemed likely to form crystals.

The breakthrough in ribosome crystallization was based on the assumption that the higher the sample homogeneity, the better the crystals, and secondly, that the preferred conformation is that of the functionally active ribosome. Consequently, highly active ribosomes of bacterial species that grow under robust conditions were selected and conditions for optimization and maintenance of their activity (Vogel et al., 1970; Zamir et al., 1971) were maintained throughout purification and crystallization. The first three-dimensional micro-crystals of ribosomal particles, treated as "powder samples" diffracted to relatively high resolution (3.5 Å) and had impressive internal order (Figure 1), were obtained from the large ribosomal subunit of a thermophilic bacterium, *Bacillus stearothermophilus* (B50S), at the beginning of the eighties (Yonath et al., 1980). A thorough screening of about 25,000 different crystallization conditions, performed by careful monitoring of the nucleation of crystalline region (Yonath et al., 1982a), was accompanied by a systematic search for parameters favoring crystallization (Yonath et al., 1982b). At the beginning of the eighties, *B. stearothermophilus* as an extremophile was considered to be exotic. Therefore frequent doubts about its suitability as a representative of "normal" bacterial ribosomes, such as those of *E. coli*, were expressed, despite the very high sequence homology between them. Nevertheless, the preliminary successes of the crystallization stimulated the use of ribosomes from other robust bacteria. Consequently, a few years later, three-dimensional crystals were obtained from the large ribosomal subunits of the halophilic archaeon, *Haloarcula marismortui*, that lives in the Dead Sea (Shevack et al., 1985). In 1987, seven years after the first crystallization of the large ribosomal subunits, parallel efforts led to the growth of crystals of the small ribosomal subunit (Yonath et al., 1988; Yusupov et al., 1988) and of the 70S ribosome (Trakhanov et al., 1987) from the extreme thermophilic bacterium, *Thermus thermophilus*.

At that time it was widely assumed that the three-dimensional structure of the ribosome may never be determined, as it was clear that alongside the improvement of the crystals, the determination of the structure would require the introduction and development of innovative methodologies. For instance, because of the weak diffraction power of the ribosome crystals even the most advanced rotating anode generators were not powerful enough to yield suitable diffraction pat-

terns. Similarly, only a few diffraction spots could be recorded (Yonath et al., 1984) (Figure 1), even when irradiating extremely large crystals (~2 mm in length) by synchrotron radiation, which at the time was in its early stage. In parallel to the advances in growing ribosomal crystals of several forms (Yonath and Wittmann, 1988), the synchrotron facilities and the detection methods underwent constant (albeit rather slow) improvement. However, even when more suitable beamlines became available, the radiation sensitivity of the ribosomal crystals caused extremely fast crystal decay. Hence, pioneering data collection at cryogenic temperature (Hope et al., 1989) became crucial. Once established, this method became routine worldwide, and, although when using very bright X-ray beam, decay was observed even at cryo-temperature, interpretable diffraction patterns were ultimately obtained from the extremely thin crystals. Additionally, multi-heavy atom clusters suitable for phasing were identified (Thygesen et al., 1996). One of these clusters, originally used for providing anomalous phasing power, was found to play a dual role in the determination of the structure of the small ribosomal subunit from *T. thermophilus* (T30S). Post-crystallization treatment with minute amounts of these clusters dramatically increased the resolution from the initial 7–9 Å to 3 Å (Schluenzen et al., 2000), presumably by minimizing the internal flexibility of the particle (Bashan and Yonath, 2008a).

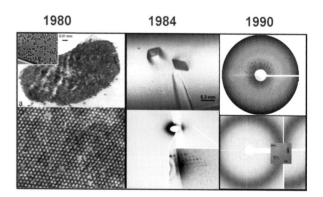

Fig. 1 From micro-crystals to three-dimensional crystals yielding useful diffraction. Left, top: The first microcrystals of B50S (Yonath at al., 1980). Left, bottom: a negatively stained section of the microcrystals, viewed by electron microscopy. Middle, top: the tip of a ~2 mm-long crystal of B50S. Middle, bottom: its diffraction pattern, obtained at 4°C in 1984 using the EMBL/DESY/Hamburg beam line. Right, top: The diffraction pattern from crystals of H50S, obtained at ID13 beamline at ESRF, Grenoble, at 95K; the diffraction extends to 2.8Å. Right, bottom: The crystals decayed completely after collecting about 0.3% of the data.

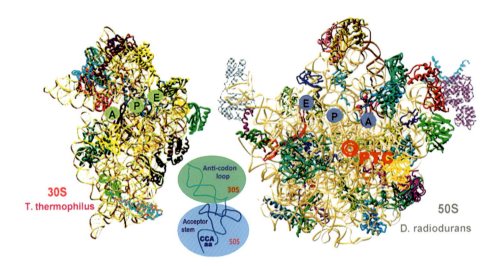

Fig. 2 The three-dimensional structures of the two ribosomal subunits from the bacteria *Deinococcus radiodurans* and *T. thermophilus*. The interface faces are facing the reader. The rRNA is shown in brownish colors, and each of the r-proteins is painted in a different color. The approximate site of the PTC is marked in red. Insert: the backbone of a tRNA molecule. The circles designate the regions interacting with each of the ribosomal subunits.

Continuous efforts were aimed at improving crystals, including the assessment of the influence of the relative concentrations of mono- and divalent ions (von Bohlen et al., 1991) on crystal properties. These efforts led to dramatic improvements in the quality of the crystals of the large ribosomal subunits from *H. marismortui* (H50S). In addition, constant refinements of bacterial growth protocols (Auerbach-Nevo et al., 2005) alongside a thorough investigation of crystallization conditions (Zimmerman and Yonath, 2009), indicated a noteworthy correlation between the conditions at which the ribosome functions and the crystal quality. Along these lines it is worth mentioning that flexible ribosomal regions were detected in electron-density maps obtained from crystals of ribosomal particles that were obtained under conditions that supported optimal functional activity (Harms et al., 2001), whereas the same regions were significantly more disordered in crystals obtained under non-physiological conditions (Ban et al., 2000).

An alternative strategy for improvement of crystal quality was the crystallization of complexes of ribosomes with substrates, inhibitors and/or factors, presumably because these factors trap the ribosomal particles in preferred conformations. Indeed, the initial diffracting crystals of the 70S ribosome from *T. thermophilus* (T70S) with mRNA and tRNA molecules (Hansen et al., 1990) a decade later led to impressive advances in resolution from crystals of functional ribosome complexes (Yusupov et al., 2001; Korostelev et al., 2006b). Importantly,

these techniques also enabled the structural analysis of snapshots of ribosomes trapped in specific conformations, albeit not necessarily functional ones (Schuwirth et al., 2005).

3. The ribosome is a polymerase

The crystal structures of bacterial ribosomes showed that the interface surfaces of both ribosomal subunits and their active sites (the decoding center and the peptidyl transferase center) are rich in RNA (Figure 2). These observations verified previous biochemical suggestions that the ribosome is a ribozyme. Particularly, as seen below, the striking architecture of the ribosome governs all tasks related to nascent protein elongation: namely the formation of peptide bonds, the processivity of this reaction, and the detachment of the growing polypeptide chain from the P-site tRNA. Thus, in addition to factor-assisted movements, i.e. the entrance and exit of the tRNA molecules and the progression of the mRNA, amino acid polymerization requires several major motions including peptidyl-tRNA translocation from the A to the P site, entry of the growing chain into the ribosome exit tunnel (Figure 3), passage of the deacylated tRNA molecule from the P to the E site and its subsequent release. However, it should be kept in mind that, although single peptide bonds can be produced by mixtures containing mainly rRNA and traces of r-proteins (Noller et al., 1992), the production

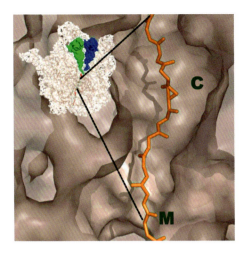

Fig. 3 The ribosomal exit tunnel. The entire large subunit, viewed from its interface surface with A- and P-site tRNAs (blue and green, respectively) and with polyalanine (orange) modeled in the tunnel, is shown in the top left panel. The main view is a zoom into the upper end of the tunnel. C denotes a crevice where co-translational initial folding may occur (Amit et al., 2005), and M shows the tunnel constriction, add.

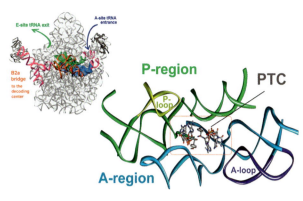

Fig. 4 The symmetrical region within the ribosome. Left: The symmetrical region within the ribosome and its details. The A-region is shown in blue, the P-region in green, and the non-symmetrical extensions are depicted in magenta. The bridge to the small subunit is shown in light brown. Right: Zoom into the symmetrical region, highlighting the basic structure that can form the active-site pocket and the loops that accommodate the C74 of the 3' ends of the A- and P-site tRNAs.

of peptide bonds by pure ribosomal RNA has not yet been demonstrated (Anderson et al., 2007).

The PTC is situated within a highly conserved universal symmetrical region that is embedded in the otherwise asymmetric ribosome structure (Figure 4). This region provides the machinery required for peptide bond formation, for the translocation of the A-site tRNA 3' end, for the detachment of the free substrate after peptide bond formation, and for the entry of the growing chain into the ribosome tunnel. This region is composed of 180 nucleotides, the fold, but not the sequence, of which is related by an internal pseudo two-fold symmetry. This region contains the two conserved nucleotides, G2552 and G2553, which form symmetrical Watson-Crick G-C base-pairs with the universally conserved CCA termini of the P and A-site tRNAs, respectively (Samaha et al., 1995; Kim and Green, 1999), and has been identified in all known ribosome structures, regardless of the source or functional state of the ribosomes (Bashan et al., 2003a; Zarivach et al., 2004; Agmon et al., 2005; Baram and Yonath, 2005). More specifically, the same sub-structure was identified in the cores of ribosomes from mesophilic, thermophilic, radiophilic and halophilic bacteria and archaea, regardless of their functional state or the ligands bound to them (Agmon et al., 2005). It is conceivable that the central location of the symmetrical region allows it to serve as the central signaling feature between all the

functional regions involved in protein biosynthesis that are located remote from one another (up to 200 Å away), but must communicate during elongation (Uemura et al., 2007).

The PTC is built as an arched void located at the bottom of a V-shaped cavity that hosts the helical portion of the acceptor stems and the 3' ends (Figure 2) of both A- and P-site tRNAs (Figure 5). In the crystal structure of the large ribosomal subunit from *Deinococcus radiodurans* (D50S) in complex with a substrate analog that mimics the acceptor stem and the 3' end of the A-site tRNA (called acceptor stem mimic, ASM), the acceptor stem interacts extensively with the walls of the cavity (Bashan et al., 2003b), forming an elaborate network of interactions (Figure 5). These interactions dictate a specific orientation that facilitates the processivity of peptide bond formation. Thus, although the PTC has some tolerance in the positioning of "fragment reaction substrates" (Hansen et al., 2002), the interactions of the tRNA acceptor stem seem to be crucial for accurate substrate positioning in the PTC in a configuration allowing for peptide bond formation (Yonath, 2003). This structural observation supports the finding that the tRNA core region contributes to interactions with the ribosome (Pan et al., 2006). The linkage between the elaborate architecture of the symmetrical region and the position of the A-site tRNA suggests that the translocation of the tRNA 3' end within the PTC is related to the overall tRNA/mRNA translocation, assisted by EF-G, which is performed by a combination of two synchronized motions: a side-

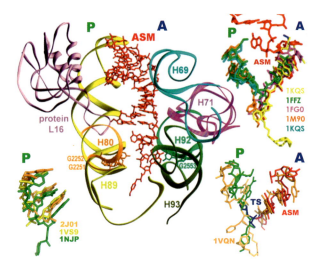

Fig. 5 Various substrates bound at the PTC. The central image shows the PTC and the cavity leading to it. The position of a mimic of A-site tRNA acceptor stem and 3' end (ASM, red) is shown. The components of the base pairs between the PTC upper rim (one at the A site and two in the P site) are highlighted. The RNA helices and the nucleotides located at the walls of the cavity are identified using *E. coli* numbering. Bottom left: Superposition of the 3' ends of P-site tRNAs: two experimentally determined (yellow and orange) and one derived (by rotation) moiety, in green. PDB entries are indicated by numbers. Top right: Positions and orientations of ASM and of various "minimal fragments" (puromycin derivatives) in the PTC; a dipeptide (yellow), produced by a minimal fragment that could not rotate into the P site because of its minimal size and its specific orientation, is shown in the A site. Bottom right: Superposition of ASM, the computed transition state (Gindulyte et al., 2006) (TS) and a chemically designed TS analog. Note that the extension of the chemical TS (Schmeing et al., 2005b), which is supposed to represent the nascent chain, is originating in the P site.

ways shift (the main component) and a rotatory motion of the A-site tRNA 3' end along a path confined by the PTC walls.

This rotatory motion appears to be navigated and guided by the ribosomal architecture, mainly the PTC rear wall that confines the rotatory path. Two flexible nucleotides, A2602 and U2585, seem to anchor and propel this motion. This means that the ribosomal architecture and its mobility provides all structural elements enabling ribosome function as an amino acid polymerase, including the formation of a symmetrical universal base pair between each of the tRNAs and the PTC (Bashan et al., 2003a; Agmon et al., 2005; Pan et al., 2006), alongside an additional base pair between the P-site tRNA and the PTC, a prerequisite for substrate-mediated acceleration (Weinger et al., 2004).

Importantly, all nucleotides involved in this rotatory motion have been classified as essential by a comprehensive genetic selection analysis (Sato et al., 2006).

Furthermore, the rotatory motion positions the proximal 2'-hydroxyl of A76 of the P-site tRNA in the same position and orientation as found in crystals of the entire ribosome with mRNA and tRNAs (Korostelev et al., 2006a; Selmer et al., 2006) (Figure 5) and allows for chemical catalysis of peptide bond formation by A76 of the P-site tRNA (Weinger et al., 2004).

Simulation studies of the rotatory motion indicated that during this motion the rotating moiety interacts with ribosomal components confining the rotatory path, along the "PTC rear wall" (Agmon et al., 2005; Agmon et al., 2006). Consistently, quantum-mechanical calculations, based on D50S structural data, indicated that the transition state (TS) of peptide bond formation is formed during the rotatory motion and is stabilized by hydrogen bonds with rRNA nucleotides (Gindulyte et al., 2006). Importantly, the TS location suggested by quantum-mechanics is close to the location of a designed TS analog in the crystal structure of its complex with H50S (Schmeing et al., 2005) (Figure 5).

In short, the structure of D50S in complex with a substrate analog mimicking the part of the A-site tRNA that interacts with the large subunit advanced the comprehension of peptide bond formation by showing that ribosomes position their substrates in stereochemical configurations suitable for peptide bond formation (Bashan et al., 2003a; Agmon et al., 2005). Furthermore, the ribosomal architecture that facilitates positional catalysis of peptide bond formation, promotes substrate-mediated chemical acceleration, in accord with the requirement of full-length tRNAs for rapid and smooth peptide bond formation observed by various methods, including the usage of chemical, genetic (Polacek et al., 2001; Weinger et al., 2004; Youngman et al., 2004; Beringer et al., 2005; Polacek and Mankin, 2005; Brunelle et al., 2006; Sato et al., 2006), computational (Sharma et al., 2005; Gindulyte et al., 2006; Trobro and Aqvist, 2006) and kinetic studies (Beringer et al., 2005; Wohlgemuth et al., 2006; Beringer and Rodnina, 2007; Rodnina et al., 2007).

4. Structural disorder with functional meaning

The significance of the interactions of the acceptor stem with the cavity leading to the PTC was indicated by biochemical studies and clearly demonstrated crystallographically. Thus shedding light on the differences between the binding modes of full-size tRNA to 70S ribosomes (or acceptor stem mimics to 50S subunits) and the binding modes of the various minimal substrates used for the fragment reaction. In functional experiments, the ribosome activity is determined by the reaction between substrate analogs capable of producing single peptide bonds. These "fragment reaction substrates" (Figure 5) are basically derivatives of puromycin, an A-site analog which acts as an inhibitor of the ribosomal polymerase activity. In many biochemical and structural studies, puromycin derivatives were used because they are good substrate analogs for a single peptide bond formation and because of the relative ease to detect in vitro single peptide bonds formed by them. However, caution is required when treating them as suitable to mimic the natural ribosome polymerase function. Interestingly, despite being small and consequently presumed to diffuse swiftly into its binding site within the ribosome, the rate of puromycin reaction with fMetPhe-tRNA as P-site substrate in 50S is comparable with rates of peptide bond formation with full-size tRNA (Wohlgemuth et al., 2008). It appears, therefore, that conformational rearrangements that were found to be required for productive positioning (Selmer et al., 2006) also play a role in determining the reaction rate. This idea is consistent with the biochemical finding that the peptidyl transfer reaction may be modulated by conformational changes at the active site (Youngman et al., 2004; Beringer et al., 2005; Schmeing et al., 2005; Brunelle et al., 2006; Beringer and Rodnina, 2007).

In the crystal structure of H50S with reactive fragment reaction substrates (Schmeing et al., 2002), the dipeptide resides in the A site (Figure 5) in an orientation hardly suitable for entrance into the tunnel. This complex was named "pre-translocational intermediate", meaning that in each elongation step the nascent protein has to translocate from the A to the P site. As the A to P site translocation within the PTC is performed by a 180 deg rotation, the translocation of the suggested "pre-translocational intermediate" requires that in each elongation step (15–20 times a second) such rotation is performed by the 3' end of A-site tRNA. If the entire growing chain is attached to its 3' end, it would have to rotate with it. Such motion would require a lot of space, which is probably not available in the dense PTC environment. Similarly, the energetic requirements of such a complicated operation should be very high. It appears, therefore that the dipeptide was trapped in the A site since the 3' end of the A-site tRNA did not undergo the rotatory motion because it, like all minimal substrate analogs,

is too short and hence lacks moieties that facilitate the rotatory motion. Notably, the substrate was detected near the P site in the crystal structure of H50S with a substrate designed to mimic the transition state (Schmeing et al., 2005). As mentioned above, this finding is in accord with the results of the quantum mechanical computations that placed the transition state of the peptide bond in the PTC, close to the P site (Gindulyte et al., 2006). This result and the correlation between the rotatory motion and amino acid polymerization rationalize the detection of the dipeptide at the A site (Schmeing et al., 2002).

Hence, it appears that for the ribosome's polymerase activity the A-site substrate needs to be a full-length tRNA with its 3' end accurately positioned in the active site of the ribosome. This conclusion is supported by the study of a crystal structure of H50S in complex with a tRNA "mini-helix" (similar to the ASM described above) which led to the suggestion that specific rRNA nucleotides catalyze peptide bond formation by the general acid/base reverse mechanism (Ban et al., 2000), a proposition that was challenged by a number of biochemical and mutational studies, e. g. (Polacek et al., 2001).

Notably, in this structure, only the tip of the 3'-end is resolved, whereas the entire acceptor stem is disordered, presumably because its interactions with the partially disordered cavity leading to the PTC could not be formed (Nissen et al., 2000). Importantly, the H50S crystals used for this study were obtained under far from optimal functional conditions, namely rather low KCl concentration, while it was found that a very high KCl concentration is essential for the function of the ribosomes from the halophile H. marismortui (Shevack et al., 1985; Gluehmann et al., 2001).

The observed disorder in otherwise very well ordered crystals of H50S suggest that ribosomes kept under far from the conditions allowing for their efficient activity, can form peptide bonds but may not be capable of elongating nascent chains. This finding seems to point to a natural strategy for avoiding or minimizing the formation of proteins under stressful or far from physiological conditions. Specifically, the disorder of almost all of the functional regions of H50S may reflect a natural response of the halophilic ribosomes to a salt deficient environment, potentially indicating a natural mechanism for conserving cellular resources under stressful circumstances. In support of this suggestion, structural comparisons showed that the H50S active site contains key PTC components in orientations that differ significantly from those observed in empty D50S

(Harms et al., 2001) as well as in functional complexes of T70S ribosomes (Selmer et al., 2006).The main conclusion from the analyses of all of these structures is that single peptide bond formation can be performed even when the initial substrate binding is not accurate. However, for elongating the nascent proteins, accurate positioning of the A-site tRNA 3' end is mandatory. It appears, therefore, that the choice of substrate analogs for the various studies as well as the discrepancy between the definitions of ribosomal activity (namely the mere ability to form single peptide bonds versus the requirement for elongating nascent proteins) are the main reason for different structures, and potentially for their interpretation. It is clear that accurate substrate positioning within the ribosome frame, accompanied by the P-site tRNA interactions with 23S rRNA that allow for substrate catalysis (Weinger et al., 2004), plays the key role in ribosome catalytic functions, a notion that is currently widely accepted [e. g. (Beringer et al., 2005; Beringer and Rodnina, 2007; Simonovic and Steitz, 2008; Bashan and Yonath, 2008b)].

5. On the ribosomal tunnel and initial nascent protein folding

It was widely assumed that nascent proteins advance on the surface of the ribosome until their maturation. Even after biochemical experiments indicated that nascent chains are masked (hence protected) by the ribosome (Malkin and Rich, 1967; Sabatini and Blobel, 1970) and a tunnel was visualized in EM reconstructions from two-dimensional sheets at 60 and 25 Å resolution (Milligan and Unwin, 1986; Yonath et al., 1987), the existence of a tunnel was not generally accepted (Moore, 1988). Furthermore, it was assumed that nascent proteins are not degraded during protein synthesis because all of them adopt the conformation of an alpha helix (Ryabova et al., 1988). Doubts regarding the existence of the ribosomal tunnel were removed when it was visualized by cryo electron microscopy (Frank et al., 1995; Stark et al., 1995). Remarkably, the tunnel is of variable width and shape (Figure 3), suggesting its possible involvement in the fate of the nascent chains in accord with previous observations [e. g. (Crowley et al., 1993; Walter and Johnson, 1994)]. Furthermore, results of biochemical, microscopic, and computational experiments verified the existence of the tunnel and in several cases indicated that it may participate actively in nascent chain progression and its initial compaction, as well as in translation arrest and cellular signal-

ing (Gabashvili et al., 2001; Gong and Yanofsky, 2002; Nakatogawa and Ito, 2002; Berisio et al., 2003; Gilbert et al., 2004; Johnson and Jensen, 2004; Woolhead et al., 2004; Amit et al., 2005; Ziv et al., 2005; Berisio et al., 2006; Cruz-Vera et al., 2006; Mankin, 2006; Mitra et al., 2006; Tenson and Mankin, 2006; Voss et al., 2006; Woolhead et al., 2006; Deane et al., 2007; Schaffitzel and Ban, 2007; Bornemann et al., 2008; Petrone et al., 2008; Starosta et al., 2010; Nakatogawa and Ito, 2004; Chiba et al., 2011).

While emerging from the ribosome, nascent chains may interact with chaperones that assist their folding and/or prevent their aggregation and misfolding. In bacteria the first chaperone that encounters the nascent proteins, called trigger factor, forms a shelter composed of hydrophobic and hydrophilic regions that provide an environment that can compete with the aggregation tendency of the still unfolded chains (Baram et al., 2005; Schluenzen et al., 2005; Kaiser et al., 2006). Interestingly, free trigger factor seems also to rescue proteins from misfolding and to accelerate protein folding (Martinez-Hackert and Hendrickson, 2009).

6. Antibiotics targeting the ribosome: strategies, expectations and problems

Because of the major significance of the ribosomes for cell viability many antibiotics target them. An immense amount of biochemical studies performed over four decades, alongside medical research and recent crystallographic analysis, showed that despite the high conservation of the ribosomal active sites, subtle differences facilitate their clinical relevance (Wilson, 2004; Yonath and Bashan, 2004; Polacek and Mankin, 2005; Yonath, 2005; Tenson and Mankin, 2006; Bottger, 2007; Sohmen et al., 2009). As so far there are no crystals of ribosomes from pathogenic organisms, structural information is currently obtained only from the crystallizable bacterial ribosomes that have been shown to be relevant as pathogen models, namely *E. coli, D. radiodurans,* and *T. thermophilus.* Antibiotic action on ribosomes from these bacteria, in conjunction with data obtained from the ribosomes of other organisms such as *Mycobacterium smegmatis* (a reasonable mimic of *Mycobacterium tuberculosis*), were found to be useful for determining antibiotic modes of action directly (described below) or indirectly (Pfister et al., 2005; Tu et al., 2005; Bommakanti et al., 2008; Hobbie et al., 2008).

The crystallographic analyses have shown that antibiotics targeting ribosomes exploit diverse strategies with common denominators. All antibioitcs target ribosomes at distinct locations within functionally relevant sites, mostly composed solely of rRNA. Each exerts its inhibitory action by competing with a crucial step in the biosynthetic cycle, including substrate binding, ribosomal dynamics, progression of the mRNA chain and decoding. Examples include hindering tRNA substrate accommodations at the PTC, stabilizing the tRNA in the A site in the pre-translocation state, preventing interactions of the ribosomal recycling factor and blocking the protein exit tunnel.

The identification of the various modes of action of antibiotics targeting ribosomes and an analysis of the ribosomal components comprising the binding pockets confirmed that the imperative distinction between ribosomes from eubacterial pathogens and mammalian cells hinges on subtle structural differences within the antibiotic binding pockets (Yonath and Bashan, 2004; Yonath, 2005; Pyetan et al., 2007; Auerbach et al., 2009; Auerbach et al., 2010). Apparently, fine tuning of the binding pocket can alter the binding parameters. These subtle sequence and/or conformational variations enable drug selectivity, thus facilitating clinical usage. Furthermore, the available structures illuminate features that are distinct between ribosomes from bacteria and non-pathogenic archaea that may be of crucial clinical importance.

Noteworthy are the results of comparisons between the crystal structures of different ribosomal particles complexed with the same antibiotics. Although leading to effective binding, disparities observed between the binding modes to pathogen models, namely *D. radiodurans* and *T. thermophilus*, may point to species specificity. Furthermore, comparison of the modes of antibiotic binding to ribosomal particles from the pathogen models (*D. radiodurans* and *T. thermophilus*) with modes observed in the archaeon *H. marismortui* (which shares properties with eukaryotes) indicated some variability in the binding modes, and in specific cases showed that binding is not synonymous with inhibitory activity. These comparisons highlighted the distinction between mere binding and binding leading to inhibitory activity. Specifically, for the macrolide family, these studies indicated that the identity of a single nucleotide can determine the strength of antibiotic binding, whereas proximal stereochemistry governs the antibiotic orientation within the binding pocket (Yonath and Bashan, 2004; Yonath, 2005) and consequently its therapeutic effective-

ness. This is in accord with recent mutagenesis studies showing that mutation from guanine to adenine in 25S rRNA at the position equivalent to *E. coli* A2058 does not confer erythromycin sensitivity in *Saccharomyces cerevisae* (Bommakanti et al., 2008). Thus, it was clearly demonstrated that the mere binding of an antibiotic is not sufficient for therapeutic effectiveness and that minute variations in the chemical moieties of the antibiotics can lead to significantly different binding modes. An appropriate example is the extreme difference between the modes of function of erythromycin, which competes with lankacidin binding, and lankamycin, which acts synergistically with lankacidin (Figure 6) (Auerbach et al., 2010; Belousoff et al., 2011).

In addition to rationalizing genetic, biochemical, and medical observations, the available structures have revealed unexpected inhibitory modes. Examples are the stabilization of the pre-translocation state (Stanley et al., 2010), as well as exploitation of the inherent ribosome flexibility for antibiotic synergism (Harms et al., 2004; Yonath, 2005; Auerbach et al., 2010) and for triggering an induced-fit mechanism by remote interactions that reshape the antibiotic binding pocket (Davidovich et al., 2007). Among the ribosomal antibiotics, the pleuromutilins are of special interest since they bind to the almost fully conserved PTC, yet they discriminate between bacterial and mammalian ribosomes. To circumvent the high conservation of the PTC, the pleuromutilins exploit the inherent functional mobility of the PTC and stabilize a conformational rearrangement that involves a network of remote interactions between flexible PTC nucleotides and less

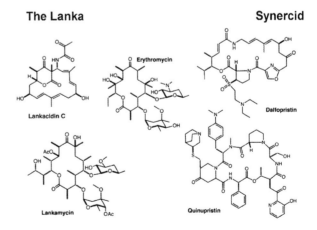

Fig. 6 The chemical compositions of antibiotic pairs acting on the ribosomal PTC and the exit tunnel. Both the lankacidin-lankamycin and the Synercid pairs display synergism. However, erythromycin (middle), which resembles lankamycin, competes with lankacidin.

conserved nucleotides residing in the PTC-vicinity (the second and third shells around the PTC). These interactions reshape the PTC contour and trigger its closure on the bound drug (Davidovich et al., 2007). The uniqueness of this pleuromutilin binding mode led to new insights as it indicated the existence of an allosteric network around the ribosomal active site. Indeed, the value of these findings is far beyond their perspective on clinical usage, as they highlight basic issues, such as the possibility of remote reshaping of binding pockets and the ability of ribosome inhibitors to benefit from the inherent flexibility of the ribosome.

Similar to the variability in binding modes, seemingly identical mechanisms of drug resistance can indeed be different, as they are dominated, directly or via cellular effects, by the antibiotics' chemical properties (Davidovich et al., 2007; Davidovich et al., 2008). The observed variability in antibiotic binding and inhibitory modes justifies expectations for structurally based improvement of the properties of existing compounds as well as for the development of novel drug classes. Detailed accounts can be found in several reviews (Auerbach et al., 2004; Yonath and Bashan, 2004; Poehlsgaard and Douthwaite, 2005; Yonath, 2005; Bottger, 2006; Tenson and Mankin, 2006; Bottger, 2007).

In short, over two dozen three-dimensional structures of ribosome complexes with antibiotics have revealed the principles allowing for clinical use, have provided unparalleled insight into the mode of antibiotic function, have illuminated mechanisms for acquiring resistance and have shown the basis for discrimination between pathogens and host cells. The elucidation of common principles of the mode of action of antibiotics targeting the ribosome combined with variability in binding modes led to the uncovering of diverse mechanisms for acquiring antibiotic resistance.

7. The ribosomal core is the optimized vestige of an ancient entity

A high level of sequence conservation in the symmetrical region has been maintained throughout ribosome evolution, even in mitochondrial ribosomes in which half the ribosomal RNA has been replaced by proteins (Mears et al., 2002; Thompson and Dahlberg, 2004; Agmon et al., 2006; Davidovich et al., 2009). This conservation and the observation that the symmetrical region provide all structural elements required for performing polypeptide elongation led us to suggest that the modern ribosome evolved by gene fusion or gene

duplication (Figure 7). We refer to this ancestral entity as the proto-ribosome (Agmon et al., 2005; Davidovich et al., 2009; Belousoff et al., 2010a; Belousoff et al., 2010b).

In particular, the preservation of the three-dimensional structure of the two halves of the ribosomal frame, independent of the sequence, emphasizes the superiority of functional requirements over sequence conservation and demonstrates the rigorous requirements of accurate substrate positioning for peptide bond formation. As mentioned above, this, as well as the universality of the symmetrical region, led to the assumption that the ancient ribosome was composed of a pocket confined by two RNA chains that formed a dimer, and that this pocket is still embedded in the heart of the modern ribosome (Figures 4, 7). In fact, as mentioned above, suggestions that proteins are made by an RNA machine have been made already in the sixties (Crick, 1968), and extensive research over the past three decades (summarized in a recent book (Yarus, 2010)) has supported the idea that nucleic acids are capable of independent replication, selection, and self splicing [e. g. (Been and Cech, 1986; Abelson, 1990; Ellington and Szostak, 1990; Tuerk and Gold, 1990; Pino et al., 2008; Costanzo et al., 2009; Lincoln and Joyce, 2009)].

In accord with these findings we have proposed (Agmon et al., 2006; Davidovich et al., 2009; Belousoff et al., 2010a; Belousoff et al., 2010b; Turk et al., 2010) that the ancient machinery that could form peptide bonds was made exclusively from RNA molecules, utilizing substituents available in the primordial soup, namely RNA chains that could acquire conformations sufficiently stable to survive evolutionary stress.

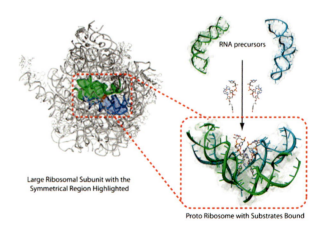

Large Ribosomal Subunit with the Symmetrical Region Highlighted

RNA precursors

Proto Ribosome with Substrates Bound

Fig. 7 The suggested proto-ribosome. Regions hosting A- and P-site tRNAs are shown in blue and green, respectively. The A-site tRNA mimic (Bashan et al., 2003) is shown in blue, and the derived P-site tRNA (by the rotatory motion) is shown in green.

These ancient RNA chains could fold spontaneously and then dimerize. The products of the dimerization yielded three-dimensional structures with a symmetrical pocket that could accommodate small substrates (e. g. amino acids conjugated with mono or oligo RNA nucleotides) in a stereochemistry suitable for spontaneous formation of various chemical bonds, including the peptide bond. These dimeric RNA complexes could have become the ancestors of the symmetrical region in the contemporary ribosome. The most appropriate pockets for promoting peptide bond formation survived.

The surviving ancient pockets became the templates for the ancient ribosomes. In a later stage, these initial RNA genes underwent optimization to produce more defined, relatively stable pockets, and when the correlation between the amino acid and the growing peptidyl sites was established, each of the two halves was further optimized for its task so that their sequences evolved differently. The entire ribosome could have evolved gradually around this symmetrical region until it acquired its final shape (Bokov and Steinberg, 2009).

The substrates of the ancient ribosomes, which were initially spontaneously produced amino acids conjugated with single or short oligo-nucleotides (Illangasekare et al., 1995; Turk et al., 2010), could have evolved in parallel to allow accurate binding, as occurs for aminoacylated CCA 3'-end. Later on, these could have been converted into longer compounds with a contour that can complement the inner surface of the reaction pocket. For increasing specificity, these short RNA segments could have been extended to larger entities by their fusion with stable RNA features, to form the ancient tRNA, presumably capable of storing, selecting and transferring instructions for producing useful proteins. The structural entity for decoding could have been combined with the structural machinery able to form a peptide bond in a single entity. Adding a feature similar to the modern anticodon loop allowed some genetic control, and could have led to the modern protein-nucleic acids world.

8. Conclusion

The currently available high-resolution structures of ribosomes and their subunits proved that the ribosome is a ribozyme. All functions of the ribosome, including decoding, peptide bond formation, protein elongation and tRNA release, are performed by ribosomal RNA while being assisted by ribosomal proteins.

A key requirement for highly efficient processivity of the ribosome's main catalytic activities hinges on accurate positioning of the ribosomal substrate, which hinges on the maintenance of the functional conformations of all ribosomal regions involved in ribosomal function, despite their high flexibility. Hence, disorder of these regions has functional meaning, and may be the result of a natural strategy to minimize cell function under hostile conditions.

The ribosomal active site, namely where the peptide bonds are being formed and where the nascent chain is elongated, is situated within a universal symmetrical region that is embedded in the otherwise asymmetric ribosome structure. This symmetrical region is highly conserved and provides the machinery required for peptide bond formation as well as for the ribosome polymerase activity. Therefore, it may be the remnant of the proto-ribosome, which seems to be a dimeric prebiotic machine that initially catalyzed prebiotic reactions, including the formation of chemical bonds, and then produced non-coded oligopeptides.

Structures of complexes of ribosomes with antibiotics revealed principles allowing for the clinical use of antibiotics, identified resistance mechanisms, and pointed at the structural basis for discriminating pathogenic bacteria from hosts. Thus, structural analyses provided valuable information for the improvement of antibiotics and for the design of novel compounds that can serve as antibiotics.

9. Future prospects

Ribosome research has undergone astonishing progress in recent years. The high-resolution structures have shed light on many of the functional properties of the translation machinery and revealed how the ribosome's striking architecture is ingeniously designed as the framework for its unique capabilities: precise decoding, substrate mediated peptide-bond formation and efficient polymerase activity. By analyzing these structures it appears that the ribosomal tasks are performed by the ribosomal RNA and may be supported by the ribosomal proteins.

Among the new findings that emerged from the structures are the intricate mode of decoding, the mobility of most of the ribosomal functional features, the symmetrical region at the core of the ribosome, the dynamic properties of the ribosomal tunnel, the interactions of the ribosome with the progressing nascent chains, the signaling between the ribosome and cellular components, and the shelter formed by the first chaperone that encounters the nascent chains (trigger factor) for preventing nascent chain aggregation and misfolding. Novel insights from these new findings include the suggestion that the translocation of the tRNA involves at least two concerted motions: sideways shift (which may be performed in a hybrid mode) and a ribosome-navigated rotation. The linkage between these findings and crystal structures of ribosomes with over two dozen antibiotics targeting the ribosome, illuminated various modes of binding and action of these antibiotics. They also deciphered mechanisms leading to resistance and identified the principles allowing for the discrimination between pathogens and eukaryotes despite high ribosome conservation. Further studies enlightened the basis for antibiotic synergism (Figure 6), indicated correlations between antibiotic susceptibility and fitness cost, and revealed a novel induced-fit mechanism exploiting inherent ribosomal flexibility for reshaping the antibiotic binding pocket by remote interactions. Thus, the high-resolution structures of the complexes of the ribosomes with the antibiotics bound to them address key issues associated with the structural basis for antibiotic resistance, synergism, and selectivity and provide unique structural tools for improving antibiotic action.

The availability of the high-resolution structures has stimulated an unpredictable expansion in ribosome research, which in turn has resulted in new insights into the translation process. An appropriate example is the study on real-time tRNA transit on single translating ribosomes at codon resolution (Uemura et al., 2010). However, despite the extensive research and the immense progress, several key issues are still unresolved, some of which are described above. Thus, it is clear that the future of ribosome research and its applications hold more scientific excitement.

Acknowledgements

Thanks are due to all members of the ribosome groups at the Weizmann Institute and at the Unit for Ribosome Research of the Max Planck Society at DESY/ Hamburg for their experimental efforts and illuminating discussion. Support was provided by the US National Inst. of Health (GM34 360), the German Ministry for Science and Technology (BMBF 05−641EA), GIF 853−2004, Human Frontier Science Program (HFSP) RGP0076/2003 and the Kimmelman Center for Macromolecular Assemblies. AY holds the Martin and Helen Kimmel Professorial Chair. X-ray diffraction data were collected the EMBL and MPG beam lines at DESY; F1/CHESS, Cornell University, SSRL/ Stanford University, ESRF/EMBL, Grenoble, BL26/PF/ KEK, Japan, and 19ID&23ID/APS/Argonne National Laboratory.

References

Abelson J (1990) Directed evolution of nucleic acids by independent replication and selection. Science 249: 488–489

Agmon I, Bashan A, Yonath A (2006) On Ribosome Conservation and Evolution. Isr J Ecol Evol 52: 359–379

Agmon I, Bashan A, Zarivach R, Yonath A (2005) Symmetry at the active site of the ribosome: structural and functional implications. Biol Chem 386: 833–844

Amit M, Berisio R, Baram D, Harms J, Bashan A, Yonath A (2005) A crevice adjoining the ribosome tunnel: hints for cotranslational folding. FEBS Lett 579: 3207–3213

Anderson RM, Kwon M, Strobel SA (2007) Toward ribosomal RNA catalytic activity in the absence of protein. J Mol Evol 64: 472–483

Auerbach-Nevo T, Zarivach R, Peretz M, Yonath A (2005) Reproducible growth of well diffracting ribosomal crystals. Acta Crystallogr D Biol Crystallogr 61: 713–719

Auerbach T, Bashan A, Yonath A (2004) Ribosomal antibiotics: structural basis for resistance, synergism and selectivity. Trends Biotechnol 22: 570–576

Auerbach T, Mermershtain I, Bashan A, Davidovich C, Rosenberg H, Sherman DH, Yonath A (2009) Structural basis for the antibacterial activity of the 12-membered-ring mono-sugar macrolide methymycin. Biotechnolog 84: 24–35

Auerbach T, Mermershtain I, Davidovich C, Bashan A, Belousoff M, Wekselman I, Zimmerman E, Xiong L, Klepacki D, Arakawa K, Kinashi H, Mankin AS, Yonath A (2010) The structure of ribosome-lankacidin complex reveals ribosomal sites for synergistic antibiotics. Proc Natl Acad Sci USA 107: 1983–1988

Ban N, Nissen P, Hansen J, Moore PB, Steitz TA (2000) The complete atomic structure of the large ribosomal subunit at 2.4 A resolution. Science 289: 905–920

Baram D, Pyetan E, Sittner A, Auerbach-Nevo T, Bashan A, Yonath A (2005) Structure of trigger factor binding domain in biologically homologous complex with eubacterial ribosome reveals its chaperone action. Proc Natl Acad Sci USA 102: 12 017–12 022

Baram D, Yonath A (2005) From peptide-bond formation to cotranslational folding: dynamic, regulatory and evolutionary aspects. FEBS Lett 579: 948–954

Bashan A, Agmon I, Zarivach R, Schluenzen F, Harms J, Berisio R, Bartels H, Franceschi F, Auerbach T, Hansen HA, Kossoy E, Kessler M, Yonath A (2003a) Structural basis of the ribosomal machinery for peptide bond formation, translocation, and nascent chain progression. Mol Cell 11: 91–102

Bashan A, Agmon I, Zarivach R, Schluenzen F, Harms J, Berisio R, Bartels H, Franceschi F, Auerbach T, Hansen HAS, Kossoy E, Kessler M, Yonath A (2003b) Structural basis of the ribosomal machinery for peptide bond formation, translocation, and nascent chain progression. Mol Cell 11: 91–102

Bashan A, Yonath A (2008a) The linkage between ribosomal crystallography, metal ions, heteropolytungstates and functional flexibility. J Mol Struct 890 289–294

Bashan A, Yonath A (2008b) Correlating ribosome function with high-resolution structures. Trends Microbiol 16: 326–335

Been MD, Cech TR (1986) One binding site determines sequence specificity of Tetrahymena pre-rRNA self-splicing, trans-splicing, and RNA enzyme activity. Cell 47: 207–216

Belousoff M, Davidovich C, Bashan A, Yonath A (2010a) in the press. Orig Life Evol Biosph

Belousoff MJ, Shapira,T, Bashan A, Zimmerman E, Rozenberg H, Arakawa K, Kinashi H, Yonath A (2011) Crystal structure of the synergistic antibiotic pair lankamycin and lankacidin in complex with the large ribosomal subunit. Proc Nat Acad USA; e-pub Feb 2011

Belousoff MJ, Davidovich C, Zimmerman E, Caspi Y, Wekselman I, Rozenszajn L, Shapira T, Sade-Falk O, Taha L, Bashan A, Weiss MS, Yonath A (2010b) Ancient machinery embedded in the contemporary ribosome. Biochem Soc Trans 38: 422–427

Beringer M, Bruell C, Xiong L, Pfister P, Bieling P, Katunin VI, Mankin AS, Bottger EC, Rodnina MV (2005) Essential mechanisms in the catalysis of peptide bond formation on the ribosome. J Biol Chem 280: 36 065–36 072

Beringer M, Rodnina MV (2007) The ribosomal peptidyl transferase. Mol Cell 26: 311–321

Berisio R, Corti N, Pfister P, Yonath A, Bottger EC (2006) 23S rRNA 2058A->G Alteration Mediates Ketolide Resistance in Combination with Deletion in L22. Antimicrob Agents Chemother 50: 3816–3823

Berisio R, Schluenzen F, Harms J, Bashan A, Auerbach T, Baram D, Yonath A (2003) Structural insight into the role of the ribosomal tunnel in cellular regulation. Nat Struct Biol 10: 366–370

Blaha G, Stanley RE, Steitz TA (2009) Formation of the first peptide bond: the structure of EF-P bound to the 70S ribosome. Science 325: 966–970

Bokov K, Steinberg SV (2009) A hierarchical model for evolution of 23S ribosomal RNA. Nature 457: 977–980

Bommakanti AS, Lindahl L, Zengel JM (2008) Mutation from guanine to adenine in 25S rRNA at the position equivalent to E. coli A2058 does not confer erythromycin sensitivity in Sacchromyces cerevisae. RNA 14: 460–464

Bornemann T, Jockel J, Rodnina MV, Wintermeyer W (2008) Signal sequence-independent membrane targeting of ribosomes containing short nascent peptides within the exit tunnel. Nat Struct Mol Biol 15: 494–499

Borovinskaya MA, Pai RD, Zhang W, Schuwirth BS, Holton JM, Hirokawa G, Kaji H, Kaji A, Cate JH (2007) Structural basis for aminoglycoside inhibition of bacterial ribosome recycling. Nat Struct Mol Biol 14: 727–732

Bottger EC (2006) The ribosome as a drug target. Trends Biotechnol 24: 145–147

Bottger EC (2007) Antimicrobial agents targeting the ribosome: the issue of selectivity and toxicity – lessons to be learned. Cell Mol Life Sci 64: 791–795

Brunelle JL, Youngman EM, Sharma D, Green R (2006) The interaction between C75 of tRNA and the A loop of the ribosome stimulates peptidyl transferase activity. RNA 12: 33–39

Carter AP, Clemons WM, Jr., Brodersen DE, Morgan-Warren RJ, Hartsch T, Wimberly BT, Ramakrishnan V (2001) Crystal structure of an initiation factor bound to the 30S ribosomal subunit. Science 291: 498–501

Chiba S, Kanamori T, Ueda T, Akiyama Y, Pogliano, K, Ito, K (2011) Recruitment of a species-specific arrest module to monitor different cellular processes. Proc Natl Acad Sci USA; e pub March 2011

Costanzo G, Pino S, Ciciriello F, Di Mauro E (2009) Generation of long RNA chains in water. J Biol Chem 284: 33 206–33 216

Crick FH (1968) The origin of the genetic code. J Mol Biol 38: 367–379

Crowley KS, Reinhart GD, Johnson AE (1993) The signal sequence moves through a ribosomal tunnel into a noncytoplasmic aqueous environment at the ER membrane early in translocation. Cell 73: 1101–1115

Cruz-Vera LR, Gong M, Yanofsky C (2006) Changes produced by bound tryptophan in the ribosome peptidyl transferase center in response to TnaC, a nascent leader peptide. Proc Natl Acad Sci USA 103: 3598–3603

Davidovich C, Bashan A, Auerbach-Nevo T, Yaggie RD, Gontarek RR, Yonath A (2007) Induced-fit tightens pleuromutilins binding to ribosomes and remote interactions enable their selectivity. Proc Natl Acad Sci USA 104: 4291–4296

Davidovich C, Bashan A, Yonath A (2008) Structural basis for cross-resistance to ribosomal PTC antibiotics. Proc Natl Acad Sci USA 105: 20 665–20 670

Davidovich C, Belousoff M, Bashan A, Yonath A (2009) The evolving ribosome: from non-coded peptide bond formation to sophisticated translation machinery. Res Microbiol 160: 487–492

Deane CM, Dong M, Huard FP, Lance BK, Wood GR (2007) Cotranslational protein folding–fact or fiction? Bioinformatics 23:i142–148

Demeshkina N, Jenner L, Yusupova G, Yusupov M (2010) Interactions of the ribosome with mRNA and tRNA. Curr Opin Struct Biol:12

Ellington AD, Szostak JW (1990) In vitro selection of RNA molecules that bind specific ligands. Nature 346: 818–822

Frank J, Zhu J, Penczek P, Li Y, Srivastava S, Verschoor A, Radermacher M, Grassucci R, Lata RK, Agrawal RK (1995) A model of protein synthesis based on cryo-electron microscopy of the E. coli ribosome. Nature 376: 441–444

Gabashvili IS, Gregory ST, Valle M, Grassucci R, Worbs M, Wahl MC, Dahlberg AE, Frank J (2001) The polypeptide tunnel system in the ribosome and its gating in erythromycin resistance mutants of L4 and L22. Mol Cell 8: 181–188

Garrett RA, Wittmann HG (1973) Structure and function of the ribosome. Endeavour 32: 8–14

Gilbert RJ, Fucini P, Connell S, Fuller SD, Nierhaus KH, Robinson CV, Dobson CM, Stuart DI (2004) Three-Dimensional Structures of Translating Ribosomes by Cryo-EM. Mol Cell 14: 57–66

Gindulyte A, Bashan A, Agmon I, Massa L, Yonath A, Karle J (2006) The transition state for formation of the peptide bond in the ribosome. Proc Natl Acad Sci USA 103: 13 327–13 332

Gluehmann M, Zarivach R, Bashan A, Harms J, Schluenzen F, Bartels H, Agmon I, Rosenblum G, Pioletti M, Auerbach T, Avila H, Hansen HA, Franceschi F, Yonath A (2001) Ribosomal crystallography: from poorly diffracting microcrystals to high-resolution structures. Methods 25: 292–302

Gong F, Yanofsky C (2002) Instruction of translating ribosome by nascent Peptide. Science 297: 1864–1867

Hansen HA, Volkmann N, Piefke J, Glotz C, Weinstein S, Makowski I, Meyer S, Wittmann HG, Yonath A (1990) Crystals of complexes mimicking protein biosynthesis are suitable for crystallographic studies. Biochim Biophys Acta 1050: 1–7

Hansen JL, Schmeing TM, Moore PB, Steitz TA (2002) Structural insights into peptide bond formation. Proc Natl Acad Sci USA 99: 11 670–11 675

Harms J, Schluenzen F, Fucini P, Bartels H, Yonath A (2004) Alterations at the peptidyl transferase centre of the ribosome induced by the synergistic action of the streptogramins dalfopristin and quinupristin. BMC Biol 2: 4;1–10

Harms J, Schluenzen F, Zarivach R, Bashan A, Gat S, Agmon I, Bartels H, Franceschi F, Yonath A (2001) High resolution structure of the large ribosomal subunit from a mesophilic eubacterium. Cell 107: 679–688

Hobbie SN, Bruell CM, Akshay S, Kalapala SK, Shcherbakov D, Bottger EC (2008) Mitochondrial deafness alleles confer misreading of the genetic code. Proc Natl Acad Sci USA 105: 3244–3249

Hope H, Frolow F, von Bohlen K, Makowski I, Kratky C, Halfon Y, Danz H, Webster P, Bartels KS, Wittmann HG, et al. (1989) Cryocrystallography of ribosomal particles. Acta Crystallogr B 45: 190–199

Illangasekare M, Sanchez G, Nickles T, Yarus M (1995) Aminoacyl-RNA synthesis catalyzed by an RNA. Science 267: 643–647

Jenner L, Ito K, Rees B, Yusupov M, Yusupova G (2007) Messenger RNA conformations in the ribosomal E site revealed by X-ray crystallography. EMBO Rep 8: 846–850

Jenner LB, Demeshkina N, Yusupova G, Yusupov M (2010) Structural aspects of messenger RNA reading frame maintenance by the ribosome. Nat Struct Mol Biol 17: 555–560

Johnson AE, Jensen RE (2004) Barreling through the membrane. Nat Struct Mol Biol 11: 113–114

Kaiser CM, Chang HC, Agashe VR, Lakshmipathy SK, Etchells SA, Hayer-Hartl M, Hartl FU, Barral JM (2006) Real-time observation of trigger factor function on translating ribosomes. Nature 444: 455–460

Kim DF, Green R (1999) Base-pairing between 23S rRNA and tRNA in the ribosomal A site. Mol Cell 4: 859–864

Korostelev A, Trakhanov S, Laurberg M, Noller HF (2006a) Crystal structure of a 70S ribosome-tRNA complex reveals functional interactions and rearrangements. Cell 126: 1065–1077

Korostelev A, Trakhanov S, Laurberg M, Noller HF (2006b) Crystal Structure of a 70S Ribosome-tRNA Complex Reveals Functional Interactions and Rearrangements. Cell 126: 1065–1077

Laurberg M, Asahara H, Korostelev A, Zhu J, Trakhanov S, Noller HF (2008) Structural basis for translation termination on the 70S ribosome. Nature 454: 852–857

Lincoln TA, Joyce GF (2009) Self-Sustained Replication of an RNA Enzyme. Science 323: 1229–1232

Malkin LI, Rich A (1967) Partial resistance of nascent polypeptide chains to proteolytic digestion due to ribosomal shielding. J Mol Biol 26: 329–346

Mankin A (2006) Antibiotic blocks mRNA path on the ribosome. Nat Struct Mol Biol 13: 858–860

Martinez-Hackert E, Hendrickson WA (2009) Promiscuous substrate recognition in folding and assembly activities of the trigger factor chaperone. Cell 138: 923–934

Mears JA, Cannone JJ, Stagg SM, Gutell RR, Agrawal RK, Harvey SC (2002) Modeling a minimal ribosome based on comparative sequence analysis. J Mol Biol 321: 215–234

Milligan RA, Unwin PN (1986) Location of exit channel for nascent protein in 80S ribosome. Nature 319: 693–695

Mitra K, Schaffitzel C, Fabiola F, Chapman MS, Ban N, Frank J (2006) Elongation arrest by SecM via a cascade of ribosomal RNA rearrangements. Mol Cell 22: 533–543

Moore PB (1988) The ribosome returns. Nature 331: 223–227

Nakatogawa H, Ito K (2004) Intraribosomal Regulation of Expression and Fate of Proteins ChemBioChem 5: 48-51

Nakatogawa H, Ito K (2002) The ribosomal exit tunnel functions as a discriminating gate. Cell 108: 629–636

Nissen P, Hansen J, Ban N, Moore PB, Steitz TA (2000) The structural basis of ribosome activity in peptide bond synthesis. Science 289: 920–930

Noller HF, Hoffarth V, Zimniak L (1992) Unusual resistance of peptidyl transferase to protein extraction procedures. Science 256: 1416–1419

Palade GE (1955) A small particulate component of the cytoplasm. J Biophys Biochem Cytol 1: 59–68

Pan D, Kirillov S, Zhang CM, Hou YM, Cooperman BS (2006) Rapid ribosomal translocation depends on the conserved 18–55 base pair in P-site transfer RNA. Nat Struct Mol Biol 13: 354–359

Petrone PM, Snow CD, Lucent D, Pande VS (2008) Side-chain recognition and gating in the ribosome exit tunnel. Proc Natl Acad Sci USA 105: 16 549–16 554

Pfister P, Corti N, Hobbie S, Bruell C, Zarivach R, Yonath A, Bottger EC (2005) 23S rRNA base pair 2057–2611 determines ketolide susceptibility and fitness cost of the macrolide resistance mutation 2058A → G. Proc Natl Acad Sci USA 102: 5180–5185

Pino S, Ciciriello F, Costanzo G, Di Mauro E (2008) Nonenzymatic RNA ligation in water. J Biol Chem 283: 36 494–36 503

Pioletti M, Schluenzen F, Harms J, Zarivach R, Gluehmann M, Avila H, Bashan A, Bartels H, Auerbach T, Jacobi C, Hartsch T, Yonath A, Franceschi F (2001) Crystal structures of complexes of the small ribosomal subunit with tetracycline, edeine and IF3. Embo J 20: 1829–1839

Poehlsgaard J, Douthwaite S (2005) The bacterial ribosome as a target for antibiotics. Nat Rev Microbiol 3: 870–881

Polacek N, Gaynor M, Yassin A, Mankin AS (2001) Ribosomal peptidyl transferase can withstand mutations at the putative catalytic nucleotide. Nature 411: 498–501

Polacek N, Mankin AS (2005) The ribosomal peptidyl transferase center: structure, function, evolution, inhibition. Crit Rev Biochem Mol Biol 40: 285–311

Pyetan E, Baram D, Auerbach-Nevo T, Yonath A (2007) Chemical parameters influencing fine-tuning in the binding of macrolide antibiotics to the ribosomal tunnel. Pure Appl Chem 79: 955–968

Ramakrishnan V (2008) What we have learned from ribosome structures. Biochem Soc Trans 36: 567–574

Rodnina MV, Beringer M, Wintermeyer W (2007) How ribosomes make peptide bonds. Trends Biochem Sci 32: 20–26

Ryabova LA, Selivanova OM, Baranov VI, Vasiliev VD, Spirin AS (1988) Does the channel for nascent peptide exist inside the ribosome? Immune electron microscopy study. FEBS Lett 226: 255–260

Sabatini DD, Blobel G (1970) Controlled proteolysis of nascent polypeptides in rat liver cell fractions. II. Location of the polypeptides in rough microsomes. J Cell Biol 45: 146–157

Samaha RR, Green R, Noller HF (1995) A base pair between tRNA and 23S rRNA in the peptidyl transferase centre of the ribosome. Nature 377: 309–314

Sato NS, Hirabayashi N, Agmon I, Yonath A, Suzuki T (2006) Comprehensive genetic selection revealed essential bases in the peptidyl-transferase center. Proc Natl Acad Sci USA 103: 15 386–15 391

Schaffitzel C, Ban N (2007) Generation of ribosome nascent chain complexes for structural and functional studies. J Struct Biol 158: 463–471

Schluenzen F, Tocilj A, Zarivach R, Harms J, Gluehmann M, Janell D, Bashan A, Bartels H, Agmon I, Franceschi F, Yonath A

(2000) Structure of functionally activated small ribosomal subunit at 3.3 angstroms resolution. Cell 102: 615–623

Schluenzen F, Wilson DN, Tian P, Harms JM, McInnes SJ, Hansen HA, Albrecht R, Buerger J, Wilbanks SM, Fucini P (2005) The Binding Mode of the Trigger Factor on the Ribosome: Implications for Protein Folding and SRP Interaction. Structure (Camb) 13: 1685–1694

Schmeing TM, Huang KS, Kitchen DE, Strobel SA, Steitz TA (2005) Structural Insights into the Roles of Water and the 2' Hydroxyl of the P Site tRNA in the Peptidyl Transferase Reaction. Mol Cell 20: 437–448

Schmeing TM, Seila AC, Hansen JL, Freeborn B, Soukup JK, Scaringe SA, Strobel SA, Moore PB, Steitz TA (2002) A pretranslocational intermediate in protein synthesis observed in crystals of enzymatically active 50S subunits. Nat Struct Biol 9: 225–230

Schmeing TM, Voorhees RM, Kelley AC, Gao YG, Murphy FVt, Weir JR, Ramakrishnan V (2009) The Crystal Structure of the Ribosome Bound to EF-Tu and Aminoacyl-tRNA. Science 326: 688–694

Schuwirth BS, Borovinskaya MA, Hau CW, Zhang W, Vila-Sanjurjo A, Holton JM, Cate JHD (2005) Structures of the Bacterial Ribosome at 3.5 A Resolution. Science 310: 827–834

Selmer M, Dunham CM, Murphy Iv FV, Weixlbaumer A, Petry S, Kelley AC, Weir JR, Ramakrishnan V (2006) Structure of the 70S Ribosome Complexed with mRNA and tRNA. Science 313: 1935–1942

Sharma PK, Xiang Y, Kato M, Warshel A (2005) What Are the Roles of Substrate-Assisted Catalysis and Proximity Effects in Peptide Bond Formation by the Ribosome? Biochemistry 44: 11 307–11 314

Shevack A, Gewitz HS, Hennemann B, Yonath A, Wittmann HG (1985) Characterization and crystallization of ribosomal particles from Halobacterium marismortui. FEBS Lett 184: 68–71

Simonetti A, Marzi S, Jenner L, Myasnikov A, Romby P, Yusupova G, Klaholz BP, Yusupov M (2009) A structural view of translation initiation in bacteria. Cell Mol Life Sci 66: 423–436

Simonetti A, Marzi S, Myasnikov AG, Fabbretti A, Yusupov M, Gualerzi CO, Klaholz BP (2008) Structure of the 30S translation initiation complex. Nature 455: 416–420

Simonovic M, Steitz TA (2008) Peptidyl-CCA deacylation on the ribosome promoted by induced fit and the O3'-hydroxyl group of A76 of the unacylated A-site tRNA. RNA 14: 2372–2378

Sohmen D, Harms JM, Schlunzen F, Wilson DN (2009) Enhanced SnapShot: Antibiotic inhibition of protein synthesis II. Cell 139: 212–212

Stanley RE, Blaha G, Grodzicki RL, Strickler MD, Steitz TA (2010) The structures of the anti-tuberculosis antibiotics viomycin and capreomycin bound to the 70S ribosome. Nat Struct Mol Biol 17: 289–293

Stark H, Mueller F, Orlova EV, Schatz M, Dube P, Erdemir T, Zemlin F, Brimacombe R, van Heel M (1995) The 70S *Escherichia coli* ribosome at 23 A resolution: fitting the ribosomal RNA. Structure 3: 815–821

Starosta AL, Karpenko VV, Shishkina AV, Mikolajka A, Sumbatyan NV, Schluenzen F, Korshunova GA, Bogdanov AA, Wilson DN (2010) Interplay between the ribosomal tunnel, nascent chain, and macrolides influences drug inhibition. Chem Biol 17: 504–514

Tenson T, Mankin A (2006) Antibiotics and the ribosome. Mol Microbiol 59: 1664–1677

Thompson J, Dahlberg AE (2004) Testing the conservation of the translational machinery over evolution in diverse environments: assaying Thermus thermophilus ribosomes and initiation factors in a coupled transcription-translation system from *Escherichia coli*. Nucleic Acids Res 32: 5954–5961

Thygesen J, Krumbholz S, Levin I, Zaytzevbashan A, Harms J, Bartels H, Schlunzen F, Hansen HAS, Bennett WS, Volkmann N, Agmon I, Eisenstein M, Dribin A, Maltz E, Sagi I, Morlang S, Fua M, Franceschi F, Weinstein S, Boddeker N, Sharon R, Anagnostopoulos K, Peretz M, Geva M, Berkovitchyellin Z, Yonath A (1996) Ribosomal crystallography – from crystal growth to initial phasing. J Cryst Growth 168: 308–323

Trakhanov SD, Yusupov MM, Agalarova SC, Garber MB, Ryazantseva SN, Tischenko SV, Shirokova VA (1987) Crystallization of 70 S ribosomes and 30 S ribosomal subunits from Thermus thermophilus. FEBS Lett 220: 319–322

Trobro S, Aqvist J (2006) Analysis of predictions for the catalytic mechanism of ribosomal peptidyl transfer. Biochemistry 45: 7049–7056

Tu D, Blaha G, Moore PB, Steitz TA (2005) Structures of MLSBK Antibiotics Bound to Mutated Large Ribosomal Subunits Provide a Structural Explanation for Resistance. Cell 121: 257–270

Tuerk C, Gold L (1990) Systematic evolution of ligands by exponential enrichment: RNA ligands to bacteriophage T4 DNA polymerase. Science 249: 505–510

Turk RM, Chumachenko NV, Yarus M (2010) Multiple translational products from a five-nucleotide ribozyme. Proc Natl Acad Sci USA 107: 4585–4589

Uemura S, Aitken CE, Korlach J, Flusberg BA, Turner SW, Puglisi JD (2010) Real-time tRNA transit on single translating ribosomes at codon resolution. Nature 464: 1012–1017

Uemura S, Dorywalska M, Lee TH, Kim HD, Puglisi JD, Chu S (2007) Peptide bond formation destabilizes Shine-Dalgarno interaction on the ribosome. Nature 446: 454–457

Unwin PN, Taddei C (1977) Packing of ribosomes in crystals from the lizard Lacerta sicula. J Mol Biol 114: 491–506

Vogel Z, Vogel T, Elson D, Zamir A (1970) Ribosome activation and the binding of dihydrostreptomycin: effect of polynucleotides and temperature on activation. J Mol Biol 54: 379–386

von Bohlen K, Makowski I, Hansen HA, Bartels H, Berkovitch-Yellin Z, Zaytzev-Bashan A, Meyer S, Paulke C, Franceschi F, Yonath A (1991) Characterization and preliminary attempts for derivatization of crystals of large ribosomal subunits from Haloarcula marismortui diffracting to 3 A resolution. J Mol Biol 222: 11–15

Voorhees RM, Weixlbaumer A, Loakes D, Kelley AC, Ramakrishnan V (2009) Insights into substrate stabilization from snapshots of the peptidyl transferase center of the intact 70S ribosome. Nat Struct Mol Biol 16: 528–533

Voss NR, Gerstein M, Steitz TA, Moore PB (2006) The geometry of the ribosomal polypeptide exit tunnel. J Mol Biol 360: 893–906

Walter P, Johnson AE (1994) Signal sequence recognition and protein targeting to the endoplasmic reticulum membrane. Annu Rev Cell Biol 10: 87–119

Watson JD (1963) Involvement of RNA in the synthesis of proteins. Science 140: 17–26

Weinger JS, Parnell KM, Dorner S, Green R, Strobel SA (2004) Substrate-assisted catalysis of peptide bond formation by the ribosome. Nat Struct Mol Biol 11: 1101–1106

Weixlbaumer A, Jin H, Neubauer C, Voorhees RM, Petry S, Kelley AC, Ramakrishnan V (2008) Insights into translational termination from the structure of RF2 bound to the ribosome. Science 322: 953–956

Wekselman I, Davidovich C, Agmon I, Zimmerman E, Rozenberg H, Bashan A, Berisio R, Yonath A (2008) Ribosome's mode of function: myths, facts and recent results. J Pept Sci 15: 122–130

Wilson DN (2004) Antibiotics and the inhibition of ribosome function. In: Nierhaus KH, Wilson DN (eds) Protein synthesis and ribosome structure. Weinheim, Germany: Wiley-VCH, pp 449–527

Wilson DN, Schluenzen F, Harms JM, Yoshida T, Ohkubo T, Albrecht R, Buerger J, Kobayashi Y, Fucini P (2005) X-ray crystallography study on ribosome recycling: the mechanism of binding and action of RRF on the 50S ribosomal subunit. Embo J 24: 251–260

Wohlgemuth I, Beringer M, Rodnina MV (2006) Rapid peptide bond formation on isolated 50S ribosomal subunits. EMBO Rep 7: 699–703

Wohlgemuth I, Brenner S, Beringer M, Rodnina MV (2008) Modulation of the rate of peptidyl transfer on the ribosome by the nature of substrates. J Biol Chem 283: 32229–32235

Woolhead CA, Johnson AE, Bernstein HD (2006) Translation arrest requires two-way communication between a nascent polypeptide and the ribosome. Mol Cell 22: 587–598

Woolhead CA, McCormick PJ, Johnson AE (2004) Nascent membrane and secretory proteins differ in FRET-detected folding far inside the ribosome and in their exposure to ribosomal proteins. Cell 116: 725–736

Yarus M (2010) Life from an RNA world: The ancestor within. Harvard University Press

Yonath A (2003) Ribosomal tolerance and peptide bond formation. Biol Chem 384: 1411–1419

Yonath A (2005) Antibiotics targeting ribosomes: resistance, selectivity, synergism, and cellular regulation. Annu Rev Biochem 74: 649–679

Yonath A, Bartunik HD, Bartels KS, Wittmann HG (1984) Some X-ray diffraction patterns from single crystals of the large ribosomal subunit from Bacillus stearothermophilus. J Mol Biol 177: 201–206

Yonath A, Bashan A (2004) Ribosomal Crystallography: Initiation, Peptide Bond Formation, and Amino Acid Polymerization are Hampered by Antibiotics. Annu Rev Microbiol 58: 233–251

Yonath A, Glotz C, Gewitz HS, Bartels KS, von Bohlen K, Makowski I, Wittmann HG (1988) Characterization of crystals of small ribosomal subunits. J Mol Biol 203: 831–834

Yonath A, Khavitch G, Tesche B, Muessig J, Lorenz S, Erdmann VA, Wittmann HG (1982a) The nucleation of crystals of the large ribosomal subunits from Bacillus stearothermophilus. Biochem Int 5: 629–636

Yonath A, Leonard KR, Wittmann HG (1987) A tunnel in the large ribosomal subunit revealed by three-dimensional image reconstruction. Science 236: 813–816

Yonath A, Muessig J, Tesche B, Lorenz S, Erdmann VA, Wittmann HG (1980) Crystallization of the large ribosomal subunit from B. stearothermophilus. Biochem Int 1: 315–428

Yonath A, Mussig J, Wittmann HG (1982b) Parameters for crystal growth of ribosomal subunits. J Cell Biochem 19: 145–155

Yonath A, Wittmann HG (1988) Crystallographic and image reconstruction studies on ribosomal particles from bacterial sources. Methods Enzymol 164: 95–117

Youngman EM, Brunelle JL, Kochaniak AB, Green R (2004) The active site of the ribosome is composed of two layers of conserved nucleotides with distinct roles in peptide bond formation and peptide release. Cell 117: 589–599

Yusupov MM, Tischenko SV, Trakhanov SD, Riazantsev SN, Garber MB (1988) A new crystallin form of 30S ribosomal subunits from Thermus thermophilus. FEBS Lett 238: 113–115

Yusupov MM, Yusupova GZ, Baucom A, Lieberman K, Earnest TN, Cate JH, Noller HF (2001) Crystal structure of the ribosome at 5.5 A resolution. Science 292: 883–896

Yusupova G, Jenner L, Rees B, Moras D, Yusupov M (2006) Structural basis for messenger RNA movement on the ribosome. Nature 444: 391–394

Zamir A, Miskin R, Elson D (1971) Inactivation and reactivation of ribosomal subunits: amino acyl-transfer RNA binding activity of the 30s subunit of Escherichia coli. J Mol Biol 60: 347–364

Zarivach R, Bashan A, Berisio R, Harms J, Auerbach T, Schluenzen F, Bartels H, Baram D, Pyetan E, Sittner A, Amit M, Hansen HAS, Kessler M, Liebe C, Wolff A, Agmon I, Yonath A (2004) Functional aspects of ribosomal architecture: symmetry, chirality and regulation. J Phys Org Chem 17: 901–912

Zhang B, Cech TR (1997) Peptide bond formation by in vitro selected ribozymes. Nature 390: 96–100

Zimmerman E, Yonath A (2009) Biological implications of the ribosome's stunning stereochemistry. ChemBioChem 10: 63–72

Ziv G, Haran G, Thirumalai D (2005) Ribosome exit tunnel can entropically stabilize {alpha}-helices. Proc Natl Acad Sci USA 102: 18 956–18 961

Structural studies on decoding, termination and translocation in the bacterial ribosome

2

Venki Ramakrishnan

1. Introduction

With the determination of the atomic structures of the ribosomal subunits in 2000, focus has shifted in the last decade to the study of functional states of the ribosome with a view to helping elucidate the mechanisms underlying the various steps of translation. Some of these studies could be carried out using crystals of the ribosomal subunits, recognition of codon-anticodon pairing by the ribosome during decoding in the 30S subunit, or studies on peptidyl transferase intermediates in the 50S subunit. Others, such as the interaction of elongation or release factors with the ribosome require high-resolution crystal forms of the intact 70S ribosome. In this chapter, we review our studies on decoding using the 30S subunit, which could use crystals of the subunit that diffracted to high resolution. However, other functional studies such as those on elongation and termination required new crystal forms of the 70S ribosome that also diffracted to high-resolution.

The structure at 5.5 Å resolution of the entire 70S ribosome complexed with mRNA and tRNA (Yusupov et al., 2001) was a major achievement and the culmination of over a decade of work that began with the crystallization of the 70S ribosome and its 30S subunit from *Thermus thermophilus* (Trakhanov et al., 1987). However, the limitation of resolution inherent to this crystal form precluded detailed structural studies on functional states of the ribosome. Another significant landmark was the determination of the structure of the *E. coli* ribosome to 3.5 Å resolution (Schuwirth et al., 2005), although, to date, this crystal form has not been able to accommodate full-length tRNAs with mRNA or most factors, thus limiting its utility.

In this context, two new crystal forms of the bacterial 70S ribosome discovered in our laboratory have proved very useful. The first, an orthorhombic crystal

form that diffracts to about 3 Å resolution (Selmer et al., 2006) was very useful in both our and several other laboratories for studies on termination (Laurberg et al., 2008, Weixlbaumer et al., 2008, Korostelev et al., 2008), recycling (Weixlbaumer et al., 2007), peptidyl transferase (Voorhees et al., 2009), various antibiotics (Stanley et al., 2010) and a factor involved in the formation of the first peptide bond (Blaha et al., 2009). However, this crystal form, like the other 70S crystal forms that preceded it (regardless of space group and species), has ribosomal protein L9 extending out from the body of the 50S subunit and interacting with a neighboring 30S subunit in the crystal lattice in such a way that it sterically occludes the binding site for translational GTPases. In our experience, the interaction is strong enough to displace even a stably bound EF-G or EF-Tu; thus the crystals containing L9 we obtained always lacked GTPase factors.

As a result, we decided to delete the portion of the gene coding for the C-terminus of L9 that protruded out from the body of the molecule. The resulting strain, MRC-MSAW1, gave crystals of ribosomes bound to EF-G (Gao et al., 2009) and EF-Tu (Schmeing et al., 2009). In both cases, L9 was not observed in the resulting structure, suggesting that the residual fragment of L9 is either too unstable and is degraded upon synthesis, or does not assemble onto 50S subunits. Nevertheless, this crystal form has paved the way for studies of GTPase factors bound to the ribosome.

We describe below our studies on decoding and other functional aspects of the ribosome, initially using the 30S subunit but subsequently with the two new crystal forms of the 70S ribosome described above. Our focus is on results obtained in our laboratory since our chapter in the last ribosome book (Brodersen et al., 2002) and is not intended to be a comprehensive description of the field.

2. Studies on decoding

The accuracy of translation is much greater than can be expected from the free-energy difference between correct and incorrect base pairs in codon-anticodon interactions (reviewed in Ogle and Ramakrishnan, 2005). The ribosome plays an active role in tRNA selection, and pre-steady-state kinetics showed that the binding of cognate but not near-cognate tRNA induces a conformational change in the ribosome that leads to the acceleration of GTP hydrolysis (reviewed in Rodnina and Wintermeyer, 2001).

2.1. Insights from studies on the 30S subunit

With the determination of the atomic structure of the 30S subunit, it became possible to try to understand decoding in structural terms. The first clue in this direction was the structure of the 30S subunit with paromomycin (Carter et al., 2000). Paromomycin was previously known to bind to an internal loop of 16S RNA in the decoding center (Fourmy et al., 1996). In the 30S subunit, paromomycin induced a change in the conformation of two universally conserved adenosines, A1492 and A1493, so that the bases were displaced from an internal loop of helix 44 in the decoding center and were in a position to interact directly with the codon-anticodon helix (Figures 1A and B; Carter et al., 2000). Subsequently, we determined the structure of the 30S subunit in complex with oligonucleotide mimics of the A-site mRNA codon and the anticodon stem-loop of tRNA (Figure 1C) (Ogle et al., 2001). The binding of cognate tRNA to the 30S subunit not only induced changes in the conformation of the two adenines that were displaced by paromomycin, but also that of G530 on the other side of the decoding center, so that all three bases made close interactions with the minor groove of the codon-anticodon helix (Figure 1C).

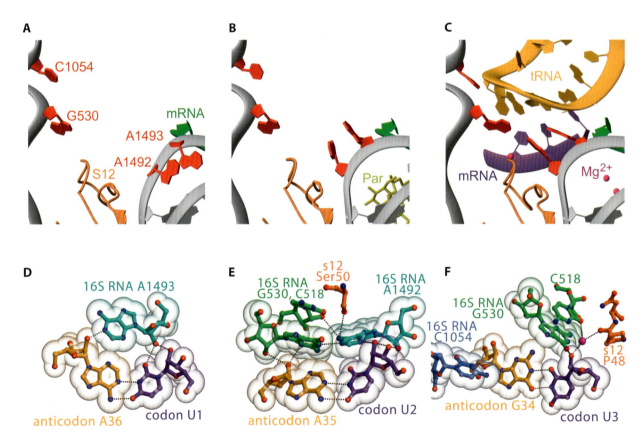

Fig. 1 Changes in the decoding center of the 30S subunit with the binding of paromomycin or tRNA. (A) The decoding center in the empty 30S subunit. (B) The decoding center in the presence of the antibiotic paromomycin. (C) The decoding center in the presence of oligonucleotide mimics of the mRNA codon and the tRNA anti-codon stem-loop. (D, E) Interactions of the three ribosomal bases with the minor groove of the first, second and third base pairs respectively between the codon and anticodon. (Figure adapted from Ogle et al., 2001)

The shape of the minor groove was closely monitored at the first two positions (Figure 1D and E), but not at the third position (Figure 1F), where a GU wobble base pair was readily accommodated. Since the shape of the minor groove is invariant for all Watson-Crick base pairs, including the location of hydrogen-bonding acceptors at identical positions (Seeman et al., 1976), this allows the ribosome to monitor Watson-Crick base pairing at the first two positions while being more tolerant at the third position, thus offering a structural basis for the wobble hypothesis. The minor groove interactions also provide binding energy for cognate tRNA in excess of that for near-cognate tRNAs. The excess free energy has been estimated at 15–20 kJ/mole (Battle and Doudna, 2002), which by itself could explain the accuracy of decoding, except that, as predicted by kinetic data, the ribosome uses much of this binding energy to drive a conformational change that leads to GTP hydrolysis and tRNA selection rather than just to preferentially stabilize cognate tRNA (Pape et al., 1999).

The second consequence of tRNA binding appears to be a global conformational change in the 30S subunit (Ogle et al., 2001). A subsequent study with near-cognate tRNAs showed that although they bound to the 30S subunit with only slightly less affinity than cognate tRNA, they failed to induce this global conformational change unless the antibiotic paromomycin was also present (Ogle et al., 2002). This study provided a structural understanding of previous kinetic data that showed that paromomycin worked primarily by helping to accelerate GTP hydrolysis for near-cognate tRNAs rather than stabilizing their binding (Pape et al., 2000).

The global conformational changes induced in the 30S subunit involve a movement of the shoulder and platform domains relative to the rest of the 30S subunit (Figure 2). The movement would involve disruption or formation of contacts between the two domains in going from the empty (open) form to the tRNA-bound (closed) form. Interestingly, an analysis of a large body of biochemical and genetic data suggests that mutations such as the *ram* mutations in S4 and S5 that would make it easier to reach the closed form increase the error rate of the ribosome, while those such as the streptomycin resistance mutations in S12 that make it more difficult to reach the closed form result in lowering the error rate. The antibiotic streptomycin stabilizes the 30S subunit in a conformation similar to the closed form, thereby increasing error rates. This led to the proposal that the closed

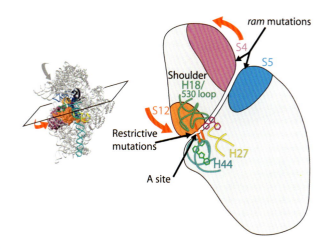

Fig. 2 Conformational changes and domain closure in the 30S ribosomal subunit effected by antibiotics or accuracy mutations of the ribosome. On the left is an overview of the 30S subunit showing the directions of the conformational changes in the domains. On the right is a schematic cross-section of the 30S subunit (as defined by the plane cutting the 30S subunit on the left), in the region of the decoding centre and proteins S4 (violet), S5 (blue) and S12 (orange). G530 and A1492/3 are represented by red bars; helices H44, H27 and H18 (with the G530-loop) are cyan, yellow and turquoise, respectively. The rotation of the shoulder domain (red arrows) during the transition to the closed 30S conformation disrupts an interface between S4 and S5, while the H18/530-loop/S12 region forms new contacts to H27 and H44. Mutations in these regions either increase or decrease mRNA misreading. Paromomycin (dark green rings) and streptomycin (dark pink rings) induce translational errors by facilitating domain closure. (Figure reproduced from Ogle et al., 2003)

form of the 30S subunit is required for tRNA selection (Ogle et al., 2002).

In conjunction with cryo-EM structures of EF-Tu bound to the ribosome (Stark et al., 1997) (Valle et al., 2002), it could be seen that domain closure upon codon recognition involved a movement of the shoulder domain of the 30S subunit towards the ternary complex of EF-Tu and tRNA. The tRNA in the cryo-EM structure was shown to be bent in the anticodon stem (Valle et al., 2002). This led to an integrated model for decoding which proposed that the additional energy from the interaction with the minor groove of the Watson-Crick codon-anticodon base pairs would be used to induce a domain closure in the 30S subunit that would be essential to reach the state required for GTP hydrolysis by EF-Tu (Ogle et al., 2002). After GTP hydrolysis and release of EF-Tu from the ribosome, the distorted tRNA would then relax into the peptidyl transferase center during accommodation.

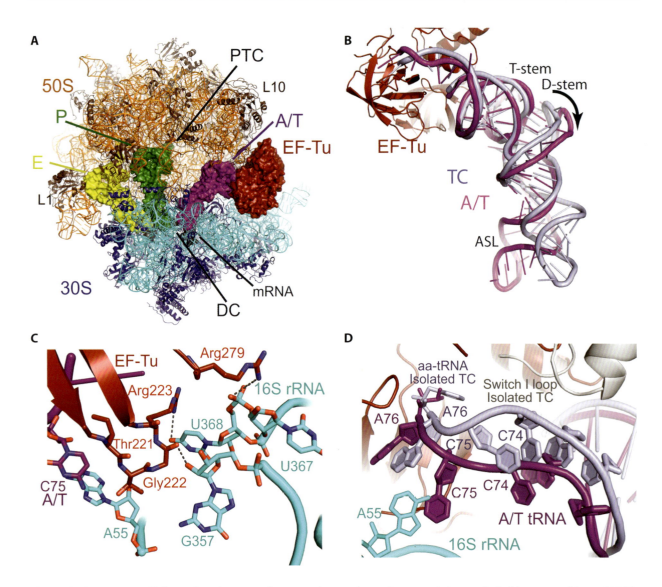

Fig. 3 The kirromycin-stalled complex of EF-Tu and aminoacyl tRNA with the ribosome. (A) Overview of the structure showing the three tRNAs and EF-Tu, the peptidyl transferase center (PTC) and the decoding center (DC). (B) Comparison of the ternary complex in the ribosome (purple) with that in isolation (slate blue), showing the distortion in the anticodon and D stems. (C) Detail showing how domain closure of the shoulder of the 30S subunit and a conformational change of a highly conserved loop of EF-Tu stabilize an interaction between EF-Tu and 16S RNA. (D) Detail showing how a stacking interaction between A55 of 16S RNA and C75 of tRNA stabilize a 3' end of tRNA that is displaced by as much as 6 Å relative to its conformation in the isolated ternary complex (gray). (Figure reproduced from Schmeing et al., 2009)

2.2. Structures of the ribosome with elongation factor Tu and aminoacyl tRNA

It became clear that despite increasingly higher resolution models of the EF-Tu-tRNA-ribosome complex determined by cryo-EM, the molecular details of decoding would require a high-resolution structure, currently reachable only by crystallography. Crystallization attempts of this complex over several years were unsuccessful. However, the L9-deletion mutant ribosomes described earlier were amenable to crystallization of GTPase factors with the ribosome (Gao et al., 2009) and were used to crystallize and solve the structure of a kirromycin-stalled complex of EF-Tu with aminoacyl-tRNA and the ribosome (Figure 3A) (Schmeing et al., 2009). Kirromycin prevents the large conformational change in EF-Tu between the GTP and GDP forms, thus trapping EF-Tu on the ribosome after GTP hydrolysis.

The detailed structure of the EF-Tu-tRNA-ribosome complex clarified many issues. The distortion in the tRNA was visible in detail and involved an untwisting accompanied by a widening of the groove in the anticodon-stem as well as a movement of the D loop (Figure 3B). The movement of the shoulder domain of the 30S subunit predicted from earlier studies on the subunit structure was accompanied by a movement of a highly conserved loop in EF-Tu that would stabilize the conformation of EF-Tu on the ribosome (Figure 3C). A striking observation was that the 3' end of tRNA was pulled away by almost 6 Å compared to the structure of the ternary complex in the absence of the ribosome. This altered conformation was stabilized by a stacking interaction with a base in the 30S subunit (Figure 3D), and would disrupt interactions between the 3' end and the switch I helix, which was disordered in the structure.

The structure shed light on how codon recognition could lead to GTP hydrolysis on EF-Tu and thus influence tRNA selection. The energy from the minor-groove interactions of residues of the decoding center with the codon-anticodon helix was used to stabilize the ternary complex on the ribosome, which consisted of the distorted form of tRNA as well as changes in EF-Tu. The distortion at the 3' end of the tRNA would destabilize the switch I loop, which according to previous suggestions would lead to exposure of the gamma phosphate of GTP to water and subsequent hydrolysis. After phosphate release, the GDP form of EF-Tu would have a significantly altered conformation, resulting in loss of many of its contacts with the ribosome, leading to its release. After EF-Tu release, there would be nothing to stabilize the tRNA in the distorted form; however since it would still be bound tightly to the decoding center through its anticodon loop, its 3' end would relax into the peptidyl transferase center thus initiating peptidyl transfer.

Because the kirromycin-stalled structure represented the state after GTP hydrolysis, it did not reveal much about the detailed mechanism of catalysis. Indeed, the catalytic histidine was in the "inactive" conformation, facing away from the GDP and making an interaction with the sarcin-ricin loop (SRL). The role of the SRL in catalysis was unclear.

The mechanism of GTP hydrolysis has more recently been investigated by a structure of the EF-Tu-tRNA complex bound with a GTP analogue, GDPCP, in complex with the ribosome (Voorhees et al., 2010). This structure captures the catalytic histidine of EF-Tu in the activated form (Figure 4A), where it coordinates the catalytic water molecule that interacts with the gamma phosphate of GDPCP (Figure 4B, middle panel). The other nitrogen of the histidine interacts with the phosphate oxygen of A2662 of the SRL. This is precisely the nucleotide that is the target of the nuclease toxin α-sarcin. An unexpected observation is that switch I is not disordered in the structure even though the 3' end of tRNA is distorted as in the kirromycin-stalled complex. Thus loss of contacts with the 3' end of tRNA is not sufficient for switch I to become disordered. Also, it appears that the disordering of switch I to open a "hydrophobic gate" is not necessary for the repositioning of the catalytic histidine and water molecule as has been previously proposed (Vogeley et al., 2001) (Sengupta et al., 2008, Villa et al., 2009, Schuette et al., 2009) (Schmeing et al., 2009). Rather, subtle movements of elements of EF-Tu around the GTP (Figure 4A) may facilitate the positioning of His84 and the water to coordinate with the gamma phosphate. The role of the SRL appears to be to hold the catalytic histidine in the active conformation. Interestingly, the conformation of the SRL itself is essentially identical to that found in factor-free structures of the ribosome. The structure rationalizes the universal importance of the SRL and a common catalytic mechanism for all ribosomal GTPase factors. The difference in their action can be attributed to the differences in the conformation of the ribosome in various states, so that each state would allow only a particular factor to be positioned with the SRL in exactly the right position to activate the catalytic histidine.

3. Studies on termination

The three stop codons are decoded by the so-called "class-I" release factors. In bacteria, there are two factors, RF1 and RF2, which have overlapping specificity: RF1 recognizes UAG, RF2 recognizes UGA, whereas both recognize UAA. The high-resolution crystal form of the 70S ribosome (Selmer et al., 2006) made it possible to obtain high-resolution structures of the ribosome complexed with either RF1 (Laurberg et al., 2008) or RF2 (Weixlbaumer et al., 2008, Korostelev et al., 2008, Korostelev et al., 2010). Taken together, these structures provided a structural basis for understanding the specificity of stop-codon recognition by RF1 and RF2. However, the structures did not make it entirely clear why the substitution of a tripeptide motif would switch the specificity of RF1 and RF2 (Ito et al., 2000), nor why a single mutation at a location that did not interact directly with the stop codon would cause a release factor to read all three stop codons (Ito et al., 1998). Thus although the basis of specificity is much better understood, more work is still required.

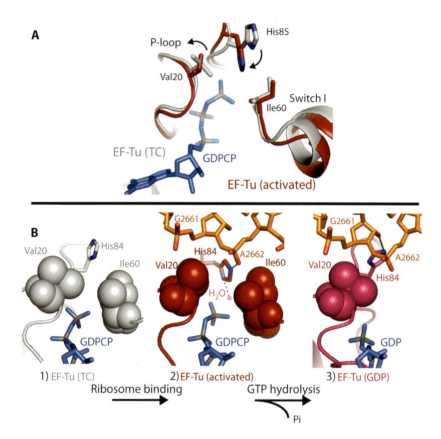

Fig. 4 The catalytic site of GTP hydrolysis in the structure of a complex of EF-Tu and aminoacyl tRNA bound to the ribosome in the presence of the nonhydrolyzable analog GDPCP. (A) Details of the catalytic site. A comparison with the isolated ternary complex (gray) shows that only small changes occur in Switch I and the P loop on binding to the ribosome, but the catalytic histidine has nevertheless assumed its active conformation. (B) Comparison of the catalytic site in (1) the isolated ternary complex, (2) the complex with GDPCP representing the state just before GTP hydrolysis, and (3) the kirromycin-stalled complex after GTP hydrolysis.

The highly conserved GGQ motif of class-I release factors has been implicated in catalysis (Frolova et al., 1999, Song et al., 2000). As expected from earlier cryo-EM and low-resolution crystal structures (Klaholz et al., 2003, Rawat et al., 2003, Petry et al., 2005), this motif is found in the heart of the peptidyl transferase center (PTC). The conservation of the glycines is rationalized because only glycine would allow the loop containing the GGQ motif to adopt the required conformation. However, the role of the glutamine was not entirely clear; it was proposed to coordinate the catalytic water molecule that acted as the nucleophile for cleavage of the ester bond connecting the peptide chain to the tRNA in peptidyl-tRNA (Song et al., 2000). Although highly conserved, the glutamine can be mutated with only a modest effect on the rate of catalysis (Mora et al., 2003). However, mutation of the glutamine specifically affects the activity of water as compared to other nucleophiles (Shaw and Green, 2007). Molecular dynamics calculations suggest that when glutamine is mutated to an alanine (Trobro and Aqvist, 2007) a second water molecule can adopt the role of the glutamine side chain. These studies suggest that despite the mutational data, the glutamine plays an important role in catalysis, explaining its conservation in all kingdoms of life.

The initial structures of class-I release factors on the ribosome all represent the state after catalysis and peptide release, in which the P-site tRNA was deacylated. More recently, the structure of RF2 bound to the ribosome with peptidyl-tRNA containing an amide-linked amino acid was determined (Jin et al., 2010). This complex mimicked the substrate complex and allowed a snapshot of the state just before hydrolysis. A water molecule was observed in the catalytic site but not positioned for catalysis, perhaps because the amide linkage is chemically different from the ester linkage of the true substrate. The tight packing of the GGQ motif with the substrates in the catalytic site supported the role of the glutamine in specifically activating water over other nucleophiles, as previously suggested (Shaw and Green, 2007).

Both the post-release states and the substrate complex confirmed that conformational changes induced by RF2 binding in the peptidyl transferase center expose the peptidyl ester for attack by a water molecule (Weixlbaumer et al., 2008, Jin et al., 2010). This is exactly as predicted on the basis of studies of A-site tRNA binding to the ribosome (Schmeing et al., 2005).

4. The peptidyl transferase center in the 70S ribosome

Although most studies on peptidyl transferase have been carried out in the context of the 50S subunit, questions have been raised about whether the PTC has a different conformation in the 70S ribosome compared to the 50S subunit (Schuwirth et al., 2005, Korostelev et al., 2006). No protein was observed within 18 Å of a transition state analogue in the PTC of the archaeal *Haloarcula marismortui* 50S subunit (Nissen et al., 2000). However, in bacteria, protein L27 was shown to crosslink to the 3' end of both A- and P-site tRNAs (Wower et al., 2000). Subsequently, it was shown that deletion of L27 or even just the N-terminal three residues lowered peptidyl transferase activity (Maguire et al., 2005).

The high-resolution structure of the 70S ribosome with mRNA and tRNAs showed weak density for the backbone of L27 that reached all the way into the peptidyl transferase center (Selmer et al., 2006), where it could potentially interact with both A- and P-site tRNAs, consistent with previous biochemical data. In that structure, the 3' end of the A-site tRNA was disordered owing to a combination of its being deacylated and a low concentration of Mg^{2+} ions in the cryoprotection buffer. Subsequently, aminoacylated tRNAs were used to study intermediates in peptidyl transfer (Voorhees et al., 2009). These structures were the first high-resolution structures of the ribosome containing full-length tRNA in both P and A sites, and showed that the conformation of the PTC in both the 70S bacterial ribosome and the archaeal 50S subunit was identical within error. Moreover, the induced fit in the PTC observed upon A-site substrate binding in the 50S subunit (Schmeing et al., 2005) was observed in the context of the entire 70S ribosome using full-length tRNAs. Finally, the N-terminal tail of L27 was well-ordered, and details of its interactions with tRNAs in the PTC could be seen. These interactions suggested that L27 stabilized the tRNA substrates in

the PTC and thus increased the rate of peptidyl transfer as previously observed.

In the presence of A-site tRNA, protein L16 becomes better ordered and makes interactions with the backbone of the elbow of the tRNA. These interactions occur between conserved residues in L16 and the backbone of the tRNA and are thus possible for all tRNAs. They rationalize previous biochemical data suggesting a role for L16 in promoting A-site tRNA binding and peptidyl transferase activity (Moore et al., 1975, Kazemie, 1976).

5. Structure of a post-translocation complex of the ribosome with elongation factor G

Translocation involves the movement of tRNAs and mRNA through the ribosome during the elongation cycle, to position a new codon in the A site and enable the ribosome to accept the next incoming aminoacyl tRNA. It is now widely accepted that translocation proceeds via an intermediate of the ribosome involving hybrid states (Bretscher, 1968) in which the tRNAs have moved with respect to the 50S but not the 30S subunit (Moazed and Noller, 1989). Translocation is completed by the movement of the tRNAs and mRNA with respect to the 30S subunit, and this movement requires the action of the GTPase elongation factor G (EF-G) (Moazed and Noller, 1989). Although slow translocation occurs without GTP hydrolysis, or indeed even without factors under certain conditions (Gavrilova et al., 1976), pre-steady-state kinetic studies have shown that GTP hydrolysis by EF-G accelerates translocation (Rodnina et al., 1997). The formation of the hybrid states to which EF-G presumably binds is accompanied by a ratcheting of the two subunits relative to each other by about 6 degrees (Frank and Agrawal, 2000). The antibiotic fusidic acid traps EF-G on the ribosome, presumably by inhibiting a conformational change in EF-G after GTP hydrolysis and thus preventing its release from the ribosome, in a manner analogous to the way kirromycin traps EF-Tu on the ribosome. Using the new monoclinic crystal form of the L9 deletion mutant 70S ribosomes, we were able to crystallize the EF-G-fusidic acid complex on the ribosome with mRNA and P-site tRNA (Figure 6) (Gao et al., 2009).

This structure reveals details of interactions of EF-G with the ribosome in the post-translocation state. Unlike kirromycin for EF-Tu, fusidic acid has

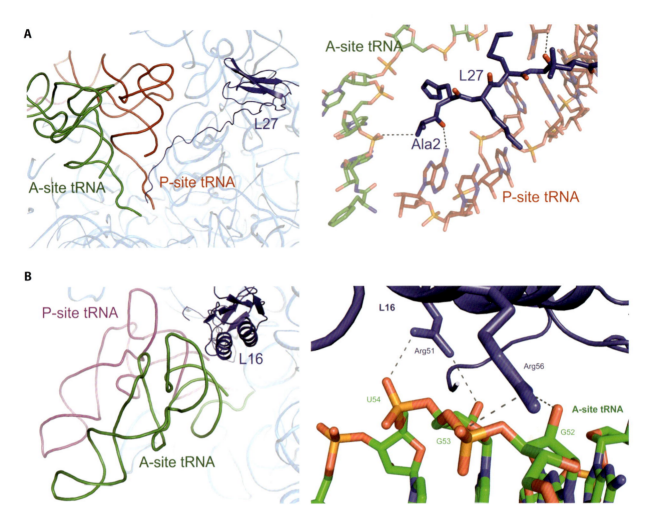

Fig. 5 Proteins in the PTC. (A) Overview (left) and details (right) of the interaction of L27 with the tRNA substrates in the PTC. (B) Overview (left) and details (right) of the interaction of L16 with A-site tRNA. (Figure adapted from Voorhees et al., 2009)

very low affinity for free EF-G, so the structure also reveals for the first time the location and interactions of fusidic acid with the factor. Domain IV of EF-G is in a conformation different from the GTP or GDP forms of isolated EF-G; its orientation may be the result of fusidic acid locking EF-G in the post-translocation state. Apart from its global orientation, the local conformation of the tip of domain IV is also altered, allowing it to interact with the minor groove of the P-site codon and anticodon. The domain does not make interactions with the A-site codon. This is not surprising, since before translocation, the A-site codon would be in the entrance tunnel in the 30S subunit and not necessarily accessible to the factor. Moreover, the P-site anticodon and codon would have been in the A site prior to translocation. One possibility is that domain IV interacts with the codon and anticodon initially in the A site, and these contacts are maintained through

their translocation into the P site. If so, domain IV would have to displace the ribosomal bases A1492, A1493 and G530 from the minor groove of the codon-anticodon helix at the decoding center. This may explain why certain antibiotics that stabilize those bases in the minor groove also prevent translocation (Peske et al., 2004).

Although fusidic acid could not be placed unambiguously, the observed difference Fourier density clarified its location and restricted its orientation to a limited number of possibilities. The interactions of fusidic acid with EF-G are completely consistent with mutational data on fusidic acid resistance and hypersensitivity (Laurberg et al., 2000, Hansson et al., 2005). The main effect of fusidic acid appears to be to lock switch II of EF-G in a conformation similar to the GTP form even after GTP hydrolysis. As a result, transmission of conformational changes to domains III and IV is prevented, trapping EF-G on the ribosome.

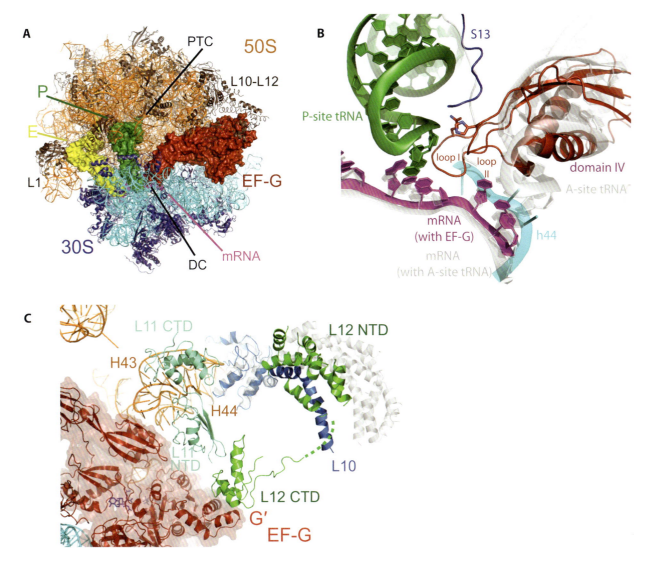

Fig. 6 The complex of EF-G and fusidic acid with the ribosome in the post-translocational state. (A) Overview of the structure. (B) Details showing the interaction of the tip of domain IV with the minor groove of the P-site codon-anticodon helix. A superposition of the A-site tRNA and mRNA (gray) is also shown. (C) A view of the interaction of EF-G with elements of the ribosome implicated in GTPase activity, including the N-terminal domain (NTD) of L11, and the C-terminal domain (CTD) of L12. A comparison with the isolated L10-(L12)$_4$ stalk (gray, Diaconu et al., 2005) shows that the stalk is shifted towards EF-G in the structure. (Figure reproduced from Gao et al., 2009)

Finally, the L10-(L12)$_6$ heptameric stalk can be visualized in the structure. In previous crystal structures of the ribosome, this stalk is almost totally disordered. The base of the stalk, consisting of L10 and several copies of the N-terminal domain of L12, can be seen in maps calculated to low resolution. The stalk appears to be bent towards EF-G as compared to the structure of the isolated stalk (Diaconu et al., 2005). A copy of the C-terminal domain of L12 can also be seen, but, surprisingly, in a location different from that seen in the 50S subunit bound with micrococcin (Harms et al., 2008). The significance of these two distinct locations of the C-terminal domain and the precise role of the domain in promoting GTP hydrolysis by EF-G remains to be determined.

6. A structure of a ribosome recycling factor

After termination by release factors, the ribosome is left with mRNA and a deacylated tRNA in the P site. This complex is disassembled by ribosome release factor (RRF) acting in concert with EF-G (Hirashima and Kaji, 1973). We have determined the structure of RRF in the ribosome in the presence of mRNA and the anticodon stem-loop of P-site tRNA (Weixlbaumer et al., 2007). This structure showed an orientation of RRF similar to that seen in crystal structures with the *E. coli* ribosome (Borovinskaya et al., 2007) or the 50S subunit (Wilson et al., 2005), as well as in previous cryo-EM studies (Agrawal et al., 2004, Gao et al., 2005). Although the 70S structures are all of an unratcheted ribosome, none of them is compatible with the presence of a tRNA in the 50S P site even though a deacylated tRNA is present after peptide release. Thus it is not clear that any of the structures address what is happening *in vivo*. On the other hand, if physiologically RRF binds to a state of the ribosome in which the subunits are ratcheted (or rotated relative to each other) so that the deacylated tRNA is in a P/E hybrid state that would help explain why EF-G is required in recycling. However, to date, no such structure has been shown, and it is clear that we do not have a structural understanding of recycling despite several studies.

7. Conclusions

Breakthroughs in the crystallization of functional complexes of the ribosome have led to the structures of a number of important states in the translational pathway. Nevertheless, there are still no high-resolution structures of initiation complexes, or those of the ratcheted ribosome in the pre-translocational state or during recycling These remain important goals in the structural biology of bacterial translation.

Acknowledgements

Our work is supported by the Medical Research Council (UK), the Wellcome Trust, the Agouron Institute and the Louis-Jeantet Foundation. In the past, I have also been generously supported by the U. S. National Institutes of Health. I am grateful to several generations of students, postdoctoral fellows, and research assistants without whom there would be nothing to report from my laboratory.

References

Agrawal RK, Sharma MR, Kiel MC, Hirokawa G, Booth TM, Spahn CM, Grassucci RA, Kaji A, Frank J (2004) Visualization of ribosome-recycling factor on the *Escherichia coli* 70S ribosome: functional implications. Proc Natl Acad Sci USA 101: 8900–8905

Battle DJ, Doudna JA (2002) Specificity of RNA-RNA helix recognition. Proc Natl Acad Sci 99: 11676–11681

Blaha G, Stanley RE, Steitz TA (2009) Formation of the first peptide bond: the structure of EF-P bound to the 70S ribosome. Science 325: 966–970

Borovinskaya MA, Pai RD, Zhang W, Schuwirth BS, Holton JM, Hirokawa G, Kaji H, Kaji A, Cate JH (2007) Structural basis for aminoglycoside inhibition of bacterial ribosome recycling. Nat Struct Mol Biol 14: 727–732

Bretscher MS (1968) Translocation in protein synthesis: a hybrid structure model. Nature 218: 675–677

Brodersen DE, Carter AP, Clemons WM, Jr., Morgan-Warren RJ, Murphy FVIV, Ogle JM, Tarry MJ, Wimberly B, Ramakrishnan V (2002) Atomic structures of the 30S subunit and its complexes with ligands and antibiotics. Cold Spring Harb Symp Quant Biol 66: 17–32

Carter AP, Clemons WM, Jr., Brodersen DE, Morgan-Warren RJ, Wimberly BT, Ramakrishnan V (2000) Functional insights from the structure of the 30S ribosomal subunit and its interactions with antibiotics. Nature 407: 340–348

Diaconu M, Kothe U, Schlunzen F, Fischer N, Harms JM, Tonevitsky AG, Stark H, Rodnina MV, Wahl MC (2005) Structural basis for the function of the ribosomal L7/12 stalk in factor binding and GTPase activation. Cell 121: 991–1004

Fourmy D, Recht MI, Blanchard SC, Puglisi JD (1996) Structure of the A site of *Escherichia coli* 16S ribosomal RNA complexed with an aminoglycoside antibiotic. Science 274: 1367–1371

Frank J, Agrawal RK (2000) A ratchet-like inter-subunit reorganization of the ribosome during translocation. Nature 406: 319–332

Frolova LY, Tsivkovskii RY, Sivolobova GF, Oparina NY, Serpinsky OI, Blinov VM, Tatkov SI, Kisselev LL (1999) Mutations in the highly conserved GGQ motif of class 1 polypeptide release factors abolish ability of human eRF1 to trigger peptidyl-tRNA hydrolysis. RNA 5: 1014–1020

Gao N, Zavialov AV, Li W, Sengupta J, Valle M, Gursky RP, Ehrenberg M, Frank J (2005) Mechanism for the disassembly of the posttermination complex inferred from cryo-EM studies. Mol Cell 18: 663–674

Gao YG, Selmer M, Dunham CM, Weixlbaumer A, Kelley AC, Ramakrishnan V (2009) The structure of the ribosome with elongation factor G trapped in the posttranslocational state. Science 326: 694–699

Gavrilova LP, Kostiashkina OE, Koteliansky VE, Rutkevitch NM, Spirin AS (1976) Factor-free ("non-enzymic") and factor-dependent systems of translation of polyuridylic acid by *Escherichia coli* ribosomes. J Mol Biol 101: 537–552

Hansson S, Singh R, Gudkov AT, Liljas A, Logan DT (2005) Structural insights into fusidic acid resistance and sensitivity in EF-G. J Mol Biol 348: 939–949

Harms JM, Wilson DN, Schluenzen F, Connell SR, Stachelhaus T, Zaborowska Z, Spahn CM, Fucini P (2008) Translational regulation via L11: molecular switches on the ribosome turned on and off by thiostrepton and micrococcin. Mol Cell 30: 26–38

Hirashima A, Kaji A (1973) Role of elongation factor G and a protein factor on the release of ribosomes from messenger ribonucleic acid. J Biol Chem 248: 7580–7587

Ito K, Uno M, Nakamura Y (1998) Single amino acid substitution in prokaryote polypeptide release factor 2 permits it to terminate translation at all three stop codons. Proc Natl Acad Sci USA 95: 8165–8169

Ito K, Uno M, Nakamura Y (2000) A tripeptide 'anticodon' deciphers stop codons in messenger RNA. Nature 403: 680–684

Jin H, Kelley AC, Loakes D, Ramakrishnan V (2010) Structure of the 70S ribosome bound to release factor 2 and a substrate analog provides insights into catalysis of peptide release. Proc Natl Acad Sci USA 107: 8593–8598

Kazemie M (1976) Binding of aminoacyl-tRNA to reconstituted subparticles of Escherichia coli large ribosomal subunits. Eur J Biochem 67: 373–378

Klaholz BP, Pape T, Zavialov AV, Myasnikov AG, Orlova EV, Vestergaard B, Ehrenberg M, van Heel M (2003) Structure of the Escherichia coli ribosomal termination complex with release factor 2. Nature 421: 90–94

Korostelev A, Zhu J, Asahara H, Noller HF (2010) Recognition of the amber UAG stop codon by release factor RF1. EMBO J 29: 2577–2585

Korostelev A, Asahara H, Lancaster L, Laurberg M, Hirschi A, Zhu J, Trakhanov S, Scott WG, Noller HF (2008) Crystal structure of a translation termination complex formed with release factor RF2. Proc Natl Acad Sci USA 105: 19684–19689

Korostelev A, Trakhanov S, Laurberg M, Noller HF (2006) Crystal structure of a 70S ribosome-tRNA complex reveals functional interactions and rearrangements. Cell 126: 1065–1077

Laurberg M, Kristensen O, Martemyanov K, Gudkov AT, Nagaev I, Hughes D, Liljas A (2000) Structure of a mutant EF-G reveals domain III and possibly the fusidic acid binding site. J Mol Biol 303: 593–603

Laurberg M, Asahara H, Korostelev A, Zhu J, Trakhanov S, Noller HF (2008) Structural basis for translation termination on the 70S ribosome. Nature 454: 852–857

Maguire BA, Beniaminov AD, Ramu H, Mankin AS, Zimmermann RA (2005) A protein component at the heart of an RNA machine: the importance of protein l27for the function of the bacterial ribosome. Mol Cell 20: 427–435

Moazed D, Noller HF (1989) Intermediate states in the movement of transfer RNA in the ribosome. Nature 342: 142–148

Moore VG, Atchison RE, Thomas G, Moran M, Noller HF (1975) Identification of a ribosomal protein essential for peptidyl transferase activity. Proc Natl Acad Sci USA 72: 844–848

Mora L, Heurgue-Hamard V, Champ S, Ehrenberg M, Kisselev LL, Buckingham RH (2003) The essential role of the invariant GGQ motif in the function and stability in vivo of bacterial release factors RF1 and RF2. Mol Microbiol 47: 267–275

Nissen P, Hansen J, Ban N, Moore PB, Steitz TA (2000) The structural basis of ribosome activity in peptide bond synthesis. Science 289: 920–930

Ogle JM, Brodersen DE, Clemons WM, Jr., Tarry MJ, Carter AP, Ramakrishnan V (2001) Recognition of cognate transfer RNA by the 30S ribosomal subunit. Science 292: 897–902

Ogle JM, Carter AP, Ramakrishnan V (2003) Insights into the decoding mechanism from recent ribosome structures. Trends Biochem Sci 28: 259–266

Ogle JM, Murphy FV, Tarry MJ, Ramakrishnan V (2002) Selection of tRNA by the ribosome requires a transition from an open to a closed form. Cell 111: 721–732

Ogle JM, Ramakrishnan V (2005) Structural Insights into Translational Fidelity. Ann Rev Biochem 74.: 129–177

Pape T, Wintermeyer W, Rodnina M (1999) Induced fit in initial selection and proofreading of aminoacyl-tRNA on the ribosome. Embo J 18: 3800–3807

Pape T, Wintermeyer W, Rodnina MV (2000) Conformational switch in the decoding region of 16S rRNA during aminoacyl-tRNA selection on the ribosome. Nat Struct Biol 7: 104–107

Peske F, Savelsbergh A, Katunin VI, Rodnina MV, Wintermeyer W (2004) Conformational changes of the small ribosomal subunit during elongation factor G-dependent tRNA-mRNA translocation. J Mol Biol 343: 1183–1194

Petry S, Brodersen DE, Murphy FVt, Dunham CM, Selmer M, Tarry MJ, Kelley AC, Ramakrishnan V (2005) Crystal structures of the ribosome in complex with release factors RF1 and RF2 bound to a cognate stop codon. Cell 123: 1255–1266

Rawat UB, Zavialov AV, Sengupta J, Valle M, Grassucci RA, Linde J, Vestergaard B, Ehrenberg M, Frank J (2003) A cryo-electron microscopic study of ribosome-bound termination factor RF2. Nature 421: 87–90

Rodnina MV, Savelsbergh A, Katunin VI, Wintermeyer W (1997) Hydrolysis of GTP by elongation factor G drives tRNA movement on the ribosome. Nature 385: 37–41

Rodnina MV, Wintermeyer W (2001) Fidelity of aromanoacyl-tRNA selection on the ribosome: kinetic and structural mechanisms. Annu Rev Biochem 70: 415–435

Schmeing TM, Huang KS, Strobel SA, Steitz TA (2005) An induced-fit mechanism to promote peptide bond formation and exclude hydrolysis of peptidyl-tRNA. Nature 438: 520–524

Schmeing TM, Voorhees RM, Kelley AC, Gao YG, Murphy FVt, Weir JR, Ramakrishnan V (2009) The crystal structure of the ribosome bound to EF-Tu and aminoacyl-tRNA. Science 326: 688–694

Schuette JC, Murphy FVt, Kelley AC, Weir JR, Giesebrecht J, Connell SR, Loerke J, Mielke T, Zhang W, Penczek PA, Ramakrishnan V, Spahn CM (2009) GTPase activation of elongation factor EF-Tu by the ribosome during decoding. Embo J 28: 755–765

Schuwirth BS, Borovinskaya MA, Hau CW, Zhang W, Vila-Sanjurjo A, Holton JM, Cate JH (2005) Structures of the bacterial ribosome at 3.5 A resolution. Science 310: 827–834

Seeman NC, Rosenberg JM, Rich A (1976) Sequence-specific recognition of double helical nucleic acids by proteins. Proc Natl Acad Sci 73: 804–808

Selmer M, Dunham CM, Murphy FVt, Weixlbaumer A, Petry S, Kelley AC, Weir JR, Ramakrishnan V (2006) Structure of the 70S ribosome complexed with mRNA and tRNA. Science 313: 1935–1942

Sengupta J, Nilsson J, Gursky R, Kjeldgaard M, Nissen P, Frank J (2008) Visualization of the eEF2–0S ribosome transition-state complex by cryo-electron microscopy. J Mol Biol 382: 179–187

Shaw JJ, Green R (2007) Two distinct components of release factor function uncovered by nucleophile partitioning analysis. Mol Cell 28: 458–467

Song H, Mugnier P, Das AK, Webb HM, Evans DR, Tuite MF, Hemmings BA, Barford D (2000) The crystal structure of human eukaryotic release factor eRF1– mechanism of stop codon recognition and peptidyl-tRNA hydrolysis. Cell 100: 311–321

Stanley RE, Blaha G, Grodzicki RL, Strickler MD, Steitz TA (2010) The structures of the anti-tuberculosis antibiotics viomycin and capreomycin bound to the 70S ribosome. Nat Struct Mol Biol

Stark H, Rodnina MV, Rinke-Appel J, Brimacombe R, Wintermeyer W, van Heel M (1997) Visualization of elongation factor Tu on the Escherichia coli ribosome. Nature 389: 403–406

Trakhanov SD, Yusupov MM, Agalarov SC, Garber MB, Ryazantsev SN, Tischenko SV, Shirokov VA (1987) Crystallization of

70 S ribosomes and 30S ribosomal subunits from *Thermus thermophilus*. FEBS Lett 220: 319–322

Trobro S, Aqvist J (2007) A model for how ribosomal release factors induce peptidyl-tRNA cleavage in termination of protein synthesis. Mol Cell 27: 758–766

Valle M, Sengupta J, Swami NK, Grassucci RA, Burkhardt N, Nierhaus KH, Agrawal RK, Frank J (2002) Cryo-EM reveals an active role for aminoacyl-tRNA in the accommodation process. Embo J 21: 3557–3567

Villa E, Sengupta J, Trabuco LG, LeBarron J, Baxter WT, Shaikh TR, Grassucci RA, Nissen P, Ehrenberg M, Schulten K, Frank J (2009) Ribosome-induced changes in elongation factor Tu conformation control GTP hydrolysis. Proc Natl Acad Sci USA 106: 1063–1068

Vogeley L, Palm GJ, Mesters JR, Hilgenfeld R (2001) Conformational change of elongation factor Tu (EF-Tu) induced by antibiotic binding. Crystal structure of the complex between EF-Tu. GDP and aurodox. J Biol Chem 276: 17149–17155

Voorhees RM, Weixlbaumer A, Loakes D, Kelley AC, Ramakrishnan V (2009) Insights into substrate stabilization from snapshots of the peptidyl transferase center of the intact 70S ribosome. Nat Struct Mol Biol 16: 528–533

Voorhees RM, Schmeing TM, Kelley AC, Ramakrishnan V. (2010) The mechanism for activation of GTP hydrolysis on the ribosome. Science 330: 835–838

Weixlbaumer A, Jin H, Neubauer C, Voorhees RM, Petry S, Kelley AC, Ramakrishnan V (2008) Insights into translational termination from the structure of RF2 bound to the ribosome. Science 322: 953–956

Weixlbaumer A, Petry S, Dunham CM, Selmer M, Kelley AC, Ramakrishnan V (2007) Crystal structure of the ribosome recycling factor bound to the ribosome. Nat Struct Mol Biol 14: 733–737

Wilson DN, Schluenzen F, Harms JM, Yoshida T, Ohkubo T, Albrecht R, Buerger J, Kobayashi Y, Fucini P (2005) X-ray crystallography study on ribosome recycling: the mechanism of binding and action of RRF on the 50S ribosomal subunit. Embo J 24: 251–260

Wower J, Kirillov SV, Wower IK, Guven S, Hixson SS, Zimmermann RA (2000) Transit of tRNA through the *Escherichia coli* ribosome. Cross-linking of the 3' end of tRNA to specific nucleotides of the 23S ribosomal RNA at the A, P, and E sites. J Biol Chem 275: 37887–37894

Yusupov MM, Yusupova GZ, Baucom A, Lieberman K, Earnest TN, Cate JH, Noller HF (2001) Crystal structure of the ribosome at 5.5 A resolution. Science 292: 883–896

Structural studies of complexes of the 70S ribosome

3

C. Axel Innis, Gregor Blaha, David Bulkley, Thomas A. Steitz

1. Introduction

During the first decade of our studies of the structural basis of ribosome functions, we concentrated our efforts on the *Haloarcula marismortui* (*H. ma.*) 50S ribosomal subunit. This resulted in our obtaining its atomic structure in 2000 from a 2.4 Å resolution map (Ban et al., 2000) and the structures of many complexes with substrate analogues (Nissen et al., 2000; Hansen et al., 2002b; Schmeing et al., 2002; Schmeing et al., 2005b) and antibiotics (Hansen et al., 2002a; Hansen et al., 2003; Tu et al., 2005b) in the subsequent five years. More recently our focus has turned to studies of the 70S ribosome and its complexes with either protein factors, tRNAs, antibiotics or a peptidyl-tRNA with a peptide sequence that stalls protein synthesis; these 70S ribosome structures will be the primary results discussed here.

Our structural studies of the *H. ma.* 50S subunit complexes with substrate analogues as well as mutational, biochemical and kinetic studies by others (Schmeing et al., 2002; Dorner et al., 2003; Weinger et al., 2004) have led to a detailed understanding of the mechanism by which the ribosome, functioning as a ribozyme, is able to catalyze peptide bond formation (Schmeing et al., 2005a; Schmeing et al., 2005b; Steitz, 2008). In the absence of an A-site substrate the ribosome protects the peptidyl-CCA from hydrolysis by preventing the access of water to the peptidyl ester link. Binding of a CC-hydroxypuromycin substrate analogue to the A site produces a conformational change in the peptidyl transferase center (PTC) that repositions and deprotects the carbonyl carbon of the ester-linked P-site substrate, thereby enabling a nucleophilic attack by the alpha-amino group. The 2'OH group of A76 of the P-site tRNA forms a hydrogen bond with the attacking alpha-amino group and acts as a proton shuttle to remove a proton from the attacking alpha-amino group and transfer a proton to the leaving 3'-OH. Structures of transition state analogues show that catalysis is also enhanced by interactions of the oxyanion with a ribosome-positioned water molecule, thereby stabilizing the transition state (Schmeing et al., 2005a; Schmeing et al., 2005b).

The suggestion has been made that, since these studies were carried out using lower than physiological salt concentrations, our crystals therefore consisted of inactive 50S subunits (Bashan et al., 2001). However, assays of the catalytic activity of crystals of *H. ma.* 50S subunit showed that the crystals are nearly as active as the subunit in solution and that the activity does not vary between 1.6 M (where the crystals are grown) and 3M NaCl concentrations (Schmeing et al., 2002). Another criticism of our structural studies of the mechanism of peptide bond formation was that we used fragment substrate analogues on the isolated 50S subunit. Noller and colleagues concluded from a model they built of the 70S ribosome complex with an A-site tRNA derived from a 3.7 Å resolution map (Korostelev et al., 2006) that the orientations and interactions of the CCA when a full tRNA and 70S ribosome were used were different from those observed when CCA analogues were bound to the 50S subunit. However, subsequent re-refinement using averaging of the Ramakrishnan and Noller 70S ribosome data (Simonovic and Steitz, 2008) as well as two 3.0 Å resolution structures of the 70S ribosome complexed with tRNAs in the A and P sites from our lab (Stanley et al., 2010) and from the Ramakrishnan lab (Selmer et al., 2006; Voorhees et al., 2009) showed that the structures of the PTC and the CCA ends bound to the A and P sites were identical to those seen in the *H. ma.* studies.

The structures of many complexes of the *H. ma.* 50S subunit with antibiotics from numerous families bound to the PTC showed how they inhibit translation and in many cases explained the mechanism of resistance produced by mutations (Hansen et al., 2002a; Hansen et al., 2003; Tu et al., 2005b; Schroeder et al., 2007; Blaha et al., 2008; Gurel et al., 2009a; Gurel et al., 2009b). In spite of *H. ma.* being an archaeon and closer to eukaryotes than bacteria, several macrolide antibiotics, including azithromycin, could be bound to the *H. ma.* 50S subunit using high antibiotic concentrations. The orientations of their macrolide rings were nearly identical to each other in our structures, but differed dramatically from that proposed for erythromycin bound to crystals of the *Deinococcus radiodurans* (*D. ra.*) subunit (Schlunzen et al., 2001). The question then arose – was this difference due to incorrect modeling into modest resolution density maps, or was it due to species-related differences between bacteria and archaea? To address this issue, we mutated G2099 to A in the *H. ma.* subunit, which enabled erythromycin to bind with approximately 10^4-fold higher affinity (Tu et al., 2005a). In the structure of the complex, the lactone ring of erythromycin was seen to be oriented exactly as observed in the *H. ma.* complexes with other macrolides, but nearly orthogonal to the *D. ra.* model (Tu et al., 2005b). These results were consistent with our view that the differences between our antibiotic results and those from the *D. ra.* studies were a consequence of the insufficient resolution of the maps in the latter studies, but doubts persisted.

In order to study the structural bases of processes carried out only by the whole ribosome and to address any remaining concerns about our conclusions from our studies of the *H. ma.* 50S subunit substrate and antibiotic complexes, we turned our attention to the 70S ribosomes from *Thermus thermophilus* (*T. th.*) and *Escherichia coli*, which are the subjects of this review. We have found that the lactone rings of erythromycin and azithromycin bound to the *T. th.* 70S ribosome exhibit positions, orientations and conformations that are nearly identical to those seen in equivalent complexes of the G2099A mutant *H. ma.* 50S subunit (Bulkley et al., 2010); furthermore, chloramphenicol binds with a completely different orientation from that proposed for the *D. ra.* complex (Bulkley et al., 2010). The elongation factor EF-P is seen to bind between the bound P-site tRNA[fMet] and the E site, making non-sequence specific interactions with the tRNA[fMet] and presumably stabilizing it for the first step in peptide bond formation (Blaha et al., 2009). Also, the structure of the 70S ribosome complexed with a short mRNA and tRNAs bound in the A, P and E sites have the CCA

ends of their A- and P-site tRNAs oriented identically to the CCA moieties bound to the A and P sites of the *H. ma.* 50S subunits (Stanley et al., 2010), thereby validating the general relevance of the previous substrate analogue complex structures with the *H. ma.* 50S subunit. Finally, a cryo-electron microscopy (cryo-EM) reconstruction of the 70S ribosome containing a P-site peptidyl-tRNA with a stalling sequence shows an extended polypeptide making interactions all the way down the tunnel (Seidelt et al., 2009).

2. Something old, something new: Antibiotics and the ribosome

Antibiotics targeting the ribosome are indispensable both as therapeutic agents and as tools for basic research. Since the determination of the structures of the two individual ribosomal subunits at atomic resolution (Ban 2000; Wimberley 2000), a plethora of structures of antibiotics in complex with both large and small subunits have become available (reviewed in (Wilson, 2009)). More recently, high resolution models of 70S ribosomes from the bacteria *E. coli* (Schuwirth et al., 2005) and *T. th.* (Selmer et al., 2006) have made it possible to gain a structural understanding of translation inhibition within the context of the entire ribosome. Recent work carried out in this area by our group is reviewed below.

2.1. The macrolides and chloramphenicol revisited

Among the antibiotics that bind to the large ribosomal subunit, macrolides and macrolide-derived compounds comprise one of the most common classes of antimicrobial agents. It is known that macrolide antibiotics bind to regions of the 23S rRNA lining the walls of the ribosomal exit tunnel in close proximity to the P site of the PTC (Figure 1) (Moazed and Noller, 1987). Due to the existence of conflicting crystallographically-derived structural models for some ribosome-antibiotic complexes, the specific interactions established between macrolide, ketolide or azalide antibiotics and the ribosome remained the subject of significant controversy for many years. In the first models of erythromycin, azithromycin and telithromycin bound to a bacterial large ribosomal subunit (Schlunzen et al., 2001; Berisio et al., 2003; Schlunzen et al., 2003), the lactone rings of the antibiotics were all modeled with

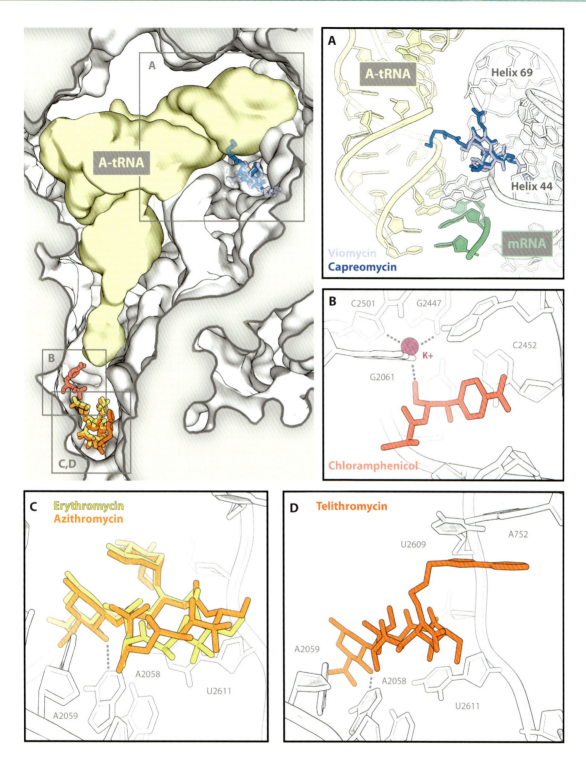

Fig. 1 The structures of several antibiotics in complex with a bacterial 70S ribosome. The top left panel shows a cross section of the 70S ribosome from *T. th.* with an A-site tRNA (Stanley et al., 2010) in pale yellow and various antibiotics bound near the decoding center (inset panel A), in the A-site crevice (inset panel B) or near the constriction of the exit tunnel (inset panels C, D). (A) Close-up view of the viomycin (light blue; PDB codes: 3KNH, 3KNI, 3KNJ and 3KNK) and capreomycin (dark blue; PDB codes: 3KNL, 3KNM, 3KNN and 3KNO) binding site (Stanley et al., 2010). The codon-anticodon base pairing between the A-site tRNA (pale yellow) and the mRNA (green) can be seen in the lower part of the panel. (B) Close-up view of the chloramphenicol (pink) binding pocket in the *T. th.* 70S, with a key potassium ion highlighted in purple (Bulkley et al., 2010). (C) Interactions of erythromycin (yellow) and azithromycin (light orange) with the *T. th.* ribosome (Bulkley et al., 2010). (D) Interactions between telithromycin (orange) and the *T. th.* ribosome. The alkyl-aryl moiety of the drug can be seen stacking with the bases of U2609 and A752 (Bulkley et al., 2010). The structures of chloramphenicol, erythromycin, azithromycin and telithromycin in complex with *T. th.* ribosomes have not yet been assigned a PDB identifier. (Figures reproduced from the references, as indicated)

different conformations and orientations with respect to the surface of the ribosomal exit tunnel. In general, the lactone ring of each antibiotic was bent away from the base of A2058[1] and the adjacent tunnel wall, though for each antibiotic a different orientation of the lactone ring was reported. However, crystallographic work published subsequently and carried out at higher resolution with crystals of either wild type (Hansen et al., 2002a) or G2099A (*H. ma.* numbering) mutant (Tu et al., 2005b) 50S ribosomal subunits from *H. ma.* told a different story: the lactone rings of erythromycin, azithromycin and telithromycin all shared approximately the same conformation and in each case were shown to be parallel to the surface formed by the bases of the 23S rRNA residues U2611, A2058 and A2059 (Tu et al., 2005b). This later structural work therefore suggested that all macrolides bind the ribosome in essentially the same manner.

In spite of this, some skepticism remained regarding the universality of this conclusion. In particular, it was suggested that the structures of macrolides bound to the G2099A mutant 50S subunit may not be representative of macrolide binding in bacteria in general and that the discrepancies between the two sets of experiments may have arisen from differences in the mode of binding of macrolide antibiotics to ribosomes from different kingdoms of life (Yonath, 2005). In order to address these concerns, we generated complexes between the aforementioned antibiotics and the 70S ribosome from the bacterium *T. th.* (Bulkley et al., 2010). The structures of the complexes that we have obtained are entirely consistent with those obtained with *H. ma.* complexes. In particular, the lactone rings of erythromycin, azithromycin and telithromycin are all oriented equivalently in complex with either the 70S ribosome of *T. th.* or the 50S ribosomal subunit of *H. ma.* (Figure 2). Minor differences were observed for the antibiotic telithromycin, whose alkyl-aryl moiety has distinct modes of interaction with ribosomes from each of the three species studied (Figure 2K).

While the macrolides act primarily by impeding the progress of the nascent polypeptide chain through the ribosomal exit tunnel, chloramphenicol directly competes with the aminoacyl moiety of an incoming tRNA for binding to the A-site crevice of the PTC (Moazed and Noller, 1987). To date, only one structural model of chloramphenicol bound to a bacterial ribosome is available (Schlunzen et al., 2001). In order to evaluate

the accuracy of that model, we solved the structure of this antibiotic in complex with the 70S bacterial ribosome from *T. th.* and discovered that the antibiotic binds in a radically different way from what had been proposed (Bulkley et al., 2010). Using an unbiased F_o-F_c difference map, the electron density that we observe suggests a model for chloramphenicol in which the antibiotic is rotated essentially 180° relative to the earlier model. Further, the interactions between the ribosomal RNA and chloramphenicol that we observe are entirely different from what had been proposed earlier and a single, well-characterized potassium ion (rather than the two putative magnesium ions from the previous model) is seen to be stabilizing the antibiotic in the ribosomal A-site crevice (Figure 1B). Interestingly, the position and orientation of the chloramphenicol moiety bound to the bacterial ribosome that we observe is very similar to that of the eukaryotic-specific antibiotic anisomycin complexed with the archaeal ribosome from *H. ma.* (Blaha et al., 2008). The structure of the present complex of chloramphenicol with the 70S ribosome explains the species specificity for bacterial rather than archaeal ribosomes. Most importantly, only this corrected structure of the chloramphenicol complex will be enabling for the design of new antibiotics.

To summarize, we conclude that the overall mode of macrolide binding is conserved across species and that the quality and interpretation of electron density, rather than species-specificity, is responsible for the differences between models derived from macrolide complexes with the bacterial ribosome. Although differences between species are likely to modulate the finer details of some antibiotic-ribosome interactions, drugs from the same class can be expected to interact with their target in a more or less identical fashion, be it a *bona fide* bacterial ribosome or an archaeal ribosome turned bacterial through mutagenesis. This notion is further supported by the observation that even in the case of drugs as disparate as choloramphenicol and anisomycin, which act on different kingdoms of life, common themes emerge in the way that certain binding pockets on the ribosome can be targeted by small molecules to inhibit protein synthesis. The antibiotics that we chose to reexamine are either actively used in a clinical setting or constitute potential leads for the development of new drugs. Consequently, accurate structures of the corresponding antibiotic-ribosome complexes are an essential prerequisite for the rational design of therapeutics seeking to inhibit multidrug resistant pathogens.

---

1 The numbering system for the *E. coli* 23S and 16S rRNAs will be used throughout this review, unless specified otherwise.

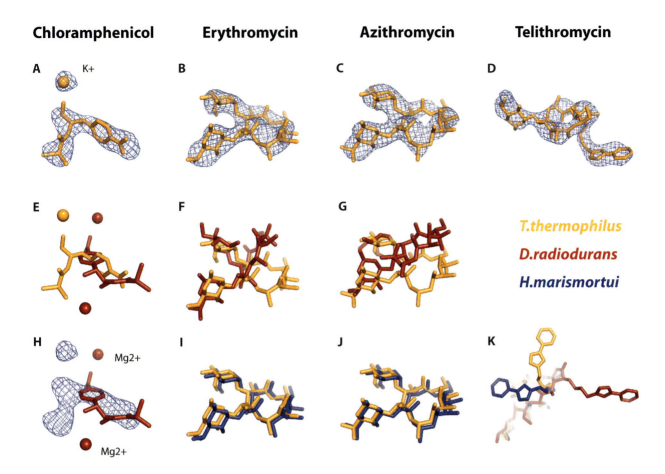

Fig. 2 Comparison of the *T. th.*, *D. ra.* and *H. ma.* antibiotic complex structures. Structures of the antibiotics chloramphenicol (A,E,H), erythromycin (B,F,I), azithromycin (C,G,J) and telithromycin (D,K) in complex with 70S ribosomes from *T. th.* (orange) or isolated 50S subunits from *D. ra.* (red) and *H. marismortui* (blue). Unbiased Fo-Fc difference electron density maps contoured at +3s calculated using amplitudes from *T. th.* 70S ribosome crystals soaked with the different antibiotics are shown as a blue mesh (A,B,C,D,H). Overlays of the various antibiotic models were obtained by superimposing phosphate atoms from equivalent 23S rRNA residues of the *T. th.*/ *D. ra.* (E,F,G,K) and *T. th.* / *H. ma.* (H,I,J,K) structures. The PDB codes corresponding to the structures shown are as follows: 1K01 (*D. ra.* 50S with chloramphenicol (Schlunzen et al., 2001)), 1JZY (*D. ra.* 50S with erythromycin (Schlunzen et al., 2001)), 1NWY (*D. ra.* 50S with azithromycin (Schlunzen et al., 2003)), 1P9X (*D. ra.* 50S with telithromycin (Berisio et al., 2003)), 1YI2 (*H. ma.* 50S with erythromycin (Tu et al., 2005b)), 1M1K (*H. ma.* 50S with azithromycin (Hansen et al., 2002a)), 1YIJ (*H. ma.* 50S with telithromycin (Tu et al., 2005b)). The structures of antibiotics in complex with *T. th.* 70S have not yet been assigned a PDB identifier. (Figure reproduced, in part, from Stanley et al. 2010 by permission.)

2.2. The tuberactinomycins

While many antibiotics target the large or the small ribosomal subunit individually, some are known to interact with both subunits. One such group of drugs – the tuberactinomycins – is comprised of several closely related compounds containing the same macrocyclic peptide core and differing only in the modification of their amino acid side chains (Thomas et al., 2003). Members of this family are particularly important from a clinical standpoint, since they are among the most effective antibiotics in use today for the treatment of multidrug-resistant tuberculosis.

We have determined the co-crystal structures of the 70S ribosome from *T. th.* complexed with three full-length tRNAs and a short piece of mRNA as well as one of two of the tuberactinomyins: viomycin or capreomycin (Stanley et al., 2010). The three tRNA molecules are bound in the classical A, P and E sites in this complex with the 70S ribosome. The anticodons of both the A-site and the P-site tRNAs form base pairs with the corresponding codons of the mRNA. As observed earlier (Ogle et al., 2001), the bases of A1492 and A1493 make stabilizing A-minor interactions with the two base pairs between the codon and anticodon of the A-site tRNA and mRNA. Both drugs bind to a site that lies between the large and small subunit in a cleft formed between helix 44 of the 16S rRNA (h44) and the tip of Helix 69 of the 23S rRNA (H69) (Figure 1A).

The macrocycle of both drugs lies within hydrogen bonding distance of the ribose-phosphate backbone of the binding pocket, which is formed by residues A1913 and C1914 of the 23S rRNA and residues A1493 and G1494 of the 16S rRNA. The orientation of the macrocycle is determined by stacking interactions with bases G1491 and G1494 from the 16S rRNA, which bring the six-membered ring of the drug into the vicinity of bases of A1492 and A1493 from the same subunit. The guanidinium moiety of this six-membered ring effectively locks the antibiotics into place by engaging in a salt bridge to the backbone phosphate of A1493. Previous biochemical work suggests that the steric restriction imposed by this ring on the orientation of A1492 and A1493 contributes to the inhibitory effect of viomycin and capreomycin (Nomoto and Shiba, 1977).

Mutations of nucleotides in both H69 and h44 that surround the antibiotic binding site confer resistance to capreomycin and viomycin (Monshupanee et al., 2008). Of particular interest is the 23S rRNA mutation A1913U, which was identified in *T. th*. In the structure of the *apo* 70S ribosome the base of A1913 is swung out of the loop at the base of H69 and inserted into h44 (Schuwirth et al., 2005), a position which partially overlaps with the binding site for both capreomycin and viomycin. The A1913U mutation may alleviate this potential clash between A1913 and the antibiotic, thereby conferring resistance to these drugs. This also suggests that bound tuberactinomycins inhibit translocation by preventing A1913 from moving into the position it occupies in the structure of the *apo* 70S-ribosome.

Classes of antibiotics that affect translocation of the tRNAs on the ribosome, such as the tuberactinomycins or aminoglycosides (Reviewed in (Shoji et al., 2009)), either stabilize the tRNA in the pre-translocation state or interfere with the conformational changes of the ribosome or tRNA required for translocation (Peske et al., 2004). Paromomycin is an aminoglycoside that binds into the major groove of h44 and stabilizes A1492 and A1493 of the 16S rRNA in the flipped-out conformation adopted upon readout of the mRNA (Ogle et al., 2001). Because paromomycin stabilizes the binding of both cognate and near-cognate tRNAs to the A site, it not only inhibits translocation but also promotes miscoding (Carter et al., 2000). Interestingly, although hygromycin B also binds to h44, it causes only the base of A1493 to adopt a unique orientation,

in which the base interferes with the movement of the tRNA from the A site to the P site (Borovinskaya et al., 2008).

Viomycin and capreomycin not only affect the positioning of the bases of A1492 and A1493, but they also appear to affect the position of A1913 of the 23S rRNA. The base of A1913 adopts a position where it forms hydrogen bonds with the A-site tRNA. Consistent with our results, biochemical experiments show that viomycin increases the affinity of tRNA to the A site by 1000-fold (Peske et al., 2004) and promotes the back-translocation of the complex of tRNA with mRNA on the ribosome (Szaflarski et al., 2008). Therefore we propose that the tuberactinomycins inhibit translocation by stabilizing the tRNA in the A site.

3. Insights into the regulation of bacterial protein synthesis

The availability of crystal forms of the entire bacterial ribosome diffracting to high resolution now enables the study of translation factors and modulators of ribosome function, with much effort being dedicated to obtaining snapshots of the various stages of the translation cycle through structures of 70S ribosomal complexes. One aspect of our work is thus concerned with elucidating some of the molecular mechanisms by which the bacterial protein synthetic machinery is able to meet the needs of rapidly dividing cells and in some cases directly sense and respond to a number of environmental stimuli these cells may become exposed to. Recent results from our group not only provide some insights into how the translation elongation factor P (EF-P) is likely to facilitate the formation of the first peptide bond, but also paint a preliminary picture of an as yet poorly understood phenomenon, namely the regulation of protein synthesis *in cis* by nascent polypeptide chains.

3.1. Regulation at the level of initiation: EF-P and the first peptide bond

EF-P is conserved in all bacteria and, though it is not required in a minimal *in vitro* translation system (Shimizu et al., 2001), it has been shown to have a stimulatory effect *in vitro* on the formation of the first peptide bond (Aoki et al., 1997b). EF-P binds stoichiometrically to each of the ribosomal subunits as well as to the 70S ribosome. Consistent with an involvement

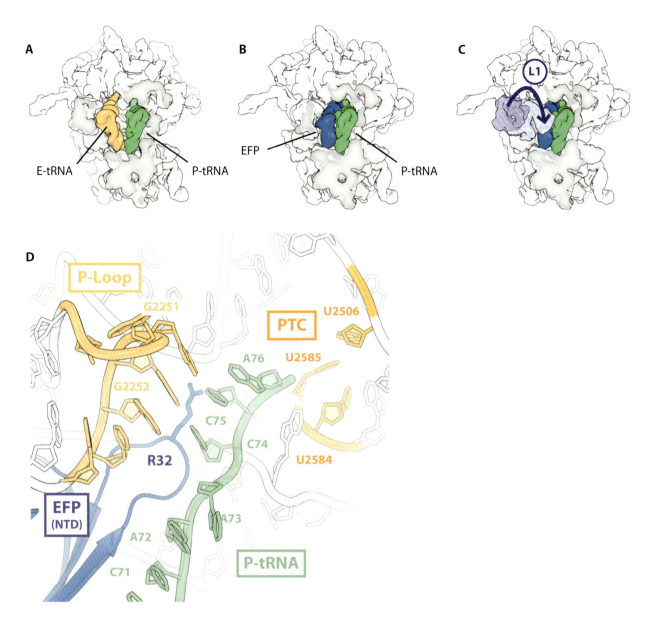

Fig. 3 The structure of EF-P bound to the ribosome. Schematic overview of (A) E- (orange) and P-site (green) tRNAs bound to the 70S ribosome (Selmer et al., 2006; PDB codes: 2J00, 2J01, 2J02 and 2J03), and of (B) EF-P (blue) and a P-site tRNA (green) in complex with the T. th.70S (Blaha et al., 2009; PDB codes: 3HUW, 3HUX, 3HUY and 3HUZ). (C) Same as in (B), but with the movement of the L1 stalk observed upon binding of EF-P indicated by an arrow. The stalk of a 70S ribosome occupied with all three tRNAs is shown in dark purple, while the stalk from the EF-P complex is depicted in light purple. (D) Interactions of the N-terminal domain of EF-P near the PTC of the large ribosomal subunit. The 23S rRNA is colored in white, except for the P loop (light orange) and the PTC (orange). EF-P is shown in blue and the acceptor arm of the P-site tRNA is shown in green.

in the initial stages of protein synthesis, the ratio of bound EF-P declines with the increasing size of polysomes (Aoki et al., 2008). In the cell, one copy of EF-P is present per ten ribosomes, a ratio comparable to that observed for the translational initiation factors (An et al., 1980; Cole et al., 1987).

Archaea and eukarya possess a factor known as initiation factor 5A (eIF-5A, previously called eIF-4D) that shares sequence and structural similarity with the first two domains of EF-P (Hanawa-Suetsugu et al., 2004). Like EF-P, eIF-5A stimulates the formation of peptide bonds between initiator tRNA and puromycin *in vitro* (Benne et al., 1978). eIF-5A is the only protein known to contain hypusine, an amino acid derived from lysine through post-translational modification (Zanelli and Valentini, 2007) and which is important for the activity of eIF-5A in translation (Saini et al., 2009). A recent mass spectrometric analysis of endogenous EF-P from *E. coli*, indicates a possible post-translational modification (Aoki et al., 2008). This modification has been identified as a lysyl modification of the position corresponding to the eIF5A hypusine residue in EF-P (Yanagisawa et al., 2010).

In order to shed light on the mechanism by which EF-P facilitates the formation of the first peptide bond, we determined the crystal structure of EF-P bound to the complex of the *T. th.* 70S ribosome, mRNA, and initiator tRNA at a resolution of 3.5 Å (Blaha et al., 2009). In our model, EF-P spans both ribosomal subunits and binds between the P- and E-tRNA binding sites (Hanawa-Suetsugu et al., 2004), contacting the initiator tRNA near the anticodon stem-loop on the 30S subunit, the D loop and the acceptor stem on the 50S subunit (Figure 3A,B). The L1 stalk also undergoes a major conformational change that positions ribosomal protein L1 in the E site to interact with EF-P (Figure 3C). Domain I of EF-P binds next to the acceptor stem of the initiator tRNA in the P site and domain III of EF-P binds adjacent to the anticodon stem loop of the P-site tRNA, partially overlapping with the E site. Domain II of EF-P is sandwiched between domains I and II of ribosomal protein L1.

The initiator tRNA bound to the P site displays the same conformation as the P-site tRNA of the 2.8 Å-resolution structure of the *T. th.* 70S ribosome with bound mRNA and tRNAs (Selmer et al., 2006), but is distorted compared to the crystal structure of the unliganded yeast tRNAPhe. Thus, even though EF-P binds adjacent to the tRNA, it does not induce any significant conformational changes.

A loop of domain I of EF-P makes numerous interactions with the CCA end of the acceptor stem of the initiator tRNA, allowing the semi-conserved Arg/Lys residue (position 32) at its tip to reach into the PTC, though not deep enough to directly participate in peptide bond formation (Figure 3D). The modification of the corresponding Lys residue with a 4-amino-2-hydroxybutyl group to hypusine in eukaryotes and with a lysyl group in *E. coli* would extend the side chain, reaching even closer to the active site. We cannot exclude the possibility that a similar modification of Arg32 occurs in *T. th*, since the *T. th.* EF-P used in our study was overexpressed in *E. coli*. In our current structure, neither the initiator tRNA nor the PTC undergoes a conformational change in the presence of EF-P, suggesting that EF-P's role may be to correctly position or stabilize the initiator tRNA in the P site.

Premature movement of the initiator tRNA to the E site may also be prevented by domain III of EF-P. Residues Tyr180 and Arg183 of domain III stabilize the A- minor interactions between two G·C base pairs in the anticodon stem-loop of the initiator tRNA and the bases of residues A1339 and G1338 of the 16S rRNA. A1339 and G1338 have been proposed to function as a "gate" between P and E sites, because their A-minor interactions with the P-site tRNA have to be broken during translocation (Selmer et al., 2006). By stabilizing these interactions, EF-P may strengthen this gate and stabilize the fMet-tRNA^{fMet} in the P site. Since domain III contacts only the small ribosomal subunit and is missing from eIF-5A, the latter must bind specifically to the 60S subunit and not span both ribosomal subunits. We cannot exclude the possibility that in eukaryotes its function is provided by another protein.

The L1 stalk is a highly dynamic component of the 50S subunit, which presumably accounts for it being disordered in most of the high resolution structures of the ribosome. Only in the *T. th.* 70S structures with bound tRNAs is the majority of the stalk ordered, seemingly because the interaction between the rRNA of the stalk and the tRNA bound in the E site stabilizes the position of the stalk (Yusupov et al., 2001; Korostelev et al., 2006; Selmer et al., 2006). The structure of the 70S ribosome with EF-P bound exhibits the largest movement of the L1 stalk relative to its position in the *apo* ribosome seen in any crystal structure. Without any disruption of the protein-RNA interface the L1 stalk is twisted along and rotated at helix H76 to reposition the ribosomal protein L1 into the E site. This motion results in a larger movement of the stalk than the previously observed 30° rotation toward the E site upon binding of a tRNA into the E site (Figure 3C; reviewed in (Korostelev and Noller, 2007)).

The structure of EF-P bound to the ribosome suggests that EF-P helps to correctly position the P-site tRNA and to restrict the mobility of the aminoacyl acceptor arm in order to facilitate peptide bond formation, a function similar to that attributed to ribosomal proteins L16 and L27 for the correct positioning of the A-site tRNA (Voorhees et al., 2009). Although ribosomes lacking L16 have severely impaired peptidyl transferase activity (Moore et al., 1975), formation of the first peptide bond can still be stimulated by EF-P (Aoki et al., 1997a). Moreover, the presence of L27 stimulates the reaction of puromycin with fMet-tRNA^{fMet} by the same magnitude as EF-P (Aoki et al., 1997a; Wower et al., 1998), though unlike EF-P, L27 also affects all subsequent peptidyl transfer reactions. By analogy, the stabilization of the CCA tail of the P-site tRNA by EF-P could explain the increased puromycin reactivity of fMet-tRNA^{fMet}. Whether this stabilization function is subsequently taken up by the nascent polypeptide chain is unknown. Interestingly, a recent publication suggests that eIF-5A not only stimulates the first but also subsequent peptidyl transfer reactions during protein synthesis (Saini et al., 2009). This is consistent with our observation that in the structure of the ribosome complex with bound EF-P all the interactions between EF-P and the fMet-tRNA^{fMet} are not specific for the initiator tRNA, which is also corroborated by previous biochemical data (Ganoza and Aoki, 2000). However, the positions of domains 1 and 2 of EF-P, that are homologous to eIF-5A, exclude the simultaneous binding of eIF-5A and a tRNA to the E site. While only the initiator tRNA occupies the P site and no deacylated tRNA is bound in the E site during initiation, the involvement of EF-P in subsequent steps of peptide elongation implies the existence of multiple and as of yet unknown additional steps in each cycle of peptide elongation.

Initiation of translation in bacteria is a multistep process that involves the formation of several intermediate complexes with different compositions and conformations (Simonetti et al., 2009). Structures of initiation complexes derived from cryo-EM studies have revealed that, during the process of initiation, fMet-tRNA^{fMet} adopts several different conformations on both the 30S subunit and the 70S ribosome before finally reaching its proper position in the P/P state (Allen et al., 2005; Simonetti et al., 2008; Simonetti et al., 2009). By stabilizing the P/P state of the initiator tRNA, EF-P could shift initiation towards the first elongation step of protein translation.

The structure of the EF-P complex with the 70S ribosome reveals the detailed interactions between EF-P and the 70S ribosome, initiator tRNA and the ribosomal protein L1. The essential role of EF-P in the cell may be to correctly position the fMet-tRNA^{fMet} in the P site for the first step of peptide bond formation by making several interactions with the backbone of the tRNA. Since eIF5A shows high structural similarity to EF-P, the conclusions drawn for EF-P likely also apply to eIF5A's role in the rate enhancement of the formation of the first peptide bond in eukarya.

3.2. Nascent chains as regulators of protein synthesis: The TnaC leader peptide

Following initiation of protein synthesis, the nascent peptide begins its journey through the ribosomal exit tunnel. Progression of the nascent chain down the exit tunnel is thought to be a relatively straightforward and uneventful process, with the exception of some well documented nascent peptides that can influence their own rate of translation by interacting with components of the 23S rRNA and ribosomal proteins lining the walls of the tunnel (Lovett and Rogers, 1996; Tenson and Ehrenberg, 2002; Beringer, 2008; Ramu et al., 2009; Ito et al., 2010). In these few instances, translation can come to a complete halt, and cells have evolved to exploit the unusual properties of these peptides as a means of regulating a variety of physiological processes. Although several arresting peptides have been relatively well characterized from a biochemical perspective (Nakatogawa and Ito, 2002; Woolhead et al., 2006; Cruz-Vera and Yanofsky, 2008; Vazquez-Laslop et al., 2008; Yang et al., 2009; Yap and Bernstein, 2009), biophysical techniques have so far failed to provide us with a detailed mechanistic understanding of the underlying structural processes. Over the past few years, our laboratory has begun to investigate the way in which specific nascent peptide sequences can stall translation in *cis*, our ultimate goal being to dissect the molecular details governing this process using X-ray crystallography.

The nascent chain that we initially focused on is TnaC, a 24-residue peptide encoded by the leader region of the tryptophanase (*tna*) operon of *E. coli* (Gong and Yanofsky, 2002). Stalling of ribosomes on the *tna* transcript occurs in the presence of inducing levels of free tryptophan. This in turn leads to the expression of the downstream *tnaA* and *tnaB* genes and to the clearance of cytoplasmic tryptophan until non-inducing levels are restored, thus making TnaC a *de facto* tryptophan sensor for certain species of bacteria. In order to render the process of nascent peptide-mediated translational stalling amenable to structural studies by X-ray crystallography, a major technical hurdle had to be overcome: stalled ribosome-nascent chain (RNC) complexes carrying a nascent peptide of defined length and sequence had to be obtained in pure form and in milligram quantities. Like other groups interested in the structural aspects of nascent-chain related processes, we modified existing cell-free protein synthesis protocols geared towards the production of milligram quantities of soluble protein by multiple-turnover protein synthesis (Calhoun and Swartz, 2006) to obtain RNC preparations featuring an enriched monosomal fraction with high peptidyl-tRNA occupancy (Evans et al., 2005; Schaffitzel and Ban, 2007). By designing a peptide sequence with calmodulin-binding peptide linked to the N terminus of TnaC via a flexible linker, stalled TnaC-70S ribosome complexes generated using our cell-free protein synthesis system could then be purified to near homogeneity by affinity chromatography on a calmodulin-sepharose matrix. Yields of up to 2 mg of pure complex could be obtained from 10 ml of translation reaction, making it both possible and practical to perform crystallization trials on this complex (Seidelt et al., 2009).

While crystallization experiments were under way, we began to collaborate with the group of Roland Beckmann at the University of Munich, hopeful that we may learn more about the homogeneity and occupancy of the TnaC-70S complex through cryo-electron microscopy (cryo-EM). To everyone's surprise, the first set of cryo-EM data collected for the stalled TnaC-70S complexes rapidly yielded a 5.8 Å-resolution single-particle reconstruction, in which not only a P-site tRNA was visible, but an extended nascent peptide could be seen spanning the entire length of the ribosomal exit tunnel (Figure 4A) (Seidelt et al., 2009). The high quality of the map revealed distinct points of contact between the nascent chain and ribosomal components of the exit tunnel. At the peptidyl transferase center of the ribosome, the universally conserved A2602 and U2585 were shown to adopt conformations that restrict the access of the catalytic GGQ loop of the termination release factors to the PTC. Through the use of the Molecular Dynamics Flexible Fitting (MDFF) method developed in the group of Klaus Schulten (University of Illinois at Urbana-Champaign) (Trabuco et al., 2008), a high-resolution structure of the entire ribosome was fitted into the experimental electron-density map. Using the fitted structure as the basis for interpreting the cryo-EM data, a model was proposed to explain how the coordinated jamming of the peptide inside the ribosomal exit tunnel ultimately leads to the complete inactivation of the PTC and to the shutdown of translation (Figure 4B, C).

While a clearer understanding of the mechanism by which the TnaC peptide causes translational stalling in the presence of tryptophan will require structural data at a resolution of 3–4 Å or better, our results prompt us to reconsider the notion of a predominantly hydrophilic, "Teflon-like" tunnel, where a lack of significantly large hydrophobic patches seemingly leaves

little or no possibility for interaction with hydrophobic sequences within translating nascent chains (Nissen et al., 2000). Indeed, we have shown that even regions of the nascent chain that are not implicated in translational arrest can make specific interactions with components of the exit tunnel (Seidelt et al., 2009). This being the case, we can speculate that many, if not all, nascent chains have the potential to modulate translation rates *in cis* by interacting with the ribosomal exit tunnel. This concept is further substantiated by the work of other laboratories, which have put under scrutiny the idea that the exit tunnel is merely a passive conduit through which nascent chains must pass on their way out of the ribosome. For instance, the movement of nascent chains containing charged amino acids down the exit tunnel has been likened to a dynamic wave of electrostatic potential, capable in principle of altering the chemical and structural environment of the tunnel (Lu et al., 2007; Lu and Deutsch, 2008). In agreement with this proposal, biochemical studies have shown that protein sequences containing positively charged residues may indeed cause transient pauses during elongation (Lu and Deutsch, 2008). Moreover, molecular dynamics simulations looking at the interaction between tunnel components and various chemical probes point to the existence of discrete binding sites and free-energy barriers for amino acids and ions within the exit tunnel. Thus, an isolated tryptophan side chain faces a considerable free energy barrier at the tunnel constriction, while an aspartate side chain is predicted to interact favorably with positive residues in protein L22 (Petrone et al., 2008).

It also appears, however, that the sequence of the nascent chain is not alone in establishing the intricate dialogue between nascent peptides and the ribosomal tunnel. Evidence for local zones of secondary structure induction or stabilization within the exit tunnel (Lu and Deutsch, 2005; Bhushan et al., 2010) raises the possibility that ribosomal elements lining these zones monitor not only the sequence of the nascent chain, but also its local structure. This is in agreement with biochemical observations made for several stalling and non-stalling nascent chains, which highlight the importance of ribosomal residues within these zones in mediating translational arrest or pausing (Nakatogawa and Ito, 2002; Cruz-Vera et al., 2005; Cruz-Vera et al., 2007; Lawrence et al., 2008) and provide backing for nascent chain compaction near the PTC (Woolhead et al., 2006). Our observation that nascent chains follow distinct paths during their journey through the ribosomal exit tunnel is therefore likely to gain new mean-

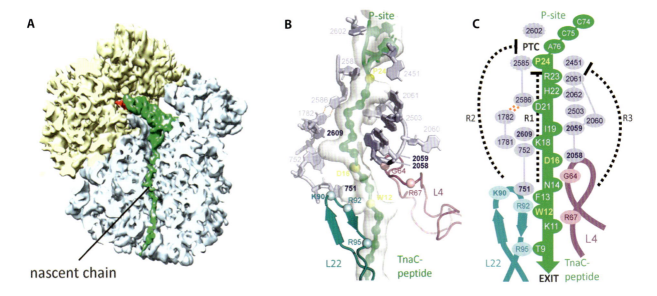

Fig. 4 Translational stalling mediated by the TnaC peptide. (A) Cryo-EM reconstruction of the TnaC-70S complex at a resolution of 5.8 Å (FSC=0.5), with the small and large subunits colored yellow and blue, respectively ((Seidelt et al., 2009); EMDB accession code: 1657)). Density corresponding to the P-site tRNA and to the TnaC nascent chain is shown in green and the mRNA is depicted in red. (B,C) Relay model for PTC silencing. Ribosomal elements that are likely to play a role in a relay mechanism to inactivate the PCT are shown, with residues important for stalling highlighted in bold. The TnaC nascent chain is in green, with residues critical for stalling in yellow. Density for the nascent chain is shown as a transparent surface in B. Potential relay pathways from Trp12 of TnaC to the PTC (R1-R3) are indicated in C. (Figure adapted from Seidelt et al. (2009))

ing as we unravel the general and specific features employed by different arresting sequences to cause ribosome stalling, as well as the more subtle language by which a broader range of nascent polypeptides influence the process of translation.

In the future, we seek to obtain a full atomic model of a stalled RNC complex by X-ray crystallography in order to establish the structural mechanism by which the stalling of polypeptide synthesis is achieved. Nascent chains of interest include TnaC and more recently ErmCL, an arresting peptide that leads to ribosome stalling in the presence of the macrolide erythromycin (Vazquez-Laslop et al., 2008). When work on the ribosome nascent chain project was initiated in our laboratory, it was not known whether nascent peptides could be visualized inside a translating ribosome. By providing the first direct evidence that this is indeed possible, our latest findings suggest that a complete molecular understanding of this process is now within reach.

Acknowledgements

The work reviewed in this chapter was funded by NIH grant GM022778 awarded to T. A. S. and by the Howard Hughes Medical Institute

References

Allen GS, Zavialov A, Gursky R, Ehrenberg M, Frank J (2005) The cryo-EM structure of a translation initiation complex from *Escherichia coli*. Cell 121: 703–712

An G, Glick BR, Friesen JD, Ganoza MC (1980) Identification and quantitation of elongation factor EF-P in *Escherichia coli* cell-free extracts. Can J Biochem 58: 1312–1314

Aoki H, Adams SL, Turner MA, Ganoza MC (1997a) Molecular characterization of the prokaryotic efp gene product involved in a peptidyltransferase reaction. Biochimie 79: 7–11

Aoki H, Dekany K, Adams SL, Ganoza MC (1997b) The gene encoding the elongation factor P protein is essential for viability and is required for protein synthesis. J Biol Chem 272: 32254–2259

Aoki H, Xu J, Emili A, Chosay JG, Golshani A, Ganoza MC (2008) Interactions of elongation factor EF-P with the *Escherichia coli* ribosome. FEBS J 275: 671–681

Ban N, Nissen P, Hansen J, Moore PB, Steitz TA (2000) The complete atomic structure of the large ribosomal subunit at 2.4 A resolution. Science 289: 905–920

Bashan A, Agmon I, Zarivach R, Schluenzen F, Harms J, Pioletti M, Bartels H, Gluehmann M, Hansen H, Auerbach T, Franceschi

F, Yonath A (2001) High-resolution structures of ribosomal subunits: initiation, inhibition, and conformational variability. Cold Spring Harb Symp Quant Biol 66: 43–56

Benne R, Brown-Luedi ML, Hershey JW (1978) Purification and characterization of protein synthesis initiation factors eIF-1, eIF-4C, eIF-4D, and eIF-5from rabbit reticulocytes. J Biol Chem 253: 3070–3077

Beringer M (2008) Modulating the activity of the peptidyl transferase center of the ribosome. RNA 14: 795–801

Berisio R, Harms J, Schluenzen F, Zarivach R, Hansen HA, Fucini P, Yonath A (2003) Structural insight into the antibiotic action of telithromycin against resistant mutants. J Bacteriol 185: 4276–4279

Bhushan S, Gartmann M, Halic M, Armache JP, Jarasch A, Mielke T, Berninghausen O, Wilson DN, Beckmann R (2010) alpha-Helical nascent polypeptide chains visualized within distinct regions of the ribosomal exit tunnel. Nat Struct Mol Biol 17: 313–317

Blaha G, Gurel G, Schroeder SJ, Moore PB, Steitz TA (2008) Mutations outside the anisomycin-binding site can make ribosomes drug-resistant. J Mol Biol 379: 505–519

Blaha G, Stanley RE, Steitz TA (2009) Formation of the first peptide bond: the structure of EF-P bound to the 70S ribosome. Science 325: 966–970

Borovinskaya MA, Shoji S, Fredrick K, Cate JH (2008) Structural basis for hygromycin B inhibition of protein biosynthesis. RNA 14: 1590–1599

Bulkley D, Innis CA, Blaha G, Steitz TA (2010) Revisiting the structures of several antibiotics bound to the bacterial ribosome. To be published

Calhoun KA, Swartz JR (2006) Total amino acid stabilization during cell-free protein synthesis reactions. J Biotechnol 123: 193–203

Carter AP, Clemons WM, Brodersen DE, Morgan-Warren RJ, Wimberly BT, Ramakrishnan V (2000) Functional insights from the structure of the 30S ribosomal subunit and its interactions with antibiotics. Nature 407: 340–348

Cole JR, Olsson CL, Hershey JW, Grunberg-Manago M, Nomura M (1987) Feedback regulation of rRNA synthesis in *Escherichia coli*. Requirement for initiation factor IF2. J Mol Biol 198: 383–392

Cruz-Vera LR, New A, Squires C, Yanofsky C (2007) Ribosomal features essential for tna operon induction: tryptophan binding at the peptidyl transferase center. J Bacteriol 189: 3140–3146

Cruz-Vera LR, Rajagopal S, Squires C, Yanofsky C (2005) Features of ribosome-peptidyl-tRNA interactions essential for tryptophan induction of tna operon expression. Mol Cell 19: 333–343

Cruz-Vera LR, Yanofsky C (2008) Conserved residues Asp16 and Pro24 of TnaC-tRNAPro participate in tryptophan induction of Tna operon expression. J Bacteriol 190: 4791–4797

Dorner S, Panuschka C, Schmid W, Barta A (2003) Mononucleotide derivatives as ribosomal P-site substrates reveal an important contribution of the 2'-OH to activity. Nucleic Acids Res 31: 6536–6542

Evans MS, Ugrinov KG, Frese MA, Clark PL (2005) Homogeneous stalled ribosome nascent chain complexes produced in vivo or in vitro. Nat Methods 2: 757–762

Ganoza MC, Aoki H (2000) Peptide bond synthesis: function of the efp gene product. Biol Chem 381: 553–559

Gong F, Yanofsky C (2002) Instruction of translating ribosome by nascent peptide. Science 297: 1864–1867

Gurel G, Blaha G, Moore PB, Steitz TA (2009a) U2504 determines the species specificity of the A-site cleft antibiotics: the structures of tiamulin, homoharringtonine, and bruceantin bound to the ribosome. J Mol Biol 389: 146–156

Gurel G, Blaha G, Steitz TA, Moore PB (2009b) Structures of triacetyloleandomycin and mycalamide A bind to the large ribosomal subunit of Haloarcula marismortui. Antimicrob Agents Chemother 53: 5010–5014

Hanawa-Suetsugu K, Sekine S, Sakai H, Hori-Takemoto C, Terada T, Unzai S, Tame JR, Kuramitsu S, Shirouzu M, Yokoyama S (2004) Crystal structure of elongation factor P from Thermus thermophilus HB8. Proc Natl Acad Sci USA 101: 9595–9600

Hansen JL, Ippolito JA, Ban N, Nissen P, Moore PB, Steitz TA (2002a) The structures of four macrolide antibiotics bound to the large ribosomal subunit. Mol Cell 10: 117–128

Hansen JL, Moore PB, Steitz TA (2003) Structures of five antibiotics bound at the peptidyl transferase center of the large ribosomal subunit. J Mol Biol 330: 1061–1075

Hansen JL, Schmeing TM, Moore PB, Steitz TA (2002b) Structural insights into peptide bond formation. Proc Natl Acad Sci USA 99: 11670–11675

Ito K, Chiba S, Pogliano K (2010) Divergent stalling sequences sense and control cellular physiology. Biochem Biophys Res Commun 393: 1–5

Korostelev A, Noller HF (2007) The ribosome in focus: new structures bring new insights. Trends Biochem Sci 32: 434–441

Korostelev A, Trakhanov S, Laurberg M, Noller HF (2006) Crystal structure of a 70S ribosome-tRNA complex reveals functional interactions and rearrangements. Cell 126: 1065–1077

Lawrence MG, Lindahl L, Zengel JM (2008) Effects on translation pausing of alterations in protein and RNA components of the ribosome exit tunnel. J Bacteriol 190: 5862–5869

Lovett PS, Rogers EJ (1996) Ribosome regulation by the nascent peptide. Microbiol Rev 60: 366–385

Lu J, Deutsch C (2005) Folding zones inside the ribosomal exit tunnel. Nat Struct Mol Biol 12: 1123–1129

Lu J, Deutsch C (2008) Electrostatics in the ribosomal tunnel modulate chain elongation rates. J Mol Biol 384: 73–86

Lu J, Kobertz WR, Deutsch C (2007) Mapping the electrostatic potential within the ribosomal exit tunnel. J Mol Biol 371: 1378–1391

Moazed D, Noller HF (1987) Chloramphenicol, erythromycin, carbomycin and vernamycin B protect overlapping sites in the peptidyl transferase region of 23S ribosomal RNA. Biochimie 69: 879–884

Monshupanee T, Gregory ST, Douthwaite S, Chungjatupornchai W, Dahlberg AE (2008) Mutations in conserved helix 69 of 23S rRNA of Thermus thermophilus that affect capreomycin resistance but not posttranscriptional modifications. J Bacteriol 190: 7754–7761

Moore VG, Atchison RE, Thomas G, Moran M, Noller HF (1975) Identification of a ribosomal protein essential for peptidyl transferase activity. Proc Natl Acad Sci USA 72: 844–848

Nakatogawa H, Ito K (2002) The ribosomal exit tunnel functions as a discriminating gate. Cell 108: 629–636

Nissen P, Hansen J, Ban N, Moore PB, Steitz TA (2000) The structural basis of ribosome activity in peptide bond synthesis. Science 289: 920–930

Nomoto S, Shiba T (1977) Chemical studies on tuberactinomycin. XIII. Modification of beta-ureidodehydroalanine residue in tuberactinomycin N. J Antibiot (Tokyo) 30: 1008–1011

Ogle JM, Brodersen DE, Clemons WM, Jr., Tarry MJ, Carter AP, Ramakrishnan V (2001) Recognition of cognate transfer RNA by the 30S ribosomal subunit. Science 292: 897–902

Peske F, Savelsbergh A, Katunin VI, Rodnina MV, Wintermeyer W (2004) Conformational changes of the small ribosomal subunit during elongation factor G-dependent tRNA-mRNA translocation. J Mol Biol 343: 1183–1194

Petrone PM, Snow CD, Lucent D, Pande VS (2008) Side-chain recognition and gating in the ribosome exit tunnel. Proc Natl Acad Sci USA 105: 16549–16554

Ramu H, Mankin A, Vazquez-Laslop N (2009) Programmed drug-dependent ribosome stalling. Mol Microbiol 71: 811–824

Saini P, Eyler DE, Green R, Dever TE (2009) Hypusine-containing protein eIF5A promotes translation elongation. Nature 459: 118–121

Schaffitzel C, Ban N (2007) Generation of ribosome nascent chain complexes for structural and functional studies. J Struct Biol 158: 463–471

Schlunzen F, Harms JM, Franceschi F, Hansen HA, Bartels H, Zarivach R, Yonath A (2003) Structural basis for the antibiotic activity of ketolides and azalides. Structure 11: 329–338

Schlunzen F, Zarivach R, Harms J, Bashan A, Tocilj A, Albrecht R, Yonath A, Franceschi F (2001) Structural basis for the interaction of antibiotics with the peptidyl transferase centre in eubacteria. Nature 413: 814–821

Schmeing TM, Huang KS, Kitchen DE, Strobel SA, Steitz TA (2005a) Structural insights into the roles of water and the 2' hydroxyl of the P site tRNA in the peptidyl transferase reaction. Mol Cell 20: 437–448

Schmeing TM, Huang KS, Strobel SA, Steitz TA (2005b) An induced-fit mechanism to promote peptide bond formation and exclude hydrolysis of peptidyl-tRNA. Nature 438: 520–524

Schmeing TM, Seila AC, Hansen JL, Freeborn B, Soukup JK, Scaringe SA, Strobel SA, Moore PB, Steitz TA (2002) A pre-translocational intermediate in protein synthesis observed in crystals of enzymatically active 50S subunits. Nat Struct Biol 9: 225–230

Schroeder SJ, Blaha G, Tirado-Rives J, Steitz TA, Moore PB (2007) The structures of antibiotics bound to the E site region of the 50 S ribosomal subunit of Haloarcula marismortui: 13-deoxytedanolide and girodazole. J Mol Biol 367: 1471–1479

Schuwirth BS, Borovinskaya MA, Hau CW, Zhang W, Vila-Sanjurjo A, Holton JM, Cate JH (2005) Structures of the bacterial ribosome at 3.5 A resolution. Science 310: 827–834

Seidelt B, Innis CA, Wilson DN, Gartmann M, Armache JP, Villa E, Trabuco LG, Becker T, Mielke T, Schulten K, Steitz TA, Beckmann R (2009) Structural insight into nascent polypeptide chain-mediated translational stalling. Science 326: 1412–1415

Selmer M, Dunham CM, Murphy FVt, Weixlbaumer A, Petry S, Kelley AC, Weir JR, Ramakrishnan V (2006) Structure of the 70S ribosome complexed with mRNA and tRNA. Science 313: 1935–1942

Shimizu Y, Inoue A, Tomari Y, Suzuki T, Yokogawa T, Nishikawa K, Ueda T (2001) Cell-free translation reconstituted with purified components. Nat Biotechnol 19: 751–755

Shoji S, Walker SE, Fredrick K (2009) Ribosomal translocation: one step closer to the molecular mechanism. ACS Chem Biol 4: 93–107

Simonetti A, Marzi S, Jenner L, Myasnikov A, Romby P, Yusupova G, Klaholz BP, Yusupov M (2009) A structural view of translation initiation in bacteria. Cell Mol Life Sci 66: 423–436

Simonetti A, Marzi S, Myasnikov AG, Fabbretti A, Yusupov M, Gualerzi CO, Klaholz BP (2008) Structure of the 30S translation initiation complex. Nature 455: 416–420

Simonovic M, Steitz TA (2008) Cross-crystal averaging reveals that the structure of the peptidyl-transferase center is the same in the 70S ribosome and the 50S subunit. Proc Natl Acad Sci USA 105: 500–505

Stanley RE, Blaha G, Grodzicki RL, Strickler MD, Steitz TA (2010) The structures of the anti-tuberculosis antibiotics viomycin and capreomycin bound to the 70S ribosome. Nat Struct Mol Biol 17: 289–293

Steitz TA (2008) A structural understanding of the dynamic ribosome machine. Nat Rev Mol Cell Biol 9: 242–253

Szaflarski W, Vesper O, Teraoka Y, Plitta B, Wilson DN, Nierhaus KH (2008) New features of the ribosome and ribosomal inhibitors: non-enzymatic recycling, misreading and back-translocation. J Mol Biol 380: 193–205

Tenson T, Ehrenberg M (2002) Regulatory nascent peptides in the ribosomal tunnel. Cell 108: 591–594

Thomas MG, Chan YA, Ozanick SG (2003) Deciphering tuberactinomycin biosynthesis: isolation, sequencing, and annotation of the viomycin biosynthetic gene cluster. Antimicrob Agents Chemother 47: 2823–2830

Trabuco LG, Villa E, Mitra K, Frank J, Schulten K (2008) Flexible fitting of atomic structures into electron microscopy maps using molecular dynamics. Structure 16: 673–683

Tu D, Blaha G, Moore PB, Steitz TA (2005a) Gene replacement in Haloarcula marismortui: construction of a strain with two of its three chromosomal rRNA operons deleted. Extremophiles 9: 427–435

Tu D, Blaha G, Moore PB, Steitz TA (2005b) Structures of MLSBK antibiotics bound to mutated large ribosomal subunits provide a structural explanation for resistance. Cell 121: 257–270

Vazquez-Laslop N, Thum C, Mankin AS (2008) Molecular mechanism of drug-dependent ribosome stalling. Mol Cell 30: 190–202

Voorhees RM, Weixlbaumer A, Loakes D, Kelley AC, Ramakrishnan V (2009) Insights into substrate stabilization from snapshots of the peptidyl transferase center of the intact 70S ribosome. Nat Struct Mol Biol 16: 528–533

Weinger JS, Parnell KM, Dorner S, Green R, Strobel SA (2004) Substrate-assisted catalysis of peptide bond formation by the ribosome. Nat Struct Mol Biol 11: 1101–1106

Wilson DN (2009) The A-Z of bacterial translation inhibitors. Crit Rev Biochem Mol Biol 44: 393–433

Woolhead CA, Johnson AE, Bernstein HD (2006) Translation arrest requires two-way communication between a nascent polypeptide and the ribosome. Mol Cell 22: 587–598

Wower IK, Wower J, Zimmermann RA (1998) Ribosomal protein L27 participates in both 50 S subunit assembly and the peptidyl transferase reaction. J Biol Chem 273: 19847–19852

Yanagisawa T, Sumida T, Ishii R, Takemoto C, Yokoyama S (2010) A paralog of lysyl-tRNA synthetase aminoacylates a conserved lysine residue in translation elongation factor P. Nat Struct Mol Biol 17: 1136–1143

Yang R, Cruz-Vera LR, Yanofsky C (2009) 23S rRNA nucleotides in the peptidyl transferase center are essential for tryptophanase operon induction. J Bacteriol 191: 3445–3450

Yap MN, Bernstein HD (2009) The plasticity of a translation arrest motif yields insights into nascent polypeptide recognition inside the ribosome tunnel. Mol Cell 34: 201–211

Yonath A (2005) Antibiotics targeting ribosomes: resistance, selectivity, synergism and cellular regulation. Annu Rev Biochem 74: 649–679

Yusupov MM, Yusupova GZ, Baucom A, Lieberman K, Earnest TN, Cate JH, Noller HF (2001) Crystal structure of the ribosome at 5.5 A resolution. Science 292: 883–896

Zanelli CF, Valentini SR (2007) Is there a role for eIF5A in translation? Amino Acids 33: 351–358

Interaction of bacterial ribosomes with mRNA and tRNA as studied by X-ray crystallographic analysis

4

Lasse B. Jenner, Natalia Demeshkina, Gulnara Yusupova, Marat Yusupov

1. Introduction

Protein synthesis, i. e. the translation of the genetic information encoded in messenger RNA into a polypeptide chain, is carried out by the ribosomes. These giant ribonucleoprotein assemblies, of particle weights of about 2.5 MDa in bacteria and up to 4 MDa in higher organisms, are composed of many different proteins (about 55 in bacteria) and long RNA chains (comprising about 4600 nucleotides in bacteria). Although the ribosome and its basic functions were discovered more than 50 years ago, the large size and the highly dynamic nature of this supramolecular assembly have made it difficult to obtain useful crystals. However, recent advances in ribosome crystallization and X-ray crystallography have led to great progress in structural studies of the ribosome. Over the last decade, X-ray structures of small (30S) and large (50S) subunits as well as of entire (70S) bacterial and archaeal ribosomes have vastly increased our understanding of the workings of this macromolecular machine.

This chapter summarizes recent studies of the ribosome structure performed by the X-ray crystallographic approach, focusing primarily on recent work from our laboratory on prokaryotic functional ribosome complexes. On the one hand, these structures aim at understanding how the ribosome maintains the reading frame on the mRNA during translation. On the other hand, our crystallographic studies provide novel, detailed structural information on the interaction of both cognate and near-cognate tRNA with the A site of the ribosome.

2. How does the ribosome maintain the reading frame of messenger RNA during protein synthesis?

One of the key questions in protein biosynthesis is how the ribosome couples mRNA and tRNA movements in order to prevent the disruption of the weak codon-anticodon interactions and the loss of the reading frame during translocation. The nature of tRNA-mRNA interactions on the ribosome has been structurally investigated over the course of almost a decade with continuously improved crystal diffraction and ribosome complex quality. While there is a substantial amount of crystallographic information available related to the interactions of the A and P-site codons of the mRNA with the tRNA anticodons (Ogle et al., 2001; Yusupov et al., 2001; Korostelev et al., 2006; Selmer et al., 2006), much less is known about the interactions of the ribosome with the mRNA up- and downstream of those two codons. The path of the mRNA has been visualized by X-ray crystallography of ribosome complexes that represent different states of translation (Yusupova et al., 2001; Jenner et al., 2005; Yusupova et al., 2006) at moderate resolution (4.5–7 Å). However, atomic-level information about the complete path of mRNA on the ribosome was missing.

To address these issues, we recently determined the high-resolution crystal structures of *Thermus thermophilus* ribosomes complexed with two different mRNA constructs and naturally modified tRNAs. The structures of these complexes, modeling the initiation and elongation states of translation, were determined at resolutions of 3.1 Å and 3.5 Å, respectively. Details of the crystallization and data analysis have been published elsewhere (Jenner et al., 2010). Crystals of the ribosome modeling the elongation state were obtained from ribosome complexes prepared with a

60 nucleotides long poly(U) mRNA containing a Shine-Dalgarno (SD) sequence, UUU (Phe) codons in A and P sites, and tRNA[Phe]. The ribosome complex modeling the initiation state was prepared with a 27 nucleotide long mRNA comprising the SD sequence with AUG (Met) codon and initiator tRNA[fMet].

2.1. Overview of the mRNA path through the ribosome

We were able to model the mRNA in the elongation state from positions –18 to +12 which delimit the boundaries of the ribosome (Figures 1A, B). The 5'-end of the mRNA upstream of the E-site codon along with the 3'-terminal tail of 16S rRNA form the SD duplex (Shine and Dalgarno, 1974) located on the platform of the 30S subunit. The A, P and E-site codons interact with the respective tRNAs on the interface between the ribosomal subunits, where modified nucleotides of the tRNAs such as ms²i⁶A37 can enhance the strength of interaction between tRNA anticodon and mRNA codon. Downstream of the A-site codon, the mRNA goes through a tunnel formed by a layer of 16S rRNA which can contract and form a tight network around the mRNA (mRNA nucleotides from +7 to +9). This is followed by a second layer of the downstream tunnel composed of proteins S3, S4 and S5 (mRNA nucleotides from +10 to +12), before the mRNA emerges on the solvent side of the 30S subunit. Although the density for the mRNA becomes rather weak in the latter region, it indicates that sequence-independent interactions between the mRNA and ribosomal proteins S3, S4, and S5 are mediated by the ribose-phosphate backbone of the mRNA and side chains of basic amino acids. This was not seen before, as previous high-resolution structures have only shown mRNA positions from –4 to +7 (Selmer et al., 2006; Weixlbaumer et al., 2007; Korostelev et al., 2008; Laurberg et al., 2008; Weixlbaumer et al., 2008).

The comparison of the two ribosome structures reveals that, upon transition from the initiation to the elongation state, the 30S subunit undergoes a conformational change whereupon helices 15, 16, 17 and 18 from the body of the 30S subunit contract towards the 30S neck, whereas the rest of the body and most of the head of the 30S subunit remain immobile (Figure 1C). This domain closure results in a contraction of the mRNA tunnel immediately downstream of the A-site codon in the elongation state by 1 to 2 Å (Figure 1D). The domain closure is most likely caused by the bind-

ing of the cognate tRNA to the A site, similar to what was seen in the isolated 30S subunit with an anticodon stem-loop bound in the A site (Ogle et al., 2002).

2.2. Network of interactions of the ribosome with the mRNA downstream of the A-site codon

In the initiation complex, the third nucleotide of the A-site codon interacts with the base of C1397 (h28) from the neck region of 16S rRNA, which nearly intercalates between the mRNA bases at positions +6 and +7 (Jenner et al., 2010a). C1397 is on the side of the mRNA tunnel and seems to be able to adopt different conformations, depending on the presence or absence of tRNA in the A site (elongation and initiation complexes, respectively). In the initiation complex, the sugar moiety of mRNA nucleotide +9 forms a hydrogen bond with Gln162 of protein S3 (Jenner et al., 2010a). However, in the elongation complex it becomes apparent that the contraction of the downstream mRNA tunnel triggers the formation of an intricate network of interactions between 16S RNA, protein S3, and the mRNA adjacent to the A-site codon. Nucleotides +8 and +9 of the mRNA are held in place by a combination of hydrogen bonding and continuous aromatic base stacking with Gln162 of protein S3 and nucleotides U1196 and C1054 from helix 34 of 16S rRNA. Finally, C1054 interacts with G34 of A-tRNA (Ogle et al., 2001) (Figure 2A, B). This network of interactions between mRNA and the head of the 30S most likely stabilizes the decoding center. Additionally, it may align the mRNA immediately downstream of the A-site codon before its movement into the A site, such that the codon approaching the decoding center is pre-oriented for the interaction with the tRNA. We suggest that the ribosome preserves this network of interactions during translocation in order to strongly and accurately safeguard the mRNA. Thus, the mRNA reading frame is maintained not only by codon-anticodon interactions, but also by this network in the downstream tunnel during swiveling of the 30S head in the course of the ratchet-like movement of the small ribosomal subunit relative to the large ribosomal subunit that accompanies translocation (Valle et al., 2003; Spahn et al., 2004; Zhang et al., 2009). After translocation, the mRNA interactions with h34 of 16S rRNA and protein S3 must be disrupted, and the 30S subunit head returns to its initial position.

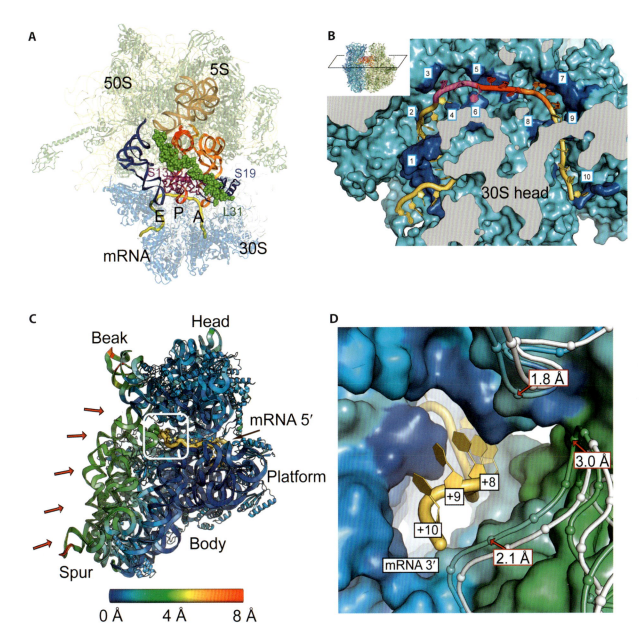

Fig. 1 Overview of the mRNA path through the ribosome. (A) Top view of the 70S elongation complex. A, P and E tRNA^{Phe} shown in orange, red and blue, respectively, and the 60-mer mRNA (position –18 to +12 visible) shown in gold. New intersubunit bridge, formed by protein L31 is shown as spheres. (B) Cross-section of the ribosome at the level of the mRNA showing interactions between mRNA and the following ribosomal elements: (1) Shine-Dalgarno sequence of the 3' of 16S rRNA; (2) Ribosomal proteins S11 and S18; (3) Loop of helix 23b (16S rRNA); (4) A1507 of 16S rRNA; (5) Interaction with modified nucleotide 37 of the P site tRNA through $Mg(H_2O)_6^{2+}$ and stabilization of the mRNA kink between the P and A-site codons via interactions with nucleotides from h44 (16S rRNA); (6) Stacking of the base of mRNA position –1 with G926 from h28 (16S rRNA); (7) the mRNA A codon interactions with nucleotides G530, A1492 and A1493 (16S rRNA); (8) C1397 from 16S RNA; (9) Aromatic stacking network between mRNA and U1196 and C1054 (16S rRNA); (10) Ribosomal proteins S3, S4 and S5. (C) Overall conformational changes of the 70S ribosome seen from the subunit interface (50S subunit and tRNAs are removed).

The 30S structure is colored according to the difference between phosphate and Cα positions in the initiation and elongation complexes, ranging from blue (0 Å difference) to red (8 Å difference). Arrows indicate the direction of movement of the domain closure during transition from the initiation to elongation state. The downstream mRNA tunnel has been marked with a white outline. From the superposition it is clear that only the shoulder of the 30S subunit moves, whereas the other parts of the 30S subunit remain immobile, and that the resulting movement leads to a contraction of the mRNA tunnel downstream of the A-site codon. (D) Detailed view of the RNA-layer part of the downstream mRNA tunnel seen from the solvent side of the 30S subunit. The RNA chains with the largest movements are shown in white (initiation) and color (elongation) with difference vectors marking the changes in position. The contraction of the downstream mRNA tunnel leads to a narrowing of the tunnel diameter by 1–3 Å, tightening the ribosome grip on the mRNA. (Figure reproduced from Jenner et al. (2010a) Nat Struc Mol Biol 17: 555–560 with permission of the publisher)

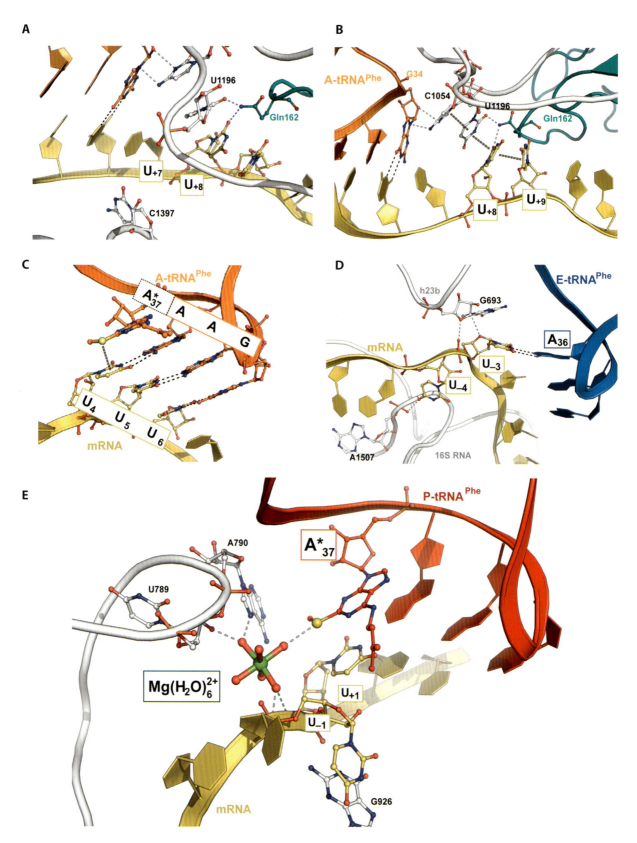

Fig. 2 mRNA-ribosome interactions in the elongating ribosome. (A and B) View of the interaction downstream of the A-site codon. (C) Stabilization of codon-anticodon interaction by cross-strand stacking of ms^{2}i^{6}A37 (A*37). (D) mRNA-ribosome interactions at the E-site codon. (E) Anchoring of tRNA in the P site. (Figure reproduced from Jenner et al. (2010a) Nat Struc Mol Biol 17: 555–560 with permission of the publisher)

2.3. Role of tRNA modifications in stabilizing mRNA-tRNA interactions on the ribosome. mRNA kink between P- and E-site codons

In contrast to missense errors, which are not necessarily destructive to proteins, practically all frame-shift errors are injurious to the synthesis of a functional protein, since the amino acid sequence in most cases will be entirely incorrect, and frequently premature stop codons are encountered. The error frequency associated with frameshift events has been estimated to be not higher than one event per 30,000 amino acids incorporated (Jorgensen and Kurland, 1990). Biochemical and genetic studies have shown that natural posttranscriptional modifications of tRNAs play a large role in maintaining the correct reading frame and decreasing the frequency of frameshifting (Bouadloun et al., 1986; Urbonavicius et al., 2001; Konevega et al., 2006; Gustilo et al., 2008). At present more than 80 different modified nucleosides have been characterized (Rozenski et al., 1999). Although they are present in tRNAs from all organisms and located at different positions within the tRNA, the majority is located in the anticodon region, especially at positions 34 (the wobble position) and 37 (3' adjacent to the anticodon) with the type of modification varying with codon specificity (Rozenski et al., 1999). Crystals of isolated 30S subunits carrying an anticodon stem-loop in the A site indicated that a modification at position 34 (uridine 5-oxyacetic acid) not only increases the stability of codon-anticodon interaction, but also expands the decoding capacity of the tRNA (Weixlbaumer et al., 2007a).

The present elongation complex was formed with *E. coli* tRNA^{Phe} which contains the hypermodified nucleotide 2-methylthio-N6-isopentenyl adenosine (ms^{2}i^{6}A37) in the anticodon loop. ms^{2}i^{6}A37 is present in almost all *E. coli* tRNAs which read codons with uridine in the first position. We found that the sulfur atom of ms^{2}i^{6}A37 in the A, P and E sites stabilizes the codon-anticodon duplex by cross-strand stacking with the base of the first nucleotide of the mRNA codon, thereby increasing the normal triplet interaction strength in A and P sites to nearly a quadruplet interaction (Figure 2C, D).

The presence of the methylthio-group of ms^{2}i^{6}A37 in P-site tRNA^{Phe} allows the formation of a network of interactions around the first nucleotide of the P-site codon and the third nucleotide of the E-site codon involving a fully hydrated magnesium ion, which is not found in the initiation complex. The magnesium ion is coordinated by the sulfur atom of ms^{2}i^{6}A37 of the P-site tRNA, the phosphates of the mRNA nucleotides U+1 and U-1, and the bases of nucleotides A790 and U789 of 16S RNA (Fig. 2E). Nucleotide U-1 flips out of the A-helical conformation of the E-site codon and stacks directly on top of nucleotide G926 of 16S rRNA in the neck-helix 28. Such non-specific anchoring of the mRNA probably increases the resistance against peptidyl-tRNA-slippage and mRNA frameshift.

Translational errors due to modification deficiencies of ms^{2}i^{6}A37 in tRNAs have been studied extensively *in vivo* and *in vitro*, but the molecular mechanism of the ms^{2}i^{6}A37 contribution to reading frame maintenance has remained unclear (Wilson and Roe, 1989; Urbonavicius et al., 2001; Urbonavicius et al., 2003).

On the basis of our present structural results, we can now rationalize previous biochemical and genetic data indicating that the absence of the methylthio group of ms^{2}i^{6}A37 significantly decreased the efficiency of tRNA^{Phe} binding to the ribosome (Menichi and Heyman, 1976; Hoburg et al., 1979) and increased the frameshift frequency, since slippage of tRNA in the P site in the direction of the E site (−1) or the A site (+1) is enhanced when the additional stabilization of the P and E-site codon region by ms^{2}i^{6}A37 is missing.

2.4. Interactions of the ribosome with the mRNA region upstream of the P-site codon

In previous ribosome structures where the nature of codon-anticodon pairing in the E site was examined (Selmer et al., 2006; Yusupova et al., 2006; Jenner et al., 2007), the mRNAs used were in a tense conformation, with a minimal distance (seven nucleotides) between the core adenosine (−8) of the SD sequence and the first nucleotide of the P-site codon (+1) (Schurr et al., 1993; Ma et al., 2002). The E-site codons in those complexes and in the present initiation complex were found in a conformation not favorable for codon-anticodon interaction, which fully agrees with our knowledge of the initiation of translation, since deacylated tRNA should not appear in the E site until after the first translocation event. In the initiation complex, we observed that the presence of the SD duplex causes strong anchoring of the 5'-end of the mRNA onto the platform of the 30S subunit, with the upstream region taking a position close to protein S11 and S18 on the platform (Figure 3A).

To study the ribosome in the elongation state we used a 60 nucleotides long poly(U) mRNA containing a SD sequence, which was designed without a specific

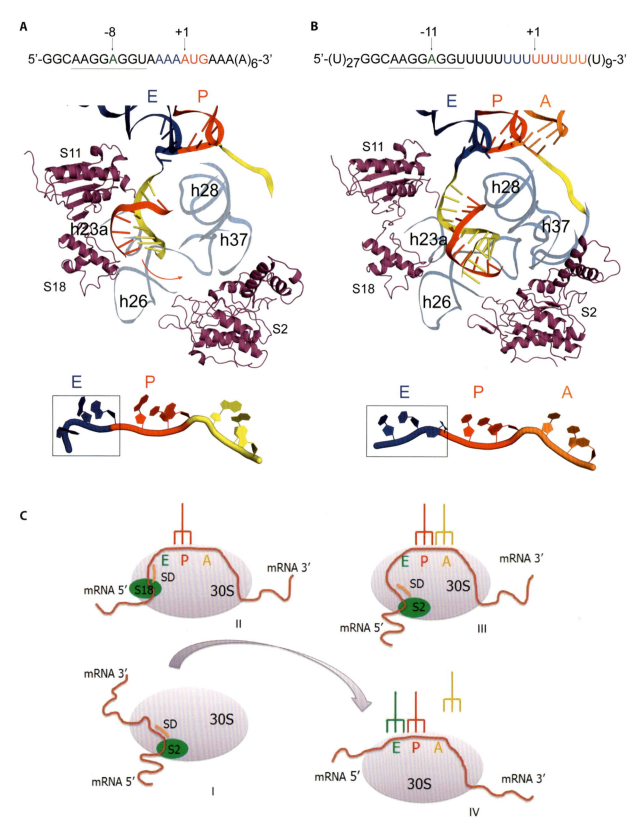

Fig. 3 mRNA-ribosome interactions upstream of the P-site codon (A, B) The ribosomal environment of the SD helix (30S proteins in magenta and 16S rRNA in light blue with the anti-SD sequence highlighted in red) changes as the SD helix moves upon transition from the initiation (A) to the elongation (B) state. The mRNA se- quences are given with the SD region underlined. Conformations of the E-site codon (framed) are shown at the bottom of panels A and B. (C) Simplified scheme for mRNA motion on the 30S sub- unit. (Figure reproduced from Demeshkina et al. (2010) Curr Opin Struct Biol 20: 325–332 © 2010 with permission of Elsevier)

codon to fix the reading frame so that the ribosome could freely choose the P-site codon (Fig. 3B). Compared to the initiation state described above, in the elongation state, with A and P sites occupied by elongator tRNA^{Phe}, the SD duplex assumes a different position on the platform where it contacts ribosomal protein S2. The mRNA is in a relaxed conformation where the distance between the SD core adenosine (−11) and the P-site codon (+1) has increased from seven to ten nucleotides. The observed SD duplex has been extended to 12 nucleotides, although classical Watson-Crick base-paring in the double helix occurs in only nine pairs. Thus, almost all nucleotides (from 1531 to 1542) of the single-stranded 3' end of 16S rRNA are involved in the formation of the SD duplex. As a consequence of the relaxation, the E-site codon adopts a conformation that is closer to the classic A-helical conformation (Fig. 3B), resulting in the formation of a base pair between the first position of the E-site codon and position 36 of the cognate E-site tRNA (Figure 2D). The interaction is stabilized by G693 of 16S rRNA and the 2-methylthio group of ms^{2}i^{6}A37 of the tRNA. Whether the observed codon-anticodon interaction in the E site would exist without stabilization by the modification remains uncertain. The second and third positions of the E-site codon are too distant from positions 35 and 34 of the E-site tRNA to form standard Watson-Crick base pairs. In the relaxed conformation immediately upstream of the E-site codon, the base of mRNA nucleotide −4 interacts with A1507 of 16S rRNA (Figure 2D).

The SD duplexes in the tense and relaxed conformations of the mRNA may delimit the extent of the movement that the mRNA upstream region can undergo on the platform of the ribosome, implying that the SD helix can take intermediate positions between these two extremes. In support of this notion, crystal structures of the 30S subunit (Kaminishi et al., 2007) and the 70S ribosome (Korostelev et al., 2006) report positions of the SD duplex that are between or near the ones observed here for the initiation and elongation states.

2.5. mRNA movement on the ribosome

Based on our crystal structures, we suggest the following simplified scheme for the movement of mRNA at different stages of translation (Yusupova et al. 2006; Jenner et al. 2007; Jenner et al. 2010a; Demeshkina et al. 2010) (Figure 3C). In the first step of initiation complex formation, the mRNA binds to the ribosome and establishes the SD duplex between the 5'-untranslated region

and the single-stranded 3' end of 16S rRNA. Subsequently, the SD duplex interacts with the 30S platform and is oriented towards ribosomal protein S2 (presumably as seen in the ribosomal complex with the nonanucleotide SD, see above), while the rest of the mRNA molecule is still unbound (Figure 3C, stage I) (Yusupova et al., 2006). The following step is a simultaneous positional adjustment of the initiation codon in the P site, where it is stabilized by codon-anticodon interaction with initiator tRNA^{fMet}, and a rearrangement of the SD duplex towards ribosomal protein S18 (Figure 3C, stage II). At this stage of initiation, the mRNA is present in a tense conformation and is precisely positioned on the 30S subunit, so that the start codon will be read first and in the correct frame. Our findings of mRNA adjustments during the initiation process are in good agreement with earlier cross-linking results (Canonaco et al., 1989; Rinke-Appel et al., 1994; La Teana et al., 1995). Immediately after initiation, when one or several codons have already been translated by the ribosome but the SD interaction is still intact, a simultaneous movement of the complete mRNA in the 5'-end direction and a lengthening of the SD helix take place (Figure 3C, stage III). The SD helix is once again shifted towards ribosomal protein S2. Further translation of mRNA by the ribosome is accompanied by movement in the 3'- to- 5' direction and leads to melting of the SD interaction. Eventually, at some stage during elongation, the 5' end of mRNA will no longer interact with the ribosome at all (Figure 3C, stage IV) and will become accessible for other ribosomes, which can lead to formation of a polysome.

3. Stabilization of tRNA in the A site of the 70S ribosome

During initial selection, aminoacyl-tRNA in a complex with elongation factor Tu (EF-Tu) and GTP is delivered to the ribosome. In the decoding site of the small ribosomal subunit, cognate aminoacyl-tRNA is recognized by a network of interactions between the codon-anticodon helix with residues of 16S rRNA (Ogle et al., 2001; Schmeing and Ramakrishnan, 2009; Schmeing et al., 2009a). After GTP hydrolysis and dissociation from EF-Tu, aminoacyl-tRNA is fully accommodated on the large ribosomal subunit and stabilized by a network of interactions with ribosomal proteins and 23S rRNA (Yusupov et al., 2001; Voorhees et al., 2009; Jenner et al., 2010b).

The comparison of two high-resolution structures of ribosomes with either empty A site or with the A

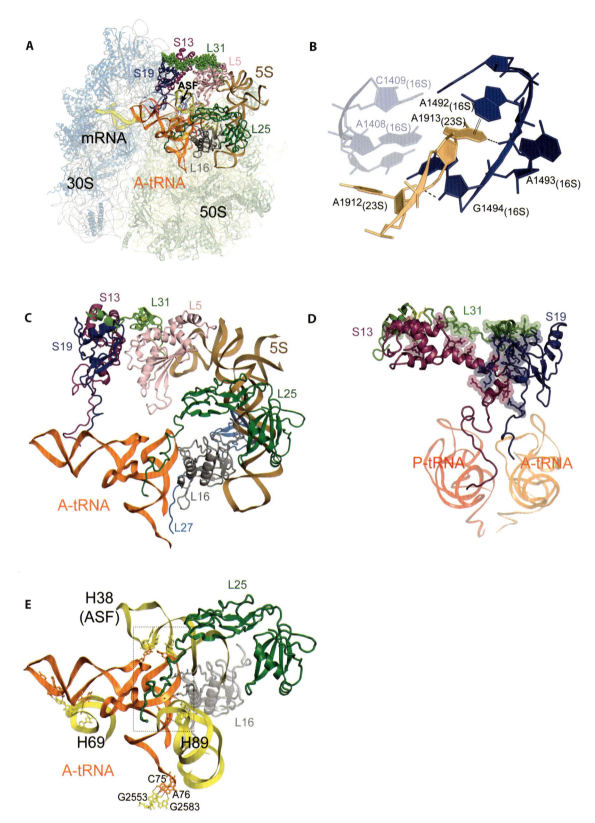

Fig. 4 Protein environment of the A-site tRNA. (A) Overall view of the ribosome. Ribosomal elements around the A-site tRNA are highlighted in colors. Protein L31 is represented as spheres. (B) Decoding center without tRNA. Stacking between A1492 of 16S rRNA and A1913 of 23S rRNA is indicated by the double line. (C) Circle of interactions around cognate tRNA composed of proteins of the small (S) and large (L) ribosomal subunits. (D) Intersubunit bridge formed by protein L31. Interactions between L31, S13 and S19 are indicated by transparent spheres. (E) Stabilization of the elbow region of cognate tRNA by the C-terminus of protein L25 (framed). (Figure reproduced from Jenner et al. 2010b with permission)

site occupied by cognate deacylated tRNA (Figure 4A) revealed new structural details of how tRNA is accommodated on the ribosome (Jenner et al., 2010b). According to our data, rearrangements of the universally conserved A1492 and A1493 of 16S rRNA, which monitor the correct codon-anticodon duplex in the A site of the small subunit (Ogle et al., 2001), are accompanied by changes in 23S rRNA of the large subunit. Thus, when the A site is empty, A1492 of 16S rRNA is held inside helix 44 (h44) of 16S rRNA by stacking with A1913 of helix 69 (H69) of 23S rRNA (Figure 4B), while A1493 protrudes from h44. This arrangement of A1492, A1493 of 16S and A1913 of 23S is stabilized by hydrogen bonds to the neighboring G1494 of 16S rRNA and A1912 of 23S rRNA. This orientation of A1493 and A1492 of 16S rRNA as well as of A1913 of 23S rRNA is apparently induced by contacts of the full-length tRNA in the P site with both h44 and H69 of rRNA. Binding of cognate tRNA to the A site does not greatly affect the position of A1493 of 16S rRNA, which approaches the first base-pair of the codon-anticodon duplex, but results in rearrangements of A1492 of 16S and A1913 of 23S rRNA. The base of A1913 (H69) swings more than 180° around the N-glycosidic bond and A1492 flips out from h44 and interacts with the second position of the codon-anticodon duplex in the A site. G530 of 16S rRNA, which was implicated in decoding as well, changes its conformation from syn to anti (Ogle et al., 2001). As a result of these rearrangements, the entire anticodon loop of tRNA becomes closely monitored by residues from both 16S and 23S rRNA, with the additional involvement of magnesium ions in the stabilization network (Jenner et al., 2010b).

We also observed substantial changes in several ribosomal proteins, conserved among bacteria, that approach the A-site tRNA by their flexible tails, which were either built differently or not modeled in previous crystal structures (Figure 4C). The full-length *T. thermophilus* protein S19 comprises 93 amino acids. None of the previous high-resolution structures contains the entire protein, as density for the last 12 amino acid was missing. In the present elongation complex, we traced the main chain of S19 up to Ala89 and found that the C-terminal tail of S19 comes close to the anticodon stem of the A-site tRNA (Figure 4C, D). Protein S19 interacts with protein S13 (Brodersen et al., 2002), whose C-terminus, according to our data, is quite different from the known crystal structures of this region of S13, where the C-terminal tail resides roughly between the anticodon stem-loops of the tRNAs in A and P sites (Selmer et al., 2006; Schmeing et al., 2009b;

Voorhees et al., 2009). In our model, the positively charged C-terminal tail of S13 is stabilized by the sugar-phosphate backbone of the anticodon loop of the P-site tRNA (Figure 4D).

Proteins S13 and S19 are coupled by an intersubunit bridge formed by protein L31 (Figure 4A, D), which is conserved among bacteria and has not been described before (Jenner et al., 2010b). The existing models of L31 are incomplete – they comprise either the N-terminal zinc-binding region only (Korostelev et al., 2008a) or the N-terminal region together with a part of the C-terminal loop. The structure of the latter is similar to our structure, but is largely out of register because it was modeled as an unstructured loop (Schmeing et al., 2009b). Protein L31 links the labile head domain of the 30S subunit through S13 and S19 with protein L5 and 5S rRNA located in the central protuberance of the large subunit (Figure 4C). Such a location of L31 and the observation that proteins S13 and S19, upon binding of tRNA in the A site, are displaced 2–3 Å toward the P site suggest that protein L31 has a regulatory role. Thus, L31, through proteins S13 and S19, may modulate the initial swiveling movements of the 30S head domain, which were described to accompany the ratchet-like movement of the small subunit relative to the large one (Frank and Agrawal, 2000; Spahn et al., 2004). This way, L31 may be involved in signaling, through 5S rRNA, the correct tRNA accommodation to the base and tip of the L7/L12 stalk.

The binding of the elbow region of tRNA in the A site on the ribosome is stabilized by interactions with ribosomal proteins and 23S rRNA (Figure 4E). The previous X-ray structures demonstrated that the elbow region of cognate tRNA is stabilized by protein L16 and H38 (A-site finger, ASF) with H89 of 23S rRNA (Yusupov et al., 2001; Voorhees et al., 2009). The quality of the electron density of the elongation complex permitted us to model the C-terminal tail of L25 up to residue 200 (out of 206) (Figure 4E), which is absent in other X-ray structures. This extremity of L25 follows the cleft formed by protein L16, the ASF, the elbow region of the A-site tRNA, and H89 of 23S rRNA. It also approaches H43 (L7/L12 stalk region) and H95 of 23S rRNA (sarcin-ricin loop), suggesting that L25 stabilizes the contacts between these components. The stabilization of the elbow region of the tRNA on the large subunit seems to induce further rearrangements in and around the peptidyl transferase center (PTC), which are similar to those reported for an aminoacylated tRNA fragment and for full tRNA in the A site (Schmeing et al., 2005; Voorhees et al., 2009). The coordinated shift of H89 and L16 that is induced by the fixation of the elbow region of the A-site tRNA may lead to re-arrangements of the contacts between the flexible loop of L16 and the N-terminus of protein L27 (Figure 4C), which approaches the PTC. The N-terminal tail of L27, which appears disordered in the initiation complex, is aligned between the acceptor ends of tRNAs in the P and A sites (Jenner et al., 2010b), as in a previous complex with aminoacylated tRNAs (Voorhees et al., 2009). In the latter structure, the first residue of L27 (Ala2) stabilizes the base of H89, where A2451 is located in the PTC (Lang et al., 2008).

Overall, the resulting structural framework suggests that the accommodated A- site tRNA forms part of a circular layer which is rich in proteins, centered on the ASF (Figure 4A), and comprises about half of the diameter of the ribosome. Therefore, our structures provide the view that the induced-fit mechanism of the accommodation step is accomplished by surrounding the cognate tRNA with dynamic layers of rRNA and proteins from both ribosomal subunits. In such a closed system of interactions, signals can be transmitted efficiently in order to increase the accuracy during the accommodation step and, at the same time, ensure flexibility for the ensuing translocation.

References

Bouadloun F, Srichaiyo T, Isaksson LA, Bjork GR (1986) Influence of modification next to the anticodon in tRNA on codon context sensitivity of translational suppression and accuracy. J Bacteriol 169: 1022–1027

Brodersen DE, Clemons WM, Jr., Carter AP, Wimberly BT, Ramakrishnan V (2002) Crystal structure of the 30 S ribosomal subunit from Thermus thermophilus: structure of the proteins and their interactions with 16 S RNA. J Mol Biol 316: 725–768

Canonaco MA, Gualerzi CO, Pon CL (1989) Alternative occupancy of a dual ribosomal binding site by mRNA affected by translation initiation factors. Eur J Biochem 182: 501–506

Demeshkina N, Jenner L, Yusupova G, Yusupov M (2010) Interactions of the ribosome with mRNA and tRNA. Curr Opin Struc Biol 20: 325–332

Gustilo EM, Vendeix FA, Agris PF (2008) tRNA's modifications bring order to gene expression. Curr Opin Microbiol 11: 134–140

Hoburg A, Aschhoff HJ, Kersten H, Manderschied U, Gassen HG (1979) Function of modified nucleosides 7-methylguanosine, ribothymidine, and 2-thiomethyl-N6-(isopentenyl)adenosine in procaryotic transfer ribonucleic acid. J Bacteriol 140: 408–414

Jenner L, Romby P, Rees B, Schulze-Briese C, Springer M, Ehresmann C, Ehresmann B, Moras D, Yusupova G, Yusupov M (2005) Translational operator of mRNA on the ribosome: how repressor proteins exclude ribosome binding. Science 308: 120–123

Jenner L, Rees B, Yusupov M, Yusupova G (2007) Messenger RNA conformations in the ribosomal E site revealed by X-ray crystallography. EMBO Rep 8: 846–850

Jenner L, Demeshkina N, Yusupova G, Yusupov M (2010a) Structural aspects of messenger RNA reading frame maintenance by the ribosome. Nat Struc Mol Biol 17: 555–560

Jenner L, Demeshkina N, Yusupova G, Yusupov M (2010b) Structural rearrangements of the ribosome at the tRNA proofreading step. Nat Struc Mol Biol 17: 1072–1078

Jorgensen F, Kurland CG (1990) Processivity errors of gene expression in Escherichia coli. J Mol Biol 215: 511–521

Jukes TH (1973) Possibilities for the evolution of the genetic code from a preceding form. Nature 246: 22–26

Kaminishi T, Wilson DN, Takemoto C, Harms JM, Kawazoe M, Schluenzen F, Hanawa-Suetsugu K, Shirouzu M, Fucini P, Yokoyama S (2007) A snapshot of the 30S ribosomal subunit capturing mRNA via the Shine-Dalgarno interaction. Structure 15: 289–297

Konevega AL, Soboleva NG, Makhno VI, Peshekhonov AV, Katunin VI (2006) [The effect of modification of tRNA nucleotide-37 on the tRNA interaction with the P- and A-site of the 70S ribosome Escherichia coli]. Mol Biol (Mosk) 40: 669–683

Korostelev A, Trakhanov S, Laurberg M, Noller HF (2006) Crystal structure of a 70S ribosome-tRNA complex reveals functional interactions and rearrangements. Cell 126: 1065–1077

Korostelev A, Asahara H, Lancaster L, Laurberg M, Hirschi A, Zhu J, Trakhanov S, Scott WG, Noller HF (2008a) Crystal structure of a translation termination complex formed with release factor RF2. Proc Natl Acad Sci USA 105: 19 684–19 689

Korostelev A, Asahara H, Lancaster L, Laurberg M, Hirschi A, Zhu J, Trakhanov S, Scott WG, Noller HF (2008b) Crystal structure of a translation termination complex formed with release factor RF2. Proc Natl Acad Sci USA 105: 19 684–19 689

La Teana A, Gualerzi CO, Brimacombe R (1995) From stand-by to decoding site. Adjustment of the mRNA on the 30S ribosomal subunit under the influence of the initiation factors. RNA 1: 772–782

Laurberg M, Asahara H, Korostelev A, Zhu J, Trakhanov S, Noller HF (2008) Structural basis for translation termination on the 70S ribosome. Nature 454: 852–857

Ma J, Campbell A, Karlin S (2002) Correlations between Shine-Dalgarno sequences and gene features such as predicted expression levels and operon structures. J Bacteriol 184: 5733–5745

Menichi B, Heyman T (1976) Study of tyrosine transfer ribonucleic acid modification in relation to sporulation in Bacillus subtilis. J Bacteriol 127: 268–280

Ogle JM, Brodersen DE, Clemons WM, Jr., Tarry MJ, Carter AP, Ramakrishnan V (2001) Recognition of cognate transfer RNA by the 30S ribosomal subunit. Science 292: 897–902

Ogle JM, Murphy FV, Tarry MJ, Ramakrishnan V (2002) Selection of tRNA by the ribosome requires a transition from an open to a closed form. Cell 111: 721–732

Rinke-Appel J, Junke N, Brimacombe R, Lavrik I, Dokudovskaya S, Dontsova O, Bogdanov A (1994) Contacts between 16S ribosomal RNA and mRNA, within the spacer region separating the AUG initiator codon and the Shine-Dalgarno sequence; a site-directed cross-linking study. Nucleic Acids Res 22: 3018–3025

Rozenski J, Crain PF, McCloskey JA (1999) The RNA Modification Database: 1999 update. Nucleic Acids Res 27: 196–197

Schmeing TM, Huang KS, Strobel SA, Steitz TA (2005) An induced-fit mechanism to promote peptide bond formation and exclude hydrolysis of peptidyl-tRNA. Nature 438: 520–524

Schmeing TM, Ramakrishnan V (2009) What recent ribosome structures have revealed about the mechanism of translation. Nature 461: 1234–1242

Schmeing TM, Voorhees RM, Kelley AC, Gao YG, Murphy FVt, Weir JR, Ramakrishnan V (2009a) The crystal structure of the ribosome bound to EF-Tu and aminoacyl-tRNA. Science 326: 688–694

Schurr T, Nadir E, Margalit H (1993) Identification and characterization of *E. coli* ribosomal binding sites by free energy computation. Nucleic Acids Res 21: 4019–4023

Selmer M, Dunham CM, Murphy FVt, Weixlbaumer A, Petry S, Kelley AC, Weir JR, Ramakrishnan V (2006) Structure of the 70S ribosome complexed with mRNA and tRNA. Science 313: 1935–1942

Shine J, Dalgarno L (1974) The 3'-terminal sequence of *Escherichia coli* 16S ribosomal RNA: complementarity to nonsense triplets and ribosome binding sites. Proc Natl Acad Sci USA 71: 1342–1346

Spahn CM, Gomez-Lorenzo MG, Grassucci RA, Jorgensen R, Andersen GR, Beckmann R, Penczek PA, Ballesta JP, Frank J (2004) Domain movements of elongation factor eEF2 and the eukaryotic 80S ribosome facilitate tRNA translocation. EMBO J 23: 1008–1019

Urbonavicius J, Qian Q, Durand JM, Hagervall TG, Bjork GR (2001) Improvement of reading frame maintenance is a common function for several tRNA modifications. EMBO J 20: 4863–873

Urbonavicius J, Stahl G, Durand JM, Ben Salem SN, Qian Q, Farabaugh PJ, Bjork GR (2003) Transfer RNA modifications that alter +1 frameshifting in general fail to affect −1 frameshifting. RNA 9: 760–768

Valle M, Zavialov A, Sengupta J, Rawat U, Ehrenberg M, Frank J (2003) Locking and unlocking of ribosomal motions. Cell 114: 123–134

Voorhees RM, Weixlbaumer A, Loakes D, Kelley AC, Ramakrishnan V (2009) Insights into substrate stabilization from snapshots of the peptidyl transferase center of the intact 70S ribosome. Nat Struct Mol Biol 16: 528–533

Weixlbaumer A, Murphy FVt, Dziergowska A, Malkiewicz A, Vendeix FA, Agris PF, Ramakrishnan V (2007a) Mechanism for expanding the decoding capacity of transfer RNAs by modification of uridines. Nat Struct Mol Biol 14: 498–502

Weixlbaumer A, Petry S, Dunham CM, Selmer M, Kelley AC, Ramakrishnan V (2007b) Crystal structure of the ribosome recycling factor bound to the ribosome. Nat Struct Mol Biol 14: 733–737

Weixlbaumer A, Jin H, Neubauer C, Voorhees RM, Petry S, Kelley AC, Ramakrishnan V (2008) Insights into translational termination from the structure of RF2 bound to the ribosome. Science 322: 953–956

Wilson RK, Roe BA (1989) Presence of the hypermodified nucleotide N6-(delta 2-isopentenyl)-2-methylthioadenosine prevents codon misreading by *Escherichia coli* phenylalanyl-transfer RNA. Proc Natl Acad Sci USA 86: 409–413

Yusupov MM, Yusupova GZ, Baucom A, Lieberman K, Earnest TN, Cate JH, Noller HF (2001) Crystal structure of the ribosome at 5.5 A resolution. Science 292: 883–896

Yusupova G, Jenner L, Rees B, Moras D, Yusupov M (2006) Structural basis for messenger RNA movement on the ribosome. Nature 444: 391–394

Yusupova GZ, Yusupov MM, Cate JH, Noller HF (2001) The path of messenger RNA through the ribosome. Cell 106: 233–241

Zhang W, Dunkle JA, Cate JH (2009) Structures of the ribosome in intermediate states of ratcheting. Science 325: 1014–1017

Genetic and crystallographic approaches to investigating ribosome structure and function

5

Steven T. Gregory, Hasan Demirci, Jennifer F. Carr, Riccardo Belardinelli, Jill R. Thompson, Dale Cameron, Daniel Rodriguez-Correa, Frank Murphy, Gerwald Jogl and Albert E. Dahlberg

1. Introduction

The past decade has seen tremendous advances in our understanding of the mechanism of protein synthesis, due in part to the solution of ribosome structures by X-ray crystallography. These structures have clarified our view of the decoding process so that it can now be understood in stereochemical terms, and have demonstrated that the ribosome is a ribozyme, catalyzing peptide bond formation using RNA's capacity to adopt complex three-dimensional arrangements. While the ribosome structure solutions represent fundamental technical achievements, perhaps their most important contribution is that they explain some four decades of genetic and biochemical studies of the ribosome. It could perhaps be said with only slight exaggeration that little about the ribosome makes sense except in the light of its three-dimensional structure.

Genetics has played a central role in developing our understanding of ribosome function, and geneticists studying the ribosome have long desired to interpret the phenotypic behavior of their mutants in structural terms. While it is tempting to use structures of wild-type ribosomes to infer the molecular basis of various mutant phenotypes, such interpretations are intrinsically difficult to authenticate and many are destined to be incorrect. It would be more efficacious to determine directly the X-ray crystal structures of mutant ribosomes. This is not a trivial task, as it requires a single organism to be amenable to both genetics and X-ray crystallography. One approach is to achieve crystallization of ribosomes from the geneticist's bacterium of choice, *Escherichia coli*. The Cate laboratory has had remarkable success in just such an endeavor (Vila-Sanjurjo et al., 2003; Schuwirth et al., 2005; Zhang et al., 2009) making the vast resource of pre-existing *E. coli* ribosomal mutants accessible via X-ray crys-

tallography. Our laboratory has taken a parallel approach, developing the ribosomal genetics of *T. thermophilus*, whose 30S subunit and 70S ribosome X-ray crystal structures have been repeatedly solved at high resolution and in various complexes with substrates and protein factors (reviewed by Schmeing and Ramakrishnan, 2009; Korostelev and Noller, 2007). Here we summarize our work on ribosomal genetics in *T. thermophilus* and describe some of our early efforts to determine the structures of mutant ribosomes.

2. Thermus thermophilus as a model system

2.1. Genetics of T. thermophilus

T. thermophilus exhibits a number of properties which make it appealing to the bacterial geneticist (for a recent review, see Cava et al., 2009). Most importantly, it is naturally competent for transformation with chromosomal DNA, allowing the genetic mapping of tightly linked mutations by homologous recombination. Other gene transfer methods that would allow mapping of more widely separated markers, such as generalized transduction and mating, have yet to be developed, although the existence of a number of *Thermus*-specific phages (Sakaki and Oshima, 1975; Yu et al., 2006) and an Hfr-like mechanism (Ramírez-Arcos et al., 1998) have both been reported. Putative insertion sequences (IS elements) have been identified in the *T. thermophilus* genome and the demonstration of active transposition of an IS element, IS*Tth7*, in *T. thermophilus* HB8 (Gregory and Dahlberg, 2008) suggests that composite transposons could be constructed. Three antibiotic resistance genes, conferring resistance to kanamycin, hygromycin B or bleomycin, have been experimentally evolved to function at high temperature and serve as selectable markers for the construction of gene knockout alleles (Cameron et al., 2004a; Gregory and Dahlberg, 2009) and a variety of shuttle vectors. *T. thermophilus* is sensitive to almost all antibiotics that target the ribosome, with the exceptions of spectinomycin and kasugamycin, thus providing numerous selections for ribosomal mutants.

One important advantage of working with *T. thermophilus* is that it has not been subjected to extensive laboratory manipulations and chemical mutagenesis and has not been adapted to laboratory growth conditions. The *T. thermophilus* strains used for genetics and X-ray crystallography are therefore very close to, if not identical with, the original natural isolates. Genome sequences have been completed for two Japanese strains of *T. thermophilus*, HB27 (Henne et al., 2004) and the closely related HB8 (Masui et al., 2005) which is the primary source of proteins and ribosomes for X-ray crystallography. These sequences allow the rapid identification of mutations by sequencing. Our genetic studies have utilized both HB8 and the Icelandic strain IB-21 interchangeably, with efficient and straightforward transfer of alleles from one strain to the other by natural transformation.

2.2. Ribosomal genes and antibiotic-resistance mutations

The ribosomal genes of *T. thermophilus* are organized into rRNA and r-protein operons as in other bacterial species, but with some interesting idiosyncrasies. Two *rrs* genes encoding 16S rRNA are transcribed independently of two *rrl-rrf-glyT* operons encoding 23S rRNA, 5S rRNA and tRNA^{Gly}. There are thus four separate genetic loci for rRNA in *T. thermophilus*. Our own analyses indicate that these loci are genetically unlinked, consistent with the large physical distances between them in the genome. Ribosomal protein operons are similar to those of other bacteria in their gene order, although the *str*, S10, and *spc* operons, which are each transcribed from their own promoters in *E. coli*, are fused into a single large transcription unit in *T. thermophilus* (Pfeiffer et al., 1995).

Because of the high frequency with which homologous recombination occurs in *T. thermophilus*, mutations in one rRNA gene arising in a genetic selection are often rapidly transferred to the second locus by gene conversion (Gregory et al., 2005). Thus, strains producing homogeneous populations of mutant ribosomes can be generated without prior strain engineering and differ from the original natural isolates at only two base-pairs in the entire genome. That homogeneity results from gene conversion rather than two independent mutational events is indicated by the fact that all double mutants we have ever characterized have the same base substitution in both rRNA genes. We have found this to be the case for *rrl* mutations conferring resistance to thiostrepton (Cameron et al., 2004b), tylosin, lincomycin, sparsomycin or chloramphenicol, and *rrs* mutations conferring resistance to kanamycin, neomycin, gentamicin, apramycin, hygromycin B or capreomycin (Gregory et al., 2005). Where this has not been the case is in selections for erythromycin-resistance, which produced heteroallelic *rrlA* mutants (Gregory et al., 2001a); with streptomycin-resistance mutations in *rrs* genes, which are genetically recessive (Gregory and Dahlberg, 2009); and with capreomycin-resistance mutations in *rrlA* (Monshupanee et al., 2008). Similarly, dominant kirromycin-resistance mutations appear to arise preferentially in *tufB* rather than in *tufA* (our unpublished observation).

Simple antibiotic selections have therefore enabled us to target all the major functional centers of the ribosome and to isolate in *T. thermophilus* many of the same ribosomal mutations that have been accumulated with several decades of work in *E. coli* and other bacteria. The similarity of mutations isolated in *T. thermophilus* and in other organisms supports the validity of *T. thermophilus* as a model system for ribosomal genetics and provides a foundation for the identification and characterization of novel mutations.

2.3. Streptomycin resistance and dependence

The aminoglycoside antibiotic streptomycin and mutations affecting the cell's response to it have historically played important roles in deciphering the mechanism of tRNA selection and we continue to use them to examine this process in detail. Streptomycin causes translational misreading by uncoupling codon recognition from GTP hydrolysis by EF-Tu (Gromadski and Rodnina, 2004). Streptomycin's bactericidal action, and the attending physiological effects, such as breakdown of membrane potential and leakage of small molecules, are ultimately the consequences of aberrant protein synthesis. As such, all these effects are eliminated by bacteriostatic inhibitors of protein synthesis, such as chloramphenicol (Jawetz et al., 1952), or by mutations in the ribosome (Gale et al., 1981).

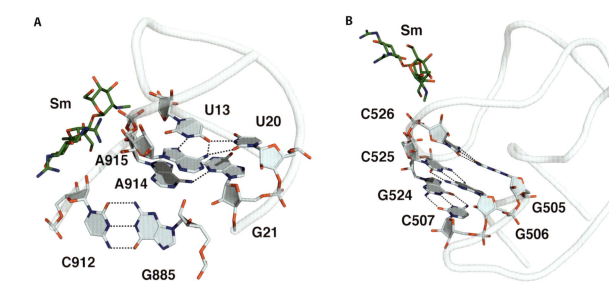

Fig. 1 (A) The central pseudoknot of 16S rRNA. Base substitutions U13C, U20G, C912A, A914G or A915G confer streptomycin resistance. (B) The helix 18 pseudoknot of 16S rRNA. Base substitutions C507A or G524U confer streptomycin dependence. Sm, streptomycin. Both figures derived from pdb entry 1fjg (Carter et al., 2000).

The connection between misreading and membrane damage is integral to the bactericidal action of streptomycin (Davis et al., 1986). Streptomycin resistance was observed by Lederberg to be a genetically recessive trait in *E. coli* (Lederberg, 1951) which, in light of streptomycin's bactericidal action, can be interpreted as resulting from misreading by the sensitive fraction of ribosomes in merodiploid strains.

Streptomycin resistance and dependence appear to be recessive in *T. thermophilus* as well, since our initial selections led exclusively to mutations in *rpsL* encoding ribosomal protein S12 (Gregory et al., 2001b). These included many of the "classical" *rpsL* mutations identified in other bacteria. The most commonly observed streptomycin-dependence mutations in all organisms are those that substitute arginine or leucine for the conserved Pro90 of ribosomal protein S12. To examine the basis for this specificity, we assessed the phenotypes of several substitutions at this position introduced by site-directed mutagenesis and gene replacement (Carr et al., 2005). The phenotypic behavior of these mutants was revealing, with bulky R groups such as arginine, glutamic acid, leucine, methionine, or tryptophan resulting in dependence and smaller R groups such as alanine, glycine, or cysteine resulting only in resistance. These observations can be interpreted readily in the context of the induced-fit model for decoding in which cognate-codon recognition favors a closed conformation of the 30S subunit (Ogle et al., 2002). This closed conformation is stabilized by interaction of ribosomal protein S12 with 16S rRNA helix 44; replacement of Pro90 with bulkier side chains could interfere with this interaction to such an extent that streptomycin is required to achieve the closed conformation (Carr et al., 2005).

Upon deleting the 16S rRNA gene *rrsA*, we were able to isolate streptomycin-resistance mutations in the single remaining intact 16S rRNA gene, *rrsB* (Gregory and Dahlberg, 2009). Five individual base substitutions were identified in the central pseudoknot of 16S rRNA, which forms part of the streptomycin binding site (Figure 1). Given that streptomycin interacts with 16S rRNA exclusively through backbone contacts, it seems probable that these base substitutions act by perturbing the conformation of the pseudoknot, a conclusion that is supported by base-specific chemical probing.

Using the *rrsA* deletion strain, we also identified two streptomycin-dependence mutations in *rrsB* affecting 16S rRNA helix 18 (Figure 1). This helix contains a pseudoknot formed by the three consecutive Watson-Crick base pairs G505-C526, G506-C525 and C507-G524. Either C507A or G524U, generating A•G or C•U mismatches, respectively, confers dependence, while combining these two substitutions and restoring Watson-Crick pairing reverses dependence. Secondary mutations relieving the streptomycin dependence phenotypes of the *rpsL*-P90L, *rrsB*-C507A or *rrsB*-G524U mutations arose readily. While most of these produce base substitutions in 16S rRNA in close proximity to the original substitution in the three-dimensional structure of the 30S subunit, others occur some distance away. These cannot be explained simply in terms of local static conformational effects, but seem more likely to act by influencing the conformational dynamics of the 30S subunit during tRNA selection.

2.4. Communication between ribosomal protein S12 and EF-Tu

From genetic selections for streptomycin-dependent mutants, we identified several *rpsL* double mutants including *rpsL*-P90L H76R. The *rpsL*-P90L mutation on its own has been selected in a number of organisms for conferring streptomycin dependence, and, while Timms and Bridges described P90L H76R double mutants of *E. coli*, the phenotype produced by the H76R mutation alone was not determined (Timms and Bridges 1993). Using site-directed mutagenesis and gene replacement, we constructed the *rpsL*-H76R single mutant of *T. thermophilus* and found it to be resistant to streptomycin (Gregory et al., 2009). This was unexpected since His76 is some 30 Å away from the streptomycin binding site (Carter et al., 2000). Calorimetry experiments also indicated that H76R mutant ribosomes retain their ability to bind streptomycin, ruling out a long-range effect on ribosomal protein S12 conformation. His76 is also not near any contact sites with 16S rRNA, but instead is oriented toward the intersubunit space.

As stated earlier, streptomycin abolishes tRNA discrimination by uncoupling codon recognition from GTPase activation of EF-Tu, such that GTP hydrolysis rates become similar for both cognate and near-cognate substrates (Gromadski and Rodnina, 2004). These observations suggest that mutations influencing the coupling of codon-recognition and GTPase activation could themselves either generate or suppress streptomycin-resistance. Such mutations need not alter residues proximal to the streptomycin binding site, but instead could affect residues engaging in contacts important for signaling codon recognition to EF-Tu. His76 is part of a conserved QEH triplet that protrudes from the surface of S12 and contacts the acceptor helix of aminoacyl-tRNA in the A/T state, bound to EF-Tu and the ribosome (Schmeing et al., 2009) (Figure 2). The QEH triplet is therefore in a position suitable for signaling codon recognition to EF-Tu via aminoacyl-tRNA. This suggests a hypothesis in which substitutions at His76 generate streptomycin-resistance by impinging upon the contribution of the QEH triplet to GTPase activation.

To further establish the connection between the S12 QEH triplet and EF-Tu, we performed a genetic selection for spontaneous kirromycin-resistant derivatives of the *rpsL*-H76R mutant and screened for ones whose streptomycin-resistance phenotype was masked by the kirromycin-resistance mutation. The use of kirromycin allowed us to target the *tuf* genes encoding EF-Tu even if it restricted our search to a subclass of all possible suppressors. One suppressor we identified had a *tufB* mutation, producing an A375T amino acid substitution in EF-Tu. This mutant allele is identical to *tufA*$_R$ of *E. coli* (Duisterwinkel et al., 1981) and *tufA8* of *Salmonella typhimurium* (Tubulekas et al., 1991). Interestingly, the *tufA*$_R$ allele in *E. coli* also suppresses the streptomycin-dependence phenotype of the *rpsL*-P90L mutation (Zuurmond et al., 1998) and the streptomycin-resistance phenotype of an *rpsL*-R53L mutation (Tubulekas et al., 1991). We found that the *T. thermophilus* mutant *tufB* allele suppresses the streptomycin-resistance phenotypes of the mutations *rpsL*-R37C and *rpsL*-K53E (Gregory et al., 2009). Further, *rpsL*-K53E in *T. thermophilus* and *rpsL*-R53L in *E. coli* each, on their own, confer weak kirromycin-resistance phenotypes. Pro90, Arg37 and Lys53 do not contact the ternary complex, nor do they contact streptomycin, but instead are on the face of S12 that contacts 16S rRNA helix 44 upon codon recognition and domain closure. The similarity in behavior of the EF-Tu mutation toward both sets of S12 mutations suggests a functional relationship between domain closure induced by codon recognition and the contact between S12 and aminoacyl-tRNA in the A/T state.

Streptomycin is thought to stabilize codon-anticodon interaction for near-cognate aminoacyl-tRNA by shifting the conformational equilibrium of the 30S subunit toward domain closure. The contact involving the QEH triplet is one of the few direct interactions between the ribosome and aminoacyl-tRNA in the A/T state, and loss of this contact could restore discrimination in the presence of streptomycin by increasing the relative influence of codon-anticodon interaction. Alternatively, the QEH contact may contribute to the bending of aminoacyl-tRNA in the A/T state, with disruption of this contact making bending more difficult. Either of these effects is expected to produce an error-restrictive phenotype, opposite that of the Hirsh

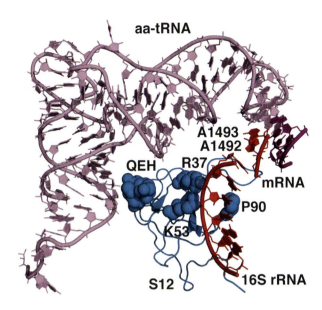

Fig. 2 Spatial relationship between the QEH triplet of ribosomal protein S12, the ternary complex, and 16S rRNA. Critical residues A1492 and A1493 of 16S rRNA are shown in red, as are 16S rRNA residues C1409-G1415. Based on pdb entry 2wrn (Schmeing et al., 2009).

suppressor mutation in the D arm of tRNA^{Trp}, which may facilitate GTPase activation by reducing the thermodynamic barrier to tRNA bending (Cochella and Green, 2005; Schmeing et al., 2009). While the precise mechanism by which the *rpsL*-H76R mutation confers streptomycin resistance remains somewhat speculative, the sum of the data point to a previously unrecog-

nized role for ribosomal protein S12 in GTPase activation of EF-Tu.

3. X-ray crystallography of mutant 30S subunits

3.1. Probing the structural basis for streptomycin dependence

Crystallization of 30S subunits bearing the G524U substitution produced a novel crystal form. These monoclinic crystals have an asymmetric unit cell with eight subunits, rather than the single subunit in the tetragonal crystals formed by wild-type subunits, creating some challenges in structure determination and resulting in a slightly lower resolution (4.5 Å). Remarkably, tetragonal crystals were obtained with G524U mutant subunits in the presence of streptomycin or with double mutant subunits containing G524U and a second-site mutation that relieves streptomycin dependence. Thus, the crystallization behavior of these mutants mirrors their phenotype.

Some functional consequences of the G524U substitution are implicit in the mutant X-ray crystal structure. The most notable effect is on the anticodon stem loop (ASL)-mRNA complex in the A site (Figure 3). Electron density for much of the ASL and mRNA is very weak, despite the presence of the antibiotic paromomycin, which stabilizes ASL binding to wild-type

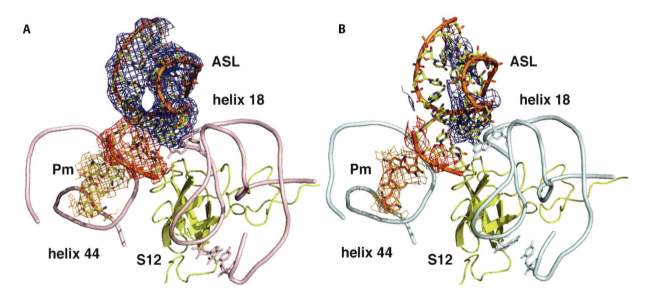

Fig. 3 Destabilization of codon-anticodon interaction by the streptomycin-dependence mutation G524U. Shown are electron densities for the wild-type 30S subunit (pdb entry 1fjg, left) and for the G524U mutant (right).

30S subunit crystals (Ogle et al., 2001). Destabilization of the ASL-mRNA interaction could result in a diminished capacity of cognate-codon recognition to stimulate GTPase activation and may be intrinsic to the streptomycin-dependence phenotype.

Amino acid substitutions located at the interface of ribosomal proteins S4 and S5 typically exhibit a *ram* (ribosomal *am*biguity, or error-prone) phenotype. However, the Hughes laboratory has described *Salmonella* mutants with substitutions in S4 at this interface that do not behave as expected. Surprisingly, these mutations have error-restrictive phenotypes and confer streptomycin resistance, but nevertheless relieve streptomycin dependence (Björkman et al., 1999). Combining these error-restrictive alleles of *rpsD* (encoding ribosomal protein S4) with error-restrictive alleles of *rpsL* actually raised error frequencies to near wild-type levels. Similarly, the error-restrictive streptomycin-resistance substitution C912U in 16S rRNA suppresses the streptomycin-dependence phenotype of the P90L mutation in *rpsL* (Vila-Sanjurjo et al., 2007). Such behavior complicates interpretation of *ram* and error-restrictive mutations in the context of the domain closure model (or any model for that matter) and suggests that streptomycin dependence is not simply an extreme form of error restriction as originally envisioned (Gorini et al., 1967). It may also indicate that additional conformational states of the 30S subunit exist besides the two thus far identified crystallographically (Ogle et al., 2002). Unfortunately, there are as yet no direct biochemical or biophysical assays for domain closure and therefore no direct experimental evidence that the classical *ram* and error-restrictive mutations behave in the manner predicted by the domain closure model (Ogle et al., 2002). Such assays for domain closure are therefore needed to complement ongoing crystallographic studies and to clarify the relationship between error proneness and 30S subunit conformational equilibria.

In light of these considerations, it is worth noting the locations of some of the base substitutions relieving streptomycin-dependence that we have identified. Several are in close proximity to the C507-G524 base pair (Figure 4), and among these is A10G in helix 1 of 16S rRNA. Helices 1 and 18 interact with one another directly and exclusively through contacts via the 2′ OH groups of the A10-U24 and C507-G524 base pairs. Thus, as with amino acid substitutions at the S4-S5 interface, base substitutions affecting the helix 1-helix 18 interface can either generate or relieve streptomycin dependence, in a manner not yet explained within the currently accepted theoretical framework.

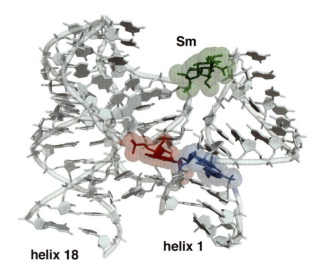

Fig. 4 Helix packing interaction at the 16S rRNA helix 1-helix 18 interface mediating streptomycin dependence. The C507-G524 base pair in helix 18 is shown as red spheres and plates. The A10-U24 base pair in helix 1 is shown as blue spheres and plates. Streptomycin is shown as green sticks and spheres. Based on pdb entry 1fjg (Carter et al., 2000).

3.2. Restructuring of the 30S subunit by rRNA modification

Translational accuracy is influenced by posttranscriptional modifications of rRNA and one of the most conserved of these is the dimethylation of two adenosines in the GGAA tetraloop of 16S rRNA helix 45. Mutations in the *ksgA* gene, encoding the KsgA methyltransferase, confer resistance to the antibiotic kasugamycin. Kasugamycin inhibits protein synthesis initiation and suppresses decoding errors during elongation (van Buul et al., 1984). Conversely, *ksgA* mutants are weakly error prone (van Buul et al., 1984) and exhibit increased initiation from non-AUG codons (O'Connor et al., 1997). Thus, modification by the KsgA methyltransferase influences the accuracy of codon recognition in both A and P sites.

To investigate the structural basis for both the pleiotropic phenotype of *ksgA* mutants and the functional role of this modification, we constructed a *ksgA* deletion mutant of *T. thermophilus* (Demirci et al., 2009) and determined the X-ray crystal structure of its 30S subunit. Comparison of the *ksgA* mutant structure with that of wild-type 30S subunits indicates that dimethylation by KsgA disrupts the canonical GGAA tetraloop fold, and thereby allows the formation of a hydrogen bonding network between helices 44 and 45 (Figure 5). The absence of this helix-packing interaction in the

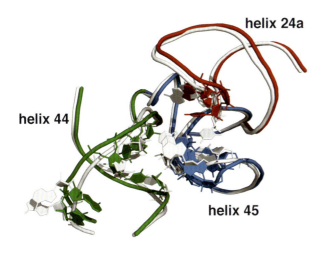

Fig. 5 Influence of the KsgA-dependent methylations on tertiary interactions in the 30S subunit. Dimethylation of A1518 and A1519 destabilizes the GGAA tetraloop of helix 45 and causes a bend that allows the formation of hydrogen bonding interactions with helix 44. The fully methylated structure is shown in white.

ksgA mutant impacts the surrounding structure, with small displacements in the decoding center, suggesting a basis for the *ram* phenotype, and shifts in the P site potentially explaining both kasugamycin resistance and effects on start codon recognition. The conformation of the helix 6 spur, which inserts into the P site of an adjacent subunit in the tetragonal crystal form, is also distorted, giving some indication of how the absence of methylation might perturb fMet-tRNA[fMet] interaction with the ribosome. Taken together, these data suggest that methylation by KsgA serves to optimize the conformation of the 30S subunit to engage in accurate protein synthesis. This conclusion is consistent with the notion of KsgA as a checkpoint in 30S subunit assembly (Xu et al., 2008), but goes further in implicating methylation itself as having a role in optimizing ribosome structure at a late stage of assembly.

4. Conclusions

Our efforts to combine genetics and X-ray crystallography have allowed us to amass a collection of ribosomal mutants for structural studies. While the range of genetic tools available for manipulation of *T. thermophilus* remains somewhat limited, we have nevertheless succeeded in recapitulating many of the ribosomal mutants identified by a number of laboratories over the past several decades. In addition, our genetic experiments have provided clues to the role of ribosomal protein S12 in coupling codon recogni-

tion and GTPase activation of EF-Tu. X-ray crystal structures of streptomycin-dependent and streptomycin-independent 30S subunits are beginning to produce insights into the mechanism of tRNA selection. However, these data, together with a reevaluation of earlier genetic data, indicate that a complete understanding of the structural basis for mutations affecting decoding accuracy lay sometime in the future. An X-ray crystal structure of ribosomes from a *ksgA* mutant also reveals the role of a conserved posttranscriptional modification in establishing the optimal conformation of the 30S subunit during assembly. We expect that further development of the genetics of *T. thermophilus* will allow even more sophisticated approaches to mutational studies of the ribosome, culminating in a structural interpretation of their effects on protein synthesis.

References

Björkman J, Samuelsson P, Andersson DI, Hughes D (1999) Novel ribosomal mutations affecting translational accuracy, antibiotic resistance and virulence of Salmonella typhimurium. Mol Microbiol 31: 53–58

Cameron DM, Gregory ST, Thompson J, Dahlberg AE (2004a) Thermus thermophilus L11 methyltransferase, PrmA, is dispensable for growth and preferentially modifies free ribosomal protein L11 prior to ribosome assembly. J Bacteriol 186: 5819–5825

Cameron DM, Thompson J, Gregory ST, March PE, Dahlberg AE (2004b) Thiostrepton-resistant mutants of Thermus thermophilus. Nucleic Acids Res 32: 3220–3227

Carr JF, Gregory ST, Dahlberg AE (2005) Severity of the streptomycin resistance and streptomycin dependence phenotypes of ribosomal protein S12 of Thermus thermophilus depends on the identity of highly conserved amino acid residues. J Bacteriol 187: 3548–3550

Carter AP, Clemons WM, Brodersen DE, Morgan-Warren RJ, Wimberly BT, Ramakrishnan V (2000) Functional insights from the structure of the 30S ribosomal subunit and its interactions with antibiotics. Nature 407: 340–348

Cava F, Hidalgo, A, Berenguer J (2009) Thermus thermophilus as biological model. Extremophiles 13: 213–231

Cochella L, Green R (2005) An active role for tRNA in decoding beyond codon:anticodon pairing. Science 308: 1178–1180

Davis BD, Chen L, Tai PC (1986) Misread protein creates membrane channels: an essential step in the bactericidal action of aminoglycosides. Proc Natl Acad Sci USA 83: 6164–6168

Demirci H, Belardinelli R, Seri E, Gregory ST, Gualerzi C, Dahlberg AE, Jogl G (2009) Structural rearrangements in the active site of the Thermus thermophilus 16S rRNA methyltransferase KsgA in a binary complex with 5'-methylthioadenosine. J Mol Biol 388: 271–282

Duisterwinkel FJ, de Graaf JM, Kraal B, Bosch L (1981) A kirromycin resistant elongation factor EF-Tu from Escherichia coli contains a threonine instead of an alanine residue in position 375. FEBS Lett 131: 89–93

Gale EF, Cundliffe E, Reynolds PE, Richmond MH, Waring MJ (1981) The molecular basis of antibiotic action. Wiley & Sons, London

Gorini L, Rosset R, Zimmermann RA (1967) Phenotypic masking and streptomycin dependence. Science 157: 1314–1317

Gregory ST, Cate JH, Dahlberg AE (2001a) A spontaneous, erythromycin-resistance mutation in a 23S rRNA gene, *rrlA*, of the extreme thermophile Thermus thermophilus. IB-21. J Bacteriol 183: 4382–4385

Gregory ST, Cate JH, Dahlberg AE (2001b) Streptomycin-resistant and streptomycin-dependent mutants of the extreme thermophile Thermus thermophilus. J Mol Biol 309: 333–338

Gregory ST, Carr JF, Rodriguez-Correa D, Dahlberg AE (2005) Mutational analysis of 16S and 23S rRNA genes of Thermus thermophilus. J Bacteriol 187: 4804–4812

Gregory ST, Dahlberg AE (2008) Transposition of an insertion sequence, ISTth7, in the genome of the extreme thermophile Thermus thermophilus HB8. FEMS Microbiol Lett 289: 187–192

Gregory ST, Dahlberg AE (2009) Genetic and structural analysis of base substitutions in the central pseudoknot of Thermus thermophilus 16S ribosomal RNA. RNA 15: 215–223

Gregory ST, Carr JF, Dahlberg AE (2009) A signal relay between ribosomal protein S12 and elongation factor EF-Tu during decoding of mRNA. RNA 15: 208–214

Gromadski KB, Rodnina MV (2004) Streptomycin interferes with conformational coupling between codon recognition and GTPase activation on the ribosome. Nat Struct Mol Biol 11: 316–322

Henne A, Brüggemann H, Raasch C, Wiezer A, Hartsch T, Liesegang H, Johann A, Lienard T, Gohl O, Martinez-Arias R, Jacobi C, Starkuviene V, Schlenczeck S, Dencker S, Huber R, Klenk HP, Kramer W, Merkl R, Gottschalk G, Fritz HJ (2004) The genome sequence of the extreme thermophile *Thermus thermophilus*. Nat Biotechnol 22: 547–553

Jawetz E, Gunnison JB, Bruff JB, Coleman VR (1952) Studies on antibiotic synergism and antagonism. Synergism among seven antibiotics against various bactera in vitro. J Bacteriol 64: 29–39

Korostelev A, Noller HF (2007) The ribosome in focus: new structures bring new insights. Trends Biochem Sci 32: 434–441

Lederberg J (1951) Streptomycin resistance; a genetically recessive mutation. J Bacteriol 61: 549–550

Masui R, Kurokawa K, Nakagawa N, Tokunaga F, Koyama Y, Shibata T, Oshima T, Yokoyama S, Yasunaga T, Kuramitsu S, NCBI (2005) Complete genome sequence of *Thermus thermophilus* HB8. *http://www. ncbi. nlm. nih. gov/sites/entrez?db=genome &cmd=Retrieve&dopt=Overview&list_uids=530*

Monshupanee T, Gregory ST, Douthwaite S, Chungjatupornchai W, Dahlberg AE (2008) Mutations in conserved helix 69 of 23S rRNA of *Thermus thermophilus* that affect capreomycin resistance but not posttranscriptional modifications. J Bacteriol 190: 7754–7761

O'Connor M, Thomas CL, Zimmermann RA, Dahlberg AE (1997) Decoding fidelity at the ribosomal A and P sites: influence of mutations in three different regions of the decoding domain in 16S rRNA. Nucleic Acids Res 25: 1185–1193

Ogle JM, Brodersen DE, Clemons WM Jr, Tarry MJ, Carter AP, Ramakrishnan V (2001) Recognition of cognate transfer RNA by the 30S ribosomal subunit. Science 292: 897–902

Ogle JM, Murphy FV, Tarry MJ, Ramakrishnan V (2002) Selection of tRNA by the ribosome requires a transition from an open to a closed form. Cell 111: 721–732

Pfeiffer T, Jorcke D, Feltens R, Hartmann RK (1995) Direct linkage of *str-*, S10- and spc-related gene clusters in Thermus thermophilus HB8, and sequences of ribosomal proteins L4 and S10. Gene 167: 141–145

Ramírez-Arcos S, Fernández-Herrero LA, Marín I, Berenguer J (1998) Anaerobic growth, a property horizontally transferred by an Hfr-like mechanism among extreme thermophiles. J Bacteriol 180: 3137–3143

Sakaki Y, Oshima T (1975) Isolation and characterization of a bacteriophage infectious to an extreme thermophile, Thermus thermophilus HB8. J Virol 15: 1449–1453

Schmeing TM, Ramakrishnan V (2009) What recent ribosome structures have revealed about the mechanism of translation. Nature 461: 1234–1242

Schmeing TM, Voorhees RM, Kelley AC, Gao YG, Murphy FV 4th, Weir JR, Ramakrishnan V (2009) The crystal structure of the ribosome bound to EF-Tu and aminoacyl-tRNA. Science 326: 688–694

Schuwirth BS, Borovinskaya MA, Hau CW, Zhang W, Vila-Sanjurjo A, Holton JM, Cate JH (2005) Structures of the bacterial ribosome at 3.5 Å resolution. Science 310: 827–834

Timms AR, Bridges BA (1993) Double, independent mutational events in the *rpsL* gene of *Escherichia coli*: an example of hypermutability? Mol Microbiol 9: 335–342

Tubulekas I, Buckingham RH, Hughes D (1991) Mutant ribosomes can generate dominant kirromycin resistance. J Bacteriol 173: 3635–3643

van Buul CP, Visser W, van Knippenberg PH (1984) Increased translational fidelity caused by the antibiotic kasugamycin and ribosomal ambiguity in mutants harbouring the *ksgA* gene. FEBS Lett 177: 119–124

Vila-Sanjurjo A, Ridgeway WK, Seymaner V, Zhang W, Santoso S, Yu K, Cate JH (2003) X-ray crystal structures of the WT and a hyper-accurate ribosome from *Escherichia coli*. Proc Natl Acad Sci USA 100: 8682–8687

Vila-Sanjurjo A, Lu Y, Aragonez JL, Starkweather RE, Sasikumar M, O'Connor M (2007) Modulation of 16S rRNA function by ribosomal protein S12. Biochimica et Biophysica Acta 1769: 462–471

Xu Z, O'Farrell HC, Rife JP, Culver GM (2008) A conserved rRNA methyltransferase regulates ribosome biogenesis. Nat Struct Mol Biol 15: 534–536

Yu MX, Slater MR, Ackermann HW (2006) Isolation and characterization of Thermus bacteriophages. Arch Virol 151: 663–679

Zhang W, Dunkle JA, Cate JH (2009) Structures of the ribosome in intermediate states of ratcheting. Science 325: 1014–1017

Zuurmond A-M, Zeef LAH, Kraal B (1998) A kirromycin-resistant EF-Tu species reverses streptomycin dependence of *Escherichia coli* strains mutated in ribosomal protein S12. Microbiol 144: 3309–3316

The packing of ribosomes in crystals and polysomes

Jack A. Dunkle and Jamie H. D. Cate

6

1. Introduction

Many structures of the bacterial ribosome have been determined at high resolution by x-ray crystallography over the past few years. These structures span a number of steps in the protein elongation and termination cycle, and include a wide variety of conformations of the ribosome. While these structures are landmarks in our understanding of protein synthesis, the very fact that ribosomes crystallize at all is quite striking. Notably, bacterial ribosomes pack in most of the crystal forms by recapitulating the organization of polysomes. While not all bacterial polysomes adopt the same overall architecture, the fact that ribosome crystals recapitulate common forms of polysomes has structural implications for how these polysomes function. We show that ribosome packing in crystals can be correlated with ribosome organization in active polysomes if ribosome dynamics are considered. The packing arrangement also suggests possible means for polysome regulation that may be used in bacteria.

2. Organization of polysomes

2.1. Polysomes in bacteria

In bacteria, transcription and translation are coupled (Miller et al., 1970). As transcription of mRNA proceeds, ribosomes initiate protein synthesis and form an "assembly line" of multiple ribosomes per mRNA. Recent experiments have shown that translation and transcription can directly influence each other in fundamental ways. For example, the rate of transcription is directly dependent on the rate of translation (Proshkin et al., 2010). In addition, termination of transcription is controlled by the translational status of an mRNA

transcript (Burmann et al., 2010). These results indicate that ribosomes studied in isolation do not fully recapitulate their regulatory functions in cells.

The spacing of ribosomes along mRNAs can be quite variable (Arava et al., 2003; Arava et al., 2005), even along a particular mRNA (Ingolia et al., 2009). Furthermore, mutations in the translational machinery can have dramatic effects on polysome function. For example, mutation of nucleotide A2451 in 23S rRNA in the peptidyl transferase center does not block ribosome entry into polysomes, yet causes a dominant lethal phenotype in *Escherichia coli* (Thompson et al., 2001). Aberrant ribosome recycling can also confer a dominant lethal phenotype (Yamami et al., 2005). Recently, a new method of translational profiling has provided a wealth of information on ribosome spacing along eukaryotic mRNAs (Ingolia et al., 2009), but the underlying mechanisms responsible for the spacing are not well understood. Further, the spacing of translating bacterial ribosomes on mRNAs remains to be determined. A quantitative description of polysomes is in the early stages of development (Underwood et al., 2005; Zouridis and Hatzimanikatis, 2007), but the structural implications of their organization have not been probed in detail.

2.2. Ribosome positioning in polysomes

The organization of ribosomes in bacterial polysomes was recently imaged for the first time at high enough resolution to discern the positioning of neighboring ribosomes. Using cryoelectron tomography, the Hartl and Baumeister groups found that polysomes in *E. coli* translation extracts or spheroplasts adopt compact configurations, predominantly staggard or superhelical arrays (Brandt et al., 2009). In these configurations,

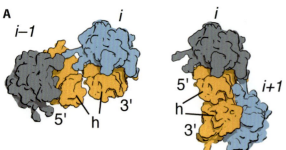 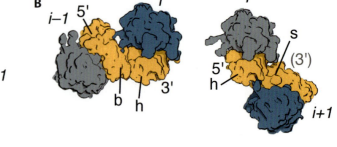

Fig. 1 Organization of polysomes in bacteria. (A) "Top-to-top" polysomes (t-t). Ribosomes progress in the 5' to 3' direction, from ribosome *i*–1 to ribosome *i*, and then to ribosome *i*+1, as indicated. 30S subunits are in gold, 50S subunits in blue or grey. Domains in the 30S subunit are marked: h, head; b, body; s, shoulder. (B) "Top- to-bottom" polysomes (t-b), labelled as in panel A. The 3' end of mRNA after the *i*+1 ribosome is not visible in this view. (Figure adapted from Brandt et al. (2009) Cell 136: 261–271. © 2009 with permission from Elsevier).

the mRNA resides in the interior of the array, and the peptide exit tunnel points towards the outside. The most common polysome arrays are pseudo-helical, and contain 30S subunits in the interior that contact each other in a "top-to-top" (t-t) arrangement. This arrangement positions the 30S subunit head domains next to each other. It also locates the mRNA entry tunnel of one ribosome (i. e., the solvent face of the 30S subunit body domain) near to the mRNA exit site on the 30S platform of the 3'-downstream ribosome (Figure 1A). On average, this would separate ribosomal P sites by about 70–80 nucleotides in adjacent ribosomes (Brandt et al., 2009). Other features of the ribosome in t-t polysomes include positioning of protein L9 and the L1 stalk of the 3'-downstream ribosome, where E-site tRNA leaves the ribosome, near to the mRNA decoding site of the 5'-upstream ribosome.

The next most common polysome configuration, termed "top-to-bottom" (t-b), positions the 30S head domain of one ribosome closer to the base of the 30S body domain of its neighbor (Figure 1B). In these polysomes, the path of the mRNA between ribosomes follows a more sinusoidal pattern. As opposed to the t-t polysomes, the position of the neighboring ribosomes in t-b polysomes does not seem to be as highly constrained (Brandt et al., 2009). Other polysome arrangements were also observed, with shorter stretches of t-t and t-b polysomes interspersed with variable ribosome arrangements.

Although polysomes in eukaryotes have not yet been imaged to the same resolution as those in bacteria (Madin et al., 2004; Kopeina et al., 2008), distinct arrangements of ribosomes have been observed in polysomes that form on the rough endoplasmic reticu-lum (Christensen et al., 1987; Christensen and Bourne, 1999). Associated with protein secretion, these polysomes are constrained to lie near the plane of the membrane, presumably with the exit tunnels of the large (60S) subunits affixed to translocons. They have been observed to form circles, spirals and hairpins (Christensen et al., 1987; Christensen and Bourne, 1999), providing strong evidence for the circularization of mRNAs in eukaryotes (Wells et al., 1998). Whether cytoplasmic ribosomes not involved in secretion organize in the same arrangements as seen in bacteria remains to be determined.

2.3. Evidence for inter-ribosome regulation

Some elements of translation regulation in bacteria and in eukaryotes suggest that ribosomes influence each other in polysomes. For example, a genome-wide analysis of apparent ribosome density on mRNAs in yeast revealed that ribosomes tend to be more densely packed near the 5' end of the open reading frame (ORF) in yeast (Ingolia et al., 2009). Furthermore, codon bias in yeast suggests that substrate channeling between neighboring ribosomes may be prevalent (Cannarozzi et al., 2010). In both bacteria and eukaryotes, translation re-initiation is likely to be influenced by polysome dynamics (Underwood et al., 2005) or structure (Christensen et al., 1987). Some bacterial mRNAs exploit ribosome stalling on leader peptide ORFs to regulate downstream ORF expression (Ramu et al., 2009). Although these regulatory events are suggestive of inter-ribosome regulation, no direct evidence for inter-ribosomal interactions is available to date.

3. Packing of ribosomes in crystals

Crystal forms of the bacterial ribosome have been obtained in a number of different space groups. Notably, most of the crystal forms involve super-helical arrays that closely resemble the arrangement of ribosomes in t-t polysomes observed in *E. coli*. This is true of many crystal forms of both the *E. coli* and the *Thermus thermophilus* 70S ribosome, in a number of functional states. All of these functional states pack in crystals with at least pseudo-four-fold helical symmetry, but with varying unit cell dimensions of five general classes (Petry et al., 2005; Schuwirth et al., 2005; Berk et al., 2006; Selmer et al., 2006; Zhang et al., 2009). Other crystal forms adopt a four-fold planar symmetry (Cate et al., 1999; Korostelev et al., 2006). In all of these crystal forms, as in t-t ribosomes, the 30S subunits form the inside of the array, with the mRNA entry and exit sites juxtaposed. Packing in the crystals is slightly more compact than in polysomes, with the distance between the mRNA entry and exit sites of neighboring ribosomes about 6 nm apart (Jenner et al., 2010), as opposed to about 7 nm apart in translating polysomes (Brandt et al., 2009). Neighboring ribosomes in the crystals would span only slightly fewer than the 70–80 nucleotides predicted between P sites in neighboring ribosomes (Brandt et al., 2009), likely setting the minimum distance between ribosomes that could be achieved. Notably, the density of ribosomes observed in the bacterial ribosomes and in crystals is consistent with that of more densely-packed yeast ribosomes (Arava et al., 2003; Arava et al., 2005).

One drawback to the packing of the ribosome in the above crystal forms involves an inter-ribosome interaction that blocks access of translation factors to their binding site at the subunit interface. In the crystal packing, ribosomal protein L9 from one ribosome projects into the inter-subunit space of the 5'-upstream ribosome, contacting the body of the small subunit (Petry et al., 2005; Schuwirth et al., 2005; Berk et al., 2006; Selmer et al., 2006; Zhang et al., 2009) (Figure 2). This packing arrangement hampered attempts to determine structures of the ribosome in complexes with translation factors such as elongation factor Tu and elongation factor G. The Ramakrishnan lab overcame this problem by simply deleting L9 from *T. thermophilus*, since it is not essential for growth (Herr et al., 2000; Lieberman et al., 2000; Herr et al., 2001; Gao et al., 2009). The resulting ribosomes crystallized in a new packing arrangement that resulted in the first high-resolution structures of EF-Tu and EF-G

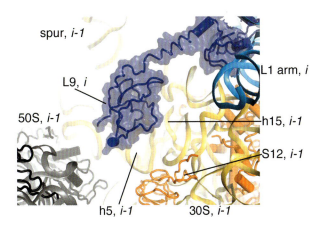

Fig. 2 Position of protein L9 in crystals of the 70S ribosome. The view is a zoom-in of that shown in Fig. 1A. Ribosomes are colored and indicated as in Fig. 1. Helices in 16S rRNA of ribosome *i-1* near to protein L9 from ribosome *i* are labeled.

bound to the ribosome (Gao et al., 2009; Schmeing et al., 2009).

Notably, the packing of ribosomes in the new crystal form seems to resemble those of t-b polysomes. In the crystals, the 30S subunits pack against each other, with the head of on 30S subunit adjacent to the 30S subunit body of the neighboring ribosome, as observed in t-b polysomes (Figure 1B). Although the packing between the 30S subunits is rotated nearly 180° relative to that seen in t-t polysomes, the mRNA exit and entry tunnels of neighboring ribosomes are still juxtaposed. Ribosomes in these crystals approach each other with the approximate distance between adjacent mRNA entry and exit sites of about 10 nm (Gao et al., 2009; Jenner et al., 2010). In translating t-b polysomes, the distance is about 7 nm (Brandt et al., 2009). Interestingly, in t-b polysomes, the arrangement of neighboring ribosomes is more variable, and often is interspersed with t-t ribosome packing (Brandt et al., 2009). This variance in orientation may be required to shorten the distance between mRNA entry and exit sites relative to that observed in an ordered crystal lattice.

4. Reconciling ribosome packing in crystals and polysome function

4.1. Translation initiation in bacteria

Translation initiation in bacteria generally relies on two elements, the ribosome binding site (RBS) within mRNAs, or Shine-Dalgarno interaction, and ribosomal protein S1 (Laursen et al., 2005; Simonetti et al.,

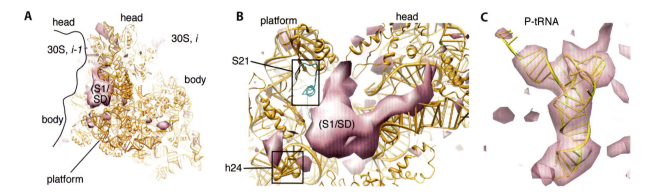

Fig. 3 Binding site of protein S1 in 70S ribosomes. Protein S1 likely extends towards the neighboring 30S subunit, beyond the EM density, which accounts for protein S1, S21 (Sengupta et al., 2001), the SD helix (Sengupta et al., 2001; Yusupova et al., 2006), and phylogenetic differences in 16S rRNA (Cannone et al., 2002). (A) An overview of the 30S subunit in ribosome *i* viewed from the solvent side of the molecule. Difference density corresponding to a portion of protein S1 and the Shine-Dalgarno helix (SD) is indicated (Sengupta et al., 2001; Yusupova et al., 2006). The approximate position of ribosome *i-1* is also shown in outline to the left. (B) Difference electron density calculated by subtracting the filtered volume of the *E. coli* 70S ribosome coordinates (Berk et al., 2006) from EM maps of the 70S ribosome (Gabashvili et al., 2000; Sengupta et al., 2001). Protein S1 likely extends towards the neighboring 30S subunit, beyond the EM density, which accounts for part of protein S1 and SD helix (Sengupta et al., 2001; Yusupova et al., 2006). In the published EM difference density map (Sengupta et al., 2001), the density map also included protein S21 (Berk et al., 2006) and the tip of 16S rRNA helix h24 (Cannone et al., 2002), due to the use of *T. thermophilus* 30S subunit coordinates (Sengupta et al., 2001). (C) The difference map in B was contoured such that adequate density was observed for P-site tRNA, which is present in the EM experiment (Gabashvili et al., 2000; Sengupta et al., 2001), but is absent in the subtracted 70S ribosome coordinates used here.

2009). In *E. coli*, ribosomal protein S1 is essential for growth. It contributes to translation initiation on most mRNAs. Based on experiments that depleted protein S1 *in vivo*, it was proposed that protein S1 does not exchange between ribosomes in the cell, and ribosomes lacking protein S1 are inactive in translation elongation (Sorensen et al., 1998). The requirement for protein S1 implies that all of the recent crystal structures of the bacterial ribosome may involve inactive ribosomes, since protein S1 has been removed in all of them. Indeed, the packing arrangement of ribosomes in the crystals that mimic t-t polysomes would likely preclude any possibility of S1 binding, due to steric hindrance (Figure 3) (Sengupta et al., 2001; Petry et al., 2005; Schuwirth et al., 2005; Berk et al., 2006; Selmer et al., 2006; Zhang et al., 2009).

Although t-t polysomes are slightly less compact when compared to ribosomes in crystals, as discussed above, several lines of evidence indicate that protein S1 is not directly required for translation elongation in bacteria. First, the genetic experiments supporting the importance of protein S1 in protein elongation (Sorensen et al., 1998) could be due to the fact that ribosomes not engaged in translation are often sequestered from the translating pool and may not be inherently inactive. For example, during cold shock, stationary phase, or other stress conditions, ribosomes are sequestered as 70S monosomes or higher-order

structures (Ortiz et al.; Wilson and Nierhaus, 2007). In addition, it is unlikely that the function of protein S1 in *E. coli* is conserved in many bacteria. For example, protein S1 is not essential in *Bacillus subtilis* (Sorokin et al., 1995). Furthermore, bacteria can be divided into at least three groups with respect to their regulation of translation initiation. Ribosomes from Gram-positive bacteria with low to moderate G+C content can be distinguished from those with high G+C content, and from Gram-negative bacteria such as *E. coli*, in their ability to initiate protein synthesis (Farwell et al., 1992). The distinction depends on the source of ribosomes, as well as the source of protein S1, which varies in its ability to stably bind to the ribosome (Farwell et al., 1992). Future experiments are needed to determine whether t-t polysomes occur in a wide variety of bacteria, include Gram-positive bacteria.

4.2. A functional role for protein L9 in polysomes

The position of ribosomal protein L9 near the neighboring 5'-upstream ribosome in t-t polysomes suggests that it might contribute to inter-ribosome regulation in some cases. Ribosomal protein L9 is not essential in *E. coli* or in *T. thermophilus* (Herr et al., 2000; Lieberman et al., 2000; Herr et al., 2001; Gao et al., 2009), yet it is conserved across bacteria (Gasteiger et al., 2003). Protein L9 binds to 23S rRNA in the large ribosomal subunit at the base of the L1 stalk (Adamski et al., 1996). It has a bi-lobed structure (Hoffman et al., 1994) and binds the large ribosomal subunit through both its N-terminal and C-terminal domains (Adamski et al., 1996; Matadeen et al., 1999). The interaction of the C-terminal domain with the large ribosomal subunit is weaker, as shown biochemically (Lieberman et al., 2000) and by its extended conformation in x-ray crystal structures of the ribosome (Cate et al., 1999; Petry et al., 2005; Schuwirth et al., 2005; Berk et al., 2006; Korostelev et al., 2006; Selmer et al., 2006; Zhang et al., 2009).

At present, there is little data to explain the functional role of protein L9 in bacterial translation. Although the protein is not essential, defects in L9 lead to mRNA slippage on the ribosome (Herr et al., 2000; Herr et al., 2001). One striking effect involves translation of a topoisomerase subunit encoded by *E. coli* bacteriophage T4 that is essential to its life cycle. The gene for this subunit, gene *60*, contains a discontinuous open reading frame (ORF) with a 50-nucleotide insertion that is efficiently bypassed during translation (Huang et al., 1988). The insertion encodes stop codons in all three reading frames, as well as sequence and structural elements important for its bypass (Wills et al., 2008) (Figure 4A). This mRNA insertion element, in conjunction with the nascent peptide's sequence and/ or 5' mRNA sequence, promotes bypass in about half of the ribosomes translating the ORF (Maldonado and Herr, 1998; Wills et al., 2008). Remarkably, accurate

bypass depends on the presence of ribosomal protein L9 (Herbst et al., 1994; Adamski et al., 1996). Codon-anticodon pairing in the ribosomal P site, which must be disrupted in the bypass mechanism, occurs on the 30S subunit nearly 90 Å away from the binding site for protein L9 on the large subunit. The mechanism for bypass is therefore not obvious from the structure of the ribosome (Herr et al., 2001; Wills et al., 2008).

One model for how L9 influences bypass is that it controls the dynamics of the L1 stalk in the large subunit (Wills et al., 2008). The L1 stalk, which can adopt many different conformational states in the ribosome (Blaha et al., 2009), is quite dynamic (Fei et al., 2008; Cornish et al., 2009; Fei et al., 2009; Schmeing and Ramakrishnan, 2009; Munro et al., 2010a; Munro et al., 2010b), and is coupled to P-site tRNA movement into the E site, as well as E-site tRNA release (Schmeing and Ramakrishnan, 2009). In the proposed model, the structure and dynamics of the mRNA in the A site is somehow influenced by tRNA gating in the E site by the L1 stalk (Wills et al., 2008). Recent results using single-molecule methods, however, suggest that E-site tRNA dissociation is uncoupled from A-site mRNA decoding in normal translation (Uemura et al., 2010). Under physiological conditions, i.e. in the crowded conditions of a cell (Ellis, 2001), it remains to be determined whether coupling between E-site tRNA release and A-site mRNA decoding occurs.

Alternatively, or in addition, protein L9 may function in phage T4 gene *60* bypass based on its position in t-t polysomes (Herr et al., 2001). In t-t polysomes, as seen in the crystal packing, the C-terminal domain

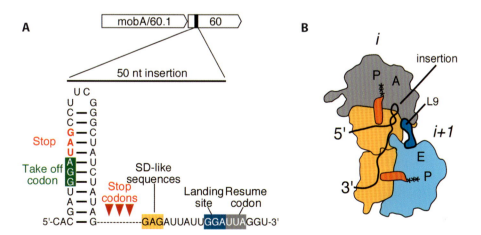

Fig. 4 Translation of bacteriophage T4 gene *60*. (A) Elements in mRNA required for bypass of the 50-nucleotide insertion in the ORF. The 50-nucleotide insertion, shown in white, separates codons 46 and 47 in the ORF of gene *60* (Wills et al., 2008). Features thought to be important for ribosome bypass of the insertion are colored and labeled. (B) Model for L9 role in polysome-mediated bypass. L9 of ribosome *i+1* is located immediately adjacent to the insertion element in the A site of ribosome *i*.

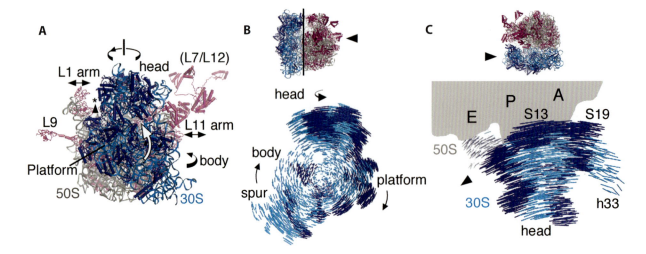

Fig. 5 Conformational rearrangements in the ribosome during translation. (A) Conformational dynamics of the intact ribosome. Arrows describe the directions of motions of the 30S subunit, including head rotation, body closure, subunit rotation, and tRNA groove opening (arrow with asterisk), along with motions of mobile 50S elements, L1 and L11. Proteins L7/L12 are highly dynamic and generally not visible in crystal structures. The ribosome is color-coded as follows: 16S rRNA, light blue; 30S proteins, dark blue; 23S and 5S rRNA, grey; 50S proteins, magenta. (B) Motions within the 30S subunit during translocation. Arrows, along with positional difference vectors, show the extent of motions that occur in the 30S subunit, as viewed from the subunit interface. The icon above shows the orientation of the view. (C) Extent of motions of the 30S head domain during translation. Difference vectors are shown as in B. A silhouette of the 50S subunit is shown with the A, P and E sites marked. The icon above shows the orientation of the view.

of protein L9 protrudes into the ribosomal A site of the preceding ribosome in the polysome, where it contacts the body of the 30S subunit near 16S rRNA helices h5 and h15 (Figure 2). Notably, mutations in the C-terminal domain of protein L9 have the largest effect in bypass rescue assays (Herbst et al., 1994; Adamski et al., 1996). Given the specific requirements for mRNA structure and unwinding for bypass to occur (Wills et al., 2008), the location of L9 in the A site of a neighboring 5'-upstream ribosome could play an important role. For example, protein L9 of the 3'-downstream ribosome (one that has just bypassed the insertion element) might influence inter-subunit rotation of the target ribosome's 30S subunit in such a way that mRNA unwinding and repositioning is favored (Figure 4B).

Even if protein L9 plays a functional role in certain circumstances at the polysome level, i. e. in bacteriophage T4 gene *60* translation, its position in t-t polysomes, as seen in crystal packing, would seemingly preclude elongation factor binding (Gao et al., 2009; Schmeing et al., 2009). The position of L9 must therefore change on a rapid timescale to allow elongation factors to bind, i. e. by transiently engaging with the base of the L1 stalk (Matadeen et al., 1999; Lieberman et al., 2000). The extended conformation of protein L9 always observed in ribosome crystal structures may reflect the intrinsic helical propensity of the central helix

of L9 under crystallization conditions. Stabilization of a central alpha helix is not unprecedented in crystallography, and was most prominently observed in the case of calmodulin (Wilson and Brunger, 2000; Chang et al., 2003).

5. Conformational dynamics of ribosomes within polysomes

During translation, the ribosome undergoes a number of global conformational changes. These include inter-subunit rotation, rotation and opening of the 30S subunit head domain, and opening and closing of the 30S subunit body (Dunkle and Cate, 2010). A minimal number of rearrangements thought to occur along the mRNA and tRNA translocation pathway have been termed ratcheting states R_0 to R_F, and involve inter-subunit rotations of ~10° and head rotations of up to 14° (Zhang et al., 2009; Dunkle and Cate, 2010) (Figure 5A). In addition, elements of the 50S subunit are quite dynamic, including the L1 arm, the L11 arm, and proteins L7/L12 in the stalk (Figure 5B). Although the total number of conformational states of the ribosome remain to be determined (Munro et al., 2009; Dunkle and Cate, 2010; Fischer et al., 2010), known conformational rearrangements can lead to movement of parts of the ribosome by over 20–30 Å at its periphery. How these conformational changes can be accommodated within tightly packed t-t polysomes (Brandt et al., 2009) is an open question.

Based on the available ribosome crystal forms and tomograms of t-t polysomes (Petry et al., 2005; Schuwirth et al., 2005; Berk et al., 2006; Selmer et al., 2006; Brandt et al., 2009; Zhang et al., 2009), inter-ribosome contacts in the superhelical array involve predominantly contacts between 30S subunits, the L1 arm, and protein L9 of the 50S subunit. As noted above, protein L9 likely fluctuates between extended and compact arrangements within the ribosome, indicating that its contacts to the preceding ribosome in a polysome may constrain ribosome dynamics transiently. Such transient contacts could become more significant as a function of temperature or stress or due to ribosome pausing signals (Brandt et al., 2009). The L1 arm contacts the head domain of the 30S subunit in the 5'-preceding ribosome and could couple to 30S subunit dynamics through this interaction. Since the position of the 30S head domain is thought to control tRNA positioning on the ribosome during translocation (Spahn et al., 2004; Schuwirth et al., 2005), it will be interesting to test whether the L1 arm of a 3'-downstream ribosome can affect translocation on the neighboring 5'-upstream ribosome. Contacts between neighboring 30S subunit body and shoulder domains may also influence translocation on neighboring ribosomes.

6. Conclusions

Ribosomes within bacterial polysomes have been shown to adopt two major configurations. Crystals of the bacterial ribosome recapitulate this organization, particularly favoring the t-t polysome. In actively translating polysomes, the ribosome must sample a wide range of conformational states, a requirement that would seemingly be at odds with crystal formation. Notably, crystals of the ribosome that recapitulate t-t polysomes are able to accommodate the full range of ribosome conformations that are now known, including the fully ratcheted state R_F (our unpublished results). The t-t polysome organization also suggests that ribosomal protein S1 would be excluded after translation initiation is complete. Ribosomal protein L9 and the L1 stalk, which bridge neighboring ribosomes in the crystals, as well as 30S subunit interactions in polysome arrays, may contribute to translational control

at the polysome level. Future experiments with ribosomes from all kingdoms of life will be required to unravel this phenomenon.

Acknowledgements

We thank G. M. Alushin for help with calculating difference maps comparing EM reconstructions with 70S ribosome coordinates. This work was funded by the NIH (grant R01-GM65050).

References

Adamski FM, Atkins JF, Gesteland RF (1996) Ribosomal protein L9 interactions with 23 S rRNA: the use of a translational bypass assay to study the effect of amino acid substitutions. J Mol Biol 261: 357–371

Arava Y, Boas FE, Brown PO, Herschlag D (2005) Dissecting eukaryotic translation and its control by ribosome density mapping. Nucleic Acids Res 33: 2421–2432

Arava Y, Wang Y, Storey JD, Liu CL, Brown PO, Herschlag D (2003) Genome-wide analysis of mRNA translation profiles in Saccharomyces cerevisiae. Proc Natl Acad Sci USA 100: 3889–3894

Berk V, Zhang W, Pai RD, Cate JH (2006) Structural basis for mRNA and tRNA positioning on the ribosome. Proc Natl Acad Sci USA 103: 15830–15834

Blaha G, Stanley RE, Steitz TA (2009) Formation of the first peptide bond: the structure of EF-P bound to the 70S ribosome. Science 325: 966–970

Brandt F, Etchells SA, Ortiz JO, Elcock AH, Hartl FU, Baumeister W (2009) The native 3D organization of bacterial polysomes. Cell 136: 261–271

Burmann BM, Schweimer K, Luo X, Wahl MC, Stitt BL, Gottesman ME, Rosch P (2010) A NusE:NusG complex links transcription and translation. Science 328: 501–504

Cannarozzi G, Schraudolph NN, Faty M, von Rohr P, Friberg MT, Roth AC, Gonnet P, Gonnet G, Barral Y (2010) A role for codon order in translation dynamics. Cell 141: 355–367

Cannone JJ, Subramanian S, Schnare MN, Collett JR, D'Souza LM, Du Y, Feng B, Lin N, Madabusi LV, Muller KM, Pande N, Shang Z, Yu N, Gutell RR (2002) The comparative RNA web (CRW) site: an online database of comparative sequence and structure information for ribosomal, intron, and other RNAs. BMC Bioinformatics 3: 2

Cate JH, Yusupov MM, Yusupova GZ, Earnest TN, Noller HN (1999) X-ray crystal structures of 70S ribosome functional complexes. Science 285: 2095–2104

Chang SL, Szabo A, Tjandra N (2003) Temperature dependence of domain motions of calmodulin probed by NMR relaxation at multiple fields. J Am Chem Soc 125: 11379–11384

Christensen AK, Bourne CM (1999) Shape of large bound polysomes in cultured fibroblasts and thyroid epithelial cells. Anat Rec 255: 116–129

Christensen AK, Kahn LE, Bourne CM (1987) Circular polysomes predominate on the rough endoplasmic reticulum of somatotropes and mammotropes in the rat anterior pituitary. Am J Anat 178: 1–10

Cornish PV, Ermolenko DN, Staple DW, Hoang L, Hickerson RP,

Noller HF, Ha T (2009) Following movement of the L1 stalk between three functional states in single ribosomes. Proc Natl Acad Sci USA 106: 2571–2576

Dunkle JA, Cate JH (2010) Ribosome structure and dynamics during translocation and termination. Annu Rev Biophys 39: 227–244

Ellis RJ (2001) Macromolecular crowding: obvious but underappreciated. Trends Biochem Sci 26: 597–604

Farwell MA, Roberts MW, Rabinowitz JC (1992) The effect of ribosomal protein S1 from *Escherichia coli* and Micrococcus luteus on protein synthesis in vitro by *E. coli* and Bacillus subtilis. Mol Microbiol 6: 3375–3383

Fei J, Bronson JE, Hofman JM, Srinivas RL, Wiggins CH, Gonzalez RL, Jr (2009) Allosteric collaboration between elongation factor G and the ribosomal L1 stalk directs tRNA movements during translation. Proc Natl Acad Sci USA 106: 15702–15707

Fei J, Kosuri P, MacDougall DD, Gonzalez RL, Jr (2008) Coupling of ribosomal L1 stalk and tRNA dynamics during translation elongation. Mol Cell 30: 348–359

Fischer N, Konevega AL, Wintermeyer W, Rodnina MV, Stark H (2010) Ribosome dynamics and tRNA movement by time-resolved electron cryomicroscopy. Nature 466: 329–333

Gabashvili IS, Agrawal RK, Spahn CM, Grassucci RA, Svergun DI, Frank J, Penczek P (2000) Solution structure of the *E. coli* 70S ribosome at 11.5 A resolution. Cell 100: 537–549

Gao YG, Selmer M, Dunham CM, Weixlbaumer A, Kelley AC, Ramakrishnan V (2009) The structure of the ribosome with elongation factor G trapped in the posttranslocational state. Science 326: 694–699

Gasteiger E, Gattiker A, Hoogland C, Ivanyi I, Appel RD, Bairoch A (2003) ExPASy: The proteomics server for in-depth protein knowledge and analysis. Nucleic Acids Res 31: 3784–3788

Herbst KL, Nichols LM, Gesteland RF, Weiss RB (1994) A mutation in ribosomal protein L9 affects ribosomal hopping during translation of gene 60 from bacteriophage T4. Proc Natl Acad Sci USA 91: 12525–12529

Herr AJ, Gesteland RF, Atkins JF (2000) One protein from two open reading frames: mechanism of a 50 nt translational bypass. EMBO J 19: 2671–2680

Herr AJ, Nelson CC, Wills NM, Gesteland RF, Atkins JF (2001) Analysis of the roles of tRNA structure, ribosomal protein L9, and the bacteriophage T4 gene 60 bypassing signals during ribosome slippage on mRNA. J Mol Biol 309: 1029–1048

Hoffman DW, Davies C, Gerchman SE, Kycia JH, Porter SJ, White SW, Ramakrishnan V (1994) Crystal structure of prokaryotic ribosomal protein L9: a bi-lobed RNA-binding protein. EMBO J 13: 205–212

Huang WM, Ao SZ, Casjens S, Orlandi R, Zeikus R, Weiss R, Winge D, Fang M (1988) A persistent untranslated sequence within bacteriophage T4 DNA topoisomerase gene 60. Science 239: 1005–1012

Ingolia NT, Ghaemmaghami S, Newman JR, Weissman JS (2009) Genome-wide analysis in vivo of translation with nucleotide resolution using ribosome profiling. Science 324: 218–223

Jenner LB, Demeshkina N, Yusupova G, Yusupov M (2010) Structural aspects of messenger RNA reading frame maintenance by the ribosome. Nat Struct Mol Biol 17: 555–560

Kopeina GS, Afonina ZA, Gromova KV, Shirokov VA, Vasiliev VD, Spirin AS (2008) Step-wise formation of eukaryotic double-row polyribosomes and circular translation of polysomal mRNA. Nucleic Acids Res 36: 2476–2488

Korostelev A, Trakhanov S, Laurberg M, Noller HF (2006) Crystal structure of a 70S ribosome-tRNA complex reveals functional interactions and rearrangements. Cell 126: 1065–1077

Laursen BS, Sorensen HP, Mortensen KK, Sperling-Petersen HU (2005) Initiation of protein synthesis in bacteria. Microbiol Mol Biol Rev 69: 101–123

Lieberman KR, Firpo MA, Herr AJ, Nguyenle T, Atkins JF, Gesteland RF, Noller HF (2000) The 23 S rRNA environment of ribosomal protein L9 in the 50 S ribosomal subunit. J Mol Biol 297: 1129–1143

Madin K, Sawasaki T, Kamura N, Takai K, Ogasawara T, Yazaki K, Takei T, Miura K, Endo Y (2004) Formation of circular polyribosomes in wheat germ cell-free protein synthesis system. FEBS Lett 562: 155–159

Maldonado R, Herr AJ (1998) Efficiency of T4 gene 60 translational bypassing. J Bacteriol 180: 1822–1830

Matadeen R, Patwardhan A, Gowen B, Orlova EV, Pape T, Cuff M, Mueller F, Brimacombe R, van Heel M (1999) The *Escherichia coli* large ribosomal subunit at 7.5 A resolution. Structure Fold Des 7: 1575–1583

Miller OL, Jr., Hamkalo BA, Thomas CA, Jr (1970) Visualization of bacterial genes in action. Science 169: 392–395

Munro JB, Altman RB, Tung CS, Cate JH, Sanbonmatsu KY, Blanchard SC (2010a) Spontaneous formation of the unlocked state of the ribosome is a multistep process. Proc Natl Acad Sci USA 107: 709–714

Munro JB, Altman RB, Tung CS, Sanbonmatsu KY, Blanchard SC (2010b) A fast dynamic mode of the EF-G-bound ribosome. Embo J 29: 770–781

Munro JB, Sanbonmatsu KY, Spahn CM, Blanchard SC (2009) Navigating the ribosome's metastable energy landscape. Trends Biochem Sci 34: 390–400

Ortiz JO, Brandt F, Matias VR, Sennels L, Rappsilber J, Scheres SH, Eibauer M, Hartl FU, Baumeister W. Structure of hibernating ribosomes studied by cryoelectron tomography in vitro and in situ. J Cell Biol 190: 613–621

Petry S, Brodersen DE, Murphy FVt, Dunham CM, Selmer M, Tarry MJ, Kelley AC, Ramakrishnan V (2005) Crystal structures of the ribosome in complex with release factors RF1 and RF2 bound to a cognate stop codon. Cell 123: 1255–1266

Proshkin S, Rahmouni AR, Mironov A, Nudler E (2010) Cooperation between translating ribosomes and RNA polymerase in transcription elongation. Science 328: 504–508

Ramu H, Mankin A, Vazquez-Laslop N (2009) Programmed drug-dependent ribosome stalling. Mol Microbiol 71: 811–824

Schmeing TM, Ramakrishnan V (2009) What recent ribosome structures have revealed about the mechanism of translation. Nature 461: 1234–1242

Schmeing TM, Voorhees RM, Kelley AC, Gao YG, Murphy FVt, Weir JR, Ramakrishnan V (2009) The crystal structure of the ribosome bound to EF-Tu and aminoacyl-tRNA. Science 326: 688–694

Schuwirth BS, Borovinskaya MA, Hau CW, Zhang W, Vila-Sanjurjo A, Holton JM, Cate JH (2005) Structures of the bacterial ribosome at 3.5 A resolution. Science 310: 827–834

Selmer M, Dunham CM, Murphy FVt, Weixlbaumer A, Petry S, Kelley AC, Weir JR, Ramakrishnan V (2006) Structure of the 70S ribosome complexed with mRNA and tRNA. Science 313: 1935–1942

Sengupta J, Agrawal RK, Frank J (2001) Visualization of protein S1 within the 30S ribosomal subunit and its interaction with messenger RNA. Proc Natl Acad Sci USA 98: 11991–11996

Simonetti A, Marzi S, Jenner L, Myasnikov A, Romby P, Yusupova G, Klaholz BP, Yusupov M (2009) A structural view of translation initiation in bacteria. Cell Mol Life Sci 66: 423–436

Sorensen MA, Fricke J, Pedersen S (1998) Ribosomal protein S1 is required for translation of most, if not all, natural mRNAs in *Escherichia coli* in vivo. J Mol Biol 280: 561–569

Sorokin A, Serror P, Pujic P, Azevedo V, Ehrlich SD (1995) The Bacillus subtilis chromosome region encoding homologues of the *Escherichia coli* mssA and rpsA gene products. Microbiology 141 (Pt 2):311–319

Spahn CM, Gomez-Lorenzo MG, Grassucci RA, Jorgensen R, Andersen GR, Beckmann R, Penczek PA, Ballesta JP, Frank J (2004) Domain movements of elongation factor eEF2 and the eukaryotic 80S ribosome facilitate tRNA translocation. EMBO J 23: 1008–1019

Thompson J, Kim DF, O'Connor M, Lieberman KR, Bayfield MA, Gregory ST, Green R, Noller HF, Dahlberg AE (2001) Analysis of mutations at residues A2451 and G2447 of 23S rRNA in the peptidyltransferase active site of the 50S ribosomal subunit. Proc Natl Acad Sci USA 98: 9002–9007

Uemura S, Aitken CE, Korlach J, Flusberg BA, Turner SW, Puglisi JD (2010) Real-time tRNA transit on single translating ribosomes at codon resolution. Nature 464: 1012–1017

Underwood KA, Swartz JR, Puglisi JD (2005) Quantitative polysome analysis identifies limitations in bacterial cell-free protein synthesis. Biotechnol Bioeng 91: 425–435

Wells SE, Hillner PE, Vale RD, Sachs AB (1998) Circularization of mRNA by eukaryotic translation initiation factors. Mol Cell 2: 135–140

Wills NM, O'Connor M, Nelson CC, Rettberg CC, Huang WM, Gesteland RF, Atkins JF (2008) Translational bypassing without peptidyl-tRNA anticodon scanning of coding gap mRNA. EMBO J 27: 2533–2544

Wilson DN, Nierhaus KH (2007) The weird and wonderful world of bacterial ribosome regulation. Crit Rev Biochem Mol Biol 42: 187–219

Wilson MA, Brunger AT (2000) The 1.0 A crystal structure of Ca(2+)-bound calmodulin: an analysis of disorder and implications for functionally relevant plasticity. J Mol Biol 301: 1237–1256

Yamami T, Ito K, Fujiwara T, Nakamura Y (2005) Heterologous expression of Aquifex aeolicus ribosome recycling factor in *Escherichia coli* is dominant lethal by forming a complex that lacks functional co-ordination for ribosome disassembly. Mol Microbiol 55: 150–161

Yusupova G, Jenner L, Rees B, Moras D, Yusupov M (2006) Structural basis for messenger RNA movement on the ribosome. Nature 444: 391–394

Zhang W, Dunkle JA, Cate JH (2009) Structures of the ribosome in intermediate states of ratcheting. Science 325: 1014–1017

Zouridis H, Hatzimanikatis V (2007) A model for protein translation: polysome self-organization leads to maximum protein synthesis rates. Biophys J 92: 717–730

Crystal structure of the eukaryotic 80S ribosome

7

Adam Ben-Shem, Lasse B. Jenner, Gulnara Yusupova, Marat Yusupov

1. Introduction

The macromolecular assembly called the ribosome is responsible for protein biosynthesis following genetic instructions in all organisms, and composed of two unequal subunits, a small and a large subunit (40S and 60S in eukaryotes; 30S and 50S in prokaryotes). The full eukaryotic ribosome has a sedimentation coefficient of 80S and a minimal mass of ~3.3 MDa (yeast, plants). The large 60S subunit consists of three ribosomal RNA (rRNA) molecules (28S, 5.8S, 5S) and 46 proteins whereas the small 40S subunit includes only one rRNA chain (18S) and harbors 33 proteins.

Although the eukaryotic ribosome was identified first, most structural information has been obtained from prokaryotic systems. The wealth of information gained from recent high-resolution crystal structures of prokaryotic ribosomes (for reviews see (Schmeing and Ramakrishnan, 2009; Demeshkina et al., 2010) obviously enhance our knowledge of prokaryotic protein synthesis; however, our understanding of the eukaryotic ribosome is sparse. Despite the evolutionary conservation of ribosomal RNA (rRNA) and ribosomal proteins, there are substantial differences between prokaryotic and eukaryotic ribosomes. Owing to the presence of 20 to 30 additional proteins and more than 50 additional nucleotide sequences, called RNA expansion segments, that are inserted into the evolutionary conserved rRNA core, eukaryotic ribosomes have a roughly 40% larger mass than prokaryotic ones. The current knowledge of the structural organization of eukaryotic ribosomes is based on fitting high-resolution X-ray structures of prokaryotic ribosomes into medium-resolution cryo-EM single-particle reconstructions. Presently these cryo-EM studies provide models at resolutions of 6–15 Å and provide the first pictures of how the eukaryotic ribosome may function

(Spahn et al., 2001; Spahn et al., 2004; Becker et al., 2009). Nevertheless, the elucidation of X-ray structures of the eukaryotic ribosome will enable us to discern the relationships between structure and function at the atomic level.

The first indication that the eukaryotic ribosome can be crystallized was provided in 1966 when two-dimensional ribosome crystals were discovered in the tissues of hypothermic chick embryos (Byers, 1966). However, until now, no three-dimensional crystals of eukaryotic ribosomes or ribosomal subunits were reported. We recently succeeded in crystallizing the complete 80S ribosome from *Saccharomyces cerevisiae* and were able to determine its structure at 4.15 Å resolution. The results of this study are presented in this chapter.

2. Overall view of the 80S yeast ribosome

We have obtained crystals of the 80S ribosome from *S. cerevisiae* diffracting beyond 3 Å. The space group was determined to be $P2_1$ with cell parameters a = 437 Å, b = 288 Å, c = 307 Å, (β = 99°, with two 80S molecules per asymmetric unit. A complete data set has been collected to 4.0 Å (I/σI = 2.0 at 4.15 Å).

The model of the 80S ribosome contains the entire rRNA moiety except for a single flexible expansion segment in 25S rRNA (ES27) and a small part of ES7. Our model contains also the backbone of all proteins with homologues in prokaryotic ribosome X-ray structures, including in most cases their eukaryotic-specific segments. The model includes many additional α-helices and β-strands that belong to eukaryote-specific proteins. With the exception of several proteins, we refrained from assigning these secondary structure elements to individual proteins, since biochemical data

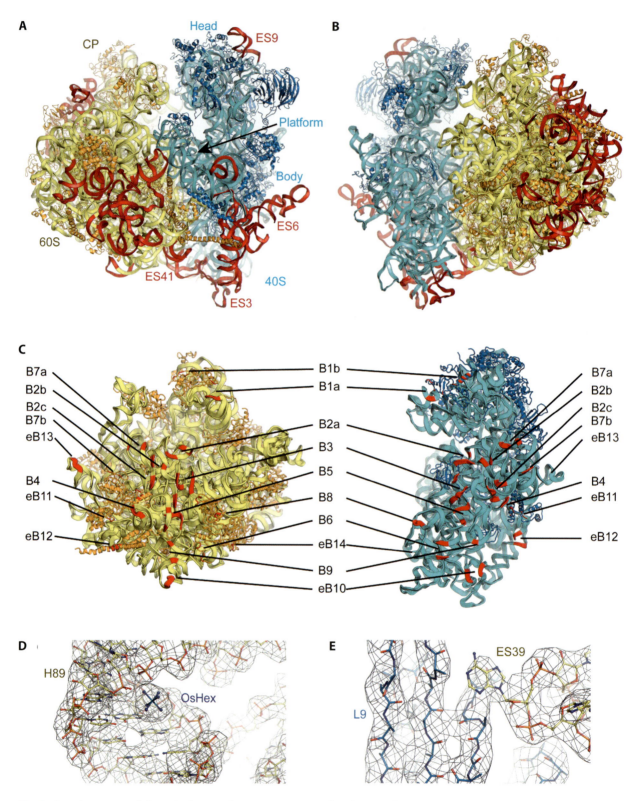

Fig. 1 Crystal structure of the 80S ribosome from *S. cerevisiae*. (A) View from the E site. Proteins and rRNA in the 40S subunit are colored in dark and light blue, respectively, and in dark and pale yellow, respectively, in the 60S subunit (this color scheme is maintained in all following figures unless otherwise indicated). Expansion segment are colored in red. (B) View from the A site. (C) Interface views of the 60S and 40S subunits with bridges num-bered essentially as previously (Yusupov et al., 2001), and colored in red. (D and E) Electron density maps calculated with SAD/MR combined phases and contoured at 1.5 σ, showing (D) H89 of 25S with an osmium hexamine molecule bound in the major grove of H89, and (E) the interaction between expansion segment 39 and protein L9. (Figure reproduced from Ben-Shem et al., 2010)

that may confirm their position are lacking. Figure 1A and B show the basic architecture of full 80S ribosome viewed from the exit and entrance sites.

The overall structure of the eukaryotic ribosome reveals a considerably larger assembly than its prokaryotic counterpart, but the basic architecture is similar with common recognizable landmarks. The rRNA expansion elements are located predominantly on the solvent-exposed sides at the periphery of both subunits. Figure 1C shows the intersubunit contacts viewed from the interface of the 60S and 40S. The intersubunit bridges of the yeast 80S ribosome were first visualized in low-resolution cryo-EM studies (Spahn et al., 2001; Spahn et al., 2004b). Our model at 4.15 Å provides a more detailed view of the molecular components involved in these contacts between ribosomal subunits. The importance of the bridges is evident as they maintain communication pathways between the two subunits during protein synthesis. The evolutionary conservation of intersubunit bridges at the core of the ribosome is noteworthy – for all intersubunit bridges that have been observed in the crystal structure of the bacterial ribosome (Yusupov et al., 2001) there is a corresponding bridge in the eukaryotic ribosome. Figure 1D demonstrates the quality of the electron density. Density for the phosphates was usually clear and guided modeling of the rRNA parts.

3. Ribosomal domain movements in the ratcheted state

Our crystals capture the ribosome in the so-called ratcheted conformation. It was postulated more than 40 years ago that the translocation of mRNA and tRNA during protein synthesis is coupled to inter-subunit movements (Bretscher, 1968; Spirin, 1969). More recently, cryo-EM studies suggested that translocation is facilitated by large-scale movements involving a rotation of the small subunit relative to the large subunit (Frank and Agrawal, 2000). As confirmed by various studies, this ratchet-like inter-subunit reorganization of the ribosome is essential for translocation (Horan and Noller, 2007). Recently, structures of the *Escherichia coli* ribosome trapped in intermediate states of ratcheting were described (Zhang et al., 2009); however, an X-ray structure of the ribosome in the fully ratcheted state was lacking.

Our model of the eukaryotic ribosome, when compared to the structure of the unratcheted prokaryotic ribosome (Jenner et al., 2010a), shows a 5° counterclockwise rotation of the 40S subunit body relative to the 60S subunit and a swiveling of the 40S head domain by 14° in the direction of the exit (E)-site tRNA (Figure 2A and B). These are the characteristics of the ratcheted state, in accordance with the observation, made by cryo-EM, that yeast ribosomes devoid of ligands assume the same ratcheted conformation as the one stabilized by the binding of eukaryotic elongation factor 2 (eEF2) (Spahn et al., 2004). Indeed, our X-ray model fits well into the cryo-EM maps of the 80S-eEF2 complex (data not shown).

The large-scale movements implicated in ratcheting result in significant alterations in the bridges, or contacts, between the head domain of the small subunit and the large subunit, in comparison with the unratcheted prokaryotic ribosome (Jenner et al., 2010a) (Figure 2C and D). In the latter, the first bridge between the small-subunit head domain and the large subunit, bridge B1a, is formed by the A-site finger (H38 of 23S) and protein S13. Since head swiveling displaces components at the periphery of the head domain by as much as 25 Å, this bridge is rearranged in our model (Figure 2C and E). We find that ratcheting brings residues 1239–1241 at the tip of h33 (a component of the beak of 40S) as well as protein S15 (prokaryotic homolog, S19p), in the proximity of the tip of H38 which bends significantly in order to form interactions with these partners. Conformational changes are also observed at the base of H38 where it contacts the central protuberance. In the second bridge between the head domain of 40S and the 60S subunit, B1b (Figure 1C), the large shift in the position of protein S18 (S13p) places its largest helix, instead of the N-terminal loop, in contact with protein L11 (L5p) of the central protuberance (Figure 2C and F). Similar rearrangements were observed in intermediates of the partially ratcheted *E. coli* 70S ribosome (Zhang et al., 2009). In addition, residues from loop 65–75 in S15 (S19p) may also interact with L11 (L5p) in the ratcheted state. The prokaryotic homologues of the two proteins that are so strongly shifted, S15 (S19p) and S18 (S13p), were shown in the non-ratcheted state to monitor the occupancy of the A and P sites respectively (Jenner et al., 2010b, a). The direct interaction between these two proteins is probably stronger in eukaryotes due to the involvement of additional residues.

The entire central protuberance of the large subunit, dominated by 5S rRNA and the proteins enveloping it, undergoes considerable structural rearrange-

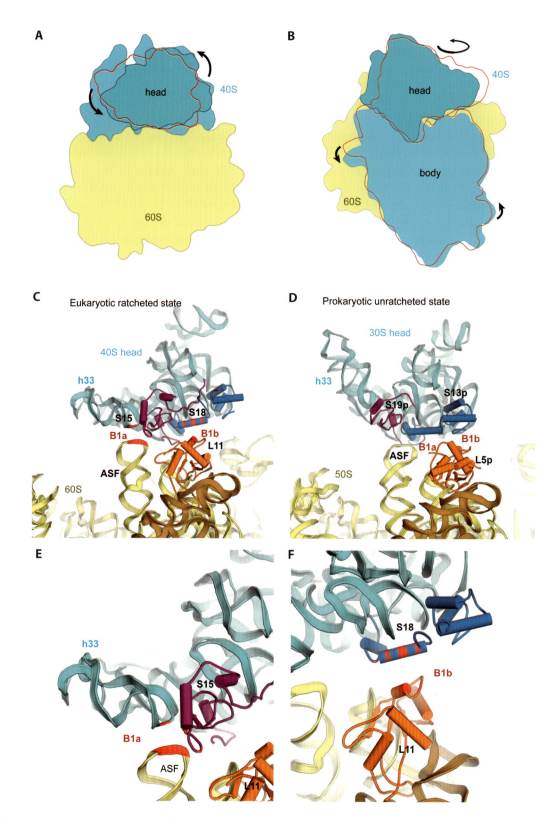

Fig. 2 Ratcheted state of the eukaryotic 80S ribosome. (A and B) Schematic representation of the motion from the unratcheted to the ratcheted state. The red line indicates the outline of the 40S in the unratcheted state, with arrows indicating the trajectory. (A) Top-view of the yeast 80S ribosome. (B) View from the solvent side of 40S. (C) Bridge B1 in the ratcheted 80S. (D) Bridge B1 of the non-ratcheted prokaryotic ribosome. (E) Close-up view of bridge B1a. The tip of the A-site finger (ASF-H38) from 25S RNA forms interactions (colored in red) with the head of the 40S subunit including protein S15 (magenta). (F) View of bridge B1b formed between proteins S18 and protein L11. Residues thought to interact are indicated in red. (Figure adopted from Ben-Shem et al., 2010)

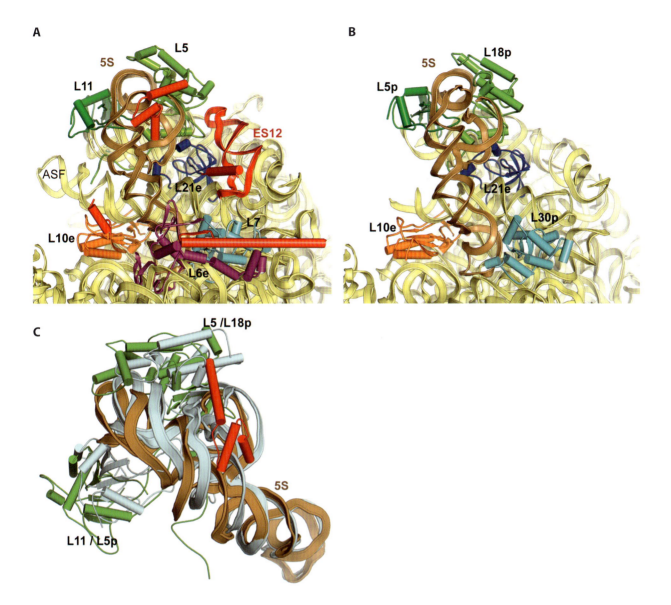

Fig. 3 Ratcheted state of the eukaryotic 80S ribosome (continued). (A and B) The central protuberance (CP) of (A) *S. cerevisiae* 60S (eukaryote-specific elements marked in red) and (B) the archaeal *H. marismortui* 50S. (C) Representation of the CP of the 50S prokaryotic ribosome (light blue) superimposed on the eukaryotic ratcheted 60S (5S rRNA in brown, proteins in green and eukaryote-specific elements in red). Only 5S rRNA and proteins L5 (L18p) and L11 (L5p) are shown for clarity. (Figure reproduced from Ben-Shem et al., 2010)

ment. These alterations involve: tilting of 5S helices, shift in the position of L11 (L5p) elements, including the conserved first and last helices, and displacement by up to 7 Å of all L5 (L18p) domains except the N' tail and first helix (Figure 3).

Their plasticity in response to ratcheting and their strategic position in the large subunit suggest that the central protuberance and the A-site finger may coordinate changes in different sites of the 60S structure with the 40S head rotation and mRNA translocation, notably the L1 stalk, the GTPase center and the peptidyl-transferase center. This proposed role for the central protuberance and the required plasticity might underlie the need for keeping 5S as a separate rRNA chain.

The central protuberance in eukaryotes is larger than its prokaryotic counterpart, forming additional interactions with other regions in the ribosome. Comparing our model to the archaeal large ribosomal subunit (Figure 3A and B) (Ban et al., 2000) we find that these additional features include: (i) expansion segment ES12 forming an RNA helix that runs in parallel to 5S rRNA and has its base interacting with the

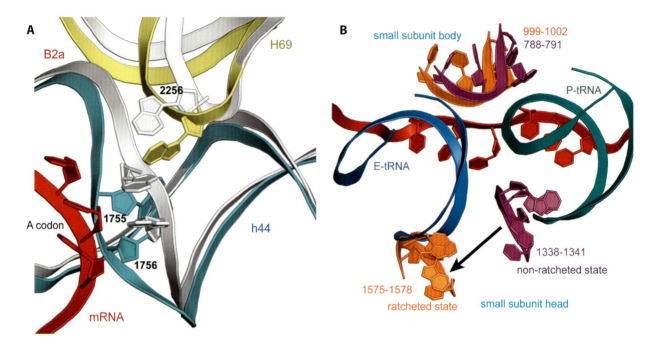

Fig. 4 Rearrangement of functional sites of the ribosome upon ratcheting. (A) Superposition of the prokaryotic non-ratcheted ribosome with the ratcheted yeast ribosome. View of the decoding region with prokaryotic 16S and 23S rRNA in grey, A-site codon in red, and eukaryotic 18S and 25S RNA in blue and yellow, respectively. (B) Top view of the superposition of the 70S ribosome containing P-tRNA (green), E-tRNA (blue) and mRNA (red) with the 80S ratcheted ribosome. The ridge consisting of nucleotides 1575–1578 (1338–1341 in prokaryotes) forming the steric block between P-tRNA and E-tRNA and the region 999–1002 (788–791 in prokaryotes) from the platform of the small subunit. (Figure reproduced from Ben-Shem et al., 2010)

eukaryote-specific C-terminal helix of L21e and its tip in contact with L5 (L18p); (ii) an extra helix at the C-terminus of protein L10e (L16p); (iii) an additional domain at the N-terminus of L5 (L18p); (iv) the eukaryote-specific protein L6e that binds 5S rRNA and is embedded in a region rich in other eukaryote-specific elements (protein L14e, the N-terminal domain of L7 (L30p) and ES7).

The numerous interactions between the head and the large subunit may serve to limit or determine the extent of the head's rotation. It is important to note that all these interactions are relatively weak and entail large distances. Hence, multiple weak interactions facilitate a wide but precise ratcheting movement. Flexibility or plasticity of the interacting partners, as observed here, is probably crucial for constantly adjusting the bridges as the ratcheting movement progresses.

4. Rearrangement of ribosome functional sites upon ratcheting

In structures of the prokaryotic ribosome in putative intermediate states along the ratcheting pathway (Zhang et al., 2009), the central bridges are almost in-distinguishable from those observed in the unratcheted state. We find that in our fully ratcheted state some of these bridges undergo considerable alterations. Bridge B2a is of special interest as it is formed by the base of the penultimate stem (h44) of 18S and H69 of 25S. In this universally conserved region, residues 1755/1756 (1492/1493 in *E. coli*) of 18S and 2256 (1913 in *E. coli*) of 25S rRNA play a key role in decoding (Ogle et al., 2001). We find that in the ratcheted yeast ribosome nucleotides 1755/1756 (1492/1493) and the region of 18S rRNA encompassing these decoding elements undergo a local conformation change. The 25S rRNA strand that encompasses residues 1754–1758 (1491–1495) is twisted and pushed away from the mRNA pathway in the direction of the P site by up to 9 Å. This position would preclude interactions of the rRNA backbone with the mRNA-tRNA complex (Figure 4A).

In addition, the tip of H69 bends so as to maintain the bridge and several residues involved in the bridge assume different orientation than in empty non-ratcheted ribosome. We suggest that the rotation of the small subunit results in breaking, or at least considerably loosening, the interaction of the 40S body with the mRNA-tRNA complex so that this complex would be free to follow the rotation of the head. These findings concur with the notion that ratcheting is a multistep process and suggest that rotation of the body and rearrangement of bridge B2a occur prior to the rotation of the head (Zhang et al., 2009).

Other instances where ratcheting breaks interactions that may impede translocation include: the bending of the A site finger, which removes its interaction with the elbow of A-tRNA, and the breaking of the contact between A1001 of 18S (A790) and the P-tRNA by a rotation of the 40S platform. However, the swiveling of the head alone can account only for a movement of the mRNA-tRNA complex by 11 Å, instead of the required 20 Å (one codon) distance. Perhaps the local 9 Å shift at the base of h44, following the rotation of the head, plays a role in the further translocation of the mRNA-tRNA complex (Taylor et al., 2007). In any case, to complete the translocation step and proceed to the non-ratcheted post-translocation state, further conformational changes that would result in breaking the bonds between the head and the mRNA-tRNA complex must ensue (Yusupov et al., 2001; Selmer et al., 2006; Jenner et al., 2010a). Our model suggests that these changes would entail large structural rearrangement in the head domain. Considering, for example, the steric block between the P-tRNA and the E site formed by the universally conserved ridge of residues 1575–1578 in 18S (1338–1341 in *E. coli*) (Schuwirth et al., 2005). In our model this ridge has rotated with the head in a trajectory that passed through the E site, suggesting that the forward swiveling of the head keeps the position of the ridge with respect to the P-tRNA (Figure 4B). Hence, the ridge creates a large physical barrier for further forward movement and release of the tRNA, while the tRNA physically prevents back-ratcheting. A large conformational change, perhaps driven by GTP hydrolysis (Rodnina et al., 1997), therefore seems to be required to remove this barrier. The nature and timing of events that complete the translocation of the mRNA-tRNA complex are unknown.

5. Concluding remarks

The current knowledge of the relationship between structure and function of eukaryotic ribosomes, which are structurally, biochemically, and functionally much more complex than their prokaryotic counterparts, has so far been provided only by cryo-EM studies. We believe that our crystal structure of the eukaryotic 80S ribosome at 4.15 Å resolution will offer a strong foundation for future higher-resolution X-ray studies of the eukaryotic ribosome in different functional complexes, which will unveil the features of the eukaryotic ribosomal machinery. A thorough understanding of the mechanisms that govern eukaryotic translation at the atomic level will herald a new era of medical therapies.

References

Ban N, Nissen P, Hansen J, Moore PB, Steitz TA (2000) The complete atomic structure of the large ribosomal subunit at 2.4 A resolution. Science 289: 905–920

Ben-Shem A, Jenner L, Yusupova G, Yusupov M (2010) Crystal structure of the eukaryotic ribosome. Science (in press)

Bretscher MS (1968) Translocation in protein synthesis: a hybrid structure model. Nature 218: 675–677

Byers B (1966) Ribosome crystallization induced in chick embryo tissues by hypothermia. J Cell Biol 30: C1–6

Frank J, Agrawal RK (2000) A ratchet-like inter-subunit reorganization of the ribosome during translocation. Nature 406: 318–322

Horan LH, Noller HF (2007) Intersubunit movement is required for ribosomal translocation. Proc Natl Acad Sci USA 104: 4881–4885

Jenner L, Demeshkina N, Yusupova G, Yusupov M (2010a) Structural aspects of messenger RNA reading frame maintenance by the ribosome. Nat Struc Mol Biol 17: 555–560

Jenner L, Demeshkina N, Yusupova G, Yusupov M (2010b) Structural rearrangements of the ribosome at the tRNA proofreading step. Nat Struct Mol Biol 17: 1072–1078

Ogle JM, Brodersen DE, Clemons WM, Jr., Tarry MJ, Carter AP, Ramakrishnan V (2001) Recognition of cognate transfer RNA by the 30S ribosomal subunit. Science 292: 897–902

Rodnina MV, Savelsbergh A, Katunin VI, Wintermeyer W (1997) Hydrolysis of GTP by elongation factor G drives tRNA movement on the ribosome. Nature 385: 37–41

Schuwirth BS, Borovinskaya MA, Hau CW, Zhang W, Vila-Sanjurjo A, Holton JM, Cate JH (2005) Structures of the bacterial ribosome at 3.5 A resolution. Science 310: 827–834

Selmer M, Dunham CM, Murphy FVt, Weixlbaumer A, Petry S, Kelley AC, Weir JR, Ramakrishnan V (2006) Structure of the 70S ribosome complexed with mRNA and tRNA. Science 313: 1935–1942

Spahn CM, Gomez-Lorenzo MG, Grassucci RA, Jorgensen R, Andersen GR, Beckmann R, Penczek PA, Ballesta JP, Frank J (2004) Domain movements of elongation factor eEF2 and the eukaryotic 80S ribosome facilitate tRNA translocation. EMBO J 23: 1008–1019

Spirin AS (1969) A model of the functioning ribosome: locking and unlocking of the ribosome subparticles. Cold Spring Harb Symp Quant Biol 34: 197–207

Yusupov MM, Yusupova GZ, Baucom A, Lieberman K, Earnest TN, Cate JH, Noller HF (2001) Crystal structure of the ribosome at 5.5 A resolution. Science 292: 883–896

Zhang W, Dunkle JA, Cate JH (2009) Structures of the ribosome in intermediate states of ratcheting. Science 325: 1014–1017

8

Structure and function of organellar ribosomes as revealed by cryo-EM

Rajendra K. Agrawal, Manjuli R. Sharma, Aymen Yassin, Indrajit Lahiri, and Linda L. Spremulli

1. Introduction

During the last decade ground-breaking progress was made in resolving the structure of ribosomes from several bacterial and archaeal species. A number of high-resolution X-ray crystallographic structures of bacterial ribosomes, including some of their functional complexes, and cryo-electron microscopic (cryo-EM) structures of both prokaryotic and eukaryotic cytoplasmic ribosomes in various functional states were resolved. More recently, the first X-ray crystallographic structures of eukaryotic ribosomes, the yeast 80S ribosome (see chapter by Jenner and coworkers in this volume) and a protozoan 40S ribosomal subunit in complex with eukaryotic initiation factor 1 (Rabl et al., 2011) at ~4 Å resolution, have been obtained. Those studies have helped us tremendously in understanding some of the key functions of cytoplasmic ribosomes. Certain organelles of the cell, such as mitochondria and chloroplasts, have their own translational machineries, including ribosomes (Harris et al., 1994; O'Brien, 2002), for the synthesis of proteins that are involved in oxidative phosphorylation and photosynthesis, respectively. Structural studies of organellar ribosomes have

Table 1 Comparison between bacterial, mammalian mitochondrial, protistan mitochondrial, and chloroplast ribosomes

Ribosome source → Properties ↓	Bacteria *Escherichia coli*	Mammalian mitochondria *Bos taurus*	Protistan mitochondria *Leishmania tarentolae*	Chloroplast *Spinacea oleracea*
Molecular mass	2.3 MDa	2.7 MDa	2.2 MDa	2.6 MDa
Diameter	~260 Å	~320 Å	~245 Å	~265 Å
Sedimentation coefficient	70S	55S	50S	70S
RNA: protein ratio	~2:1	~1:2	~1:3	~3:2
Subunits	30S + 50S	28S + 39S	(28S – 30S) + 40S	30S + 50S
Small subunit composition	16S rRNA (1542 nt) + 21 proteins	12S rRNA (950 nt) + 29 proteins (15)	9S rRNA (610 nt) + 56 proteins* (46)	16S rRNA (1491 nt) + 25 proteins (4)
Large subunit composition	23S rRNA and 5S rRNA (total 3024 nt) + 34 proteins	16S rRNA (1560 nt) + 50 proteins (20)	12S rRNA (1173 nt) + 77 proteins* (66)	23S rRNA, 5S rRNA, and 4.8S rRNA (total 3033 nt) + 33 proteins (2)
Number of inter-subunit bridges	13	15	9	13

* The estimate of the number of proteins in the *L. tarentolae* ribosomal subunits is based on an exhaustive mass spectrometric analysis of *Trypanosoma brucei* (Zíková et al., 2008), the closest known relative of *Leishmania*.
The number of organellar ribosomal proteins that do not have bacterial homologs are shown in brackets.
nt, nucleotide(s)

lagged behind studies of cytoplasmic ribosomes. In this article, we describe the current state of the structural information available for the organellar ribosomes, which was obtained by using the techniques of single-particle cryo-EM and three-dimensional image processing in our laboratory.

At the present time, we have determined cryo-EM structures of three different organellar ribosomes, including a mammalian mitochondrial ribosome (mitoribosome), a chloroplast ribosome (chloro-ribosome), and a protistan mitoribosome. The goal of these studies has been to understand these structures from the perspective of their structural and functional evolution, by comparing their structures with the structures of the widely studied bacterial and eukaryotic cytoplasmic ribosomes. Since chloroplasts and mitochondria are thought to have evolved from endosymbiotic primitive bacteria in eukaryotic host cells (Margulis, 1970; Gray et al., 2001), it is expected that the structure of organellar ribosomes will be more similar to that of bacterial ribosomes rather than to that of cytoplasmic ribosomes of eukaryotes. However, as described below, we observe a number of distinct structural features of organellar ribosomes which differ significantly from their bacterial counterpart.

Before describing the structures of individual organellar ribosomes, an understanding of their composition is essential. Thus, we will first provide an overall comparison of the constituents of bacterial (*Escherichia coli*) and the above-mentioned three organellar ribosomes (Table 1), the secondary structure diagrams of the ribosomal RNAs (rRNAs; Figure 1), and their overall cryo-EM structures (Figure 2). A more detailed description of the proteins of the mammalian mitoribosome was presented in a recent review by Koc and coworkers (2010).

2. The structure of the mammalian mitochondrial ribosome

The mammalian mitoribosome is responsible for the synthesis of 13 membrane proteins that form components of complexes involved in oxidative phosphorylation (Attardi, 1985, Chomyn et al., 1986). Apart from these 13 polypeptides, mammalian mitochondrial DNA encodes 2 rRNAs and 22 tRNAs. The mRNAs almost completely lack 5' and 3'-untranslated nucleotides, and the initiation codon is generally located within three nucleotides from the 5' end of the mRNA (Anderson et al., 1982; Montoya et al., 1981). There is no Shine-

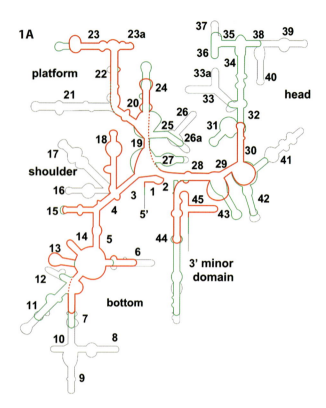

Fig. 1A Secondary structures of SSU RNAs from bacteria (16S, grey), mammalian mitochondria (12S, green), and *Leishmania tarentolae* mitochondria (9S, red). (Adapted from Suppl. Information in Sharma et al. 2009)

Dalgarno sequence and no cap structure. The mechanism by which the mitoribosome recognizes the start site on the mRNA is unknown. Animal mitochondrial tRNAs (tRNA$_{mt}$) are generally shorter (59–75 nucleotides in length) than bacterial or eukaryotic cytoplasmic tRNAs. Many tRNAs$_{mt}$ lack the D loop/T loop interactions in the elbow region that stabilize the L-shape in the "more conventional" tRNAs (Steinberg and Cedergren, 1994; Steinberg et al., 1994; Yokogawa et al., 1991; Watanabe et al., 1994; Zagryadskaya et al., 2004). The first structure of a tRNA$_{mt}$, revealed in our cryo-EM study of the mammalian mitoribosome (Sharma et al., 2003), showed an L-shape, but with a "caved-in" elbow region. Since then, the resolution of the mammalian mitoribosome in our laboratory has significantly improved (currently at 7.0 Å resolution; Sharma et al., in preparation). This structure shows several new features that are better defined; however, the basic structural organization, including overall distributions of rRNAs and proteins within the map, is essentially the same as reported earlier (Sharma et al., 2003). Therefore, for the present comparison with other organellar ribosomes, the structures of which were determined at

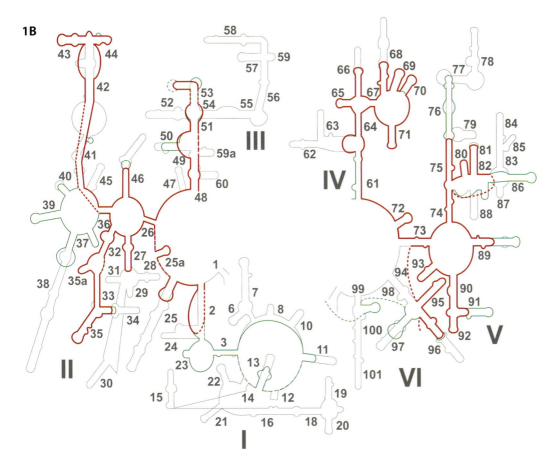

Fig. 1B Secondary structures of LSU RNAs from bacteria (23S, grey), mammalian mitochondria (16S, green), and *L. tarentolae* mitochondria (12S, red). Dashed lines indicate unassigned segments of rRNA.

9–14 Å resolution, we will use an 11 Å resolution map of the mammalian mitoribosome.

The mammalian mitoribosomes are substantially different from bacterial ribosomes. Their mass ratio of protein to RNA is almost inversed (61:39, as compared to 33:67; O'Brien, 2002), and with a molecular weight of 2.71MDa and a diameter of ~ 320 Å, mitoribosomes are larger in size (O'Brien, 1971; Sharma et al., 2003; Table 1). However, due to its porous nature, mitoribosomes exhibit a sedimentation coefficient of 55S, lower than that of bacterial 70S ribosomes. Like all ribosomes, the 55S mitoribosome is composed of two unequal subunits, the 28S small subunit (SSU) and the 39S large subunit (LSU). Overall, the mammalian mitoribosome SSU is elongated (by ~ 70 Å) and narrower (by ~ 15 Å) in its mid-body region, while the mammalian mitoribosome LSU possesses larger structures in the L1 stalk, L7/L12 stalk, and the central protuberance regions (Sharma et al., 2003). The mammalian mitoribosome SSU is composed of 12S rRNA and 29 proteins, while its LSU is composed of 16S rRNA and

about 50 proteins (Koc et al., 2001, Suzuki et al, 2001). This is in sharp contrast to the bacterial 70S ribosome, which has its 30S SSU composed of 16S rRNA and 21 proteins and its 50S LSU composed of 23S rRNA, 5S rRNA, and 33 proteins (Wittmann-Liebold, 1985). Fourteen out of the 29 SSU mitochondrial ribosomal proteins (MRPs) have bacterial homologues, while 15 proteins are classified as mito-specific. Likewise, 22 out of the 50 LSU MRPs have bacterial homologues, while the remaining 28 are classified as specific for mitochondria. Thus, almost half of the 79 MRPs are unique to the mammalian mitochondrial system and have no equivalent counterparts in bacteria. Most of these MRPs are situated in peripheral regions and mostly on the solvent sides of the two ribosomal subunits. Thus, unlike bacterial ribosomes, rRNAs in the mammalian mitoribosomes are largely covered by MRPs (Sharma et al., 2003). A few mito-specific MRPs are present on the interface side and participate in the formation of the protein-protein intersubunit bridges characteristic for mitoribosomes.

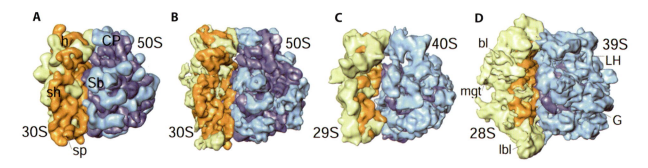

Fig. 2 Segmented cryo-EM maps of ribosomes from (A) bacteria (70S, *E. coli*), (B) chloroplasts (70S, *S. oleracea*), (C) protistan mitochondria (50S, *L. tarentolae*), and (D) mammalian mitochondria (55S, *B. taurus*). Proteins of small and large ribosomal subunits are colored yellow and blue, respectively, and rRNAs orange and ma-genta, respectively. Landmarks of the small subunit: bl, beak lobe; h, head; lbl, lower body lobe; mgt, mRNA gate; sh, shoulder; sp, spur. Landmarks of the large subunit: CP, central protuberance; G, gap due to a shorter rRNA segment in the mitoribosome; LH, large subunit handle; Sb, L7/L12 stalk base.

2.1. The small subunit of mammalian mitochondrial ribosomes

The mitoribosomal RNAs of both the SSU and LSU show significant deletion of rRNA segments, compared to the bacterial rRNAs (Figure 1). In the mammalian mitochondrial SSU, many 12S rRNA segments, corresponding to bacterial 16S rRNA helices 6, 8–10, 12, 13, 16, 17, 21, 33, 37 and 39–41, as well as the rRNA segment bearing the anti-Shine-Dalgarno sequence, are absent altogether, while several other segments of the mitochondrial rRNA are shorter, compared to bacterial 16S rRNA. There is little sequence conservation between mitochondrial 12S rRNA and bacterial 16S rRNA, and the base composition is quite different (18% G residues and 37% A residues, compared to 32% G residues and 25% A residues in bacterial rRNA). In addition, only about 50% of the residues in 12S rRNA are engaged in base-pairing, as compared to 63% in bacterial 16S rRNA (Mears et al., 2006). Furthermore, only about 19% of the missing segments in the mito SSU rRNA are compensated by additional and enlarged MRPs, even though 12 of the 14 SSU MRPs that show sequence homology with bacterial counterparts are significantly (4–25 kDa) larger, while two other MRPs, S6 and S12, are shorter.

Differences in the lengths of the rRNA helices and the protein composition between the mammalian mitochondrial and bacterial SSUs lead to the special morphological characteristics observed in the mitoribosome SSU (Sharma et al., 2003). Due to the absence of the bacterial 16S rRNA helices 33 and 33a, the mitoribosomal SSU lacks the beak-like feature of the head. Instead there is a large globular mass, apparently formed by one or more MRPs, referred to as beak lobe,

in the mitoribosome SSU, that emerges from the solvent side of the head and extends to the position where bacterial helix 33a would be situated (Sharma et al., 2003). Due to the absence of bacterial rRNA helices 16 and 17, the mitoribosomal SSU has a narrower shoulder and body regions, and a more open factor-binding region. The absence of bacterial helices 12 and 21 results in a channel-like feature in the center of the mitoribosome SSU body, and the lack of helices 39 and 41 results in another channel-like feature in the head region, connecting the interface side of the subunit with the solvent side. Helix 6 of the 16S rRNA that gives rise to the spur-like feature in the bacterial SSU is absent in the mitoribosomal SSU and appears to be partially compensated by a MRP mass designated as lower body finger (Sharma et al., 2003). An elongated MRP mass spans the platform and the body on the solvent side of the mitoribosome SSU and compensates for the missing bacterial rRNA helix 21. The lower body of the mitoribosome SSU is composed entirely of MRPs. The RNA-binding bacterial proteins S4, S8 and S20 are absent in the mitoribosome SSU, as is protein S13, a main component of the intersubunit bridges B1a and B1b in the bacterial ribosome. Mito-specific MRPs compensate for the role of S13 by forming mitochondria-specific bridges B1a-B1d with the central protuberance of the mitoribosome LSU (Sharma et al., 2003). Proteins S8, S19 and S20 are also absent in the mitoribosome SSU with no apparent compensation by MRPs.

One of the unique features of the mitoribosome SSU is the presence of a triangular gate-like structure surrounding the mRNA entry site (Sharma et al., 2003). This gate is formed by the extended MRPS2, which is larger in the mitoribosome, and mito-specific MRPs that compensate for the absence of bacterial proteins S3 and S4. The gate may play a direct role in recruiting mitochondrial mRNAs, which usually lack a 5'-UTR (Temperley et al., 2010). There are 11 mitochondrially-encoded mRNAs, nine monocistronic and two dicistronic with overlapping reading frames, that are responsible for the synthesis of 13 polypeptides.

2.2. The large subunit of mammalian mitochondrial ribosomes

Similar to the SSU, the mitochondrial LSU lacks many rRNA elements characteristic of the bacterial 50S LSU (Figure 1B). Approximately 28% of the missing rRNA elements appear to be compensated by MRPs. 5S rRNA has not been setected in the mitoribosomal LSU, and about 50% of its mass is replaced by mito-specific MRPs, making the central protuberance of LSU appear almost twice as big as that of its bacterial counterpart. A mass of protein(s) connects the central protuberance to the rest of LSU through a unique cylindrical feature designated as LSU handle (LH) (Figure 2D), whose long axis is situated such that it would be parallel and close to helices 4 and 5 of the 5S rRNA in the bacterial LSU. Based on their location and structural features, it is conceivable that the MRPs involved in the formation of LSU handle in the mitoribosome play the role of 5S rRNA in the bacterial ribosome. A comparison of the secondary structure elements of mitochondrial 16S rRNA and bacterial 23S rRNA, which is divided into six structural domains, reveals the absence of most of domain I and a significant portion of domain III in the mitochondrial 16S rRNA (Figure 1B). The absence of a major portion of domain I leads to the formation of a significant gap (area marked as "G" in Figure 2D) in the lower back (solvent side) region of the mitochondrial LSU. However, some parts of domain III are replaced by mito-specific MRPs. One of the distinct features of the mitochondrial LSU is the absence of helix 38 which, together with protein S13, forms bridge B1a in bacterial ribosomes. This particular bridge is replaced by protein-protein interaction in the mitoribosome. Several other 23S rRNA helices, which are known to contribute in the formation of bridges in bacterial ribosome, are also absent or truncated, including helices 34, 62 and 68, which are elements of bacterial intersubunit bridges B4, B5/B6 and B7a, respectively. In the mitoribosome, bridge B4 has an altered configuration, bridge B6 is missing, while both positions of bridges B5 and B7a are preserved. The 23S rRNA helices H77 and H78, which form the L1 stalk in bacterial ribosomes,

are absent in the mitoribosome. However, a prominent analogous structural feature, which appears to be formed by a larger L1 MRP homolog and other MRPs, is present. Interestingly, the 23S rRNA helices 11 and 68 and the loop formed between helices 76 and 77 are absent in mitochondrial 16S rRNA (Figure 1B). These rRNA elements are known to be directly involved in tRNA binding at the bacterial ribosomal E site (tRNA exit site) (Yusupov et al., 2001). Thus, their absence in the mitoribosome results in very weak interactions for tRNA in the E-site region (see also Mears et al., 2006).

A finger-like structure (P-site finger) is formed by a mito-specific MRP that extends from the central protuberance of the mito LSU and contacts the P-site tRNA at its T-loop side. This extended protein mass may stabilize and regulate tRNA binding and movement during protein synthesis. The stabilization of the P-site tRNA and the absence of any tRNA mass at the E site in the mitoribosome might help explain the consistent co-purification of a P-site tRNA along with the mitoribosome, in sharp contrast to the bacterial ribosome that invariably co-purifies with a tRNA bound at its E site. More recently, it has been shown that the unstructured C-terminus of bacterial ribosomal protein L25, the major globular portion of which is located on the solvent side, extends into the intersubunit face and interacts with the T-loop side of the A-site tRNA (Jenner et al., 2010). These findings reveal that direct interactions between the ribosomal protein extensions and tRNAs are more common than previously suggested. However, the functional implication of such an interaction is unknown.

Following the first cryo-EM reconstruction of the mammalian mitoribosome (Sharma et al., 2003), our laboratory has begun exploring some of the functional complexes of the mammalian mitoribosome, in order to understand the functional roles of mito-specific MRPs and the significant alterations that has been observed in the structural organizations of the translational factors of the mammalian mitochondria (Spremulli et al., 2004). In the following section, we briefly describe the mammalian mitochondrial initiation factors and elongation factor G.

2.3. Mammalian mitochondrial translation initiation factors

One of the striking differences between bacterial and mitochondrial initiation is the number of initiation factors involved. In contrast to bacterial systems, where there are three initiation factors, only two initiation factors have been identified in mammalian mitochondria (Koc and Spremulli, 2002; Liao and Spremulli, 1990, 1991; Ma and Spremulli, 1995, 1996; Ma et al., 1995). Notably, a homolog of bacterial IF1 is not present in mammalian mitochondria. In bacteria, IF1 binds in the region of the A site of the 30S subunit, where it is postulated to prevent premature binding of initiator- or aminoacyl-tRNAs to the A site (Carter et al., 2001; Moazed et al., 1995). Mitochondrial IF2 (IF2$_{mt}$, 74 kDa) stimulates the binding of fMet-tRNA to the small subunit of the mitoribosome, while mitochondrial IF3 (IF3$_{mt}$, 29 kDa) promotes the dissociation of the 55S mitoribosome, thereby providing a supply of small subunits for initiation complex formation. IF2$_{mt}$ has 4 major domains. The G domain (or domain IV) shows significant structural similarity to the G domains of other GTP-binding proteins (Brock et al., 1988; Lee et al., 1999). Domain V is structurally similar to domain II of the bacterial translation elongation factors and is involved in the binding of IF2 to the small ribosomal subunit (Allen et al., 2005; Moreno et al., 1998, 1999). Domain VI C1 is viewed as a "link" between Domains V and VI C2. The X-ray crystallographic structure of aIF5B, an IF2 homolog in an archaeon, *Methanobacterium thermoautotrophicum*, is known (Roll-Mecak et al., 2000). A homology model of IF2 shows that Domain VI C2 is a compact globular structure that would interact with the initiator tRNA (Allen et al., 2005). Interestingly, IF2$_{mt}$ lacks a significant portion (~240 amino acids) of the polypeptide chain from the N-terminus of domain III, as compared to *E. coli* IF2 (Spencer and Spremulli, 2005). However, IF2$_{mt}$ has a small domain of 37 amino acids inserted into domain VI C1. Mutation of several conserved basic residues of the insertion domain reduces the ability of IF2$_{mt}$ to bind 28S subunits, implying that this region makes important contacts with the small subunit of the mitoribosome (Spencer and Spremulli, 2005). A previous biochemical study has shown that the insertion domain of IF2$_{mt}$ substitutes functionally for IF1 in bacteria (Gaur et al., 2008). IF2$_{mt}$ was shown to be capable of replacing bacterial IF2 in a strain with IF1 and IF2 gene knockouts. However, the deletion of the insertion domain from IF2$_{mt}$ necessitates the presence of IF1 in *E. coli*, suggesting that the insertion domain in IF2$_{mt}$ functionally mimics bacterial IF1 (Gaur et al., 2008). Our cryo-EM map of the 70S ribosome in complex with initiator tRNA and IF2$_{mt}$ indicates that the density corresponding to the insertion domain indeed occupies the ribosomal A site (Yassin et al., in preparation).

As in the case of IF2$_{mt}$, also IF3$_{mt}$ is structurally different from bacterial IF3. IF3$_{mt}$ shares a central conserved region with bacterial IF3, for which two X-ray crystallographic structures, one each for the C- and N-terminal domains (Biou et al., 1995), and an NMR structure for the C-terminal domain (Garcia et al., 1995), are known. In addition, IF3$_{mt}$ possesses unique ~30 amino acids long extensions at both N and C termini (Haque and Spremulli, 2008). Biochemical studies have shown that these extensions are important for the dissociation of IF3$_{mt}$ from the 28S subunit during the formation of the 55S mitoribosome (Haque et al., 2008). Our preliminary cryo-EM study of the mito 28S SSU-IF3$_{mt}$ complex indicates that the C-terminal domain (CTD) of IF3$_{mt}$ interacts with the rim of the 28S SSU platform, while the density assigned to the N-terminal domain (NTD) of IF3$_{mt}$ enters the decoding region (Lahiri et al., unpublished).

2.4. Elongation factor G of mammalian mitochondria

The basic steps in the translation elongation cycle in mammalian mitochondria appear to be the same as those taking place in bacteria (Agrawal et al., 2000; Schmeing and Ramakrishnan, 2009). The main function of EF-G is to catalyze the critical step of tRNA translocation from the ribosomal A and P sites to the P and E sites, respectively. Like bacterial EF-G (Ævarsson et al., 1994; Czworkowski et al., 1994), mitochondrial EF-G (EF-G$_{mt}$) folds into five domains (Bhargava et al., 2004). In most organisms, two forms of EF-G$_{mt}$ are present (Hammarsund et al., 2001) which vary slightly in their amino-acid sequences. In comparison to bacterial EF-G, form 1 (EF-G1$_{mt}$) has a 10 amino acid extension at the C-terminus, while form 2 (EF-G2$_{mt}$) has an insertion of 25 amino acid within domain II. Mutants of the yeast EF-G1$_{mt}$ are impaired in mitochondrial protein synthesis, while mutations in EF-G2$_{mt}$ fail to show any clear phenotype (Winzeler et al., 1999). EF-G1$_{mt}$ (~80 kDa) is active on *E. coli* ribosomes, whereas *E. coli* EF-G is not active on mitoribosomes (Eberly et al., 1985). Mammalian EF-G2$_{mt}$ (77 kDa) works exclusively in conjunction with mitochondrial ribosome recycling factor (RRF$_{mt}$, 26 kDa) to facilitate the process of ribosome disassembly after the translation termination step (Tsuboi et al., 2009). Our cryo-EM study of several functional complexes of 55S mitoribosome and EF-G1$_{mt}$ (Sharma et al., in preparation) reveals unique structural features of tRNA translocation in the mammalian mitochondria, including novel interactions between mito-specific MRPs and EF-G1$_{mt}$, and less pronounced movements of the mitoribosomal SSU head upon EF-G1$_{mt}$ binding to the mitoribosome (cf. Agrawal et al., 1999; Frank and Agrawal, 2000; Valle et al., 2003).

3. Structure of a protistan mitochondrial ribosome

Leishmania is a protist that belongs to a protozoan parasite genus known to cause a spectrum of diseases in humans. *Leishmania* contains a protein translational machinery in its single, large kinetoplast mitochondrion (Vickerman et al., 1976). Furthermore, the *Leishmania* mitoribosome (Lmr) represents one of the smallest rRNA-containing ribosomes. The *L. tarentolae* mitoribosomes are classified as protein-rich 50S particles (Table 1) that have a dramatically reduced mitochondrial rRNA, resulting in an unstable ribosomal structure, and are tightly associated with the mitochondrial membrane (Maslov et al., 2006, 2007).

Our cryo-EM analysis has revealed that Lmr is composed of two unequal subunits, the ~ 28S -30S SSU and a ~ 40S LSU (Maslov et al., 2006, 2007). The Lmr SSU is comprised of only a 610 nucleotide-long 9S rRNA while its LSU contains a 1,173-nucleotide-long 12S rRNA (de la Cruz et al., 1985, 1985a). When compared to bacterial and mammalian ribosomal subunits, this accounts for ~ 61 % and 28 % less rRNA content for the SSU and LSU, respectively (Figure 1). Nevertheless, the functionally important regions of the rRNAs that are required for peptide bond formation and subunit interactions have been maintained. The cryo-EM map shows that the Lmr possesses most of the features that are typical for bacterial ribosomes, but the Lmr has a highly porous structure with an overall diameter of ~ 245 Å (Figure 2C). Thus, the diameter of Lmr is ~ 15 Å smaller than that of its bacterial (~ 260 Å) counterpart and ~ 75 Å smaller than that of the mammalian mitoribosome (~ 320 Å). Similar to the situation in mammalian mitoribosomes, the rRNAs of the Lmr are largely shielded by proteins.

in the 9S rRNA (Figure 1A), several stem-loops are either reduced in size or eliminated altogether, accounting for ~ 61 % overall reduction of rRNA content. Even with such a low rRNA content, the Lmr SSU map displays all the typical features of a ribosomal SSU, relating closely to its bacterial counterpart. Particularly noteworthy is its beak-like feature, which is present despite the complete absence of the stem-loop corresponding to helix h33a of bacterial 16S rRNA, highlighting the importance of the head's overall architecture for ribosome function (Figure 2C). The 9S Lmr rRNA is comprised of only 21 of 45 bacterial rRNA helices (Wimberly et al., 2000; Sharma et al., 2009). The missing 24 rRNA helices are positionally replaced in parts by Lmr-specific proteins. These missing helices are h8-h12, h16, h17, h21, h25, h26, h26a, h31-h43, and a large part of h44. The feature of compensation of missing rRNA segments by proteins in Lmr is much more extensive in Lmr compared to the mammalian mitoribosome, where only ~19 % of deleted rRNA segments are positionally replaced by MRPs (Sharma et al., 2003). Although structurally the Lmr SSU maintains the head structure, the lack of h35-h43 leads to a large tunnel that runs from top of the head of the subunit to the neck region (Sharma et al., 2009). The critical nucleotides that are directly involved in the decoding process are retained in h44, though a dramatically reduced size of h44 creates a large gap in the lower inter-subunit region (Figure 2C). As far as the ribosomal proteins are concerned, the Lmr SSU possesses more than 50 proteins (Table 1). This estimate is based on an exhaustive mass spectrometric analysis of mitoribosomes from both *Leishmania* and its close relative *Trypanosoma* (see Zíková et al., 2008). Of these proteins only 10 are homologous to bacterial SSU proteins, including proteins S5, S6, S8, S9, S11, S12 and S15–S18. In addition to occupying the majority of the solvent exposed side of the subunit, proteins occupy most of interface side of the Lmr SSU head, thereby significantly altering the composition of mRNA and tRNA paths in the Lmr (Sharma et al., 2009) as compared to those in bacterial and other organellar ribosome structures.

3.1. Small subunit of the *Leishmania* mitoribosome

Comparing the secondary structure of the Lmr SSU 9S rRNA to its bacterial counterpart reveals that, although all four domains of the 16S rRNA are present

3.2. Large subunit of the *Leishmania* mitoribosome

The topology of the Lmr LSU is characterized by features that include the central protuberance and two stalk-like features on either side of the central protu-

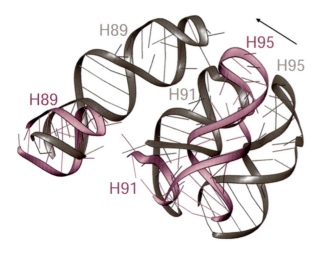

Fig. 3 Spatial shift of helix 95 (H95, SRL) in the *Leishmania* mitoribosomal LSU. Most of bacterial 23S rRNA helix 91 (H91) and a major portion of helix 89 (H89) are absent in Lmr (pink) (Figure 1B). These deletions allow the spatial shift of the SRL and bring it closer to the peptidyl-transferase center in the Lmr. (Figure modified from Suppl. Information in Sharma et al. 2009 with permission)

berance. The greatly reduced rRNA content here is mainly due to the absence of domain III and numerous stem-loop structures in domains I and II (Sharma et al., 2009; Figure 1B). The critical role of H69 in forming the central inter-subunit bridge, B2 a, and to provide a platform for A- and P-site tRNA binding during the translation elongation cycle and the ribosome recycling step has been maintained.

A modified rRNA model from the mammalian mito-LSU (Mears et al., 2006), i. e., after the removal of rRNA segments that are known to be absent in the Lmr LSU (Figure 1B), fits extremely well into the Lmr LSU cryo-EM map. The structurally most conserved regions of the LSU rRNA are domain V, which encompasses the peptidyl transferase center (PTC), and domain VI, which is comprised of the universally conserved α-sarcin/ricin stem-loop (SRL). In this region of the Lmr, there is a significant spatial shift of the SRL towards the PTC, possibly due to the truncation of helices H89 and H91 in the Lmr 12S rRNA (Figure 3). Thus, the SRL and the PTC in the Lmr LSU are closer by ~25 Å, as compared to that in the bacterial ribosome. These findings suggest that there are substantial alterations in size and structural organization in the translation elongation factors of *Leishmania* mitochondria. Like the Lmr SSU, the Lmr LSU is also estimated to possess a large number, ~77, of ribosomal proteins. Of these only 21 proteins are homologous to bacterial LSU proteins, including L2–L4, L9, L11–L17,

L20–L24, L27–L30 and L33. Among these, the locations of L16, L30 and L33 appear to be slightly shifted in the Lmr LSU, compared to their respective positions in the bacterial and mammalian mito LSUs. Similarly to Lmr SSU, additional proteins in Lmr LSU also occupy some of the functional regions on the interface side, including the tRNA corridor at the ribosomal A and P sites (Sharma et al., 2009).

4. Structure of a chloroplast ribosome

We have studied the structure of 70S ribosomes from spinach (*Spinacea oleracea*), whose chloroplasts have a genome of 146 genes, of which 98 code for proteins and 45 for tRNAs and rRNAs (Schmitz-Linneweber et al., 2001). The majority of the chloroplast-encoded proteins are targeted to the thylakoid membranes and include components of the ATP synthase, cytochrome b/f and, especially, photosystem I and II complexes. In addition, chloroplast ribosomes synthesize NADH dehydrogenase, the large subunit of RuBisCO, four subunits of RNA polymerase, and a subset of ribosomal proteins, i.e., 12 of the 30S SSU proteins and 8 of the 50S LSU proteins. Similar to the situation in mitochondria, the bulk of chloroplast proteins are encoded in the nucleus and imported into the chloroplast.

The chloroplast rRNAs are similar in length and exhibit relatively few differences when compared with their *E. coli* counterparts. The 16S rRNA of the chloroplast 30S subunit is slightly smaller and contains 1491 nucleotides. While the chloroplast 50S subunit is composed of three rRNAs, rather than two, as in *E. coli*, the combined size of chloroplast 50S subunit rRNAs is slightly larger (3033 nucleotides) than the *E. coli* counterpart (3024 nucleotides) (Table 1). Chloro-ribosomes possess a total of 58 plastid ribosomal proteins (PRPs); 25 and 33 PRPs in the 30S and 50S subunits, respectively (Yamaguchi and Subramanian, 2000; Yamaguchi et al., 2000), and have orthologs to nearly all *E. coli* ribosomal proteins, with the exception of L25 and L30. PRPs are generally larger than their *E. coli* counterparts, containing extensions at the N- and C-termini. In addition, the chloro-ribosome possesses six non-orthologous proteins, which are termed "plastid-specific ribosomal proteins" (or PSRPs): Four (PSRP1–4) are associated with the 30S subunit and two (PSRP5 and 6) with the 50S subunit. The larger size of the PRPs as well as the presence of six PSRPs leads to a protein to RNA mass-ratio of 2/3 in the case of chloroplast ribosomes, compared to 1/2 for *E. coli* (Table 1). The presence of PSRPs has prompted the suggestion that chloroplasts have evolved specific proteins that play unique functional roles during chloroplast translation, such as the light-dependent stimulation of protein synthesis (Yamaguchi and Subramanian, 2003).

The three-dimensional cryo-EM map of the 70S chloro-ribosome (Figure 2B) reveals known structural features of the bacterial ribosome (Figure 2A), such as the body, head, and platform of the 30S subunit, and the central protuberance, and L1 and L7/L12 stalks, of the 50S subunit. However, the chloro-ribosome is larger in overall size (by ~ 10 Å along the longest diameter) compared to the *E. coli* ribosome (Gabashvili et al., 2000). Compositions of inter-subunit bridges are mostly conserved between the bacterial (Gabashvili et al., 2000; Selmer et al., 2006; Schuwirth et al., 2005; Yusupov et al., 2001) and chloroplast (Sharma et al., 2007) ribosomes.

4.1. Small subunit of the chloroplast ribosome

The overall shape and size of the chloro-ribosomal 30S subunit is very similar to that of its *E. coli* counterpart. However, there is a striking difference in size and shape of their foot or "spur" feature, which distinguishes the chloro-ribosomal 30S subunit from its *E. coli* counterpart (Figure 2B, see also Sharma et al., 2007). The secondary structure maps of the 16S rRNA indicate that the major differences are the absence of nucleotides in this region, specifically, in helix 6, which forms the spur, and helices 10 and 17, which lie in the immediate vicinity (see Wimberly et al., 2000). Most of the orthologous PRPs of the chloro-ribosomal 30S subunit are larger, compared to their bacterial counterparts, generally possessing N- (NTE) or C-terminal extensions (CTE) (Yamaguchi and Subramanian, 2000). Among those, the CTE in PRP S21, which is positioned to interact directly with the 5'-untranslated region of the mRNA (Yusupova et al., 2006), appear to change the overall topology of the mRNA path in the chloro-ribosomes as compared to that in the bacterial ribosome (Sharma et al., 2007).

All four PSRPs of the chloro-ribosomal SSU have been localized within the cryo-EM map. The PSRP1 is located at the neck of the 30S subunit (Sharma et al., 2010), interacting with 16S rRNA helices 18 and 44 from the 30S body; helices 30 and 34, and PRPs S9 and S13 from the 30S head; and is within 10 Å distance of PRP S12. This position of PSRP1 suggests that PSRP1 is not in fact a *bona fide* ribosomal protein, but rather a regulatory factor that has homology to the cold-shock protein pY, the binding site of which is located in the corresponding region on *E. coli* ribosomes (Vila-Sanjuro et al., 2004). PSRP2 and PSRP3 interact with one another and with 16S rRNA helices 6 and 10, respectively. The combined densities of PSRP2 and PSRP3 span a large portion of the bottom of the chloro-30S SSU. Thus, like in the mitoribosomes (Sharma et al., 2003; Sharma et al., 2009), the bottom of the SSU in the chloro-ribosome is especially protein-rich, as compared to its bacterial counterparts. PSRP4, the most basic protein among PRPs (pI 11.79; Yamaguchi and Subramanian, 2000), is buried between the 16S rRNA helices 30, 41, 41a, 42 and 43 within the 30S head. The NTD of PSRP4 shows homology with the Thx protein that is present in the *Thermus thermophilus* 70S ribosome (Selmer et al., 2006), but absent in the *E. coli* 70S ribosome (Schuwirth et al., 2005). Thx and PSRP4 appear to share a similar function in stabilizing the arrangement of rRNA helices in the 30S head region.

4.2. Large subunit of the chloroplast ribosome

The overall shape of the chloro-ribosomal 50S subunit is similar to its bacterial counterpart, except that the PRP L1 stalk and the L7/L12-stalk are larger (Sharma et al., 2007). Almost all of the orthologous PRPs of the 50S LSU are larger in size than their respective *E. coli* counterparts, due to NTEs and CTEs, such that PRP L21 and L22 are almost twice the size of the respective bacterial proteins (Yamaguchi and Subramanian, 2000). The extensions of most PRPs are located on the solvent side of the subunit (Sharma et al., 2007). Extensions of PRPs L13, L21 and L22 apparently converge into an elongated mass localized between PRP L11 and the polypeptide exit tunnel.

The 50S subunit contains two small PSRPs, PSRP5 (6.6–9.2 kDa) and PSRP6 (7.4 kDa). The position of PSRP5 has been tentatively identified near the tRNA exit, or E site, in a groove formed between 23S rRNA helices 68 and 88 (Sharma et al., 2007). This position of PSRP5 suggests a possible role of this protein in the ejection of the deacylated tRNAs from the chloro-ribosome. Since a distinct mass of density that can be assigned to PSRP6 was absent in our cryo-EM map, it is likely that PSRP6 is relatively loosely associated with the ribosome.

Since most of the polypeptides synthesized in cell organelles are targeted to the organelles' inner mem-

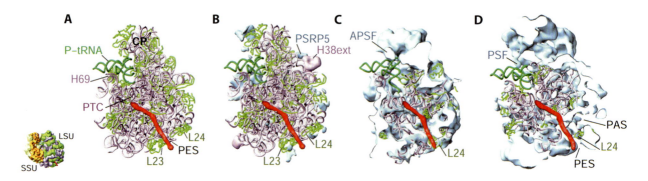

Fig. 4 Topology of the polypeptide exit tunnel, shown in cut-away views of the LSU. A portion of the LSU structure has been computationally removed (white surfaces in panels C and D correspond to the cutting planes) from the side toward the viewer, to reveal the tunnel topology. An α-helical polypeptide chain (red) has been docked into the tunnel and low-pass filtered to match the resolution of the cryo-EM map. (A) Bacterial *(E. coli)* 50S LSU, (B) chloroplast *(S. oleracea)* 50S LSU, (C) protistan *(L. tarentolae)* mitochondrial 40S LSU, and (D) mammalian *(B. taurus)* mitochondrial 39S LSU. Positions of conserved bacterial rRNA segments (pink) and proteins (light green) were derived by docking coordinates of the LSU from *E. coli* (Schuwirth et al., 2005), chloroplasts (Sharma et al., 2007) and mitochondria (Mears et al., 2006) into the cryo-EM maps. In panel B, blue densities correspond to chloroplast-specific protein masses, including PSRPs, and pink solid masses correspond to insertions, as compared to the *E. coli* sequence, in the chloroplast 23S rRNA. In panels C and D, blue densities correspond to mito-specific protein masses in the protistan and mammalian mitoribosomal LSUs, respectively. Landmarks: PTC, peptidyl-transferase center; P-tRNA (dark green), P-site tRNA; APSF, A- and P-site finger (see Sharma et al., 2009); PSF, P-site finger; PAS, nascent polypeptide accessible site; PES, conventional polypeptide exit site. Thumbnail to the left of panel A depicts the overall orientation of the ribosome.

branes, it is likely that organellar ribosomes are closely and dynamically associated with those membranes. In the remainder of this article, we will provide a comparison of topologies of nascent polypeptide exit tunnels among three organellar ribosomes described above.

5. The polypeptide exit tunnels of the organellar ribosomes

Newly synthesized nascent polypeptide chains emerge from the ribosome through a tunnel-like feature, the polypeptide exit tunnel. This tunnel connects the peptidyl-transferase center (PTC), the site of peptide-bond formation, to the exterior of the ribosome, where nascent chains emerge through the polypeptide exit site (PES, Figure 4). The structural organization of the exit tunnel in cytoplasmic ribosomes has been studied by both cryo-EM (Frank et al., 1995; Beckmann et al., 2001; Gabashvili et al., 2001) and X-ray crystallography (Nissen et al., 2000). However, in contrast to the variety of possible destinations for the polypeptides synthesized by cytoplasmic and chloroplast ribosomes, all of the polypeptides synthesized by mitoribosomes are inserted into the mitochondrial inner membrane (mtIM). Although it is not yet clear whether polypeptides are inserted co-translationally into the membrane, current evidence indicates this is most likely the case (see Jia et al., 2009).

In the chloroplast ribosome, the majority of the polypeptides are inserted into the thylakoid membrane, therefore it might be expected that chloroplasts have evolved a more specialized polypeptide exit tunnel for efficient post-translational protein export. The cryo-EM map of the chloro-ribosome shows that while the overall topography of the tunnel exit is similar to that observed in bacterial ribosomes, with ribosomal proteins L22, L23, L24 and L29 encircling the tunnel exit site (Figure 5A, lower panel), a number of differences are evident. All these PRPs are larger in chloro-ribosomes (Figure 5B). Extensions of the PRP L23, L24 and L29 may help establish additional contacts with signal recognition particle (SRP), which lacks an RNA moiety in the chloroplast.

The polypeptide exit tunnel in the Lmr differs from its bacterial and chloroplast counterparts (Figures 4A–C and 5A–C). The tunnel in the Lmr LSU has two openings on the solvent side, one corresponding to the conventional polypeptide exit site (PES) and the second upwards of the tunnel at a distance of ~ 25 Å from the PES. This site of premature exposure of the nascent polypeptide chain is located closer to the peptidyl-transferase center and allows free access to the solvent. Therefore, we designate this opening as the polypeptide-accessible site (PAS, Figure 5C, upper panel, also see Sharma et al., 2009). Possibly, some of the nascent chains that are being synthesized on the mitoribosome emerge through this accessible site, rather than being

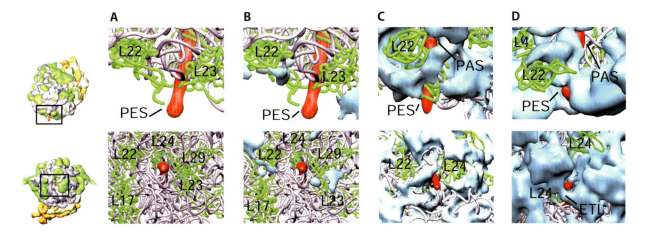

Fig. 5 Topography of the two putative sites of polypeptide exit in the LSUs of ribosomes from (A) bacteria *(E. coli)*, (B) chloroplasts *(S. oleracea)*, (C) protistan *(L. tarentolae)* mitochondria, and (D) mammalian *(B. taurus)* mitochondria. In the upper panels, the LSUs are shown from the L1-protein side to reveal a large opening (PAS) in the tunnel of mitoribosomes (C and D) distant from the PES. In the lower panels, the polypeptide exit site (PES) is shown from the bottom of the LSUs. The overall orientation of the ribosomes, with the boxed area that has been enlarged in panels A-D, is shown by thumbnails to the left of panel A. Landmarks: PAS, polypeptide-accessible site; ETL, exit tunnel lid; segments of LSU proteins (green) present in the immediate vicinity are identified.

routed all the way through to the conventional exit site. Thus, two pathways may exist by which nascent chains can emerge from the mitoribosome, perhaps allowing alternative interactions with different classes of chaperones involved in the assembly of the large oligomeric complexes of which they are components. The solvent-side openings of both PES and PAS are predominantly encircled by Lmr-specific proteins. The area between the two exit sites is also dominated by mito-specific MRPs, which may be involved in the adherence of the mitoribosome to the mtIM through the PES, or remain embedded in the mtIM such that the PAS and the PES are partially exposed at opposite sides of the mtIM.

Like the polypeptide exit tunnel in the Lmr, the tunnel of the mammalian mitoribosome is distinctly different from the tunnels in the cytoplasmic and chloroplast ribosomes. Like in Lmr, the mammalian mitoribosome also has a wide opening (PAS) in the tunnel, well before the conventional PES (Figures 4D, 5D, upper panels). As noted above, the majority of domains I and III of the bacterial 23S rRNA are absent in the mitoribosome (Figure 1B). In addition, a clear homolog of bacterial protein L29 is lacking in mitoribosomes. These are critical components of the polypeptide exit tunnel in the bacterial ribosome (Nissen et al., 2000). The other LSU proteins, L22, L23, and L24, which together with L29 surround the exit of the tunnel in the bacterial ribosome, are represented by much larger homologs in the mammalian mitoribosome (Koc et al., 2001; Suzuki et al., 2001).

Because of the absence of 23S rRNA domains I and III, domains which together form the inner lining of most of the solvent side of the tunnel in the bacterial ribosome, the structure of the lower two-thirds of the tunnel in both types of mitoribosome structures presented here is drastically different from their bacterial counterparts, and is composed mainly of mitoribosome-specific proteins (Figures 4C, D and 5C, D). Since the majority of the rRNA domain I and a significant portion of domain III (Figure 1B) are not replaced by proteins, the PAS is formed, making the tunnel more open to the solvent. In the mammalian mitoribosome, the conventional PES is partially covered by a lid-like structure (exit-tunnel lid, or "ETL" in Figure 5D, lower panel), which appears to be formed by a protrusion of a large protein that replaces bacterial L29.

The protein-rich nature of the tunnel exit sites observed in organellar ribosomes suggests that this feature may be necessary for the efficient insertion of the mitochondrial translation products into mtIM. The PAS has not been observed in any of the cytoplasmic or the chloroplast ribosomes; it appears to be characteristic of mitoribosomes and may be necessary for the co-translational insertion of polypeptides into the mtIM.

6. Conclusions

This article provides a side-by-side comparison of the cryo-EM structures of the bacterial ribosome with those of selected organellar ribosomes. We find that chloroplast ribosomes are very similar to, and mitoribosomes are significantly different from, their bacterial counterparts. However, several features are unique to organellar (both mitochondrial and chloroplast) ribosomes as a group. These features include the presence of a protein-rich base in the small subunit body and the exit of the polypeptide tunnel. Both of these regions could be involved in the attachment of organellar ribosomes to the membrane. Furthermore, there is additional structural variability among mitoribosomes from different sources. Through molecular analysis of the cryo-EM maps, we find that the truncated regions of the rRNA are only partially compensated by the presence of additional and enlarged proteins in the mammalian mitoribosome, while such compensation is more pronounced in the case of a protistan mitoribosome. Despite its strikingly small complement of rRNA, the latter is similar to its bacterial counterpart in overall size and morphology. The protist's mitoribosomal proteins that have bacterial homologs retain segments crucial for their incorporation into the ribosome, suggesting that all components required to build a functional translational machine are present in this highly divergent form of the ribosome. Although mitoribosomes differ dramatically in structure between mammalian and protistan organisms, they evidently have several common characteristic features including a protein-rich mRNA path, a P-site finger, a tunnel in the small subunit body, and a polypeptide-accessible site (PAS). The comparison presented here suggests that the maintenance of a certain minimum size and the retention of key architectural elements have underpinned a notably conserved basic functioning of the ribosome, despite compositional changes during long structural evolution. Further structural and functional characterization of the protein components present at key functional sites will be critical in establishing detailed mechanisms for the recruitment of leaderless mRNAs to mitoribosomes and insertion of the nascent chains into the mitochondrial inner membrane.

Acknowledgements

This work was supported by a grant from the National Institutes of Health, GM61576 (to R. K. A.). A. Y. would like to acknowledge his affiliation with the Department of Microbiology and Immunology, Faculty of Pharmacy, Cairo University, Egypt.

References

Ævarsson A, Brazhnikov E, Garber M, Zheltonosova J, Chirgadze Yu, Al-Karadaghi S, Svensson LA, Liljas A (1994) Three-dimensional structure of the ribosomal translocase: elongation factor G from *Thermus thermophilus*. EMBO J 13: 3669–3677

Agrawal RK, Heagle AB, Penczek P, Grassucci RA, Frank J (1999) Elongation factor-G-dependent GTP hydrolysis induces translocation accompanied by large conformational changes in the 70S ribosome. Nature Struct Biol 6: 643–647

Agrawal RK, Spahn CM, Penczek P, Grassucci RA, Nierhaus KH, Frank J (2000) Visualization of tRNA movements on the *Escherichia coli* 70S ribosome during the elongation cycle. J Cell Biol 150: 447–460

Allen GS, Zavialov A, Gursky R, Ehrenberg M, Frank J (2005) The cryo-EM structure of a translation initiation complex from *Escherichia coli*. Cell 121: 703–712

Anderson S, de Brujin M, Coulson A, Eperon I, Sanger F, Young I (1982) Complete sequence of bovine mitochondrial DNA: Conserved features of the mammalian mitochondrial genome. J Mol Biol 156: 683–717

Attardi G (1985) Animal mitochondrial DNA: an extreme example of genetic economy. Int Rev Cytol 93: 93–145

Beckmann R, Spahn CM, Eswar N, Helmers J, Penczek PA, Sali A, Frank J, Blobel G (2001) Architecture of the protein-conducting channel associated with the translating 80S ribosome. Cell 107: 361–372

Bhargava K, Templeton PD, Spremulli LL (2004) Expression and characterization of isoform 1 of human mitochondrial elongation factor G. Protein Exp & Purifi 37: 368–376

Biou V, Shu F, Ramakrishnan V (1995) X-ray crystallography shows that translational initiation factor IF3 consists of two compact alpha/beta domains linked by an alpha-helix. EMBO J 14: 4056–4064

Brock S, Szkaradkiewicz K, Sprinzl M (1998) Initiation factors of protein biosynthesis in bacteria and their structural relationship to elongation and termination factors. Mol Microbiol 29: 409–417

Carter AP, Clemons WM, Jr. Brodersen DE, Morgan-Warren RJ, Hartsch T, Wimberly BT, Ramakrishnan V (2001) Crystal structure of an initiation factor bound to the 30S ribosomal subunit. Science 291: 498–501

Chomyn A, Cleeter MW, Ragan CI, Riley M, Doolittle RF, Attardi G (1986) URF6, last unidentified reading frame of human mtDNA, codes for an NADH dehydrogenase subunit. Science 234: 614–618

Czworkowski J, Wang J, Steitz TA, Moore PB (1994) The crystal structure of elongation factor G complexed with GDP, at 2.7Å. EMBO J 13: 3661–3668

de la Cruz VF, Lake JA, Simpson AM, Simpson L (1985) A minimal ribosomal RNA: Sequence and secondary structure of the 9S kinetoplast ribosomal RNA from *Leishmania tarentolae*. Proc Natl Acad Sci USA 82: 1401–1405

de la Cruz VF, Simpson AM, Lake JA, Simpson L (1985a) Primary sequence and partial secondary structure of the 12S kinetoplast (mitochondrial) ribosomal RNA from *Leishmania tarentolae*: Conservation of peptidyl-transferase structural elements. Nucl Acids Res 13: 2337–2356

Eberly SL, Locklear V, Spremulli LL (1985) Bovine mitochondrial ribosomes. Elongation factor specificity. J Biol Chem 260: 8721–8725

Frank J, Agrawal RK (2000) A ratchet-like inter-subunit reorganization of ribosome during translocation. Nature 406: 318–322

Frank J, Zhu J, Penczek P, Li Y, Srivastava S, Verschoor A, Radermacher M, Grassucci R, Lata KR, Agrawal RK (1995) A model of protein synthesis based on cryo-electron microscopy of the *E. coli* ribosome. Nature 376: 441–444

Gabashvili IS, Agrawal RK, Spahn CM, Grassucci R, Svergun D, Frank J, Penczek P (2000) Solution structure of the *E. coli* 70S ribosome at 11.5 Å resolution. Cell 100: 537–549

Gabashvili IS, Gregory ST, Valle M, Grassucci R, Worbs M, Wahl MC, Dahlberg AE, Frank J (2001) The polypeptide tunnel system in the ribosome and its gating in erythromycin resistance mutants of L4 and L22. Mol Cell 8: 181–188

Garcia C, Fortier PL, Blanquet S, Lallemand JY, Dardel F (1995) Solution structure of the ribosome-binding domain of *E. coli* translation initiation factor IF3. Homology with the U1A protein of the eukaryotic spliceosome. J Mol Biol 254: 247–259

Gaur R, Grasso D, Datta PP, Krishna PD, Das G, Spencer A, Agrawal RK, Spremulli L, Varshney U (2008) A single mammalian mitochondrial translation initiation factor functionally replaces two bacterial factors. Mol Cell 29: 180–190

Gray MW, Burger G, Lang BF (2001) The origin and early evolution of mitochondria. Genome Biol 2: 1018.1–1018.5

Hammarsund M, Wilson W, Corcoran M, Merup M, Einhorn S, Grander D, Sangfelt O (2001) Identification and characterization of two novel human mitochondrial elongation factor genes, hEFG2 and hEFG1, phylogenetically conserved through evolution. Hum Genet 109: 542–550

Haque ME, Spremulli LL (2008) Roles of the N- and C-terminal domains of mammalian mitochondrial initiation factor 3 in protein biosynthesis. J Mol Biol 384: 929–940

Haque ME, Grasso D, Spremulli LL (2008) The interaction of mammalian mitochondrial translational initiation factor 3 with ribosomes: evolution of terminal extensions in IF3$_{mt}$. Nucleic Acids Res 36: 589–597

Harris EH, Boynton JE, Gillham NW (1994) Chloroplast ribosomes and protein synthesis. Microbiol Rev 58: 700–754

Jenner L, Demeshkina N, Yusupova G, Yusupov M (2010) Structural rearrangements of the ribosome at the tRNA proofreading step. Nat Struct Mol Biol 17: 1072–1078

Jia L, Kaur J, Stuart RA (2009) Mapping of the *Saccharomyces cerevisiae* Oxa1-mitochondrial ribosome interface and identification of MrpL40, a ribosomal protein in close proximity to Oxa1 and critical for oxidative phosphorylation complex assembly. Eukaryot Cell 8: 1792–802

Koc EC, Spremulli LL (2002) Identification of mammalian mitochondrial translational initiation factor 3 and examination of its role in initiation complex formation with natural mRNAs. J Biol Chem 277: 35541–35549

Koc EC, Burkhart W, Blackburn K, Moyer MB, Schlatzer DM, Moseley A, Spremulli LL (2001) The large subunit of the mammalian mitochondrial ribosome. Analysis of the complement of ribosomal proteins present. J Biol Chem 276:43958–43969

Koc EC, Haque ME, Spremulli LL (2010) Current views of the structure of the mammalian mitochondrial ribosome. Israel J Chem 50:45–59

Lee JH, Choi SK, Roll-Mecak A, Burley SK, Dever TE (1999) Universal conservation in translation initiation revealed by human and archaeal homologs of bacterial translation initiation factor IF2. Proc Natl Acad Sci USA 96:4342–4347

Liao HX, Spremulli LL (1990) Identification and initial characterization of translational initiation factor 2 from bovine mitochondria. J Biol Chem 265:13618–13622

Liao HX, Spremulli LL (1991) Initiation of protein synthesis in animal mitochondria: Purification and charac-

terization of translational initiation factor 2. J Biol Chem 266:20714–20719

Ma L, Spremulli LL (1995) Cloning and sequence analysis of the human mitochondrial translational initiation factor 2 cDNA. J Biol Chem 270: 1859–1865

Ma J, Spremulli LL (1996) Expression, purification and mechanistic studies of bovine mitochondrial translational initiation factor 2. J Biol Chem 271: 5805–5811

Ma J, Farwell M, Burkhart W, Spremulli LL (1995) Cloning and sequence analysis of the cDNA for bovine mitochondrial translational initiation factor 2. Biochim Biophys Acta 1261: 321–324

Margulis L (1970) Origin of Eukaryotic Cells. Yale Univ. Press, New Haven

Maslov DA, Sharma MR, Butler E, Falick AM, Gingery M, Agrawal RK, Spremulli LL, Simpson L (2006) Isolation and characterization of mitochondrial ribosomes and ribosomal subunits from *Leishmania tarentolae*. Mol Biochem Parasitol 148: 69–78

Maslov DA, Spremulli LL, Sharma MR, Bhargava K, Grasso D, Falick AM, Agrawal RK, Parker CE, Simpson L (2007) Proteomics and electron microscopic characterization of the unusual mitochondrial ribosome-related 45S complex in *Leishmania tarentolae*. Mol Biochem Parasitol 152(2):203–212

Mears JA, Sharma MR, Gutell RR, McCook AS, Richardson PE, Caulfield TR, Agrawal RK, Harvey SC (2006) A structural model for the large subunit of the mammalian mitochondrial ribosome. J Mol Biol 358: 193–212

Moazed D, Samaha RR, Gualerzi C, Noller HF (1995) Specific protection of 16S rRNA by translational initiation factors. J Mol Biol 248: 207–210

Montoya J, Ojala D, Attardi G (1981) Distinctive features of the 5'-terminal sequences of the human mitochondrial mRNAs. Nature 290: 465–470

Moreno JMP, Kildsgaard J, Siwanowicz I, Mortensen KK, Sperling-Petersen HU (1998) Binding of *E. coli* initiation factor IF2 to 30S ribosomal subunits: a functional role for the N-terminus of the factor. Biochem Biophys Res Comm 252: 465–471

Moreno JMP, Dyrskjøtersen L, Kristensen J, Mortensen K, Sperling-Petersen, H (1999) Characterization of the domains of *E. coli* initiation factor IF2 responsible for recognition of the ribosome. FEBS Lett 455: 130–134

Nissen P, Hansen J, Ban N, Moore PB, Steitz TA (2000) The structural basis of ribosome activity in peptide bond synthesis. Science 289: 920–930

O'Brien TW (1971) The general occurrence of 55 S ribosomes in mammalian liver mitochondria. J Biol Chem 246: 3409–3417

O'Brien TW (2002) Evolution of a protein-rich mitochondrial ribosome: implications for human genetic disease. *Gene* 286: 73–79

Rabl J, Leibundgut M, Ataide SF, Haag A, Ban N (2011) Crystal structure of the eukaryotic 40S ribosomal subunit in complex with initiation factor 1. Science 331:730–736

Roll-Mecak A, Cao C, Dever TE, Burley SK (2000) X-Ray structures of the universal translation initiation factor IF2/eIF5B: conformational changes on GDP and GTP binding. Cell 103: 781–792

Schmeing TM, Ramakrishnan V (2009) What recent ribosome structures have revealed about the mechanism of translation. Nature 461: 1234–1242

Schmitz-Linneweber C, Maier RM, Alcaraz JP, Cottet A, Herrmann RG, Mache R (2001) The plastid chromosome of spinach (Spinacia oleracea): complete nucleotide sequence and gene organization. Plant Mol Biol 45: 307–315

Schuwirth BS, Borovinskaya MA, Hau CW, Zhang W, Vila-Sanjurjo A, Holton JM, Cate JH (2005) Structures of the bacterial ribosome at 3.5 Å resolution. Science 310: 827–834

Selmer M, Dunham CM, Murphy FV 4 th, Weixlbaumer A, Petry S, Kelley AC, Weir JR, Ramakrishnan V (2006) Structure of the 70S ribosome complexed with mRNA and tRNA. Science 313: 1935–1942

Sharma MR, Koc EC, Datta PP, Booth, TM, Spremulli LL, Agrawal RK (2003) Structure of the mammalian mitochondrial ribosomes reveals an expanded role for its component proteins. Cell 115: 97–108

Sharma MR, Wilson DN, Datta PP, Barat C, Schluenzen F, Fucini P, Agrawal RK (2007) Cryo-EM study of the spinach chloroplast ribosome reveals the structural and functional roles of plastid-specific ribosomal proteins. Proc Natl Acad Sci USA 104:19315–19320

Sharma MR, Booth TM, Simpson L, Maslov DA, Agrawal RK (2009) Structure of a mitochondrial ribosome with minimal RNA. *Proc* Natl Acad Sci USA 106: 9637–9642

Sharma MR, Dönhöfer A, Barat C, Marquez V, Datta PP, Fucini P, Wilson DN, Agrawal RK (2010) PSRP1 is not a ribosomal protein, but a ribosome-binding factor that is recycled by the ribosome-recycling factor (RRF) and elongation factor G (EF-G). J Biol Chem 285: 4006–4014

Spencer AC, Spremulli LL (2005) The interaction of mitochondrial translational initiation factor 2 with the small ribosomal subunit. Biochim Biophys Acta 1750: 69–81

Spremulli LL, Coursey A, Navratil T, Hunter SE (2004) Initiation and elongation factors in mammalian mitochondrial protein biosynthesis. Prog Nucleic Acid Res Mol Biol 77: 211–261

Steinberg S, Cedergren R (1994) Structural compensation in atypical mitochondrial tRNAs. Nat Struct Biol 1: 507–510

Steinberg S, Gautheret D, Cedergren R (1994) Fitting the structurally diverse animal mitochondrial tRNAs[Ser] to common three-dimensional constraints. J Mol Biol 236: 982–989

Suzuki T, Terasaki M, Takemoto-Hori C, Hanada T, Ueda T, Wada A, Watanabe K (2001) Structural compensation for the deficit of rRNA with proteins in the mammalian mitochondrial ribosome. Systematic analysis of protein components of the large ribosomal subunit from mammalian mitochondria. J Biol Chem 276:21724–21736

Temperley RJ, Wydro M, Lightowlers RN, Chrzanowska-Lightowlers ZM (2010) Human mitochondrial mRNAs-like members of all families, similar but different. Biochim Biophys Acta 1797: 1081–1085

Tsuboi M, Morita H, Nozaki Y, Akama K, Ueda T, Ito K, Nierhaus KH, Takeuchi N (2009) EF-G2$_{mt}$ is an exclusive recycling factor in mammalian mitochondrial protein synthesis. Mol Cell 35: 502–510

Valle M, Zavialov A, Sengupta J, Rawat U, Ehrenberg M, Frank J (2003) Locking and unlocking of ribosomal motions. Cell 114: 123–134

Vickerman K, Preston TM (1976) Comparative Cell Biology of the Kinetoplastid Flagellates. In: Lumsden WHR, Evans DA (eds) Biology of the Kinetoplastida. Academic Press, London, pp 35–130

Vila-Sanjurjo A, Schuwirth BS, Hau CW, Cate JHD (2004) Structural basis for the control of translation initiation during stress. Nat Struct Mol Biol 11: 1054–1059

Watanabe Y-I, Kawai G, Yokogawa T, Hayashi N, Kumazawa Y, Ueda T, Nishikawa K, Hirao I, Miura K-I, Watanabe K (1994) Higher-order structure of bovine mitochondrial tRNA[ser] UGA: chemical modification and computer modeling. Nucleic Acids Res 22: 347–353

Wimberly BT, Brodersen DE, Clemons WM, Morgan-Warren RJ, Carter AP, Vonrhein C, Hartsch T, Ramakrishnan VR (2000) Structure of the 30S ribosomal subunit. Nature 407: 327–339

Winzeler EA, Shoemaker DD, Astromoff A, Liang H, Anderson K, Andre B, Bangham R, Benito R, Boeke JD, Bussey H, Chu AM, Connelly C, Davis K et al. (1999) Functional characterization of the S. cerevisiae genome by gene deletion and parallel analysis. Science 285: 901–906

Wittmann-Liebold B (1985) Ribosomal proteins: their structure and evolution, In: Hardesty B, Kramer G (eds) Structure, Function, and Genetics of Ribosomes, Springer, New York, pp 326–361

Yamaguchi K, Subramanian AR (2000) The plastid ribosomal proteins. Identification of all the proteins in the 50 S subunit of an organelle ribosome (chloroplast). J Biol Chem 275: 28466–28482

Yamaguchi K, Subramanian AR (2003) Proteomic identification of all plastid-specific ribosomal proteins in higher plant chloroplast 30S ribosomal subunit. Eur J Biochem 270: 190–205

Yamaguchi K, von Knoblauch K, Subramanian AR (2000) The plastid ribosomal proteins. Identification of all the proteins in the 30 S subunit of an organelle ribosome (chloroplast). J Biol Chem 275: 28455–28465

Yassin AS, Haque ME, Datta PP, Elmore K, Banavali NK, Spremulli LL, Agrawal RK (2011) Insertion domain within mammalian mitochondrial translation initiation factor 2 serves the role of eubacterial initiation factor 1. Proc Natl Acad Sci USA, DOI: 10.1073/pnas.1017425108 (in press)

Yokogawa T, Watanabe YI, Kumazawa Y, Ueda T, Hirao I, Miura KI, Watanabe K (1991) A novel cloverleaf structure found in mammalian mitochondrial tRNA[ser](UCN). Nucleic Acids Res 19: 6101–6105

Yusupov MM, Yusupova GZ, Baucom A, Lieberman K, Earnest TN, Cate JH, Noller HF (2001) Crystal structure of the ribosome at 5.5 Å resolution. Science 292:883–896

Yusupova G, Jenner, L, Rees B, Moras D, Yusupov N (2006) Structural basis for messenger RNA movement on the ribosome. Nature 444: 391–394

Zagryadskaya EI, Kotlova N, Steinberg SV (2004) Key Elements in Maintenance of the tRNA L-shape. J Mol Biol 340: 435–444

Zíková A, Panigrahi AK, Dalley RA, Acestor N, Anupama A, Ogata Y, Myler PJ, Stuart K. (2008) Trypanosoma brucei mitochondrial ribosomes: affinity purification and component identification by mass spectrometry. Mol Cell Proteomics 7: 1286–1296

Modifications of ribosomal RNA: From enzymes to function

9

Petr V. Sergiev, Anna Y. Golovina, Irina V. Prokhorova, Olga V. Sergeeva, Ilya A. Osterman, Mikhail V. Nesterchuk, Dmitry E. Burakovsky, Alexey A. Bogdanov, Olga A. Dontsova

1. Introduction

Modified nucleosides are present in all kinds of stable RNA molecules, tRNAs being particularly rich in them (Auffinger and Westhof, 1998). Ribosomal RNA (rRNA) from all organisms contains modifications, and there is a correlation between the overall complexity of an organism and the number of modified nucleosides in its rRNA. The rRNA of the most primitive bacteria, such as some *Mycoplasma* species, may possess only 14 modified nucleosides (de Crécy-Lagard et al., 2007). In *Escherichia coli*, there are 36 modified nucleosides in rRNA (Table I). Yeast ribosomes possess about one hundred rRNA modifications, human rRNA over two hundred (Ofengand and Fournier, 1998; Decatur and Fournier, 2002). Eukaryotes and archaea use snoRNA guided rRNA modification mechanism. This mechanism allows archaea and eukarya to use a limited number of modification enzymes, mainly pseudouridine synthase and 2'-O-methyltransferase to introduce the majority of their rRNA modifications (Decatur and Fournier, 2002). By contrast, bacteria have developed specific enzymes for each one of the (fewer) modifications they have. Nevertheless, there are many different rRNA modifications in bacteria. Despite intensive study for several decades, many open questions remain regarding the functional role of modified rRNA nucleosides. In this review we will focus on rRNA modifications in *E. coli* and discuss their possible functions.

Even before high-resolution structures of ribosomal subunits (Ban et al., 2000; Schluenzen et al., 2000; Wimberly et al., 2000) or the entire ribosome (Yusupov et al., 2001; Schuwirth et al., 2005) and its functional complexes (Ogle et al., 2002; Yusupova et al., 2006; Schmeing et al., 2009; Gao et al., 2009) were determined, it was known that modified nucleosides are concentrated in the functional centers of the ribosome (Brimacombe et al., 1993). Now, the positions of rRNA modifications can be precisely mapped on the three-dimensional structure of the ribosome (Figure 1), and it is clearly seen that modified nucleosides are concentrated around the mRNA, the tRNAs in A and P sites, the peptidyl transferase center, the peptide exit tunnel, and on both sides of intersubunit bridges. The concentration of modified nucleosides in the functionally most important regions of the ribosome indicate their important functional role (Brimacombe et al., 1993).

Since reconstitution techniques for both subunits of bacterial ribosomes were established, it was possible

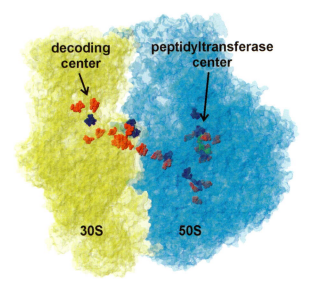

Fig. 1 Spatial distribution of modified nucleosides in the bacterial ribosome. The molecular surface of small and large ribosomal subunits are shown in yellow and blue, respectively and signed. Methylated nucleosides are depicted in red, pseudouridines in blue. Other modified bnucleosides are depicted in green. Ribosomal functional centers are marked. The structure model (Jenner et al., 2010) was created with the help of SwissPDBviewer programm (Guex and Peitsch, 1997).

Table 1 Modified nucleosides in *E. coli* rRNA and modifying enzymes

Nucleotide	Enzyme	Reference
16S rRNA		
Ψ516	RsuA (YejD[#])	Wrzesinski et al., 1995a
m⁷G527	RsmG (GidB)	Okamoto et al., 2007
m²G966	RsmD (YhhF)	Lesnyak et al., 2007
m⁵C967	RsmB (YhdB)	Tscherne et al., 1999a; Gu et al., 1999
m²G1207	RsmC (YjjT)	Tscherne et al., 1999b
m⁴Cm1402	RsmI (YraL) RsmH (MraW)	Kimura and Suzuki, 2010
m⁵C1407	RsmF (YebU)	Andersen and Douthwaite, 2006
m³U1498	RsmE (YggJ)	Basturea et al., 2006
m²G1516	RsmJ	unknown
m⁶₂A1518	RsmA (KsgA)	Helser et al., 1972; Poldermans et al., 1979
m⁶₂A1519	RsmA (KsgA)	Helser et al., 1972; Poldermans et al., 1979
23S rRNA		
m¹G745	RlmAI (RrmA,YebH)	Gustafsson and Persson, 1998
Ψ746	RluA (YabO)	Wrzesinski et al., 1995b
m⁵U747	RlmC (YbjF, RumB)	Madsen et al., 2003
Ψ955	RluC (YceC)	Conrad et al., 1998; Huang et al., 1998
m⁶A1618	RlmF (YbiN)	Sergiev et al., 2008
m²G1835	RlmG (YgjO)	Sergiev et al., 2006
Ψ1911	RluD (YfiI)	Huang et al., 1998; Raychaudhuri et al., 1998
m³Ψ1915	RluD (YfiI) RlmH (YbeA)	Huang et al., 1998; Raychaudhuri et al., 1998 Purta et al., 2008a; Ero et al., 2008
Ψ1917	RluD (YfiI)	Huang et al., 1998; Raychaudhuri et al., 1998
m⁵U1939	RlmD (YgcA, RumA)	Agarwalla et al., 2002; Madsen et al., 2003
m⁵C1962	RlmI (YccW)	Purta et al., 2008b
m⁶A2030	RlmJ	unknown
m⁷G2069	RlmK	unknown
Gm2251	RlmB (YjfH)	Lovgren and Wikstrom, 2001
m²G2445	RlmL (YcbY)	Lesnyak et al., 2006
D2449	RldA	unknown
Ψ2457	RluE (YmfC)	Del Campo et al., 2001
Cm2498	RlmM (YgdE)	Purta et al., 2009
*C2501	RltA	Unknown modification, Andersen et al., 2004
m²A2503	RlmN (YfgB)	Toh et al., 2008
Ψ2504	RluC (YceC)	Conrad et al., 1998; Huang et al., 1998
Um2552	RlmE (FtsJ, RrmJ)	Caldas et al., 2000a; Bugl et al., 2000
Ψ2580	RluC (YceC)	Conrad et al., 1998; Huang et al., 1998
Ψ2604	RluF (YjbC)	Del Campo et al., 2001
Ψ2605	RluB (YciL)	Del Campo et al., 2001

[#] Previous designations of open reading frames are given in brackets

to test the functional activity of bulk natural modifications. *In-vitro* reconstitution of ribosomal subunits using *in-vitro* transcribed, unmodified rRNA can help to elucidate the functional role of rRNA modifications. It was demonstrated that small (30S) subunits assembled on unmodified 16S rRNA exhibit reduced, but significant functional activity (Krzyzosiak et al., 1987). In contrast, complete lack of all 23S rRNA modifications leads to non-functional large (50S) subunits (Green and Noller, 1996). To test the role of individual rRNA modifications both *in vitro* and *in vivo*, it was necessary to obtain strains devoid of rRNA modification enzymes (Table 1). This work was pioneered by Björk and Isaksson (1970), who obtained the first *E. coli* strains lacking particular rRNA methylations. After the invention of the PCR-based *E. coli* gene knockout technique (Datsenko and Wanner, 2000) and the availability of the gene-knockout "Keio" collection (Baba et al., 2006), the identification of genes coding for rRNA modifying enzymes became much easier. The advent of the MALDI-MS analysis also played an important role in the burst of the number of genes for rRNA methylation enzymes that were identified (Douthwaite and Kirpekar, 2007). By now, almost all such genes are identified (Table 1), and it is expected that in a few years the entire collection of rRNA modification enzymes will have been revealed.

The surprising outcome of the characterization of rRNA modification gene knockouts was that not a single one of them was found essential for viability. Phenotypes of rRNA methyltransferase and pseudouridine synthase knockout strains ranged from severe growth retardation (Huang et al., 1998; Raychaudhuri et al., 1998) to complete lack of any phenotype (Lovgren and Wikstrom, 2001). It became a common opinion that modifications of rRNA are necessary for "fine-tuning" the translation apparatus rather than for any "core" functional role. In order to better define "fine-tuning", ideas on the exact roles of particular rRNA modifications in ribosome assembly, structure, and function will be discussed in this review.

2. Modifications of rRNA stabilize ribosome structure

Modified nucleosides could have a role in stabilization of particular "functional" conformation of rRNA. Pseudouridine is the most frequent modified nucleoside in rRNA (Ofengand et al., 1995, Ofengand and Fournier 1998; Ofengand et al., 2001). While it shares with uri-

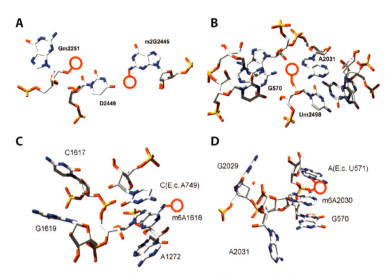

Fig. 2 Structural context of representative modified nucleotides. Relevant nucleotides are signed, methyl groups which are absent in the corresponding PDB file are marked by red circles. (A) Modified 23S rRNA nucleotides cluster Gm2251-m²G2445-D2449 in vicinity of peptidyltransferase center (Jenner et al., 2010). (B) Environment of 2'-O methylated U2498 of the 23S rRNA (Jenner et al., 2010). (C) Structural context of m⁶A1618 in the large ribosomal subunit (Jenner et al., 2010). (D) Location of modified nucleotides m⁶A2030 of *E. coli* 23S rRNA and m¹A628 of *H. marismortui* 23S rRNA (Ban et al., 2000).

dine the Watson-Crick pairing ability, it possesses an additional hydrogen bond donor (N-H) at position 5, where uridine has a C-H group (Charette and Gray, 2000). Potentially, additional hydrogen bonds could contribute to the structural stability of the ribosome. However, the analysis of crystal structures of *E. coli* ribosomes (Schuwirth et al., 2005) and higher-resolution *T. thermophilus* ribosomes (Gao et al., 2009; Schmeing et al., 2009; Jenner et al., 2010) did not reveal hydrogen bonds involving the additional N-H group of any pseudouridine. On the other hand, all pseudouridine residues are involved in stacking interactions within rRNA, consistent with the observed stabilization of base stacking by pseudouridines in model systems (Davis, 1995; Yarian et al., 1999; Desaulniers et al., 2008).

Methylated nucleosides in rRNA can be C-, O-, and N-methylated. *E. coli* rRNA contains four 2'-O-methylated nucleosides, one in 16S rRNA and three in the 23S rRNA. Methylation of a hydroxyl group removes one potential hydrogen bond donor and creates the possibility for hydrophobic contacts. Furthermore, the methyl group is bulkier than hydrogen, introducing potential steric effects. Thus, the structural role of methylation could be to "create a (hydrophobic) contact" or to "avoid a contact".

It is interesting to examine the cluster of modified nucleosides, Gm2251-m²G2445-D2449, that is located close to the peptidyl transferase center (Figure 2A). The three modified nucleosides directly contact each other. Dihydrouridine, in contrast to uridine, does not have an aromatic ring and adopts a non-

planar conformation, while the two methyl groups fill the empty space around the dihydrouridine residue. One may hypothesize that the three modifications are necessary to create a specific structure of this functionally important rRNA region (Lesnyak et al., 2006). Among these three modified nucleosides, m²G2445 is the most important, since knockout of the *ycbY* gene that codes for the corresponding modifying enzyme results in the most severe phenotype, especially in M9 medium (Lesnyak et al., 2006). The modification on Gm2251 gives no advantage to the cells grown at laboratory conditions (Lovgren and Wikstrom, 2001), despite the fact that residue 2251 pairs with C75 of the P site-bound tRNA (Samaha et al., 1995; Nissen et al., 2000). The enzyme necessary for the formation of D2449 has not been characterized yet, but the substitution of T to C in the corresponding region of rDNA, which precludes the formation of D2449, does not lead to a strong phenotype (O'Connor et al., 2001).

The 2'-O-methylated residue Cm2498 is located in the peptidyltransferase loop. The enzyme responsible for this modification was discovered recently (Purta et al., 2009), and only a small growth defect has been ascribed to the loss of this methylation. Nevertheless, it can be noted that the addition of a methyl group to the ribose at position 2498 could fill the void in the contact site between several parts of rRNA, including both sides of the peptidyl transferase loop, hairpin 72 around nucleotide A2031, and hairpin 26 around nucleotide 570 (Figure 2B), thereby stabilizing that region.

Similar to 2'-O-methylations, base methylations could introduce additional hydrophobic contacts. In the *E. coli* ribosome, there are three m^5C residues (m^5C967, m^5C1407 in 16S rRNA, and m^5C1962 in 23S rRNA), two m^5U residues (m^5U747 and m^5U1939 in 23S rRNA) and one m^2A residue (m^2A2503 in 23S rRNA). The methyl group in m^5C967 may help in stabilizing helix 31 in that the methyl group could stack between the base of the neighboring m^2G966 and the ribose moiety of A968. However, direct measurements of the stability of hairpin 31 revealed that modification of C967 moderately destabilizes its structure (Abeydeera and Chow, 2009). This argues for a functional rather than structural role for this modification which will be discussed later. The two other m^5C modifications are also unlikely to serve a mere structural role. Modified m^5U747 could form a hydrophobic contact with C2612 and Lys3 of protein L32 (Jenner et al., 2010), while m^5U1939 is unlikely to form hydrophobic interactions.

Base methylations at nitrogen atoms usually substitute for hydrogen, thus removing a potential hydrogen bond donor group. Methylation of N7 of guanosine removes a potential hydrogen bond acceptor and creates a positive charge. However, the Watson-Crick base pairing potential is not affected by the N7 methylation of guanosine. Both, m^7G527 in 16S rRNA and m^7G2069 in 23S rRNA are involved in G-C basepairs (Cannone et al., 2002). The most common N-methylated base is m^2G (Sergiev et al., 2007), being present at five positions in the *E. coli* ribosome (m^2G966, m^2G1207, and m^2G1516 in 16S rRNA; m^2G1835 and m^2G2445 in 23S rRNA). N6-methylated adenosine is found at four positions (m^6_2A1518 and m^6_2A1519 in 16S rRNA; m^6A1618 and m^6A2030 in 23S rRNA). m^3U1498 and $m^4Cm1402$ are found once and are both located in the 30S decoding center only 4 Å apart from one another. It was suggested that the methyl groups of these residues could be involved in hydrophobic interactions (Kimura and Suzuki, 2010).

In another unique modification – m^1G745 of 23S rRNA – the Watson-Crick base-pairing potential is lost. Although the formation of alternative structures in unmethylated rRNA has not been examined directly, this modification would certainly favor the formation of the sheared base pair m^1G745-A750 that has been observed (Schuwirth et al., 2005; Jenner et al., 2010).

The tetraloop of helix 45 in 16S rRNA is particularly rich in modified bases. Two N6-dimethylated adenosines and one m^2G are located in the tetraloop and could form hydrophobic interactions. Unlike monomethylated m^6A and m^2G, dimethylated m^6_2A can not engage in base-pairing and thus could influence secondary structure formation. On the basis of experiments with oligonucleotide analogus of helix 45, it was suggested that m^6_2A1518 and m^6_2A1519 might be needed to prevent unproductive secondary structure formation (Micura et al., 2001). Recently, the crystal structure of *T. thermophilus* ribosomes devoid of methylated nucleosides was solved (Dahlberg AE, personal communication). The comparison with the structure of native ribosomes suggests that the methylations influence the structure of helix 45 loop, making it possible to dock it onto the helix 44 side.

Monomethylation of guanosine and adenosine at the exocyclic amino group does not prevent Watson-Crick base-pairing, since one amino hydrogen is still available to form a hydrogen bond. The respective methyl groups then are positioned in the minor and major grooves of the double helices formed. This situation applies for m^2G1207 of 16S rRNA, as well as for m^2G1835 and m^2G2445 of 23S rRNA. The monomethylated adenosines m^6A1618 and m^6A2030 are not engaged in base-pairing. The two residues form the same type of structure in that the methylated nucleotide is flipped away from the chain formed by neighboring nucleotides, and the base stacks between bases distant in the primary structure (Figures 2C, D).

In *Haloarcula marismortui*, an unmodified cytidine is located at a position equivalent to *E. coli* m^6A2030. At the same time, *H. marismortui* A628, which is equivalent to *E. coli* U571, is methylated (Kirpekar et al., 2005). It appears that the two methyl groups in *E. coli* and *H. marismortui* rRNA would occupy approximately same position in the tertiary structure, while being attached to bases that are far apart in the primary structure of 23S rRNA (Figure 2D).

3. Modifications in rRNA enhance ribosome interaction with ligands

Modified nucleotides are clustered in ligand binding sites of the ribosome (Brimacombe et al., 1993). Most of the modified nucleotides are located close to the P-site codon and the P-site tRNA (Figure 3). The P-site codon is in contact with nucleotides $m^4Cm1402$ and m^3U1498 of 16S rRNA (Korostelev et al., 2006) which are likely to form a hydrophobic contact with each other (Kimura and Suzuki, 2010). Nucleoside $m^4Cm1402$ is involved in an interaction with the phosphate between the second and third codon. Similarly,

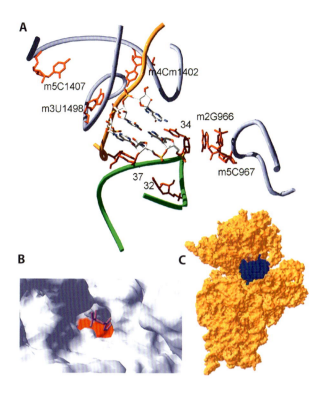

Fig. 3 (A) Modified nucleosides of 16S rRNA contacting the codon-anticodon duplex in the P site. The backbone of 16S rRNA is depicted in blue, the mRNA and tRNA backbones in yellow and green, respectively. Modified nucleosides are colored red (Jenner et al., 2010). (B) Amino acid binding pocket in the large ribosomal subunit (Nissen et al., 2000). A tyrosine sidechain (purple) is shown in the pocket, the surface formed by the modified nucleosides m^2A2503-$\Psi2504$ of 23S rRNA is colored red. (C) Model of the RsmD complex with the 30S subunit (Lesnyak et al., 2007). RsmD is depicted in blue.

m^3U1498 is close to the 3'-4'-5' side of the second-nucleotide ribose of the P-site codon.

Another group of modified nucleosides forms contacts with the anticodon loop of the P-site tRNA. m^2G966 of 16S rRNA forms a stacking interaction with the ribose of nucleotide 34 of P site-bound tRNA (Korostelev et al., 2006). The other side of m^2G966 is stacked on m^5C967, and m^2G966 is protected from kethoxal modification by P site-bound tRNA (Moazed and Noller, 1986). The functional role of the 966/967 modifications was examined in several mutagenesis experiments. Early data revealed no major phenotype of mutations of G966/C967, except for the deletion of C967 which was lethal (Jemiolo et al., 1991). Later work, using a specialized 30S system, demonstrated that various G966 substitutions lead to decreased translation activity (Abdi and Fredrick, 2005). The inhibition caused by the m^2G966A substitution could be releaved by an additional G1338A mutation (Abdi and Fredrick, 2005) which was demonstrated to strengthen

fMet-tRNA binding to the P site in the presence of excess IF3 (Lancaster and Noller, 2005). Exhaustive mutagenesis of helix 31 in a specialized ribosome system revealed that substitutions of m^2G966 and m^5C967 produced a hyperactive phenotype which could be suppressed by IF3 overexpression. Overexpression of IF2 makes ribosomes with substitutions of m^2G966 and m^5C967 even less active than the wild type (Saraiya et al., 2008). To reveal the exact role of those modifications, the respective knockout strains are to be studied.

Modifications in h44 and h31 could influence the adjustment of the P-site codon and affect the fidelity of initiation. At the stage of initiation, fMet-tRNA binds to the initiation codon in the P site. In *E. coli*, several initiation codons are used. AUG predominates, while GUG and UUG are less frequent (Blattner et al., 1997). In rare cases, even an AUU codon is used for initiation. This is observed in the gene *infC* coding for IF3 initiation factor (Sacerdot et al., 1982). Expression of IF3 is under feedback regulation since one of the functions of IF3 is to restrict initiation to "allowed" initiation codons (Butler et al., 1987; Sacerdot et al., 1996).

An influence of 16S rRNA methylation on translation initiation fidelity was found in a genetic screen using engineered fGln-tRNA (Das et al., 2008). Mutation in the *folD* gene, which leads to a reduction of the intracellular level of the methyl donor S-adenosyl-L-methionine (SAM), caused relaxation in initiator tRNA selection. It appeared that the SAM decrease reduced the modification of rRNA. Examination of individual rRNA methyltransferase knockout strains revealed that in *ksgA* and *rsmD* knockouts the selection of initiator tRNA is relaxed. A knockout of *rsmH*, the gene coding for N4-C1402 methyltransferase, resulted in increased AUU usage as initiator codon (Kimura and Suzuki, 2010). A double-knockout of *rsmB/rsmD*, the genes responsible for m^2G966/m^5C967 formation, in contrast, reduces both AUU usage *in vivo* and initiation efficiency *in vitro* (Prokhorova et al., manuscript in preparation). Clearly, modified nucleosides in the P site of the 30S subunit influence both efficiency and fidelity of translation initiation, most likely by altering the ribosome´s interactions with the P-site tRNA and codon.

4. Modifications of rRNA unify ribosome interaction with ligands

The ribosome has to interact with several dozens of different tRNA species and should keep these interactions relatively uniform (Fahlman et al., 2004). The interaction of EF-Tu with aminoacyl-tRNA is uniform in that the affinities of EF-Tu for the aminoacyl moiety and and tRNA body are balanced (LaRiviere et al., 2001). One may hypothesize that modifications of 16S rRNA nucleosides that are in contact with the codon-anticodon duplex in the decoding site could serve to balance the affinity of different tRNA species to the P site. For example, there are various modified nucleosides at position 34 in the anticodon of most tRNAs. Nucleoside 34 is in contact with m^2G966 of 16S rRNA (Figure 3A), and this hydrophobic contact could strengthen tRNA binding.

Similar reasons could be found for rRNA nucleoside modifications in the peptidyl transferase center of the large ribosomal subunit. A cluster of modified residues m^2A2503-$\Psi2504$ forms part of the A-site pocket and of the peptide tunnel entrance (Figure 3B). This binding pocket should be able to accommodate every amino acid with comparable affinity. It is possible that rRNA modifications in this area are needed for the modulation of the interactions of various aminoacyl residues with the peptidyl transferase center of ribosome, although experimental validation of these ideas is needed.

5. Modification of rRNA as an "assembly checkpoint"

Ribosome assembly is a highly coordinated multi-step process (Bunner et al., 2010a,b). The process is initiated co-transcriptionally, suggesting that rRNA is "never" protein-free in the living cell. Modification enzymes that are capable of modifying protein-free RNA *in vitro* could utilize a whole range of early assembly intermediates *in vivo*. Likewise, enzymes that modify assembled subunits *in vitro* are likely to use late assembly intermediates *in vivo*. Thus, there seem to be specific "windows" in the assembly process, where specific methyltransferases act. One example of such a "window" was provided by Ofengand and colleagues (Weitzmann et al., 1991). The methyltransferases RsmB and RsmD modify the adjacent bases G966 and C967 of 16S rRNA. Interestingly, RsmB modifies 16S rRNA only prior to the binding of both proteins S7 and S19, whereas the specificity of RsmD is the opposite – it is active only after the binding of both S7 and S19. The binding of S19 causes protection of single-stranded 16S rRNA regions around residues G966/C967 from both base-specific reagents (Powers et al., 1988) and hydroxyl radicals (Powers and Noller, 1995).

Little structural information is available for assembly intermediates, such as the earliest substrate of RsmD methyltransferase which contains only 16S rRNA and proteins S7 and S19. Since the assembly of ribosomal subunits is believed to follow 5'-to-3' order of transcription (Powers et al., 1993; Dutca and Culver, 2008; Sykes and Williamson, 2009), it is likely that at the time when S7 and S19 bind to the 3' part of 16S rRNA, forming the "head" domain of the small subunit, transcription of the "body" domain of 16S rRNA is nearly completed. The challenge for all rRNA modification enzymes is to precisely recognize a single (or, in some cases, a few) nucleoside among thousands of nucleosides. Many modification enzymes have diverse RNA recognition domains fused to a catalytic domain (Sergiev et al., 2007). RsmD is unusual in consisting of a catalytic domain only, albeit with uncompromised specificity towards G966 of 16S rRNA (Lesnyak et al., 2007). The target nucleoside of RsmD, G966, is located in a deep cleft of the small subunit, which, during translation, is occupied by P site-bound tRNA (Korostelev at al., 2006; Jenner et al., 2010). RsmD can modify G966 in the completely assembled small ribosomal subunit (Weitzmann et al., 1991; Lesnyak et al., 2007), so it should be able to fit into the P-site cleft (Figure 3C). Thus, RsmD can use almost its entire surface for precise substrate recognition. In contrast to the usual case where a substrate fits into an active-site cleft of an enzyme, RsmD fits into a cleft of its substrate.

Modification enzymes that act on protein-free rRNA or early assembly intermediates should not only precisely recognize their target, but also do it rapidly, before potentially inhibitory ribosomal proteins can bind and prevent further modification. It is, therefore, surprising that modification is usually almost 100% efficient in the presence of competing ribosomal proteins. The mechanisms by which modification enzymes win the competition with ribosomal proteins are unclear. After successful modification, the modification enzymes should make available the binding sites for ribosomal proteins for assembly to proceed. Pseudouridine synthase RsuA, which is responsible for $\Psi516$ formation in 16S rRNA, contains an S4-like recognition domain (Conrad et al., 1999). The fact that the S4 binding site is close to the RsuA modification site (Stern at al., 1986) then would suggest that the two proteins compete for the same, or an overlapping, binding site.

The RsmB methyltransferase, whose target is nucleotide C967 of 16S rRNA, contains a NusB-like recognition domain (Foster et al., 2003). The substrate of RsmB is protein-free 16S rRNA or a very early assembly intermediate, formed prior to S7 and S19 binding (Weitzmann et al., 1991). The closest homolog of the RsmB RNA-recognition domain is the transcription antitermination factor NusB (Luo et al., 2008). This relation raises the possibility of an interaction of early-acting rRNA modification enzymes with RNA polymerase while it is transcribing rDNA operons.

After methylating the rRNA, the methyltransferase should dissociate, to allow further ribosome assembly and function in mRNA translation. The trigger for dissociation presumably is the completion of the modification process itself. For rRNA methyltransferases, it is likely that the exchange of S-adenosyl-L-homocysteine, the product of methyl transfer, for SAM, the methyl donor, on the enzyme leads to a steric clash between the methyl group incorporated into the RNA and the methyl group of SAM. This may trigger enzyme dissociation or impair re-binding. It is even possible that the only purpose of the methylation reaction is to induce methyltransferase dissociation (Connoly et al., 2008; Mangat and Brown, 2008). This hypothesis assumes that the primary role of modification enzymes is to assist assembly at certain stages and/or to prevent conformation changes or the binding of certain ligands that would lead to dead-ends of the assembly pathway (Connoly et al., 2008; Mangat and Brown, 2008; Xu et al., 2008).

This hypothesis is supported by the observation that modification activity and activity in ribosome assembly can be separated. For instance, in an *rlmA* knockout *E. coli* strain an as yet unidentified secondary mutation restores nearly wild-type growth, but not the modification of the target nucleoside, G745 in 23S rRNA (Liu et al., 2004). Furthermore, transposon inactivation of the *rluD* gene that codes for a pseudouridine synthase acting on helix 69 of 23S rRNA was found to result in a slow-growth phenotype (Raychaudhuri et al., 1998). Second-site mutations close to the *rluD* gene partially restored growth, and the effect could be enhanced by transformation of the strain with a plasmid coding for catalytically inactive RluD (Gutgsell et al., 2001). A subsequent study, where the *rluD* gene knockout was performed more precisely, demonstrated, however, that only catalytically functional RluD could restore cell growth (Gutgsell et al., 2005). Nevertheless, second-site mutations in some as yet unidentified genetic loci are able to restore growth without restoring the formation of $\Psi1911$, $\Psi1915$, $\Psi1917$ (Gutgsell et al., 2001; Gutgsell et al., 2005), indicating that at least some modification enzymes act as "assembly chaperones".

An "assembly checkpoint" function has been documented for the universally conserved methyltransferase KsgA (Connolly et al., 2008; Mangat and Brown, 2008; Xu et al., 2008). The enzyme dimethylates the exocyclic amino groups of A1518 and A1519 of 16S rRNA (Helser et al., 1972; Poldermans et al., 1979). In yeast, the knockout of the homologous *dim1* gene is lethal (Lafontaine et al., 1995), and growth could be restored by introducing a catalytically inactive form of Dim1p (Lafontaine et al., 1998). In *E. coli*, the knockout of the *ksgA* gene leads to retarded growth and the accumulation of 16S rRNA precursors (Connolly et al., 2008), in keeping with the observation that KsgA acts on late assembly intermediates (Desai and Rife, 2006). The absence of the very last assembly protein, S21, is sufficient to make such "almost assembled" particles a substrate for KsgA (Desai and Rife, 2006). Loss of S21 is accompanied with the compromised ability to recognize SD regions in mRNA (Backendorf et al., 1981; Van Duin and Wijnands, 1981). Protein S21 is also required to restore the functional activity of the 30S subunit after inactivation in cold low-magnesium buffer (Backendorf et al., 1981). Later experiments showed that these inactivated 30S subunits are substrates for the methyltransferase KsgA (Desai and Rife, 2006). Binding of KsgA to 30S assembly intermediates prevents the involvement of such intermediates in translation. Methylation of target nucleotides triggers the dissociation of KsgA·30S complex. Expression of the catalytically inactive KsgA(E66A) mutant is deleterious for the cell, since mutant KsgA is not released from the 30S particle (Connolly et al., 2008). It might be concluded that KsgA functions as a protein switch that blocks 30S particles until assembly is completed successfully (Mangat and Brown, 2008).

6. Modification of rRNA as a "quality mark" in ribosome assembly

Related to the ribosome assembly checkpoint function is the function of rRNA modification as a quality mark in ribosome assembly. In the "assembly checkpoint" scenario, the modification itself, after dissociation of the modification enzyme, has no further function. According to "quality mark" hypothesis, modification can take place only after the completion of certain assembly stages and prevents rRNA degradation (Song and Nazar, 2002). If a ribosomal particle fails to be modified, it is degraded. An ideal candidate for illustrating the "quality mark" scenario is the formation of $\Psi1911$,

Ψ1915, Ψ1917 in 23S rRNA by RluD. These modifications take place late, even after the completion of 50S subunit assembly (Leppik et al., 2007; Vaidyanathan et al., 2007), and independent of each other (Leppik et al., 2007). Deletion of the RluD gene leads to a slow-growth phenotype (Huang et al., 1998; Raychaudhuri et al., 1998; Gutgsell et al., 2005), arguably the most severe growth defect caused by any of rRNA modification deficiency. Inactivation of RluD alters the ribosome sedimentation profile dramatically (Gutgsell et al., 2005). The fraction of functional 70S particles is reduced, while "lighter" particles, such as 62S and 39S, appear (Gutgsell et al., 2005). At first glance, the accumulation of 50S subunit precursor (39S) and their inclusion into 70S-like particles (62S) argues in favor of an involvement of RluD in the 50S subunit assembly pathway as "assembly checkpoint" factor. However, not only 23S rRNA precursors, but also 16S rRNA precursors accumulate in cells deficient in RluD (Gutgsell et al., 2005). It is also possible that 39S and 62S particles result not from incomplete assembly, but from breakdown of 50S and 70S particles, respectively (Gutgsell et al., 2005). Rapid turnover of 23S rRNA would result in increased transcription of rDNA operons to compensate for the loss of 23S rRNA. This, in turn, would lead to an unbalanced accumulation of 16S rRNA, which may not be processed in time. Second-site suppressors of the *rluD* knockout that restore normal growth and ribosome profiles (Gutgsell et al., 2005) could affect specific "disassembly" proteins or RNases acting on unmodified 23S rRNA. The phenomenon of rapid degradation of unmodified rRNA was documented for yeasts (Song and Nazar, 2002), but similar processes in bacteria still await experimental verification.

Another rRNA modification enzyme with high impact on 50S subunit biogenesis is the methyltransferase RrmJ. The enzyme modifies the 2'-OH group of U2552 in the A loop of 23S rRNA (Caldas et al., 2000a; Bugl et al., 2000). Knockout of *rrmJ* leads to very slow growth (Caldas et al., 2000b) and the accumulation of 40S particles containing 23S rRNA (Hager et al., 2002). The idea that RrmJ is necessary for 50S assembly was challenged by several findings. If 40S particles were assembly intermediates "waiting" for RrmJ to bring about the formation of a certain assembly stage, one would expect them to be preferred substrates for RrmJ. Surprisingly, however, RrmJ does not methylate 40S particles (Hager et al., 2002; Hager et al., 2004) accumulated in a strain deficient for RrmJ. In contrast, RrmJ readily methylates completely assembled 50S subunits (Hager et al., 2002; Hager et al., 2004). It seems contradictory

that, on the one hand, the methyltransferase acts on assembled subunits, and, on the other, is necessary for subunit assembly. This apparent contradiction could be resolved if the Um2552 modification had a "quality mark" function. According to this point of view, 40S particles are breakdown products, rather than assembly intermediates.

An alternative explanation, favoring the "assembly checkpoint" model, is that 40S particles accumulating in the RrmJ-deficient strain are dead-end assembly products. Normally, RrmJ should bind and prevent their formation until assembly is completed. Then the RrmJ methylation activity might be triggered and lead to enzyme release. This model is supported by the ability of the 50S subunit-dependent GTPases Obg and EngA to suppress a *rrmJ* deficiency (Tan et al., 2002). At the cost of GTP hydrolysis, the GTPases could resolve the assembly dead-end product, which subsequently could have another chance to fold properly.

Ribosome modification as a "quality mark" assumes the existence of ribonucleases whose activity is restricted by RNA modification. Such hypothetic system could resemble restriction-modification systems. Likely candidates for such RNases are stress-induced toxin proteins from toxin-antitoxin pairs. One of the best-known among them is MazF (Zhang et al., 2003), which currently is considered an mRNA-specific nuclease. Induction of the MazF system leads to the overexpression of at least one rRNA modification enzyme, RsuA (Amitai et al., 2009), which is responsible for Ψ516formation in 16S rRNA (Wrzesinski et al., 1995a). Deletion of the *rsuA* gene compromises cell survival upon MazF induction (Amitai et al., 2009). One may assume that RsuA protects 16S rRNA from MazF cleavage, but the RsuA modification site does not contain readily recognizable MazF cleavage consensus ACA (Zhang et al., 2003). rRNA modification could also be imagined to protect cells from exogenous RNases, similar to various ribonucleolytic colicins. For several tRNA species, 2'-O-methylation was demonstrated to protect RNA from colicin cleavage (Chan et al., 2009). Colicin E3, which cleaves 16S rRNA after nucleotide 1493 (Senior and Holland, 1971; Bowman et al., 1971), is not inhibited by any of the rRNA modifications in *E. coli*. However, it is still possible that other colicin-like RNases are inhibited by modifications in *E. coli* rRNA.

7. rRNA modification as antibiotic resistance mechanism

Apart from "house-keeping" rRNA modifications, a large group of rRNA modifications is known to protect bacterial species from antibiotics (Poehlsgaard and Douthwaite, 2005). The existence of multiple copies of rDNA operons in the majority of bacterial species makes it difficult to acquire resistance by mutations. This is perhaps the main reason behind the observed prevalence of ribosome inhibitors among antibacterial agents. Natural antibiotic producers use rRNA methylation for protecting their own ribosomes from inhibition by the antibiotic they produce. Such resistance mechanisms are documented for the producers of various aminoglycosides (Holmes and Cundliffe, 1991), macrolides (Zalacain and Cundliffe, 1989), thiazole-containing peptides (Cundliffe, 1978) and orthosomycins (Treede et al., 2003). The genes coding for rRNA methyltransferases that are responsible for antibiotic resistance are rapidly spreading among clinically important pathogens, making bacterial infections more and more dangerous. Rapid spreading of antibiotic resistance among bacteria during the past decades was promoted by the excessive use of antimicrobials (Kunin, 1985; Ferber, 1998). However, antibiotics were naturally used in competition between bacteria for billions of years. Perhaps, during this time some antibiotics became inactive towards the majority of bacterial species due to global spread of resistance.

In turn, later evolved steps in antibiotic biosynthesis could have counteracted such resistance mechanisms. Co-evolution of antibiotic biosynthesis and corresponding resistance mechanisms might have resulted in the global spread of rRNA methyltransferase genes protecting bacteria from antibacterials which were in use by bacteria billions of years ago.

It should be noted that modified nucleosides in rRNA are located close to the binding sites of antibiotics. This could be a coincidence since both modified nucleotides and antibiotic binding sites are located in functional centers of the ribosome. However, there are several cases where antibiotics require rRNA modifications for efficient inhibition. One case is kasugamycin, an antibiotic acting on translation initiation (Helser et al., 1972; Poldermans et al., 1979). This antibiotic requires the m_2^6A1518-m_2^6A1519 modifications in 16S rRNA to inhibit translation. Similarly, the lack of the m^7G527 modification in 16S rRNA leads to moderate streptomycin resistance (Okamoto et al., 2007). Antimycobacterial capreomycin depends on 2'-O-methylation of C1409 in 16S rRNA and C1920 in 23S rRNA (Johansen et al., 2006).

Modern antibiotics could be the products of evolution driven by the development of resistance caused by methylation. Examples are provided by tetracycline (Figure 4A), which binds close to m^5C967 (Brodersen et al., 2000), streptomycin (Figure 4B), which binds close to m^7G527 (Carter et al., 2000), paromomycin (Figure 4C), which binds close to m^5C1407, and several

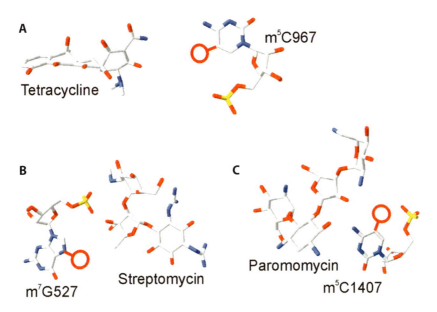

Fig. 4 Modified nucleosides in 16S rRNA contacting antibiotics bound to the small ribosomal subunit. Methyl groups which are absent in the corresponding PDB file are marked by red circles.

(A) Interaction of tetracycline with m^5C967 (Brodersen et al., 2000). (B) Interaction of streptomycin with m^7G527 (Carter et al., 2000). (C) Interaction of m^5C1407 with paromomycin (Carter et al., 2000).

other antibiotics (Carter et al., 2000). In present-time ribosomes, these methyl groups do not prevent these antibiotics from binding. Modification of m⁷G527 even enhances streptomycin action (Okamoto et al., 2007). However, the hypothetic "ancient" antibiotics may have had additional functional groups located at positions that are nowadays occupied by methyl groups. Later, enzymes like RsmB, RsmG and RsmF appeared as resistance methyltransferases against these "ancient" antibiotics. Spreading of the respective methyltransferase genes made "ancient" antibiotics inefficient and "present-day" antibiotics were invented to counteract resistances caused by RsmB, RsmG and RsmF.

Even if some methyltransferases have never been resistance enzymes, they could form a reserve for the evolution of new resistance methyltransferases. The potential involvement of a single rRNA modification enzyme in both ribosome function and antibiotic resistance was demonstrated. Inactivation of the *rluC* gene, which codes for a pseudouridine synthase, leads to hypersensitivity against thiamulin, clindamycin and linezolid. It was also demonstrated that only pseudouridine formation at position 2504 of 23S rRNA, but not at other target sites of RluC, caused the hypersensitivity phenomenon (Toh and Mankin, 2008).

8. rRNA modification and regulation of gene expression

The existence of different ribosome species and appearance of "altered" ribosomes is well documented for eukaryotes (Ramagopal, 1992). For instance, plasmodium species practice exchange of rRNA upon transition between vertebrate and mosquito hosts (Li et al., 1994). Ribosomal protein modification alters translation efficiency. A good example of such regulation is rpS6 phosphorylation (Meyuhas, 2008). In yeast, ribosomal proteins encoded in the paralog genes were shown to be specialized for mRNA-specific translation (Komili et al., 2007). Bacteria also have exchangeable ribosomal proteins. In E. coli ribosomal proteins L31 and L36 contain Zn²⁺ ribbons. Upon Zn²⁺ starvation these proteins are dismissed by their paralogues devoid of Zn²⁺ binding site (Makarova et al., 2001). This regulation apparently plays a role in saving Zn²⁺ for the incorporation into enzymes that strictly require Zn²⁺ for activity.

Pseudouridine synthase RluC, whose products are the 23S rRNA nucleosides Ψ955, Ψ2504 and Ψ2580, is involved in periplasmic stress response. Inactivation of the rluC gene suppresses the phenotype of bipA knockout (Krishnan and Flower, 2008). Deletion of bipA, a gene coding for a ribosome-dependent GTPase, leads to cold sensitivity and inability to cope with periplasmic stress (Pfennig and Flower, 2001). All three RluC target nucleotides are essential for the suppression (Krishnan and Flower, 2008). Although RluC apparently is involved in regulation of envelope stress response, it would be important to demonstrate that rluC expression is by itself regulated.

Inactivation of the RsmB/RsmD methyltransferases, two enzymes that are responsible for the base methylations at G966 and C967, respectively, alter the proteome of the cell (Prokhorova et al., manuscript in preparation). Particularly, the level of initiation factor 3 is decreased when the modification is inhibited.

RrmJ is co-expressed with FtsH, a membrane-bound protease (Caldas et al., 2000a; Bugl et al., 2000). Both genes are induced by heat-shock (Richmond et al., 1999). RrmJ, which modifies U2552 of 23S rRNA, is possibly involved in the heat-shock response, although U2552 is methylated constitutively, independent of heat shock (Caldas et al., 2000a).

Depending on the particular modified residue, different possibilities can apply: structure stabilization, enhancement and unification of ribosome interaction with ligands, "assembly checkpoint", "quality mark", protection from unidentified antibacterial RNases or small molecule binding and regulation of translation.

References

Abdi NM, Fredrick K (2005) Contribution of 16S rRNA nucleotides forming the 30S subunit A and P sites to translation in Escherichia coli. RNA 11: 1624–1632

Abeydeera ND, Chow CS (2009) Synthesis and characterization of modified nucleotides in the 970 hairpin loop of Escherichia coli 16S ribosomal RNA. Bioorg Med Chem 17: 5887–5893

Agarwalla S, Kealey JT, Santi DV, Stroud RM (2002) Characterization of the 23S ribosomal RNA m^{5}U1939 methyltransferase from Escherichia coli. J Biol Chem 277: 8835–8840

Amitai S, Kolodkin-Gal I, Hananya-Meltabashi M, Sacher A, Engelberg-Kulka H (2009) Escherichia coli MazF leads to the simultaneous selective synthesis of both "death proteins" and "survival proteins". PLoS Genet 5: e1 000 390

Andersen NM, Douthwaite S (2006) YebU is a m^{5}C methyltransferase specific for 16S rRNA nucleotide 1407. J Mol Biol 359: 777–786

Andersen TE, Porse BT, Kirpekar F (2004) A novel partial modification at C2501 in Escherichia coli 23S ribosomal RNA. RNA 10: 907–913

Atherly AG (1974) Ribonucleic acid regulation in amino acid-limited cultures of Escherichia coli grown in a chemostat. J Bacteriol 120: 1322–1330

Auffinger P, Westhof E (1998) Location and distribution of modified nucleotides in tRNA. In: Grosjean H, Benne R (eds) Modification and editing of RNA. ASM Press, Washington, pp 569–576

Baba T, Ara T, Hasegawa M, Takai Y, Okumura Y, Baba M, Datsenko KA, Tomita M, Wanner BL, Mori H (2006) Construction of Escherichia coli K-12 in-frame, single-gene knock-out mutants the Keio collection. Mol Syst Biol 2: 2006–2008

Backendorf C, Ravensbergen CJ, Van der Plas J, van Boom JH, Veeneman G, Van Duin J (1981) Basepairing potential of the 3' terminus of 16S RNA: dependence on the functional state of the 30S subunit and the presence of protein S21. Nucleic Acids Res 9: 1425–1444

Ban N, Nissen P, Hansen J, Moore PB, Steitz TA (2000) The complete atomic structure of the large ribosomal subunit at 2.4 Å resolution. Science 289: 905–920

Basturea GN, Rudd KE, Deutscher MP (2006) Identification and characterization of RsmE, the founding member of a new RNA base methyltransferase family. RNA 12: 426–434

Basturea GN, Deutscher MP (2007) Substrate specificity and properties of the Escherichia coli 16S rRNA methyltransferase, RsmE. RNA 13: 1969–1976

Björk GR, Isaksson LA (1970) Isolation of mutants of Escherichia coli lacking 5-methyluracil in transfer ribonucleic acid or 1-methylguanine in ribosomal RNA. J Mol Biol 51: 83–100

Blattner FR, Plunkett G 3rd, Bloch CA, Perna NT, Burland V, Riley M, Collado-Vides J, Glasner JD, Rode CK, Mayhew GF, Gregor J, Davis NW, Kirkpatrick HA, Goeden MA, Rose DJ, Mau B, Shao Y (1997) The complete genome sequence of Escherichia coli K-12. Science 277: 1453–1462

Bowman CM, Dahlberg JE, Ikemura T, Konisky J, Nomura M (1971) Specific inactivation of 16S ribosomal RNA induced by colicin E3 in vivo. Proc Natl Acad Sci USA 68: 964–968

Brimacombe R, Mitchell P, Osswald M, Stade K, Bochkariov D (1993) Clustering of modified nucleotides at the functional center of bacterial ribosomal RNA. FASEB J 7: 161–167

Brodersen DE, Clemons WM Jr, Carter AP, Morgan-Warren RJ, Wimberly BT, Ramakrishnan V (2000) The structural basis for the action of the antibiotics tetracycline, pactamycin, and hygromycin B on the 30S ribosomal subunit. Cell 103: 1143–1154

Bugl H, Fauman EB, Staker BL, Zheng F, Kushner SR, Saper MA, Bardwell JCA, Jakob U (2000) RNA methylation under heat shock control. Mol Cell 6: 349–360

Bunner AE, Nord S, Wikström PM, Williamson JR (2010a) The effect of ribosome assembly cofactors on in vitro 30S subunit reconstitution. J Mol Biol 398: 1–7

Bunner AE, Beck AH, Williamson JR (2010b) Kinetic cooperativity in Escherichia coli 30S ribosomal subunit reconstitution reveals additional complexity in the assembly landscape. Proc Natl Acad Sci USA 107: 5417–5422

Butler JS, Springer M, Grunberg-Manago M (1987) AUU-to-AUG mutation in the initiator codon of the translation initiation factor IF3 abolishes translational autocontrol of its own gene (infC) in vivo. Proc Natl Acad Sci USA 84: 4022

Caldas T, Binet E, Bouloc P, Costa A, Desgres J, Richarme G (2000a) The FtsJ/RrmJ heat shock protein of Escherichia coli is a 23S ribosomal RNA methyltransferase. J Biol Chem 275: 16414–16419

Caldas T, Binet E, Bouloc P, Richarme G (2000b) Translational defects of Escherichia coli mutants deficient in the Um(2552) 23S ribosomal RNA methyltransferase RrmJ/FTSJ. Biochem Biophys Res Commun 271: 714–718

Cannone JJ, Subramanian S, Schnare MN, Collett JR, D'Souza LM, Du Y, Feng B, Lin N, Madabusi LV, MÜller KM, Pande N, Shang Z, Yu N, Gutell RR (2002) The Comparative RNA Web (CRW) Site: An Online Database of Comparative Sequence and Structure Information for Ribosomal, Intron, and Other RNAs. BioMed Central Bioinformatics 3: 2

Carter AP, Clemons WM, Brodersen DE, Morgan-Warren RJ, Wimberly BT, Ramakrishnan V (2000) Functional insights from the structure of the 30S ribosomal subunit and its interactions with antibiotics. Nature 407: 340–348

Chan CM, Zhou C, Huang RH (2009) Reconstituting bacterial RNA repair and modification in vitro. Science 326: 247

Charette M, Gray MW (2000) Pseudouridine in RNA: what, where, how, and why. IUBMB Life 49: 341–351

Connolly K, Rife JP, Culver G (2008) Mechanistic insight into the ribosome biogenesis functions of the ancient protein KsgA. Mol Microbiol 70: 1062–1075

Conrad J, Sun D, Englund N, Ofengand J (1998) The rluC gene of Escherichia coli codes for a pseudouridine synthase that is solely responsible for synthesis of pseudouridine at positions 955, 2504, and 2580 in 23S ribosomal RNA. J Biol Chem 273: 18562–18566

Cundliffe E (1978) Mechanism of resistance to thiostrepton in the producing-organism Streptomyces azureus. Nature 272: 792–795

de Crécy-Lagard V, Marck C, Brochier-Armanet C, Grosjean H (2007) Comparative RNomics and modomics in Mollicutes: prediction of gene function and evolutionary implications. IUBMB Life 59: 634–658

Das G, Thotala DK, Kapoor S, Karunanithi S, Thakur SS, Singh NS, Varshney U (2008) Role of 16S ribosomal RNA methylations in translation initiation in Escherichia coli. EMBO J 27: 840–851

Datsenko KA Wanner BL (2000) One-step inactivation of chromosomal genes in Escherichia coli K-12 using PCR products. Proc Natl Acad Sci USA 97: 6640–6655

Davis DR (1995) Stabilization of RNA stacking by pseudouridine. Nucleic Acids Res 23: 5020–5026

Decatur WA, Fournier MJ (2002) rRNA modifications and ribosome function. Trends Biochem Sci 27: 344–351

Del Campo M, Kaya Y, Ofengand J (2001) Identification and site of action of the remaining four putative pseudouridine synthases in Escherichia coli. RNA 7: 1603–1615

Desai PM, Rife JP (2006) The adenosine dimethyltransferase KsgA recognizes a specific conformational state of the 30S ribosomal subunit. Arch Biochem Biophys 449: 57–63

Desaulniers JP, Chang YC, Aduri R, Abeysirigunawardena SC, SantaLucia J Jr, Chow CS (2008) Pseudouridines in rRNA helix 69 play a role in loop stacking interactions. Org Biomol Chem 6: 3892–3895

Douthwaite S, Kirpekar F (2007) Identifying modifications in RNA by MALDI mass spectrometry. Meth Enzymol 425: 1–20

Dutca LM, Culver GM (2008) Assembly of the 5' and 3' minor domains of 16S ribosomal RNA as monitored by tethered probing from ribosomal protein S20. J Mol Biol 376: 92–108

Ero R, Peil L, Liiv A, Remme J (2008) Identification of pseudouridine methyltransferase in *Escherichia coli*. RNA 14: 2223–2233

Fahlman RP, Dale T, Uhlenbeck OC (2004) Uniform binding of aminoacylated transfer RNAs to the ribosomal A and P sites. Mol Cell 16: 799–805

Ferber, D (1998) New hunt for roots of resistance. Science 280: 27

Foster PG, Nunes CR, Greene P, Moustakas D, Stroud RM (2003) The first structure of an RNA m^{5}C methyltransferase, Fmu, provides insight into catalytic mechanism and specific binding of RNA substrate. Structure 11: 1609–1620

Gao YG, Selmer M, Dunham CM, Weixlbaumer A, Kelley AC, Ramakrishnan V (2009) The structure of the ribosome with elongation factor G trapped in the posttranslocational state. Science 326: 694–699

Green R, Noller HF (1996) In vitro complementation analysis localizes 23S rRNA posttranscriptional modifications that are required for Escherichia coli 50S ribosomal subunit assembly and function. RNA 2: 1011–1021

Gu XR, Gustafsson C, Ku J, Yu M, Santi DV (1999) Identification of the 16S rRNA m^{5}C967 methyltransferase from *Escherichia coli*. Biochemistry 38: 4053–4057

Guex N, Peitsch MC (1997) SWISS-MODEL and the Swiss-Pdb-Viewer: an environment for comparative protein modeling. Electrophoresis 18: 2714–2723

Gustafsson C, Persson BC (1998) Identification of the rrmA gene encoding the 23S rRNA m^{1}G745 methyltransferase in *Escherichia coli* and characterization of an m^{1}G745-deficient mutant. J Bacteriol 180: 359–365

Gutgsell NS, Del Campo M, Raychaudhuri S, Ofengand J (2001) A second function for pseudouridine synthases: A point mutant of RluD unable to form pseudouridines 1911, 1915, and 1917 in *Escherichia coli* 23S ribosomal RNA restores normal growth to an RluD-minus strain. RNA 7: 990–998

Gutgsell NS, Deutscher MP, Ofengand J (2005) The pseudouridine synthase RluD is required for normal ribosome assembly and function in *Escherichia coli*. RNA 11: 1141–1152

Hager J, Staker BL, Bugl H, Jakob U (2002) Active site in RrmJ, a heat shock-induced methyltransferase. J Biol Chem 277: 41978–41986

Hager J, Staker BL, Jakob U (2004) Substrate binding analysis of the 23S rRNA methyltransferase RrmJ. J Bacteriol 186: 6634–6642

Helser TL, Davies JE, Dahlberg JE (1972) Mechanism of kasugamycin resistance in *Escherichia coli*. Nat New Biol 235: 6–9

Holmes DJ, Cundliffe E (1991) Analysis of a ribosomal RNA methylase gene from *Streptomyces tenebrarius* which confers resistance to gentamicin. Mol Gen Genet 229: 229–237

Huang L, Ku J, Pookanjanatavip M, Gu X, Wang D, Greene PJ, Santi DV (1998) Identification of two Escherichia coli pseudouridine synthases that show multisite specificity for 23S RNA. Biochemistry 37: 15951–15957

Jemiolo DK, Taurence JS, Giese S (1991) Mutations in 16S rRNA in Escherichia coli at methyl-modified sites: G966, C967, and G1207. Nucleic Acids Res 19: 4259–4265

Jenner LB, Demeshkina N, Yusupova G, Yusupov M (2010) Structural aspects of messenger RNA reading frame maintenance by the ribosome. Nat Struct Mol Biol 17: 555–560

Johansen SK, Maus CE, Plikaytis BB, Douthwaite S (2006) Capreomycin binds across the ribosomal subunit interface using tlyA-encoded 2'-O-methylations in 16S and 23S rRNAs. Mol Cell 23: 173–182

Kimura S, Suzuki T (2010) Fine-tuning of the ribosomal decoding center by conserved methyl- modifications in the Escherichia coli 16S rRNA. Nucleic Acids Res 38: 1341–1352

Kirpekar F, Hansen LH, Rasmussen A, Poehlsgaard J, Vester B (2005) The archaeon *Haloarcula marismortui* has few modifications in the central parts of its 23S ribosomal RNA. J Mol Biol 348: 563–573

Komili S, Farny NG, Roth FP, Silver PA (2007) Functional specificity among ribosomal proteins regulates gene expression. Cell 131: 557–571

Korostelev A, Trakhanov S, Laurberg M, Noller HF (2006) Crystal structure of a 70S ribosome-tRNA complex reveals functional interactions and rearrangements. Cell 126: 1065–1077

Krishnan K, Flower AM (2008) Suppression of Δ*bipA* phenotypes in *Escherichia coli* by abolishment of pseudouridylation at specific sites on the 23S rRNA. J Bacteriol 190: 7675–7683

Krzyzosiak W, Denman R, Nurse K, Hellmann W, Boublik M, Gehrke CW, Agris PF, Ofengand J (1987) *In vitro* synthesis of 16S ribosomal RNA containing single base changes and assembly into a functional 30S ribosome. Biochemistry 26: 2353–2364

Kunin CM (1985) The responsibility of the infectious disease community for the optimal use of antimicrobial agents. J Infect Dis 151: 388–398

Lafontaine D, Vandenhaute J, Tollervey D (1995) The 18S rRNA dimethylase Dim1p is required for pre-ribosomal RNA processing in yeast. Genes Dev 9: 2470–2481

Lafontaine DL, Preiss T, Tollervey D (1998) Yeast 18S rRNA dimethylase Dim1p: a quality control mechanism in ribosome synthesis? Mol Cell Biol 18: 2360–2370

Lancaster L, Noller HF (2005) Involvement of 16S rRNA nucleotides G1338 and A1339 in discrimination of initiator tRNA. Mol Cell 20: 623–632

LaRiviere FJ, Wolfson AD, Uhlenbeck OC (2001) Uniform binding of aminoacyl-tRNAs to elongation factor Tu by thermodynamic compensation. Science 294: 165–168

Leppik M, Peil L, Kipper K, Liiv A, Remme J (2007) Substrate specificity of the pseudouridine synthase RluD in Escherichia coli. FEBS J 274: 5759–5766

Lesnyak DV, Osipiuk J, Skarina T, Sergiev PV, Bogdanov AA, Edwards A, Savchenko A, Joachimiak A, Dontsova OA (2007) Methyltransferase that modifies guanine 966 of the 16S rRNA: functional identification and tertiary structure. J Biol Chem 282: 5880–5887

Lesnyak DV, Sergiev PV, Bogdanov AA, Dontsova OA (2006) Identification of *Escherichia coli* m^{2}G methyltransferases. I. The ycbY gene encodes a methyltransferase specific for G2445 of the 23S rRNA. J Mol Biol 364: 20–25

Li J, McConkey GA, Rogers MJ, Waters AP, McCutchan TR (1994) Plasmodium: the developmentally regulated ribosome. Exp Parasitol 78: 437–441

Liu M, Novotny GW, Douthwaite S (2004) Methylation of 23S rRNA nucleotide G745 is a secondary function of the RlmAI methyltransferase. RNA 10: 1713–1720

Lovgren JM, Wikstrom PM (2001) The rlmB gene is essential for formation of Gm2251 in 23S rRNA but not for ribosome maturation in Escherichia coli. J Bacteriol 183: 6957–6960

Luo X, Hsiao HH, Bubunenko M, Weber G, Court DL, Gottesman ME, Urlaub H, Wahl MC (2008) Structural and functional analysis of the *E. coli* NusB-S10 transcription antitermination complex. Mol Cell 32: 791–802

Madsen CT, Mengel-Jorgensen J, Kirpekar F, Douthwaite S (2003) Identifying the methyltransferases for m5U747 and m5U1939 in 23S rRNA using MALDI mass spectrometry. Nucleic Acids Res 31: 4738–4746

Makarova KS, Ponomarev VA, Koonin EV (2001) Two C or not two C: recurrent disruption of Zn-ribbons, gene duplication, lineage-specific gene loss, and horizontal gene transfer in evolution of bacterial ribosomal proteins. Genome Biol 2: 0033

Mangat CS, Brown ED (2008) Ribosome biogenesis; the KsgA protein throws a methyl-mediated switch in ribosome assembly. Mol Microbiol 70: 1051–1053

Meyuhas O (2008) Physiological roles of ribosomal protein S6: one of its kind. Int Rev Cell Mol Biol 268: 1–37

Micura R, Pils W, Höbartner C, Grubmayr K, Ebert MO, Jaun B (2001) Methylation of the nucleobases in RNA oligonucleotides mediates duplex-hairpin conversion. Nucleic Acids Res 29: 3997–4005

Moazed D, Noller HF (1986) Transfer RNA shields specific nucleotides in 16S ribosomal RNA from attack by chemical probes. Cell 47: 985–994

Nissen P, Hansen J, Ban N, Moore PB, Steitz TA (2000) The structural basis of ribosome activity in peptide bond synthesis. Science 289: 920–930

O'Connor M, Lee WM, Mankad A, Squires CL, Dahlberg AE (2001) Mutagenesis of the peptidyltransferase center of 23S rRNA: the invariant U2449 is dispensable. Nucleic Acids Res 29: 710–715

Ofengand J, Bakin A, Wrzesinski J, Nurse K, Lane BG (1995) The pseudouridine residues of ribosomal RNA. Biochem Cell Biol 73: 915–4

Ofengand J, Fournier MJ (1998) The pseudouridine residues of rRNA: number, location, biosynthesis, and function. In: Grosjean H, Benne R (eds) Modification and editing of RNA. ASM Press, Washington, pp 229–253

Ofengand J, Malhotra A, Remme J, Gutgsell NS, Del Campo M, Jean-Charles S, Peil L, Kaya Y (2001) Pseudouridines and pseudouridine synthases of the ribosome. Cold Spring Harb Symp Quant Biol 66: 147–59

Ogle JM, Murphy FV, Tarry MJ, Ramakrishnan V (2002) Selection of tRNA by the ribosome requires a transition from an open to a closed form. Cell 111: 721–732

Okamoto S, Tamaru A, Nakajima C, Nishimura K, Tanaka Y, Tokuyama S, Suzuki Y, Ochi K (2007) Loss of a conserved 7-methylguanosine modification in 16S rRNA confers low-level streptomycin resistance in bacteria. Mol Microbiol 63: 1096–1106

Pfennig PL, Flower AM (2001) BipA is required for growth of *Escherichia coli* K12 at low temperature. Mol Genet Genomics 266: 313–317

Poehlsgaard J, Douthwaite S (2005) The bacterial ribosome as a target for antibiotics. Nat Rev Microbiol 3: 870–881

Poldermans B, Roza L, Van Knippenberg PH (1979) Studies on the function of two adjacent N⁶,N⁶-dimethyladenosines near the 3' end of 16S ribosomal RNA of *Escherichia coli*. III. Purification and properties of the methylating enzyme and methylase-30S interactions. J Biol Chem 254: 9094–9100

Powers T, Changchien LM, Craven GR, Noller HF (1988) Probing the assembly of the 3' major domain of 16 S ribosomal RNA. Quaternary interactions involving ribosomal proteins S7, S9 and S19. J Mol Biol 200: 309–319

Powers T, Daubresse G, Noller HF (1993) Dynamics of in vitro assembly of 16 S rRNA into 30 S ribosomal subunits. J Mol Biol 232: 362–74

Powers T, Noller HF (1995) Hydroxyl radical footprinting of ribosomal proteins on 16S rRNA. RNA 1: 194–209

Purta E, Kaminska KH, Kasprzak JM, Bujnicki JM, Douthwaite S (2008a) YbeA is the m³Ψ methyltransferase RlmH that targets nucleotide 1915 in 23S rRNA. RNA 14: 2234–2244

Purta E, O'Connor M, Bujnicki J, Douthwaite S (2008b) *YccW* is the m⁵C methyltransferase specific for 23S rRNA nucleotide 1962. J Mol Biol 383: 641–651

Purta E, O'Connor M, Bujnicki JM, Douthwaite S (2009) YgdE is the 2'-O-ribose methyltransferase RlmM specific for nucleotide C2498 in bacterial 23S rRNA. Mol Microbiol 72: 1147–1158

Ramagopal S (1992) Are eukaryotic ribosomes heterogeneous? Affirmations on the horizon. Biochem Cell Biol 70: 269–272

Raychaudhuri S, Conrad J, Hall BG, Ofengand J (1998) A pseudouridine synthase required for the formation of two universally conserved pseudouridines in ribosomal RNA is essential for normal growth of Escherichia coli. RNA 4: 1407–1417

Richmond CS, Glasner JD, Mau R, Jin H, Blattner FR (1999) Genome wide expression profiling in Escherichia coli K-12. Nucleic Acids Res 27: 3821–3835

Sacerdot C, Fayat G, Dessen P, Springer M, Plumbridge JA, Grunberg-Manago M, Blanquet S (1982) Sequence of a 1.26-kb DNA fragment containing the structural gene for E. coli initiation factor IF3: presence of an AUU initiator codon. EMBO J 1: 311–315

Sacerdot C, Chiaruttini C, Engst K, Graffe M, Milet M, Mathy N, Dondon J, Springer M (1996) The role of the AUU initiation codon in the negative feedback regulation of the gene for translation initiation factor IF3 in Escherichia coli. Mol Microbiol 21: 331–346

Samaha RR, Green R, Noller HF (1995) A base pair between tRNA and 23S rRNA in the peptidyl transferase centre of the ribosome. Nature 377: 309–14

Saraiya AA, Lamichhane TN, Chow CS, SantaLucia J Jr, Cunningham PR (2008) Identification and role of functionally important motifs in the 970 loop of Escherichia coli 16S ribosomal RNA. J Mol Biol 376: 645–657

Schluenzen F, Tocilj A, Zarivach R, Harms J, Gluehmann M, Janell D, Bashan A, Bartels H, Agmon I, Franceschi F, Yonath A (2000) Structure of functionally activated small ribosomal subunit at 3.3 angstroms resolution. Cell 102: 615–623

Schmeing TM, Voorhees RM, Kelley AC, Gao YG, Murphy FV 4 th, Weir JR, Ramakrishnan V (2009) The crystal structure of the ribosome bound to EF-Tu and aminoacyl-tRNA. Science 326: 688–694

Schuwirth BS, Borovinskaya MA, Hau CW, Zhang W, Vila-Sanjurjo A, Holton JM, Cate JH (2005) Structures of the bacterial ribosome at 3.5 Å resolution. Science 310: 827–834

Senior BW, Holland IB (1971) Effect of colicin E3 upon the 30S ribosomal subunit of Escherichia coli. Proc Natl Acad Sci USA 68: 959–963

Sergiev PV, Bogdanov AA, Dontsova OA (2007) Ribosomal RNA guanine-(N2)-methyltransferases and their targets. Nucleic Acids Res 35: 2295–2301

Sergiev PV, Lesnyak DV, Bogdanov AA, Dontsova OA (2006) Identification of Escherichia coli m(2)G methyltransferases: II. The ygjO gene encodes a methyltransferase specific for G1835 of the 23 S rRNA. J Mol Biol 364: 26–31

Sergiev PV, Serebryakova MV, Bogdanov AA, Dontsova OA (2008) The *ybiN* gene of Escherichia coli encodes adenine-N⁶ methyltransferase specific for modification of A1618 of 23S ribosomal RNA, a methylated residue located close to the ribosomal exit tunnel. J Mol Biol 375: 291–300

Song X, Nazar RN (2002) Modification of rRNA as a 'quality control mechanism' in ribosome biogenesis. FEBS Lett 523: 182–186

Stern S, Wilson RC, Noller HF (1986) Localization of the binding site for protein S4 on 16 S ribosomal RNA by chemical and enzymatic probing and primer extension. J Mol Biol 192: 101–110

Sykes MT, Williamson JR (2009) A complex assembly landscape for the 30S ribosomal subunit. Annu Rev Biophys 38: 197–215

Tan J, Jakob U, Bardwell JC (2002) Overexpression of two different GTPases rescues a null mutation in a heat-induced rRNA methyltransferase. J Bacteriol 184: 2692–2698

Toh SM, Mankin AS (2008) An indigenous posttranscriptional modification in the ribosomal peptidyl transferase center confers resistance to an array of protein synthesis inhibitors. J Mol Biol 380: 593–597

Toh S-M, Xiong L, Bae T, Mankin AS (2008) The methyltransferase YfgB/RlmN is responsible for modification of adenosine 2503 in 23S rRNA. RNA 14: 98–106

Treede I, Jakobsen L, Kirpekar F, Vester B, Weitnauer G, Bechthold A, Douthwaite S (2003) The avilamycin resistance determinants AviRa and AviRb methylate 23S rRNA at the guanosine 2535 base and the uridine 2479 ribose. Mol Microbiol 49: 309–318

Tscherne JS, Nurse K, Popienick P, Michel H, Sochacki M, Ofengand J (1999a) Purification, cloning, and characterization of the 16S RNA m⁵C967 methyltransferase from *Escherichia coli*. Biochemistry 38: 1884–1892

Tscherne JS, Nurse K, Popienick P, Ofengand J (1999b) Purification, cloning, and characterization of the 16S RNA m2G1207 methyltransferase from Escherichia coli. J Biol Chem 274: 924–929

Vaidyanathan PP, Deutscher MP, Malhotra A (2007) RluD, a highly conserved pseudouridine synthase, modifies 50S subunits more specifically and efficiently than free 23S rRNA. RNA 13: 1868–1876

Van Duin J, Wijnands R (1981) The function of ribosomal protein S21 in protein synthesis. Eur J Biochem 118: 615–619

Weitzmann C, Tumminia SJ, Boublik M, Ofengand J (1991) A paradigm for local conformational control of function in the ribosome: binding of ribosomal protein S19 to Escherichia coli 16S rRNA in the presence of S7 is required for methylation of m²G966 and blocks methylation of m⁵C967 by their respective methyltransferases. Nucleic Acids Res 19: 7089–7095

Wimberly BT, Brodersen DE, Clemons WM Jr, Morgan-Warren RJ, Carter AP, Vonrhein C, Hartsch T, Ramakrishnan V (2000) Structure of the 30S ribosomal subunit. Nature 407: 327–339

Wrzesinski J, Bakin A, Nurse K, Lane BG, Ofengand J (1995a) Purification, cloning, and properties of the 16S RNA pseudouridine 516 synthase from Escherichia coli. Biochemistry 34: 8904–8913

Wrzesinski J, Nurse K, Bakin A, Lane BG, Ofengand J (1995b) A dual-specificity pseudouridine synthase: an *Escherichia coli* synthase purified and cloned on the basis of its specificity for Ψ746 in 23S RNA is also specific for Ψ32 in tRNA[phe]. RNA 1: 437–448

Xu Z, O'Farrell HC, Rife JP, Culver GM (2008) A conserved rRNA methyltransferase regulates ribosome biogenesis. Nat Struct Mol Biol 15: 534–536

Yarian CS, Basti MM, Cain RJ, Ansari G, Guenther RH, Sochacka E, Czerwinska G, Malkiewicz A, Agris PF (1999) Structural and functional roles of the N1- and N3-protons of psi at tRNA's position 39. Nucleic Acids Res 27: 3543–3549

Yusupov MM, Yusupova GZ, Baucom A, Lieberman K, Earnest TN, Cate JH, Noller HF (2001) Crystal structure of the ribosome at 5.5 Å resolution. Science 292: 883–896

Yusupova G, Jenner L, Rees B, Moras D, Yusupov M (2006) Structural basis for messenger RNA movement on the ribosome. Nature 444: 391–394

Zalacain M, Cundliffe E (1989) Methylation of 23S rRNA caused by tlrA (ermSF), a tylosin resistance determinant from Streptomyces fradiae. J Bacteriol 171: 4254–4260

Zhang Y, Zhang J, Hoeflich KP, Ikura M, Qing G, InouyeM (2003) MazF cleaves cellular mRNAs specifically at ACA to block protein synthesis in Escherichia coli. Mol Cell 12: 913–923

Section II Recruiting the ribosome for translation

Insights into translation initiation and termination complexes and into the polysome architecture

10

Angelita Simonetti, Stefano Marzi, Alexander G. Myasnikov,
Jean-François Ménétret and Bruno P. Klaholz

1. Structure and function of bacterial translation initiation complexes

Translation initiation is the most strongly regulated phase of protein synthesis during which the synthesis of a given protein is decided on. Initiation is the least conserved step of translation, since bacteria, archaea and eukarya have distinct and very different ways to initiate translation, and many different *trans*-acting factors are involved in the process. In bacteria, translation initiation comprises the consecutive formation of three major intermediary initiation complexes that are assembled *via* a multi-step process and that differ in composition and in conformation. At the end of the initiation process, an active 70S ribosomal initiation complex (70S IC) has formed which can enter peptide bond formation.

The process starts with the assembly of the "30S pre-initiation complex" (30S PIC) in which the start codon of the mRNA (usually AUG) does not yet interact with the anticodon of the initiator tRNA in the peptidyl (P) site. In order to form an active "30S initiation complex" (30S IC), a conformational change of the 30S PIC is required that leads to the accommodation of the mRNA on the 30S subunit and the formation of specific interactions between the anticodon of fMet-tRNA^{fMet} and the start codon on the mRNA. The joining of the large ribosomal subunit (50S) to the 30S IC leads to the formation of a third initiation complex, the "70S initiation complex" (70S IC), which is ready to enter elongation. During this transition, the adjustment of fMet-tRNA^{fMet} in the P site and the release of all initiation factors (IFs) that are assisting the initiation process are coupled with the hydrolysis of GTP on IF2 (for reviews see Simonetti et al., 2009; Myasnikov et al., 2009). This latter step marks the irreversible transition from the initiation to the elongation phase (Figure 1).

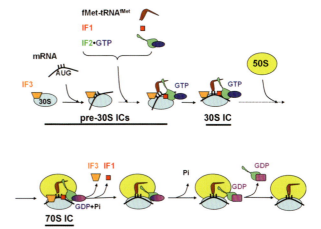

Fig. 1 Simplified scheme of bacterial protein synthesis initiation. IF3 (orange)-bound 30S subunit (blue) recruits mRNA, initiation factors IF1 (red) and IF2 (green), and fMet-tRNA^{fMet} (brown), leading to the formation of the 30S PIC and 30S IC. Upon interaction of the 30S IC with the 50S subunit (yellow), IF2 hydrolyzes GTP, the 70S IC forms, and initiation factors IF3 and IF1 are released. After IF2·GDP release, the 70S IC is ready for entering elongation. (Figure adapted from Myasnikov et al. 2009 with permission from Elsevier)

Cryo-electron microscopy (cryo-EM) and X-ray crystallography have provided structural and thereby mechanistic details of several aspects of protein synthesis on the ribosome, but for many years information on the initiation step was lagging behind. One of the reasons is probably that the assembly of the translation machinery during the initiation is a highly dynamic process. Thus, the binding of IFs, mRNA, fMet-tRNA^{fMet}, and the 50S subunit as well as GTP hydrolysis by IF2 (the largest of the initiation factors) alters the structure of the different 30S complexes. It was therefore particularly important to determine the structure of initiation intermediates and describe the landscape of the translation initiation process in time and space. Recently, the structures of three types of complexes have been investigated by single particle cryo-EM, providing snapshots of intermediates during the bacterial translation initiation phase: the 30S PIC with bound mRNA, the 30S IC with mRNA, initiator tRNA and two of the three initiation factors (IF1 and IF2), and the 70S IC of the assembled ribosome containing IF2 in presence of GDP or a non-hydrolyzable GTP-analogue.

The structure of a ribosome-bound auto-regulatory mRNA blocked on the 30S PIC has revealed the mechanism by which the ribosome recruits mRNAs and can be blocked transiently to regulate translation (Marzi et al., 2007). The study of the 30S and 70S initiation complexes has shown the important role that IF2 plays during bacterial translation initiation, illustrated by the localization and conformation of IF2 in different functional states (Allen et al., 2005; Myasnikov et al., 2005; Simonetti et al., 2008). These studies have revealed the mechanism by which the initiator tRNA is stabilized by IF2 in a functional conformation that favors the association of the ribosomal subunits (Simonetti et al., 2008), and they have disclosed the conformational transition of the factor and of the ribosome taking place upon GTP hydrolysis (Myasnikov et al., 2005).

1.1. Pre-initiation: mRNA binding and adaptation onto the 30S subunit

One of the most critical phases of the initiation process is the binding of the mRNA to the 30S subunit. This process is the target of many different mechanisms of translation regulation (for a review see Marzi et al., 2008) that ensure a rapid response to a variety of stimuli allowing a fast and transient adaptation. Moreover, as the translation of a specific mRNA directly influences its degradation rate, an irreversible response can also be achieved (Deana and Belasco, 2005).

The formation of the 30S IC can be viewed as a two-step process according to the "stand-by model" originally proposed by de Smit and van Duin (2003), and then confirmed experimentally by fast kinetics experiments (Studer and Joseph, 2006) and structural work (Marzi et al., 2007). The initial recognition of the mRNA by the 30S subunit leads to the formation of the inactive 30S PIC that is followed by the formation of the active 30S IC, when the mRNA is fully adapted into the mRNA channel and the A- and P-site codons

are exposed for the decoding. Both steps are regulated by structural elements present on the mRNA as well as by the binding to the mRNA of trans-acting factors (proteins or metabolites). A characteristic mechanism of translation inhibition, the so-called competition mechanism, is based on the sequestration of the Shine-Dalgarno (SD) sequence, a sequence of four to six nucleotides often present upstream of the start codon in bacterial mRNAs that binds to the 3' end of the 16S rRNA, thereby anchoring the mRNA to the 30S subunit. Alternatively, hindering the mRNA accommodation in the channel can result in transient translation inhibition through an entrapment mechanism.

While several crystal structures show the pathway of the mRNA in its channel (Yusupova et al., 2001; Yusupova et al., 2006), only recently it became possible to visualize, by cryo-EM, the mRNA in the 30S PIC (Marzi et al., 2007). That work shows that the platform of the 30S subunit accommodates the complete 5' UTR and the beginning of the translated region of the *rpsO* mRNA, which encodes for ribosomal protein S15. This mRNA region of 130 nucleotides is folded into a stable hairpin-loop-pseudoknot structure that is stabilized by the repressor protein S15. The binding of S15 to the pseudoknot stabilizes the 30S PIC, an otherwise rapidly resolved transient complex, thus allowing structural analysis. The mRNA in this conformation is located completely outside of the mRNA channel and, despite the presence of the initiator fMet-tRNA in the P site, there is no codon-anticodon interaction. Upon repressor release, the mRNA unfolds and adopts the classical path in the mRNA channel, leading to the formation of an active translation initiation complex.

The platform of the 30S subunit is well suited for the docking of structured mRNAs. Several ribosomal proteins located at the platform (S2, S7, S11, S18, and S21) are either in contact with, or in close proximity to, the folded mRNA (Figure 2). A systematic structure and sequence analysis revealed that conserved residues of these proteins form patches of positive charges on the surface of the platform close to the trapped mRNA. This nest of positive charges can allocate different folded mRNAs regardless of their sequence. When the mRNA is localized on the platform it may then be further stabilized using the SD sequence, which can form a double helix by binding to the anti-SD sequence of 16S rRNA just in the center of the platform.

Strikingly, structured 5' UTRs of several bacterial and eukaryotic mRNAs have been found to partially overlap with some of those conserved residues. This is the case of *Escherichia coli rpsO* (Marzi et al., 2007)

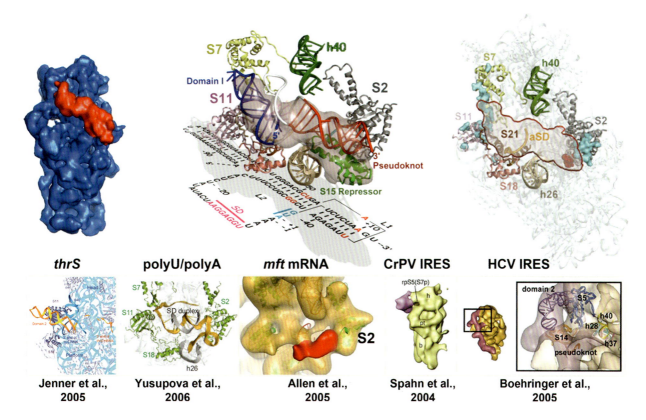

Fig. 2 Binding of folded mRNAs to the platform of the 30S subunit. Upper left panel: Global view of the cryo-EM structure of the 30S PIC at 10 Å resolution (Figure adapted from Marzi et al., 2007) with permission from Elsevier) visualizes the mRNA prior to the mRNA adjustment in the 30S mRNA channel; the mRNA (red) interacts with the platform of the 30S subunit (proteins S2, S7, S11, S18, and S21). Upper middle panel: Detailed interpretation of the cryo-EM structure after fitting atomic resolution structures into the electron densities (Figure adapted from Marzi et al. (2007) with permission from Elsevier). Upper right panel: Universally conserved residues of these proteins (highlighted in cyan) that are exposed on the surface of the 30S platform; these residues participate in the binding of different folded mRNA both in bacteria and eukaryotic ribosomes (Figure reproduced from Marzi et al., 2007; with permission from Elsevier). Bottom panels: Examples of folded mRNAs bound to the platform (Figures adapted from the following references with permission (Elsevier, Macmillan Publishers; AAAS): Jenner et al., 2005; Yusupova et al., 2006; Allen et al., 2005 (red, our coloring of density which is probably due to a short helix at the 5' end of the SD sequence of the mRNA); Spahn et al., 2004; Boehringer et al., 2005).

and *thrS* mRNAs (Jenner et al., 2005), and of the IRESs of the hepatitis C (HCV) virus (Spahn et al., 2001; Boehringer et al., 2005) and the cricket paralysis virus (CrPV) (Spahn et al., 2004). Rather unexpectedly, also a 5' poly(A)- or poly(U)-rich extension upstream the SD sequence can form stable stem-loop structures on the platform of the 30S subunit that are in contact with the ribosomal protein S2 (Yusupova et al., 2006). This suggests the existence of a common docking site for structured mRNAs on the 30S platform entered during pre-initiation, subsequently followed by the adaptation of the mRNA into the mRNA channel. The time delay between docking and adaptation reflects the stability of the mRNA structures that need to be eventually melted to promote the codon-anticodon interaction. Thus, ligands (proteins, metabolites, noncoding RNAs, etc.) that stabilize the folded state of the

mRNA can block the ribosome at the pre-initiation stage by preventing the initiator codon from reaching the decoding site inside the ribosome. Moreover, the platform is a wide and accessible open space giving the possibility for regulatory ligands to interact with the mRNA on the platform (such as the repressor protein S15 with *rpsO* mRNA). This underlines the importance of the pre-initiation mRNA binding site in timing and regulating translation. An interesting question to address is how mRNAs unfold while bound to the 30S PIC, especially when they can bind to the ribosome only in the folded state. One example of such mRNAs is the *rpsO* mRNA, which exposes its SD sequence in a large loop of a pseudoknot structure (while the mis-folded alternative three-hairpin-loop state does not bind to the ribosome; Philippe et al., 1993). It is likely that components of the platform

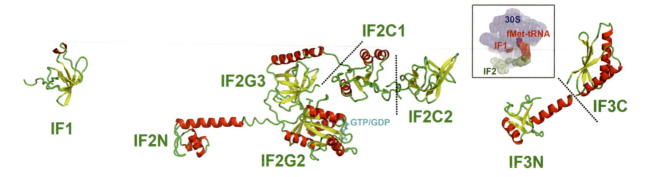

Fig. 3 Structures of bacterial translation initiation factors. IF1: *E. coli* IF1 NMR structure (Sette et al., 1997) (PDB ID 1AH9.pdb). IF2: The model of *T. thermophilus* IF2·GDP has been obtained using the 12 Å-resolution cryo-EM map of the 30S IC (Simonetti et al., 2008); G2 and G3 domains are from the recently determined crystal structure of *T. thermophilus* IF2 (A. Simonetti, S. Marzi, A. Urzhumtsev, L. Jenner, M. Yusupov, C. O. Gualerzi, B. P. Klaholz, unpublished data), the C1 and C2 domains from the NMR structure of *B. stearothermophilus* IF2 (1Z9B. pdb) (Wienk et al., 2005) and (1D1N. pdb) (Meunier et al., 2000), respectively. IF3: The structure is modeled from the crystal structures of the isolated N- and C-terminal domains (Biou et al., 1995) (1TIF. pdb and 1TIG. pdb). The factors are represented oriented as they appear on the 30S subunit as visualized in the inset (IF1 and IF2, Simonetti et al., 2008; IF3, Fabbretti et al., 2007, and McCutcheon et al., 1999).

binding site are involved in the unfolding process. One potential candidate on the platform binding site is the ribosomal protein S1, as it possesses helix-unwinding properties (Kolb et al., 1977). From its binding site on the platform, the mRNA will then adapt into the 30S mRNA channel through a process known as mRNA accommodation.

1.2. Translation initiation factors

Three initiation factors (IF1, IF2, and IF3; Figure 3) kinetically assist the formation of the specific interactions between the anticodon of fMet-tRNA^{fMet} and the start codon on the mRNA, once it has unfolded from the 30S PIC (Figures 2 and 4A) and entered the P site of the 30S subunit (Gualerzi and Pon, 1990), forming the 30S IC. These IFs have distinct, but coordinated functions that enhance the rate of the formation of the various initiation complexes and ensure translation accuracy in different ways.

IF3, which is composed of two domains that are connected through a flexible, lysine-rich linker (Fortier et al., 1994; Kycia et al., 1995; Moreas et al., 1997), maintains the two ribosomal subunits dissociated and thus provides a pool of free 30S subunits for initiation. IF3 also stimulates the P-site codon-anticodon interaction between fMet-tRNA and mRNA. Moreover, during the transition from the 30S IC to the 70S IC, IF3 acts as a fidelity factor that discriminates against incorrect complexes by favoring subunit dissociation (Grigoriadou et al., 2007) when incorrect complexes,

such as pseudo-initiation, non-canonical, or leaderless initiation complexes, are present (Petrelli et al., 2001; Grigoriadou et al., 2007; Milon et al., 2008)

The structures of the two individual domains of IF3 have been determined by X-ray crystallography and NMR (Figure 3) revealing a globular α/β fold and a two layer α/β sandwich for the N- and C-terminal domain, respectively (Fortier et al., 1994, Garcia et al., 1995). However, the precise localization of IF3 on the 30S subunit remains controversial, despite of a series of studies using immuno-EM (Stoffler et al., 1984), cryo-EM (McCucheon et al., 1999), chemical probing (Fabbretti et al., 2007), site-directed chemical probing (Dallas, 2001), or soaking 30S subunit crystals into IF3 solution (Pioletti et al., 2001).

IF1 promotes a more efficient binding of IF3 and IF2 to the 30S subunit, thus stimulating their activities, and it cooperates with the latter two factors to ensure the correct location of the initiator tRNA in the P-site. The crystal structure of the 30S·IF1 complex (Carter et al., 2001) showed IF1, the smallest of the three IFs, bound in a niche created by the ribosomal protein S12, the penultimate stem-loop (530 loop), and helix 44 of 16S RNA in the vicinity of the A site (Figure 4B). As illustrated by the solution structure of *E. coli* IF1 (Sette 1997; Figure 3), this factor is characterized by a rigid five-stranded β-barrel flanked by flexible extremities and comprises an OB-fold motif that is present in several RNA-binding proteins, including ribosomal proteins S1, S17, L2, and L17.

The largest among the initiation factors, IF2 (63.1 kDa in *Thermus thermophilus*), belongs to the fam-

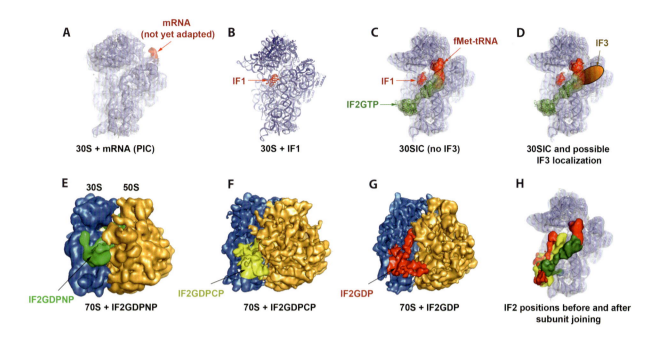

Fig. 4 Structures of translation initiation complexes. (A) Cryo-EM structure of the pre-initiation complex obtained by trapping the folded mRNA on the 30S platform (Marzi et al., 2007). (B) Crystal structure of 30S subunit in complex with IF1 obtained by soaking IF1 directly into the 30S subunits crystals (Carter et al., 2001). (C) Cryo-EM structure of the 30S IC with an mRNA molecule adapted into the decoding channel, IF1, IF2·GTP, and fMet-tRNA^{fMet} (Simonetti et al., 2008). (D) Tentative localization of IF3 on the 30S IC (Simonetti et al., 2009; adapted from Myasnikov et al., 2009 with permission from Elsevier). (E) Cryo-EM structure of a 70S IC with IF2·GDPNP bound (Allen et al., 2005). (F) Cryo-EM structure of a 70S IC with IF2·GDPCP (Myasnikov et al., 2005). (G) Cryo-EM structure of a 70S IC with IF2·GDP (Myasnikov et al., 2005). (H) IF2 positions before and after joining of the ribosomal subunits representing the states before (IF2·GTP·30S, green; IF2·GDPCP·70S, yellow) and after (IF2·GDP·70S, red) GTP hydrolysis (Myasnikov et al., 2005; Simonetti et al., 2008).

ily of GTP-binding proteins (Bourne et al., 1991), together with the elongation factors EF-Tu and EF-G and the termination factor RF3. The main function of IF2 is the stimulation of the fMet-tRNA^{fMet} binding to the ribosome. The specific interaction of IF2 with the initiator tRNA represents the most important contact occurring during translation initiation which results in an acceleration of codon-anticodon base pairing with the initiation triplet of the mRNA at the P site. In fact, IF2 recognizes specifically the formyl-methionine bound to the acceptor stem of the initiator tRNA (Guenneugues et al., 2000), discriminating against all other aminoacylated tRNAs and determining the accuracy of selecting the correct initiation site of both leadered (containing a 5' UTR with a SD sequence) and leaderless mRNAs (Grill et al., 2000). Moreover, IF2 promotes the formation of the 70S IC, thereby driving the formation of the first peptide bond (initiation dipeptide), which marks the transition from the initiation to the elongation phase (La Teana et al., 1996; Grigoriadou et al., 2007b). Comprising five domains, IF2 presents an N terminal domain (subdivided to N1, N2, and G1 subdomains) that varies in sequence and length among bacteria, followed by the G domain (G2) that contains the conserved GTP-binding elements, the G3 domain, that is homologous to domain II of EF-Tu and EF-G, and, finally, the conserved C-terminal region of IF2 that comprises two sub-domains, C1 and C2.

Compared to the other initiation factors for which NMR and crystal structures are available, the available knowledge of the structure of IF2 is scarce. In fact, only the structures of the two C-terminal domains (Meunier et al., 2000; Wienk et al., 2005) and of a small fragment of the N-terminal domain (Laursen et al., 2003) of IF2 have been solved by NMR. While IF2C1 is disk-shaped due to a central beta sheet that is surrounded by three α-helices, the C2 module that contains all the molecular determinants necessary for the recognition of the fMet-tRNA^{fMet} has a β-barrel structure. Finally, only three α-helices compose the 57 amino-acid fragment at the beginning of the N domain. To the present day, the knowledge of the spatial organization of the domains of IF2 is limited to homology models derived from the crystal structure of the archaeal factor eIF5B from *Methanobacterium thermoautotrophicum* (Roll-

Mecak et al., 2000). Bacterial and archaeal IF2/eIF5B proteins have slightly different functional characteristics (Simonetti et al., 2009), and the archaeal factor lacks the N domain.

The lack of high-resolution structural information on IF2 and its dynamics during the transition from the 30S IC to the 70S IC have so far limited the exhaustive interpretation of these structures and the understanding of the molecular mechanism of action of IF2. Indeed, the X-ray structures of isolated components comprising IF2 can be used in combination with Molecular Dynamics Flexible Fitting (MDFF) to achieve a quasi-atomic interpretation of the cryo-EM reconstructions of intermediates of the various phases of initiation, which is important for understanding the mechanism of initiation.

Recently, we have obtained crystals of *T. thermophilus* IF2 that diffract to 1.9 Å resolutions. A first attempt to determine the structure by molecular replacement using the available IF2/eIF5B crystal structure failed, suggesting that the domains of bacteria IF2 are arranged differently, consistent with the medium degree of sequence conservation between the two proteins (38 % sequence identity). However, using multi-wavelength anomalous dispersion (MAD) phasing, we could determine the atomic structure of *T. thermophilus* IF2 comprising the first 358 residues that constitute three subdomains (N/G1, G2/G3, and C1). Indeed, while the structure lacks the tRNA-interacting C-terminal domain (C2), it provides important insights into the domains involved in the interaction with the ribosome and in GTP hydrolysis, and it reveals the fold of the N-terminal domain which is specific to IF2 and not present in other translational GTPases, such as elongation factors Tu and G.

The comparison of the structures of bacterial IF2 and archaeal IF2/eIF5B shows that the G2 domain and G3 domain are smaller and more compact in the bacterial IF2. The structure of bacterial IF2 also provides first insights into the nucleotide binding pocket which is formed by four conserved sequence patches, specific residues of the G1 (P loop), G2, G3, and G4 loops which provide hydrogen bonds and hydrophobic contacts for the recognition and binding of the nucleotide. In combination with the NMR structure of the C-terminal domain of bacterial IF2, the present IF2 crystal structure now allows to build a model for the full-length bacterial IF2 (Figure 3) that is much more precise than homology models based on the IF2/eIF5B crystal structure and that can also be used for a detailed interpretation of the cryo-EM maps of 30S IC and 70S IC.

1.3. The structure of translation initiation complexes

In order to understand the mechanism of translation initiation, we have determined the cryo-EM structure of the 30S IC containing the *T. thermophilus* 30S subunit, fMet-tRNA[fMet], *T. thermophilus* initiation factors IF1 and IF2·GTP and an mRNA with 27 nucleotides (mk27) that contains an extended SD sequence of 8 nucleotides for enhanced base-pairing with the 3' end of the 16S RNA (typical for *T. thermophilus* mRNA) (Figure 4C). The resulting cryo-EM structure describes the architecture of the 30S IC and reveals how the initiator tRNA is stabilized by IF2, thus providing molecular details of the interaction between the 30S and its partners at the initiation step (Simonetti et al., 2008).

Despite the optimization of the buffer conditions to form a stable 30S IC, the complex still proved to be of limited homogeneity. Nevertheless, the combination of cryo-EM and three-dimensional (3D) reconstruction with a new method for particle separation based on three-dimensional statistical analysis allowed to resolve multiple states within a heterogeneous sample. The approach is based on a random selection of small subsets of particle images for 3D reconstruction, followed by 3D multi-variate statistical analysis and classification (3D-SC, 3D re-sampling and classification; Simonetti et al., 2008). The method does not rely on any external references and allows determining many structural states from a single sample. Applied to the heterogenous 30S IC data set, this 3D-SC procedure yielded five sets of 3D structures at resolutions around 10–12 Å that differed in composition or conformation. While IF1 and the fMet-tRNA[fMet] are visible in all sets, IF2 is absent in two of them. A clear density for the helix formed by the SD/anti-SD base pairs is seen in all complexes except in two sets. The different states of 30S·tRNA·IF1 complexes in the presence and absence of IF2 or mRNA are observed in a single sample in which all the complexes are in equilibrium with each other. In conclusion, the new particle-separation approach, applied here to study the structure of the 30S IC, has proven to be a powerful tool for the characterization of multiple states within a conformationally heterogeneous sample and allowed to obtain more information regarding the dynamics of a multi-component macromolecular assembly.

To analyze the molecular interaction within the 30S IC, structures of its individual components were fitted into the experimental map of set 1 that contains the fully occupied and best resolved complex. In this structure, electron density that fits with the size of IF1 is visible close to the decoding site, consistent with the 30S·IF1 crystal structure (Carter et al., 2001) (Figures 4B and 4C). IF2, in its GTP-bound state, is seen to be organized in two main modules corresponding in size to domains G2/G3 and C1/C2, while density for the N-terminal domain is not visible in the map (Figure 4C). Notably, the topology of the IF2 domains suggests that, when IF2 is bound to the 30S subunit, domain C1 appears to be very close to domain C2, distinguishing bacterial IF2 from the archaeal one, where the two domains are separated by a very long (40 Å) rigid α-helix (Roll-Mecak et al., 2000).

Furthermore, the 30S IC structure reveals how the initiator tRNA is stabilized by IF2 in a functional conformation that favors the association of the ribosomal subunits. Two contact points exist between the IF2·fMet-tRNA^{fMet} subcomplex and the 30S subunit on the surface facing the 50S subunit in the 70S ribosome. The first contact is between domain G3 of IF2 and helices h5 and h14 of 16S rRNA. The contact consists in the interaction of the decoding stem of the fMet-tRNA^{fMet} with the neck of the 30S subunit. The decoding stem bends towards the initiation codon of the mRNA, thereby promoting the codon-anticodon interaction. This way, IF2 appears to stabilize fMet-tRNA^{fMet} on the 30S subunit, even in absence of the 50S subunit.

Another important contact for the stabilization of the 30S IC is provided by the interaction between the C-terminal domain of IF2 and the acceptor end of the initiator tRNA. This interaction induces a kink of the 3' end of the acceptor stem that expands the interface between IF2 and fMet-tRNA^{fMet}. Altogether, these interactions affect the position of the fMet-tRNA^{fMet} that has its acceptor stem lying between the P and E sites (30S P/I state, slightly distinct from the P/I in the 70S IC containing IF1 and IF2 in that the tRNA moves out of the "translocation plane"; Allen et al., 2005). Since in the 70S IC the fMet-tRNA^{fMet} is accommodated in the P/P state, it is possible that the tRNA undergoes a conformational change upon subunit association. The 30S IC structure also provides insights into the mechanism of 70S IC assembly. In fact, the correct positioning of IF2 and the fMet-tRNA^{fMet} in the 30S IC plays a strategic role for the successful docking of this complex to the 50S subunit. While IF2 holds fMet-tRNA^{fMet} in a position ready for its insertion into the 50S subunit, it also increases (by about 25%) the surface of the 30S IC available for interacting with the 50S subunit. Moreover, the position of fMet-tRNA^{fMet} would allow

helix H69 of 23S RNA to insert below the D loop of the tRNA and form one of the key inter-subunit bridges (B2a) with the top of helix h44 (16S RNA). Remarkably, the largest part of IF2 is complementary in shape to the 50S surface. Finally, the 30S IC structure rationalizes the rapid activation of GTP hydrolysis triggered on 30S IC joining with the 50S subunit (Tomsic et al., 2000) by showing that the GTP-binding domain of IF2 would directly face the GTPase-associated center of the 50S subunit. The localization of IF3 on the 30S subunit is still a controversial issue, because no structure of a whole 30S IC (with the three IFs, fMet-tRNA^{fMet}, and mRNA) has been obtained yet (Figure 4D; for more details see Myasnikov et al., 2009)

Cryo-EM allowed to localize IF2 also in the 70S IC (Allen et al., 2005; Myasnikov et al., 2005). The cryo-EM reconstruction of the 70S IC by Allen et al. (2005) has been obtained in the presence of all initiation factors; nevertheless, the obtained 3D structure shows density for IF2-GDPNP, but no clear density for IF3 (Figure 4E). The complexes used for the two cryo-EM structures obtained by Myasnikov et al. (2005) were assembled by incubating *T. thermophilus* 70S ribosomes with a 27 nucleotide-long mRNA, initiator fMet-tRNA^{fMet}, IF2, and either a non-hydrolyzable GTP-analogue (GDPCP) or GDP. These structures showed IF2 in two different conformations representing the latest steps of the translation initiation (Myasnikov et al., 2005). In both structures, IF2 is positioned in the inter-subunit cleft of the 70S ribosome and is in contact with both subunits. Interestingly, these structures show no interaction between IF2 and fMet-tRNA^{fMet}, which is located in the P site ready to enter the first round of elongation. Furthermore, the cryo-EM structures show that the transition from the GTP-bound (GDPCP) to the GDP-bound state of IF2 involves substantial conformational changes of both IF2 and the ribosome. While in the GTP-bound state, IF2 interacts mostly with the 30S subunit, extending its C-terminal domains towards the elbow of the initiator tRNA in the P site, in the GDP-bound state IF2 moves away and assumes a "ready to leave" conformation (Figure 4F and 4G).

The combined structural analysis of the 30S IC and 70S IC has contributed to the comprehension of the mechanism of translation initiation in bacteria, providing molecular and quasi-atomic details regarding the two intermediate ribosomal complexes that characterize initiation phase of translation. In combination with kinetic data, this allows a four-dimensional analysis of the whole process. Indeed, the comparison between the three structures (Myasnikov et al., 2005; Simonetti et al., 2008) in combination with kinetic data (Tomsic et al., 2000; Grigoriadou et al., 2007b) illustrates that IF2 adopts at least three different conformations (Figure 4H) during the initiation process: (i) the GTP-bound state on the 30S subunit, in which it interacts through its C-terminal domain with the initiator tRNA in the P/I state; (ii) the GDP·Pi-bound state, in which IF2 releases the tRNA to move into the P site; (iii) the GDP-bound state, which forms after the release of inorganic phosphate. In the transition from state (ii) to state (iii), the interaction surface between IF2 and both the tRNA and the ribosome is reduced, thereby enhancing the dissociation of IF2.

2. The three-dimensional architecture of bacterial and eukaryotic polyribosomes

During protein synthesis, several ribosomes may bind to an mRNA molecule and form clusters called polyribosomes or polysomes. Polysomes harbor mostly translating ribosomes in the elongation phase, but also some in the initiation and termination phases. In eukaryotes, the initiation is thought to proceed through a scanning mechanism: the small (40S) ribosomal subunit is recruited to the capped 5' end of the mRNA from which it scans along the 5' UTR in search of a start codon where initiation takes place. When several ribosomes are bound to an mRNA molecule, translation takes place on adjacent ribosomal particles simultaneously. Early on, it was observed that ribosomes form double rows, as for example visualized for eukaryotic polysomes in the mid 1960's (Warner et al., 1962; Kuff et al., 1966; Dallner et al., 1966). In the cellular context, these double rows may assume different morphological arrangements, such as hairpins or spirals, in the free and membrane-bound states (Christensen et al., 1999). While pro- and eukaryotic polysomes share the general feature of double rows, until recently little was known about the three-dimensional organisation and topology of ribosomes within polysomes, i. e., the orientation of the subunits, the mode of ribosome interactions within the polysome chain and between the chains, the mRNA path, or the position of the 5' and 3' ends of the mRNA. It was also unclear whether the 3D architecture of pro- and eukaryotic polysomes and the mRNA path therein are similar.

These questions have been recently addressed in two studies using cryo-electron tomography (CET). CET is a powerful technique in which image series are acquired while the sample is being tilted incrementally. As for single-particle cryo-EM, the sample is preserved under native conditions in the hydrated state, i. e. without artifacts introduced by staining or spreading. However, rather than averaging many different particles (observed under different viewing angles) into a 3D reconstruction, a single object is analyzed and reconstructed in 3D from tilt series images, which allows to investigate single objects with non-uniform structures that cannot be averaged across different particles. Since in CET the averaging is limited to a relatively small amount of tilt images due to the high sensitivity of biological specimens against electron irradiation, the resolution is not yet as good as in single-particle cryo-EM, where very large numbers of particles can be averaged and classified. However, CET allows to address the general architecture of large molecular assemblies such as polyribosomes.

Using CET, two groups have recently determined the 3D structures of polysomes in prokaryotes (Brandt et al., 2009) and in eukaryotes (A. G. Myasnikov, Z. A. Afonina, J-F. Ménétret, V. A. Shirokov, A. S. Spirin and B. P. Klaholz, unpublished data), in both cases making use of cell-free translation systems. For both pro- and eukaryotic polysomes, arrangements were observed in which neighboring ribosomes exhibit preferred orientations within the double rows, with the small ribosomal subunit oriented inwards the double row and the polypeptide exit sites of the large ribosomal subunits facing the outside of the assembly (Figure 5). However, the tomographic reconstructions also reveal some differences. The prokaryotic polysomes show a unidirectional (parallel) orientation of the ribosomes with a staggered (zig-zag) or pseudo-helical organization along the mRNA path, which alternates between the rows. In order to comply with this parallel arrangement some of the 70S ribosomes are flipped upside-down within a row, and the 30S subunit head is oriented either up- or downwards, depending on the row while pointing into one common direction (Brandt et al., 2009). In contrast, eukaryotic ribosomes appear to adopt both anti-parallel and parallel organizations of the two rows, forming a double row layer (Figure 5), and the 40S subunits' heads are oriented towards one face of the layer. In some cases the double rows form a higher-order structure where ribosomal particles are arranged as a supramolecular helix. In the case of an anti-parallel organization, the 5' and 3' ends of the mRNA would be close to each other (in keeping with the known circularization of the mRNA in eukaryotes), while in a parallel organization they would not. Intermolecular polysome circulariz-

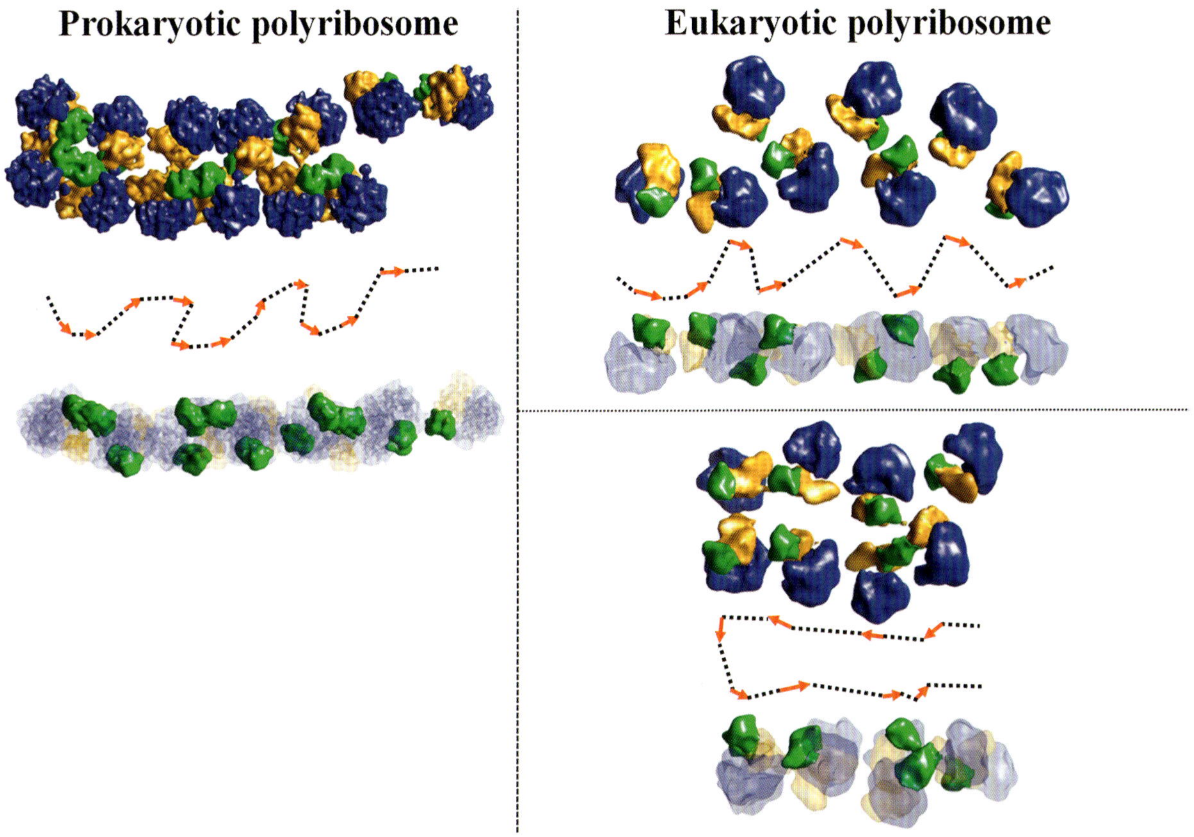

Prokaryotic polyribosome   **Eukaryotic polyribosome**

Fig. 5 Comparison of the structural organization of eukaryotic and prokaryotic polysomes. Prokaryotic (left; Brandt et al., 2009) and eukaryotic polysomes (right; A. G. Myasnikov, Z. A. Afonina, J-F. Ménétret, V. A. Shirokov, A. S. Spirin and B. P. Klaholz, unpublished data; parallel organisation upper panel, antiparallel - lower panel). For details, see text.

ation may, in turn, entropically favor the intramolecular mRNA 5'-3' looping and thus re-initiation (Alekhina et al., 2007) possibly through the known interaction of the 5'-bound initiation factor 4E (eIF4E) with the 3'-bound poly(A)-binding protein (PABP) (Tarun et al., 1995). Interestingly, removal or replacement of 5' and 3' UTRs affects the kinetics of translation initiation, but does not prevent the formation of the double-row polysomes during translation (Kopeina et al., 2008). In contrast, in bacteria the observed unidirectional, non-circular but rather parallel organization of ribosomes in polysomes is in agreement with transcription and translation being coupled and therefore directionally linked. In eukaryotes, these processes are compartmentally separated and do not require to be directional.

At present, a number of questions remain to be addressed concerning polysomes: (i) Can the mRNA be visualized directly (it is not currently directly visible because of the limited resolution in the tomographic reconstructions and the mRNA path is derived indirectly from the orientation of the individual ribosomes)? (ii) What is the mechanism of polysome formation? (iii)

What is the molecular basis for the processes of initiation, scanning, termination, or re-initiation, (iv) how do polysomes look like when bound to membrane structures such as the endoplasmic reticulum? These and other aspects will provide a detailed view into the mechanism of protein synthesis *in vivo*.

3. Structure of bacterial translation termination complexes

3.1. Bacterial termination complexes with RF1 and RF2

The basis for molecular recognition between the stop codon and the release factors (RF) has remained unknown for a long time. The RFs not only recognize the stop codons (UAA, UAG or UGA), but also catalyze the hydrolysis of the bond between the nascent peptide and the tRNA in the P site. The RFs which recognize the stop codons in the A site and induce release of the nascent peptide are called class-I RFs, while

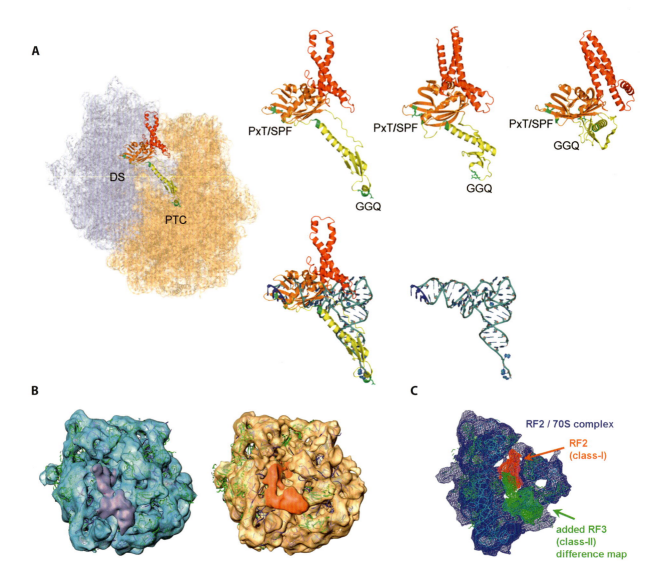

Fig. 6 Mechanism of translation termination in bacteria. (A) Structure of RF2 as observed by different approaches: by crystallography (first panel, more detailed view in the second upper panel), by cryo-EM of the 70S complex (third upper panel) and in the isolated state (fourth upper panel); second lower panel: overlay of an A-site tRNA and RF2 as seen in the crystal structure of the 70S complex; third lower panel, A-site tRNA. (B) Cryo-EM structure of the 70S/RF3 complex observed in two states as sorted out from the sample by local statistical analysis (Figure reproduced from Klaholz et al. (2004) with permission from Macmillan Publshers); the RF3 density is highlighted in magenta or orange. (C) Model of RF2 and RF3 bound to the 70S ribosome. The model is derived from a 70S/RF2 cryo-EM structure (Klaholz et al., 2003) and the RF3 part of a 70S/RF3 cryo-EM structure (Klaholz et al., 2004) suggesting that class-I and class-II RFs interact with each other on the ribosome. For clarity, only the front part of the ribosome is shown.

class-II RFs are GTPases that in bacteria accelerate the dissociation of the class-I RFs from the ribosome after peptide release. In bacteria, two RFs recognize the three stop codons with different, but overlapping specificity: RF1 recognizes UAA and UAG, while RF2 recognizes UAA and UGA (Capecchi 1967; Caskey et al., 1968; Scolnick et al., 1968). RF1 and RF2 contain a conserved GGQ amino-acid motif involved in peptidyl-tRNA hydrolysis (Frolova et al., 1999; Seit-Nebi et al., 2001; Mora et al., 2003; Shaw et al., 2007), while

distinct residues are important for recognition of different stop codons through a PxT (PAT or PVT) and an SPF motif in RF1 and RF2 respectively (Ito et al., 2000). However, the individual molecular mechanisms behind these processes and the functional connection between stop-codon recognition and peptidyl-tRNA hydrolysis remained to be addressed through structural approaches (Klaholz, 2011).

First structure-function insights were provided by the crystal structure of one of the release factors (*E. coli*

RF2, Vestergaard et al., 2001; and later also *E. coli* RF1, Shin et al., 2004, and *T. thermophilus* RF2, Zoldak et al., 2007), revealing the fold of the proteins and the location of functionally important, conserved residues. The GGQ motif was found to be in a loop of domain 3 (Figure 6), while the SPF motif is in a loop between two ß-strands of domain 2 which folds together with domain 4 to form the decoding domain. In principle, these two active sites of RF2 were expected to contact the decoding center (for the SPF motif) and the peptidyl transferase center (PTC, for the GGQ motif) respectively, but they were found in rather close proximity (23 Å) in the crystal structures.

The cryo-EM structure of the 70S/RF2 complex then showed that, when bound to the ribosome, RF2 has an extended, open conformation (Klaholz et al., 2003; Rawat et al., 2003). A small-angle X-ray diffraction study showed that isolated RF1 also has an open conformation in solution (Vestergaard et al., 2005), similar to the ribosome-bound state, also different from the isolated crystalline state. This indicates that binding of RF2 to the ribosome probably involves less conformational changes than the differences with the crystal structure of the isolated factor would suggest. For comparison, the eukaryotic counterpart of RF1/2, eRF1, shows an extended conformation both in solution and in the crystal (Song et al., 2000; Kononenko et al., 2004), but in complex with eRF3 it is more bent (Cheng et al., 2009). The structure of a eukaryotic ribosomal termination complex is unknown.

The cryo-EM structures of the *E. coli* 70S·RF2 complex with a UAA stop codon and peptidyl-tRNA in the P site (Klaholz et al., 2003; Rawat et al., 2003) and later of the 70S·RF1 complex (Rawat et al., 2006) revealed the location of the three domains of the factor when bound to the ribosome: (i) the three-helix bundle of domain 1 is close to the 30S subunit head and in the vicinity of protein L11 on the 50S subunit; (ii) domains 2 and 4 were found in the decoding site of the 30S subunit between ribosomal protein S12 and helices h18, h44 and H69, with the SPF motif next to the mRNA; and (iii) domain 3 was seen to extend through its long helix towards the PTC, next to the P-site tRNA carrying the nascent peptide. The RFs thus directly connect the decoding center with the PTC, and link stop-codon recognition with hydrolysis of the peptidyl-tRNA ester bond and release of the nascent peptide (Figure 6A). The binding site of the RFs on the ribosome suggested by the cryo-EM data is consistent with earlier cross-linking and site-directed hydroxyl radical probing data (Tate et al., 1990; Wilson et al., 2000).

The general location on the ribosome of the three RF domains and of the conserved motifs therein, as well as the concept of connecting decoding and peptidyl transferase centers revealed by cryo-EM, has been confirmed by the crystal structures of the *T. thermophilus* 70S·RF1 and 70S·RF2 complexes with deacylated tRNA at 5.9 and 6.7 Å resolution (Petry et al., 2005). The latter represents the first crystal structure of a ribosome complex with an RF and suggested that parts of the protein other than the tripeptide motif are involved in stop-codon recognition, but left unclear the fine details of stop-codon recognition and ester bond hydrolysis. The high-resolution crystal structures of *T. thermophilus* 70S ribosomes with RF1 or RF2 then provided the required details on stop-codon recognition and gave insights into the role of the GGQ motif (Laurberg et al., 2008; Weixlbaumer et al., 2008; Korostelev et al., 2008; Jin et al, 2010; Korostelev et al., 2010). These studies cover the four functional complexes in which RF1 or RF2 recognize either the common UAA or the specific UAG and UGA stop codons:

– 70S/*RF1* with UAA codon (3.2 Å resolution; Laurberg et al., 2008),
– 70S/RF2 with UGA codon (3.5 Å resolution; Weixlbaumer et al., 2008),
– 70S/RF2 with UAA codon (3 Å resolution; Korostelev et al., 2008), and
– 70S/*RF1* with UAG codon (3.6 Å resolution; Korostelev et al., 2010).

The four structures reveal that stop-codon recognition is based on a combination of specific interactions between the nucleotides and amino acids, but also on a conformational adaptation of the 30S decoding region and of the third codon base which take place upon RF binding. The specific recognition of the stop codons relies on hydrogen-bond patterns and van der Waals packing/stacking of residues. The first base, U1, packs against the tip of helix 5 of RF1 (steric discrimination against the larger purines), and forms main-chain hydrogen bonds with two glycines (and the hydroxyl moiety of Thr186 in RF1). The hydrogen-bond pattern provides selectivity for U in the first position and discriminates against C (which has the opposite hydrogen bond donor/acceptor characteristics).

A2 at the second position of the UAA and UAG codons lies between U1 and His193 and contacts the side chains of Pro184 (PxT motif) and Glu119; an additional hydrogen bond is formed with the hydroxyl moiety of Thr186 (PxT motif). Taken together, the PxT motif interacts with the first two stop-codon bases, rather than with the second and third positions as proposed earlier (Ito et al., 2000; Nakamura et al., 2002). G2 at the second position of the UGA codon contacts the hydroxyl moiety of Ser193 (first residue in the SPF motif of *E. coli* RF2; Ser206 in *T. thermophilus*). RF2 can recognize both UAA and UGA because the SPF serine can adapt equally well to either guanine or adenine in this position by switching its hydrogen bond donor/acceptor activity (Korostelev et al., 2008 and 2010).

Interestingly, the third position (A3) is rotated away from its position seen in the presence of cognate tRNA, because RF1 (and also RF2) inserts a loop (residues 193/194) between nucleotides 2 and 3, such that A3 finally stacks on G530. A3 forms hydrogen bonds with the side chains of Gln181 and Thr194 of RF1 and stacks against G530. The stacking provides the basis for discrimination against pyrimidines in the third codon position, whereas both A and G are compatible and therefore allowed in the third position (i. e. UAA or UAG codons for RF1). In addition, hydrogen-bonding with A or G is appears to be enabled in RF1 through the rotation of the side chain amide group of Gln181 (inversion of the hydrogen bond donor/acceptor potential; Korostelev et al., 2010). An interesting feature is that the key residues involved in tRNA recognition, the universally conserved G530, A1492 and A1493, are also used by the RFs, albeit in a different manner: in the presence of the RFs only A1492 flips out, whereas both A1492 and A1493 of helix 44 in 16S rRNA are flipped in upon binding of cognate tRNA. Flipping out of A1493 would prevent RF binding while it promotes tRNA binding (Youngman et al., 2007). These subtle molecular differences highlight the mechanism of the very high discrimination between regular codons and stop codons exerted by class-I RFs.

Taken together, the crystal structures have revealed that the initially identified PxT and SPF motif are only partially involved in direct codon recognition: the hydroxyl moiety of the RF1 threonine recognizes the first two bases, and the RF2 serine recognizes the second base; this comes from the fact that the hydroxyl groups in RF1 and RF2 are shifted by approximately one base with respect to the stop codon. The hydrophobic residues of the motifs are involved in not-specific van der Waals contacts. Thus, the concept of "tripeptide decoding" (Ito et al., 2000; Nakamura et al., 2002), though compelling as a hypothesis, was not confirmed by the structures.

The mechanism of peptidyl-tRNA hydrolysis depends to a large extent on the GGQ motif that is present in class-I RFs of all organisms. The GGQ motif is lo-

cated in domain 3 of RF1 and RF2 with the two glycines providing a backbone conformation that places the motif into the PTC next to A76 at the CCA-end of the P-site tRNA. The side-chain of Gln230 is oriented away from the scissile ester bond; consistently, mutation of the glutamine has only a modest effect on catalysis (Seit-Nebi et al., 2001; Shaw et al., 2007; Dincbas-Renqvist et al., 2000). The backbone amide of the glutamine is in hydrogen-bonding distance to the 3' hydroxyl group of A76 and could serve to coordinate and stabilize the transition state and the deacylated tRNA product. In comparison to the crystal structures of the isolated RFs, there is a conformational change of the linker between domains 3 and 4, in that a switch loop forms a helical structure extending helix h7 when the RFs are bound to the ribosome (Laurberg et al., 2008; Korostelev et al., 2008). The switch loop appears important for positioning the GGQ motif (located at the other extremity of helix h7) into the PTC in response to a stop codon in the decoding centre which rearranges upon stop codon recognition as mentioned above.

As revealed by the cryo-EM and crystal structures, the class-I RFs seem to mimic an A-site tRNA molecule in that they link the two main active sites of the ribosome. The RFs are functionally related to the tRNA with respect to codon recognition in the A site, but the recognition at molecular level is entirely distinct and the activity in the PTC switches from peptide bond formation to ester bond hydrolysis. Also, the RFs have no part corresponding to the T loop region of the tRNA and thus also no corresponding interactions in this area with ribosomal residues in the A site, while the N-terminal domain 1 extends to an area on the ribosome where tRNAs do not bind.

3.2. Bacterial termination complexes with the class-II release factor RF3

The bacterial class-II release factors are GTPases that promote the dissociation of the class-I RFs once the nascent polypeptide has been released (Freistroffer et al., 1997; Grentzmann et al., 1998; Kisselev et al., 2003). RF3 in the GDP-bound state binds to the ribosome that has a class-I RF bound in response to a stop codon, followed by an exchange of GDP for GTP, release of the class-I RF, and dissociation of RF3 from the ribosome which is triggered by GTP hydrolysis (Zavialov et al., 2001). RF3 is a member of the family of translational GTPases, such as initiation factor IF2 and elongation factor EF-G, which share a highly conserved G domain that contains the GDP/GTP binding pocket. Until now, structural information on RF3 is limited to two studies which we will discuss here, which use single particle cryo-EM for two related RF3-ribosome complexes (Klaholz et al., 2004; Gao et al., 2007) and X-ray crystallography for the investigation of the isolated factor (Gao et al., 2007). The cryo-EM structure of the *E. coli* ribosome with bound RF3 was trapped using the non-hydrolyzable GTP analog GDPNP (Klaholz et al., 2004; Figure 6B), thus blocking the dissociation of RF3 from the ribosome. This complex showed significant conformational heterogeneity and prompted us to develop novel image processing tools for particle separation according to their structural state. The separation was based on a two-step classification of the particles: a first classical classification that groups particles according to their orientation, and a second sub-classification of the particles from a given orientational group according to their structural differences (Suppl. Mat. in Klaholz et al., 2004). In the second classification, the attention of the examination (based on multivariate statistical analysis, MSA) can be concentrated on typical areas such as the beak, the toe, or the L1 area of the ribosome which showed structural variability (a procedure called local MSA or focused classification). This concept of particle separation has been developed further since and now can be applied directly on sets of 3D structures (see section 1.3 on the 30S initiation complex, Simonetti et al., 2008). In the case of the 70S-RF3 complex, the analysis revealed that the ribosome complex can assume two different conformational states in a single sample: in state 1 RF3 is pre-bound to the ribosome in an open conformation, whereas in state 2 RF3 contacts the GTPase center of the large subunit of the ribosome in a more closed conformation (Klaholz et al., 2004). Importantly, in state 2 the tRNA molecule is found translocated from the P to the E site and the L1 stalk has moved inwards and stabilizes the E-site tRNA. The translocation of the tRNA is associated with a large conformational rearrangement of the ribosome in which different areas move in a cooperative manner and the subunits rotate by 6° with respect to each other. These conformational changes are reminiscent of the rotational movement between the subunits, as described for ribosomes which had EF-G bound (Agrawal et al., 1999), indicating related functions of the two GTPases EF-G and RF3. This study revealed (i) the binding site of RF3 on the ribosome and the global domain assignment of RF3, (ii) the conformational changes of the ribosome induced by RF3, and (iii) indicated that the RF3 position (state-1) would

be compatible with the simultaneous presence of RF2 or RF1 (see model in Figure 6C). As the position and detailed molecular mechanism of action of the class-I RFs is known from the series of studies described in the previous section, it was proposed that the release mechanism could be based on a hinge movement of the domains with respect to the GTP-binding domain, induced by GDP-GTP exchange (Klaholz et al., 2004).

While RF3 is sequence-related to EF-G, there are also significant differences (domain V and the first part of domain IV are absent in RF3) which puts limitations to a more detailed interpretation of the 70S-RF3 complex using a homology model of RF3 (Klaholz et al., 2004) based on the EF-G crystal structure (Laurberg et al., 2000). The crystal structure of RF3 from *E. coli* in the GDP-bound state (Gao et al., 2007) showed that it is composed of three domains: domain I, a G domain that binds the nucleotide and has an insertion related to that found in EF-G (the G' domain); domain II, a ß-barrel domain as in EF-G and EF-Tu; and domain III, a ß-barrel flanked by two α-helices. The structural similarity between RF3 and EF-G or EF-Tu is limited to domains I and II. In this cryo-EM study (Gao et al., 2007), the crystal structure of RF3-GDP was fitted into the cryo-EM map of the ribosome complex with RF3-GDPNP; this required a flexible fitting of domains II and III with respect to the G domain. The resulting localization of the domains is as described in our previous study, including the interaction of helix 5 of 16S rRNA with domains II/III, the proximity to RF3 of protein S12, and the interactions of RF3 with the α-sarcin-ricin loop and protein L6 on the large ribosomal subunit (Klaholz et al., 2004). However, the crystal structure now allows to localize the GDP/GTP-binding pocket and the associated switch 2 region more precisely in the vicinity of the α-sarcin-ricin loop on the large subunit which is functionally important for triggering GTP hydrolysis. The complex was apparently more homogeneous, with the deacylated tRNA in a single hybrid P/E-state rather than in either pre- or post-translocation states. In comparison with the class I RF-ribosome complexes (see previous section), a rotational movement of the subunits was also observed and related to release of RF1/2 from the ribosome, as supported by the effects of mutations at the interface of domains II and III of RF3. Finally, GTP hydrolysis would induce RF3 release such that the ribosome switches back to a conformation similar to that prevailing prior to class-I RF binding (Gao et al., 2007).

For many of the structure-function studies on translation termination complexes discussed in this chapter, the synergy between crystallography and cryo-EM has been instrumental for a detailed interpretation of the structures in conjunction with functional studies. Crystallography has provided outstanding details on the molecular basis of stop-codon recognition and is on the way to address details of peptidyl-tRNA ester bond hydrolysis. Now that ribosomal complexes with EF-Tu or EF-G have been crystallized (Gao et al., 2009; Schmeing et al., 2009) it may also be envisaged to tackle the crystal structure of a ribosome-RF3 complex, as the accessibility of the factor-binding site relies on a permissive crystal packing for which crystallization conditions are now well-established.

References

Agrawal RK, Heagle AB, Penczek P, Grassucci RA, Frank J (1999) EF-G-dependent GTP hydrolysis induces translocation accompanied by large conformational changes in the 70S ribosome. Nat Struct Biol 6: 643–647

Alekhina OM, Vassilenko KS, Spirin AS. (2007) Translation of non-capped mRNAs in a eukaryotic cell-free system: acceleration of initiation rate in the course of polysome formation. Nucleic Acids Res. 35:6547–6559

Allen GS, Zavialov A, Gursky R, Ehrenberg M, Frank J (2005) The cryo-EM structure of a translation initiation complex from *Escherichia coli*. Cell 121: 703–712

Biou V, Shu F, Ramakrishnan V (1995) X-ray crystallography shows that translational initiation factor IF3 consists of two compact a/b domains linked by an a-helix. EMBO J 14: 4056–4064

Boehringer D, Thermann R, Ostareck-Lederer A, Lewis JD, Stark H (2005) Structure of the hepatitis C virus IRES bound to the human 80S ribosome: remodelling of the HCV IRES. Structure 13: 1695–17064

Bourne HR, Sanders DA, McCormick F (1991) The GTPase superfamily: conserved structure and molecular mechanism. Nature 349: 117–127

Brandt F, Etchells SA, Ortiz JO, Elcock AH, Hartl FU, Baumeister W (2009) The native 3D organization of bacterial polysomes. Cell 136: 261–271

Brenner S, Stretton AO, Kaplan S (1965) Genetic code: the 'nonsense' triplets for chain termination and their suppression. Nature 206: 994–998

Capecchi MR (1967) Polypeptide chain termination in vitro: isolation of a release factor. PNAS 58: 1144–1151

Carter AP, Clemons WM, Brodersen Jr DE, Morgan-Warren RJ, Hartsch T, Wimberly BT, Ramakrishnan V (2001) Crystal structure of an initiation factor bound to the 30S ribosomal subunit. Science 291: 498–501

Caskey CT, Tompkins R, Scolnick E, Caryk T, Nirenberg M (1968) Sequential translation of trinucleotide codons for the initiation and termination of protein synthesis. Science 162: 135–138

Cheng Z, Saito K, Pisarev AV, Wada M, Pisareva VP, Pestova TV, Gajda M, Round A, Kong C, Lim M, Nakamura Y, Svergun DI, Ito K, Song H (2009) Structural insights into eRF3 and stop codon recognition by eRF1. Genes Dev 23:1106–1118

Christensen AK, Bourne CM (1999) Shape of large bound polysomes in cultured fibroblasts and thyroid epithelial cells. Anat Rec 255: 116–129

Dallner G, Siekevitz P, Palade GE (1966) Biogenesis of endoplasmic reticulum membranes. II. Synthesis of constitutive microsomal enzymes in developing rat hepatocyte. J Cell Biol 30: 97–117

Deana A, Belasco JG (2005) Lost in translation: the influence of ribosomes on bacterial mRNA decay. Genes Dev 19: 2526–2533

de Smit MH, van Duin J (2003) Translational standby sites: how ribosomes may deal with the rapid folding kinetics of mRNA. J Mol Biol 331: 737–743

Dincbas-Renqvist V, Engström A, Mora L, Heurgué-Hamard V, Buckingham R, Ehrenberg M. (2000) A post-translational modification in the GGQ motif of RF2 from *Escherichia coli* stimulates termination of translation. Embo J 19: 6900–6907

Fabbretti A, Pon CL, Hennelly SP, Hill WE, Lodmell JS, Gualerzi CO (2000) Real-time dynamics of ribosome-ligand interaction by time-resolved chemical probing methods. Mol Cell 25: 285–296

Fortier PL, Schmitter JM, Garcia C, Dardel F (1994) The N-terminal half of initiation factor IF3 is folded as a stable independent domain. Biochimie 76: 376–383

Freistroffer DV, Pavlov MY, MacDougall J, Buckingham RH, Ehrenberg M (1997) Release factors RF3 in *E. coli* accelerates the dissociation of release factors RF1 and RF2 from the ribosome in a GTP-dependent manner. EMBO J 16: 4126–4133

Frolova LY, Tsivkovskii RY, Sivolobova GF, Oparina NY, Serpinsky OI, Blinov VM, et al. (1999) Mutation in the highly conserved GGQ motif of class 1 polypeptide release factors abolish ability of human eRF1 to trigger peptidyl-tRNA hydrolysis. RNA 5: 1014–1020

Frolova L, Seit-Nebi A, Kissev L (2002) Highly conserved NIKS tetrapeptide is functionally essential in eukaryotic translation termination factor eRF1. RNA 8: 129–136

Gao H, Zhou Z, Rawat U, Huang C, Bouakaz L, Wang C et al. (2007) RF3 induces ribosomal conformational changes responsible for dissociation of class I release factors. Cell 129: 929–941

Gao YG, Selmer M, Dunham CM, Weixlbaumer A, Kelley AC, Ramakrishnan V (2009) The structure of the ribosome with elongation factor G trapped in the posttranslocational state. Science 326: 694–699

Garcia C, Fortier PL, Blanquet S, Lallemand JY, Dardel F (1995) ¹H and ¹⁵N resonance assignments and structure of the N-terminal domain of *Escherichia coli* initiation factor 3. J Mol Biol 228: 395–402

Grentzmann G, Kelly PJ, Laalami S, Shuda M, Firpo MA, Cenatiempo Y, Kaji A (1998) Release factors RF-3 GTPase activity acts in disassembly of the ribosome termination complex. RNA 8: 973–983

Grigoriadou C, Marzi S, Pan D, Gualerzi CO, Cooperman BS (2007a) The translational fidelity function of IF3 during transition from the 30 S initiation complex to the 70 S initiation complex. J Mol Biol. 373: 551–561

Grigoriadou C, Marzi S, Kirillov S, Gualerzi CO, Cooperman BS (2007b) A quantitative kinetic scheme for 70 S translation initiation complex formation. J Mol Biol 373: 562–572

Grill S, Gualerzi CO, Londei P, Bläsi U (2000) Selective stimulation of translation of leaderless mRNA by initiation factor 2: evolutionary implications for translation. EMBO J 19: 4101–10

Gualerzi CO, Pon CL (1990) Initiation of mRNA translation in prokaryotes. Biochemistry 29: 5881–889

Guenneugues M, Caserta E, Brandi L, Spurio R, Meunier S, Pon CL, Boelens R, Gualerzi CO (2000) Mapping the fMet-tRNA(fMet) binding site of initiation factor IF2. EMBO J 19: 5233–5240

Ito K, Uno M, Nakamura Y (2000) A tripeptide 'anticodon' deciphers stop codons in messenger RNA. Nature 403: 680–684

Jenner L, Romby P, Rees B, Schulze-Briese C, Springer M, Ehresmann C et al. (2005) Translational operator of mRNA on the ribosome: how repressor proteins exclude ribosome binding. Science 308: 120–123

Jin H, Kelly AC, Loakes D, Ramakrishnan V (2010) Structure of the 70S ribosome bound to release factor 2 and a substrate analog provides insights into catalysis of peptide release. Proc Natl Acad Sci USA 107: 8593–8598

Kisselev L, Ehrenberg M, Frolova L (2003) Termination of translation: interplay of mRNA, rRNA and release factors? EMBO J 22: 175–182 (Review)

Klaholz BP (2011) Molecular recognition and catalysis in translation termination complexes. Trends Biochem Sc., in press

Klaholz BP, Pape T, Zavialov AV, Myasnikov AG, Orlova EV, Vestergaard B et al. (2003) Structure of the *Escherichia coli* ribosomal termination complex with release factor 2. Nature 421: 90–94

Klaholz BP, Myasnikov AG, Van Heel M (2004) Visualization of release factor 3 on the ribosome during termination of protein synthesis. Nature 427: 862–865

Kolb A, Hermoso JM, Thomas JO, Szer W (1977) Nucleic acid helix-unwinding properties of ribosomal protein S1 and the role of S1 in mRNA binding to ribosomes. Proc Natl Acad Sci USA 74: 2379–2383

Konecki DS, Aune KC, Tate W, Caskey CT (1977) Characterization of reticulocyte release factor. J Biol Chem 252: 4514–1420

Kononenko AV, Dembo KA, Kiselev LL, Volkov VV (2004) Molecular morphology of eukaryotic class I transaltion termination factor eRF1 in solution (in Russian). Mol Biol (Mosk) 38: 303–311

Kopeina GS, Afonina ZA, Gromova KV, Shirokov VA, Vasiliev VD, Spririn AS (2008) Step-wise formation of eukaryotic double-row polyribosomes and circular translation of polysomal mRNA. Nucleic Acids Res 36: 2476–2488

Korostelev A, Asahara H, Lancaster L, Laurberg M, Hirschi A, Zhu J, Trakhanov S et al. (2008) Crystal structure of a translation termination complex formed with release factor RF2. Proc Natl Acad Sci USA 150: 19684–19689

Korostelev A, Zhu J, Asahara H, Noller HF (2010) Recognition of the amber UAG stop codon by release factor RF1. EMBO J [Epub ahead of print]

Kuff EL, Hymer WC, Shelton E, Roberts NE (1966) The in vivo protein synthetic activities of free versus membrane-bound ribonucleoprotein in a plasma-cell tumor of the mouse. J Cell Biol 29: 63–75

Kycia JH, Biou V, Shu F, Gerchman SE, Graziano V, Ramakrishnan V (1995) Prokaryotic translation initiation factor IF3 is an elongated protein consisting of two crystallisable domains. Biochemistry 34: 6183–6187

La Teana A, Pon CL, Gualerzi CO (1996) Late events in translation initiation. Adjustment of fMet-tRNA in the ribosomal P-site. J Mol Biol 256: 667–675

Laursen BS, Mortensen KK, Sperling-Petersen HU, Hoffman DW (2003) A conserved structural motif at the N terminus of bacterial translation initiation factor IF2. J Biol Chem 278: 16320–16328

Laurberg M, Kristensen O, Martemyanov K, Gudkoy AT, Nagaev I, Hughes D et al. (2000) Structure of a mutant EF-G reveals domain III and possibly the fusidic acid binding site. J Mol Biol 303: 593–603

Laurberg M, Asahara H, Korostelev A, Zhu J, Trakhanov S, Noller HF (2008) Structural basis for translation termination on the 70S ribosome. Nature 454: 852–857

Marzi S, Myasnikov AG, Serganov A, Ehresmann C, Romby P, Yusupov M et al. (2007) Structured mRNA regulate translation initiation by binding to the platform of the ribosome. Cell 130: 1019–1031

Marzi S, Fechter P, Chevalier C, Romby P, Geissmann T (2008) RNA switches regulate initiation of translation in bacteria. Biol Chem 389: 585–589

McCutcheon JP, Agrawal R. K., Philips S. M., Grassucci R. A., Gerchman S. E., Clemons W. M., Ramakrishnan V. and Frank J. (1999)) Location of translational initiation factor IF3 on the small ribosomal subunit. Proc. Natl. Acad. Sci. USA 96: 4301–4306

Merkulova TI, Frolova LY, Lazar M, Camonis J, Kisselev LL (1999) C-terminal domains of human translation termination factors eRF1 and eRF3 mediate their in vivo interaction. FEBS Lett 443: 41–47

Meunier S, Spurio R, Czisch M, Wechselberger R, Geunneugues M, Gualerzi CO, Boelens R (2000) Structure of the fMet-tRNA^fMet-binding domain of *B. stearothermophilus* initiation factor IF2. EMBO J 19: 1918–1926

Milon P, Konevega AL, Gualerzi CO, Rodnina MV (2008) Kinetic checkpoint at a late step in translation initiation. Mol Cell. 30: 712–20

Mora L, Heurgué-Hamard V, Champ S, Ehrenberg M, Kisselev LL, Buckingham RH (2003) The essential role of the invariant GGQ motif in the function and stability in vivo of bacterial release factors RF1 and RF2. Mol Microbiol 47: 267–275

Moreau M, de Cock E, Fortier PL, Garcia C, Albaret C, Blanquet S, Lallemand JY, Dardel F (1997) Heteronuclear NMR studies of *E. coli* translation initiation factor IF3. Evidence that the inter-domain region is disordered in solution. J Mol Biol 266: 15–22

Myasnikov AG, Marzi S, Simonetti A, Giuliodori AM, Gualerzi CO, Yusupova G, Yusupov M, Klaholz BP (2005) Conformational transition of initiation factor 2 from the GTP- to GDP-bound state visualized on the ribosome. Nat Struct Mol Biol 12: 1145–1149

Myasnikov AG, Simonetti A, Marzi S, Klaholz BP (2009) Structure-function insights into prokaryotic and eukaryotic translation initiation. Curr Opin Struct Biol 19: 300–309

Nakamura Y, Ito K (2002) A tripeptide discriminator for stop codon recognition. FEBS Lett 514: 30–33

Petrelli D, LaTeana A, Garofalo C, Spurio R, Pon CL, Gualerzi CO. (2001) Translation initiation factor IF3: two domains, five functions, one mechanism? EMBO J 20: 4560–4569

Petry S, Brodersen DE, Murphy FV 4th, Dunham CM, Selmer M, Tarry MJ et al. (2005) Crystal structure of the ribosome in complex with release factors RF1 and RF2 bound to a cognate stop codon

Philippe C, Eyermann F, Bénard L, Portier C, Ehresmann B, Ehresmann C. (1993) Ribosomal protein S15 from *Escherichia coli* modulates its own translation by trapping the ribosome on the mRNA initiation loading site. Proc Natl Acad Sci USA 15: 4394–4398

Rawat UB, Zavialov AV, Sengupta J, Valle M, Grassucci RA, Linde J et al. (2003) A cryo-electron microscopic study of ribosome-bound termination factor RF2. Nature 421: 87–90

Rawat U, Gao H, Zavialov A, Gursky R, Ehremberg M, Frank J (2006) Interactions of the release factor RF1 with the ribosome as revealed by cryo-EM. J Mol Biol 357: 1144–1153

Roll-Mecak A, Cao C, Dever TE, Burley SK (2000) X-ray structures of the universal translation initiation factor IF2/eIF5B: conformational changes on GDP and GTP binding complex. Cell 103: 781–792

Schmeing TM, Voorhees RM, Kelley AC, Gao YG, Murphy FV 4 th, Weir JR et al. (2009) The crystal structure of the ribosome bound to EF-Tu and aminoacyl-tRNA. Science 326: 688–694

Scolnick E, Tompkins R, Caskey CT, Nirenberg M (1968) Release factors differing in specificity for terminator codons. PNAS 61: 768–774

Seit-Nebi A, Frolova L, Justesen J, Kisselev L (2001) Class-1 translation termination factors: GGQ minidomain is essential for release activity and ribosome binding but not for stop codon recognition. Nucleic Acids Res 29: 3982–3987

Sette M, van Tilborg P, Spurio R, Kaptein R, Paci M, Gualerzi CO, Boelens R. (1997) The structure of the translational initiation factor IF1 from *E. coli* contains an oligomer-binding motif. EMBO J 16: 1436–1443

Shaw JJ, Green R (2007) Two distinct components of release factor function uncovered by nucleophile partitioning analysis. Mol Microbiol 47: 267–275

Shin DH, Brandsen J, Jancarik J, Yokota H, Kim R, Kim SH (2004) Structural analyses of peptide release factor 1 from Thermotoga maritima reveal domain flexibility required for its interaction with the ribosome. J Mol Biol 341: 227–239

Simonetti A, Marzi S, Myasnikov AG, Fabbretti A, Yusupov M, Gualerzi CO, Klaholz BP (2008) Structure of the 30S translation initiation complex. Nature 455: 416–420

Simonetti A, Marzi S, Jenner L, Myasnikov AG, Romby P, Yusupova G, Klaholz BP, Yusupov M (2009) A structural view of translation initiation in bacteria. Cell Mol Life Sci 66: 423–436

Song H, Mugnier P, Das AK, Webb HM, Evans DR, Tuite MF et al. (2000) The crystal structure of human eukaryotic release factor eRF1–mechanism of stop codon recognition and peptidyl-tRNA hydrolysis. Cell 100: 311–321

Spahn CM, Kieft JS, Grassucci RA, Penczek PA, Zhou K, Doudna JA et al. (2001) Hepatitis C virus IRES RNA-induced changes in the conformation of the 40Ss ribosomal subunit. Science 291: 1959–1962

Spahn CM, Jan E, Mulder A, Grassucci RA, Sarnow P, Frank J (2004) Cryo-EM visualization of a viral internal entry site bound to human ribosomes: the IRES functions as an RNA-based translation factor. Science 118: 465–475

Studer S. M. and Joseph S. (2006) Unfolding of mRNA secondary structure by the bacterial translation initiation complex. Mol Cell 22: 105–115

Tarun SZ, Sachs AB (1995) A common function for mRNA 5' and 3' ends in translation initiation in yeast. Gene Develop 9: 2997–3007

Tate W, Greuer B, Brimacombe R (1990) Codon recognition in polypeptide chain termination: site directed crosslinking of termination codon to *Escherichia coli* release factor 2. Nucleic Acids Res 18: 6537–6544

Tomsic J, Vitali LA, Daviter T, Savelsbergh A, Spurio R, Striebeck P, Wintermeyer W, Rodnina MV, Gualerzi CO. (2000) Late events of translation initiation in bacteria: a kinetic analysis. EMBO J 19: 2127–2136

Trobro S, Aqvist J (2007) A model for how ribosomal release factors induce peptidyl-tRNA cleavage in termination of protein synthesis. Mol Cell 27: 758–766

Vestergaard B, Van LB, Andersen GR, Nyborg J, Buckingham RH, Kjeldgaard M (2001) Bacterial polypeptide release factor RF2 is structurally distinct from eukaryotic eRF1 Mol Cell 8: 1375–1372

Vestergaard B, Sanyal S, Roessle M, Mora L, Buckingham RH, Kastrup JS et al. (2005) A cryo-electron microscopic study of ribosome-bound termination factor RF2. Mol Cell 20: 929–938

Warner JR, Rich A, Hall CE (1962) Electron Microscope Studies of Ribosomal Clusters Synthesizing Hemoglobin. Science 138: 1399–1403

Weixlbaumer A, Jin H, Neubauer C, Voorhees RM, Petry S, Kelly AC, Ramakrishnan V (2008) Insights into translational termination from the structure of RF2 bound to the ribosome. Science 322: 953–956

Wienk H, Tomaselli S, Bernard C, Spurio R, Picone D, Gualerzi CO, Boelens R (2005) Solution structure of the C1-subdomain of *Bacillus stearothermophilus* translation initiation factor IF2. Protein Sci 14: 2461–2468

Wilson KS, Ito K, Noller HF, Nakamura Y (2000) Functional sites of interaction between release factor RF1 and the ribosome. Nat Struct Biol 7: 866–870

Youngman EM, Brunelle JL, Kochaniak AB, Green R (2004) The active site of the ribosome is composed of two layers of conserved nucleotides with distinct role in peptide bound formation and peptide release. Cell 117: 589–599

Youngman EM, He SL, Nikstad LJ, Green R (2007) Stop codon recognition by release factors induces structural rearrangement of the ribosomal decoding center that is productive for peptide release. Mol Cell 28: 533–543

Yusupova, G. Z., Yusupov, M. M., Cate, J. H. and Noller, H. F. (2001) The path of the messenger RNA throught the ribosome. Cell 106: 233–241

Yusupova, G., Jenner, L., Rees, B., Moras, D., and Yusupov, M. (2006) Structural basis for messenger RNA movement on the ribosome. Nature 444: 391–394

Zavialov AV, Buckingham RH, Ehrenberg M (2001) A posttermination ribosomal complex is the guanine nucleotide exchange factor for peptide release factor RF3. Cell 107: 115–124

Zoldák G, Redecke L, Svergum DI, Konarev PV, Voertler CS, Dobbek H et al. (2007) Release factor 2 from Thermus thermophilus: structural, spectroscopic and microcalorimetric studies. Nucleic Acids Res 35: 1343–1353

Initiation of bacterial protein synthesis with wild type and mutated variants of initiation factor 2

11

Michael Y. Pavlov, Suparna Sanyal and Måns Ehrenberg

1. Introduction

In-frame translation of code triplets (codons) of mRNAs into the peptide chains that define the cell's proteome requires accurate positioning of the mRNA start codon in the ribosomal P site during initiation of protein synthesis. Initiation in bacteria begins on the small (30S) ribosomal subunit. It is generated through dissociation of the bacterial ribosome into its small and large (50S) subunits. Ribosome dissociation is promoted by ribosomal recycling factor, RRF, and elongation factor G (EF-G) (Pavlov et al., 1997; Karimi et al., 1999; Peske et al., 2005; Pavlov et al., 2008; Savelsbergh et al., 2009) following termination of protein synthesis by a class-1 release factor (RF1/2) (Freistroffer et al., 2000) and RF1/2 recycling by the G-protein RF3 (Freistroffer et al., 1997; Zavialov et al., 2001).

Initiation starts with the binding of mRNA to the 30S subunit, after which the start codon, AUG, is identified among all possible base triplets of mRNA and placed in the P site of the 30S subunit (Ringquist et al., 1992; Studer and Joseph, 2006). The daunting problem of finding the correct AUG codon for initiation among a manifold of in-frame or out-of-frame AUG triplets in an mRNA sequence is solved with help of the Shine and Dahlgarno (SD) sequence in the mRNA (Shine and Dalgarno, 1974; Londei, 2005). This sequence interacts with the anti-SD sequence in the 16S rRNA of the 30S subunit, which positions AUG codon closest to the SD sequence in the neighborhood of the P site of the 30S subunit. Precise P-site positioning is accomplisheD by the formylated version of Met-charged initiator tRNA, fMet-tRNA$_i$ (fMet-tRNA$_{CAU}^{fMET}$), through its high affinity to the P site of the 30S subunit *and* to the AUG codon (Steitz and Jakes, 1975; Londei, 2005). As a result, the P site becomes programmed with the AUG codon and the A site with the second codon in

the open reading frame (ORF) of the protein to be synthesized (Hartz et al., 1989). Subsequent docking of the 50S subunit to the mRNA-programmed and fMet-tRNA$_i$-containing 30S subunit leads to formation of the 70S initiation complex (70S IC), which completes initiation of translation.

Addition of fMet-tRNA$_i$ and mRNA with proper spacing between SD sequence and initiation codon (Ringquist et al., 1992) to the empty 30S subunit correctly positions initiator tRNA and initiation codon in the P site of the 30S·mRNA·fMet-tRNA$_i$ pre-initiation complex (30S PIC) (Hartz et al., 1989). Subsequent addition of the 50S subunit to the 30S PIC leads to formation of an elongation competent 70S IC (Hartz et al., 1989). However, addition of a mixture of elongator tRNAs and fMet-tRNA$_i$ to mRNA-programmed 30S subunits results in a heterogeneous set of 30S PICs complexes containing initiator and elongator tRNAs paired to their cognate codons in the P site, generating a multiplicity of mRNA reading frames (Hartz et al., 1989). Therefore, this minimal system for initiation does not provide accurate selection of fMet-tRNA$_i$ into the 30S PIC and, hence, fails to correctly position the start codon of mRNA on the ribosome. The required selectivity is provided by the three initiation factors (IF1, IF2 and IF3) (Hartz et al., 1989). These factors use features of initiator fMet-tRNA$_i$, which separate it from all elongator tRNAs (RajBhandary, 1994; Schmitt et al., 1996), to ensure binding of fMet-tRNA$_i$, but not other tRNAs, including the AUG-reading elongator tRNA^{Met}, to 30S subunits containing mRNA, IF1, IF2•GTP, and IF3 (Wintermeyer and Gualerzi, 1983; Pon and Gualerzi, 1984; Canonaco et al., 1986; Antoun et al., 2006a).

The roles of initiation factors in selective binding of fMet-tRNA$_i$ into the 30S PIC have been under extensive study during several decades (reviewed in

(Gualerzi and Pon, 1990; Gualerzi et al., 2000; Boelens and Gualerzi, 2002; Marintchev and Wagner, 2004; Laursen et al., 2005). In contrast, the effects of initiation factors on the kinetics of 50S subunit docking to the 30S PIC for 70S IC formation have received little attention in the past (Grunberg-Manago et al., 1975; Chaires et al., 1981). Stopped-flow experiments with detection of scattered light were, however, used in pioneering work to demonstrate that complex formation between vacant 30S subunits and 50S subunits follows the law of mass action (Wishnia et al., 1975). More recently, we used this technique to study the effects of initiation factors, GTP and fMet-tRNA$_i$ on the kinetics of subunit joining (Antoun et al., 2003; Antoun et al., 2004; Antoun et al., 2006b; Antoun et al., 2006a).

2. Roles of initiation factors in rapid and accurate formation of the 30S PIC

The roles of initiation factors for rapid and selective binding of fMet-tRNA$_i$ to the 30S subunit and subsequent rapid docking of the 50S subunit have been systematically studied in (Antoun et al., 2006b; Antoun et al., 2006a). One conclusion from this work is that all three initiation factors, and, in particular, IF2 greatly favor the preferential binding of fMet-tRNA$_i$ into the 30S PIC, in line with previous results from a model system for initiation in which N-acetyl-Phe-tRNA^{Phe} was used instead of fMet-tRNA$_i$ (Gualerzi et al., 1979; Wintermeyer and Gualerzi, 1983; Canonaco et al., 1986; Gualerzi and Wintermeyer, 1986). Interestingly, we have found that IF3 effectively destabilizes the binding of *all* tRNAs, including fMet-tRNA$_i$, to the mRNA-programmed 30S subunit by increasing the rate constant for tRNA dissociation from the 30S subunit much more than the association rate constant (Antoun et al., 2006a). In particular, IF3 alone increases the rate of fMet-tRNA$_i$ association to 30S subunits programmed with an unstructured mXR7 mRNA (encoding an MFTI tetrapeptide) with a strong SD sequence about 5-fold but increases its dissociation rate about 100-fold, implying a 20-fold reduction in fMet-tRNA$_i$ affinity (association equilibrium constant) to the mRNA-programmed 30S subunit (Antoun et al., 2006a). Surprisingly, IF3-induced reduction of the elongator Phe-tRNA^{Phe} affinity to an mXR7 mRNA-programmed 30S subunit was somewhat smaller, about 10-fold (Antoun et al., 2006a). These results apparently contradict earlier findings that IF3 reduces the 30S subunit-affinity of initiator tRNA much less than the affinity of

elongator tRNAs (Risuleo et al., 1976). We note, however, that selective recognition of initiator tRNA by IF3 seems to depend on three consecutive G-C base pairs in its anticodon stem (Hartz et al., 1989; RajBhandary, 1994; O'Connor et al., 2001; Hoang et al., 2004; Lancaster and Noller, 2005) and that tRNA^{Phe} from *E. coli* contains two out of these three G-C base pairs. It is therefore possible that IF3 reduces the affinity of other elongator tRNAs, lacking all three G-C base-pairs in the anticodon stem, much more than the affinity of initiator tRNA and tRNA^{Phe} (Hartz et al., 1989; Hoang et al., 2004; Lancaster and Noller, 2005). It has been suggested that selection of G-C containing tRNAs depends on IF3-induced removal of the S9 protein tail from the P site of the 30S PIC, which allows for interaction of these G-C base-pairs with the G1338/A1339 bases of 16S rRNA (Dallas and Noller, 2001; Lancaster and Noller, 2005). The IF3-dependent de-positioning of the S9 tail may both interrupt the non-specific interaction of S9 with tRNA in the P site (Hoang et al., 2004) and broaden the passage to the P site, which could explain the increased dissociation and association rate constants for tRNA interaction with the 30S subunit in the presence of IF3 (Antoun et al., 2006a). It has also been suggested that in the absence of IF3 all tRNAs interact with S9 instead of G1338/A1339, which could give them similar and high affinity to the 30S P site, ensuring high processivity and uniform rate of protein elongation (Antoun et al., 2006a).

IF3-induced acceleration of tRNA association to and dissociation from the 30S subunit is amplified by the addition of IF1 (Antoun et al., 2006a), in line with previous observations (Pon and Gualerzi, 1984; Gualerzi and Pon, 1990). IF1 binding to the A site alters global conformation of the 30S subunit (Carter et al., 2001). It seems therefore that both IF1 and IF3 affect the kinetics of tRNA binding to the 30S PIC by changing the conformation of the 30S subunit, instead of making direct contacts with the P site-bound tRNA (Gualerzi and Pon, 1990; Carter et al., 2001; Dallas and Noller, 2001).

Addition of IF2 to the 30S PIC containing IF1, IF3, and mXR7 mRNA increases about 60-fold the rate constant for fMet-tRNA$_i$ association and decreases about 3-fold the dissociation rate constant. At the same time, IF2 addition has little effect on these rate constants for non-formylated Met-tRNA$_i$ and elongator Phe-tRNA^{Phe} (Antoun et al., 2006a). These results underline the importance of the formyl group of fMet-tRNA$_i$ as its major structural element recognized by IF2 (Gualerzi and Pon, 1990; RajBhandary, 1994;

Schmitt et al., 1996). Accordingly, addition of all three initiation factors accelerates fMet-tRNA$_i$ association to and dissociation from the 30S subunit about 400- and 300-fold, respectively, leaving the affinity of initiator tRNA binding virtually unaltered (Antoun et al., 2006a). In contrast, the three initiation factors increase the association and dissociation rates of elongator Phe-tRNA^{Phe} by about 5- and 300- fold, respectively, reducing by 60-fold the affinity of this tRNA to the 30S PIC. This strong reduction of the 30S binding affinity for tRNAs other than fMet-tRNA$_i$ by initiation factors accounts for a major part of the preferential selection of fMet-tRNA$_i$ into the 30S PIC (Antoun et al., 2006a).

3. Roles of initiation factors and initiator tRNA in subunit joining

The presence of fMet-tRNA$_i$ on the 30S subunit strongly affects the rate of subunit joining. In the absence of IF3 and presence of only IF2 or IF2 and IF1, addition of fMet-tRNA$_i$ accelerates the effective rate, k_c, of subunit docking 60- or 30-fold, respectively. These results suggest that fMet-tRNA$_i$ may provide an additional complementary surface to increase the rate of binding of the 50S subunit. Alternatively, fMet-tRNA$_i$ may drive IF2 on the 30S subunit into a conformation that is optimal for rapid subunit joining (Antoun et al., 2006a; Simonetti et al., 2008). The latter scenario is in line with our recent results as discussed below in Section 8. In the absence of IF3, k_c is directly proportional to the 50S subunit concentration, implying a rate constant for subunit dissociation close to zero (Antoun et al., 2006a). Under these conditions the overall rate of dipeptide formation after mixing 30S PICs with 50S subunits and ternary complexes composed of elongation factor EF-Tu, GTP, and aminoacyl-tRNA (aa-tRNA) cognate to the second codon of XR7 mRNA is 27 s^{-1} and significantly smaller than the rate of subunit joining (42 s^{-1}) measured by light scattering (Antoun et al., 2006b). The delay between subunit joining and peptide bond formation is due to the kinetic steps that separate these two events, which include hydrolysis of GTP on IF2, release of IF2•GDP from the 70S IC and the ternary complex binding to the 70S IC (Antoun et al., 2003; Antoun et al., 2006b).

In the presence of IF1, IF2, *and* IF3, the rate and extent of subunit joining are virtually zero in the absence of fMet-tRNA$_i$, but increase to 3 s^{-1} and near 100% 70S formation, respectively, by initiator tRNA addition (Antoun et al., 2006b). The rate of 3 s^{-1} corresponds well to the rate of dipeptide formation (3.6 s^{-1}) measured in a parallel quench-flow experiment, indicating that in the presence of IF3, subunit joining is rate limiting for di-peptide formation (Antoun et al., 2006b). Increasing IF3 concentration reduces the rate of subunit joining, suggesting that dissociation of IF3 precedes 70S complex formation (Antoun et al., 2006b). Addition of IF3 decreases the affinity of fMet-tRNA$_i$ to the 30S PIC about 20-fold implying (from the detailed balance constraint) that fMet-tRNA$_i$ similarly reduces the affinity of IF3 to the 30S subunit, thereby promoting rapid dissociation of the factor (Antoun et al., 2006b). Hence, fMet-tRNA$_i$ binding to the 30S PIC accelerates subunit joining by two distinct mechanisms. Firstly, fMet-tRNA$_i$ drives IF2 into a conformation, which promotes subunit joining, and, secondly, fMet-tRNA$_i$ alleviates inhibition of subunit joining by IF3 (Antoun et al., 2006b).

The kinetic model depicted in Figure 1 and quantified by Eq. 1 explains how the average time, $1/k_c$, for subunit joining depends on the association rate constant for 50S subunit docking, k_{50S}, the 50S concentration, $[50S]$, the IF3 concentration, $[IF3]$, and the rate constants q_{IF3} and k_{IF3} for IF3 dissociation from and association to the 30S PIC, respectively:

$$\frac{1}{k_c} = \frac{1}{q_{IF3}}\left(1 + \frac{k_{IF3}}{k_{50S}}\frac{[IF3]}{[50S]}\right) + \frac{1}{k_{50S}}\frac{1}{50S} = \frac{1}{k_g} + \frac{1}{k_{50S}}\frac{1}{50S} \quad (1)$$

This model predicts a linear relation between $1/k_c$ and $[IF3]$. It also predicts a minimal subunit joining time in the limit of large $[50S]$-values, equal to the average time ($1/q_{IF3}$) of IF3 release from the 30S PIC, as observed experimentally (Antoun et al., 2006b).

Removal of IF1 from the 30S PIC increases the subunit docking rate about 3-fold, possibly due to reduced affinity of IF3 to the 30S PIC by an increased value of q_{IF3} (Antoun et al., 2006b). Such an increase in q_{IF3} due to the IF1 removal was later observed by another group (Milon et al., 2008). IF1 may also favor a 50S-docking-inactive conformation of the 30S PIC (Milon et al., 2008). These two explanations of the adverse effect of IF1 on subunit docking rate are not mutually exclusive, since the putative, IF1-induced inactive 30S PIC conformation (Milon et al., 2008) may have higher affinity to IF3 than the active form of the subunit.

It has been suggested (Milon et al., 2008) that the kinetics of subunit docking and, in particular, the impact of IF1 on this kinetics depends on the strength of the SD-antiSD interaction (Milon et al., 2008). Here, we have compared the kinetics of 50S subunit-dock-

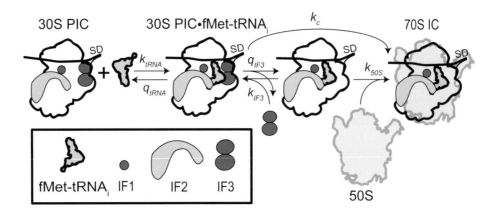

Fig. 1 Kinetic scheme of initiation of protein synthesis starting from fMet-tRNA$_i$, 50S subunits, and tRNA-free 30S PICs containing three initiation factors and mRNA. fMet-tRNA$_i$ binds to 30S PIC with the second order rate constant k_{tRNA} and dissociates with rate constant q_{tRNA}. Binding of fMet-tRNA$_i$ to 30S PIC destabilized IF3 binding increasing q_{IF3}. The presence of IF1 stabilizes IF3 binding decreasing q_{IF3}. 50S subunit associates with 30S PIC containing tRNA with the second order rate constant k_{50S}. In the absence of IF3, $k_c = k_{50S} \cdot [50S]$.

ing to 30S PICs programmed with mRNAs containing strong (XR7) or weak (SD022) SD-sequence and found the impacts of initiation factors and GTP on the rate of subunit docking for the two types of mRNA to be very similar. We found, in particular, that addition of IF1 and/or IF3 to the 30S subunit alters the subunit docking kinetics similarly for the strong and weak SD-mRNAs (Table 1; Figure 2A, B). When, furthermore, the IF3 concentration increases in the presence of IF1 and IF2, the subunit docking rate considerably decreases for both mRNA types (Table 1). When GTP is

replaced by GDP, the rate of 50S subunit docking to 30S PICs programmed with our weak-SD mRNA, and containing fMet-tRNA$_i$ along with all three initiation factors, decreases 100-fold (Figure 2D, Table 1). All these results are perfectly in line with previous observations with the strong-SD XR7 mRNA (Antoun et al., 2004; Antoun et al., 2006b). The biphasic kinetics of subunit joining with our weak-SD mRNA (Figure 2B) is similar to that previously observed in the presence of IF1, also with a weak SD-mRNA (Milon et al., 2008).

In summary, the effects of initiation factors on the rate of subunit joining observed here (Figure 2; Table 1) and in previous (Antoun et al., 2006a; Antoun et al., 2006b) experiments are similar to those observed by Milon et al (2008) for their weak-SD mRNAs. The major difference between the results of the two groups is the 200-fold reduction in subunit joining rate upon IF1 addition they observe for 30S PICs programmed with strong SD-mRNA, contrasting the three-fold reduction observed by us for both types of mRNA (Figure 2; Table 1). A temperature shift from 37 °C, routinely used by us, to 20 °C, used by (Milon et al., 2008), does not qualitatively alter the impact of IF1 and/or IF3 on the kinetics of subunit joining with our strong-SD mRNA (Figure 2C, Table 1), implying that the different results are not due to different temperatures. The discrepancies may, we suggest, be caused by different structural features of the mRNAs, unrelated to their SD sequences or, possibly, by different buffer conditions.

Table 1 Effective rates, k_c, of the 50S subunits docking to 30S PICs assembled with fMet-tRNA$_i$, IF2, and different combinations of IF1 and IF3

Conditions	IF2	IF2 and IF1	IF2 and IF3	IF2, IF1 and IF3 or IF3[a]
mXR7, 37°C, GTP	37 s^{-1}	40 s^{-1}	22 s^{-1}	8.5 s^{-1}/4.7 s^{-1} [a]
mXR7, 20°C, GTP	21 s^{-1}	24 s^{-1}	12 s^{-1}	3.6 s^{-1}/ 2.1 s^{-1} [a]
mSD022, 37°C, GTP	30 s^{-1}	39 s^{-1}	17 s^{-1}	5.1 s^{-1}/3.3 s^{-1} [a]
mSD022, 37°C, GDP	1.5 s^{-1}	1.5 s^{-1}	0.12 s^{-1}	0.05 s^{-1}

30S subunits were programmed with mXR7 (strong SD) or mSD022 (weak SD) mRNAs. [a]IF3 was added in 0.4 μM (IF3) or in 0.8 μM (IF3[a]) concentrations. The effective rate k_c is defined here as the inverse of the time at which 50% of the 70S initiation complexes have been formed after the subunit mixing (Zorzet et al., 2010).

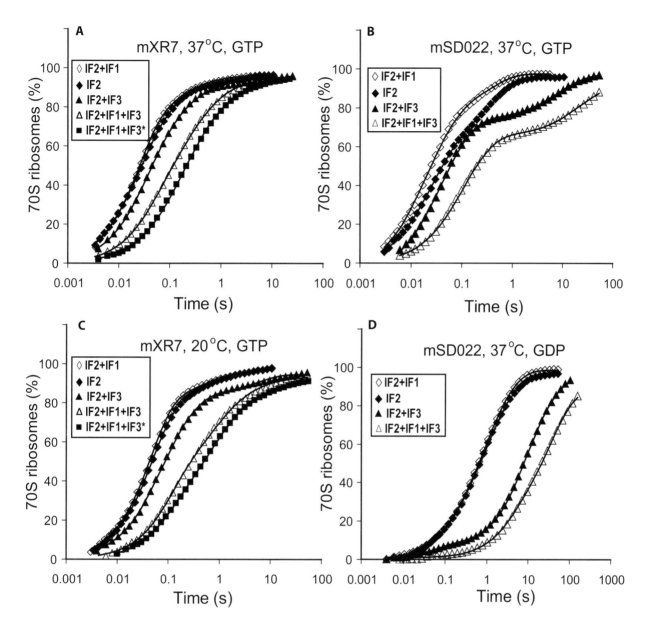

Fig. 2 Kinetics of 70S initiation complex formation after rapid mixing of 50S subunits with 30S PICs containing fMet-tRNAi, IF2, and different combinations of IF1 and IF3. (A) 0.37 μM 50S subunits were mixed in a stopped-flow apparatus with a 30S PIC mixture containing 0.31 μM 30S subunits, 0.6 μM fMet-tRNA, 0.7 μM XR7 (strong SD) mRNA supplemented with 1 μM IF2 (◆); with 0.5 μM IF2 and 0.5 μM IF1 (◇); with 1 μM IF2 and 0.4 μM IF3 (▲); with 0.5 μM IF2, 0.5 μM IF1 and 0.4 μM IF3 (△) or with 0.5 μM IF2, 0.5 μM IF1 and 0.8 μM IF3 (■) as indicated in the figure. After the mixing, light scattering at 430 nm was monitored to follow subunit association. The experiment was conducted at 37 °C in a polymix-like buffer, LS4, contained 95 mM KCl, 5 mM NH₄Cl, 0.5 mM CaCl², 8 mM putrescine, 1 mM spermidine, 30 mM HEPES pH 7.5, 1 mM DTE, 1 mM GTP, 1 mM ATP, 6 mM Mg(OAc)₂, and 2 mM PEP, supplemented with 1 μg/ml pyruvate kinase (PK) and 0.1 μg/ml myokinase (MK). Since each ATP or GTP molecule chelate one Mg^{2+} cation, the free Mg^{2+} concentration in the L4 buffer is about 4 mM (Pavlov et al., 2008). (B) The same as (A) but 30S PICs contained 0.8 μM mSD022 (weak SD) mRNA. (C) The same as (A) but the experiment was conducted at 20 °C. (D) The same as (B) but the LS4 buffer contained 1 mM GDP instead of GTP, and PK /MK enzymes were excluded. All concentrations are after mixing.

4. Roles of GTP and GTP hydrolysis in initiation of mRNA translation

An important question in initiation of protein synthesis concerns the role played by GTP in the function of IF2 that belongs to the same family of translational GTPases as EF-Tu, EF-G, RF3, and SelB (Bourne et al., 1991; Hauryliuk et al., 2008; Paleskava et al., 2010). It was early concluded that IF2•GTP is essential for initiation of mRNA translation and that GTP hydrolysis is required for release of IF2 from the 70S IC to allow for binding of ternary complex and formation of the first peptide bond (Benne et al., 1973; Fakunding and Hershey, 1973; Grunberg-Manago et al., 1975). The importance of GTP hydrolysis for fast release of IF2 from the ribosome has later been questioned by Pon et al (1985), who from binding experiments proposed that the GDP- and GTP-bound forms of IF2 have similar and high affinities to the 70S ribosome (Pon et al., 1985). However, their experimental setup did not exclude hydrolysis of GTP on ribosome bound IF2 (Luchin et al., 1999; Grigoriadou et al., 2007a), implying that the reported affinity of IF2·GTP reflects the association rate of IF2·GTP to the 70S ribosome, while the dissociation rate is more complex, involving GTP hydrolysis and dissociation of either IF2·GTP or IF2·GDP. The authors (Pon et al., 1985) do not discuss the possibility that, due to rapid GTP hydrolysis (Luchin et al., 1999; Grigoriadou et al., 2007a), their experiment might have been insensitive to any difference in dissociation rate constants for IF2·GTP and IF2·GDP. It is, in other words, possible, that their data report the IF2·GDP affinity to the 70S ribosome in both the GTP and the GDP case. Moreover, the high IF2·GDP affinity to the 70S ribosome (Pon et al., 1985) could be due to their particular experimental conditions. Indeed, the overall rate of 27 s^{-1}, measured for dipeptide formation upon mixing 30S PICs with 50S subunits and ternary complexes (Antoun et al., 2006b), provides a lower limit for the rate of IF2·GDP dissociation from the 70S IC. This suggests that under physiological conditions (37 °C and low free Mg^{2+} concentration) the affinity of IF2·GDP to the 70S ribosome is low.

The pivotal importance of GTP and its hydrolysis for rapid formation of the elongation competent 70S IC from a 30S PIC and 50S subunit has been demonstrated by stopped-flow experiments with monitoring light scattering (Antoun et al., 2003). The subunit docking rate decreases 20-fold when GDP replaces GTP but only two-fold when the non-hydrolyzable GTP analogue GDPNP replaces GTP on the naturally

truncated -form of IF2 in the 30S PIC. In subsequent work (Antoun et al., 2004), we used the full-length -form of IF2 to demonstrate a 60-fold decrease in the rate of docking of the 50S subunit to the 30S PIC when GTP was swapped for GDP and, as before, a two-fold rate decrease when GTP was swapped for GDPNP. This shows that GTP, but not GTP hydrolysis, is required for rapid subunit docking. Moreover, the 70S IC formed either with IF2 and GDPNP or with a small amount of IF2·GTP following incubation with IF2 and GDPNP is virtually inactive in dipeptide formation. From these results we concluded that the GTP (GDPNP) form of IF2 dissociates very slowly from the 70S IC, thereby blocking the entry of ternary complex to the A site of the ribosome (Antoun et al., 2003), in line with earlier conclusions (Dubnoff et al., 1972; Benne et al., 1973).

In summary, our results demonstrate that rapid 70S IC formation requires GTP, but not GTP hydrolysis, and that GTP hydrolysis leads to rapid release of IF2•GDP from the 70S IC swiftly followed by dipeptide formation.

5. Alternative models of subunit joining and 70S IC formation

A recent study with *E. coli* components, complemented with IF2 from *B. stearothermophilus*, has confirmed that both fMet-tRNA$_i$ and IF2 are required on the 30S subunit for its rapid docking to the 50S subunit and that subunit joining rates with GTP and its non-hydrolysable analogue (GDPCP) are similar (Grigoriadou et al., 2007a; Grigoriadou et al., 2007b). The dipeptide formation rate of 0.2 s^{-1}, measured upon mixing 30S PICs with 50S subunits and ternary complexes, was much smaller than the rate of subunit association due to the small rate of dissociation of thermophilic IF2•GDP from *E. coli* 70S IC (Grigoriadou et al., 2007a). Such slow dissociation of thermophilic IF2·GDP may also explain the earlier results by Tomsic et al. (2000), who found very similar and small (0.2 s^{-1}) rates of peptide bond formation after mixing 50S subunits with 30S PICs containing either IF2•GDP or IF2•GTP. They concluded that neither the presence of GTP nor GTP hydrolysis is important for formation of the first peptide bond in a nascent protein (Tomsic et al., 2000). However, slow dissociation of *B. stearothermophilus* IF2 from the *E. coli* 70S IC could have been rate-limiting for peptide bond formation, thereby masking any difference in subunit joining rate between the GTP and GDP cases.

Recent experiments have led to the proposal of a novel kinetic model of subunit joining (Grigoriadou et al., 2007a). According to this model, docking of the 50S subunit to the 30S PIC leads to an unstable 70S IC, which initially can undergo rapid dissociation into subunits but eventually becomes stable along a complex kinetic pathway. No effect of IF3 on the rate of formation of 70S IC was observed in these experiments (Grigoriadou et al., 2007b). In contrast, more recent (Milon et al., 2008) and our previous (Antoun et al., 2006b) observations show that, in the absence of IF3, ribosomal subunits do not dissociate after the initial joining, and that addition of IF3 to the 30S PIC greatly reduces the rate of subunit docking. These contrasting results put in question details of the kinetic model suggested by (Grigoriadou et al., 2007a; Grigoriadou et al., 2007b), as further discussed below in section 5.1.

It was concluded from stopped-flow data based on fluorescence detection that IF3 dissociates from the 70S IC *after* subunit joining (Milon et al., 2008) and not *before*, as previously suggested (Antoun et al., 2006b). IF3 consists of globular N- and C-domains connected by a long, helical linker (Dallas and Noller, 2001; Petrelli et al., 2001). When the isolated C-domain of IF3 is bound to the 30S subunit, its docking with the 50S subunit is inhibited (Petrelli et al., 2001). From these observations, we speculate that spontaneous dissociation of the C-domain of IF3 from the 30S PIC may allow for 70S IC formation, even though its N-domain remains transiently bound on the 30S subunit. Such transient IF3 binding after C-domain dissociation and 70S complex formation may explain why IF3 dissociation appears slower than subunit joining (Milon et al., 2008). The strong dependence of the rate of subunit joining on IF3-concentration (Antoun et al., 2006b) (Figure 2; Table 1) could then be explained on the further assumption that the C-domain dissociation permits binding of a *second* molecule of IF3 to the vacated C-domain binding site on the 30S subunit.

5.1. Roles of GTP hydrolysis and Pi release in subunit joining

It is an established fact that the GTP molecule on IF2 is hydrolyzed during initiation of translation. However, the timing of GTP hydrolysis on IF2 and subsequent Pi release from IF2 in relation to subunit joining has become controversial. That is, it was recently claimed that stable subunit association occurs *after* GTP hydrolysis and that phosphate release follows GTP hy-

drolysis after a long (>0.1s) lag period (Grigoriadou et al., 2007a; Grigoriadou et al., 2007b). Our experiments at high 50S subunit concentration in Figure 3 show, in contrast, that subunit docking is almost complete at the onset of GTP hydrolysis and that Pi release very rapidly follows GTP hydrolysis. Indeed, the time corresponding to the rapid phase of subunit joining is about 6 ms, while the times for GTP hydrolysis and Pi release after subunit mixing are 25 ms and 32 ms, respectively (Figure 3).

That stable subunit joining does not require GTP hydrolysis has been shown more directly in very recent experiments (Huang et al., in preparation), where the rate of subunit joining was monitored with the GTPase impaired IF2-mutants V400G and H448E (Luchin et al., 1999). These mutants have much slower rates of GTP hydrolysis and Pi release than wild type IF2, while the rates of subunit docking promoted by the mutated and wild type variants of IF2 are virtually the same. The rates of subunit joining are, furthermore, strictly proportional to the 50S subunit concentration for wild

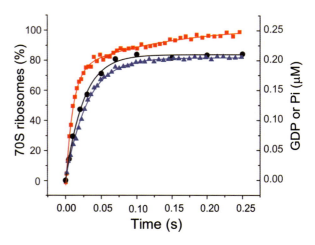

Fig. 3 Comparison of kinetics of subunit association (red trace), GTP hydrolysis (, black trace) and Pi release (blue trace) at large excess of 50S subunits over 30S PICs. Mix A containing 0.5 μM 30S subunit, 1 μM MLI mRNA, 1 μM fMet-tRNA^fMet, 1 μM IF1, 1 μM IF2 and 1μM GTP, pre-incubated in 37 °C for 5 minutes was mixed rapidly with pre-incubated mix B containing 2.5 μM 50S subunit in a stopped-flow apparatus and light scattering at 430 nm was monitored to follow subunit association at 37 °C. MLI mRNA coding for Met-Leu-Ile tripeptide had the sequence identical with XR7 mRNA outside the coding region (Huang et al., 2010). In a parallel setup, 10 μM of MDCC labeled phosphate binding protein (PBP-MDCC) was added in the mix B (as described in Huang et al. 2010) and Pi release was measured by the increase of fluorescence at 464 nm (l_Ex = 425 nm). For GTP hydrolysis, 20 μM of [³H]GTP was used in mix A in otherwise identical reaction condition in quench flow. The amount of GDP produced in every time point was estimated by separating GTP and GDP in TLC or in a Mono-Q column attached to FPLC.

type and mutants of IF2, implying negligible dissociation of the 70S IC into subunits before GTP hydrolysis (Huang et al., in preparation).

The reasons why our previous (Antoun et al., 2003; Antoun et al., 2006b) and present (Huang et al., in preparation; Figure 3) observations differ qualitatively from those by (Grigoriadou et al., 2007a; Grigoriadou et al., 2007b) are unclear. The experiments were carried out at different temperatures (37° vs. 20 °C) and had different buffer conditions. Furthermore, homologous IF2 variants from *E. coli* were used in one case (Antoun et al., 2003; Antoun et al., 2006b) (Huang et al., in preparation), while heterologous IF2 variants from *B. stearothermophilus* were used in the other case (Grigoriadou et al., 2007a; Grigoriadou et al., 2007b), but also other explanations for the discrepancies are conceivable.

5.2. IF2 interaction with the L12 protein of the 50S subunit promotes fast subunit joining

It has been suggested that ribosomal stalk proteins (L7/L12) are involved in the recruitment of translational GTPases, EF-Tu and EF-G to the ribosome and in acceleration of GTP hydrolysis (Savelsbergh et al., 2000;

Mohr et al., 2002; Diaconu et al., 2005). The ribosomal stalk in *E. coli* is composed of four copies of flexible L12 proteins (Chandra Sanyal and Liljas, 2000; Mulder et al., 2004), which interact with translation factors using their C-terminal domains (Diaconu et al., 2005; Helgstrand et al., 2007). Here, we have studied the role of the stalk proteins in IF2 recognition during initiation of mRNA translation (Huang et al., 2010).

We have compared the docking rates of 30S PICs or 30S subunits to native and L12-deficient 50S subunits. The rates of 50S subunit docking to vacant 30S subunits or to 30S PICs lacking IF2 were unaffected by L12 depletion (Figure 4A). However, in the presence of IF2 in the 30S PIC the docking rate was 40 times smaller for L12 depleted than for native 50S subunits (Figure 4B), demonstrating that L12 speeds up subunit joining in an IF2-dependent manner. We have also monitored the extents of GTP hydrolysis and Pi release of IF2-containing 30S PICs as they dock with native or L12 deficient 50S subunits. We found that the time spans between a subunit-docking event and GTP hydrolysis or Pi release are very similar for native and L12 depleted 50S subunits (Huang et al., 2010). These data suggest that L12 is essential for rapid docking of the 50S subunit to an IF2•GTP-containing 30S PIC, but not for subsequent GTP hydrolysis and Pi release.

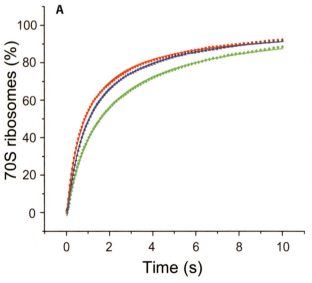
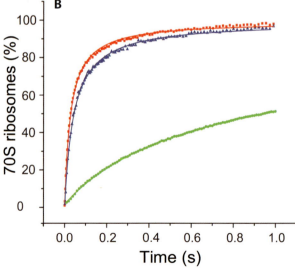
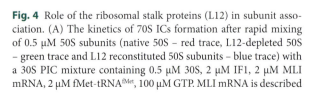

Fig. 4 Role of the ribosomal stalk proteins (L12) in subunit association. (A) The kinetics of 70S ICs formation after rapid mixing of 0.5 μM 50S subunits (native 50S – red trace, L12-depleted 50S – green trace and L12 reconstituted 50S subunits – blue trace) with a 30S PIC mixture containing 0.5 μM 30S, 2 μM IF1, 2 μM MLI mRNA, 2 μM fMet-tRNA^{fMet}, 100 μM GTP. MLI mRNA is described in the caption of Fig. 3 After the mixing in a stopped flow apparatus, light scattering at 430 nm was monitored to follow subunit association at 37 °C. (B) the same as (A), but the 30S PIC mixture contained 2 μM IF2. Individual mixtures were pre-incubated at 37 °C for 10 minutes before mixing

Thus, despite its involvement in stimulation of GTP hydrolysis on the elongation factors EF-G and EF-Tu (Savelsbergh et al., 2000; Mohr et al., 2002; Diaconu et al., 2005), L12 is not directly involved in stimulation of GTP hydrolysis on IF2 (Huang et al., 2010).

6. Accuracy of initiator tRNA selection in initiation of mRNA translation

In addition to a large rate of initiation of mRNA translation, the accuracy of fMet-tRNA$_i$ selection in the initiation process is crucial for rapid growth and high fitness of bacterial populations (Zorzet et al., 2010). Light scattering data on 50S subunit docking to 30S PICs, containing fMet-tRNA$_i$, Met-tRNA$_i$, tRNA$_i$ or Phe-tRNA^{Phe}, in combination with kinetics and equilibrium data on the binding of these tRNAs into the 30S PIC have been used to estimate the accuracy of initiator tRNA selection into the 70S IC (Antoun et al., 2006a). This means, in particular, estimation of k_{cat}/K_m parameters for the incorporation of any one of these tRNAs in the 70S IC starting from 50S subunits, free tRNAs and tRNA-free 30S PICs (Figure 1). The k_{cat}/K_m parameter for the kinetic scheme in Figure 1 is in each case the association rate constant, k_{tRNA}, for binding of a tRNA to the 30S PIC, multiplied by the probability, P_{tRNA}, that the tRNA ends up in the 70S IC, rather than dissociates from the 30S PIC:

$$\frac{k_{cat}}{K_M} = k_{tRNA}P_{tRNA} \approx k_{tRNA}\frac{k_c}{q_{tRNA}+k_c} \qquad (2)$$

Here, q_{tRNA} is the rate constant for dissociation of tRNA from the 30S PIC and k_c is a compounded rate constant for the docking of 50S subunits to tRNA-containing 30S PICs (see Figure 1). Equation (1) defines k_c in terms of elemental rate constants in the kinetic scheme of Figure 1. The accuracy, A, of initiation is the ratio of the k_{cat}/K_M parameters for cognate and non-cognate tRNAs:

$$A = \frac{\left(\frac{k_{cat}}{K_m}\right)_c}{\left(\frac{k_{cat}}{K_m}\right)_{nc}} = \frac{k_{tRNA}^c P_{in}^c}{k_{tRNA}^{nc} P_{in}^{nc}} \approx \frac{k_{tRNA}^c}{k_{tRNA}^{nc}}\frac{1+\frac{q_{tRNA}^{nc}}{k_c^{nc}}}{1+\frac{q_{tRNA}^c}{k_c^c}} \qquad (3)$$

The value of A gives the ratio between the number of 70S complexes formed with cognate (fMet-tRNA$_i$) and non cognate (e. g., Phe-tRNA^{Phe}) tRNAs at equal

free concentrations. A "non-cognate tRNA" here refers to a tRNA other than fMet-tRNA$_i$, either bound to the AUG initiation codon (e. g. Met-tRNA$_i$ or tRNA$_i$) or to another codon (e. g., Phe-tRNA^{Phe} bound to its cognate UUC codon adjacent to AUG). It follows from Equation (3) that in the limit of rapid subunit docking (i. e., $k_c \gg q_{tRNA}$), $A=k_{tRNA}^c/k_{tRNA}^{nc}$ is only determined by the association rate constants for fMet-tRNA$_i$ and its non-cognate competitor, implying low accuracy. In the limit of slow subunit docking (i. e. $k_c \ll q_{tRNA}$), $A=(k_{tRNA}^c/k_{tRNA}^{nc})\times(q_{tRNA}^{nc}/q_{tRNA}^c)\times(k_c^c/k_c^{nc})$ implying high accuracy (Antoun et al., 2006a). The later relation for A explains why IF2 plays such a pivotal role in the accuracy of initiation: firstly, IF2 selectively increases the affinity of formylated initiator tRNA to the 30S PIC in relation to non-formylated initiator and elongator tRNAs and, secondly, IF2 selectively increases the rate of 50S docking to 30S PICs containing formylated initiator tRNA (Antoun et al., 2006a).

Equation (3) predicts the accuracy of initiator tRNA selection to increase when the dissociation rate constants for initiator and non-initiator tRNAs increase by the very same factor (Antoun et al., 2006a). IF1 and IF3 greatly increase the accuracy of initiation by such a purely kinetic mechanism, i. e. by uniformly increasing the rate constant (q_{tRNA}) for tRNA dissociation from the 30S subunit. Moreover, IF1 and IF3 also greatly reduce the rate constant (k_c) of subunit docking for all tRNAs, which according to Equation (3), would further increase accuracy of initiation (Antoun et al., 2006a).

In summary, our results show that the presence of all three initiation factors in the 30S PIC is required to achieve the highest accuracy of initiation (Antoun et al., 2006a). The accuracy enhancement by IF1/IF3, achieved by the purely kinetic mechanism described above, may be further increased by IF3-induced preferential dissociation of elongator tRNAs lacking G-C base-pairs in their anticodon stem from the 30S PIC (Risuleo et al., 1976; Gualerzi et al., 1979; Hoang et al., 2004). In addition to their roles in enhancing the accuracy of initiation, IF1/IF3 also increase the rate of tRNA binding to the 30S PIC, thereby considerably increasing the rate and extent of formation of fMet-tRNA$_i$-containing 70S ICs from 50S subunits, fMet-tRNA$_i$ and tRNA-free 30S PICs (Antoun et al., 2006b; Antoun et al., 2006a).

Table 2 Kinetic parameters of 70S IC formation from 50S subunits, tRNA-free 30S PICs and Met-tRNA$_i$ (or fMet-tRNA$_i$) with different IF2s in the 30S PIC

	$k_{cat, Met-tRNAi}$	$(k_{cat}/K_M)_{Met-tRNAi}$	$(k_{cat}/K_M)_{fMet-tRNAi}$	Accuracy
WT IF2	1 s^{-1}	0.55 µM^{-1}s^{-1}	20 µM^{-1}s^{-1}	36
B-type IF2	2 s^{-1}	0.9 µM^{-1}s^{-1}	20 µM^{-1}s^{-1}	22
A-type IF2	4 s^{-1}	1.65 µM^{-1}s^{-1}	20 µM^{-1}s^{-1}	12

7. Initiation with non-formylated Met-tRNA$_i$

Formylation deficiency in bacteria is normally caused by loss-of-function mutations in the formyl-methionyl-transferase (FMT) gene or other genes involved in single-carbon metabolism and leads to slow growth phenotype (Guillon et al., 1992; Steiner-Mosonyi et al., 2004). In line with this, Met-tRNA$_i$ binds much more slowly to the 30S PIC than fMet-tRNA$_i$ and subsequent docking to the 50S subunit is slower for the Met-tRNA$_i$-containing than for the fMet-tRNA$_i$-containing 30S PICs (Antoun et al., 2006a). Formylation deficient bacteria often acquire secondary mutations that increase their growth rate (Apfel et al., 2001; Nilsson et al., 2006). A previously unknown class of such mutations was recently identified (Zorzet et al., 2010). Here, point mutations distal to the fMet-tRNA$_i$ binding domain IV, and thus unrelated to tRNA binding, display strong (A-type, mutations located in domain III of IF2) or weak (B-type, mutations located in domains III, G or in the N-terminus of IF2) growth complementation. The A-type mutations in IF2 increase the effective rate, (k_{cat}/K_m), of Met-tRNA$_i$ binding to the 30S PIC by 3-fold. They also increase the maximal rate (k_{cat}) of docking of Met-tRNA$_i$-containing 30S PIC to the 50S subunit by four-fold. Such an increased in k_{cat}/K_m and k_{cat} parameters by mutations (Table 2) accounts for the growth complementary phenotype of the A- and B-type IF2 mutants in a formylation-deficient background (Zorzet et al., 2010). In contrast, k_{cat}/K_m for fMet-tRNA$_i$ incorporation into 70S complex was not affected by mutations in IF2 (Table 2). Accordingly, it follows from k_{cat}/K_m values in Table 2 and Eq. 3 that the accuracy of initiation with wild type IF2 (A=36) is three-fold higher than with A-type IF2 (A=12). Consequently, these mutations increase the speed of initiation with Met-tRNA$_i$ at the cost of reduced accuracy in fMet-tRNA$_i$ over Met-tRNA$_i$ selection. Furthermore, the overall rate of aberrant initiation with an elongator tRNA increased even more, five-fold, by

the A-type mutations (Zorzet et al., 2010). Such an excessive formation of aberrant 70S ICs with elongator tRNAs may explain why A-type IF2 mutations have a considerable fitness cost in the formylation proficient background (Zorzet et al., 2010).

We have found that the rate, k_c, of subunit docking with wild type IF2 was 9 s^{-1} with fMet-tRNA$_i$, 1.3 s^{-1} with Met-tRNA$_i$ and 0.18 s^{-1} with tRNA$_i$, whereas with A-type IF2 the corresponding docking rates were 10 s^{-1}, 6.6 s^{-1} and to 2.2 s^{-1}, respectively. These results imply that A-type IF2 is much less sensitive to removal of the formyl group and methionine from initiator tRNA than wild-type IF2 (Zorzet et al., 2010). The docking rate reduction upon formyl group removal was here about seven-fold in contrast to a 50-fold reduction in our previous work (Antoun et al., 2006a). This discrepancy is, we suggest, caused by deacylation of Met-tRNA$_i$ to tRNA$_i$ in our previous study (Antoun et al., 2006a). The much larger k_c value for subunit docking with Met-tRNA$_i$ estimated here implies that the accuracy of fMet-tRNA$_i$ selection over Met-tRNA$_i$ should be close to the ratio k_{tRNA}^c/k_{tRNA}^{nc} of the association rates of fMet-tRNA$_i$ and Met-tRNA$_i$ to the 30S PIC. This estimates A as $\approx$ 46 using the k_{tRNA} parameters measured in (Antoun et al., 2006a), a value reasonably close to A $\approx$ 36 calculated here from directly measured k_{cat}/K_m parameters (Table 2).

8. IF2 mutants active in subunit joining in the absence of tRNA or GTP

The subunit joining properties of A-type IF2 were also studied in the absence of IF3 (Pavlov et al., 2011). In a particular experiment the subunit docking rate was about 40 s^{-1} for all IF2 variants in the presence of fMet-tRNA$_i$ and IF1. Removal of fMet-tRNA$_i$ reduced the docking rate 20-fold (from 37 to 1.9 s^{-1}) for wild-type IF2, but only by 40% (from 41 to 25 s^{-1}) for A-type IF2 (Table 3). This result shows that rapid docking is restored after fMet-tRNA$_i$ removal by the A-

Table 3 Effects of A-type mutations in IF2 on the effective rate, k_c, of the 50S subunit docking to 30S PICs assembled with fMet-tRNA$_i$ and either GTP, or GDP, or no G-nucleotide, or with only GTP without tRNA. The effective rate k_c is defined here as the inverse of the time at which 50% of the 70S ICs have been formed after the subunit mixing (Zorzet et al., 2010)

	fMet-tRNA$_i$ GTP	fMet-tRNA$_i$ GDP	fMet-tRNA$_i$ no G-nucleotide	no tRNA GTP
WT IF2	37 s^{-1}	0.6 s^{-1}	0.5 s^{-1}	1.9 s^{-1}
A-type IF2	41 s^{-1}	26 s^{-1}	1.8 s^{-1}	25 s^{-1}

type mutations, implying that fMet-tRNA$_i$ *per se* is not required for fast subunit docking. Its role is, we suggest, to induce an IF2 conformation which promotes rapid subunit docking (Simonetti et al., 2008). From this follows that the phenotype of the A-type and B-type mutations is explained by their ability to shift the equilibrium between an inactive and an active form of IF2 towards the active conformation of the factor. In line with this interpretation, we have found that when GTP is swapped for GDP in the presence of fMet-tRNA$_i$ the subunit joining rate is reduced 60-fold (from 37 to 0.6 s^{-1}) with wild type IF2 but less than two-fold (from 41 to 26 s^{-1}) with A-type IF2 in the 30S PIC (Table 3). This implies that in the presence of fMet-tRNA$_i$ a large fraction of A-type, but not wild- type, IF2 is driven into the active docking conformation in the presence of GDP. In the absence of guanine nucleotide, the rate of subunit docking was slow even with the A-type IF2 mutant (Table 3), implying that addition of GDP to the apo-form of IF2 shifts the conformational equilibrium towards the active form, albeit to a lesser extent than GTP. In summary, these results mean that activation of wild type IF2 requires the presence of both GTP and fMet-tRNA$_i$, while activation of A-type IF2 requires the presence of only one of these activating ligands. In fact, activation of IF2 and its mutants provide a striking example of conditional activation of a GTP-binding protein by GTP and other ligands (Hauryliuk et al., 2008).

tants that increase the speed of initiation with non-formylated tRNA are over-activated in the sense that they can promote fast subunit joining in the presence of only one activating ligand: either GTP or fMet-tRNA$_i$, whereas wild-type IF2 requires the presence of both these ligands for its full activation. Moreover, these IF2-activating mutations decrease the accuracy of authentic initiation with formylated fMet-tRNA$_i$ and reduce the fitness of otherwise wild-type bacteria, explaining why evolution has optimized wild-type IF2 for full activation by the simultaneous presence of GTP and fMet-tRNA$_i$. A deeper understanding of IF2-function in initiation may be eventually provided by structurally based explanations of how GTP *and* fMet-tRNA$_i$ activate IF2 on the 30S subunit and how IF2 mutants complementing formylation deficiency can by-pass the two-ligand requirement for activation.

Acknowledgements

The experiments presented in Figures 3 and 4 were conducted by Chenhui Huang and Chandra S. Mandava. We also thank Chenhui Huang for preparation of Figures 3 and 4. The research grants from the Swedish Research Council (project support to M. E. and S. S. and Linne support (URRC) to M. E.) and NIH/USA to M. E. and S. S., and from Göran Gustafsson Stiftelse, Carl Tryggers Stiftelse and Wenner-Gren Stiftelse to S. S. are thankfully acknowledged.

9. Conclusion

It is now generally accepted that IF2•GTP promotes selective binding of fMet-tRNA$_i$ into the 30S PIC and that the rapid docking of the 50S subunit requires the presence of both fMet-tRNA$_i$ and IF2•GTP in the 30S PIC. Accordingly, IF2•GTP favors fMet-tRNA$_i$ selection into the 70S IC at two steps of the initiation process, thereby ensuring high accuracy of translation initiation. Importantly, our results show that IF2 mu-

References

Antoun A, Pavlov MY, Andersson K, Tenson T, Ehrenberg M (2003) The roles of initiation factor 2 and guanosine triphosphate in initiation of protein synthesis. Embo J 22: 5593–5601

Antoun A, Pavlov MY, Lovmar M, Ehrenberg M (2006a) How initiation factors maximize the accuracy of tRNA selection in initiation of bacterial protein synthesis. Mol Cell 23: 183–193

Antoun A, Pavlov MY, Lovmar M, Ehrenberg M (2006b) How initiation factors tune the rate of initiation of protein synthesis in bacteria. Embo J 25: 2539–2550

Antoun A, Pavlov MY, Tenson T, Ehrenberg MM (2004) Ribosome formation from subunits studied by stopped-flow and Rayleigh light scattering. Biol Proced Online 6: 35–54

Apfel CM, Locher H, Evers S, Takacs B, Hubschwerlen C, Pirson W, Page MG, Keck W (2001) Peptide deformylase as an antibacterial drug target: target validation and resistance development. Antimicrob Agents Chemother 45: 1058–1064

Benne R, Naaktgeboren N, Gubbens J, Voorma HO (1973) Recycling of initiation factors IF-1, IF-2 and IF-3. Eur J Biochem 32: 372–380

Boelens R, Gualerzi CO (2002) Structure and function of bacterial initiation factors. Curr Protein Pept Sci 3: 107–119

Bourne HR, Sanders DA, McCormick F (1991) The GTPase superfamily: conserved structure and molecular mechanism. Nature 349: 117–127

Canonaco MA, Calogero RA, Gualerzi CO (1986) Mechanism of translational initiation in prokaryotes. Evidence for a direct effect of IF2 on the activity of the 30 S ribosomal subunit. FEBSL ett 207: 198–204

Carter AP, Clemons WM, Jr., Brodersen DE, Morgan-Warren RJ, Hartsch T, Wimberly BT, Ramakrishnan V (2001) Crystal structure of an initiation factor bound to the 30S ribosomal subunit. Science 291: 498–501

Chaires JB, Pande C, Wishnia A (1981) The effect of initiation factor IF-3 on Escherichia coli ribosomal subunit association kinetics. J Biol Chem 256: 6600–6607

Chandra Sanyal S and Liljas A (2000) The end of the beginning: structural studies of ribosomal proteins. Curr Opin Struct Biol 10: 633–636

Dallas A, Noller HF (2001) Interaction of translation initiation factor 3 with the 30S ribosomal subunit. Mol Cell 8: 855–864

Diaconu M, Kothe U, Schlunzen F, Fischer N, Harms JM, Tonevitsky AG, Stark H, Rodnina MV, Wahl MC (2005) Structural basis for the function of the ribosomal L7/12 stalk in factor binding and GTPase activation. Cell 121: 991–1004

Dubnoff JS, Lockwood AH, Maitra U (1972) Studies on the role of guanosine triphosphate in polypeptide chain initiation in Escherichia coli. JB iol Chem 247: 2884–2894

Fakunding JL, Hershey JW (1973) The interaction of radioactive initiation factor IF-2 with ribosomes during initiation of protein synthesis. JB iol Chem 248: 4206–4212

Freistroffer DV, Kwiatkowski M, Buckingham RH, Ehrenberg M (2000) The accuracy of codon recognition by polypeptide release factors. Proc Natl Acad Sci USA 97: 2046–2051

Freistroffer DV, Pavlov MY, MacDougall J, Buckingham RH, Ehrenberg M (1997) Release factor RF3 in E. coli accelerates the dissociation of release factors RF1 and RF2 from the ribosome in a GTP-dependent manner. Embo J 16: 4126–4133

Grigoriadou C, Marzi S, Kirillov S, Gualerzi CO, Cooperman BS (2007a) A quantitative kinetic scheme for 70s translation initiation complex formation. J Mol Biol 373: 562–572

Grigoriadou C, Marzi S, Pan D, Gualerzi CO, Cooperman BS (2007b) The Translational Fidelity Function of IF3 During Transition from the 30 SI nitiation Complex to the 70 S Initiation Complex. J Mol Biol 373: 551–561

Grunberg-Manago M, Dessen P, Pantaloni D, Godefroy-Colburn T, Wolfe AD, Dondon J (1975) Light-scattering studies showing the effect of initiation factors on the reversible dissociation of Escherichia coli ribosomes. J Mol Biol 94: 461–478

Gualerzi C, Risuleo G, Pon C (1979) Mechanism of the spontaneous and initiation factor 3-induced dissociation of 30 S. aminoacyl-tRNA. polynucleotide ternary complexes. J Biol Chem 254: 44–49

Gualerzi C, Wintermeyer W (1986) Prokaryotic initiation factor 2 acts at the level of the 30 S ribosomal subunit. A fluorescence stopped-flow study. FEBS Lett 202: 1–6

Gualerzi CO, Brandi L, Caserta E, LaTeana A, Spurio R, Tomsic J, Pon C (2000) Translation initiation in bacteria. In Garett RA, Liljas A, Matheson AT, Moore PB, Noller HF (eds.), The Ribosome: Structure, Function, Antibiotics and Cellular Interactions. ASM Press, Washington DC, pp 477–494

Gualerzi CO, Pon CL (1990) Initiation of mRNA translation in prokaryotes. Biochemistry 29: 5881–5889

Guillon JM, Mechulam Y, Schmitter JM, Blanquet S, Fayat G (1992) Disruption of the gene for Met-tRNA(fMet) formyltransferase severely impairs growth of Escherichia coli. J Bacteriol 174: 4294–4301

Hartz D, McPheeters DS, Gold L (1989) Selection of the initiator tRNA by Escherichia coli initiation factors. Genes Dev 3: 1899–1912

Hauryliuk V, Hansson S, Ehrenberg M (2008) Co-factor dependent conformational switching of GTPases. Biophys J 95: 1704–1715

Helgstrand M, Mandava CS, Mulder FA, Liljas A, Sanyal S, Akke M (2007) The ribosomal stalk binds to translation factors IF2, EF-Tu, EF-G, RF3 via a conserved region of the L12 C-terminal domain. J Mol Biol 365: 468–479

Hoang L, Fredrick K, Noller HF (2004) Creating ribosomes with an all-RNA 30S subunit P site. Proc Natl Acad Sci USA 101: 12 439–12 443

Huang C, Mandava CS, Sanyal S (2010) The ribosomal stalk plays a key role in IF2-mediated association of the ribosomal subunits. J Mol Biol 399: 145–153

Karimi R, Pavlov MY, Buckingham RH, Ehrenberg M (1999) Novel roles for classical factors at the interface between translation termination and initiation. Mol Cell 3: 601–609

Lancaster L, Noller HF (2005) Involvement of 16S rRNA nucleotides G1338 and A1339 in discrimination of initiator tRNA. Mol Cell 20: 623–632

Laursen BS, Sorensen HP, Mortensen KK, Sperling-Petersen HU (2005) Initiation of protein synthesis in bacteria. Microbiol Mol Biol Rev 69: 101–123

Londei P (2005) Evolution of translational initiation: new insights from the archaea. FEMS Microbiol Rev 29: 185–200

Luchin S, Putzer H, Hershey JW, Cenatiempo Y, Grunberg-Manago M, Laalami S (1999) In vitro study of two dominant inhibitory GTPase mutants of Escherichia coli translation initiation factor IF2. Direct evidence that GTP hydrolysis is necessary for factor recycling. J Biol Chem 274: 6074–6079

Marintchev A, Wagner G (2004) Translation initiation: structures, mechanisms and evolution. Q Rev Biophys 37: 197–284

Milon P, Konevega AL, Gualerzi CO, Rodnina MV (2008) Kinetic checkpoint at a late step in translation initiation. Mol Cell 30: 712–720

Mohr D, Wintermeyer W, Rodnina MV (2002) GTPase activation of elongation factors Tu and G on the ribosome. Biochemistry 41: 12520–12528

Mulder FA, Bouakaz L, Lundell A, Venkataramana M, Liljas A, Akke M, Sanyal S (2004) Conformation and dynamics of ribosomal stalk protein L12 in solution and on the ribosome. Biochemistry 43: 5930–5936

Nilsson AI, Zorzet A, Kanth A, Dahlstrom S, Berg OG, Andersson DI (2006) Reducing the fitness cost of antibiotic resistance by amplification of initiator tRNA genes. Proc Natl Acad Sci USA 103: 6976–6981

O'Connor M, Gregory ST, Rajbhandary UL, Dahlberg AE (2001) Altered discrimination of start codons and initiator tRNAs by mutant initiation factor 3. RNA 7: 969–978

Paleskava A, Konevega AL, Rodnina MV (2010) Thermodynamic and kinetic framework of selenocysteyl-tRNASec recognition by elongation factor SelB. J Biol Chem 285: 3014–3020

Pavlov MY, Zorzet A, Andersson DI, Ehrenberg M (2011) Activation of initiation factor 2 by ligands and mutations for rapid docking of ribosomal subunits. Embo J 30: 289–301

Pavlov MY, Antoun A, Lovmar M, Ehrenberg M (2008) Complementary roles of initiation factor 1 and ribosome recycling factor in 70S ribosome splitting. Embo J 27: 1706–1717

Pavlov MY, Freistroffer DV, MacDougall J, Buckingham RH, Ehrenberg M (1997) Fast recycling of *Escherichia coli* ribosomes requires both ribosome recycling factor (RRF) and release factor RF3. Embo J 16: 4134–4141

Peske F, Rodnina MV, Wintermeyer W (2005) Sequence of steps in ribosome recycling as defined by kinetic analysis. Mol Cell 18: 403–412

Petrelli D, LaTeana A, Garofalo C, Spurio R, Pon CL, Gualerzi CO (2001) Translation initiation factor IF3: two domains, five functions, one mechanism? Embo J 20: 4560–4569

Pon CL, Gualerzi CO (1984) Mechanism of protein biosynthesis in prokaryotic cells. Effect of initiation factor IF1 on the initial rate of 30 S initiation complex formation. FEBS Lett 175: 203–207

Pon CL, Paci M, Pawlik RT, Gualerzi CO (1985) Structure-function relationship in *Escherichia coli* initiation factors. Biochemical and biophysical characterization of the interaction between IF-2 and guanosine nucleotides. J Biol Chem 260: 8918–8924

Raj Bhandary UL (1994) Initiator transfer RNAs. J Bacteriol 176: 547–552

Ringquist S, Shinedling S, Barrick D, Green L, Binkley J, Stormo GD, Gold L (1992) Translation initiation in *Escherichia coli*: sequences within the ribosome-binding site. Mol Microbiol 6: 1219–1229

Risuleo G, Gualerzi C, Pon C (1976) Specificity and properties of the destabilization, induced by initiation factor IF-3, of ternary complexes of the 30-S ribosomal subunit, aminoacyl-tRNA and polynucleotides. Eur J Biochem 67: 603–613

Savelsbergh A, Mohr D, Wilden B, Wintermeyer W, Rodnina MV (2000) Stimulation of the GTPase activity of translation elongation factor G by ribosomal protein L7/12. J Biol Chem 275: 890–894

Savelsbergh A, Rodnina MV, Wintermeyer W (2009) Distinct functions of elongation factor G in ribosome recycling and translocation. RNA 15: 772–780

Schmitt E, Guillon JM, Meinnel T, Mechulam Y, Dardel F, Blanquet S (1996) Molecular recognition governing the initiation of translation in *Escherichia coli*. A review. Biochimie 78: 543–554

Shine J, Dalgarno L (1974) The 3'-terminal sequence of *Escherichia coli* 16S ribosomal RNA: complementarity to nonsense triplets and ribosome binding sites. Proc Natl Acad Sci USA 71: 1342–1346

Simonetti A, Marzi S, Myasnikov AG, Fabbretti A, Yusupov M, Gualerzi CO, Klaholz BP (2008) Structure of the 30S translation initiation complex. Nature 455: 416–420

Steiner-Mosonyi M, Creuzenet C, Keates RA, Strub BR, Mangroo D (2004) The Pseudomonas aeruginosa initiation factor IF-2 is responsible for formylation-independent protein initiation in P. aeruginosa. J Biol Chem 279: 52 262–52 269

Steitz JA, Jakes K (1975) How ribosomes select initiator regions in mRNA: base pair formation between the 3' terminus of 16S rRNA and the mRNA during initiation of protein synthesis in *Escherichia coli*. Proc Natl Acad Sci USA 72: 4734–4738

Studer SM, Joseph S (2006) Unfolding of mRNA secondary structure by the bacterial translation initiation complex. Mol Cell 22: 105–115

Tomsic J, Vitali LA, Daviter T, Savelsbergh A, Spurio R, Striebeck P, Wintermeyer W, Rodnina MV, Gualerzi CO (2000) Late events of translation initiation in bacteria: a kinetic analysis. Embo J 19: 2127–2136

Wintermeyer W, Gualerzi C (1983) Effect of *Escherichia coli* initiation factors on the kinetics of N-Acphe-tRNAPhe binding to 30S ribosomal subunits. A fluorescence stopped-flow study. Biochemistry 22: 690–694

Wishnia A, Boussert A, Graffe M, Dessen PH, Grunberg-Manago M (1975) Kinetics of the reversible association of ribosomal subunits: stopped-flow studies of the rate law and of the effect of Mg2+. J Mol Biol 93: 499–515

Zavialov AV, Buckingham RH, Ehrenberg M (2001) A post-termination ribosomal complex is the guanine nucleotide exchange factor for peptide release factor RF3. Cell 107: 115–124

Zorzet A, Pavlov MY, Nilsson AI, Ehrenberg M, Andersson DI (2010) Error-prone initiation factor 2 mutations reduce the fitness cost of antibiotic resistance. Mol Microbiol 75: 1299–1313

Translation initiation at the root of the cold-shock translational bias

12

Claudio O. Gualerzi, Anna Maria Giuliodori, Anna Brandi, Fabio Di Pietro, Lolita Piersimoni, Attilio Fabbretti and Cynthia L. Pon

1. Background

Research carried out in the last two decades has shown that all living organisms, from bacteria to mammals, have evolved mechanisms to cope with the effects caused by a sudden temperature downshift (cold-shock). Following cold-stress, the mesophilic bacterium *Escherichia coli* enters an acclimation phase during which cell growth stops for 3–6 hours, while bulk gene expression is drastically reduced, and a set of at least 26 well characterized cold-shock genes is selectively and transiently expressed (Yamanaka, 1999; Gualerzi et al., 2003). The proteins synthesized during the cold-acclimation phase are somewhat artificially classified into early and late cold-shock proteins, and likewise early and late cold-adapted proteins accumulate in the lag phase that precedes the resumption of cell division and growth at low temperature (Figure 1A). Overall, the present perception is that the main purposes of the proteins synthesized during cold adaptation and in cold-adapted cells are: (i) to deal with unfavorable secondary structures of nucleic acids induced/stabilized by the cold, which are expected to hinder basic functions such as transcription, ribosome assembly, and translation; (ii) to oppose the cold-shock-induced decrease in membrane fluidity; (iii) to accumulate sugars which are protective against low temperature, such as trehalose; (iv) to assist protein folding at low temperatures (Graumann and Marahiel, 1998; Phadtare et al., 1999; Gualerzi et al., 2003; Weber and Marahiel, 2003).

At the end of the '80s, when several global response networks were fairly well-known in bacteria, only a couple of studies had been published on this particular type of stress (Herendeen et al., 1979; Jones et al., 1987). Analysis of ^{35}S-labeled proteins synthesized following an abrupt temperature downshift of an *E. coli* culture had evidenced the selective expression of a handful of proteins which were identified by their electrophoretic Cartesian coordinates but whose actual nature was still unknown. Among the first cold-shock proteins identified were translation initiation factor IF2 (Jones et al., 1987), CspA defined as "major cold-shock protein" (Goldstein et al., 1990), and H-NS (La Teana et al., 1991), an abundant nucleoid-associated protein, which is both an architectural protein of the nucleoid and a transcriptional regulator (inhibitor) (Pon et al., 1988; 2005). Using extracts of cold-shocked cells, it was possible to reproduce *in vitro* the selective cold-shock stimulation of *hns* expression and to detect the presence in cold-shocked cells of a transcriptional activator of *hns* which was identified as being CspA (La Teana et al., 1991).

The finding that CspA is responsible for starting the cascade of events which characterize the cold-stress response and the reprogramming of cell gene expression during the cold adaptation phase opened a number of questions, such as: a) what is the mechanism/signal that activates *cspA* itself; b) what is the mechanism by which CspA activates cold-shock gene expression, and c) what is the structure of CspA.

The three-dimensional structure of *E. coli* CspA (71 residues) and of its *Bacillus subtilis* homologue CspB was elucidated by crystallographic and NMR studies (Schindelin et al., 1993, 1994; Newkirk et al., 1994; Feng et al., 1998) and revealed an astonishing resemblance to that of translation initiation factor IF1 (Figure 1B). This similarity apparently is not restricted to structure, but also pertains to function; in fact, it was later demonstrated that overexpression of *E. coli* IF1 can suppress *in vivo* the phenotype caused by the deletion of *cspB* and *cspC* in *B. subtilis* (Weber et al., 2001). Further investigations demonstrated that *E. coli* contains a large number of proteins similar to CspA, as it was shown that *cspA* is just one of nine paralogous

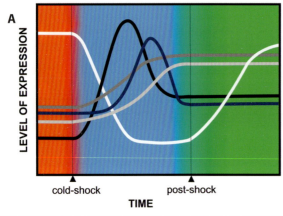

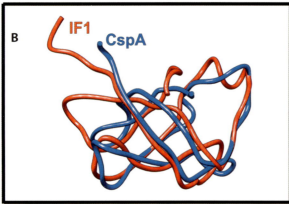

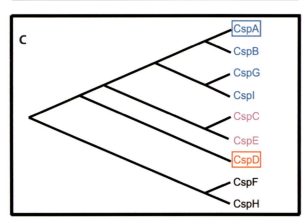

Fig. 1 (A) Expression of cellular proteins during the acclimation and recovery phases following cold-shock. The colored regions represent the time before cold-stress (red), during cold-acclimation (blue), and growth in the cold (green). The tracings represent an idealized time-dependent variation of the levels of: bulk (white), early cold-induced (black), late cold-induced (blue), early cold-adapted (dark gray), and late cold-adapted (light gray) proteins. (B) Comparison of the three-dimensional structures in solution of *E. coli* CspA (blue) and translation initiation factor IF1 (red) as determined by Newkirk et al., (1994) and Sette et al., (1997), respectively. (C) Relatedness-tree depicting the relationship between CspA and the products of its paralogous genes based on the multiple alignment of the *E. coli* Csp family members generated with the Clustal W algorithm (Thompson et al., 1994). The colors indicate proteins which are cold-shock (blue), non-cold-shock (red), constitutively expressed (magenta), and those whose regulation is unknown (gray).

genes which clearly result from extensive gene duplication events (Figure 1C). As seen from the figure, not all these genes are cold-shock-induced, and the specific functions of their individual products remain rather elusive.

As to the mechanism/signal that activates *cspA* itself, the scant literature on the subject initially available was committed to indicating transcription as the regulatory level responsible for induction of *cspA* expression after cold-shock (e. g. Tanabe et al., 1992). Thus, the finding that, after by-passing its natural promoter by replacing it with an inducible promoter, the expression of *cspA* still responded to cold-stress came as a surprise (Brandi et al., 1996). It was demonstrated that, in spite of the presence of the *cspA* mRNA transcribed from the inducible promoter, the cells incubated at 37 °C expressed CspA rather inefficiently, unlike those which had been incubated at 10 °C (Figure 2A). Furthermore, when cell-free extracts prepared from cells subjected to cold-shock and from control cells were programmed with *in-vitro* transcribed *cspA* mRNA, the extracts of cold-shock cells, particularly those derived from cells exposed to cold-shock for 90 min, proved to be more proficient in translating this mRNA than the control extract (Figure 2B). A subsequent analysis revealed that the contribution of transcriptional control in the regulation of *cspA* expression is small, if there is any, but that the regulation is instead due to two post-transcriptional phenomena: a) an enormous cold-shock induced stabilization of the *cspA* transcript (Brandi et al., 1996, Goldenberg et al., 1996) and b) the existence of a cold-shock translational bias (Brandi et al., 1996; Goldenberg et al., 1997) whereby the cold-shock transcripts are efficiently translated during the cold-acclimation phase, while the non-cold-shock mRNAs, even if present in the cells, are not efficiently translated until the cells have completed cold adaptation phase (Figure 2 C).

2. Nature and molecular basis of the cold-shock translational bias

Since it had been seen that at least some cold-shock transcripts are translated better by cell-free extracts from cold-shocked cells than from control cells, a number of mRNAs was selected which, for their characteristics, could be regarded as typical cold-shock and non-cold-shock mRNAs. At least in one case, the choice was straightforward. While *cspA* mRNA could be regarded, beyond any doubt, the most typical repre-

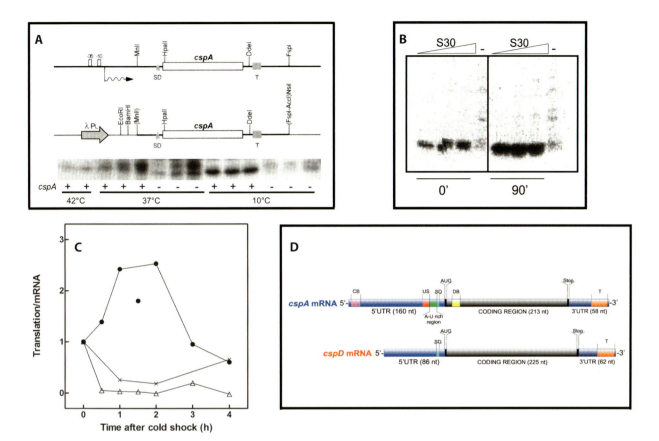

Fig. 2 (A) Cold-shock-induction of *cspA* with different promoters. *CspA* with its natural promoter as found in the *E. coli* chromosome (upper scheme) and *cspA* placed under the control of the λPL promoter (lower scheme). Autoradiogram of the electrophoretic gel showing the CspA produced *in vivo* after induction of λPL promoter and incubation at 37 °C and 10 °C (lower panel). The lanes denoted + and – refer to cells transformed with the expression vector containing (+) and not containing (–) the *cspA* gene insert. (Figure modified from Brandi et al. 1996) (B) Synthesis of CspA by 3 μl, 5 μl and 8 μl of cell-free S30 extracts from control cells grown at 37 °C (0', left panel) and from cells subjected to 90 min cold-shock at 10 °C (90′, right panel). The extracts were normalized for their ribosome content and programmed with *in-vitro* transcribed *cspA* mRNA. (C) *In-vivo* translational efficiency of cold-shock *cspA* (•), and non-cold-shock *cat* (D), *lac Z* (x) mRNAs as a function of time elapsed after the cold-stress. Translational efficiency is presented as the ratio between the steady-state level of the mRNAs determined at each time after cold-shock and the amount of protein synthesized during the corresponding period. The mRNAs were transcribed *in vivo* under the control of the *cspA* promoter. (Figure modified from Goldenberg et al. 1997) (D) Schematic comparison of the paradigm cold-shock *cspA* (upper) and non-cold-shock *cspD* (lower) mRNAs highlighting their main features indicated as 5'-(5'UTR) and 3'-(3'UTR) untranslated regions, coding regions, T (rho-independent transcription terminator), SD (Shine-Dalgarno sequence), AUG (initiation codon), stop (termination codon). Additional features of the *cspA* mRNA suggested to be important for cold-shock induction are: CB (cold box), US (upstream sequence), DB (downstream box), A-U-rich region.

sentative of the cold-shock mRNAs, the choice of the paradigm non-cold-shock mRNA fell on the transcript of the *cspD* gene, one of the aforementioned paralogous genes of *cspA* (Figure 1C). It seemed difficult to identify two genes which were at the same time more similar and more different than *cspA* and *cspD*. Indeed, they are similar because of their size, for being monocistronic, for the extensive sequence similarity of their products (50% identical, 64% homologous), and for the presence of rather long 5'- and 3'-UTRs in their transcripts (Figure 2D). However, they differ in the regulation of their expression and in the cellu-lar localization and, consequently, in the function of their respective products (Giangrossi et al., 2001). Of the additional features suggested to be important for cold-shock induction (Figure 2D), only the AU-rich region, a likely S1-binding site, has been shown to play a relevant role for the expression of *cspA* (Giuliodori et al., 2010), while neither the DB (La Teana et al., 2000, Giuliodori et al., 2010) nor the CB (Giuliodori et al., 2010) boxes seem to be relevant.

In addition to *cspA* and *cspD* mRNA, which were chosen as paradigm cold-shock and non-cold- shock mRNAs, two additional transcripts, those of *hns* (La

Teana et al., 1991) and of *hupA* (Giangrossi et al., 2002), were selected as typical cold-shock and non-cold-shock mRNAs, respectively. A third class of transcripts, which consisted of the three alternative transcripts arising from the three promoters found in the *hupB* gene, was denoted as "cold-tolerant" mRNAs because, although not having the typical behavior of the cold-shock mRNAs, they are nevertheless translated with fairly high efficiency at low temperature (Giangrossi et al., 2002). Translation experiments carried out *in vitro* with normalized cell-free extracts from both control and cold-shocked cells not only confirmed the possibility of reproducing *in vitro* the cold-shock translational bias, but gave indications that this phenomenon is due to both cis-acting and trans-acting factors and that the bias goes both ways: positive with respect to the cold-shock mRNAs and negative with the non-cold-shock mRNAs (Giuliodori et al., 2004).

2.1. Cis-acting elements

When the ratio between maximum level of *in vitro* translation of various types of mRNAs obtained at 15 °C and 37 °C with a control cell-free extract was measured, it became clear that this ratio was around one or higher for cold-shock mRNAs and much lower than one for non-cold-shock mRNAs; the cold-tolerant mRNAs yielded somewhat intermediate ratios (Figure 3A). This result suggested that cis-acting elements present in the mRNAs make them more or less prone to translation at low temperature. To elucidate the nature of these cis-acting elements, we undertook the task of determining the structure of the two paradigm mRNAs, *cspA* and *cspD*, at 37 °C as well as at the cold-shock temperatures of 10 °C and 20 °C. While no temperature-dependent difference in the structure of *cspD* mRNA could be detected (not shown), the *cspA* mRNA was clearly present in two completely different conformations, one at 37 °C and one at the two cold-shock temperatures (Giuliodori et al., 2010) (Figure 3B).

Since there is evidence that *cspA* is always transcribed in the cell, regardless of temperature and phase of growth, the structural difference was interpreted as resulting from an alternative folding of the transcript occurring at low temperature. It was postulated that an otherwise thermodynamically unstable folding intermediate is stabilized at the cold-shock temperature. When transcription occurs at 37 °C, on the other hand, the unstable intermediate structure yields to a

more stable one formed through basepairing of a portion of the 5'-UTR with the coding sequence. Another difference between the cold-shock and the 37 °C structure of *cspA* mRNA is the presence in the former but not in the latter of a pseudoknot involving the 5'-UTR of the transcript (Figure 3C). It is likely that the presence of this pseudoknot might stabilize the cold-shock structure of *cspA*. Furthermore, it was shown that the structural differences between the 37 °C and the cold-shock structure of *cspA* mRNA have a functional bearing. In fact, *cspA* mRNA in the cold-shock conformation binds better to the ribosome and is also translated better (Figure 4A). These differences can be attributed to a higher exposure of the Shine-Dalgarno (SD) sequence and to the greater availability of the initiation triplet, since this is included in a less stable helix at the cold-shock temperature compared to 37 °C (Figure 3B). Finally, the overall good translational activity of *cspA* mRNA, regardless of temperature, can be attributed to the presence in its 5'UTR of an AU-rich S1-binding region (see the toeprinting experiment in Figure 4B); indeed, using a reporter gene system, Yamanaka et al., (1999) were able to show that deletion of part of that sequence causes a strong reduction of *in vivo* translation capacity at the cold-shock temperature and, even more so, at 37 °C (Figure 4B).

2.2. Trans-acting elements

To detect the possible existence of cold-shock-induced trans-acting factors which would contribute to the cold-shock translational bias in combination with the cis-acting elements just described, a translational experiment similar to that shown in Figure 3A was performed. Also in this case, the maximum level of translation of cold-shock, cold-tolerant and non-cold-shock mRNAs was measured at 15 °C and 37 °C. However, unlike in the experiment of Figure 3A, in this case the maximum level of translation at 15 °C was measured in extracts of cells subjected to cold-shock, whereas the level of translation at 37 °C was measured in extracts of control cells exponentially growing at 37 °C. As seen from the results presented in the histogram of Figure 5A, the combined effects of cis- and trans-acting factors increase to a large extent the degree of translational discrimination between cold-shock and non-cold-shock mRNAs that has been defined as translational bias. In fact, the 15 °C/37 °C translational ratio is now > 1 for the cold-shock mRNAs, it increases only slightly for the cold-tolerant mRNAs, while for the non-cold-

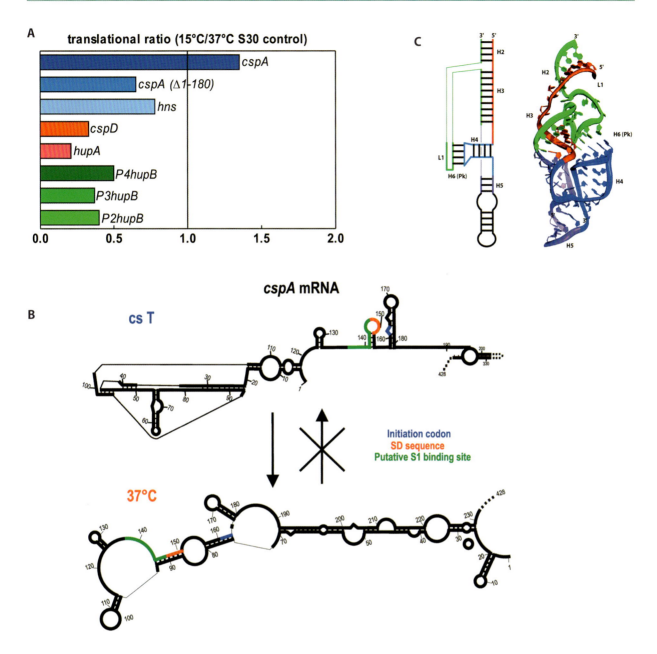

Fig. 3 (A) Translational efficiency of cold-shock (different shades of blue), non-cold-shock (different shades of red) and cold-tolerant (different shades of green) mRNAs at 37 °C and 15 °C in a cell-free extract obtained from cells growing exponentially at 37 °C. The histograms show the ratio between maximum levels of translation attained with each mRNA at 15 °C and 37 °C (Figure reproduced from Giuliodori et al. (2004)) (B) Structure of *cspA* mRNA at the cold-shock temperatures (csT) 10 °C and 20 °C (upper structure) and at 37 °C (lower structure) as determined by chemical and enzymatic probing (Giuliodori et al., 2010). The TIR elements relevant for translational efficiency are: initation codon (blue), SD sequence (red), AU-rich putative S1 binding site (green). (C) Secondary structure of the pseudoknot present at low temperature in *cspA* mRNA (left) with colors indicating the elements of the 5'UTR region giving rise to the three-dimensional structure of the pseudoknot (right). (Figure modified from Giuliodori et al., 2010)

shock mRNAs it remains essentially unmodified or becomes lower (for *cspD* mRNA) (Figure 5A). These results were obtained with an extract of cells cold-shocked for 90 min. It should be noted, however, that the extent of stimulation of cold-shock mRNA translation varies as a function of the length of exposure of the cells from which the extracts were prepared in a way that clearly indicates that the trans-acting factors are not present at the onset of the cold adaptation but accumulate in the cells as a function of the time elapsed since the temperature-downshift (Giuliodori et al., 2004).

A

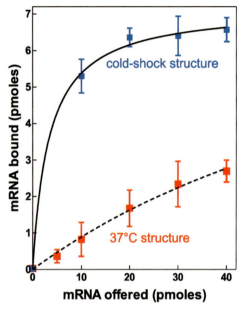

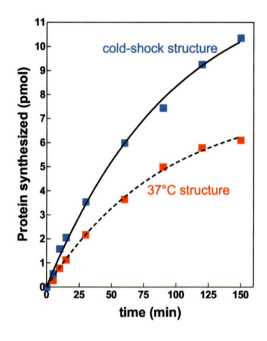

B

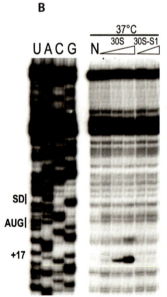

Type of mutation and change(s) in the TIR	Predicted Structure of the TIR at 37°C	Relative Translation Efficiency (-fold)	
		37°C	**15°C**
cspA wt		1	1
cspA Δ118–143 Deletion of ss AU-rich region		<<0.1	0.1

Fig. 4 (A) The cold-shock conformation of *cspA* mRNA (blue) is more active than the 37 °C conformation (red) in binding to the 30S ribosomal subunit (left panel) and translation *in vitro* at 15 °C by a cell-free system from control cells (right panel). (Figure from Giuliodori et al. 2010) (B) Role of S1 and of its RNA binding site in *cspA* mRNA translation. Toeprinting analysis (left) of the translation initiation complex formed at 37 °C with *cspA* mRNA, tRNA[fMet] and increasing amounts of 30S ribosomal subunits containing or lacking S1. Lanes U, A, C and G are sequencing ladders in which start codon AUG, SD sequence and nucleotide +17 from the AUG are indicated; lane N contains *cspA* mRNA alone. The predicted structural and functional consequences of the partial deletion (*cspA* D118–143 mutant) of the putative S1 binding region of the *cspA* mRNA 5'UTR are also shown (right). The functional activity is measured from the translational efficiency of a *cspA::lacZ* chimeric mRNA containing wt or a mutant 5'UTR lacking most of the AU-rich region constructed by Yamanaka et al., (1999). The normalized relative activities detected *in vivo* at 37 °C and 15 °C (Yamanaka et al., 1999) are reported.

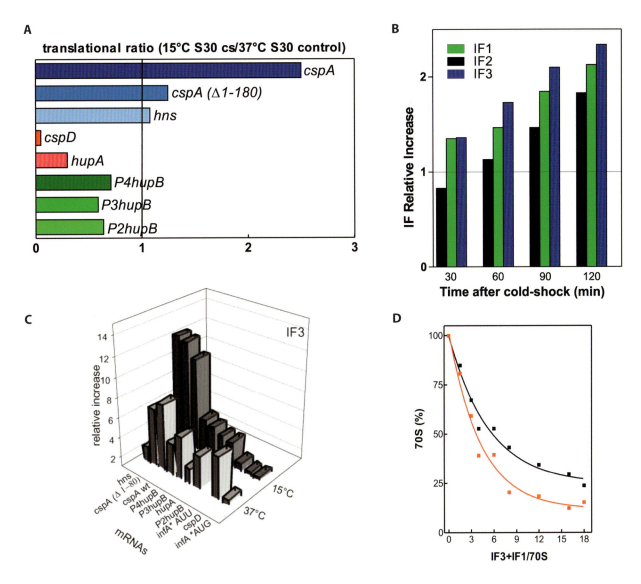

Fig. 5 (A) Translational activity of cold-shock (different shades of blue), non-cold-shock (different shades of red) and cold tolerant (different shades of green) mRNAs at 37 °C in a control cell-free extract (i. e. cells growing exponentially at 37 °C) and at 15 °C in an extract from cells exposed to cold-shock for 90 min. The histogram shows the ratios between maximum translational level attained for each mRNA at 15 °C and at 37 °C. (Figure taken from Giuliodori et al. 2004) (B) Imbalance of the stoichiometric ratios between initiation factor IF1 (green), IF2 (black), IF3 (blue) and ribosomes caused by cold-shock. The values are normalized setting the ratio detected prior to the cold-stress to 1. The data presented in the histogram refer to two hours of cold-shock, but the extent of the imbalance continues to increase up to at least 4 hours reaching a value ≥ 3 (not shown). (Figure taken from Giuliodori et al. 2004) (C) Effect of IF3 on the translation of the indicated mRNAs at 37 °C and 15 °C. The ordinate presents the relative increase in the translational activity of each mRNA obtained in the presence of IF3 with respect to the translational activity in the absence of IF3. (Figure modified from Gualerzi et al. 2003) (D) Effect of increasing IF3+IF1/70S stoichiometric ratio on the amount of 70S monomers remaining undissociated at 37 °C (red) and 15 °C (black). (Figure taken from Giangrossi et al. 2007)

2.3. Multiple origin of the cold-shock translational bias

Taken together, the results presented in sections 2.1 and 2.2 indicate that both specific cis-acting features within cold-shock mRNAs, like *cspA* mRNA, and trans-acting factors, which are synthesized and ac-cumulate in the cold stressed cells, are responsible for generating a positive translational bias in favor of cold-shock mRNAs and a negative bias which reduces the translational capacity of non-cold-shock RNAs. The cis-acting elements identified in *cspA* mRNA confer upon this RNA the properties of an unconventional temperature-sensing RNA (Giuliodori et al.,

2010) insofar as conventional thermosensor RNAs, contrary to *cspA* mRNA, become more exposed and more active as they unfold at higher temperature (Breaker, 2010).

In light of these conclusions, the components of the extract from cold-shock cells were analyzed and compared to those from control cells to identify the nature of the cold-shock trans-acting factors and to determine whether positive and negative translational bias might be due to the same or to a different set of factors.

2.4. Factors responsible for positive translational bias

Since a major involvement of the ribosomes of cold-shocked cells in causing the translational bias was ruled out, the analysis was focused on the soluble proteins that were selectively present or absent in the extracts of cold-shocked cells. Thus, both bulk incorporation of radioactive precursors and pulse-chase experiments were performed to identify the proteins synthesized after the stress and selectively present in the extracts of cold-shocked cells by electrophoretic analysis followed by semi-quantitative immunoblotting with antibodies directed against likely candidates. Among the proteins identified in this way and suspected to cause the translational bias were CspA and the three translation initiation factors IF1, IF2 and IF3. Quite striking was the finding that, as a function of the time elapsed after cold-shock, the levels of IF1 and IF3, in addition to that of IF2, drastically increased with respect to the level of the ribosomes. As a result of this increase, the IFs/ribosomes ratio, which is normally kept constant at about 0.15, increases to about 0.30 after 2 hours (Figure 5B) and to about 0.45 after 3–4 hours cold-shock (not shown). This finding is in line with the premise that expression of the three IFs is subject to a common regulatory mechanism so that their relative levels remain always constant; however, it contradicts the long-held belief that the level of the IFs is maintained constant, under all growth conditions, relative to that of the ribosomes (Howe and Hershey, 1983), and it opened a series of fundamental questions: (i) what is the mechanism responsible for generating the imbalance? (ii) do the IFs play any specific role in the process of cold acclimation? (iii) if yes, what is this role, since the three IFs have been shown to perform more than one function? (iv) are the IFs implicated in the cold-shock translational bias? (v) is there any function of the

IFs which would benefit from a higher IFs/ribosomes stoichiometric ratio at the cold-shock temperature?

To identify the function requiring an increased level of IFs, translational tests with cold-shock, cold-tolerant and non-cold-shock mRNAs were carried out at 37 °C and 15 °C, and the dependence of the translational level upon the presence of the individual aforementioned factors was determined. As seen from the histogram of Figure 5C in which the capacity of IF3 to stimulate the level of translation of various types of mRNAs at high (37 °C) and low (15 °C) temperature is displayed, the IF3-dependence does not indicate any preference for cold-shock mRNAs when translation is carried out at 37 °C, whereas a completely different behavior is observed at 15 °C. In fact, IF3 hardly stimulates translation of non-cold-shock mRNAs and has only a modest effect on the translation of cold-tolerant mRNA, while it selectively and greatly stimulates translation of the cold-shock mRNAs at low temperature. This is precisely what one would expect for a factor producing a positive cold-shock translational bias favoring cold-shock mRNA translation. On the other hand, when similar tests were carried out with IF1, IF2 and CspA, none of these factors displayed any low-temperature-specific influence on cold-shock mRNA translation. In particular, IF2 was found to have a greater beneficial effect on mRNA translation at 37 °C than under cold-shock conditions, while IF1 and CspA were found to stimulate 2–2.5-fold the translation of all types of mRNAs at both low and high temperature. On the other hand, the behavior of IF1 and CspA would justify the attribution to both proteins of an ancillary role in the cold-shock translational bias insofar as they potentiate the effects of IF3. Thus, while the discrimination in favor of cold-shock mRNAs is selectively operated by IF3, both IF1 and CspA would provide a further stimulation of the translational efficiency of those mRNAs which are actively translated at cold-shock temperature. Another function of IF1 and IF3, which was found to require an increased factor/ribosome stoichiometric ratio in the cold, was identified as the subunit anti-association activity of these factors. As seen from Figure 5D, twice as much IF1 and IF3 is needed to obtain a net dissociation of 50 % of the 70S ribosomes at 15 °C compared to 37 °C, most likely to compensate for the greater stability of the ribosomes at the lower temperature.

2.5. Factors responsible for negative translational bias

As seen above, cold-shock not only determines a high level of translation of the transcripts of cold-shock genes, but also causes an overall reduction of the translational activity of non-cold-shock mRNAs, even if these are present in the cells as a result of transcription from an inducible promoter. Translation tests carried out with the paradigm non-cold-shock *cspD* mRNA at high temperature (e. g. 60–70 °C) in cell extracts obtained from thermophilic bacteria gave a clear indication that translation of this mRNA is severely limited by its secondary/tertiary structure (unpublished results). Nevertheless, when the structure of *cspD* mRNA was determined, using the same methods which were used to elucidate the *cspA* mRNA structure, no differences were detected between the structures at 37 °C and 10 °C/20 °C, which are the temperatures relevant for the cold-stress induction. This result indicates that the unfavorable 15 °C/37 °C translational ratio seen in both control (Figure 3A) and cold-shock (Figure 5A) cell extracts is not, or only marginally, due to a direct effect of cis-acting elements, but is mediated by the presence of trans-acting factors that diminish its translational capacity. In a separate set of experiments aimed at identifying the nature of the inhibitory factor(s), the same factors which had been tested for their possible role in a positive translational bias were also tested on the translational activity of *cspD*

mRNA. These experiments clearly indicated that IF3, in addition to favoring the translation of cold-shock mRNAs, is capable of reducing the translational activity of *cspD* mRNA. When the individual steps of the translation initiation pathway were analyzed, using translation systems programmed with *cspA* or *cspD* mRNAs, it was found that, in the presence of *cspA* mRNA, IF3 stimulated the formation of the 30S initiation complex at low temperature, but did not cause any specific inhibition in a system programmed with *cspD* mRNA. Indeed, the initial step of fMet-tRNA binding during formation of the 30S initiation complex was stimulated by IF3 to the same extent with either *cspA* or *cspD* mRNA. It is instead on the subsequent step, consisting in a temperature-dependent locking of the 30S initiation complex (Wintermeyer and Gualerzi, 1983), that IF3 produces a different effect at 15 °C, depending on the nature of the template. While IF3 stimulates locking when the template is *cspA* mRNA, it has no effect when the template is *cspD* mRNA. Furthermore, fMet-tRNA is clearly placed in a non-productive position when bound to the ribosome at low temperature and in the presence of IF3 in response to the non-cold-shock mRNA. In fact, unlike the case of fMet-tRNA bound to the ribosome in the presence of IF3 in response to *cspA* mRNA, which proved to be puromycin-reactive at all temperatures, that bound in response to *cspD* mRNA was found to be productive in the formation of fMet-puromycin only at 37 °C, while at 15 °C it was to a large extent non-productive (Figure

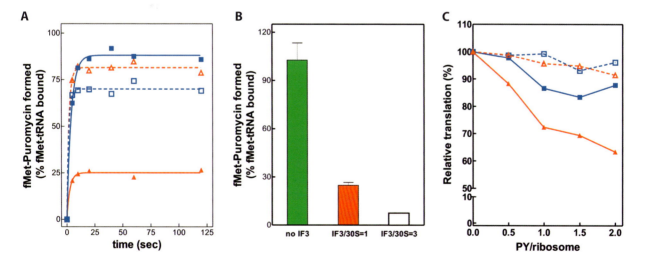

Fig. 6 Trans-acting factors causing a negative bias towards non-cold-shock mRNAs. (A) Effect of IF3 on the time course of fMet-puromycin formation on ribosomes programmed with *cspA* mRNA (blue tracings) and *cspD* mRNA (red tracings) at 37 °C (□, △) and 15 °C (■, ▲). (B) Effect of increasing amounts of IF3 on the puro-mycin reactivity of fMet-tRNA bound to ribosomes programmed with *cspD* mRNA. (Figure taken from Giuliodori et al. 2007) (C) Effect of increasing amounts of PY protein on the translation of a typical cold-shock (blue) and non-cold-shock (red) mRNA at 37 °C (open symbols) and 15 °C (closed symbols).

6A). This temperature-dependent inhibitory effect of IF3 is clearly amplified when IF3 is present in excess relative to ribosomes (Figure 6B).

The experiment presented in Figure 6C strongly suggests that another cold-shock-induced protein which may play a role in inducing a negative cold-shock-induced bias is the ribosome-binding protein PY (Agafonov et al., 1999; Vila-Sanjurjo et al., 2004). Translation of both cold-shock and non-cold-shock mRNAs at 37 °C is essentially unaffected by the addition of PY, while a clear inhibition can be detected at 15 °C. Although translation of cold-shock mRNAs is also affected by PY, the translation of a non-cold-shock mRNA is definitely more sensitive to the presence of PY.

3. Origin of the IFs/ribosome stoichiometric imbalance

Among the several possible mechanisms which could alter the IFs/ribosomes ratio after cold-shock, we examine first the simplest explanation, namely that ribosome synthesis and/or assembly stops or slows down after cold-shock, while synthesis of IFs continues. Experiments carried out to analyze the synthesis of ribosomal components, i. e. rRNA and r-proteins, and ribosomal subunit assembly after temperature-downshift show that this premise is essentially correct (Marchi, 2006; Piersimoni, 2010). After cold-stress, only those ribosomal subunits whose synthesis and assembly are already well advanced have a chance to be completed, while it takes a few hours before newly transcribed rRNA and newly synthesized r-proteins are found in active ribosomes. Also rRNA maturation slows down considerably after the cold-stress. On the other hand, it was demonstrated that the three IFs are actively synthesized during cold-adaptation and that their synthesis is not due to a cold-shock-induced stabilization of pre-existing mRNAs, but that the temperature-stress induces *de novo* transcription of the IF genes through the activation of promoters which are marginally used under optimal growth conditions (Giangrossi et al., 2007; Giuliodori et al., 2007; Brandi, A., unpublished data). In conclusion, the data indicate that the transient stoichiometric imbalance between IFs and ribosomes following cold-shock can be attributed to the mechanism depicted in Figure 7A. As seen in section 2.4, in contrast to IF1 and IF3, no specific role in translation after cold-shock could be detected for IF2. An important clue as to a possible role of this

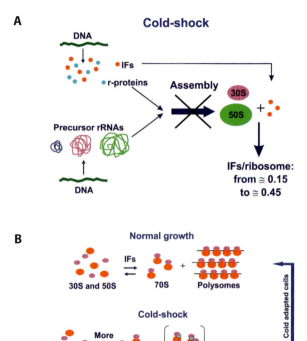

Fig. 7 (A) Mechanisms generating the cold-shock IFs/ribosome imbalance. (B) Postulated rationale for the seemingly paradoxical simultaneous, transient increase after cold-shock of the levels of proteins that favor (PY) or counteract (IF1 and IF3) subunit association. Further details can be found in the text. (Figure modified from Giuliodori et al. 2007)

factor in cold-shock came from experiments aimed at analyzing ribosome assembly after applying cold-stress. In fact, mass-spectrometric analysis of the proteins selectively associated with the ribosomal subunits after cold-stress demonstrated that, unlike the extra copies of IF1 and IF3 found in the soluble fraction of the cell, excess IF2 is associated with the ribosomal subunits. This finding suggests that, whatever role IF2 plays, it is exerted at the ribosome level. It is tempting to speculate that IF2, like other GTPases, may play a role in ribosome assembly at the cold-shock temperature and/or that its protein chaperone activity (Caldas et al., 2000) might be used to assist protein folding at low temperature.

The increased level of IF3 and IF1 relative to the ribosome was interpreted as the consequence of the need of the cell to maintain a sufficient pool of dissociated subunits at low temperature. However, this interpretation seems to clash logically with the finding that cold-shock, as mentioned above, also induces the expression of PY, a ribosome-binding protein that favors subunit association (Agafonov et al., 1999; Vila-Sanjurjo et al., 2004). An explanation capable of reconciling these apparently conflicting results is presented in the model shown in Figure 7B. The model postulates that, during growth under optimal conditions, a large proportion of the ribosomes is actively involved in translation and, therefore, sequestered in polysomes, such that the dynamic equilibrium between subunit association and dissociation applies only to the (small) fraction of ribosomes which are not engaged in polysomes. Following temperature downshift, the number of ribosomes actively synthesizing proteins and, consequently, being associated with polysomes drops considerably. It is suggested that a large proportion of the ribosomes that would have become free at the lower temperature is now stored and possibly preserved from degradation as translationally incompetent, PY-associated 70S ribosomes. Thus, the pool of free 70S ribosomes in equilibrium with their dissociated subunits would not be drastically different in cold-shocked or actively growing cells. Under these conditions, the presence of increased levels of IF1 and IF3 could be essential to counteract the increased affinity between ribosomal subunits caused by lowering the temperature and therefore to ensure that a sufficiently large pool of dissociated subunits is available to start new rounds of translation also under cold-stress conditions.

References

Agafonov DE, Kolb VA, Nazimov IV, Spirin AS (1999) A protein residing at the subunit interface of bacterial ribosome. Proc Natl Acad Sci USA 96: 12 345–12 349

Brandi A, Pietroni P, Gualerzi CO, Pon CL (1996) Post-transcriptional regulation of CspA expression in *Escherichia coli*. Mol Microbiol 19: 231–240

Breaker R (2010) RNA switches out in the cold. Mol Cell 37: 1–2

Caldas T, Laalami S, Richarme G (2000) Chaperone properties of bacterial elongation factor EF-G and initiation factor IF2. J Biol Chem 275: 855–860

Feng W, Tejero R, Zimmerman DE, Inouye M, Montelione GT (1998) Solution NMR structure and backbone dynamics of the major cold-shock protein (CspA) from *Escherichia coli*: evidence for conformational dynamics in the single-stranded RNA binding site. Biochemistry 37: 10 881–10 896

Giangrossi M, Giuliodori AM, Gualerzi CO, Pon CL (2002) Selective expression of the β-subunit of the nucleoid-associated protein HU during cold shock in *Escherichia coli*. Mol Microbiol 44: 205–216

Giangrossi M, Exley RM, LeHegarat F, Pon CL (2001) Different in vivo localization of the *Escherichia coli* proteins CspD and CspA. FEMS Microbiol Lett 202: 171–176

Giangrossi M, Brandi A, Giuliodori AM, Gualerzi CO, Pon CL (2007) Cold-shock-induced de novo transcription and trans-

lation of infA and role of IF1 during cold adaptation. Mol Microbiol 64: 807–821

Giuliodori AM, Brandi A, Gualerzi CO, Pon CL (2004) Preferential translation of cold shock mRNAs during cold adaptation. RNA 10: 265–276

Giuliodori AM, Brandi A, Giangrossi M, Gualerzi CO, Pon CL (2007) Cold-stress-induced de novo expression of infC and role of IF3 in cold-shock translational bias. RNA 13: 1355–1365

Giuliodori AM, Di Pietro F, Marzi S, Masquida B, Wagner R, Romby P, Gualerzi CO, Pon CL (2010) The cspA mRNA is a thermosensor that modulates translation of the cold shock protein CspA. Mol Cell 15: 21–33

Goldenberg D, Azar I, Oppenheim AB (1996) Differential mRNA stability of the cspA gene in the cold-shock response of *Escherichia coli*. Mol Microbiol 19: 241–248

Goldenberg D, Azar I, Oppenheim AB, Brandi A, Pon CL, Gualerzi CO (1997) Role of *Escherichia coli* cspA promoter sequences and adaptation of translational apparatus in the cold shock response. Mol Gen Genet 256: 282–290

Goldstein J, Pollitt NS, Inouye M (1990) Major cold shock protein of *Escherichia coli*. Proc Natl Acad Sci USA 87: 283–287

Graumann PL, Marahiel MA (1998) A superfamily of proteins that contain the cold-shock domain. Trends Biochem Sci 23: 286–290

Gualerzi CO, Giuliodori AM, Pon CL (2003) Transcriptional and post-transcriptional control of cold shock genes. J Mol Biol 331: 527–539

Herendeen SL, VanBogelen RA, Neidhardt FC (1979) Levels of major proteins of *Escherichia coli* during growth at different temperatures. J Bacteriol. 139: 185–194

Howe JG, Hershey JW (1983) Initiation factor and ribosome levels are co-ordinately controlled in *Escherichia coli* growing at different rates. J Biol Chem 258: 1954–1959

Jones PG, VanBogelen RA, Neidhardt FC (1987) Induction of proteins in response to low temperature in *Escherichia coli*. J Bacteriol 169: 2092–2095

La Teana A, Brandi A, Falconi M, Spurio R, Pon CL, Gualerzi CO (1991) Identification of a cold shock transcriptional enhancer of the *Escherichia coli* gene encoding nucleoid protein H-NS. Proc Natl Acad Sci USA 88: 10 907–10 911

La Teana A, Brandi A, O'Connor M, Freddi S, Pon CL (2000) Translation during cold adaptation does not involve mRNA-rRNA base pairing through the downstream box. RNA 6: 1393–1402

Marchi P (2006) Ribosomal subunits synthesis and assembly in cold-shocked and cold-adapted *Escherichia coli* cells. Doctoral thesis, University of Camerino

Newkirk K, Feng W, Jiang W, Tejero R, Emerson SD, Inouye M, Montelione GT (1994) Solution NMR structure of the major cold shock protein (CspA) from *Escherichia coli*: identification of a binding epitope for DNA. Proc Natl Acad Sci USA 91: 5114–5118

Phadtare S, Alsina J, Inouye M (1999) Cold shock response and cold-shock proteins. Curr Opin Microbiol 2: 175–180

Piersimoni L (2010) Weird facets of ribosome synthesis in stressfully chilled bacteria. Doctoral thesis, University of Camerino

Pon CL, Calogero RA, Gualerzi CO (1988) Identification, cloning, nucleotide sequence and chromosomal map location of hns, the structural gene for *Escherichia coli* DNA-binding protein H-NS. Mol Gen Genet 212: 199–202

Pon CL, Stella S, Gualerzi CO (2005) Repression of transcription by curved DNA and nucleoid protein H-NS: a mode of bacterial gene regulation. In: Ohyama T (ed) DNA conformation and transcription. Landes Bioscience, Austin, pp 52–65

Schindelin H, Marahiel MA, Heinemann U (1993) Universal nucleic acid-binding domain revealed by crystal structure of the B. subtilis major cold-shock protein. Nature 364: 164–168

Schindelin H, Jiang W, Inouye M, Heinemann U (1994) Crystal structure of CspA, the major cold shock protein of *Escherichia coli*. Proc Natl Acad Sci USA 91: 5119–5123

Sette M, Tilborg P, Spurio R, Kaptein R, Paci M, Gualerzi CO, Boelens R (1997) The structure of the translational initiation factor IF1 from *E. coli* contains an oligomer-binding motif. EMBO J 16: 1436–1443

Tanabe H, Goldstein J, Yang M, Inouye M (1992) Identification of the promoter region of the *Escherichia coli* major cold shock gene, cspA. J Bacteriol 174: 3867–3873

Thompson JD, Higgins DG, Gibson TJ (1994) CLUSTAL W: improving the sensitivity of progressive multiple sequence alignment through sequence weighting, position specific gap penalties and weight matrix choice. Nucleic Acids Res 22: 4673–4680

Vila-Sanjuruo A, Schuwirth BS, Hau CW, Cate JH (2004) Structural basis for the control of translation initiation during stress. Nat Struct Mol Biol 11: 1054–1059

Weber MH, Beckering CL, Marahiel MA (2001) Complementation of cold shock proteins by translation initiation factor IF1 in vivo. J Bacteriol 183: 7381–7386

Weber MH, Marahiel MA (2003) Bacterial cold shock responses. Sci Prog 86: 9–75

Wintermeyer W, Gualerzi C (1983) Effect of *Escherichia coli* initiation factors on the kinetics of N-AcPhe-tRNAPhe binding to 30S ribosomal subunits. A fluorescence stopped-flow study. Biochemistry 22: 690–694

Yamanaka K (1999) Cold shock response in *Escherichia coli*. J Mol Microbiol Biotechnol 1: 193–202

Recruiting knotty partners: The roles of translation initiation factors in mRNA recruitment to the eukaryotic ribosome

13

Sarah F. Mitchell, Sarah E. Walker, Vaishnavi Rajagopal, Colin Echeverría Aitken and Jon R. Lorsch

1. Introduction

Eukaryotic translation initiation begins with the binding of a ternary complex (TC) composed of eukaryotic initiation factor (eIF) 2, methionyl initiator tRNA (Met-tRNA$_i$) and GTP to the 40S ribosomal subunit to form the 43S pre-initiation complex (PIC) in a process that is promoted by eIFs 1, 1A, and 3 (Figure 1; for reviews of the entire pathway of translation initiation see (Jackson et al., 2010; Lorsch and Dever, 2010)). This complex is then capable of binding the mRNA near the 5' end and moving in a 5'-to-3' direction in search of the start codon. As the start codon in eukaryotic mRNA is usually the first AUG codon, it is necessary for the ribosome to be recruited to the very 5' end of the mRNA. Were it to bind 3' to the start codon, it would scan to the next AUG and make an aberrant polypeptide. This localization is achieved by the presence of a 7-methylguanosine cap on the 5' end of the mRNA that, through a number of protein-protein interactions, brings the ribosome to the very beginning of the mRNA. Protein factors are also thought to be responsible for removing secondary structure and RNA binding proteins from the RNA to create a single-stranded region for the ribosome to bind. Interaction between factors at the 5' and 3' ends of the message functionally circularizes the mRNA and allows communication between the ends. After initial recruitment of the PIC, many of these factors are also thought to be involved in the process of scanning, removing structure and proteins so that the ribosome can move forward in search of the start codon, and potentially also pushing the PIC along the mRNA or increasing the directionality of this movement.

A number of initiation factors that are involved in this process have been identified and their general activities studied both *in vivo* and *in vitro*. eIF4E,

also known as the cap-binding protein, binds the 5'-7-methylguanosine cap structure of the mRNA (Sonenberg et al., 1978). It also interacts with eIF4G, a large, multi-domain scaffolding protein. In addition to eIF4E, eIF4G binds eIF4A, a DEAD-box RNA helicase, believed to be responsible for removing structure from the 5'-untranslated region (5'-UTR) (Lawson et al., 1986; Svitkin et al., 1996). Together eIF4E, eIF4G, and eIF4A make up the eIF4F cap-binding complex. Mammalian eIF4B increases the helicase activity of eIF4A *in vitro* (Grifo et al., 1984; Ray et al., 1985). Mammalian cells also have a second protein, eIF4H, with sequence and activity similar to that of eIF4B (Richter-Cook et al., 1998). eIF3, a large, multimeric protein (five subunits in yeast, ≥ 12 in mammals), is involved in many steps in the initiation process including mRNA recruitment to the ribosome (reviewed in (Hinnebusch, 2006)). The goal of this chapter is to outline what we know about the roles of each of these proteins in mRNA recruitment to the ribosome.

All of the proteins listed above lack bacterial homologues. This is not surprising given a number of important differences between eukaryotic and bacterial translation initiation. Bacterial translation initiation begins on mRNAs as they are being transcribed. In eukaryotic systems, the mRNA is transcribed, spliced, and modified before being exported from the nucleus. This allows time for more complex structures to form in the mRNA and provides ample opportunity for the mRNA to interact with a variety of RNA binding proteins, both specific and non-specific. Differences in the process of start codon recognition also alter the way in which mRNA must be recruited to the ribosome in eukaryotes and bacteria. Because bacterial mRNAs are polycistronic and, like archaeal mRNAs, have Shine-Dalgarno sequences that identify the start codons, it is not necessary for

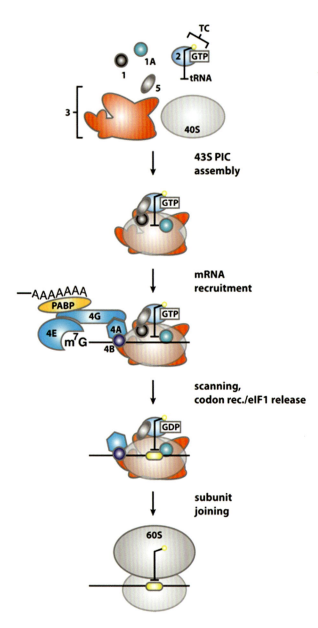

Fig. 1 Eukaryotic translation initiation. A cartoon depiction of the key steps of translation initiation as described in the introduction. In the first step, eIFs 1, 1A and 3 promote binding of a ternary complex of eIF2, initiator methionyl tRNA and GTP to the 40S ribosomal subunit. The resulting 43S pre-initiation complex (PIC) is then directed to bind to an mRNA near the 5' end by eIFs 4E, 4G, 4A and 4B. The mRNP is circularized through the interaction between eIF4G and PABP. The PIC then scans the mRNA in a 5'-to-3' direction in search of the start codon. Upon start codon recognition eIF2, and its associated GTPase activating protein eIF5, release the phosphate molecule resulting from GTP hydrolysis. The large subunit is then able to join the complex in a process facilitated by eIF5B, resulting in the final 80S initiation complex.

the ribosome to bind the 5' end of the mRNA in order to locate the correct AUG. Thus, the cap structure and cap-binding complex are found exclusively in eukaryotic organisms.

This additional level of complexity has provided eukaryotic systems with a number of opportunities to regulate translation initiation. In fact, it is generally thought that mRNA recruitment to the ribosome is the step at which the majority of translational control is exerted. At the most direct level, the structure and length of the 5'-UTR of an mRNA influences the efficiency of its translation. Short, unstructured 5'-UTR sequences correlate with more efficient initiation (Kozak, 1986; Baim and Sherman, 1988; Bettany et al., 1989; Kozak, 1989; Vega Laso et al., 1993). Additional levels of control through a variety of pathways also regulate the level of translation, either globally, or for individual mRNAs. One such system that has been particularly well studied is the mTOR pathway (Robert and Pelletier, 2009). Detailed descriptions of these mechanisms are not within the scope of this chapter, which focuses on the general pathway of mRNA recruitment (for further information on regulatory pathways see (Sonenberg and Hinnebusch, 2009)).

2. Individual initiation factors

2.1. eIF3

One of the first initiation factors to be discovered, eIF3 has been implicated in nearly every aspect of the translation initiation pathway. Composed of up to 13 nonidentical subunits, it is the largest and most complex of the initiation factors (Hinnebusch, 2006). In fact, its scale – more than 350 kDa in yeast and up to ~820 kDa in humans – approaches that of the 40S ribosomal subunit. In the yeast *Saccharomyces cerevisiae*, eIF3 is composed of five essential subunits (eIF3a, eIF3b, eIF3c, eIF3g, and eIF3i) thought to represent a core complex capable of executing the crucial functions performed by eIF3 in more complex organisms (Phan et al., 1998). A sixth, non-stoichiometric and non-essential subunit, eIF3j, is thought to modulate the activity of the core complex (Phan et al., 1998; Valasek et al., 1999; Valasek et al., 2001; Nielsen et al., 2006). It has been shown to bind near the ribosomal A site and to compete with eIF1A and mRNA for binding to the 40S subunit (Fraser et al., 2007), although the mechanistic implications of these effects are not yet clear.

During the early stages of initiation, eIF3 is thought to bind the 40S subunit, preventing its premature association with the 60S subunit (Kolupaeva et al., 2005). eIF3 promotes the association of TC with the 40S subunit to form the 43S PIC in both mammalian (Trachsel et al., 1977; Chaudhuri et al., 1999; Majumdar et al., 2003) and yeast systems (Danaie et al., 1995; Phan et al., 1998; Phan et al., 2001; Jivotovskaya et al., 2006). Although in mammals eIF3 appears to be required for productive TC recruitment to the 40S subunit, in yeast it is only stimulatory (Algire et al., 2002; Jivotovskaya et al., 2006). Recent studies have also suggested that eIF3 participates in scanning of the mRNA and recognition of the start codon by the PIC (Valasek et al., 2004; Yamamoto et al., 2005; Szamecz et al., 2008).

Mirroring its diverse roles, eIF3 interacts with many components of the initiation machinery. eIF3 has been shown to interact with the 40S subunit, ternary complex, eIF1, eIF5, eIF4G (in mammals) and the mRNA itself (Asano et al., 2000; Korneeva et al., 2000; Kolupaeva et al., 2005; Pisarev et al., 2008). Recent studies have demonstrated that in yeast eIF3 can form a higher order complex with eIF1, eIF5, and TC (Asano et al., 2000). This multifactor complex (MFC) may facilitate the formation of the 43S PIC, perhaps by cooperative binding of its constituent components (Valasek et al., 2002; Valasek et al., 2004).

Evidence that eIF3 stimulates the recruitment of mRNA to the assembled 43S complex has existed for decades. Early studies demonstrated that eIF3 is essential for binding of mRNA to the 43S PIC (Trachsel et al., 1977). This observation was confirmed by subsequent studies (Pestova and Kolupaeva, 2002; Kolupaeva et al., 2005). It was further observed that the eIF3·40S interaction is stabilized by mRNA (Kolupaeva et al., 2005). In addition, the *prt1–1* mutation in the eIF3b subunit interferes with the binding of mRNA to the 40S subunit in yeast extracts (Phan et al., 2001). This mRNA binding defect can be rescued by addition of either eIF3 or a subcomplex of the a, b, and c subunits. Consistent with this, depletion of eIF3 in yeast severely impairs mRNA recruitment to the 43S PICs *in vivo* (Jivotovskaya et al., 2006). In both of these examples, however, eIF3 mutants also resulted in decreased association of the MFC components, TC, eIF5, and eIF1, with the 40S subunit. As a consequence, these observed effects could also be partially explained as a downstream consequence of decreased TC binding to the 43S PIC; TC has previously been shown to stimulate the binding of mRNA to 40S subunits (Maag et al., 2005; Jivotovskaya et al., 2006). Owing to the participation of eIF3 in many aspects of the initiation pathway, and the difficulty of separating these processes both *in vivo* and *in vitro*, providing direct evidence for the role of eIF3 in mRNA recruitment has proved challenging.

Mammalian eIF3 has been shown to bind eIF4G in several studies (Lamphear et al., 1995; Imataka and Sonenberg, 1997; Korneeva et al., 2000; Morino et al., 2000; LeFebvre et al., 2006), leading to the suggestion that this interaction might serve as a bridge between the mRNA 5' end and the 43S PIC. In contrast, no evidence of an interaction between eIF4G and eIF3 has been obtained in yeast. Moreover, neither of the two yeast eIF4G isoforms contains the eIF3-binding domain observed in mammalian eIF4G (Marintchev and Wagner, 2005). Instead, yeast eIF3 might bind to eIF4G via its interaction with eIF5 or eIF1 (Asano et al., 2001; Phan et al., 2001). Another possibility is that eIF3 interacts with the mRNA 5' end via eIF4B (Methot et al., 1996a). In mammals, eIF3a has been observed to interact with eIF4B (Methot et al., 1996b), while in yeast there is evidence that eIF3g can interact with eIF4B (Vornlocher et al., 1999). By engaging components of the cap-binding complex, or other factors associated with the mRNA, eIF3 might serve as a hub linking the 43S PIC with the 5' end of the mRNA.

The idea that eIF3 participates in mRNA recruitment indirectly, through its interactions with other components of the initiation pathway, is consistent with the emerging proposal that eIF3 functions as a "versatile scaffold" for the initiation machinery (Hinnebusch, 2006). Through the diverse interactions of its distinct subunits, eIF3 may orchestrate connections among various initiation factors, TC, mRNA, and the 40S subunit. Consistent with this interpretation, cryo-EM studies suggested that eIF3 binds the solvent face of the 40S subunit (Srivastava et al., 1992; Siridechadilok et al., 2005). In the most recent models, eIF3 appears as a five-lobed mass, reminiscent of a starfish. Interestingly, one arm of the factor projects towards the intersubunit face of the 40S subunit, near the mRNA exit channel of the E site (Siridechadilok et al., 2005). These data are supported by *in vivo* studies in yeast, which suggest that eIF3c and the N-terminal domain of eIF3a interact with ribosomal protein Rps0A, which is predicted to be on the solvent face of the 40S subunit near the protuberance and the E site (Spahn et al., 2001; Valasek et al., 2003). This study also identified an interaction between the C-terminal domain of eIF3a and helices 16–18 of domain I of the 18S rRNA. This interaction, which is necessary and sufficient for binding of the eIF3a C-terminal domain to the 40S ribosomal subunit *in vitro*, places the C-terminal domain of eIF3a at the shoulder, near the entry channel for mRNA, and might provide it access to the intersubunit face of the 40S subunit; the simultaneous observation that the N-terminal domain of eIF3a interacts at the mRNA exit channel suggests this subunit spans the 40S solvent face. Interestingly, no density for eIF3 appears near the 40S shoulder in the cryo-EM model (Siridechadilok et al., 2005). Enzymatic probing of 43S PICs, however, indicates that G_{537} and G_{539} in helix 16 are protected when eIF3 binds, and hydroxyl radical cleavage data also suggest eIF3 interacts in the vicinity of $GC_{537-538}$, near the solvent face of the 40S shoulder (Pisarev et al., 2008). This study also confirmed the positioning of specific domains of eIF3 near the exit channel of the mRNA, as observed in *in vivo* studies in yeast (Valasek et al., 2003). Both eIF3a and eIF3d, a subunit that has not been identified in yeast, can be crosslinked in 48S PICs to 4-thiouridine residues at positions −14 and −8 to −17 of the mRNA, respectively. These observations are consistent with previous studies in which eIF3 has been crosslinked to the mRNA in 48S complexes (Nygard and Westermann, 1982; Westermann and Nygard, 1984), and suggest that either eIF3a or eIF3d (in mammals) forms the arm observed in the cryo-EM model that projects into the mRNA exit channel. Interestingly, positioning of eIF3d, as predicted by the observed crosslinking pattern, places it between eIF3a and the 40S subunit, forming part of the mRNA platform at the exit site. The absence of this subunit in yeast suggests that its role might be taken by other factors capable of mediating interactions with both eIF3a and the 40S subunit.

Recent work has linked the interaction of eIF3 near the mRNA binding platform with binding of mRNA to the 43S PIC (Szamecz et al., 2008). The observations that domains of eIF3 can be localized to both the mRNA entry and exit channels, and appear to interact with the mRNA, suggest that eIF3 may act directly in mRNA recruitment, not simply via its interactions with other factors. Consistent with this proposal, several eIF3 subunits have been shown to bind RNA (Asano et al., 1997; Block et al., 1998; Hanachi et al., 1999; Lindqvist et al., 2008).

One of the primary difficulties in dissecting the precise role of eIF3 in mRNA recruitment is the complexity of its interactions throughout the initiation pathway. Through its distinct subunits, eIF3 interacts with many of the components of the initiation machinery, thereby participating in multiple core events of the pathway. Untangling the role of eIF3 in mRNA recruitment from events both upstream and downstream has remained challenging. New *in vitro* assays that report on the overall progress of mRNA recruitment might shine light on the role of eIF3 in this process, and allow

correlation with *in vivo* observations. Notably, eIF3 is not required for stable TC recruitment in yeast, suggesting that *in vitro* approaches employing purified yeast factors might succeed in more precisely defining the role of eIF3 in mRNA recruitment in this organism. Still more interesting would be assays that can segregate the initial stages of mRNA recruitment from downstream events such as scanning and start-codon recognition. Combined with continuing *in vivo* and structural studies, these assays might finally unravel the precise role and mechanism of action of eIF3 in mRNA recruitment.

2.2. eIF4E

eIF4E was originally identified as the only initiation factor from rabbit reticulocyte lysate (RRL) to specifically crosslink to the 5'-7-methylguanosine cap on mRNAs (Sonenberg et al., 1978). Binding of eIF4E to the cap and eIF4G is thought to direct the 43S complex to the 5'-end of the mRNA so that it can scan for the first AUG codon and initiate translation at the correct position (Tarun and Sachs, 1997). eIF4E is well-conserved throughout the eukaryotes, with yeast and mammalian isoforms sharing ~33% sequence identity. Expression of murine eIF4E is able to rescue the otherwise lethal phenotype of the yeast eIF4E deletion (Altmann et al., 1989).

Several structures of eIF4E bound to m7GDP or both m7GDP and fragments of eIF4G have now been published, and reveal a conserved mode of cap binding (Marcotrigiano et al., 1997; Matsuo et al., 1997; Marcotrigiano et al., 1999; Gross et al., 2003). The structure of eIF4E resembles that of a cupped hand, which grasps the cap structure. Two conserved tryptophan residues (W58 and W104 in yeast) stack on either side of the m7G moiety in this cap-binding slot, while several positively charged residues coordinate the phosphate groups. Each of the residues involved in binding the cap is conserved in all forms of eIF4E. Binding of eIF4G to the opposite surface of eIF4E does not affect the conformation of the cap-binding pocket. eIF4G wraps around the N terminus, or wrist, of eIF4E like a bracelet, interacting with several essential and conserved residues that are also necessary for eIF4E–eIF4G interaction and translation (Ptushkina et al., 1998; Marcotrigiano et al., 1999; Gross et al., 2003).

There has been some controversy regarding the function of the eIF4E-eIF4G interaction in cap binding. Several studies initially suggested that eIF4G en-

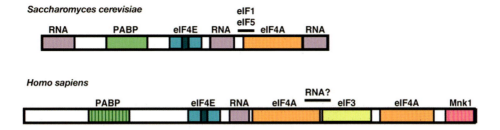

Fig. 2 Schematic of binding sites within *S. cerevisiae* and human eIF4G. Although the general positions of the PABP binding regions are similar between the two, the sequence does not appear to be conserved. The dark regions in the eIF4E binding sites denote the conserved $Y(X)_4L\Phi$ sequence.

hanced binding of eIF4E to the 5' cap in yeast, mammalian, and plant systems by conformational coupling (Haghighat and Sonenberg, 1997; Ptushkina et al., 1998; Gross et al., 2003; Michon et al., 2006), while others did not detect an effect of eIF4G on the eIF4E-cap interaction (Goss et al., 1990; Niedzwiecka et al., 2002). Moreover, the lack of a structural change in the eIF4E cap-binding slot upon eIF4G binding further suggested that eIF4E-cap interactions are not affected by eIF4G (Gross et al., 2003). A recent kinetic study established that there is no strict order of binding of cap and eIF4G to eIF4E and that interaction with eIF4G does not modulate the affinity of eIF4E for the cap (Slepenkov et al., 2008). Additionally, this study provided an explanation for the disparity in previous results by demonstrating that recombinant eIF4E purified from inclusion bodies formed inactive multimeric species. These inactive aggregates were not present in preparations from the soluble fraction and could be dissociated into active monomers upon binding to eIF4G, which increased the observed binding to cap analogs and led to the impression that the K_d for the cap-eIF4E interaction was reduced by binding to eIF4G.

It is thought that eIF4E is limiting in many cell types to maintain low translation of transcripts containing structured 5'-UTRs. eIF4E is present in approximately equimolar amounts relative to ribosomes in yeast(Von Der Haar and McCarthy, 2002), but in other cell types is expressed at much lower levels. In fact, overexpression of eIF4E in several mammalian cell lines results in malignant transformation (reviewed in (Sonenberg, 2008)). The activity of eIF4E is controlled by eIF4E-binding proteins (4E-BPs) that compete with eIF4G for binding to the factor. The activity of these 4E-BPs is regulated via phosphorylation through the TOR pathway; in mammals, hypophosphorylated 4E-BP1 binds eIF4E with increased affinity and represses cap-de-pendent initiation (Robert and Pelletier, 2009). eIF4E itself can also be phosphorylated, although the effect of this modification is not fully understood.

2.3. eIF4G

eIF4G, a large (> 900 amino acids) protein, circularizes the mRNP through interactions with eIF4E bound to the 5' cap and with PABP on the 3' poly(A) tail. Like eIF3, eIF4G is thought to act as a central interaction hub during mRNA recruitment to the PIC. In addition to binding eIF4E and PABP, eIF4G interacts with eIF4A, eIF1, eIF5 and RNA and is consequently called a scaffolding protein (He et al., 2003). As mentioned above, mammalian, but not yeast, eIF4G interacts with eIF3 (Lamphear et al., 1995). In fact, the eIF3 binding site is not present in the eIF4G of lower eukaryotes (Marintchev and Wagner, 2005).

It was initially thought that eIF4G acted primarily at the level of mRNA recruitment to the 43S complex, but recent eIF4G-degron depletion experiments suggested otherwise (Jivotovskaya et al., 2006). Depletion of eIF4G from yeast resulted in the appearance of half-mer shoulders on polysome profiles, indicating accumulation of 48S PICs. These data suggest that eIF4G function is critical for one or more steps downstream of mRNA recruitment, such as scanning, start codon recognition or joining of the large subunit. Consistent with this, yeast genetic studies have found mutations in eIF4G that reduce the fidelity of start codon recognition (He et al., 2003).

Fungal eIF4G can be divided into N-terminal and C-terminal domains, containing different segments for interactions with each of its binding partners (Figure 2). The N-terminal ~ 500 amino acids is comprised of an RNA binding site, the PABP binding site, and the eIF4E binding site, while the C-terminal domain contains the eIF4A binding site, two additional RNA binding sites, and the regions for interaction with eIF1 and eIF5 (Kessler and Sachs, 1998; Dominguez et al., 1999; Berset et al., 2003; He et al., 2003). In mammals, the C-terminal region of eIF4G has a site for eIF3 interaction and a third domain that includes a second eIF4A binding site (Imataka and Sonenberg, 1997).

The consensus eIF4E binding sequence was determined (Mader et al., 1995), and structures of this peptide from eIF4G bound to eIF4E have been published (Marcotrigiano et al., 1999; Gross et al., 2003). This consensus sequence is $Y(X)_4L\Phi$ (X = variable, $\Phi = L >> M$, F), and is conserved in several eIF4E-binding proteins (Mader et al., 1995). Although eIF4E-eIF4G interaction does not introduce major changes in eIF4E, it induces folding of this region of eIF4G (Marcotrigiano et al.,1999; Gross et al., 2003; Hershey et al., 1999).

The eIF4A binding site was mapped to the C-terminal portion of yeast eIF4G1, amino acids 542–883 (Dominguez et al., 1999). The domain is conserved in mammals and a crystal structure of the human version shows 10 alpha helices that come together to form a crescent of 5 HEAT repeats (Schutz et al., 2008). Expression of the corresponding domain in yeast inhibited cell growth, and addition of this fragment of eIF4G to eIF4A-dependent yeast extracts inhibited translation, presumably by competing with WT eIF4G for eIF4A binding (Dominguez et al., 1999). Increasing the concentration of WT eIF4A in these cells or extracts rescued cell growth and translation, respectively, suggesting that this interaction between eIF4G and eIF4A is required for translation. In mammalian eIF4G, there are two eIF4A binding sites: the central site that shares similarity to the yeast eIF4A binding site and an additional C-terminal eIF4A binding site that is not present in yeast. Despite the strong conservation of eIF4A and the central eIF4A binding site, and the ability of murine eIF4A to interact with yeast eIF4G, murine eIF4A was unable to substitute for yeast eIF4A *in vivo*, and could not support yeast translation *in vitro* (Dominguez et al., 2001). This suggests that the interaction between eIF4G and eIF4A is essential, but varies functionally between yeast and higher eukaryotes. The functional consequences of this interaction are discussed below.

The interaction of PABP and eIF4G has been addressed by genetic and biochemical studies, and captured by a co-crystal structure of a 29 amino acid segment of the human eIF4G PABP-binding site bound to the viral protein NSP3, which competes with PABP for the same binding interface on eIF4G (Kessler and

Sachs, 1998) (Tarun and Sachs, 1996; Groft and Burley, 2002). The N terminus of yeast eIF4G1 (amino acids 188–300) is responsible for interaction with PABP (Tarun et al., 1997). Although this interaction also occurs in higher eukaryotes, it is not conserved at a molecular level. The replacement of residues in yeast PABP with the corresponding human residues disrupts the interaction (Otero et al., 1999).

There are at least three RNA-binding domains in yeast eIF4G, one at the extreme N-terminus (amino acids 1–82), one in the middle (amino acids 492–539), and one at the C-terminus (amino acids 883–952). Each of these sites has roughly the same affinity for single-stranded RNA and is required for eIF4G activity in translation (Berset et al., 2003). The presence of all three binding sites confers ~100-fold greater affinity for RNA on full-length eIF4G than any of the individual RNA-binding domains has on its own (Berset et al., 2003). This RNA-binding activity may serve to stabilize interactions within the mRNP and provide additional points of contact with the mRNA or stabilize a conformation of the mRNP favorable for binding to the PIC (Berset et al., 2003).

eIF4G has multiple isoforms in eukaryotes. In *S. cerevisiae*, eIF4G is encoded by two genes, TIF4631 and TIF4632, that share ~50% sequence identity (Goyer et al., 1993) (Clarkson et al.). Although the two isoforms are thought to serve an equivalent purpose (Goyer et al., 1993) (Clarkson et al.), eIF4G1 is expressed at ~three-fold higher levels than eIF4G2. Other differences have also been noted. For example, nuclease-treated extracts lacking eIF4G1 are better able to support translation of uncapped, polyadenylated luciferase RNA than extracts lacking eIF4G2, and the eIF4G2-containing extracts are less susceptible to inhibition by cap analog (Tarun et al., 1997). This suggests that eIF4G2 may rely less on the cap for its function in translation than eIF4G1, in line with the observation that eIF4G1 copurifies with eIF4E, while eIF4G2 does not (Tarun and Sachs, 1996). Additionally, deletion of TIF4631 leads to a growth and translation defect, whereas deletion of TIF4632 does not affect growth or translation (Goyer et al., 1993; Clarkson et al., 2010). However, recent studies examining homogenic strains of yeast containing either two copies of TIF4631 or two copies of TIF4632 revealed no growth defects and similar ribosome occupancies to WT for mRNAs in both strains, suggesting the differences observed in strains in which a single copy was deleted resulted from altered levels of total eIF4G rather than differences in function (Clarkson et al., 2010). While this does indicate that mRNAs do not rely on a particular eIF4G isoform for translation, it does not rule out the possibility that conditions exist under which isoform-specific differences may be observed.

2.4. eIF4A

eIF4A is a ~50 kDa DEAD-box RNA helicase containing seven conserved sequence motifs that are important for its ATP binding and hydrolysis and RNA-unwinding activities (Schmid and Linder, 1992; Gorbalenya and Koonin, 1993). The mammalian protein exists in three isoforms: eIF4AI, eIF4AII (translation factors), and eIF4AIII (part of the exon-junction complex) (Nielsen and Trachsel, 1988). In the yeast *Saccharomyces cerevisiae*, two duplicate genes, TIF1 and TIF2, encode eIF4A (Nielsen et al., 1985).

Biochemical analyses of mammalian eIF4AI showed that it binds ssRNA in an ATP-dependent manner and possesses RNA-stimulated ATPase and helicase activities (Grifo et al., 1982; Grifo et al., 1984; Abramson et al., 1987; Lawson et al., 1989; Pause et al., 1994; Lorsch and Herschlag, 1998a; Rogers et al., 2001a). Further analysis of the RNA unwinding and ATPase activities revealed mammalian eIF4A to be a non-processive, bidirectional RNA helicase (Abramson et al., 1988; Rozen et al., 1990; Peck and Herschlag, 1999; Rogers et al., 1999). Genetic analyses with yeast eIF4A demonstrated that mutations that abrogate ATP hydrolysis, RNA unwinding and translational activity *in vitro* commensurately affect the function of the factor *in vivo* (Blum et al., 1992).

The crystal structure of eIF4A from yeast revealed that the protein is dumbbell shaped, with N- and the C-terminal domains connected by a flexible 11-residue linker (Caruthers et al., 2000) (Figure 3). All the conserved sequence motifs implicated in ATP binding/hydrolysis and RNA binding are present at the interface of the two domains. Interestingly, no spatial relationship is observed between these conserved motifs in the unliganded state of the enzyme (Caruthers et al., 2000) (Figure 3). This suggested that eIF4A exists as an extended molecule in solution, which undergoes a conformational change to couple ATP binding and/or hydrolysis to RNA binding. This conclusion was supported by limited proteolysis and kinetic and thermodynamic studies of mammalian eIF4A that demonstrated a cycle of conformational changes in the factor upon substrate binding and conversion of ATP to ADP (Lorsch and Herschlag, 1998a, b).

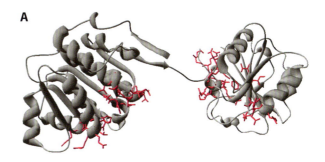

A

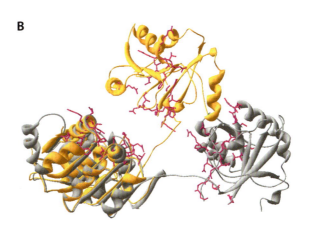

B

Fig. 3 Structures of eIF4A. (A) Crystal structure of full-length eIF4A. The non-liganded form (open form) of eIF4A (PDB ID: 1FUU) is shown here. The protein assumes an extended dumbbell shape. Side-chains of the residues implicated in ATP binding and hydrolysis and RNA binding are shown in pink. The image was rendered using SWISS-PROT/DeepView (*http://www. expasy. org/ spdbv/*). (B) Open and closed conformations of eIF4A. eIF4G binding to eIF4A induces a big change in the relative orientation of the two RecA domains of eIF4A. eIF4A in the open conformation (PDB ID: 1FUU) is shown in grey and eIF4A in the closed conformation (PDB ID: 2VSX) is shown in gold. The conserved motifs implicated in ATP binding/hydrolysis and RNA binding are shown in pink. The figure was generated by aligning the N-terminal domains of eIF4A from the two crystal structures using the "Magic fit" function of SWISS-MODEL (*http://www. expasy. org/spdbv/*).

The physical interaction between eIF4A and eIF4G (which, along with eIF4E, form the eIF4F complex) results in a large stimulation of eIF4A's RNA-stimulated ATPase and RNA unwinding activities (Abramson et al., 1987; Pause et al., 1994; Korneeva et al., 2005; Hinton et al., 2007). The structural basis of this stimulation was revealed by protein NMR studies on the human eIF4AI/eIF4GII middle domain complex (Oberer et al., 2005) and a crystal structure of the yeast eIF4A/eIF4G middle domain complex (Schutz et al., 2008). These studies revealed that eIF4G contacts both the N- and the C-terminal domains of eIF4A. This results in the two RecA domains of eIF4A becoming fixed in space such that the residues from each domain involved in ATP binding/hydrolysis and RNA binding now face each other, poised for catalysis (Oberer et al., 2005; Schutz et al., 2008) (Figure 3).

In addition to interacting physically with eIF4G, eIF4A has been shown to have functional interactions with several other initiation factors, including eIF4E, eIF4B, and eIF4H (in mammals) (Rozovsky et al., 2008; Marintchev et al., 2009). In particular, mammalian eIF4B and eIF4H both stimulate RNA unwinding by eIF4A in vitro (Richter et al., 1999; Rogers et al., 2001b; Marintchev et al., 2009).

Secondary structure in the 5'-UTR of mRNAs inhibits translation initiation (e. g., (Kozak, 1989; Vega Laso et al., 1993)). An RNA helicase could alleviate this problem and eIF4A is widely believed to serve this function (Lawson et al., 1986). Consistent with this idea, Ray and coworkers showed that mammalian eIF4F can induce a structural change in mRNA (Ray et al., 1985). More recently, the requirement of eIF4A in translation was shown to be directly proportional to the extent of secondary structure in the 5'-UTRs of mRNA (Svitkin et al., 2001). Based on these and other observations, a general model for the role of eIF4F in cap-dependent recruitment of mRNA to the ribosome was developed. In this model, eIF4E binds to the 5'-m^7G-cap, localizing the eIF4F complex to the 5' end of the mRNA. The helicase component of eIF4F, eIF4A, then unwinds any secondary structures in the 5'-UTR. Through its interactions with eIF3, eIF4G recruits the 43S PIC to the mRNA (although as eIF3 and eIF4G do not appear to interact in yeast additional mechanisms must be operative). The PIC can then scan the mRNA to identify the start codon. As mentioned above, the eIF4F complex appears to play a role in scanning and might also influence start codon recognition itself (He et al., 2003).

Although the proposal that eIF4A within the eIF4F complex removes secondary structures in the 5' end of the mRNA seems reasonably well supported at this point, it does not satisfactorily explain the unusually high concentration of eIF4A in the cell. In yeast, eIF4A is present at ~ 110,000 molecules per cell, fivefold higher than the concentration of ribosomes (Von Der Haar and McCarthy, 2002; Ghaemmaghami et al., 2003). Thus, in addition to its role in the eIF4F complex, eIF4A presumably carries out other functions. For example, it might act downstream on the mRNA to remove additional structures that are farther from the

5' end (Lindqvist et al., 2008; Rozovsky et al., 2008). It might also facilitate scanning of the mRNA by the PIC (Spirin, 2009) or alter the conformation of the ribosome or another RNP.

2.5. Other helicases in translation initiation

Several other RNA helicases have more recently been identified as translation initiation factors. In mammalian systems, the helicase DHX29 is required for translation of mRNAs with highly structured 5'-UTRs (Pisareva et al., 2008). Ded1p, another yeast DEAD-box helicase, has been implicated in translation initiation (Chuang et al., 1997; Iost et al., 1999), among other RNA-dependent processes. Much work has gone towards understanding the mechanism of helicase action by this enzyme (Yang et al.; Chen et al., 2008; Del Campo et al., 2009), but little work has addressed its role as an initiation factor. However, single- molecule studies have suggested that it may be the helicase that removes structure from the 5'-UTR as it is able to unwind more base pairs than is eIF4A (Marsden et al., 2006).

2.6. eIF4B

Early work in reconstituted mammalian systems identified eIF4B as a factor involved in the recruitment of mRNA to the PIC (Trachsel et al., 1977). Later work in yeast demonstrated that eIF4B could stimulate translation and binding of mRNA to 40S ribosomal subunits *in vitro* (Altmann et al., 1995). Work in yeast, wheat germ, and mammals demonstrated that eIF4B is particularly stimulatory for mRNAs with structured 5'-UTRs (Shahbazian et al.; Altmann et al., 1993; Dmitriev et al., 2003; Mayberry et al., 2009). *In vitro* research has branched into a number of directions, establishing a variety of functions for eIF4B and leading to several hypotheses for its mechanism of action. eIF4B from various organisms has been studied and evidence exists for the interaction of one or more of these proteins with eIF4A, eIF3, PABP, RNA, and the ribosome (Naranda et al., 1994; Methot et al., 1996b; Le et al., 1997; Vornlocher et al., 1999; Rozovsky et al., 2008). eIF4B appears to be important for the interactions controlling mRNA recruitment to the PIC, as it associates not only with both single-stranded RNA and the ribosome, but also with other factors involved in mRNA recruitment. Consequently, a confusing aspect of the biology of eIF4B is that it is not essential in yeast, although the deletion does cause significant slow growth and temperature sensitivity phenotypes (Coppolecchia et al., 1993). As can be seen from the following description of the many activities of eIF4B, further investigation will be necessary to determine which activities are involved in its ability to stimulate mRNA binding to the PIC and whether the role of eIF4B might vary among species.

Structurally, canonical eIF4B is composed of two major domains. Near the N terminus is an RNA recognition motif (RRM) (Milburn et al., 1990), the solution structure and RNA binding surface of which has been determined (Figure 4A) (Fleming et al., 2003). The C terminus of the factor contains a highly repeated stretch. In mammalian eIF4B, this region is termed the DRYG domain because it contains an abundance of aspartate, arginine, tyrosine, and glycine residues. It has been shown to be involved in the dimerization of the factor and binding to the eIF3a subunit of eIF3 (Methot et al., 1996b). In yeast this region is made up of six complete repeats and one partial repeat of 21–26 amino acids (Coppolecchia et al., 1993). This sequence is well conserved between repeats, but bears little resemblance to the DRYG sequence found in the mammalian homolog. Plant eIF4Bs contain a glycine-rich, sequence of low complexity that bears no resemblance to the C-terminal domains of either yeast or mammalian eIF4B (Metz et al., 1999). (See Figure 4B for samples of sequences from the C-terminal domains of mammalian, yeast, and plant eIF4Bs.) The N-terminal region of plant eIF4B also has little similarity to other eIF4Bs, and the putative RRM domains of plant eIF4Bs do not agree well with the canonical RRM consensus sequence (Maris et al., 2005) and are not detected as RRMs by pattern searching programs (SFM and JRL, unpublished). In fact, wheat germ eIF4B was identified by its ability to stimulate the activities of eIF4A *in vitro* rather than by sequence similarity to known eIF-4Bs (Browning et al., 1989). Prior to this discovery in 1989, wheat germ iso4G was often referred to as eIF4B, which makes early wheat germ eIF4B literature somewhat confusing to those new to the field (Browning et al., 1989). Overall, eIF4B is an unusual translation initiation factor in that it is not highly conserved among species. It is important to note this striking divergence as it may be that the eIF4Bs differ in function as well as in sequence from one species to another.

The function of eIF4B known best is its ability to stimulate the RNA binding and helicase activities of eIF4A (Ray et al., 1985; Abramson et al., 1987). Wheat

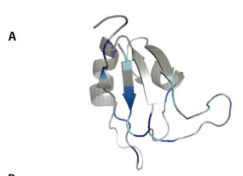

A

B

S. cerevisiae:	PRRGGGADVDWSSARGSNFQGDGRE
H. sapiens:	DDSFGDKYRDRYDSDRYRDGYRDGY
T. aestivum:	LGTGGGFRESSGGGFRESSGGGFRE

Fig. 4 (A) The structure of the RRM domain of eIF4B (PDB ID 2J76). Residues that undergo a chemical shift change when RNA is bound are shown in blue, with residues undergoing the largest change shown in a darker shade (Fleming et al., 2003). (B) Portions of the C-terminal low-complexity repeat sequences are shown for human, yeast and wheat germ eIF4B.

germ eIF4B decreases the K_m of eIF4A for ATP (Bi et al., 2000) and mammalian eIF4B binds to RNA cooperatively with eIF4A (Methot et al., 1994). Although usually a weak helicase, eIF4A gains some processivity in the presence of eIF4B (Rogers et al., 2001b). eIF4B is also thought to promote binding of eIF4A near the cap structure of mRNA, properly positioning it for preparation of the 5'-UTR (Grifo et al., 1982). Strangely, no direct physical interaction between eIF4B and eIF4A has been observed independent of RNA. However, a ternary complex of eIF4A, eIF4B and RNA has been observed with mammalian factors under conditions in which no binding of eIF4B alone was observed (Rozovsky et al., 2008).

eIF4B contains two RNA binding sites, one at the RRM and another in the C-terminal region (Naranda et al., 1994; Niederberger et al., 1998). It is possible that one or both of these sites allows eIF4B to increase the affinity of eIF4A for RNA (Pelletier and Sonenberg, 1985). Alternatively, eIF4B might increase the affinity of eIF4A for RNA by inducing a conformational change in eIF4A. It has also been suggested that eIF4B may act as the single strand binding protein for eIF4A, capturing unwound RNA to slow its re-annealing, potentially by polymerizing on the mRNA (Kapp and Lorsch, 2004; Lindqvist et al., 2008). eIF4B might also prevent back-sliding of the 43S PIC or eIF4A during scanning (Spirin, 2009).

The ability of eIF4B to both interact with the ribosome and bind single-stranded RNA non-specifically could also allow it to act as an additional linkage between the mRNA and 40S subunit (Naranda et al., 1994; Methot et al., 1996a). As mentioned above in the section on eIF3, both mammalian and yeast eIF4B interact with eIF3, though in different ways (Methot et al., 1996b; Vornlocher et al., 1999). This link could serve as a connection between eIF3 and the 5' end of the mRNA, particularly in yeast where no interaction has been observed between eIF4G and eIF3. Another possible role for eIF4B is as an RNA chaperone. eIF4B possesses both RNA annealing and strand separating activities (Altmann et al., 1995) and thus it might function directly to rearrange mRNA or rRNA by facilitating interchange of base-paired regions.

Both RNA binding domains of eIF4B have been shown to play important roles in its activities. Mutations in the RRM domain of yeast eIF4B reduced the ability of eIF4B to suppress the phenotype of temperature-sensitive mutants of eIF4A (Coppolecchia et al., 1993). The RRM has also been shown to be necessary, but not sufficient, for the strand- exchange activity of eIF4B (Niederberger et al., 1998), and to play a role in the stimulation of the helicase activity of eIF4A (Methot et al., 1994). The C-terminal region has been shown to be necessary for the suppression of the slow-growth phenotype of the eIF4B null strain (Niederberger et al., 1998). The C-terminal domains have also been implicated in the association of the factor with ribosomes and in its ability to stimulate the helicase activity of eIF4A (Methot et al., 1994; Naranda et al., 1994). That the oddly repetitive, low-complexity region of eIF4B has so many functions and plays such an important role in the activity of eIF4B is surprising. Additional structural and mechanistic studies of this region of eIF4B will be important for understanding its role in mRNA recruitment.

2.7. eIF4H

Mammalian cells possess an additional RNA-binding initiation factor that is related to eIF4B. Like eIF4B, it contains an RRM domain and increases the RNA-binding, ATPase, and helicase activities of eIF4A (Richter-Cook et al., 1998). eIF4H competes with eIF4B for interaction with eIF4A and acts much like eIF4B, forming a complex with eIF4A and RNA (Rozovsky et al., 2008). Unlike with eIF4B, a direct interaction between eIF4A and eIF4H has been detected, using surface plasmon resonance (Marintchev et al., 2009).

2.8. PABP

A conserved feature of the majority of eukaryotic mRNAs is a string of adenosines added post-transcriptionally to the 3' end. Processing of the 3' end is highly regulated and requires numerous proteins, suggesting an important role for the poly(A) tail in the cell (reviewed in (Danckwardt et al., 2008)). Although it is well-known that the poly(A) tail and its binding partner, poly(A)-binding protein (PABP), play an important role in mRNA stability (see (Mangus et al., 2003) for review), poly(A) and PABP also have roles in translation. Two major evolutionarily divergent classes of PABPs exist in eukarya: nuclear PABPs and cytoplasmic PABPs. Nuclear PABPs appear to be involved mostly in polyadenylation itself (Anderson et al., 1993; Mangus et al., 2003), whereas cytoplasmic PABPs have been implicated in mRNA export, turnover, and translation. Higher eukaryotes encode several cytoplasmic PABPs whose expression is often cell or tissue-specific, whereas unicellular eukaryotes have only one form (reviewed in (Mangus et al., 2003)). In yeast, a single essential gene, PAB1, encodes Pab1p (PABP) (Sachs et al., 1986).

Most cytoplasmic PABPs share a conserved structure of four tandem RRM domains connected to a C-terminal PABPC domain by a glycine-rich linker. Although these RRMs show considerable sequence similarity, the individual domains appear to vary in function (Kessler and Sachs, 1998). The four RRM domains possess non-specific as well as poly(A)-specific RNA-binding activity, and in Xenopus and yeast PABP, RRMs 1 and 2 are sufficient for poly(A) binding (Kessler and Sachs, 1998; Deo et al., 1999). A cocrystal structure of RRMs 1 and 2 of human PABP with poly(A)$_{11}$ shows that the two RRMs form a long narrow trough that binds poly(A). Specificity for poly(A) is provided by a combination of intramolecular and intermolecular base stacking and sandwiching of the adenine rings between aromatic and aliphatic residues (Deo et al., 1999).

PABP also binds to eIF4G, eRF3, and to a class of molecules called poly(A)-binding protein-interacting proteins (Pbp1p-Pbp6p) among others. eIF4B binds PABP in plants (Le et al., 1997), but the interaction has not been observed in yeast or mammals. The binding interface for eIF4G was localized to RRM2 opposite the RNA-binding channel in yeast and mammals (Kessler and Sachs, 1998, Groft and Burley, 2002). RRM2 was shown to be important for the ability of PABP to stimulate translation in yeast, as was eIF4G-PABP interaction

(Kessler and Sachs, 1998; Otero et al., 1999). Although the RRMs are conserved among species, the molecular mechanism for binding eIF4G is not (Otero et al., 1999).

The C-terminal domain of PABP has some similarity to ubiquitin E3 ligases (although no role for PABP in ubiquitination has been described) and is responsible for the interaction with eRF3, Pan3p (part of the poly(A) nuclease (PAN) complex), Pbp1p (PABP-binding protein 1), Pbp2p (hnRNPk homolog), Pbp3p, Pkc1p (protein kinase C homolog), and Kre6p, among others, in yeast (Mangus et al., 1998; Dunn et al., 2005). Mutations in eRF3, eRF1, or PABP result in less capped- and poly(A)-tailed mRNA in polysomes in the presence of cap analog, which could suggest that the interaction between release factors and PABP stabilizes initiation complexes (Amrani et al., 2004). However, it is unclear whether PABP binding to eRFs contributes to initiation, re-initiation, or another stage of translation.

PABP has been implicated at several stages of translation. Mutations in PABP or eIF4G that prevent their interaction cause severe growth and translation defects in yeast (Tarun and Sachs, 1996; Kessler and Sachs, 1998). Such mutations also decrease translation of uncapped, polyadenylated mRNA, indicating that eIF4G-PABP interaction is necessary for PABP to stimulate translation regardless of the presence of the cap-eIF4E interaction (Otero et al., 1999). A number of experiments have implicated the PABP-eIF4G interaction in stimulating recruitment of the 43S PIC to the mRNA (Tarun and Sachs, 1996; Preiss and Hentze, 1998; Kahvejian et al., 2005). Removal of PABP from mammalian extracts decreased the rate of translation and the efficiency of 48S and 80S complex formation (Kahvejian et al., 2005). Because PABP depletion conferred a stronger defect on 80S than on 48S formation, it was suggested that PABP is also involved in ribosomal subunit joining. However, a decreased level of 80S monosomes could also indicate defects in other steps following mRNA recruitment, such as scanning or start codon recognition, and thus the exact steps at which PABP acts remain to be fully resolved. Several lines of evidence also indicate that the poly(A) tail and PABP are most important for translation under competitive conditions, suggesting they impart a selective advantage for mRNAs rather than being essential for the core processes of initiation (Preiss and Hentze, 1998; Svitkin et al., 2009).

3. Conclusion

The fact that mRNA recruitment is an integral, highly regulated step of protein synthesis with potential clinical implications has made it a focus of study for many years. What emerges from the plethora of information about the many factors involved in mRNA recruitment is a picture in which much is known yet little is fully understood. Combined studies using new approaches *in vitro* and *in vivo* have the potential to finally organize the tangle of functions and interactions involved in mRNA recruitment into a clear model of the molecular mechanics underlying this central biological process.

References

Abramson RD, Browning KS, Dever TE, Lawson TG, Thach RE, Ravel JM, Merrick WC (1988) Initiation factors that bind mRNA. A comparison of mammalian factors with wheat germ factors. J Biol Chem 263: 5462–5467

Abramson RD, Dever TE, Lawson TG, Ray BK, Thach RE, Merrick WC (1987) The ATP-dependent interaction of eukaryotic initiation factors with mRNA. J Biol Chem 262: 3826–3832

Algire MA, Maag D, Savio P, Acker MG, Tarun SZ, Sachs AB, Asano K, Nielsen KH, Olsen DS, Phan L, Hinnebusch AG, Lorsch JR (2002) Development and characterization of a reconstituted yeast translation initiation system. RNA 8: 382–397

Altmann M, Muller PP, Pelletier J, Sonenberg N, Trachsel H (1989) A mammalian translation initiation factor can substitute for its yeast homologue in vivo. J Biol Chem 264: 12 145–12 147

Altmann M, Muller PP, Wittmer B, Ruchti F, Lanker S, Trachsel H (1993) A Saccharomyces cerevisiae Homologue of Mammalian Translation Initiation Factor 4B Contributes to RNA Helicase Activity. EMBO J 12: 3997–4003

Altmann M, Wittmer B, Methot N, Sonenberg N, Trachsel H (1995) The Saccharomyces cerevisiae translation initiation factor Tif3 and its mammalian homologue, eIF-4B, have RNA annealing activity. EMBO J 14: 3820–3827

Amrani N, Ganesan R, Kervestin S, Mangus DA, Ghosh S, Jacobson A (2004) A faux 3'-UTR promotes aberrant termination and triggers nonsense-mediated mRNA decay. Nature 432: 112–118

Anderson JT, Wilson SM, Datar KV, Swanson MS (1993) NAB2: a yeast nuclear polyadenylated RNA-binding protein essential for cell viability. Mol Cell Biol 13: 2730–2741

Asano K, Clayton J, Shalev A, Hinnebusch AG (2000) A multifactor complex of eukaryotic initiation factors, eIF1, eIF2, eIF3, eIF5, and initiator tRNA(Met) is an important translation initiation intermediate in vivo. Genes Dev 14: 2534–2546

Asano K, Kinzy TG, Merrick WC, Hershey JW (1997) Conservation and diversity of eukaryotic translation initiation factor eIF3. J Biol Chem 272: 1101–1109

Asano K, Shalev A, Phan L, Nielsen K, Clayton J, Valassek L, Donahue TF, Hinnebusch AG (2001) Multiple roles for the C-terminal domain of eIF5 in translation initiation complex assembly and GTPase activation. EMBO J 20: 2326–2337

Baim SB, Sherman F (1988) mRNA structures influencing translation in the yeast Saccharomyces cerevisiae. Mol Cell Biol 8: 1591–1601

Berset C, Zurbriggen A, Djafarzadeh S, Altmann M, Trachsel H (2003) RNA-binding activity of translation initiation factor eIF4G1 from Saccharomyces cerevisiae. RNA 9: 871–880

Bettany AJ, Moore PA, Cafferkey R, Bell LD, Goodey AR, Carter BL, Brown AJ (1989) 5'-secondary structure formation, in contrast to a short string of non-preferred codons, inhibits the translation of the pyruvate kinase mRNA in yeast. Yeast 5: 187–198

Bi X, Ren J, Goss DJ (2000) Wheat germ translation initiation factor eIF4B affects eIF4A and eIFiso4F helicase activity by increasing the ATP binding affinity of eIF4A. Biochemistry 39: 5758–5765

Block KL, Vornlocher HP, Hershey JW (1998) Characterization of cDNAs encoding the p44 and p35 subunits of human translation initiation factor eIF3. J Biol Chem 273: 31901–31908

Blum S, Schmid SR, Pause A, Buser P, Linder P, Sonenberg N, Trachsel H (1992) ATP hydrolysis by initiation factor 4A is required for translation initiation in Saccharomyces cerevisiae. Proc Natl Acad Sci USA 89: 7664–7668

Browning KS, Fletcher L, Lax SR, Ravel JM (1989) Evidence that the 59-kDa protein synthesis initiation factor from wheat germ is functionally similar to the 80-kDa initiation factor 4B from mammalian cells. J Biol Chem 264: 8491–8494

Caruthers JM, Johnson ER, McKay DB (2000) Crystal structure of yeast initiation factor 4A, a DEAD-box RNA helicase. Proc Natl Acad Sci USA 97: 13 080–13 085

Chaudhuri J, Chowdhury D, Maitra U (1999) Distinct functions of eukaryotic translation initiation factors eIF1A and eIF3 in the formation of the 40S ribosomal preinitiation complex. J Biol Chem 273: 17 975–17 980

Chen YH, Su LH, Sun CH (2008) Incomplete nonsense-mediated mRNA decay in Giardia lamblia. Int J Parasitol 38: 1305–1317

Chuang RY, Weaver PL, Liu Z, Chang TH (1997) Requirement of the DEAD-Box protein ded1p for messenger RNA translation. Science 275: 1468–1471

Clarkson BK, Gilbert WV, Doudna JA. Functional overlap between eIF4G isoforms in Saccharomyces cerevisiae. PLoS One 5: e9114

Clarkson BK, Gilbert WV, Doudna JA (2010) Functional overlap between eIF4G isoforms in *Saccharomyces cerevisiae*. PLoS One 5: e9114

Coppolecchia R, Buser P, Stotz A, Linder P (1993) A new yeast translation initiation factor suppresses a mutation in the eIF-4A RNA helicase. EMBO J 12: 4005–4011

Danaie P, Wittmer B, Altmann M, Trachsel H (1995) Isolation of a protein complex containing translation initiation factor Prt1 from Saccharomyces cerevisiae. J Biol Chem 270: 4288–4292

Danckwardt S, Hentze MW, Kulozik AE (2008) 3' end mRNA processing: molecular mechanisms and implications for health and disease. EMBO J 27: 482–498

Del Campo M, Mohr S, Jiang Y, Jia H, Jankowsky E, Lambowitz AM (2009) Unwinding by local strand separation is critical for the function of DEAD-box proteins as RNA chaperones. J Mol Biol 389: 674–693

Deo RC, Bonanno JB, Sonenberg N, Burley SK (1999) Recognition of polyadenylate RNA by the poly(A)-binding protein. Cell 98: 835–845

Dmitriev SE, Pisarev AV, Rubtsova MP, Dunaevsky YE, Shatsky IN (2003) Conversion of 48S translation preinitiation complexes into 80S initiation complexes as revealed by toeprinting. FEBS Lett 533: 99–104

Dominguez D, Altmann M, Benz J, Baumann U, Trachsel H (1999) Interaction of translation initiation factor eIF4G with eIF4A in the yeast Saccharomyces cerevisiae. J Biol Chem 274: 26 720–26 726

Dominguez D, Kislig E, Altmann M, Trachsel H (2001) Structural and functional similarities between the central eukaryotic initiation factor (eIF)4A-binding domain of mammalian eIF4G and the eIF4A-binding domain of yeast eIF4G. Biochem J 355: 223–230

Dunn EF, Hammell CM, Hodge CA, Cole CN (2005) Yeast poly(A)-binding protein, Pab1, and PAN, a poly(A) nuclease complex recruited by Pab1, connect mRNA biogenesis to export. Genes Dev 19: 90–103

Fleming K, Ghuman J, Yuan X, Simpson P, Szendroi A, Matthews S, Curry S (2003) Solution structure and RNA interactions of the RNA recognition motif from eukaryotic translation initiation factor 4B. Biochemistry 42: 8966–8975

Fraser CS, Berry KE, Hershey JW, Doudna JA (2007) eIF3j is located in the decoding center of the human 40S ribosomal subunit. Mol Cell 26: 811–819

Ghaemmaghami S, Huh WK, Bower K, Howson RW, Belle A, Dephoure N, O'Shea EK, Weissman JS (2003) Global analysis of protein expression in yeast. Nature 425: 737–741

Gorbalenya AE, Koonin EV (1993) Helicases – Amino-Acid-Sequence Comparisons and Structure-Function-Relationships. Current Opinion in Structural Biology 3: 419–429

Goss DJ, Carberry SE, Dever TE, Merrick WC, Rhoads RE (1990) Fluorescence study of the binding of m7GpppG and rabbit globin mRNA to protein synthesis initiation factors 4A, 4E, and 4F. Biochemistry 29: 5008–5012

Goyer C, Altmann M, Lee HS, Blanc A, Deshmukh M, Woolford JL, Trachsel H, Sonenberg N (1993) TIF4631 and TIF4632: Two Yeast Genes Encoding the High-Molecular-Weight Subunits of the Cap-Binding Protein Complex (Eukaryotic Initiation Factor 4F) Contain an RNA Recognition Motif-Like Sequence and Carry Out an Essential Function. Mol Cell Biol 13: 4860–4874

Grifo JA, Abramson RD, Satler CA, Merrick WC (1984) RNA-stimulated ATPase activity of eukaryotic initiation factors. J Biol Chem 259: 8648–8654

Grifo JA, Tahara SM, Leis JP, Morgan MA, Shatkin AJ, Merrick WC (1982) Characterization of eukaryotic initiation factor 4A, a protein involved in ATP-dependent binding of globin mRNA. J Biol Chem 257: 5246–5252

Groft CM, Burley SK (2002) Recognition of eIF4G by rotavirus NSP3 reveals a basis for mRNA circularization. Mol Cell 9: 1273–1283

Gross JD, Moerke NJ, von der Haar T, Lugovskoy AA, Sachs AB, McCarthy JE, Wagner G (2003) Ribosome loading onto the mRNA cap is driven by conformational coupling between eIF4G and eIF4E. Cell 115: 739–750

Haghighat A, Sonenberg N (1997) eIF4G dramatically enhances the binding of eIF4E to the mRNA 5'-cap structure. J Biol Chem 272: 21 677–21 680

Hanachi P, Hershey JW, Vornlocher HP (1999) Characterization of the p33 subunit of eukaryotic translation initiation factor-3 from Saccharomyces cerevisiae. J Biol Chem 274: 8546–8553

He H, von der Haar T, Singh CR, Ii M, Li B, Hinnebusch AG, McCarthy JE, Asano K (2003) The yeast eukaryotic initiation factor 4G (eIF4G) HEAT domain interacts with eIF1 and eIF5 and is involved in stringent AUG selection. Mol Cell Biol 23: 5431–5445

Hershey PE, McWhirter SM, Gross JD, Wagner G, Alber T, Sachs AB (1999) The Cap-binding protein eIF4E promotes folding of a functional domain of yeast translation initiation factor eIF4G1. J Biol Chem 274: 21 297–21 304

Hinnebusch AG (2006) eIF3: a versatile scaffold for translation initiation complexes. Trends Biochem Sci 31: 553–562

Hinton TM, Coldwell MJ, Carpenter GA, Morley SJ, Pain VM (2007) Functional analysis of individual binding activities of the scaffold protein eIF4G. J Biol Chem 282: 1695–1708

Imataka H, Sonenberg N (1997) Human eukaryotic translation initiation factor 4G (eIF4G) possesses two separate and independent binding sites for eIF4A. Mol Cell Biol 17: 6940–6947

Iost I, Dreyfus M, Linder P (1999) Ded1p, a DEAD-box protein required for translation initiation in Saccharomyces cerevisiae, is an RNA helicase. J Biol Chem 274: 17 677–17 683

Jackson RJ, Hellen CU, Pestova TV (2010) The mechanism of eukaryotic translation initiation and principles of its regulation. Nat Rev Mol Cell Biol 11: 113–127

Jivotovskaya AV, Valasek L, Hinnebusch AG, Nielsen KH (2006) Eukaryotic translation initiation factor 3 (eIF3) and eIF2 can promote mRNA binding to 40S subunits independently of eIF4G in yeast. Mol Cell Biol 26: 1355–1372

Kahvejian A, Svitkin YV, Sukarieh R, M'Boutchou MN, Sonenberg N (2005) Mammalian poly(A)-binding protein is a eukaryotic translation initiation factor, which acts via multiple mechanisms. Genes Dev 19: 104–113

Kapp LD, Lorsch JR (2004) The molecular mechanics of eukaryotic translation. Annu Rev Biochem 73: 657–704

Kessler SH, Sachs AB (1998) RNA recognition motif 2 of yeast Pab1p is required for its functional interaction with eukaryotic translation initiation factor 4G. Mol Cell Biol 18: 51–57

Kolupaeva VG, Unbehaun A, Lomakin IB, Hellen CU, Pestova TV (2005) Binding of eukaryotic initiation factor 3 to ribosomal 40S subunits and its role in ribosomal dissociation and anti-association. RNA 11: 470–486

Korneeva NL, First EA, Benoit CA, Rhoads RE (2005) Interaction between the NH2-terminal domain of eIF4A and the central domain of eIF4G modulates RNA-stimulated ATPase activity. J Biol Chem 280: 1872–1881

Korneeva NL, Lamphear BJ, Hennigan FL, Rhoads RE (2000) Mutually cooperative binding of eukaryotic translation initiation factor (eIF) 3 and eIF4A to human eIF4G-1. J Biol Chem 275: 41 369–41 376

Kozak M (1986) Influences of mRNA Secondary Structure on Initiation by Eukaryotic Ribosomes. Proc Natl Acad Sci USA 83: 2850–2854

Kozak M (1989) Circumstances and mechanisms of inhibition of translation by secondary structure in eucaryotic mRNAs. Mol Cell Biol 9: 5134–5142

Lamphear BJ, Kirchweger R, Skern T, Rhoads RE (1995) Mapping of Functional Domains in Eukaryotic Protein Synthesis Initiation Factor 4G (eIF4G) with Picornaviral Proteases. J Biol Chem 270: 21 975–21 983

Lawson TG, Lee KA, Maimone MM, Abramson RD, Dever TE, Merrick WC, Thach RE (1989) Dissociation of double-stranded polynucleotide helical structures by eukaryotic initiation factors, as revealed by a novel assay. Biochemistry 28: 4729–4734

Lawson TG, Ray BK, Dodds JT, Grifo JA, Abramson RD, Merrick WC, Betsch DF, Weith HL, Thach RE (1986) Influence of 5' proximal secondary structure on the translational efficiency of eukaryotic mRNAs and on their interaction with initiation factors. J Biol Chem 261: 13 979–13 989

Le H, Tanguay RL, Balasta ML, Wei CC, Browning KS, Metz AM, Goss DJ, Gallie DR (1997) Translation initiation factors eIF-iso4G and eIF-4B interact with the poly(A)-binding protein and increase its RNA binding activity. J Biol Chem 272: 16 247–16 255

LeFebvre AK, Korneeva NL, Trutschl M, Cvek U, Duzan RD, Bradley CA, Hershey JW, Rhoads RE (2006) Translation initiation factor eIF4G-1 binds to eIF3 through the eIF3e subunit. J Biol Chem 281: 22 917–22 932

Lindqvist L, Imataka H, Pelletier J (2008) Cap-dependent eukaryotic initiation factor-mRNA interactions probed by cross-linking. RNA 14: 960–969

Lorsch JR, Dever TE (2010) Molecular view of 43 S complex formation and start site selection in eukaryotic translation initiation. J Biol Chem 285: 21 203–21 207

Lorsch JR, Herschlag D (1998a) The DEAD Box Protein eIF4A. 1. A Minimal Kinetic and Thermodynamic Framework Reveals Coupled Binding of RNA and Nucleotide. Biochemistry 37: 2180–2193

Lorsch JR, Herschlag D (1998b) The DEAD Box Protein eIF4A. 2. A Cycle of Nucleotide and RNA-dependent Conformational Changes. Biochemistry 37: 2194–2206

Maag D, Fekete CA, Gryczynski Z, Lorsch JR (2005) A conformational change in the eukaryotic translation preinitiation complex and release of eIF1 signal recognition of the start codon. Mol Cell 17: 265–275

Mader S, Lee H, Pause A, Sonenberg N (1995) The translation initiation factor eIF-4E binds to a common motif shared by the translation factor eIF-4 gamma and the translational repressors 4E-binding proteins. Mol Cell Biol 15: 4990–4997

Majumdar R, Bandyopadhyay A, Maitra U (2003) Mammalian translation initiation factor eIF1functions with eIF1A and eIF3 in the formation of a stable 40 S preinitiation complex. J Biol Chem 278: 6580–6587

Mangus DA, Amrani N, Jacobson A (1998) Pbp1p, a factor interacting with Saccharomyces cerevisiae poly(A)-binding protein, regulates polyadenylation. Mol Cell Biol 18: 7383–7396

Mangus DA, Evans MC, Jacobson A (2003) Poly(A)-binding proteins: multifunctional scaffolds for the post-transcriptional control of gene expression. Genome Biol 4: 223

Marcotrigiano J, Gingras AC, Sonenberg N, Burley SK (1997) Cocrystal structure of the messenger RNA 5' cap-binding protein (eIF4E) bound to 7-methyl-GDP. Cell 89: 951–961

Marcotrigiano J, Gingras AC, Sonenberg N, Burley SK (1999) Cap-dependent translation initiation in eukaryotes is regulated by a molecular mimic of eIF4G. Mol Cell 3: 707–716

Marintchev A, Edmonds KA, Marintcheva B, Hendrickson E, Oberer M, Suzuki C, Herdy B, Sonenberg N, Wagner G (2009) Topology and regulation of the human eIF4A/4G/4H helicase complex in translation initiation. Cell 136: 447–460

Marintchev A, Wagner G (2005) eIF4G and CBP80 share a common origin and similar domain organization: implications for the structure and function of eIF4G. Biochemistry 44: 12 265–12 272

Maris C, Dominguez C, Allain FH (2005) The RNA recognition motif, a plastic RNA-binding platform to regulate post-transcriptional gene expression. FEBS J 272: 2118–2131

Marsden S, Nardelli M, Linder P, McCarthy JE (2006) Unwinding single RNA molecules using helicases involved in eukaryotic translation initiation. J Mol Biol 361: 327–335

Matsuo H, McGuire AM, Fletcher CM, Gingras AC, Sonenberg N, Wagner G (1997) Structure of translation factor eIF4E bound to m7GDP and interaction with 4E-binding protein. Nat Struct Biol 4: 717–724

Mayberry LK, Allen ML, Dennis MD, Browning KS (2009) Evidence for variation in the optimal translation initiation complex: plant eIF4B, eIF4F, and eIF(iso)4F differentially promote translation of mRNAs. Plant Physiol 150: 1844–1854

Methot N, Pause A, Hershey JW, Sonenberg N (1994) The translation initiation factor eIF-4B contains an RNA-binding region that is distinct and independent from its ribonucleoprotein consensus sequence. Mol Cell Biol 14: 2307–2316

Methot N, Pickett G, Keene JD, Sonenberg N (1996a) In vitro RNA selection identifies RNA ligands that specifically bind to eukaryotic translation initiation factor 4B: the role of the RNA remotif. RNA 2: 38–50

Methot N, Song MS, Sonenberg N (1996b) A region rich in aspartic acid, arginine, tyrosine, and glycine (DRYG) mediates eukaryotic initiation factor 4B (eIF4B) self-association and interaction with eIF3. Mol Cell Biol 16: 5328–5334

Metz AM, Wong KC, Malmstrom SA, Browning KS (1999) Eukaryotic initiation factor 4B from wheat and Arabidopsis thaliana is a member of a multigene family. Biochem Biophys Res Commun 266: 314–321

Michon T, Estevez Y, Walter J, German-Retana S, Le Gall O (2006) The potyviral virus genome-linked protein VPg forms a ternary complex with the eukaryotic initiation factors eIF4E and eIF4G and reduces eIF4E affinity for a mRNA cap analogue. FEBS J 273: 1312–1322

Milburn SC, Hershey JW, Davies MV, Kelleher K, Kaufman RJ (1990) Cloning and expression of eukaryotic initiation factor 4B cDNA: sequence determination identifies a common RNA recognition motif. EMBO J 9: 2783–2790

Morino S, Imataka H, Svitkin YV, Pestova TV, Sonenberg N (2000) Eukaryotic translation initiation factor 4E (eIF4E) binding site and the middle one-third of eIF4GI constitute the core domain for cap-dependent translation, and the C-terminal one-third functions as a modulatory region. Mol Cell Biol 20: 468–477

Naranda T, Strong WB, Menaya J, Fabbri BJ, Hershey JW (1994) Two structural domains of initiation factor eIF-4B are involved in binding to RNA. J Biol Chem 269: 14 465–14 472

Niederberger N, Trachsel H, Altmann M (1998) The RNA recognition motif of yeast translation initiation factor Tif3/eIF4B is required but not sufficient for RNA strand-exchange and translational activity. RNA 4: 1259–1267

Niedzwiecka A, Marcotrigiano J, Stepinski J, Jankowska-Anyszka M, Wyslouch-Cieszynska A, Dadlez M, Gingras AC, Mak P, Darzynkiewicz E, Sonenberg N, Burley SK, Stolarski R (2002) Biophysical studies of eIF4E cap-binding protein: recognition of mRNA 5′ cap structure and synthetic fragments of eIF4G and 4E-BP1 proteins. J Mol Biol 319: 615–635

Nielsen KH, Valasek L, Sykes C, Jivotovskaya A, Hinnebusch AG (2006) Interaction of the RNP1 motif in PRT1 with HCR1 promotes 40S binding of eukaryotic initiation factor 3 in yeast. Mol Cell Biol 26: 2984–2998

Nielsen PJ, McMaster GK, Trachsel H (1985) Cloning of eukaryotic protein synthesis initiation factor genes: isolation and characterization of cDNA clones encoding factor eIF-4A. Nucleic Acids Res 13: 6867–6880

Nielsen PJ, Trachsel H (1988) The mouse protein synthesis initiation factor 4A gene family includes two related functional genes which are differentially expressed. EMBO J 7: 2097–2105

Nygard O, Westermann P (1982) Specific interaction of one subunit of eukaryotic initiation factor eIF-3 with 18S ribosomal RNA within the binary complex, eIF-3 small ribosomal subunit, as shown by cross-linking experiments. Nucleic Acids Res 10: 1327–1334

Oberer M, Marintchev A, Wagner G (2005) Structural basis for the enhancement of eIF4A helicase activity by eIF4G. Genes Dev 19: 2212–2223

Otero LJ, Ashe MP, Sachs AB (1999) The yeast poly(A)-binding protein Pab1p stimulates in vitro poly(A)-dependent and cap-dependent translation by distinct mechanisms. EMBO J 18: 3153–3163

Pause A, Methot N, Svitkin Y, Merrick WC, Sonenberg N (1994) Dominant negative mutants of mammalian translation initiation factor eIF-4A define a critical role for eIF-4F in cap-dependent and cap-independent initiation of translation. EMBO J 13: 1205–1215

Peck ML, Herschlag D (1999) Effects of oligonucleotide length and atomic composition on stimulation of the ATPase activity of translation initiation factor eIF4A. RNA 5: 1210–1221

Pelletier J, Sonenberg N (1985) Insertion Mutagenesis to Increase Secondary Structure within the 5′ Noncoding Region of a Eukaryotic mRNA Reduces Translational Efficiency. Cell 40: 515–526

Pestova TV, Kolupaeva VG (2002) The roles of individual eukaryotic translation initiation factors in ribosomal scanning and initiation codon selection. Genes Dev 16: 2906–2922

Phan L, Schoenfeld LW, Valasek L, Nielsen KH, Hinnebusch AG (2001) A subcomplex of three eIF3 subunits binds eIF1 and eIF5 and stimulates ribosome binding of mRNA and tRNA(i) (Met) EMBO J 20: 2954–2965

Phan L, Zhang X, Asano K, Anderson J, Vornlocher HP, Greenberg JR, Qin J, Hinnebusch AG (1998) Identification of a translation initiation factor 3 (eIF3) core complex, conserved in yeast and mammals, that interacts with eIF5. Mol Cell Biol 18: 4935–4946

Pisarev AV, Kolupaeva VG, Yusupov MM, Hellen CU, Pestova TV (2008) Ribosomal position and contacts of mRNA in eukaryotic translation initiation complexes. EMBO J 27: 1609–1621

Pisareva VP, Pisarev AV, Komar AA, Hellen CU, Pestova TV (2008) Translation initiation on mammalian mRNAs with structured 5′UTRs requires DExH-box protein DHX29. Cell 135: 1237–1250

Preiss T, Hentze MW (1998) Dual function of the messenger RNA cap structure in poly(A)-tail-promoted translation in yeast. Nature 392: 516–520

Ptushkina M, von der Haar T, Vasilescu S, Frank R, Birkenhager R, McCarthy JE (1998) Cooperative modulation by eIF4G of eIF4E-binding to the mRNA 5′ cap in yeast involves a site partially shared by p20. EMBO J 17: 4798–4808

Ray BK, Lawson TG, Kramer JC, Cladaras MH, Grifo JA, Abramson RD, Merrick WC, Thach RE (1985) ATP-dependent Unwinding of Messenger RNA Structure by Eukaryotic Initiation Factors. J Biol Chem 260: 7651–7658

Richter-Cook NJ, Dever TE, Hensold JO, Merrick WC (1998) Purification and characterization of a new eukaryotic protein translation factor. Eukaryotic initiation factor 4H. J Biol Chem 273: 7579–7587

Richter NJ, Rogers GW, Jr., Hensold JO, Merrick WC (1999) Further biochemical and kinetic characterization of human eukaryotic initiation factor 4H. J Biol Chem 274: 35 415–35 424

Robert F, Carrier M, Rawe S, Chen S, Lowe S, Pelletier J (2009) Altering chemosensitivity by modulating translation elongation. PLoS One 4: e5428

Robert F, Pelletier J (2009) Translation initiation: a critical signalling node in cancer. Expert Opin Ther Targets 13: 1279–1293

Rogers GW, Jr., Lima WF, Merrick WC (2001a) Further characterization of the helicase activity of eIF4A. Substrate specificity. J Biol Chem 276: 12 598–12 608

Rogers GW, Jr., Richter NJ, Lima WF, Merrick WC (2001b) Modulation of the helicase activity of eIF4A by eIF4B, eIF4H, and eIF4F. J Biol Chem 276: 30 914–30 922

Rogers GW, Richter NJ, Merrick WC (1999) Biochemical and kinetic characterization of the RNA helicase activity of eukaryotic initiation factor 4A. J Biol Chem 274: 12 236–12 244

Rozen F, Edery I, Meerovitch K, Dever TE, Merrick WC, Sonenberg N (1990) Bidirectional RNA helicase activity of eucaryotic translation initiation factors 4A and 4F. Mol Cell Biol 10: 1134–1144

Rozovsky N, Butterworth AC, Moore MJ (2008) Interactions between eIF4AI and its accessory factors eIF4B and eIF4H. RNA 14: 2136–2148

Sachs AB, Bond MW, Kornberg RD (1986) A single gene from yeast for both nuclear and cytoplasmic polyadenylate-binding proteins: domain structure and expression. Cell 45: 827–835

Schmid SR, Linder P (1992) D-E-A-D protein family of putative RNA helicases. Mol Microbiol 6: 283–292

Schutz P, Bumann M, Oberholzer AE, Bieniossek C, Trachsel H, Altmann M, Baumann U (2008) Crystal structure of the yeast eIF4A-eIF4G complex: an RNA-helicase controlled by protein-protein interactions. Proc Natl Acad Sci USA 105: 9564–9569

Shahbazian D, Parsyan A, Petroulakis E, Topisirovic I, Martineau Y, Gibbs BF, Svitkin Y, Sonenberg N. Control of cell survival and proliferation by mammalian eukaryotic initiation factor 4B. Mol Cell Biol 30: 1478–1485

Siridechadilok B, Fraser CS, Hall RJ, Doudna JA, Nogales E (2005) Structural roles for human translation factor eIF3 in initiation of protein synthesis. Science 310: 1513–1515

Slepenkov SV, Korneeva NL, Rhoads RE (2008) Kinetic mechanism for assembly of the m7GpppG. eIF4E. eIF4G complex. J Biol Chem 283: 25 227–25 237

Sonenberg N (2008) eIF4E, the mRNA cap-binding protein: from basic discovery to translational research. Biochem Cell Biol 86: 178–183

Sonenberg N, Hinnebusch AG (2009) Regulation of translation initiation in eukaryotes: mechanisms and biological targets. Cell 136: 731–745

Sonenberg N, Morgan MA, Merrick WC, Shatkin AJ (1978) A polypeptide in eukaryotic initiation factors that crosslinks specifically to the 5'-terminal cap in mRNA. Proc Natl Acad Sci USA 75: 4843–4847

Spahn CM, Beckmann R, Eswar N, Penczek PA, Sali A, Blobel G, Frank J (2001) Structure of the 80S ribosome from Saccharomyces cerevisiae–tRNA-ribosome and subunit-subunit interactions. Cell 107: 373–386

Spirin AS (2009) How does a scanning ribosomal particle move along the 5'-untranslated region of eukaryotic mRNA? Brownian Ratchet model. Biochemistry 48: 10688–10692

Srivastava S, Verschoor A, Frank J (1992) Eukaryotic initiation factor 3 does not prevent association through physical blockage of the ribosomal subunit-subunit interface. J Mol Biol 226: 301–304

Svitkin YV, Evdokimova VM, Brasey A, Pestova TV, Fantus D, Yanagiya A, Imataka H, Skabkin MA, Ovchinnikov LP, Merrick WC, Sonenberg N (2009) General RNA-binding proteins have a function in poly(A)-binding protein-dependent translation. EMBO J 28: 58–68

Svitkin YV, Ovchinnikov LP, Dreyfuss G, Sonenberg N (1996) General RNA binding proteins render translation cap dependent. EMBO J 15: 7147–7155

Svitkin YV, Pause A, Haghighat A, Pyronnet S, Witherell G, Belsham GJ, Sonenberg N (2001) The requirement for eukaryotic initiation factor 4A (eIF4A) in translation is in direct proportion to the degree of mRNA 5' secondary structure. RNA 7: 382–394

Szamecz B, Rutkai E, Cuchalova L, Munzarova V, Herrmannova A, Nielsen KH, Burela L, Hinnebusch AG, Valasek L (2008) eIF3a cooperates with sequences 5' of uORF1 to promote resumption of scanning by post-termination ribosomes for reinitiation on GCN4 mRNA. Genes Dev 22: 2414–2425

Tarun SJ, Sachs AB (1996) Association of the yeast poly(A) tail binding protein with translation initiation factor eIF-4G. EMBO J 15: 7168–7177

Tarun SZ, Jr., Sachs AB (1997) Binding of eukaryotic translation initiation factor 4E (eIF4E) to eIF4G represses translation of uncapped mRNA. Mol Cell Biol 17: 6876–6886

Tarun SZ, Jr., Wells SE, Deardorff JA, Sachs AB (1997) Translation initiation factor eIF4G mediates in vitro poly(A) tail-dependent translation. Proc Natl Acad Sci USA 94: 9046–9051

Trachsel H, Erni B, Schreier MH, Staehelin T (1977) Initiation of mammalian protein synthesis. II. The assembly of the initiation complex with purified initiation factors. J Mol Biol 116: 755–767

Valasek L, Hasek J, Trachsel H, Imre EM, Ruis H (1999) The Saccharomyces cerevisiae HCR1 gene encoding a homologue of the p35 subunit of human translation initiation factor 3 (eIF3) is a high copy suppressor of a temperature-sensitive mutation in the Rpg1p subunit of yeast eIF3. J Biol Chem 274: 27 567–27 572

Valasek L, Mathew AA, Shin BS, Nielsen KH, Szamecz B, Hinnebusch AG (2003) The yeast eIF3 subunits TIF32/a, NIP1/c, and eIF5 make critical connections with the 40S ribosome in vivo. Genes Dev 17: 786–799

Valasek L, Nielsen KH, Hinnebusch AG (2002) Direct eIF2-eIF3 contact in the multifactor complex is important for translation initiation in vivo. EMBO J 21: 5886–5898

Valasek L, Nielsen KH, Zhang F, Fekete CA, Hinnebusch AG (2004) Interactions of eukaryotic translation initiation factor 3 (eIF3) subunit NIP1/c with eIF1 and eIF5 promote preinitiation complex assembly and regulate start codon selection. Mol Cell Biol 24: 9437–9455

Valasek L, Phan L, Schoenfeld LW, Valaskova V, Hinnebusch AG (2001) Related eIF3 subunits TIF32 and HCR1 interact with an RNA recognition motif in PRT1 required for eIF3 integrity and ribosome binding. EMBO J 20: 891–904

Vega Laso MR, Zhu D, Sagliocco F, Brown AJ, Tuite MF, McCarthy JE (1993) Inhibition of translational initiation in the yeast Saccharomyces cerevisiae as a function of the stability and position of hairpin structures in the mRNA leader. J Biol Chem 268: 6453–6462

Von Der Haar T, McCarthy JE (2002) Intracellular translation initiation factor levels in Saccharomyces cerevisiae and their role in cap-complex function. Mol Microbiol 46: 531–544

Vornlocher HP, Hanachi P, Ribeiro S, Hershey JW (1999) A 110-kilodalton subunit of translation initiation factor eIF3 and an associated 135-kilodalton protein are encoded by the Saccharomyces cerevisiae TIF32 and TIF31 genes. J Biol Chem 274: 16 802–16 812

Westermann P, Nygard O (1984) Cross-linking of mRNA to initiation factor eIF-3, 24 kDa cap binding protein and ribosomal proteins S1, S3/3a, S6 and S11 within the 48S pre-initiation complex. Nucleic Acids Res 12: 8887–8897

Yamamoto Y, Singh CR, Marintchev A, Hall NS, Hannig EM, Wagner G, Asano K (2005) The eukaryotic initiation factor (eIF) 5 HEAT domain mediates multifactor assembly and scanning with distinct interfaces to eIF1, eIF2, eIF3, and eIF4G. Proc Natl Acad Sci USA 102: 16 164–16 169

Yang Q, Del Campo M, Lambowitz AM, Jankowsky E (2007) DEAD-box proteins unwind duplexes by local strand separation. Mol Cell 28: 253–263

The mechanism of ribosome recycling in eukaryotes

14

Andrey V. Pisarev, Maxim A. Skabkin, Vera P. Pisareva, Olga V. Skabkina, Christopher U. T. Hellen, and Tatyana V. Pestova

1. Introduction

Protein synthesis is a cyclical process, consisting of initiation, elongation, termination and ribosome recycling stages. Initiation requires pools of separated small and large ribosomal subunits, and it has been known for about forty years that, after termination, eukaryotic 80S ribosomes dissociate from polysomes into 40S and 60S subunits, which then either participate in the next round of initiation, or enter a reservoir of stable, translationally inactive 80S monosomes (Adamson et al., 1969; Hogan and Korner, 1968b; Kaempfer 1969; Falvey and Staehelin, 1970; Howard et al., 1970; Henshaw et al., 1973). Although 80S monosomes accumulate in response to stresses that inhibit initiation, such as amino acid or glucose starvation, reversal of these conditions allows them to be dissociated and their subunits to re-enter the translation process (Hogan and Korner, 1968a). In bacteria, recycling of post-termination 70S ribosomal complexes (post-TCs) requires elongation factor EF-G, the dedicated ribosome recycling factor RRF (Hirashima and Kaji, 1973) and initiation factor IF3. EF-G and RRF split post-TCs in a GTP-dependent manner into free 50S subunits and mRNA- and tRNA-associated 30S subunits, after which IF3 promotes the release of tRNA, followed by spontaneous dissociation of mRNA (Karimi et al., 1999; Peske et al., 2005; Zavialov et al., 2005). RRF, comprising two domains, binds at the subunit interface, where it mainly interacts with the 50S subunit, making contact with helix 69 of 23S rRNA, which is involved in formation of the central intersubunit bridge B2a (e.g. Gao et al., 2005, 2007; Barat et al., 2007). Hydrolysis of GTP and subsequent P_i release induce conformational changes in EF-G, which were thought to trigger movement of RRF, particularly of its head domain, which promotes separation of ribos-

omal subunits by disrupting bridge B2a and possibly other bridges (Gao et al., 2005, 2007; Savelsbergh et al., 2009). However, in eukary otes, RRF homologs function exclusively in organelles such as chloroplasts and mitochondria (Rorbach et al., 2008), indicating that the factors required for recycling of cytoplasmic post-termination complexes must be different from those involved in recycling in bacteria. It has been known for decades that post-termination 'native' 40S and 60S subunits are associated with factors that prevent their re-association with complementary subunits (Henshaw et al., 1973). In early studies, ribosome dissociation activity in the ribosomal salt wash from native ribosomal subunits was resolved into two or more species that in some instances appeared to associate differentially with 40S and 60S subunits (e. g. Mizuno and Rabinowitz, 1973; Lubsen and Davies, 1974; Merrick et al., 1973; Thompson et al., 1977; Jones et al., 1980). Thus, several laboratories reported anti-association activity for eukaryotic initiation factor (eIF) 3, which binds to 40S subunits, and whose anti-association activity is augmented by eIF1A (Goumans et al., 1980; Thompson et al., 1977; Trachsel and Staehelin, 1979), and for eIF6 (Russell and Spremulli, 1979; Valenzuela et al., 1982; Raychaudhuri et al., 1984), which binds to a conserved region on the intersubunit surface of 60S subunits and prevents their association with 40S subunits, presumably by sterically impairing the formation of the intersubunit bridge B6 and the surrounding intersubunit contacts B5a, B5b, B3 and B7 (Gartmann et al., 2010). However, other reports indicated that eIF3 alone, or together with eIF1A, is unable to split 80S monosomes or prevent the re-association of 40S and 60S subunits (Ceglarz et al., 1980; Checkley et al., 1981; Chaudhuri et al., 1999), or that it can do so only in the presence of ssRNAs that have the potential to bind directly to the mRNA-binding cleft of the

40S subunit (Kolupaeva et al., 2005). Questions have even been raised as to whether eIF6 is involved in the translation process at all, at least in yeast (Si and Maitra, 1999). Additional proteins that showed monosome dissociation activity, but that were apparently unrelated to eIF3 and eIF6, have been purified but were not identified (e. g. Merrick et al., 1973; Jones et al., 1980).

Another important caveat is that the dissociation/anti-association activities of eIF1A, eIF3 and eIF6 have been assayed only using purified 'derived' (i. e. factor-free) ribosomal subunits and 80S monosomes reconstituted from them, rather than authentic ribosomal post-termination complexes (post-TCs) containing mRNA and deacylated P-site tRNA. Moreover, the properties of these factors could not account for the reported requirement of energy for eukaryotic recycling (e. g. Mizuno and Rabinowitz, 1973). Thus, despite all of these studies, both the actual mechanism of eukaryotic post-termination ribosomal recycling and the question of whether it involves the factors previously identified as having ribosome dissociation/anti-association activity remained unresolved.

The development of the methodology necessary for the purification of individual components of the translational apparatus, for their use in the sequential reconstitution *in vitro* of all preceding stages of translation (initiation, elongation and termination), and for the subsequent isolation of individual pre- and post-termination complexes has cleared the way for renewed efforts to elucidate the mechanism of eukaryotic post-termination ribosomal recycling, which we will review below.

2. Translation termination in eukaryotes

The mechanism of termination in eukaryotes (Figure 1 A) differs in significant respects from that in bacteria, and, as a result, the compositions of the respective post-TCs are also distinct. In bacteria, two class-I release factors, RF1 and RF2, recognize stop codons in the ribosomal A-site (RF1 recognizes UAG/UAA, whereas RF2 recognizes UGA/UAA codons) and mediate the hydrolysis of the ester bond of the P-site peptidyl tRNA in the peptidyltransferase center (PTC) of the large ribosomal subunit. The universally conserved GGQ motif located in domains 3 of RF1/RF2 enters the PTC and is essential for hydrolysis (Youngman et al., 2008). The bacterial class-II release factor RF3, a ribosome-dependent GTPase, is not an essential protein and does not participate in peptidyl-tRNA

hydrolysis. Rather, it accelerates the release of RF1/RF2 from post-termination ribosomes, thereby ensuring the rapid turnover of the factors, and then dissociates itself after hydrolyzing GTP (Zavialov et al., 2001, 2002). As a result of this termination mechanism, bacterial post-TCs consist of 70S ribosomes, mRNA and P-site deacylated tRNA.

In eukaryotes, all three stop codons are recognized by a single class-I release factor, eRF1, which is unrelated to RF1/RF2, except for presence of the GGQ motif (Kisselev et al., 2003). In contrast to bacterial RF1/RF2, peptide release by eRF1 alone is very inefficient and is strongly accelerated by the eukaryotic class-II release factor eRF3 (Alkalaeva et al., 2006). Another distinguishing feature of eukaryotic release factors is that, unlike bacterial class-1 and class-2 RFs, which do not exhibit binding affinity for each other, eukaryotic eRF1 and eRF3 form a stable complex by an interaction of their C-terminal domains (Frolova et al., 1998; Ito et al., 1998). eRF3/eRF1 association stabilizes eRF3•GTP binding by two orders of magnitude due to lowering of the dissociation rate constant (Hauryliuk et al. 2006; Mitkevich et al. 2006; Pisareva et al., 2006) so that eRF1, eRF3 and GTP form a long-lived complex, suggesting that eRF1 and eRF3 bind to the pre-termination complex (pre-TC), and enter the termination process as a ternary complex, eRF1•eRF3•GTP. eRF1/eRF3 association also induces conformational changes in eRF1 (Cheng et al., 2009), which probably increase its affinity to pre-TCs. Binding of eRF1•eRF3•GTP to pre-TCs induces conformational changes in the latter that are manifested as a two-nucleotide forward shift of their toe-print, which may be caused by a rotation of the head relative to the body of the 40S subunit, preventing reverse transcriptase from penetrating further (Alkalaeva et al., 2006). However, peptide release does not occur until eRF3 hydrolyzes GTP (Salas-Marco and Bedwell, 2004; Alkalaeva et al., 2006; Fan-Minogue et al., 2008). GTP hydrolysis induces further conformational changes, most likely in eRF1 (Cheng et al., 2009), that enable the GGQ loop in eRF1's middle (M) domain to enter the peptidyltransferase center of the 60S subunit and induce peptidyl-tRNA hydrolysis. Thus, whereas bacterial RF3 mediates the release of RF1/RF2 from post-TCs and its GTPase activity is required for the subsequent dissociation of RF3 itself from post-termination ribosomes, eRF3 works cooperatively with eRF1 to ensure rapid peptide release, and GTP hydrolysis is required to couple stop-codon recognition with peptide release. Importantly, in contrast to bacteria, at least one eukaryotic release factor,

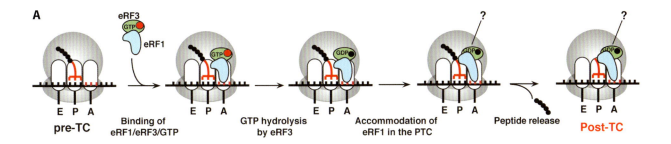

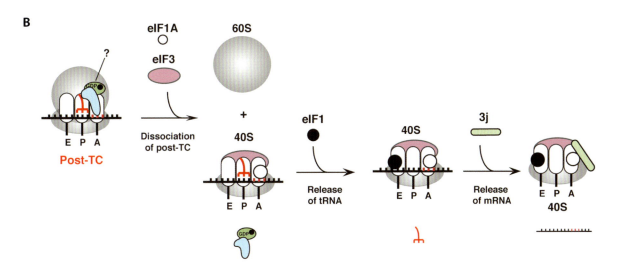

Fig. 1 (A) Model for translation termination in eukaryotes. Pre-termination complexes (pre-TCs) contain peptidyl-tRNA in the P site. eRF1 and eRF3 bind to pre-TCs in a ternary complex eRF1/eRF3/GTP. The stop codon is recognized by eRF1. After GTP hydrolysis by eRF3, eRF1 induces peptide release. At least one release factor, eRF1, remains associated with post-termination complexes (post-TCs). (B) Model for eukaryotic ribosome recycling mediated by initiation factors. eIF3, in cooperation with its 3j subunit, eIF1 and eIF1A, dissociates post-TCs into free 60S subunits and tRNA- and mRNA-bond 40S subunits. eIF1 promotes the subsequent release of tRNA, which is followed by 3j-mediated dissociation of mRNA.

eRF1, remains associated with post-TCs, accounting for the maintenance of the shifted toe-print in post-TCs after peptide release (Pisarev et al., 2007, 2010). It has not yet been established whether eRF3-GDP remains associated with ribosomal complexes throughout the entire termination process, but eRF3's high affinity for eRF1, both in the presence of GDP and in the absence of nucleotides, suggests that such a possibility cannot be excluded. As the binding sites of eRF1/eRF3 and bacterial EF-G/RRF are located in equivalent regions in eukaryotic and bacterial ribosomes, the continued association of eRF1 with post-TCs following hydrolysis of peptidyl-tRNA was another indication that the mechanism of ribosomal recycling in eukaryotes must differ from that in bacteria.

3. Recycling of mammalian post-TCs by initiation factors at low Mg²⁺ concentrations

eIF3 is able to dissociate 80S ribosomes in the presence of ssRNA that can bind directly to the mRNA-binding cleft (Kolupaeva et al., 2005) and, following eIF5-induced GTP hydrolysis by eIF2, eIF1 promotes the dissociation from 48S initiation complexes of tRNAs lacking the G-C base-pairs in the anticodon stem characteristic for initiator tRNAs (Lomakin et al., 2006). These observations suggested that eIF3 and eIF1, possibly in combination with other eIFs, might be able to mediate recycling of post-TCs. To test this hypothesis, pre-TCs were assembled on a modified β-globin mRNA encoding a tetrapeptide followed by a UAA stop codon, purified by sucrose density gradient centrifugation, and then incubated with eRF1/ eRF3 (to induce peptide release), Met-tRNA$^{Met}_i$ and eIFs 1, 1A, 2, 3, 4A, 4B and 4F. At 1 mM of free Mg^{2+}, this full set of initiation factors was indeed able to promote efficient recycling of post-TCs and formation of new 48S complexes on the initiation codon (Pisarev et al., 2007). Systematic omission of factors revealed that eIF3, its loosely associated 3j subunit, eIF1, and eIF1A are sufficient for recycling (Figure 1 B). eIF3 is the principal factor that promotes splitting of post-TCs into free 60S subunits and 40S subunits with associated tRNA and mRNA, and its dissociating activity is strongly enhanced by 3j and less so by eIF1 and eIF1A. After dissociation of post-TCs, eIF3 remains on 40S subunits, preventing their re-association with 60S subunits. Subsequent release of the P-site deacylated tRNA from the 40S subunits is induced by eIF1 and followed by 3j-mediated dissociation of mRNA. eIF3, 3j, eIF1, and eIF1A also promoted efficient recycling when peptide release was induced by puromycin instead of eRF1/eRF3, indicating that splitting of post-TCs by this mechanism was not dependent on release factors *per se*.

The 40S/eIF3 complex was modeled based on cryo-electron microscopic (cryo-EM) reconstructions of two separate complexes of the hepatitis C virus IRES with 40S subunits and of the IRES with eIF3. The model indicates that the major part of the ~800kDa five-lobed eIF3 binds to the solvent side of the 40S subunit with its left "leg" located below the platform near the 60S subunit interface and covering rpS13, which contributes to the formation of intersubunit bridge B4, and it has therefore been suggested that the ribosome-dissociating activity of eIF3 could, at least in part, be due to the disruption of that bridge (Siridechadilok et al., 2005). Binding of eIF3 most likely induces conformational changes in the 40S subunit that might also contribute to the dissociation of post-TCs. However, the visualization of such potential conformational changes will require the direct determination, by cryo-EM or X-ray crystallography, of the structure of eIF3/40S subunit complexes. The enhancement of the eIF3 association with 40S subunits by 3j (Fraser et al., 2004) could, at least in part, be responsible for the stimulation of the dissociating activity of eIF3 by 3j. However, the fact that the C-terminal domain of 3j is located in the mRNA-binding channel in the A-site area of the 40S subunit (Fraser et al., 2007) suggests that the mechanism of its action might be more complicated. eIF1 binds to the interface surface between the platform and Met-tRNA$^{Met}_i$

(Lomakin et al., 2003), whereas the structured domain of eIF1A resides in the A site and its N- and C-terminal tails extend into the P site (Yu et al., 2009). Thus, both eIF1 and eIF1A would be unable to gain access to their binding sites until the intersubunit space has been opened by eIF3, but subsequently could further enhance splitting and contribute to the prevention of subunit re-association.

The function of eIF1 in mediating the dissociation of deacylated tRNA from the P site of the 40S subunit is consistent with its well-documented activity in maintaining the fidelity of initiation codon and initiator tRNA selection (for a review see Jackson et al., 2010) and is most likely based on the conformational changes that eIF1 in cooperation with eIF1A induces in the 40S subunit (Passmore et al., 2007). The fact that even when there is no tRNA in the P site, the mRNA remains associated with 40S/eIF3 complexes is consistent with the mutual stabilization of eIF3 and mRNA on the 40S subunit (Unbehaun et al., 2004; Kolupaeva et al., 2005). The role of 3j in mediating the release of mRNA from the 40S subunit is consistent with the reported negative cooperativity between 3j and mRNA in binding to the 40S subunit (Unbehaun et al., 2004; Fraser et al., 2007). Interestingly, the yeast 3j ortholog HCR1 is not essential for viability (Valasek et al., 1999), whereas the *Drosophila melanogaster* 3j ortholog Adam is an essential protein (Goldstein et al., 2001).

The mechanisms of bacterial ribosome recycling and eIF-mediated recycling of mammalian post-TCs show some similarities, such as the splitting of post-TCs into free large subunits and tRNA/mRNA-bound small subunits, and the subsequent promotion of the dissociation of deacylated tRNA by IF3 and eIF1, respectively, factors that bind to comparable regions on small ribosomal subunits and perform equivalent roles during initiation (Lomakin et al., 2006, and references therein). However, the differences between them are much greater. Thus, dissociation of bacterial post-TCs occurs from the intersubunit space, whereas dissociation of mammalian post-TCs mostly relies on eIF3 acting from the 40S subunit's solvent side, and more importantly, the dissociation of bacterial post-TCs requires GTP hydrolysis (by EF-G), whereas the dissociation of mammalian post-TCs by eIFs occurs without energy consumption. It has therefore been suggested (Rodnina, 2010) that the recycling of eukaryotic post-TCs by eIFs 3/3j/1/1A is instead reminiscent of the IF3/IF1-mediated ribosome recycling that occurs under particular conditions in bacterial systems (Pavlov et al., 2008).

Splitting of eukarytic post-termination complexes by initiation factors without GTP hydrolysis functions at low Mg^{2+} concentrations ($\leq$1mM) only (Pisarev et al., 2010), most likely because at elevated Mg^{2+} concentrations the flexibility of ribosomal subunits is reduced and subunit association is stabilized (e.g. Shenvi et al., 2005). Although mammalian cells are thought to maintain a cytosolic level of free Mg^{2+} within a range of 0.2–1 mM, the level can increase, e.g. in response to vasoactive peptides (Grubbs, 2002; Romani, 2007). However, the investigation of recycling of preassembled post-TCs in rabbit reticulocyte lysate showed that it can occur in a wide range of Mg^{2+} concentrations (up to 6 mM total Mg^{2+}), immediately indicating the existence of another recycling mechanism operating at higher Mg^{2+} concentrations (Pisarev et al., 2010).

4. Functions of protein ABCE1

4.1. ABCE1-mediated recycling of mammalian post-TCs

Purification from rabbit reticulocyte lysate of the missing factors that could mediate recycling at elevated Mg^{2+} concentrations (Pisarev et al., 2010) yielded a ~ 65 kDa protein that was identified by mass spectrometry as ABCE1, an essential member of the ATP-binding cassette (ABC) family of proteins. ABC proteins are mostly involved in transport across membranes, but also in DNA repair and translation (Rees et al., 2009). In contrast to ABC transporters, ABCE1 lacks transmembrane domains. ABCE1 is the fourth ABC family member that is involved in eukaryotic translation. The others are: (i) ABC50, which stimulates formation of eIF2/GTP/Met-tRNA$^{Met}_i$ complexes (Paytubi et al., 2009), (ii) GCN20, which, together with GCN1, functions in activating the GCN2 eIF2(kinase (Vazquez de Aldana et al., 1995), and (iii) the fungus-specific elongation factor eEF3, which was reported to facilitate the release of deacylated tRNA from the E site of the ribosome after translocation (Andersen et al., 2006).

ABCE1 alone can promote a rather efficient dissociation into subunits of post-TCs obtained with eRF1/eRF3. However, the level of dissociation is substantially increased in the presence of eIF6 or eIFs 3/1/1A, presumably by preventing subunit re-association, indicating that the dissociation by ABCE1 is transient in nature (Figure 2 A, B). Post-TC dissociation by ABCE1 is rapid (~ 30% of post-TCs were dissociated within 30 seconds of incubation) and efficient at up to 3.5 mM

free Mg^{2+} (but marginal at 5 mM Mg^{2+}) (Figure 2 C, D). ABCE1 is most likely involved in recycling at all Mg^{2+} concentrations, accelerating the process at lower and becoming essential at higher concentrations.

After ABCE1-mediated dissociation of post-TCs, both tRNA and mRNA remain bound to the 40S subunits. Although eIF1 together with eIF1A can weaken the binding of deacylated tRNA to 40S subunits, leading to near-complete tRNA and mRNA dissociation during sucrose density gradient centrifugation, efficient tRNA and mRNA release without centrifugation also requires eIF3 and its 3j subunit. Thus, the factor requirements for tRNA and mRNA release from recycled 40S subunits obtained with ABCE1 at elevated Mg^{2+} concentrations are similar to those observed at low Mg^{2+} concentrations in the absence of ABCE1. Interestingly, it was recently shown that release of deacylated tRNA and mRNA from 40S subunits after ABCE1-mediated dissociation of post-termination ribosomes can also be promoted by ligatin and to a slightly lesser extent by MCT-1/DENR, an interacting pair of proteins that correspond to N- and C-terminal regions of ligatin, respectively (Skabkin et al., 2010). Ligatin, MCT-1, and DENR are present in all eukaryotes, and their orthologs in yeast (Sacharomyces cerevisiae) are Tma64, Tma20 and Tma22 (Prosniak et al., 1998; Deyo et al., 1998; Fleischer et al., 2006). Ligatin and Tma64 contain N-terminal DUF1947 and PUA domains and C-terminal SWIB/MDM 2 and SUI1/eIF1 domains that also occur in MCT-1/Tma20 and DENR/Tma22, respectively. The mechanism of tRNA/mRNA release from 40S subunits mediated by these proteins and the fate of 40S subunits freed this way in the translation pathway has not yet been determined. Nevertheless, the existence of these factors suggests that there may be considerable redundancy in the factors that can promote the dissociation of tRNA and mRNA from 40S subunits in eukaryotes. It will be interesting to see whether these different branches of the recycling pathway are connected to different downstream processes.

Importantly, ABCE1 promotes efficient recycling only if peptide release is triggered by both eRF1 and eRF3 (Figure 2 E). Dissociation of post-TCs obtained with eRF1 alone is less efficient, and requires concentrations of eRF1 that are higher than those needed for peptide release. Post-TCs obtained by treatment of pre-TCs with puromycin in the absence of release factors could not be dissociated by ABCE1 at all. Post-TCs obtained with eRF1/eRF3 remain bound to eRF1 (or possibly eRF1/eRF3) and are characterized by a two-nucleotide shift of their mRNA toe-print

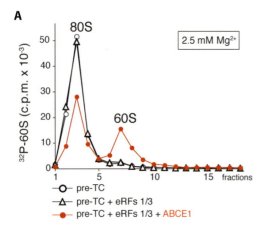

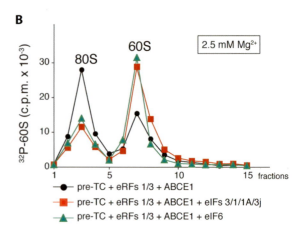

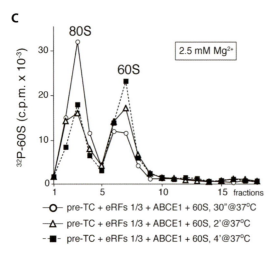

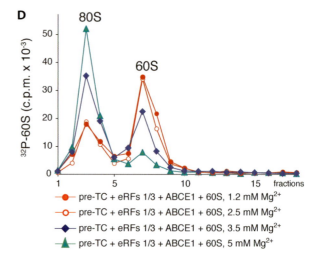

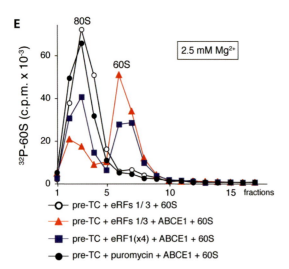

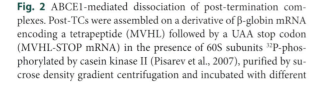

Fig. 2 ABCE1-mediated dissociation of post-termination complexes. Post-TCs were assembled on a derivative of β-globin mRNA encoding a tetrapeptide (MVHL) followed by a UAA stop codon (MVHL-STOP mRNA) in the presence of 60S subunits ^{32}P-phosphorylated by casein kinase II (Pisarev et al., 2007), purified by sucrose density gradient centrifugation and incubated with different combinations of eRFs, ABCE1, eIFs, puromycin and unlabeled 60S subunits at the indicated concentrations of free Mg^{2+}. Dissociation of post-TCs into subunits was assayed by sucrose density gradient centrifugation followed by Cerenkov counting. (Figure reproduced from Pisarev et al. (2010) Mol Cell 37: 196–210 © 2010 with permission from Elsevier)

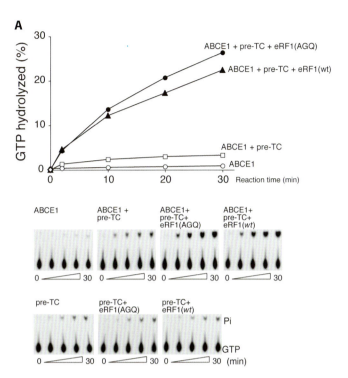

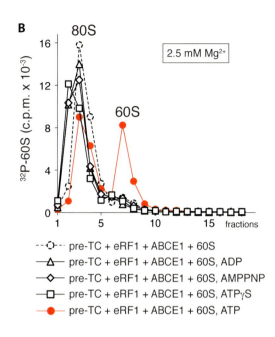

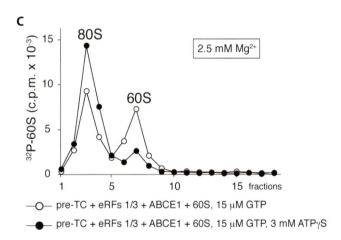

Fig. 3 NTPase activity of ABCE1 is required for its activity in ribosome recycling. (A) Time courses of GTP hydrolysis by ABCE1 in the presence/absence of pre-TC (assembled on MVHL-STOP mRNA), pre-TC/eRF1(AGQ) and pre-TC/eRF1(wt), as indicated. 15 μl reaction mixtures containing 0.5 pmol ABCE1, 0.33 μM [γ-^{32}P]GTP and combinations of 0.5 pmol pre-TC, 10 pmol eRF1(AGQ), 10 pmol eRF1(wt) and 10 pmol eIF6, were incubated at 37 °C for the times indicated. GTP hydrolysis in the presence of both ABCE1 and pre-TCs (upper panels) was corrected to take into account the intrinsic GTPase activity of pre-TCs (lower panels). (B, C) Dissociation by ABCE1 of post-TCs. Post-TCs were obtained by incubation of pre-TCs assembled on MVHL-STOP mRNA in the presence of [^{32}P]60S subunits with (B) eRF1, or (C) eRF1/eRF3, depending on the presence/absence of nucleotides as indicated, assayed by sucrose density gradient centrifugation and Cerenkov counting. (Figure reproduced from Pisarev et al. (2010) Mol Cell 37: 196–210 © 2010 with permission from Elsevier)

that probably results from conformational changes. In contrast, when peptide release is induced by eRF1 alone, eRF1 does not remain bound to post-TCs, and a weak two-nucleotide toe-print shift is apparent only at high concentrations of eRF1. Thus, the efficiency of recycling of post-TCs obtained either with eRF1 and eRF3 together or with eRF1 alone correlates with the association eRF1 with the ribosome after peptide release. The reason why eRF1 remains firmly bound to post-TCs only in the presence of eRF3 is not clear. One possibility is that eRF3 remains associated with ribosomal complexes throughout the termination process and stabilizes the binding of eRF1 after peptide release. Alternatively, the conformations of eRF1 in which it induces peptide release with and without eRF3 could differ to some extent: in the presence of eRF3, eRF1 might establish additional ribosomal contacts that would allow it to remain bound after peptide release, even if eRF3 dissociates after GTP hydrolysis.

This, in turn, poses the question of why the dissociation of post-TCs by ABCE1 requires the presence of eRF1. Consistent with previous reports (Andersen and Leevers, 2007; Dong et al., 2004; Kispal et al., 2005; Yarunin et al., 2005), we observed that ABCE1 in the ADPNP-bound form efficiently associated with 40S subunits and 43S complexes, but not with 80S ribosomes or pre-TCs, which suggests that in the latter the binding site of ABCE1 is occluded. However, it could bind stably to post-TCs and even to pre-TCs associated with the eRF1(AGQ) mutant, complexes that both exhibit the shifted toe-print mentioned above. ABCE1 and eRF1 can interact directly, and the interaction was mapped to the second NBD of ABCE1 (Khoshnevis et al., 2010; Pisarev et al., 2010). Although a direct contact between ABCE1 and eRF1 could certainly contribute to the binding of ABCE1 to 80S ribosomal complexes,

it seems more likely that the principal role of eRF1 in promoting the binding of ABCE1 could be to induce conformational changes in 80S ribosomal complexes that unmask the binding site of ABCE1. In this respect it is worth noting that the binding of eEF3 also depends on the conformational state of 80S ribosomes: in its ATP-bound form, eEF3 binds most stably to ribosomes in the post-translocation state, whereas binding to empty ribosomes is weaker, and a rotated conformation of the head in pre-translocated ribosomes is not permissive for eEF3-binding (Andersen et al., 2006). However, although eRF1 is certainly required for the association of ABCE1 with post-TCs, the function of eRF1 in ribosomal recycling might extend beyond the creation of a binding site for ABCE1 (see discussion below). Interestingly, in its GTP-bound form (i. e. with GDPNP), eRF3 prevents the association of ABCE1 with eRF1-bound ribosomes. Thus, prior to GTP hydrolysis, eRF3 either sterically blocks the association of ABCE1 with the ribosome or induces conformational changes that are unfavorable for the binding of ABCE1. To allow ABCE1 to bind, eRF3/GDP must therefore either dissociate from post-TCs, or one must assume that, unlike eRF3/GTP, eRF3/GDP does not interfere with the binding of ABCE1.

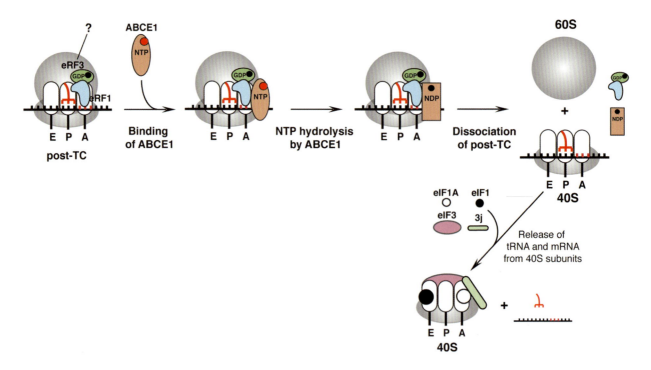

Fig. 4 Model for eukaryotic ribosome recycling mediated by ABCE1. ABCE1 associates with post-TCs that contain eRF1 (or eRF1/eRF3, if eRF3 remains associated with ribosomal complexes after GTP hydrolysis). After NTP hydrolysis by ABCE1, post-TCs dissociate into free 60S subunits and tRNA- and mRNA-bound 40S subunits. Initiation factors eIF3, 3j, eIF1 and eIF1A promote subsequent release of tRNA and mRNA from 40S subunits preparing them for the next round of initiation.

ABCE1 has low intrinsic NTPase activity and lacks nucleotide specificity, hydrolyzing ATP, GTP, CTP and UTP. The NTPase activity is weakly enhanced by vacant 80S ribosomes, but strongly stimulated by post-TCs and eRF1(AGQ)-associated pre-TCs (Figure 3A). Dissociation of post-TCs by ABCE1 is dependent on NTP hydrolysis by ABCE1 (Figures 3B and 3C), suggesting that NTP hydrolysis might be coupled to the splitting of post-TCs into subunits. The fact that stable binding of ABCE1 with ribosomal subunits was not observed in the presence of ADP (Andersen and Leevers, 2007; Pisarev et al., 2010) suggests that, after NTP hydrolysis and dissociation of post-TCs, ABCE1 is also released.

Taken together, the data described above suggest the following model for ABCE1-mediated eukaryotic ribosome recycling (Figure 4). ABCE1 binds to post-TCs containing eRF1 (or eRF1/eRF3, if eRF3 remains associated), and, after hydrolyzing NTP, promotes their dissociation into 60S subunits and tRNA- and mRNA-bound 40S subunits. eRF1 (or eRF1/eRF3) and ABCE1/NDP are released into solution. After that, eIF3, 3j, eIF1 and eIF1A mediate the release of tRNA and mRNA from 40S subunits, preparing them for the next round of initiation.

4.2. The structure of ABCE1 and its relationship to function

Mammalian ABC proteins are categorized into seven sub-families (Dean et al., 2001), two of which (ABCE and ABCF) lack membrane-spanning domains. ABCE1 is the only member of the ABCE subfamily, whereas the other three ABC family members that are involved in translation, ABC50, GCN20 and eEF3, belong to the ABCF subfamily (Kerr, 2004). ABC proteins typically contain twin ABC-type nucleotide-binding domains (NBD), which can convert chemical energy into mechanical work (Higgins and Linton, 2004; Rees et al., 2009). NBDs comprise catalytic core domains containing Walker A, Walker B and Q-loop motifs, and α-helical domains containing the "LSG-GQ" signature motif. NBDs are arranged in a head-to-tail orientation, which creates two composite nucleotide-binding sites formed by Walker A/B and Q-loop motifs from one NBD and by the signature motif from the other. Thus, the ATP molecules bind at the interface between NBDs, with the γ-phosphates located between the P loop of the Walker A motif on one NBD and the signature motif of the other. The basis for the

activity of ABC proteins is their ability to undergo cyclical conformational changes determined by changes in the relative positions of NBDs, which depend on the nucleotide-bound state of the protein: the ATP-bound state is characterized by a closed, 'dimerized' conformation of the ABC cassettes with an extensive interface between the NBDs, whereas in the ADP-bound and nucleotide-free states the separation between the NBDs is much greater. Thus, nucleotide-dependent conformational transitions between the stages of ATP binding and hydrolysis induce a tweezer-like power stroke between the NBDs that could be transmitted to associated domains and/or macromolecules.

The ~68 kDa ABCE1 consists of four domains: a cysteine-rich N-terminal domain harboring two [4Fe-4S] clusters, followed by two NBDs arranged in the typical head-to-tail orientation by a hinge domain formed by the highly conserved region connecting the NBDs, and the C terminus of the protein (Karcher et al., 2005, 2008; Figure 5). In the currently available crystal structures of ABCE1, both nucleotide-binding sites contain ADP. The hinge domain that binds along the NBD1-NBD2 interface may potentially act as a pivot point for the putative conformational changes between NTP- and NDP-bound states of the protein. ABCE1 is the only ABC enzyme that contains an iron-sulfur-cluster domain. The Fe-S domain has the $(\beta\alpha\beta)_2$ fold characteristic of bacterial type ferredoxins, with two small insertions located at suggested positions for electron entry and exit in ferredoxin domains that are active in electron transfer. Coordination by cysteines of one of the [4Fe-4S] clusters is typical of bacterial ferredoxin domains, whereas the second cluster, in which the coordinating cysteine triad has an unusual spacing $CX_4CX_{3/4}C$, is of a type unique to ABCE1 (Barthelme et al., 2007; Karcher et al., 2008). Both [4Fe-4S] clusters are surrounded by a hydrophobic core, do not interact directly with the solvent (Karcher et al., 2008), and are stable down to redox potentials of -560 mV, which argues against an electron transfer function of the Fe-S clusters (Barthelme et al., 2007). On the other hand, the presence of a conserved positively charged patch on the surface of the Fe-S domain suggests that it might be involved in the interaction of ABCE1 with nucleic acids, e.g. rRNA (Karcher et al., 2008). Mutations of conserved amino acids involved in ATP binding and hydrolysis or of structurally important residues in the Fe-S domain are lethal (Coelho et al., 2005; Dong et al., 2004; Karcher et al., 2005; Kispal et al., 2005; Barthelme et al., 2007), indicating that both NBDs and the Fe-S domain are essential for function.

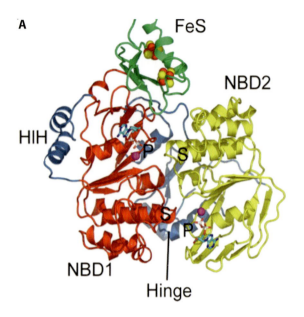

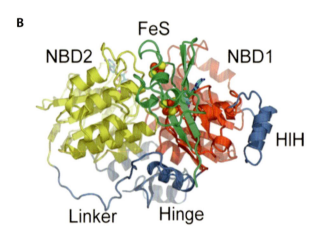

Fig. 5 Structural overview of ABCE1 from *Pyrococcus abyssi*. (A) "Top view" and (B) "Front view": ribbon representations showing the nucleotide binding domains NBD1 (red) with its helix-loop-helix insertion (HlH, blue) and NBD2 (yellow), oriented head-to-tail by the Hinge domain (light blue). The Fe-S domain (green) includes representations of the two [4Fe-4S] clusters (red spheres, iron; yellow spheres, sulphur), and binds to NBD1 at the lateral opening of the nucleotide-binding cleft. Two ADP molecules (colored stick models) and magnesium ions (magenta spheres) show the positions of the two composite binding sites (P loop/Walker A motifs (P) and the signature motifs (S)). (Figure adapted from Karcher et al. (2008) J Biol Chem 283: 7962–7971 © the American Society for Biochemistry and Molecular Biology and reprinted with permission)

The Fe-S domain binds directly to Lobe I of NBD1 at the opening of the cleft between NBD1 and NBD2. Moreover, it directly contacts the Y loop of NBD1, which is a central element of the ATP-binding site of NBD1 (Karcher et al., 2005; 2008). It has therefore been suggested that such a direct link between NBD1 and the associated Fe-S domain might provide a mechanism for potential substrate-mediated allosteric control of the ATPase activity of ABCE1 (Karcher et al., 2008). Although the Fe-S domain does not contact NDB2 in the ADP-bound form of ABCE1, modeling of the ATP-bound state of ABCE1 suggests that NBD2 would rotate by about 40° relative to the FeS-NBD1 hinge for Walker A and signature motifs of opposing NBDs to form the composite binding sites for ATP. The reorientation of NBD2 would lead to a steric clash with the Fe-S domain that could be resolved by repositioning of the latter, which, in turn, might affect the interaction of ABCE1 with components of the translation apparatus, e. g. the ribosome (Karcher et al., 2008). Mutational analyses have yielded data consistent with a functionally critical role of the suggested interface of the Fe-S and NBD2 domains (Karcher et al., 2005).

ABCE1 is extremely conserved, and it is the only human ABC protein that has orthologs in archaea (Dean and Annilo, 2005), with sequence identity ranging from 40 to 50%. Although the mechanisms of translation termination and ribosomal recycling in archaea have not yet been elucidated, it is interesting to note that, even though archaea do not encode an eRF3 homologue, archaeal aRF1 is homologous to eRF1 (Atkinson et al., 2008), which – as our data show – is required for ABCE1-mediated ribosomal recycling. Thus, it is tempting to suggest that the mechanism of ribosome recycling in archaea might be similar to that in eukaryotes.

4.3. Evidence for the involvement of ABCE1 in translation termination and initiation

Recently, ABCE1 has also been implicated in termination. In genetic experiments in yeast (Khoshnevis et al., 2010), partial down-regulation of ABCE1 (RLI1) increased stop-codon read-through, whereas overexpression partially suppressed the stop-codon read-through defect of an eRF1 mutant containing a substitution (I222S) in its middle domain (Stansfield et al., 1997). The mechanism by which ABCE1 influences termination is not known. Notably, although the site of interaction of ABCE1 with eRF1 was mapped to the NBD2 of ABCE1, the Fe-S domain was essential for the function of ABCE1 in termination (Khoshnevis et al., 2010).

Interestingly, before the identification of the function of ABCE1 in ribosome recycling and termination, an involvement of ABCE1 in translation initiation was noted. Depletion of ABCE1 in yeast (Dong et al., 2004; Kispal et al., 2005; Yarunin et al., 2005), human (Chen et al., 2006), and *Drosophila melanogaster* (Andersen and Leavers, 2007) cells resulted in a severe reduction in the level of translation as well as in polysome size and levels, and in the accumulation of mRNA-free 80S monosomes, changes that are all consistent with a defect in initiation. Co-immunoprecipitation experiments identified human eIF2(and eIF5 (Chen et al., 2006) and yeast eIF2, eIF3 and eIF5 (Dong et al., 2004) as proteins that associated with ABCE1 in cell extracts. On the other hand, more rigorous tandem affinity purification experiments identified association of ABCE1 with subunits of eIF3 in yeast (Yarunin et al., 2005) and in *Drosophila* (Andersen and Leevers, 2007), but not with other factors, suggesting that the reported association of ABCE1 with eIF2 and eIF5 may have been indirect. The particularly strong interactions of ABCE1 with the eIF3j orthologs ADAM (*Drosophila*) and HCR1 (*S. cerevisiae*) are consistent with interactions identified in two-hybrid assays (Ito et al., 2001; Kispal et al., 2005; Khoshnevis et al., 2010). Strong reductions in 40S-associated eIF2 and eIF1 in ABCE1-depleted yeast cells, as well as the observation that ABCE1 can interact with eIF3, led to the suggestion that ABCE1 promotes the assembly of 43S pre-initiation complexes (Dong et al., 2004). Since the estimated abundance in cells of ABCE1 is considerably lower than that of eIF2 or eIF3 (Ghaemmaghami et al., 2003), it was proposed that ABCE1 acts catalytically in 43S complex formation (Dong et al., 2004). However, depletion of 90% of ABCE1 in *Drosophila* cells did not influence the levels of 40S-bound eIF2 and eIF3, but resulted in a marked reduction of polysomes, arguing against the involvement of ABCE1 in the recruitment of eIF2 during 43S complex formation and suggesting that ABCE1 is instead involved in a stage of translation downstream of 43S complex formation (Andersen and Leevers, 2007). The mechanism by which ABCE1 affects translation initiation also remains to be established.

4.4. Involvement of ABCE1 in cellular processes other than translation

ABCE1 has been implicated in several cellular processes other than protein synthesis. It was first identified as an inhibitor of the endoribuclease RNase L, which is an effector of the antiviral and antiproliferative effects of interferon (hence ABCE1's original name RLI1 or RNase L inhibitor 1) (Bisbal et al., 1995). Interferon induces synthesis of 2'-5'-oligoadenylate (2−5A) during viral infection. 2−5A activates RNase L, which then cleaves cellular and viral mRNAs, preventing replication of RNA viruses and leading to inhibition of translation. As a countermeasure, some viruses induce the expression of ABCE1, which inhibits RNase L by binding to it, antagonizing its interaction with and subsequent activation by 2−5A (Martinand et al., 1997; Silverman, 2007). It is not known whether the repression of RNase L activation reflects steric hindrance by bound ABCE1 or is a consequence of ABCE1-mediated conformational changes in RNase L. Interestingly, it has also been shown that RNase L binds to eRF3, increasing stop-codon read-through (Le Roy et al., 2005).

ABCE1 is also required for HIV capsid assembly. Formation of the capsid of HIV and other retroviruses involves multimerization of thousands of gag polyprotein precursors, which is coordinated with encapsidation of genomic RNA, and is followed by HIV protease (PR)-mediated processing of gag and maturation of the immature capsids, by budding and finally by virus release. ABCE1 initially associates (via the basic residues in the viral nucleocapsid protein component of gag) with HIV capsid assembly intermediates that increase up to the size of complete, immature capsids (Zimmerman et al., 2002; Lingappa et al., 2006; Dooher et al., 2007). ABCE1 is released following maturation, which is an ATP-dependent process, consistent with a chaperone-like role for ABCE1 in a step in capsid maturation that could involve formation of the capsid shell, packaging of the RNA genome and/or coordination of the PR-mediated process of maturation and virion release.

ABCE1 is also critical for ribosome biogenesis: in addition to causing a strong defect in translation, depletion of ABCE1 in yeast leads to a defect in the nuclear export of 40S and 60S ribosomal subunits, and in late stages in the processing of pre-ribosomal RNAs (Kispal et al., 2005; Yarunin et al., 2005). ABCE1 is associated with pre-40S particles and binds particularly strongly and specifically to eIF3j (Hcr1) (Ito et al., 2001; Dong et al., 2004; Kispal et al., 2005; Yarunin et al., 2005), which plays a dual role in translation and in the final, cytoplasmic stage of maturation of 20S premRNA to 18S rRNA (Valasek et al., 2001). However, the molecular roles of ABCE1 in ribosomal biogenesis have not yet been established.

5. Alternative mechanisms for recycling of ribosomes in yeast?

It has recently been reported that model yeast post-termination complexes, obtained by puromycin treatment of polysomes isolated from growing yeast cells, can be dissociated by the fungus-specific elongation factor eEF3 in an ATP-dependent manner (Kurata et al., 2010). Interestingly, in contrast to the dissociation of bacterial post-TCs by EF-G/RRF and of mammalian post-TCs by eIFs or by ABCE1/eRF1 into free large subunits and tRNA/mRNA-associated small subunits, dissociation of yeast model post-TCs by eEF3 did not yield mRNA-associated 40S subunits, which led to the suggestion that during eEF3-mediated recycling, release of mRNA, tRNA and dissociation of ribosomes into subunits may occur simultaneously (Kurata et al., 2010). If eRF1 remains associated with post-TCs in yeast, as it is the case in higher eukaryotes, it will be important to determine whether eEF3 is able to recycle post-TCs in which peptide release was induced by release factors instead of puromycin. It is also important to emphasize that eEF3 occurs only in fungi, and any recycling mechanism involving eEF3 can thus not be generally applicable to other eukaryotes.

6. Perspectives

Despite the significant progress that has been made in the characterization of factors involved in ribosome recycling in eukaryotes, including the identification of ABCE1 as a central player, the molecular mechanism of eukaryotic recycling remains unresolved. The elucidation of the mechanism will require the concerted efforts of biochemists, structural biologists, and kineticists. The first general question concerns the composition and structure of post-TCs. eRF3 enhances ribosomal association of eRF1 after peptide release, so it will be important to determine whether eRF3 also remains bound to post-TCs, and if not, at which stage it is released. The structure of post-TCs bound to eRF1, which may reveal potential ribosomal conformational changes, would shed light on why ABCE1 associates selectively with eRF1-bound post-TCs, but not with vacant 80S ribosomes.

A second major question concerns the mechanism by which ABCE1 splits eRF1-associated post-TCs, which will require the determination of the structure of ABCE1 in the ATP-bound conformation (which will show whether binding of ATP leads to the predicted ro-

tation of NBD2, and whether that results in repositioning of the Fe-S domain), and also of the position of ABCE1 on the ribosome (which will indicate whether ABCE1 might be able to split post-TCs directly, or whether this step requires transmission and possibly amplification by eRF1 of the effect of ATP-dependent conformational changes in ABCE1, potentially in a manner analogous to the impact of EF-G on RRF). In the last scenario, eRF1 would have an active role in recycling beyond simply creating a binding site for ABCE1.

Although it is tempting to assume that the mechanical work into which ABCE1 converts the chemical energy of NTP hydrolysis is responsible for dissociation of post-TCs, the coupling of NTP hydrolysis and phosphate release with splitting of post-termination ribosomes into subunits needs to be established by kinetic experiments. The mechanism by which NTP hydrolysis by ABCE1 is induced and the mutual relation between the two NTP binding/hydrolysis sites have not been determined yet. Whereas there is a general agreement (based on crystallographic data showing symmetrical binding site occupancy; e. g. Smith et al., 2002) that both nucleotide-binding pockets must contain ATP for stable closure of the NBD dimer, the hydrolysis of ATP in the two pockets could be sequential, but take place in random order within a single cycle, leading to opening of the NBD1-NBD2 dimer interface, or occur in an alternating fashion at different stages within a working cycle (Higgins and Linton, 2004).

Finally, the elucidation of the suggested roles of ABCE1 in termination and initiation would also reveal whether ABCE1 indeed links the termination, ribosome recycling, and initiation stages of eukaryotic protein synthesis.

Acknowledgements

This work was supported by NIH Grant GM80 623 to TVP.

References

Adamson SD, Howard GA, Herbert E (1969) The ribosome cycle in a reconstituted cell-free system from reticulocytes. Cold Spring Harb Symp Quant Biol 34: 547– 554

Alkalaeva EZ, Pisarev AV, Frolova LY, Kisselev LL, Pestova TV (2006) In vitro reconstitution of eukaryotic translation reveals cooperativity between release factors eRF1 and eRF3. Cell 125: 1125–1136

Andersen CB, Becker T, Blau M, Anand M, Halic M, Balar B, Mielke T, Boesen T, Pedersen JS, Spahn CM, Kinzy TG, Andersen GR, Beckmann R (2006) Structure of eEF3 and the mechanism of transfer RNA release from the E-site. Nature 443: 663–668

Andersen DS, Leevers SJ (2007) The essential Drosophila ATP-binding cassette domain protein, pixie, binds the 40 S ribosome in an ATP-dependent manner and is required for translation initiation. J Biol Chem 282: 14752–14760

Atkinson GC, Baldauf SL, Hauryliuk V (2008) Evolution of non-stop, no-go and nonsense-mediated mRNA decay and their termination factor-derived components. BMC Evol Biol 8: 290

Barat C, Datta PP, Raj VS, Sharma MR, Kaji H, Kaji A, Agrawal RK (2007) Progression of the ribosome recycling factor through the ribosome dissociates the two ribosomal subunits. Mol Cell 27: 250–261

Barthelme D, Scheele U, Dinkelaker S, Janoschka A, Macmillan F, Albers SV, Driessen AJ, Stagni MS, Bill E, Meyer-Klaucke W, Schünemann V, Tampé R (2007) Structural organization of essential iron-sulfur clusters in the evolutionarily highly conserved ATP-binding cassette protein ABCE1. J Biol Chem 282: 14598–14607

Bisbal C, Martinand C, Silhol M, Lebleu B, Salehzada T (1995) Cloning and characterization of a RNase L inhibitor. A new component of the interferon-regulated 2–5A pathway. J Biol Chem 270: 13308–13317

Ceglarz E, Goumans H, Thomas A, Benne R (1980) Purification and characterization of protein synthesis initiation factor eIF-3 from wheat germ. Biochim Biophys Acta 610: 181–188

Chaudhuri J, Chowdhury D, Maitra U (1999) Distinct functions of eukaryotic translation initiation factors eIF1A and eIF3 in the formation of the 40 S ribosomal preinitiation complex. J Biol Chem 274: 17975–17980

Checkley JW, Cooley L, Ravel JM (1981) Characterization of initiation factor eIF-3 from wheat germ. J Biol Chem 256: 1582–1586

Chen ZQ, Dong J, Ishimura A, Daar I, Hinnebusch AG, Dean M (2006) The essential vertebrate ABCE1 protein interacts with eukaryotic initiation factors. J Biol Chem 281: 7452–7457

Cheng Z, Saito K, Pisarev AV, Wada M, Pisareva VP, Pestova TV, Gajda M, Round A, Kong C, Lim M, Nakamura Y, Svergun DI, Ito K, Song H (2009) Structural insights into eRF3 and stop codon recognition by eRF1. Genes Dev 23: 1106–1118

Coelho CM, Kolevski B, Bunn C, Walker C, Dahanukar A, Leevers SJ (2005) Growth and cell survival are unevenly impaired in pixie mutant wing discs. Development 132: 5411–5424

Dean M, Annilo T (2005) Evolution of the ATP-binding cassette (ABC) transporter superfamily in vertebrates. Annu Rev Genomics Hum Genet 6: 123–142

Deyo JE, Chiao PJ, Tainsky MA (1998) drp, a novel protein expressed at high cell density but not during growth arrest. DNA Cell Biol 17: 437–447

Dong J, Lai R, Nielsen K, Fekete CA, Qiu H, Hinnebusch AG (2004) The essential ATP-binding cassette protein RLI1 functions in translation by promoting preinitiation complex assembly. J Biol Chem 279: 42157–42168

Dooher JE, Schneider BL, Reed JC, Lingappa JR (2007) Host ABCE1 is at plasma membrane HIV assembly sites and its dissociation from Gag is linked to subsequent events of virus production. Traffic 8: 195–211

Falvey AK, Staehelin T (1970) Structure and function of mammalian ribosomes. II. Exchange of ribosomal subunits at various stages of in vitro polypeptide synthesis. J Mol Biol 53: 21–34

Fan-Minogue H, Du M, Pisarev AV, Kallmeyer AK, Salas-Marco J, Keeling KM, Thompson SR, Pestova TV, Bedwell DM (2008) Distinct eRF3 requirements suggest alternate eRF1 conform-

ations mediate peptide release during eukaryotic translation termination. Mol Cell 30: 599–609

Fleischer TC, Weaver CM, McAfee KJ, Jennings JL, Link AJ (2006) Systematic identification and functional screens of uncharacterized proteins associated with eukaryotic ribosomal complexes. Genes Dev 20: 1294–1307

Fraser CS, Berry KE, Hershey JW, Doudna JA (2007) eIF3j is located in the decoding center of the human 40S ribosomal subunit. Mol Cell 26: 811–819

Fraser CS, Lee JY, Mayeur GL, Bushell M, Doudna JA, Hershey JW (2004) The j-subunit of human translation initiation factor eIF3 is required for the stable binding of eIF3 and its subcomplexes to 40 S ribosomal subunits in vitro. J Biol Chem 279: 8946–8956

Frolova LY, Simonsen JL, Merkulova TI, Litvinov DY, Martensen PM, Rechinsky VO, Camonis JH, Kisselev LL, Justesen J (1998) Functional expression of eukaryotic polypeptide chain release factors 1 and 3 by means of baculovirus/insect cells and complex formation between the factors. Eur J Biochem 256: 36–44

Gao N, Zavialov AV, Li W, Sengupta J, Valle M, Gursky RP, Ehrenberg M, Frank J (2005) Mechanism for the disassembly of the posttermination complex inferred from cryo-EM studies. Mol Cell 18: 663–674

Gao N, Zavialov AV, Ehrenberg M, Frank J (2007) Specific interaction between EF-G and RRF and its implication for GTP-dependent ribosome splitting into subunits. J Mol Biol 374: 1345–13458

Gartmann M, Blau M, Armache JP, Mielke T, Topf M, Beckmann R (2010) Mechanism of eIF6-mediated inhibition of ribosomal subunit joining. J Biol Chem 285: 14848–14851

Ghaemmaghami S, Huh WK, Bower K, Howson RW, Belle A, Dephoure N, O'Shea EK, Weissman JS (2003) Global analysis of protein expression in yeast. Nature 425: 737–741

Goldstein ES, Treadway SL, Stephenson AE, Gramstad GD, Keilty A, Kirsch L, Imperial M, Guest S, Hudson SG, LaBell AA, O'Day M, Duncan C, Tallman, M, Cattelino A, Lim J (2001) A genetic analysis of the cytological region 46C-F containing the Drosophila melanogaster homolog of the jun proto-oncogene. Mol Genet Genomics 266: 695–700

Goumans H, Thomas A, Verhoeven A, Voorma HO, Benne R (1980) The role of eIF-4C in protein synthesis initiation complex formation. Biochim Biophys Acta 608: 39–46

Grubbs RD (2002) Intracellular magnesium and magnesium buffering. Biometals 15: 251–259

Hauryliuk V, Zavialov A, Kisselev L, Ehrenberg M (2006) Class-1 release factor eRF1 promotes GTP binding by class-2 release factor eRF3. Biochimie 88: 747–757

Henshaw EC, Guiney DG, Hirsch CA (1973) The ribosome cycle in mammalian protein synthesis. I. The place of monomeric ribosomes and ribosomal subunits in the cycle. J Biol Chem 248: 4367–4376

Higgins CF, Linton KJ (2004) The ATP switch model for ABC transporters. Nat Struct Mol Biol 11: 918–926

Hirashima A, Kaji A (1973) Role of elongation factor G and a protein factor on the release of ribosomes from messenger ribonucleic acid. J Biol Chem 248: 7580–7587

Hogan BL, Korner A (1968a) Ribosomal subunits of Landschütz ascites cells during changes in polysome distribution. Biochim Biophys Acta 169: 129–138

Hogan BL, Korner A (1968b) The role of ribosomal subunits and 80-S monomers in polysome formation in an ascites tumour cell. Biochim Biophys Acta 169: 139–149

Howard GA, Adamson SD, Herbert E (1970) Subunit recycling during translation in a reticulocyte cell-free system. J Biol Chem 245: 6237–6239

Ito K, Ebihara K & Nakamura Y (1998) The stretch of C-terminal acidic amino acids of translational release factor eRF1 is a primary binding site for eRF3 of fission yeast. RNA 4: 958–972

Ito T, Chiba T, Ozawa R, Yoshida M, Hattori M, Sakaki Y (2001) A comprehensive two-hybrid analysis to explore the yeast protein interactome. Proc Natl Acad Sci USA 98: 4569–4574

Jackson RJ, Hellen CU, Pestova TV (2010) The mechanism of eukaryotic translation initiation and principles of its regulation. Nat Rev Mol Cell Biol 11: 113–127

Jones RL, Sadnik I, Thompson HA, Moldave K (1980) Studies on native ribosomal subunits from rat liver. Evidence for a low molecular weight ribosome dissociation factor. Arch Biochem Biophys 199: 277–285

Kaempfer R (1969) Ribosomal subunit exchange in the cytoplasm of a eukaryote. Nature 222: 950–953

Karcher A, Büttner K, Märtens B, Jansen RP, Hopfner KP (2005) X-ray structure of RLI, an essential twin cassette ABC ATPase involved in ribosome biogenesis and HIV capsid assembly. Structure 13: 649–659

Karcher A, Schele A, Hopfner KP (2008) X-ray structure of the complete ABC enzyme ABCE1 from Pyrococcus abyssi. J Biol Chem 283: 7962–7971

Karimi R, Pavlov MY, Buckingham RH, Ehrenberg M (1999) Novel roles for classical factors at the interface between translation termination and initiation. Mol Cell 3: 601–609

Kerr ID (2004) Sequence analysis of twin ATP binding cassette proteins involved in translational control, antibiotic resistance, and ribonuclease L inhibition. Biochem Biophys Res Commun 315: 166–173

Khoshnevis S, Gross T, Rotte C, Baierlein C, Ficner R, Krebber H (2010) The iron-sulphur protein RNase L inhibitor functions in translation termination. EMBO Rep 11: 214–219

Kispal G, Sipos K, Lange H, Fekete Z, Bedekovics T, Janáky T, Bassler J, Aguilar Netz DJ, Balk J, Rotte C, Lill R (2005) Biogenesis of cytosolic ribosomes requires the essential iron-sulphur protein Rli1 p and mitochondria. EMBO J 24: 589–598

Kisselev L, Ehrenberg M, Frolova L (2003) Termination of translation: interplay of mRNA, rRNAs and release factors? EMBO J 22: 175–182

Kolupaeva VG, Unbehaun A, Lomakin IB, Hellen CU, Pestova TV (2005) Binding of eukaryotic initiation factor 3 to ribosomal 40S subunits and its role in ribosomal dissociation and anti-association. RNA 11: 470–486

Kurata S, Nielsen KH, Mitchell SF, Lorsch JR, Kaji A, Kaji H (2010) Ribosome recycling step in yeast cytoplasmic protein synthesis is catalyzed by eEF3 and ATP. Proc Natl Acad Sci USA 107: 10854–10859

Le Roy F, Salehzada T, Bisbal C, Dougherty JP, Peltz SW (2005) A newly discovered function for RNase L in regulating translation termination. Nat Struct Mol Biol 12: 505–512

Lingappa JR, Dooher JE, Newman MA, Kiser PK, Klein KC (2006) Basic residues in the nucleocapsid domain of Gag are required for interaction of HIV-1 gag with ABCE1 (HP68), a cellular protein important for HIV-1 capsid assembly. J Biol Chem 281: 3773–3784

Lomakin IB, Kolupaeva VG, Marintchev A, Wagner G, Pestova TV (2003) Position of eukaryotic initiation factor eIF1 on the 40S ribosomal subunit determined by directed hydroxyl radical probing. Genes Dev 17: 2786–2797

Lomakin IB, Shirokikh NE, Yusupov MM, Hellen CU, Pestova TV (2006) The fidelity of translation initiation: reciprocal activities of eIF1, IF3 and YciH. EMBO J 25: 196–210

Lubsen NH, Davies BD (1974) A ribosome dissociation factor on both native subunits in rabbit reticulocytes. Biochim Biophys Acta 335: 196–200

Martinand C, Montavon C, Salehzada T, Silhol M, Lebleu B, Bisbal C (1999) RNase L inhibitor is induced during human immunodeficiency virus type 1 infection and down regulates the 2–5A/RNase L pathway in human T cells. J Virol 73: 290–296

Merrick WC, Lubsen NH, Anderson WF (1973) A ribosome dissociation factor from rabbit reticulocytes distinct from initiation factor M3. Proc Natl Acad Sci USA 70: 2220–2223

Mitkevich VA, Kononenko AV, Petrushanko IY, Yanvarev DV, Makarov AA, Kisselev LL (2006) Termination of translation in eukaryotes is mediated by the quaternary eRF1*eRF3*GTP*Mg2+ complex. The biological roles of eRF3 and prokaryotic RF3 are profoundly distinct. Nucleic Acids Res 34: 3947–3954

Mizuno S, Rabinovitz M (1973) Factor-promoted dissociation of free ribosomes in a rabbit reticulocyte lysate system: inhibition and requirement for an energy source. Proc Natl Acad Sci USA 70: 787–791

Passmore LA, Schmeing TM, Maag D, Applefield DJ, Acker MG, Algire MA, Lorsch JR, Ramakrishnan V (2007) The eukaryotic translation initiation factors eIF1 and eIF1A induce an open conformation of the 40S ribosome. Mol Cell 26: 41–50

Pavlov MY, Antoun A, Lovmar M, Ehrenberg M (2008) Complementary roles of initiation factor 1 and ribosome recycling factor in 70S ribosome splitting. EMBO J 27: 1706–1717

Paytubi S, Wang X, Lam YW, Izquierdo L, Hunter MJ, Jan E, Hundal HS, Proud CG (2009) ABC50 promotes translation initiation in mammalian cells. J Biol Chem 284: 24061–24073

Peske F, Rodnina MV, Wintermeyer W (2005) Sequence of steps in ribosome recycling as defined by kinetic analysis. Mol Cell 18: 403–412

Pisarev AV, Hellen CU, Pestova TV (2007) Recycling of eukaryotic posttermination ribosomal complexes. Cell 131: 286–299

Pisarev AV, Skabkin MA, Pisareva VP, Skabkina OV, Rakotondrafara AM, Hentze MW, Hellen CU, Pestova TV (2010) The role of ABCE1 in eukaryotic posttermination ribosomal recycling. Mol Cell 37: 196–210

Pisareva VP, Pisarev AV, Hellen CU, Rodnina MV, Pestova TV (2006) Kinetic analysis of interaction of eukaryotic release factor 3 with guanine nucleotides. J Biol Chem 281: 40224–40235

Prosniak M, Dierov J, Okami K, Tilton B, Jameson B, Sawaya BE, Gartenhaus RB (1998) A novel candidate oncogene, MCT-1, is involved in cell cycle progression. Cancer Res 58: 4233–4237

Raychaudhuri P, Stringer EA, Valenzuela DM, Maitra U (1984) Ribosomal subunit antiassociation activity in rabbit reticulocyte lysates. Evidence for a low molecular weight ribosomal subunit antiassociation protein factor (Mr = 25,000). J Biol Chem 259: 11930–11935

Rees DC, Johnson E, Lewinson O (2009) ABC transporters: the power to change. Nat Rev Mol Cell Biol 10: 218–227

Rodnina MV (2010) Protein synthesis meets ABC ATPases: new roles for Rli1/ABCE1. EMBO Rep 11: 143–144

Romani A (2007) Regulation of magnesium homeostasis and transport in mammalian cells. Arch Biochem Biophys 458: 90–102

Rorbach J, Richter R, Wessels HJ, Wydro M, Pekalski M, Farhoud M, Kühl I, Gaisne M, Bonnefoy N, Smeitink JA, Lightowlers RN, Chrzanowska-Lightowlers ZM (2008) The human mitochondrial ribosome recycling factor is essential for cell viability. Nucleic Acids Res 36: 5787–5799

Russell DW, Spremulli LL (1979) Purification and characterization of a ribosome dissociation factor (eukaryotic initiation factor 6) from wheat germ. J Biol Chem 254: 8796–8800

Salas-Marco J, Bedwell DM (2004) GTP hydrolysis by eRF3 facilitates stop codon decoding during eukaryotic translation termination. Mol Cell Biol 24: 7769–7778

Savelsbergh A, Rodnina MV, Wintermeyer W (2009) Distinct functions of elongation factor G in ribosome recycling and translocation. RNA 15: 772–780

Shenvi CL, Dong KC, Friedman EM, Hanson JA, Cate JH (2005) Accessibility of 18S rRNA in human 40S subunits and 80S ribosomes at physiological magnesium ion concentrations – implications for the study of ribosome dynamics. RNA 11: 1898–1908

Si K, Maitra U (1999) The Saccharomyces cerevisiae homologue of mammalian translation initiation factor 6 does not function as a translation initiation factor. Mol Cell Biol 19: 1416–1426

Silverman RH (2007) Viral encounters with 2',5'-oligoadenylate synthetase and RNase L during the interferon antiviral response. J Virol 81: 12720–12729

Siridechadilok B, Fraser CS, Hall RJ, Doudna JA, Nogales E (2005) Structural roles for human translation factor eIF3 in initiation of protein synthesis. Science 310: 1513–1515

Skabkin MA, Skabkina OV, Dhote V, Komar AA, Hellen CUT, Pestova TV (2010) Activities of Ligatin and MCT-1/DENR in eukaryotic translation initiation and ribosomal recycling. Genes Dev 24: 1787–1801

Smith PC, Karpowich N, Millen L, Moody JE, Rosen J, Thomas PJ, Hunt JF (2002) ATP binding to the motor domain from an ABC transporter drives formation of a nucleotide sandwich dimer. Mol Cell 10: 139–149

Stansfield I, Kushnirov VV, Jones KM, Tuite MF (1997) A conditional-lethal translation termination defect in a sup45 mutant of the yeast Saccharomyces cerevisiae. Eur J Biochem 245: 557–563

Thompson HA, Sadnik I, Scheinbuks J, Moldave K (1977) Studies on native ribosomal subunits from rat liver. Purification and characterization of a ribosome dissociation factor. Biochemistry 16: 2221–2230

Trachsel H, Staehelin T (1979) Initiation of mammalian protein synthesis. The multiple functions of the initiation factor eIF-3. Biochim Biophys Acta 565: 305–314

Unbehaun A, Borukhov SI, Hellen CU, Pestova TV (2004) Release of initiation factors from 48S complexes during ribosomal subunit joining and the link between establishment of codon-anticodon base-pairing and hydrolysis of eIF2-bound GTP. Genes Dev 18: 3078–3093

Valásek L, Hasek J, Nielsen KH, Hinnebusch AG (2001) Dual function of eIF3j/Hcr1p in processing 20 S pre-rRNA and translation initiation. J Biol Chem 276: 43351–43360

Valásek L, Hasek J, Trachsel H, Imre EM, Ruis H (1999) The Saccharomyces cerevisiae HCR1 gene encoding a homologue of the p35 subunit of human translation initiation factor 3 (eIF3) is a high copy suppressor of a temperature-sensitive mutation in the Rpg1p subunit of yeast eIF3. J Biol Chem 274: 27567–27572

Valenzuela DM, Chaudhuri A, Maitra U (1982) Eukaryotic ribosomal subunit anti-association activity of calf liver is contained in a single polypeptide chain protein of Mr = 25,500 (eukaryotic initiation factor 6). J Biol Chem 257: 7712–7719

Vazquez de Aldana CR, Marton MJ, Hinnebusch AG (1995) GCN20, a novel ATP binding cassette protein, and GCN1 reside in a complex that mediates activation of the eIF-2 alpha kinase GCN2 in amino acid-starved cells. EMBO J 14: 3184–3199

Yarunin A, Panse VG, Petfalski E, Dez C, Tollervey D, Hurt EC (2005) Functional link between ribosome formation and biogenesis of iron-sulfur proteins. EMBO J 24: 580–588

Youngman EM, McDonald ME, Green R (2008) Peptide release on the ribosome: mechanism and implications for translational control. Annu Rev Microbiol 62: 353–373

Yu Y, Marintchev A, Kolupaeva VG, Unbehaun A, Veryasova T, Lai SC, Hong P, Wagner G, Hellen CU, Pestova TV (2009) Position of eukaryotic translation initiation factor eIF1A on the 40S ribosomal subunit mapped by directed hydroxyl radical probing. Nucleic Acids Res 37: 5167–5182

Zavialov AV, Buckingham RH, Ehrenberg M (2001) A posttermination ribosomal complex is the guanine nucleotide exchange factor for peptide release factor RF3. Cell 107: 115–124

Zavialov AV, Hauryliuk VV, Ehrenberg M (2005) Splitting of the posttermination ribosome into subunits by the concerted action of RRF and EF-G. Mol Cell 18: 675–686

Zavialov AV, Mora L, Buckingham RH, Ehrenberg M (2002) Release of peptide promoted by the GGQ motif of class 1 release factors regulates the GTPase activity of RF3. MolCell 10: 789–798

Zimmerman C, Klein KC, Kiser PK, Singh AR, Firestein BL, Riba SC, Lingappa JR (2002) Identification of a host protein essential for assembly of immature HIV-1 capsids. Nature 415: 88–92

Section III Decoding, fidelity, and peptidyl transfer

The specific interaction between aminoacyl-tRNAs and elongation factor Tu

15

Jared M. Schrader, Margaret E. Saks, and Olke C. Uhlenbeck

1. Introduction

EF-Tu couples the hydrolysis of GTP with the accurate delivery of aminoacyl-tRNAs (aa-tRNAs) into the encoded ribosomal A site. The well-studied catalytic cycle of EF-Tu (Figure 1) can be subdivided into five phases: (1) the binding of EF-Tu•GTP to elongator aa-tRNAs; (2) the binding of the resulting ternary complex to the ribosomal A/T site where codon sampling occurs; (3) a conformational change of both the ribosome and the ternary complex with subsequent hydrolysis of GTP; (4) disruption of the ternary complex with release of phosphate, accommodation of the aa-tRNA into the A site, and release of EF-Tu•GDP from the ribosome; and (5) GDP-GTP exchange catalyzed by EF-Ts. Many of these phases have been dissected into several discrete steps using a variety of biochemical and biophysical methods (Pape et al., 1998; Gromadski et al., 2002; Blanchard et al., 2004). The mechanistic details of this EF-Tu-dependent decoding pathway are the focus of several articles in this volume. Here we discuss how the thermodynamic details of the interaction between EF-Tu and aa-tRNA differ for each tRNA species. We will summarize data showing that the sequence of three base pairs in the T stem of tRNA "tunes" its affinity for EF-Tu in a way that compensates for the variable contribution of the esterified amino acid to the overall binding affinity. This ensures that any correctly aminoacylated tRNA can initially bind to EF-Tu•GTP tightly enough for delivery to the ribosome but weakly enough that it can be released from EF-Tu•GDP during decoding. It appears that this sequence-specific tuning is highly conserved in bacteria and can largely explain the complex pattern of sequence conservation in the T stems of all bacterial tRNAs.

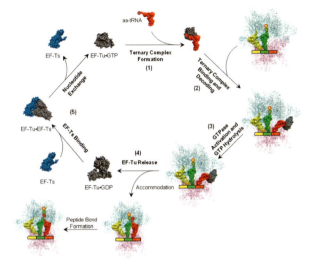

Fig. 1 The catalytic cycle of EF-Tu. A structure-based diagram of the five phases of EF-Tu function described in the text. The structures are not drawn to a uniform scale but are based on appropriate crystal structures. The tRNAs, amino acids, and codons in the E site (yellow), P site (green) and decoding site (red) are indicated.

2. Structures of the interface between EF-Tu and aa-tRNA

To understand the specificity of the initial binding of aa-tRNA to EF-Tu•GTP and the subsequent release of aa-tRNA from EF-Tu•GDP on the ribosome, the appropriate crystal structures must be scrutinized. Three structures of ternary complexes are available (Nissen et al., 1995; Nissen et al., 1999; pdb 1OB2). Although they differ with respect to the source of EF-Tu and the type of tRNA, all three are quite similar, with domains 2 and 3 of EF-Tu forming an extensive interface with the esterified amino acid and the acceptor and T stems of tRNA (Figure 2A). While the overall structures of the protein and the tRNA in the complex resemble

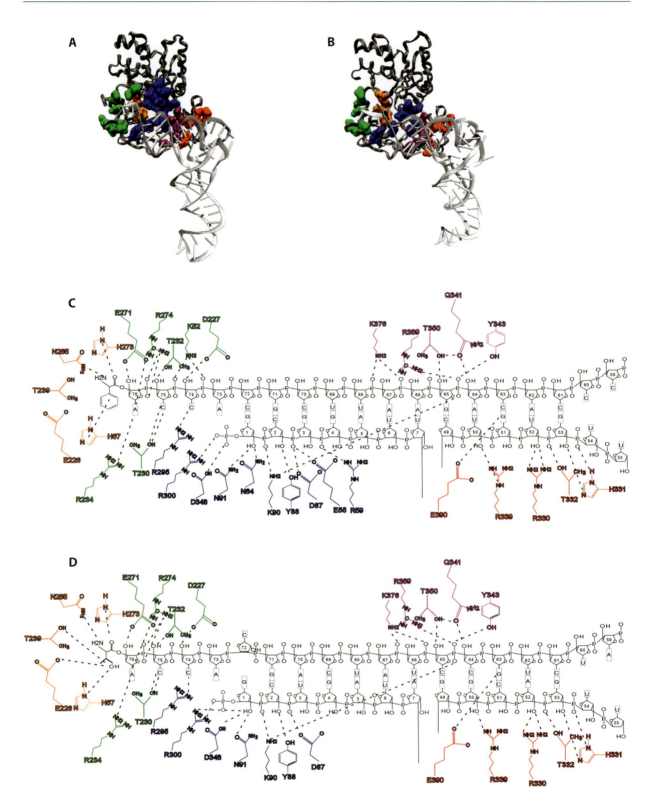

Fig. 2 The binding interface between EF-Tu and aa-tRNA. (A) The X-ray cocrystal structure of *T. thermophilus* EF-Tu•GDPNP (gray) with yeast Phe-tRNA[Phe] (white) (pdb 1TTT). EF-Tu side chains are colored by region (amino acid binding pocket – orange; region 1 – green; region 2 – blue; region 3 – purple; region 4 – red). (B) The X-ray cocrystal structure of *T. thermophilus* EF-Tu•GTP (gray) with *E. coli* Thr-tRNA[Thr] (white) bound to the 70S *T. thermophilus* ribosome stalled with kirromycin (pdb 2WRN). (C) Potential interactions between the *T. thermophilus* EF-Tu side chains and yeast Phe-tRNA[Phe]. Dotted lines indicate possible hydrogen bonds or ion pairs. (D) Potential interactions between the *T. thermophilus* EF-Tu side chains and Thr-tRNA[Thr] in the A/T site.

their corresponding free structures, there are small rearrangements in the interface that permit numerous potential hydrogen bonds and ion pairs to form. A diagram of the interface is shown in Figure 2C where the residues of EF-Tu are subdivided into five adjacent groups based upon their potential for interacting with the esterified amino acid, the 3' terminus, the 5' terminus, the inner T stem, and the outer T stem. Table 1 lists these potential contacts for both the *Thermus thermophilus* and *Escherichia coli* proteins. In general, the interface between the protein and the tRNA shows only a single interaction with a base. The remainder of the amino acids form hydrogen bonds and ion pairs with 2' hydroxyls and phosphates in the A and T helices of tRNA.

A recent crystal structure of the *T. thermophilus* ternary complex bound to 70S ribosomes in the presence of kirromycin (Figure 2B) provides a view of the complex trapped during decoding (Schmeing et al., 2009). Biochemical experiments suggest that kirromycin blocks decoding after GTPase activation and GTP hydrolysis have occurred (Rodnina et al., 1995; Blanchard et al., 2004). While the tRNA structure in this complex is distorted, the interface between EF-Tu and tRNA is significantly changed only near the 3' end of the tRNA. As diagrammed in Fig. 2D and summarized in Table 1, several residues that were close to the aa-tRNA in the ternary complex are no longer nearby and several other residues are in proximity. Thus, although the interface is slightly different on the ribosome, no clear additional sequence specificity is observed.

3. Conformity and thermodynamic compensation

Since all the different elongator aa-tRNAs must interact with EF-Tu•GTP to enter the translational machinery, it is perhaps not surprising that they all bind the protein with quite similar affinities. Experiments testing large panels of aa-tRNAs with EF-Tu•GTP from either *E. coli* (Louie et al., 1984; Louie and Jurnak, 1985; Ott et al., 1990) or *T. thermophilus* (Asahara and Uhlenbeck, 2005) reveal similar K_d or k_{off} values. While this indicates that aa-tRNAs have approximately equal access to EF-Tu, it does not necessarily mean that the protein lacks specificity for aa-tRNAs. Indeed, the small differences in K_d observed among aa-tRNAs appear to be inversely proportional to their intracellular concentration, suggesting that their affinities may be carefully adjusted to ensure that equivalent

concentrations of ternary complexes are available to bind to the ribosome (Jakubowski, 1988). In addition, since EF-Tu•GTP binds very poorly to Met-tRNA[fMet] or Sec-tRNA[Sec], the protein must have some underlying specificity (Louie and Jurnak, 1985; Forster et al., 1990).

Interestingly, aa-tRNAs also interact with the ribosome in a uniform manner. Not only do different aa-tRNAs bind with similar affinities to the ribosomal A and P sites, but ternary complexes containing different aa-tRNAs bind ribosomes and undergo decoding in a similar manner (Fahlman et al., 2004; Ledoux and Uhlenbeck, 2008). Such uniform kinetic and thermodynamic properties for the different aa-tRNAs presumably reflect selective pressure to ensure their interchangeable role in translation.

The aa-tRNAs no longer bind uniformly to *T. thermophilus* EF-Tu when they are misacylated. Initial experiments using a matrix of four *E. coli* tRNAs each acylated with four different amino acids demonstrated that misacylated tRNAs could bind EF-Tu from 60-fold weaker to 120-fold tighter than the correctly acylated counterparts (LaRiviere et al., 2001). Each esterified amino acid contributes a different but discrete amount to the overall EF-Tu binding affinity, and the sequences of their cognate tRNAs have evolved to compensate, so that uniform binding is observed. This principle of thermodynamic compensation was confirmed by (1) misacylating a large set of *E. coli* tRNAs with valine to establish a hierarchy of tRNA bodies, and (2) misacylating yeast tRNA[Phe] with multiple amino acids to establish a hierarchy of esterified amino acids (Asahara and Uhlenbeck, 2002; Dale et al., 2004). As shown in Figure 3A, the two hierarchies were inversely related so that tight-binding amino acids have weak-binding cognate tRNAs and vice versa. Analysis of the binding free energies of all the misacylated tRNAs revealed that the thermodynamic contributions of the amino acid and tRNA bodies were strictly independent, permitting accurate prediction of the free energy of binding of any misacylated tRNA (Asahara and Uhlenbeck, 2005). The large difference in $\Delta G°$ observed among tRNA bodies indicates that EF-Tu is a sequence-specific tRNA binding protein, but its specificity is masked when cognate aa-tRNAs are studied.

Table 1 Aspects of the interface between EF-Tu and aa-tRNA

Residue in *T. thermophilus* (*E. coli*)	Percent conserved in bacteria	Closest contact in ternary complex (pdb 1TTT)	$\Delta\Delta G^{oa}$ (kcal/mol) observed for mutation	Closest contact in ribosome-bound ternary complex (pdb 2WRN)
Amino Acid Binding Pocket				
H67 (H66)	100	His 3.9Å Stack Phe[b]		ND1 3.9Å OG1 Thr
H273 (F261)	1	ND1 4.2Å O77 Phe		ND1 5.2Å O77 Thr
N285 (N273)	99	NH2 4.0Å NH2 Phe		OD1 7.0Å NH2 Thr
T239 (T228)	92	OG1 3.7Å CE1 Phe		OG1 6.5Å OG1 Thr
E226 (E215)	100	OE1 3.4Å CE1 Phe		OD1 6.3Å OG1 Thr
D227 (D216)	100	OD2 5.6Å CZ Phe		OD2 9.4Å OG1 Thr
Region 1				
K52 (N51)	24	NZ 2.3Å OP1 C74	0.2	Not resolved
		NZ 4.8Å OP1 C75		
D227 (D216)	100	OD2 5.3Å O2' C74		OD2 4.6Å O2' C74
T232 (S221)	46	OG1 7.2Å O2' C75	−0.1	OG1 4.2Å N4 C75
R234 (R223)	100	NH2 4.9Å N6 A76		NH2 3.2Å N6 A76
T230 (S219)	20	OG1 7.1Å O2' C75		OG1 5.3Å O2' C75
		OG1 10.0Å N4 C75		OG1 5.3Å N4 C75
E271 (E259)	100	OE1 2.7Å O2' A76	>1.5	OE1 4.2Å O2' A76
		OE2 3.6Å N6 A76		OE2 3.0Å N6 A76
R274 (R262)	81	NH2 4.3Å OP1 A76	0.9	NH2 7.0Å OP1 A76
		NH2 4.3Å OP2 A76		NH2 6.8Å OP2 A76
Region 2				
E55 (E54)	98	OE1 3.9Å O2' C2	−2.9	Not resolved
R59 (R58)	100	NH1 6.9Å O2' C2	−0.1	Not resolved
N64 (N63)	23	OD1 3.5Å O2' A73	−0.2	Not resolved
D87 (D86)	100	OD2 4.8Å OP1 G3		OD2 3.5Å OP2 U3
Y88 (Y87)	100	OH 3.6Å O2' C2	−0.2	OH 3.7Å O2' C2
K90 (K89)	100	NZ 2.8Å OP3 G1	0.5	NZ 3.9Å OP1 C2
		NZ 5.5Å OP2 G65		NZ 2.5Å OP2 C65
N91 (N90)	100	ND2 3.7Å O2' G1	0.9	OD1 3.6Å O2' G1
R295 (R283)	79	NH2 4.8Å N6 C74		NH1 3.1Å O2 C74
R300 (R288)	100	NH1 2.6Å OP1 G1	0.2	NH1 5.5Å OP1 G1
		NH2 9.8Å O2' G1		NH2 3.5Å O2' G1
D348 (D336)	100	OD1 5.4Å OP1 G1		OD1 2.5Å OP1 G1
Region 3				
Q341 (Q329)	100	NE2 3.2Å O2' A64	1.0	NE2 3.8Å O2' C64
Y343 (Y331)	96	OH 4.6Å OP1 G65		OH 3.1Å OP1 C65
T350 (T338	99	OG2 2.5Å O2' G65	1.1	OG2 3.5Å O2' C65
K376 (H364)	12	NZ 2.3Å OP1 A67	0.1	NZ 4.0Å OP1 A67
		NZ 5.4Å OP1 A67		NZ 3.1Å OP1 U66
R389 (R377)	100	NH2 6.8Å OP1 G65	>1.5	NH2 6.9Å OP1 C65
Region 4				
R330 (R318)	99	NH1 2.7Å O2' U52	>1.5	NH1 3.2Å O2' A52
H331 (H319)	99	ND1 3.8Å OP1 5MU54	0.8	ND1 5.3Å OP1 5MU54
		NE2 4.8Å O2' G53		NE2 3.4Å O2' G53
T332 (T320)	89	OG1 2.7Å OP1 G53	1.0	OG1 4.0Å OP1 G53
R339 (R327)	95	NH1 4.5Å O2' U50	0.1	NH1 4.2Å O2' C50
E390 (E378)	100	OE2 3.4Å N2 G51	1.0	OE2 2.8Å N2 G63

a Data from Sanderson and Uhlenbeck (2007b).

b Residue numbers and atoms named using pdb nomenclature. Each entry includes the atom of the indicated EF-Tu residue, the distance, the nucleotide atom, and the tRNA residue

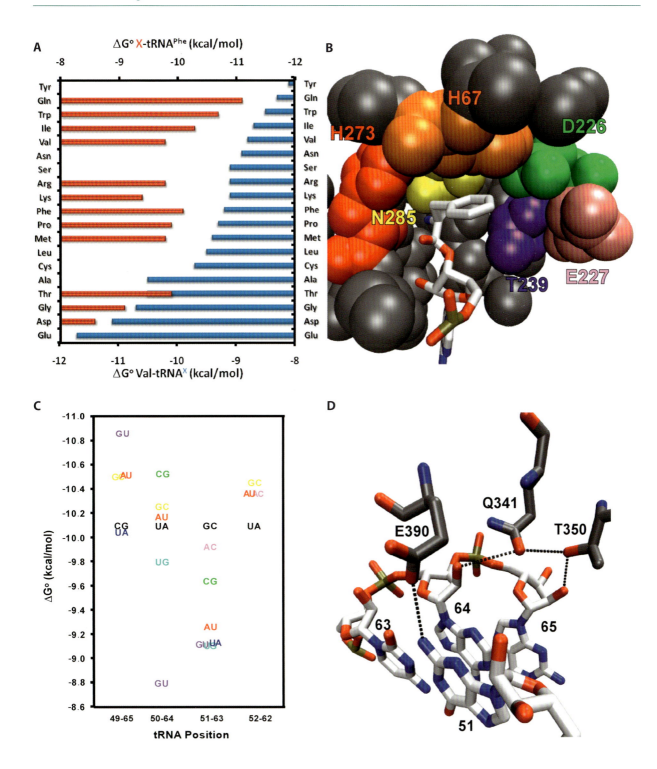

Fig. 3 The specificity of EF-Tu for esterified amino acids and tRNA bodies. (A) The free energy of binding of misacylated X-tRNA^Phe to *T. thermophilus* EF-Tu•GTP (red) (Dale et al., 2002) compared to the free energy of binding of Val-tRNA^X to *T. thermophilus* EF-Tu•GTP (blue) (Asahara and Uhlenbeck, 2002) where X indicates the different amino acids. (B) The esterified amino acid binding pocket of *T. thermophilus* EF-Tu bound to yeast tRNA^Phe (pdb1TTT). Side chains that could potentially contribute to the binding affinity and specificity of the esterified amino acid are labeled. (C) The free energy of binding of single base-pair mutations of yeast tRNA^Phe to *T. thermophilus* EF-Tu•GTP (Schrader et al., 2009). The wild-type tRNA^Phe sequence is shown in black.

4. How amino acid and tRNA specificity are achieved

The specificity of EF-Tu for different esterified amino acids is poorly understood. As shown in Figure 3B, the side chain of the esterified phenylalanine fits into a pocket between domains 1 and 2 that is large enough to fit all 20 amino acids. The top of the pocket consists of the imidazole of H67 which appears to stabilize the phenylalanine ring by a stacking interaction. The back and sides of the pocket consist of several hydrophobic side chains, as well as several main chain carbonyl and amide groups that restrict the position of the phenylalanine ring. The nearby residues D226 and E227 cause the pocket to have a net negative charge. With the exception of cysteine (Nissen et al., 1999), it is not yet known how other esterified amino acids fit into the pocket. However, the chemical environment, the net negative charge, and the presence of several hydrogen bond donors and acceptors in the pocket presumably provide opportunities for the specific interactions that lead to the range in $\Delta G°$ observed for the different amino acids. As would be expected, mutations of several of the amino acid side chains that line the pocket can alter the specificity of EF-Tu for the esterified amino acid (Dale et al., 2004; Roy et al., 2007; S. Chapman, unpublished data).

The specificity of EF-Tu for different tRNA bodies is better understood. Alanine mutagenesis of 20 interface residues in *T. thermophilus* EF-Tu revealed 11 amino acids that contribute to the overall binding affinity for yeast Phe-tRNA^{Phe} (Sanderson and Uhlenbeck, 2007b). As summarized in Table 1, these thermodynamically important residues are spread over the entire tRNA binding interface, but most are in regions 3 and 4 which contact the T stem. When this panel of EF-Tu mutants was assayed with three other aa-tRNAs that occupied different positions in the tRNA hierarchy (Figure 3A), the thermodynamic effect of the alanine mutation was different in five of the 20 positions tested (Sanderson and Uhlenbeck, 2007c). This clearly shows that the protein interacts with the four tRNAs in a thermodynamically different manner. Three of these five residues interact with a region of the T stem which differs in sequence among the tRNAs tested, implying that they make sequence-specific contacts. However, the remaining two residues contacted tRNA at conserved positions, suggesting that the overall structure or dynamics of the tRNA could also contribute to the specificity of binding.

Experiments defining the region of tRNA responsible for its specific interaction with EF-Tu were initially performed by substituting single base pairs in the acceptor and T stems of yeast tRNA^{Phe} and measuring the affinity of the mutant tRNAs for *T. thermophilus* EF-Tu (Schrader et al., 2009). Each base pair in tRNA^{Phe} was changed to every alternative pair observed at that position among bacterial tRNAs for a total of 42 mutations. While base-pair changes in the acceptor stem only showed small changes in binding affinity, mutations in all four of the non-conserved base pairs in the T stem altered the binding affinity in a sequence-specific manner (Fig. 3C). Since very few bacterial tRNAs have the U52-A62 present in yeast tRNA^{Phe} and other base pairs at this site bound similarly, the data suggested that the remaining three T-stem base pairs are primarily responsible for the sequence-specific interaction with EF-Tu. The data from two different experiments support this model. In the first, the three T-stem base pairs present in seven different *E. coli* tRNAs were transplanted into yeast tRNA^{Phe}. The resulting chimeric tRNAs bound *T. thermophilus* EF-Tu with a $\Delta G°$ that was very similar to the parental tRNA. In a second experiment, a panel of single base-pair substitutions in the T stem of the very different *E. coli* tRNA^{Leu} structure gave a qualitatively very similar, but quantitatively slightly different sequence dependence of binding to *T. thermophilus* EF-Tu. More recent experiments measuring the affinities of sets of base-pair substitutions in the T stems of *E. coli* tRNA^{Val} (Schrader et al., 2010) and *E. coli* tRNA^{Thr} (Saks et al., 2010) to *E. coli* EF-Tu showed a sequence specificity similar to the data obtained earlier with *T. thermophilus* EF-Tu.

It is striking that the three T-stem base pairs identified as sequence-specific by tRNA mutagenesis directly contact three of the five amino acids in EF-Tu, defined as sequence-specific by alanine mutagenesis. The detailed structure of this "specificity patch" of EF-Tu is shown in Figure 3D. Two of the residues (Q341 and T350) contact the 2' hydroxyls of riboses 64 and 65 and presumably sense the sequence-dependent helical structure of the T stem. The third residue, E390, directly reads the 51–63 base pair by making a hydrogen bond with the amino group of guanine in the minor groove (Sanderson and Uhlenbeck, 2007a). These three amino acids are almost universally conserved among bacteria, but are not present in archaea or eukarya. While this small "specificity patch" is undoubtedly responsible for a substantial amount of the specificity of EF-Tu for different tRNAs, it is unlikely that it fully explains the specificity for every tRNA. For example, it is clear that the mismatched 1–72 pair in tRNA^{fMet} is part of the reason that this initiator tRNA binds poorly to EF-Tu (Schulman et al., 1974; Fischer et al., 1985; Seong and RajBhandary, 1987), and there is some data suggesting that the relatively few *E. coli* elongator tRNAs with U1-A72 pairs use this site to modulate EF-Tu affinity (L. Behlen, unpublished data). In addition, the binding of tRNA^{Thr} to *E. coli* EF-Tu appears in some cases to depend on the identity of the 52–62 pair (Saks et al., 2010). Finally, the existence of the two "specificity" residues in EF-Tu that contact tRNA in conserved regions of the molecule suggests that the specificity of EF-Tu for tRNA may also be the result of indirect mechanisms that are not yet understood. For example, the small bend and twist at the junction between the acceptor and T helices could subtly depend on the junction sequence and thereby affect EF-Tu affinity.

5. Estimating binding affinities and explaining T-stem sequence variation

The successful prediction of the binding free energies of misacylated tRNAs to EF-Tu (Asahara and Uhlenbeck, 2005) suggested that a similar approach could be used to predict the free energy of EF-Tu binding to any tRNA based upon its T-stem sequence and the $\Delta G°$ values of single base-pair substitutions. Indeed, the sum of the ranges in $\Delta G°$ observed for different sequences at the three sites (Figure 3C) equals 3.2 kcal/mol which is similar to the range in $\Delta G°$ observed among tRNAs (Asahara and Uhlenbeck, 2002). Thus, if the three recognition sites were thermodynamically independent, it should be possible to predict $\Delta G°$ for any T-stem sequence by summing up the $\Delta\Delta G°$ values calculated for the single-base pair substitutions. However, considering that the three base pairs are adjacent to each other and that two of them are recognized indirectly through the backbone, it seemed quite possible that the effect of a given pair could depend upon the identity of its neighboring pairs and thus not be thermodynamically independent. Somewhat surprisingly, several different yeast tRNA^{Phe} derivatives containing two or three base-pair changes had $\Delta G°$ values nearly equal to those predicted by adding up the $\Delta\Delta G°$ values of the corresponding single base-pair substitutions (Schrader et al., 2009). A similar conclusion was reached using a series of single and multiple base-pair substitutions in *E. coli* tRNA^{Thr} binding to *E. coli* EF-Tu (Saks et al., 2010). Thus, for sets of mutations within a

given tRNA body, the three recognition sites appear to be thermodynamically independent.

An attempt to use the same approach to predict the $\Delta G°$ values for 19 different *E. coli* tRNAs to *T. thermophilus* EF-Tu was only moderately successful (Schrader et al., 2009). While both the total range and hierarchy of $\Delta G°$ values of different tRNAs were predicted generally correct, the calculated $\Delta G°$ of many tRNAs differed from the measured value by as much as 1 kcal/mol, well outside the expected experimental errors. This means either that one or more of the $\Delta\Delta G°$ values for the individual base-pair substitutions are incorrect, the additivity assumption is invalid, or, most likely, structural elements outside the specificity patch contribute to $\Delta G°$ of certain tRNAs. The tRNA molecules that are not well predicted by the specificity patch can now be studied in detail to find such additional elements and ultimately refine the model.

Despite current limitations in accurately predicting the $\Delta G°$ of different tRNAs, the data in Figure 3C provide interesting insights into the T-stem sequences present in bacterial tRNAs. Figure 4A shows a compilation of the T-stem base-pair sequences from 145 bacteria for tRNA$^{Gly}_{GCC}$, tRNA$^{Thr}_{UGU}$ and tRNA$^{Gln}_{UUG}$ which respectively have tight, intermediate and weak binding affinities to EF-Tu. In general, their T-stem sequences are consistent with the sequence specificity deduced from the tRNA^{Phe} mutagenesis. Thus, the tight-binding tRNA$^{Gly}_{GCC}$ from different bacteria contain a limited number of different T-stem sequences with combinations of the tighter-binding base pairs at all three positions, while the weak-binding tRNA$^{Gln}_{UUG}$ sequences possess different combinations of the weaker pairs. In contrast, the T-stem base pairs present in bacterial tRNA$^{Thr}_{UGU}$ are highly varied because they can achieve an intermediate binding affinity with multiple combinations of tight, weak and intermediate base pairs. A more quantitative view of this is shown in Figure 4B where the $\Delta G°$ value for every individual tRNA was calculated from its T-stem sequence using the data in Figure 3C, and the fraction of bacterial species possessing a given $\Delta G°$ was plotted. Each tRNA species has a relatively narrow distribution of calculated $\Delta G°$ values with a mean value consistent with that seen for *T. thermophilus* EF-Tu. This not only strongly suggests that bacterial T-stem sequences have primarily evolved to adjust binding of tRNAs to EF-Tu but also indicates that the sequence specificity of EF-Tu binding is similar in all bacteria.

6. EF-Tu affinity and ribosomal function

Since the interface between EF-Tu and tRNA persists during decoding, it was of interest to assess whether modifying the affinity of aa-tRNA for EF-Tu would alter the performance of the ternary complex on the ribosome. Seven T-stem mutations of *E. coli* tRNA^{Val} were created with differing binding affinities for *E. coli* EF-Tu (Schrader et al., 2010). Three bound more tightly, three bound less tightly, and one bound as well as wild-type tRNA^{Val}. tRNA^{Val} was chosen because it lacked modifications in the anticodon hairpin, and the relatively few modifications in the remainder of the tRNA did not have a detectable effect on its decoding properties. As long as the EF-Tu concentrations were adjusted to saturate the binding of the weaker tRNAs, all of the mutant ternary complexes bound to the ribosomal A/T site with affinities similar to that of wild type. In addition, the maximal rates of GTP hydrolysis of a tight and weak binding tRNA were similar to wild type. However, the rates of peptide bond formation (k_{pep}) of the three ternary complexes with tighter-binding tRNAs were significantly slower than wild type, while the remaining four ternary complexes showed

normal k_{pep} values. This suggested that the three tighter-binding tRNAs released from EF-Tu•GDP slowly enough that they limited the rate of the subsequent peptidyl transfer step (Figure 1). The three weaker-binding tRNAs and the T-stem mutant that bound as well as wild type presumably release from EF-Tu•GDP fast enough to not impact k_{pep}. To test this idea, we used the E378A mutation of EF-Tu which selectively weakens the affinity for tRNAs containing a G51-C63 base pair (Sanderson and Uhlenbeck, 2007a). With EF-Tu(E378A), the two tight-binding tRNAs that contain G51-C63 no longer showed the slow k_{pep} observed with wild-type EF-Tu. However, as expected, the slow k_{pep} for the tight-binding tRNA^{Val} that contains an A51-C63 pair was not affected by the E378A mutation. An additional control is that the k_{pep} for wild-type tRNA^{Val} which contains U51-A63 is not altered when EF-Tu(E378A) is used. These experiments support the idea that the slow k_{pep} observed with hyperstabilized tRNA mutations is the result of slow release from EF-Tu•GDP.

To confirm that enhancing the affinity of aa-tRNAs for EF-Tu leads to a reduced k_{pep}, the T-stem sequence of the tightest tRNA^{Val} mutant was introduced into four other *E. coli* tRNAs, and the rates of dissociation (k_{off}) of the resulting chimeric aa-tRNAs from EF-Tu were compared to their rates of peptide bond formation (k_{pep}). The parental tRNAs were chosen to include both tight, intermediate and weak binding bodies (Figure 3A). In each case, k_{pep} was reduced in proportion to the reduction in k_{off}. For example, since tRNA^{Ala} is a tight-binding tRNA body, the stabilizing T-stem chimera only shows a four-fold slower k_{off} and a three-fold slower k_{pep} compared to wild-type tRNA^{Ala}. In contrast, the chimera made from the weak-binding tRNA^{Tyr} shows a 60-fold slower k_{off} and a 55-fold slower k_{pep} compared to wild-type Tyr-tRNA^{Tyr}. The linear relationship between k_{off} and k_{pep} observed for all the hyperstabilized aa-tRNA mutations lends additional support to the model that the slow k_{pep} is the result of the abnormally slow release from EF-Tu•GDP. Since wild-type tRNAs and all of the weaker-binding mutations show similar k_{pep} values, it appears that native tRNA sequences have evolved such that release from EF-Tu•GDP is not rate limiting for peptide bond formation. In other words, the T-stem sequences of tRNAs have evolved to bind EF-Tu•GTP tightly enough to form ternary complex, but not too tightly to slow protein synthesis.

While the observation that the five *E. coli* tRNA chimeras containing the same stabilizing T-stem sequence had a broad range of k_{pep} values implies that the esterified amino acid can influence the rate of

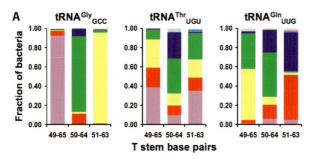

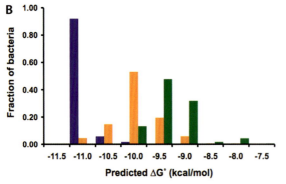

Fig. 4 (A) The frequency of base pairs at the three T-stem positions from tRNA$^{Gly}_{GCC}$, tRNA$^{Thr}_{UGU}$ and tRNA$^{Gln}_{UUG}$ present in 145 fully-sequenced bacterial genomes (Saks and Conery, 2007). G•U (purple), A-U (red), G-C (yellow), C-G (green), U-A (blue), U-G (light blue), A-C (pink), and "other" (gray). (B) Binding free energies for bacterial tRNA$^{Gly}_{GCC}$ (blue), tRNA$^{Thr}_{UGU}$ (orange) and tRNA$^{Gln}_{UUG}$ (green) calculated from adding $\Delta\Delta G°$ values obtained from single base-pair substitutions (Schrader et al., 2009).

release of aa-tRNAs from EF-Tu•GDP, it was important to confirm this directly using misacylated tRNAs. To do this, we made use of the fact that an esterified phenylalanine stabilizes binding to EF-Tu by about three-fold compared to an esterified valine (Asahara and Uhlenbeck, 2002). Phenylalanylated versions of both wild-type tRNA^{Val} and a tight-binding T-stem mutation were indeed found to have two or three-fold slower k_{pep} values than their valylated counterparts. However, a phenylalanylated weak-binding tRNA^{Val} mutation did not stabilize EF-Tu binding sufficiently to make it tighter than wild-type Val-tRNA^{Val}, and therefore gave the same k_{pep} as its valylated counterpart. In a second example, the very slow k_{pep} observed for the hyperstabilized Tyr-tRNA^{Tyr} chimera was increased 57-fold when it was esterified with alanine, consistent with the much stronger stabilizing effect of an esterified tyrosine compared to alanine (Dale et al., 2004). These experiments show that the compensating thermodynamic relationship between the esterified amino acid and tRNA body originally observed for the formation of ternary complex is also operative at the step where EF-Tu releases from aa-tRNA during decoding.

An alternative approach to explore how modifying the affinity of tRNAs for EF-Tu can affect tRNA function is to examine how selected T-stem mutations behave in living cells. This was done using a gene replacement strategy where an *E. coli* strain was created where the chromosomal copy of the single, essential tRNA$^{Thr}_{UGU}$ gene was inactivated and a wild-type copy of the gene was supplied on a plasmid containing a temperature-sensitive replicon (Saks et al., 1998). These cells were then transformed with a second, compatible low-copy plasmid containing a mutant tRNA^{Thr} gene that was of interest to test. When the cells containing the two different plasmids were grown at the non-permissive temperature, the plasmid containing the wild-type gene could not replicate, so cells could only survive if the mutant test tRNA could adequately decode the cognate ACA codons in *E. coli*. Using this approach, 24 different single and double base-pair substitutions of the 49–65, 50–64, and 51–63 base pairs in tRNA^{Thr} were individually tested. While 14 strains grew normally, seven did not support growth at all, and three showed significantly slower growth rates. These experiments clearly show that certain combinations of T-stem base pairs do not support tRNA^{Thr} function even though the same combinations are present in the T stems of other *E. coli* tRNAs. Dot blot and Northern blot experiments showed that all 24 mutant tRNA^{Thr} genes were fully processed to the correct length and were present in amounts that were similar to tRNA^{Thr} in wild-type *E. coli*.

After *in vitro* transcription and aminoacylation with threonine, the 24 T-stem tRNA^{Thr} mutations were tested for binding to *E. coli* EF-Tu. As expected, the mutations generally displayed the broad range of affinities and sequence specificities that was seen with *T. thermophilus* EF-Tu, and the effects of individual mutations were quite similar. As was the case for *T. thermophilus* EF-Tu, the thermodynamic effects of mutations at the three base pairs were essentially independent. However, it was striking that there was no clear correlation between the ability of individual base-pair mutations to support growth and the EF-Tu affinity measured *in vitro*. Although cells containing several of the tRNA^{Thr} mutations that bind poorly to EF-Tu are not viable, cells containing other weak-binding T stems grow normally. In addition, cells containing several T-stem mutations that bind to EF-Tu more tightly than wild type are not viable, although cells containing the tightest binding T stems are viable. This somewhat surprising result may indicate that assays *in vitro* do not adequately reflect the interaction between EF-Tu and aa-tRNA *in vivo*. Alternatively, the data may highlight the sequence dependence of some unappreciated aspect of decoding or some later step in translation. The system is poised for further genetic and biochemical experiments to understand these results and cement the connection between the properties of EF-Tu *in vitro* and *in vivo*.

Acknowledgement

This work was funded by National Institutes of Health Grant GM037 552 (to O. C. U.).

References

Asahara H, Uhlenbeck OC (2002) The tRNA specificity of Thermus thermophilus EF-Tu. Proc Natl Acad Sci USA 99: 3499–3504

Asahara H, Uhlenbeck OC (2005) Predicting the binding affinities of misacylated tRNAs for Thermus thermophilus EF-Tu•GTP. Biochemistry 44: 11 254–11 261

Blanchard SC, Gonzalez RL, Kim HD, Chu S, Puglisi JD (2004) tRNA selection and kinetic proofreading in translation. Nat Struct Mol Biol 11: 1008–1014

Dale T, Sanderson LE, Uhlenbeck OC (2004) The affinity of elongation factor Tu for an aminoacyl-tRNA is modulated by the esterified amino acid. Biochemistry 43: 6159–6166

Fahlman RP, Dale T, Uhlenbeck OC (2004) Uniform binding of aminoacylated transfer RNAs to the ribosomal A and P sites. Mol Cell 16: 799–805

Fischer W, Doi T, Ikehara M, Ohtsuka E, Sprinzl, M (1985) Interaction of methionine-specific tRNAs from *Escherichia coli* with immobilized elongation factor Tu. FEBS Lett 192: 151–154

Forster C, Ott G, Forchhammer K, Sprinzl M (1990) Interaction of a selenocysteine-incorporating tRNA with elongation factor Tu from *E. coli*. Nucleic Acids Res 18: 487–491

Gromadski KB, Wieden HJ, Rodnina MV (2002) Kinetic mechanism of elongation factor Ts-catalyzed nucleotide exchange in elongation factor Tu. Biochemistry 41: 162–169

Jakubowski H (1988) Negative correlation between the abundance of *Escherichia coli* aminoacyl-tRNA families and their affinities for elongation factor Tu-GTP. J Theor Biol 133: 363–370

LaRiviere FJ, Wolfson AD, Uhlenbeck OC (2001) Uniform binding of aminoacyl-tRNAs to elongation factor Tu by thermodynamic compensation. Science 294: 165–168

Ledoux S, Uhlenbeck OC (2008) Different aa-tRNAs are selected uniformly on the ribosome. Mol Cell 31: 114–123

Louie A, Ribeiro NS, Reid BR, Jurnak F (1984) Relative affinities of all *Escherichia coli* aminoacyl-tRNAs for elongation factor Tu-GTP. J Biol Chem 259: 5010–5016

Louie A, Jurnak F (1985) Kinetic studies of *Escherichia coli* elongation factor Tu-guanosine 5'-triphosphate-aminoacyl-t RNA complexes. Biochemistry 24: 6433–6439

Nissen P, Kjeldgaard M, Thirup S, Polekhina G, Reshetnikova L, Clark BF, Nyborg J (1995) Crystal structure of the ternary complex of Phe-tRNAPhe, EF-Tu, and a GTP analog. Science 270: 1464–1472

Nissen P, Thirup S, Kjeldgaard M, Nyborg J (1999) The crystal structure of Cys-tRNACys-EF-Tu-GDPNP reveals general and specific features in the ternary complex and in tRNA. Structure 7: 143–156

Ott G, Schiesswohl M, Kiesewetter S, Forster C, Arnold L, Erdmann VA, Sprinzl M (1990) Ternary complexes of *Escherichia coli* aminoacyl-tRNAs with the elongation factor Tu and GTP: thermodynamic and structural studies. Biochim Biophys Acta 1050: 222–225

Pape T, Wintermeyer W, Rodnina MV (1998) Complete kinetic mechanism of elongation factor Tu-dependent binding of aminoacyl-tRNA to the A site of the *E. coli* ribosome. EMBO J 17: 7490–7497

Rodnina MV, Fricke R, Kuhn L, Wintermeyer W (1995) Codon-dependent conformational change of elongation factor Tu preceding GTP hydrolysis on the ribosome. EMBO J 14: 2613–2619

Roy H, Becker HD, Mazauric MH, Kern D (2007) Structural elements defining elongation factor Tu mediated suppression of codon ambiguity. Nucleic Acids Res 35: 3420–3430

Saks ME, Sampson JR, Abelson J (1998) Evolution of a transfer RNA gene through a point mutation in the anticodon. Science 279: 1665–1670

Saks ME, Conery J S. (2007) Anticodon-dependent conservation of bacterial tRNA gene sequences. RNA 13: 651–660

Saks ME, Sanderson LE, Choi DS, Crosby CM, Uhlenbeck O C. (2010) Manuscript in preparation

Sanderson LE, Uhlenbeck OC (2007a) The 51–63 base pair of tRNA confers specificity for binding by EF-Tu. RNA 13: 835–840

Sanderson LE, Uhlenbeck OC (2007b) Directed Mutagenesis Identifies Amino Acid Residues Involved in Elongation Factor Tu Binding to yeast Phe-tRNA(Phe) J Mol Biol 368: 119–130

Sanderson LE, Uhlenbeck OC (2007c) Exploring the specificity of bacterial elongation factor Tu for different tRNAs. Biochemistry 46: 6194–6200

Schmeing TM, Voorhees RM, Kelley AC, Gao YG, Murphy F V. t., Weir JR, Ramakrishnan V (2009) The crystal structure of the ribosome bound to EF-Tu and aminoacyl-tRNA. Science 326: 688–694

Schrader JM, Chapman SJ, Uhlenbeck OC (2009) Understanding the sequence specificity of tRNA binding to elongation factor Tu using tRNA mutagenesis. J Mol Biol 386: 1255–1264

Schrader JM, Chapman SJ, Uhlenbeck OC (2011) The affinity of aminoacyl-tRNA to elongation factor Tu is tuned for optimal decoding. PNAS (in press)

Schulman LH, Pelka H, Sundari RM (1974) Structural requirements for recognition of *Escherichia coli* initiator and non-initiator transfer ribonucleic acids by bacterial T factor. J Biol Chem 249: 7102–7110

Seong BL, RajBhandary UL (1987) Mutants of *Escherichia coli* formylmethionine tRNA: a single base change enables initiator tRNA to act as an elongator in vitro. Proc Natl Acad Sci USA 84: 8859–886

Mechanisms of decoding and peptide bond formation

16

Marina V. Rodnina

1. Introduction

During protein synthesis, the ribosome translates the genetic information carried by the mRNA into the amino acid sequence of proteins with the help of adaptor molecules, aminoacyl-tRNAs (aa-tRNA). In each round of elongation, the ribosome selects the correct (cognate) aa-tRNA corresponding to the mRNA codon from the total cellular pool of aa-tRNAs. The delivery of aa-tRNA to the decoding site (A site), where tRNA recognition and selection takes place, is brought about by elongation factor Tu (EF-Tu). The recognition of aa-tRNA by the ribosome occurs via a series of selection steps that control the stepwise movement of aa-tRNA from EF-Tu into the A site and the accommodation of the aminoacyl end of the aa-tRNA in the peptidyl transferase center. Accommodation is followed by peptide bond formation between the A-site aa-tRNA and peptidyl-tRNA in the P site that results in the elongation of the nascent peptide chain by one amino acid. The functional centers of the ribosomes are composed mostly of rRNA. Thus, understanding decoding and peptide bond formation requires answers to several fundamental questions: (i) how does the rRNA machinery recognize tRNAs and mRNAs? (ii) how does the ribosome discriminate between very similar cognate and near-cognate tRNAs which change identity in every round of elongation? (iii) how does the ribosome balance the requirements for speed and accuracy? and (iv) how does rRNA catalyze peptide bond formation? The goal of this review is to summarize the recent progress towards answering these questions.

2. The mechanism of decoding

2.1. The kinetic scheme of decoding

The stepwise movement of aa-tRNA from EF-Tu into the ribosomal A site entails a number of intermediates (Figure 1) (Rodnina and Wintermeyer, 2001; Marshall et al., 2008). In the first step, the ternary complex of aa-tRNA with EF-Tu and GTP forms a labile initial binding complex with the ribosome (rate constants of the forward and backward reactions are defined as k_1 and k_{-1}, respectively) followed by codon reading by the anticodon of the tRNA. The latter step was identified by single-molecule FRET techniques but is not resolved from the following codon recognition; hence, no rate constants are assigned in Figure 1. Subsequent formation of the codon-recognition complex on the 30S subunit (k_2, k_{-2}) triggers the GTPase activation of EF-Tu (k_3) and GTP hydrolysis (k_{GTP}). The release of inorganic phosphate (Pi) induces the conformational transition of EF-Tu from the GTP- to the GDP-bound form (k_4), whereby the factor loses the affinity for aa-tRNA and dissociates from the ribosome (k_6). The aa-tRNA released from EF-Tu moves into the peptidyl transferase center on the 50S subunit (k_5) where it takes part in rapid peptide bond formation (k_{pep}). Alternatively, aa-tRNA may be rejected from the ribosome (k_7).

2.2. Initial binding

The initial step in the interaction of the ternary complex, EF-Tu·GTP·aa-tRNA, with the ribosome is the codon-independent binding that is determined by EF-Tu interactions with ribosomal proteins at the L7/12 stalk of the ribosome (Kothe et al., 2004; Diaconu et al.,

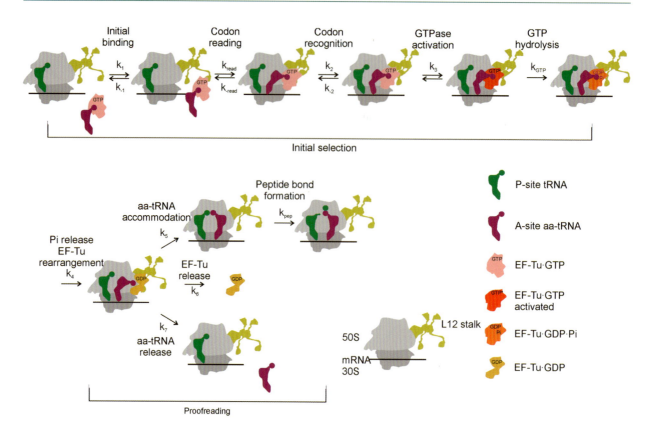

Fig. 1 Kinetic mechanism of EF-Tu-dependent aa-tRNA binding to the A site. Kinetically resolved steps are indicated by rate constants k_1–k_7 (forward reactions) and k_{-1}–k_{-3} (backward reactions). The rate of codon-reading (likely a readily reversible step) could not be determined by rapid kinetics; the values available from single-molecule FRET experiments are not comparable to the respective values presented here due to differences in buffer conditions. Rate constants of the two chemical steps that are rate-limited by the preceding step are designated k_{GTP} and k_{pep}.

2005). The selective removal of the C-terminal domain (CTD) of L7/12 impairs the binding. The mutational analysis suggested that the initial contact involves helices 4 and 5 in the CTD of L7/12 as well as helix D in the G domain of EF-Tu (Kothe et al., 2004). It is not known whether the ribosome can simultaneously bind more than one ternary complex at each L7/12 protein, of which there are four or six copies on the ribosome, depending on the organism (Diaconu et al., 2005; Ilag et al., 2005; Miyoshi et al., 2009); ribosomes from *Escherichia coli*, for instance, have four copies of L7/12. X-ray scattering and NMR measurements indicated that the L7/12 CTDs are highly mobile, although one of them is more retracted on the average (Bernado et al., 2010). The presence of multiple copies, the flexibility of the stalk, and the electrostatic properties of L7/12 may account for the high value of k_1 (about 10^8 $M^{-1}s^{-1}$ (20 °C), Table 1 (Rodnina et al., 1996; Gromadski and Rodnina, 2004a)). These observations are consistent with a model where L7/12 CTDs reach out into solution to bind ternary complex and hand it over to the factor binding site around the sarcin-ricin loop (SRL) of 23S rRNA on the 50S subunit of the ribosome.

2.3. Codon recognition

Codon recognition proceeds through a number of intermediates identified by rapid kinetics (Rodnina and Wintermeyer, 2001; Rodnina et al., 2005; Daviter et al., 2006), single molecule fluorescence (Marshall et al., 2008), and x-ray crystallography (Ogle and Ramakrishnan, 2005; Schmeing and Ramakrishnan, 2009). The most likely sequence of events is the following. First, the tRNA enters the decoding site on the 30S subunit maintaining its conformation as present in the unbound ternary complex (Rodnina et al., 1994; Schmeing et al., 2009; Schuette et al., 2009). The relative orientation of codon and anticodon in this early complex may not be optimal, but might allow transient probing of the mRNA codon (Blanchard et al., 2004), e. g. due to spontaneous tRNA fluctuations between

Table 1 Rate constants of elemental steps of decoding the cognate UUC codon by Phe-tRNA[Phe]

Rate constant	HiFi, 20°[a]	TAKM$_{10}$, 20°[b]	LoFi, 20°[c]
k_1, $\mu M^{-1}s^{-1}$	140	110	nd
k_{-1}, s^{-1}	85	25	nd
k_2, s^{-1}	190	100	32
k_{-2}, s^{-1}	0.2[d]	0.2[e]	0.005[d]
k_3, s^{-1}	260	500	>500
k_4, s^{-1}	nd	60	nd
k_5, s^{-1}	7–14	7	2
k_6, s^{-1}	nd	3	nd
k_7, s^{-1}	<0.3	<0.3	<0.3

[a] From (Gromadski and Rodnina, 2004a) and (Wohlgemuth et al., 2010). The substrates were 70S initiation complexes with fMet-tRNA[fMet] in the P site. HiFi buffer contains 3.5 mM Mg^{2+}, 0.5 mM spermidine, and 8 mM putrescine.
[b] From (Pape et al., 1998). The substrates were 70S ribosomes programmed with poly(U) and carrying AcPhe-tRNA[Phe] in the P site at 10 mM Mg^{2+}
[c] From (Gromadski et al., 2006) The substrates were 70S initiation complexes with fMet-tRNA[fMet] in the P site. LoFi buffer contains 20 mM Mg^{2+} and no polyamines.
[d] Determined using EF-Tu(H84A) defective in GTP hydrolysis (Rodnina et al., 1996; Gromadski and Rodnina, 2004a)
[e] Determined using the non-hydrolyzable GTP analog GDPNP (Pape et al., 1998)

undistorted and distorted states. Single-molecule studies suggest rapid and reversible A-site sampling by incoming cognate and near-cognate aa-tRNAs after the initial, codon-independent binding of the EF-Tu·aa-tRNA complex to L7/L12 (Marshall et al., 2008). Second, the formation of the cognate codon-anticodon duplex leads to local conformational changes of the conserved residues A1492, A1493, and G530 of 16S rRNA (Ogle et al., 2001). The bases change their positions and form A-minor interactions with the minor grove of the first two base pairs of the codon-anticodon complex in a fashion which is specific for Watson-Crick base pair geometry, but independent of the sequence. The codon-recognition complex is stabilized by purines at position 37 in the anticodon loop of the tRNA, mainly by stronger stacking interactions and binding of additional Mg^{2+} ions (Konevega et al., 2004). The local rearrangement of the decoding center is accompanied by rotations of the 30S head and shoulder domains toward the subunit interface, collectively described as domain closure (Ogle et al., 2002). This conformational change could force the anticodon stem-loop into its accommodated orientation while the acceptor stem of the tRNA is still held by the interactions with EF-Tu, thus distorting the tRNA molecule

(Schmeing et al., 2009; Schuette et al., 2009; Villa et al., 2009). The tRNA is distorted by bending at the junction of the anticodon and D stems at nucleotides 44, 45, and 26 (Stark et al., 2002; Valle et al., 2003; Schmeing et al., 2009; Schuette et al., 2009; Villa et al., 2009). The distorted structure of aa-tRNA is maintained by interactions between several elements of the ribosome, aa-tRNA, and EF-Tu. So far, the elemental rate constants of these rearrangements are not resolved. However, the kinetic analysis revealed that the overall rate of codon recognition (k_2) is very similar for cognate and near-cognate ternary complexes (Pape et al., 1999; Gromadski and Rodnina, 2004a; Gromadski et al., 2006), indicating that an early step that is common for cognate and near-cognate complexes, such as entering the decoding site, dominates the kinetics.

2.4. GTPase activation

GTPase activation of EF-Tu can be envisaged as a rearrangement of the active site that is necessary to assemble the groups involved in GTP hydrolysis. The unstimulated, intrinsic GTPase activity of EF-Tu is extremely low, because His84 in the switch II region of EF-Tu, which is critical for catalysis, is oriented away from the GTP-binding pocket and prevented from entering the active site by a "hydrophobic gate" formed by the side chains of Val20 and Ile61 (Berchtold et al., 1993). Upon GTPase activation of EF-Tu on the ribosome, His84 has to move towards the γ-phosphate, and this movement is thought to be induced only when a correct codon-anticodon complex is formed. The neighboring residue Gly83 plays an important role in both the rearrangement of the switch II region during GTPase activation, due to the conformational flexibility inherent to Gly residues, and in GTP hydrolysis itself, probably by taking part in positioning the catalytic water with a hydrogen bond to its main chain oxygen (Knudsen et al., 2001).

During the GTPase activation step, the conformational signal elicited by correct codon-anticodon interaction in the decoding site is communicated to the GTPase center on EF-Tu, accelerating rearrangement steps that precede and limit the rate of GTP hydrolysis (Rodnina and Wintermeyer, 2001). These rearrangements constitute an essential part of the induced-fit mechanism of aa-tRNA selection. Distortion of the tRNA molecule due to interactions at the decoding site on the 30S subunit results in a displacement of the aminoacyl end of the tRNA from its binding site in EF-

Tu, and the current models assume that the following steps that lead to GTPase activation originate from that rearrangement (Schmeing et al., 2009; Schuette et al., 2009; Villa et al., 2009). Cryo-EM structures suggest that there is movement of the switch I region of EF-Tu away from the nucleotide binding pocket towards the 30S subunit where it interacts with the junction of helices 8/14 at the 30S subunit shoulder (Schuette et al., 2009; Villa et al., 2009). This interaction would open one wing of the hydrophobic gate, and reorient His84 toward the nucleotide, whereas the other wing is held fixed by an interaction between the SRL and the P loop. A movement of the switch I region would be consistent with kinetic studies indicating that the environment of a fluorescence reporter group attached to the ribose of GTP (mant-GTP), located in close proximity of the switch I region, is altered upon GTPase activation, but prior to GTP hydrolysis (Rodnina et al., 1995; Daviter et al., 2003). However, deletion mutants of helix 14 retained substantial activity in GTP hydrolysis, arguing against an essential role of h14 in EF-Tu activation (McClory et al., 2010). In the crystal structure of the GTPase-activated state (Voorhees et al., 2010), the switch I region is not disordered, and GTPase activation appears to involve subtle movements of elements of EF-Tu around the GTP.

2.5. GTP hydrolysis by EF-Tu on the ribosome

His84 in the switch II region of EF-Tu is directly involved in the chemistry step of GTP hydrolysis, as the H84A mutation reduces the reaction rate by five orders of magnitude, whereas the steps preceding GTP cleavage remain unaffected (Daviter et al., 2003). The presumed catalytic role of His84 is in hydrogen-bonding to the substrates, i. e. the attacking water molecule and/or the γ-phosphate, to precisely align the groups directly involved in the reaction; this role would be similar to that of Gln which stabilizes the transition state of GTP hydrolysis in Ras-like or heterotrimeric G proteins (Vetter and Wittinghofer, 1999; Bos et al., 2007) (Figure 2). The arginine residue (Arg58) that is located in the switch I region of EF-Tu in a position homologous to that of the catalytic arginine in G_α proteins, is not essential for GTP hydrolysis (Knudsen and Clark, 1995), and extensive mutagenesis did not identify any other amino acid side chains that would be important for GTP hydrolysis in *E. coli* EF-Tu (Knudsen and Clark, 1995; Wiborg et al., 1996; Mansilla et al., 1997; Rattenborg et al., 1997).

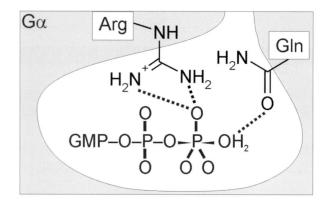

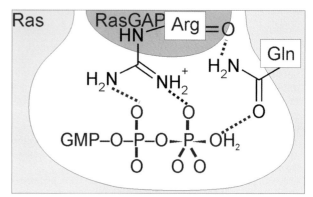

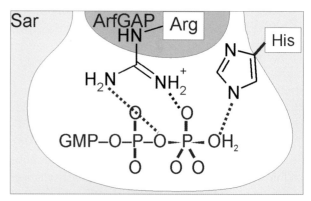

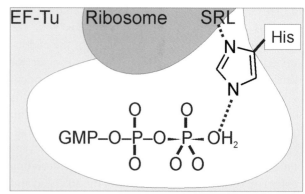

Fig. 2 Mechanism of GTP hydrolysis in the α-subunit of heterotrimeric G-proteins, in the complex Ras-RasGAP, Sar-ArfGAP (modified from (Bos et al., 2007)), and in the EF-Tu·aa-tRNA·ribosome complex. The GTP-binding proteins and the GTPase-activating proteins (GAPs) are shown in light and dark grey, respectively. GTP is shown in the transition state of hydrolysis; negative charges on phosphoryl groups are omitted for clarity.

In addition to the catalytic His84, several other contacts are important for GTP hydrolysis. The switch regions of the G domain of EF-Tu form contacts with the SRL of 23S rRNA (Stark et al., 2002; Valle et al., 2003). SRL cleavage impedes progression from the GTPase-activated state (Blanchard et al., 2004), and, as a result, GTP hydrolysis by EF-Tu is abolished. Recent crystal structures (Voorhees et al., 2010) indicate that contacts with the SRL stabilize His84 in the GTPase-activated state. Other contacts, which may contribute to GTP hydrolysis, include ribosomal proteins L7/12 and L11 as well as the L11-binding region of 23S rRNA. The interaction of EF-Tu with L7/12 accounts for a 2500-fold stimulation of GTP hydrolysis (Mohr et al., 2002), although none of the conserved amino acids in the C-terminal domain of L7/12 was specifically responsible for the observed stimulation (Kothe et al., 2004). However, since these contacts are somewhat remote from the nucleotide binding pocket, they may act indirectly by inducing conformational transitions of EF-Tu or aa-tRNA, rather than by donating catalytic groups for GTP hydrolysis.

2.6. Pi release and the conformational change of EF-Tu

GTP hydrolysis yields the reaction products, GDP and Pi, in the nucleotide-binding site of EF-Tu, and the rate of Pi release limits the rate of the conformational change of EF-Tu from the GTP- to the GDP-bound form (Kothe and Rodnina, 2006). The rate of this step is probably determined by the flexibility of residues involved in the reorientation of the switch I and II regions required for Pi release and the subsequent switch of the conformation (Knudsen et al., 2001). The release of aa-tRNA from EF-Tu, which so far has not been measured directly, probably takes place during the conformational transition, and may be affected by the nature of the amino acid in aa-tRNA (Effraim et al., 2009; Burakovsky et al., 2010; see chapter by O. Uhlenbeck, this volume).

2.7. aa-tRNA accommodation

After its release from EF-Tu, the 3' end of aa-tRNA has to move a distance of almost 70 Å within the ribosome from its binding site on EF-Tu into the peptidyl transferase center. In the case of Phe-tRNA[Phe], which has been extensively studied, accommodation limits the rate of peptide bond formation, which is intrinsically very rapid (Pape et al., 1998, 1999; Bieling et al., 2006; Wohlgemuth et al., 2010). Computer simulation of accommodation suggested a step-wise movement of the tRNA through a corridor of conserved rRNA bases, which engage in various interactions with the tRNA during this movement (Sanbonmatsu et al., 2005). The final accommodation of aa-tRNA in the A site of the peptidyl transferase center appears to be determined by the interaction between the 3'-CCA end of aa-tRNA and the A loop (helix 92) of 23S rRNA, in particular the universally conserved residues U2492, C2556, and C2573 (Kim and Green, 1999). The latter residues are proposed to act as a gate, causing the acceptor stem to pause before allowing entrance into the peptidyl transferase center (Sanbonmatsu et al., 2005). Previous mutagenesis studies (O'Connor and Dahlberg, 1995) indicated that substitutions of U2492 and U2555 decreased the fidelity of translation, supporting the view that the accommodation gate may attenuate aa-tRNA binding. However, mutations of C2573 and the neighboring A2572 did not affect aa-tRNA accommodation, peptide bond formation, or the fidelity of aa-tRNA selection, suggesting that the ribosome may allow rapid aa-tRNA accommodation in spite of defects at the accommodation gate, potentially through an alternative pathway (Burakovsky et al., 2010).

3. Fidelity of aa-tRNA selection

3.1. Error frequency

Speed and accuracy of translation determine cell growth and the quality of newly synthesized proteins. The rate of protein elongation in *E. coli* was estimated to $4-22$ s^{-1} per codon at 37 °C (Bremer and Dennis, 1987; Sorensen and Pedersen, 1991; Liang et al., 2000; Proshkin et al., 2010). Error frequencies *in vivo*, estimated by quantification of misincorporation due to different physicochemical properties of native and altered proteins or by using reporter constructs expressing proteins which gain activity upon amino acid misincorporation, range between 10^{-5} and 10^{-3}, depending on the type of measurement, concentrations and nature of tRNAs that perform misreading, as well as the mRNA context (for reviews, see (Parker, 1989; Drummond and Wilke, 2009)). Such an error frequency can be reproduced *in vitro* in reconstituted translation systems (Thompson et al., 1981; Gromadski and Rodnina, 2004a; Gromadski et al., 2006; Zaher and

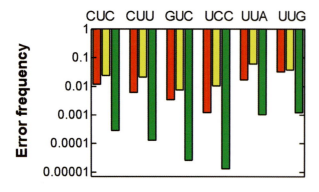

Fig. 3 Error frequency of Phe incorporation on various near-cognate codons. Error frequencies of initial selection (red bars) and proofreading (yellow bars), as well as the overall error frequency (green bars) are shown. (Figure reproduced from Daviter et al., 2006 with permission)

Green, 2009; Wohlgemuth et al., 2010) (Figure 3). The much lower *in-vitro* error frequency of 10^{-7} reported recently (Johansson et al., 2008) can now be attributed to the influence of inadvertently low Mg^{2+} concentrations on aa-tRNA interactions with cognate and near-cognate codons (Wohlgemuth et al., 2010).

The fidelity of amino acid incorporation into protein is controlled at three basic selection stages: preferential rejection of incorrect ternary complexes prior to GTP hydrolysis at the initial selection stage; preferential rejection of near-cognate aa-tRNAs at the proofreading stage after GTP hydrolysis (Thompson and Stone, 1977; Ruusala et al., 1982; Rodnina and Wintermeyer, 2001), and preferential hydrolysis of near-cognate peptidyl-tRNAs in the P site by termination factors (Zaher and Green, 2009). The missense error frequency of translation in cells depends on the combined efficacy of these three stages and on the abundance of the aa-tRNA cognate to the given codon relative to the near-cognate competitors (Kramer and Farabaugh, 2007). Assuming an error frequency of initial selection and proofreading of about 0.03 each, the cumulative error frequency of decoding is close to 10^{-3} (Gromadski and Rodnina, 2004b, a; Gromadski et al., 2006; McClory et al., 2010; Wohlgemuth et al., 2010). A single initial miscoding event is then amplified by preferential incorporation of an incorrect amino acid at the next codon, followed by premature chain termination, thereby reducing the observed error frequency of synthesis of a full-length protein about 10-fold (Zaher and Green, 2009). Thus, the overall error frequency is expected to be around 10^{-4}, which may be further modulated by the tRNA concentrations in the cell. Assuming an average protein length in *E. coli*

of about 335 amino acids (Netzer and Hartl, 1997), this means that most of the proteins (97%) in the cell are synthesized correctly. To date, error frequencies were measured for only a small number of proteins at a few different positions; it can be expected that the error frequency may vary from case to case. Assuming higher error frequencies, e. g. $3 \cdot 10^{-3}$, leads to the prediction that all proteins of the assumed length on the average will have one incorrect amino acid, and longer proteins naturally will accumulate more errors. This level of infidelity in protein synthesis may affect the fitness of the organism by reducing the amount of active proteins *(loss of function)*, producing proteins that are toxic for the cell *(gain of toxic function)*, or increasing misfolding, thereby raising the *clean-up costs* (summarized in (Drummond and Wilke, 2009)). Of course, some missense errors may be more readily tolerated than others. Indeed, the common experience in expressing mutant proteins suggests that amino acids at many positions in a protein, except the catalytic active-site residues, may be substituted without adverse effects on activity. In some cases cells can tolerate an astonishingly high degree of mistranslation where up to 10% of an expressed protein carries a missense error at the active site (Ruan et al., 2008). Quantifying the extent of mistranslation and understanding which levels of mistranslation can be tolerated and dealt with by the cellular degradation and refolding machineries is an exciting future avenue of research.

3.2. Molecular mechanism of tRNA selection

Kinetic studies have identified four elemental reactions that have different rates for cognate and near-cognate aa-tRNAs (Figure 1). The ribosome controls the differences in the stabilities of the codon-anticodon complexes (k_{-2}, k_7) and specifically accelerates the rates of GTPase activation (k_3) and accommodation (k_5) of correct substrates, implicating codon-anticodon duplex stability and induced fit as sources of selectivity. Kinetic partitioning between GTPase activation and ternary complex dissociation strongly favors the acceptance of cognate and the rejection of near-cognate ternary complexes (Figure 4). Likewise, cognate aa-tRNA is preferentially accommodated during proofreading, while near-cognate tRNA is mostly rejected.

The primary means of codon-anticodon recognition by the ribosome are A-minor interactions between A1492 and A1493 of 16S rRNA and the codon-anticodon duplex in the decoding site. Mismatches distort

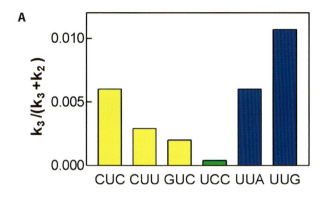

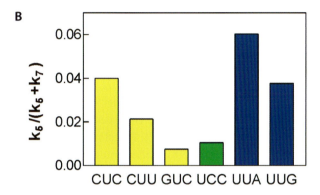

Fig. 4 Kinetic partitioning of initial selection (A) and proofreading (B). Binding of Phe-tRNA^Phe (anticodon 3'-AAG-5') to ribosomes with a near-cognate codon in the A site with one mismatch in the first (CUC, CUU, GUC; yellow), second (UCC; green) or third (UUA, UUG; blue) position (Gromadski et al., 2006). Kinetic partitioning for the cognate tRNA is close to 1. (Figure reproduced from Daviter et al., 2006 with permission)

the Watson-Crick geometry of the duplex, thereby impairing the interactions with decoding-site residues (Ogle et al., 2001). Single mismatches at any position of the codon-anticodon complex result in slower forward reactions and a uniformly 1000-fold faster dissociation of ternary complexes from the ribosome (Gromadski et al., 2006). This suggests that high-fidelity tRNA selection is achieved by a conformational switch of the decoding site between accepting and rejecting modes, independent of the thermodynamic stability of the respective codon-anticodon complexes or the position of mismatch. Furthermore, A-site binding is a non-equilibrium process that–provided the tRNA is cognate–is driven by the rapid, irreversible forward reactions of GTP hydrolysis and peptide bond formation, reducing the apparent affinity (K_M) of the cognate ternary complex for the ribosome. Due to the large differences in the rates of GTPase activation, the K_M values for cognate and near-cognate ternary complexes

are similar (Gromadski and Rodnina, 2004a; Gromadski et al., 2006; Kothe and Rodnina, 2007; McClory et al., 2010); hence, discrimination on the basis of K_M is not possible. Rather, discrimination is based on the large differences in forward reaction rates, which are high with correct and low with incorrect substrates, resulting in differences in k_{cat}/K_M values between cognate and near-cognate species that are sufficient to explain the observed levels of discrimination. Notably, the rates of GTPase activation are differently impaired by the nature and position of mismatches (Gromadski and Rodnina, 2004a; Gromadski et al., 2006). Thus, the free energy provided by the interactions of ribosomal residues with the codon-anticodon duplex at the decoding center is used for inducing conformational changes which, in turn, modulate the rates of the forward reactions.

The molecular basis for the differential GTPase activation by cognate and near-cognate codon-anticodon interactions is not known. The distortion of the tRNA is important (Piepenburg et al., 2000; Cochella and Green, 2005; Pan et al., 2008). However, fluorescence quenching data suggest that, upon formation of the codon-recognition complex, a near-cognate aa-tRNA also becomes distorted (our unpublished data), indicating the existence of alternative (or additional) mechanisms by which the signal from the decoding region is communicated. Recent structural work suggested that ribosomal proteins S13, S19, L16, L25, L27, and L31 may be actively involved, at least at the proofreading stage (Jenner et al., 2010). According to these structural observations, proofreading would begin with monitoring of the entire anticodon loop of the tRNA by nucleotides from 16S rRNA (helices 18 and 44) and 23S rRNA (helix 69) with the essential involvement of magnesium ions. In fact, mutations in h8 and h14 that compromise inter-subunit bridge B8 (which normally acts to counteract the inward rotation of the 30S shoulder domain) decrease the stringency of aa-tRNA selection (McClory et al., 2010).

The importance of local and global conformational changes at the decoding center becomes particularly clear when the effects of antibiotics are analyzed. Paromomycin is capable of inducing the conformational flip at the decoding center even in the absence of correct codon-anticodon interaction (Ogle et al., 2001; Ogle et al., 2002). Thus, binding of paromomycin stabilizes the binding of near-cognate aa-tRNA and increases the rate of GTP hydrolysis in the near-cognate ternary complex almost to the level of the cognate substrate (Pape et al., 2000; Gromadski and Rodnina,

2004b). Restricting the flexibility of the 30S subunit by the antibiotic streptomycin (Carter et al., 2000; Ogle et al., 2003), which is likely to lock the position of the 30S subunit head relative to the 30S body, affects the GTPase activation step primarily by decreasing the rate of the reaction with the cognate, and increasing the rate with the near-cognate ternary complex, resulting in an almost complete loss of selectivity (Gromadski and Rodnina, 2004b).

3.3. Trade-off between speed and accuracy

A significant trade-off between speed and accuracy is a consequence of the necessity for rapid protein synthesis in the cell. Whereas k_{cat} values for the GTPase activation and accommodation reaction are grossly different between cognate and near-cognate ternary complexes/aa-tRNAs, their K_M values are very similar, indicating that the potential for very accurate substrate selection in translation is not fully exploited in order to accelerate the speed of the overall process. One interesting question is how–given these similar affinities–the cognate ternary complex escapes excessive competition with the bulk of non- and near-cognate substrates. The rates of GTP hydrolysis and peptide bond formation measured in the presence of bulk aa-tRNA suggested that excess near- and non-cognate ternary complexes reduce the rate of GTP hydrolysis in the cognate ternary complex by a factor of 10. However, due to the large difference between the rates of GTP hydrolysis and peptide bond formation, the inhibition of GTP hydrolysis does not lead to a significant decrease in the rate of cognate peptide bond formation (Wohlgemuth et al., 2010). Thus, the high speed of GTP hydrolysis in the cognate ternary complex causes a loss in the fidelity of selection by increasing the K_M value for the cognate substrate, but at the same time precludes that the rate of cognate peptide bond formation is decreased by competition with bulk ternary complexes. Thus, the rate of GTP hydrolysis, and not that of peptide bond formation, likely governs the optimization of speed and accuracy of translation during evolution and explains why the maximum intrinsic accuracy of the system is not utilized (Wohlgemuth et al., 2010).

4. Peptide bond formation

4.1. Structures of reaction intermediates

During peptide bond formation, the α-amino group of the A-site aa-tRNA attacks the carbonyl carbon of the P-site peptidyl- tRNA to produce a new, one amino acid-longer peptidyl-tRNA in the A site and a deacylated tRNA in the P site (Figure 5A). The catalytic center for peptide bond formation is located on the 50S subunit, and the rate of peptide bond formation is the same on isolated 50S subunits and on native 70S ribosomes, provided a full-length tRNA substrate is present in the P site (Wohlgemuth et al., 2006). High-resolution crystal structures of ribosomes revealed that the peptidyl transferase center is composed of RNA only (Ban et al., 2000; Nissen et al., 2000; Schmeing et al., 2002; Schmeing et al., 2005b; Selmer et al., 2006) (Figure 5B), which lent support to the earlier hypothesis that rRNA, rather than proteins, catalyzes peptide bond formation (Noller et al., 1992). Protein L27, whose N-terminal truncation led to modest impairment of peptidyl transferase activity (Maguire et al., 2005), appears to interact with the tRNA, but does not reach into the catalytic center of the ribosome (Voorhees et al., 2009; Jenner et al., 2010). The crystal structures of *Haloarcula marismortui* 50S subunits complexed with different transition state analogs revealed that the reaction proceeds through a tetrahedral intermediate with S chirality (Schmeing et al., 2005a). The oxyanion of the tetrahedral intermediate seems to be stabilized by a water molecule that is positioned by nucleotides A2602 and U2584 (Schmeing et al., 2005a). The only atom within hydrogen-bonding distance of the α-amino group mimic was the 2'-OH of A76 of the P-site moiety of the transition state analog (Schmeing et al., 2005a). N3 of A2451 appears to be within hydrogen-bonding distance of the α-amino group of the nucleophile in pre-reaction states (see below), although this distance is increased during the course of the reaction.

The conserved residues A2451, U2506, U2585, C2452 and A2602 are located at the so-called "inner shell" of the peptidyl transferase center (Bashan et al., 2003; Schmeing et al., 2005a; Schmeing et al., 2005b). Strikingly, mutation of neither these residues nor of the adjacent G2447 resulted in defects in the rate of peptide bond formation when full-length, intact aa-tRNA was used as A-site substrate, while the reaction with puromycin was strongly impaired (Polacek et al., 2001; Thompson et al., 2001; Katunin et al., 2002; Beringer et al., 2003; Hesslein et al., 2004; Youngman et

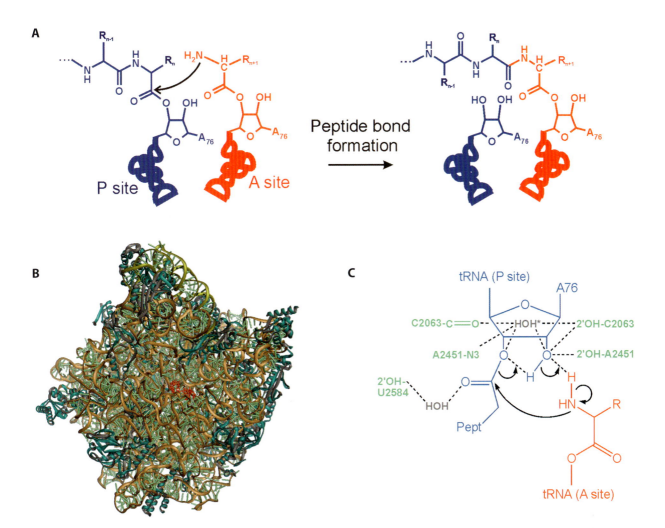

Fig.5 Peptide bond formation. (A) Schematic of the reaction. The α-amino group of aminoacyl-tRNA in the A site (red) attacks the carbonyl carbon of the peptidyl-tRNA in the P site (blue), resulting in peptidyl-tRNA in the A site and deacylated tRNA in the P site. (B) Crystal structure of the 50S subunit from *H. marismortui* with a transition state analog (red) bound to the active site (PDB entry 1VQP; (Schmeing et al., 2005a)). Ribosomal proteins are blue, the 23S rRNA backbone is brown, the 5S rRNA backbone olive, and rRNA bases are pale green. Reproduced from (Beringer and Rodnina, 2007) with permission. (C) Concerted proton shut- tle. Peptidyl-tRNA (P site) and aminoacyl-tRNA (A site) are blue and red, respectively, ribosome residues are pale green, and ordered water molecules are gray. The attack of the α-NH$_2$ group on the ester carbonyl carbon results in a six-membered transition state in which the 2'-OH group of the A-site A76 ribose moiety donates its proton to the adjacent leaving 3'-oxygen and simultaneously receives a proton from the amino group (Schmeing et al., 2005a; Trobro and Åqvist, 2005). Ribosomal residues are not involved in chemical catalysis but are part of the H-bond network that stabilizes the transition state.

al., 2004; Beringer et al., 2005). The largest decrease in the reaction rates with native aa-tRNA (200-fold) was observed with the A2450G-C2063U mutation (Hesslein et al., 2004). Most likely, however, the replacement of the ionizing A⁺-C pair with a G-U pair was not fully isosteric, as initially predicted, and disrupted the structure of the active site (Hesslein et al., 2004).

4.2. Enzymology of the peptidyl transfer reaction

The ribosome brings about a 10^7-fold rate enhancement of the peptidyl transfer reaction, relative to the second order reaction between model substrates in solution (Sievers et al., 2004). This acceleration is the result of a lowering of the entropy of activation, whereas the enthalpy of activation is the same for the reaction on the ribosome and in solution. In contrast, enzymes that employ general acid-base or covalent catalysis typically act by lowering the activation enthalpy of the catalyzed reaction. Thus, the ribosome appears to use mechanisms of catalysis that are largely entropic in origin, such as substrate positioning in the active site, desolvation, or electrostatic shielding (Sievers et al., 2004). The reaction rate is modulated by the length of the nascent peptide of the peptidyl-tRNA, the nature of its C-terminal amino acid (Wohlgemuth et al., 2008), and of the A-site aa-tRNA (Pavlov et al., 2009), suggesting a contribution of steric and charge effects.

While the small contribution of the activation enthalpy to catalysis argues against a significant role of an acid-base mechanism, it was nevertheless important to approach this question directly. If the ribosome employed acid-base catalysis to an appreciable extent, the rate of reaction should depend on pH, and from the k_{pep}/pH plots the number of ionizing groups involved in catalysis and their pK_a values could be determined. A caveat of such measurements is the fact that the α-amino group ionizes with a pK_a of about 8 (Wolfenden, 1963); as only the deprotonated form of the amino group can act as a nucleophile, the reaction rate is expected to increase with pH, even if other ionizing groups were not involved. In fact, early measurements indicated that the reaction rate increased with pH, and the increase per pH unit suggested that a single ionizing group was involved (Maden and Monro, 1968; Pestka, 1972). At that time – well before RNA catalysis was discovered – a histidine residue of a ribosomal protein was proposed to act as a catalyst, which prompted a search for catalytic histidines among ribosomal proteins (Diedrich et al., 2000). However, the conditions of those early experiments did not allow for rigorous monitoring of the chemistry step, and the observed pH dependence very likely was due to ionization of the α-amino group nucleophile. These ambiguities were ultimately circumvented by using an aa-tRNA derivative with a hydroxyl group replacing the amino group as the reactive nucleophile, e. g. by using hydroxy-Phe-tRNA (phenyl-lactyl-tRNA) (Fahnestock et al., 1970; Katunin et al., 2002; Bieling et al., 2006). Hydroxy-Phe-tRNA is readily accepted as a substrate by the ribosome, which then catalyzes the formation of an ester bond instead of a peptide bond. In that case, the reaction rate was no longer affected by pH changes (Bieling et al., 2006), indicating that catalysis by the peptidyl transferase center is intrinsically independent of pH. This observation argues against an involvement of ionizing groups of the ribosome in catalysis and indicates that the ribosome

does not utilize general acid-base catalysis to any significant extent.

Another group found within hydrogen-bonding distance of the nucleophilic group of transition state analogs is the 2'-OH of A76 of peptidyl-tRNA in the P site (Schmeing et al., 2005a; Schmeing et al., 2005b). The 2'-OH plays a crucial role in the reaction on both isolated 50S subunits (Krayevsky and Kukhanova, 1979) and 70S ribosomes (Weinger et al., 2004), although the magnitude of the effect is disputed (Koch et al., 2008). The essential role of the 2'-OH of A76 of the P-site substrate suggested that the ribosome utilizes substrate-assisted catalysis, i. e. a mechanism in which a functional group of the substrate contributes to catalysis, in addition to the contributions of substrate positioning and desolvation described above. Molecular dynamics simulations of the reactant and tetrahedral intermediate states of the peptidyl transferase center suggested a stable pre-organized hydrogen-bond network poised for catalysis (Trobro and Åqvist, 2005, 2006). The peptidyl transferase center could thus be viewed as a rather rigid environment of pre-organized dipoles that do not need major rearrangements during the reaction. According to the molecular dynamics simulations, the most favorable mechanism does not involve general acid-base catalysis by ribosomal groups (Trobro and Åqvist, 2005). As the thermodynamic data predicted (Sievers et al., 2004), the catalytic effect was found to be of entropic origin and was associated with a reduction in solvent reorganization energy rather than with substrate alignment or proximity (Trobro and Åqvist, 2005). The Brønsted coefficient of the α-amino nucleophile displayed linear free-energy relationships with slopes close to zero under conditions where chemistry was rate limiting (Kingery et al., 2008). These results indicated that, in the transition state, the nucleophile is uncharged in the ribosome-catalyzed reaction, in contrast to the substantial positive charge reported for typical uncatalyzed aminolysis reactions. This suggests that the transition state involves deprotonation to a degree that is commensurate with nitrogen-carbon bond formation, and that the ribosome-catalyzed reaction is facilitated by amine deprotonation (Kingery et al., 2008), most likely through a proton shuttle mechanism (Figure 5C).

There are several shuttle pathways that can be envisaged (Das et al., 1999; Dorner et al., 2002; Changalov et al., 2005; Schmeing et al., 2005a; Trobro and Åqvist, 2005, 2006; Kingery et al., 2008). All these models have in common that the attack of the α-amino group on the ester carbon results in a six-membered transition state, where the 2'-OH group donates its proton to the adjacent 3'-oxygen while simultaneously receiving one of the amino protons. Such a scenario would not require a pKa shift of the 2'-OH group, due to the concerted nature of the bond-forming and -breaking events. The 2'-OH of the ribose moiety of A2451 seems to be part of the intricate hydrogen bond network in the active site and to interact directly with the critical 2'-OH group of the P-site tRNA. Consistent with this view, substitution of the 2'-OH of A2451 by hydrogen resulted in impaired peptidyl transferase activity (Erlacher et al., 2006). The mechanism is also consistent with the notion that none of the rRNA bases within the inner shell act as general acids or bases in catalysis (Beringer et al., 2005; Bieling et al., 2006; Brunelle et al., 2006).

The ribosome is an ancient RNA catalyst. It is much less efficient than many protein enzymes which use chemical catalysis and accelerate reactions by up to 10^{23}-fold (Radzicka and Wolfenden, 1995). Apparently, evolutionary pressure had a much larger influence on increasing the speed and fidelity of the rate-limiting steps of protein synthesis than on the chemical step of peptide bond formation (Rodnina and Wintermeyer, 2001). This route of evolution allowed the ribosome to retain its relatively primitive catalytic strategy during the transition from a prebiotic translational ribozyme into the modern ribosome. The catalytic mechanism employed by the ribosome appears to represent a fossil of a primitive catalyst of the RNA world.

References

Ban N, Nissen P, Hansen J, Moore PB, Steitz TA (2000) The complete atomic structure of the large ribosomal subunit at 2.4 Å resolution. Science 289: 905–920

Bashan A, Agmon I, Zarivach R, Schluenzen F, Harms J, Berisio R, Bartels H, Franceschi F, Auerbach T, Hansen HA, Kossoy E, Kessler M, Yonath A (2003) Structural basis of the ribosomal machinery for peptide bond formation, translocation, and nascent chain progression. Mol Cell 11: 91–102

Berchtold H, Reshetnikova L, Reiser CO, Schirmer NK, Sprinzl M, Hilgenfeld R (1993) Crystal structure of active elongation factor Tu reveals major domain rearrangements. Nature 365: 126–132

Beringer M, Adio S, Wintermeyer W, Rodnina MV (2003) The G2447A mutation does not affect ionization of a ribosomal group taking part in peptide bond formation. RNA 9: 919–922

Beringer M, Bruell C, Xiong L, Pfister P, Bieling P, Katunin VI, Mankin AS, Bottger EC, Rodnina MV (2005) Essential mechanisms in the catalysis of peptide bond formation on the ribosome. J Biol Chem 280: 36065–36072

Beringer M, Rodnina MV (2007) The ribosomal peptidyl transferase. Mol Cell 26: 311–321

Bernado P, Modig K, Grela P, Svergun DI, Tchorzewski M, Pons M, Akke M (2010) Structure and dynamics of ribosomal protein L12: An ensemble model based on SAXS and NMR relaxation. Biophys J 98: 2374–2382

Bieling P, Beringer M, Adio S, Rodnina MV (2006) Peptide bond formation does not involve acid-base catalysis by ribosomal residues. Nat Struct Mol Biol 13: 423–428

Blanchard SC, Gonzalez RL, Kim HD, Chu S, Puglisi JD (2004) tRNA selection and kinetic proofreading in translation. Nat Struct Mol Biol 11: 1008–1014

Bos JL, Rehmann H, Wittinghofer A (2007) GEFs and GAPs: critical elements in the control of small G proteins. Cell 129: 865–877

Bremer H, Dennis PP. 1987. Modulation of chemical composition and other parameters of the cell by growth rate. In: Neidhardt FC, ed. *Escherichia coli and Salmonella typhimurium: cellular and molecular biology*. Washington, DC: American Society for Microbiology. pp 1553–1569

Brunelle JL, Youngman EM, Sharma D, Green R (2006) The interaction between C75 of tRNA and the A loop of the ribosome stimulates peptidyl transferase activity. RNA 12: 33–39

Burakovsky DE, Sergiev PV, Steblyanko MA, Kubarenko AV, Konevega AL, Bogdanov AA, Rodnina MV, Dontsova OA (2010) Mutations at the accommodation gate of the ribosome impair RF2-dependent translation termination. RNA 16: 1848–1853

Carter AP, Clemons WM, Jr., Brodersen DE, Morgan-Warren RJ, Wimberly BT, Ramakrishnan V (2000) Functional insights from the structure of the 30S ribosomal subunit and its interactions with antibiotics. Nature 407: 340–348

Changalov MM, Ivanova GD, Rangelov MA, Acharya P, Acharya S, Minakawa N, Foldesi A, Stoineva IB, Yomtova VM, Roussev CD, Matsuda A, Chattopadhyaya J, Petkov DD (2005) 2'/3'-O-peptidyl adenosine as a general base catalyst of its own external peptidyl transfer: implications for the ribosome catalytic mechanism. Chem Biochem 6: 992–996

Cochella L, Green R (2005) An active role for tRNA in decoding beyond codon:anticodon pairing. Science 308: 1178–1180

Das GK, Bhattacharyya D, Burma DP (1999) A possible mechanism of peptide bond formation on ribosome without mediation of peptidyl transferase. J theor Biol 200: 193–205

Daviter T, Gromadski KB, Rodnina MV (2006) The ribosome's response to codon-anticodon mismatches. Biochimie 88: 1001–1011

Daviter T, Wieden H-J, Rodnina MV (2003) Essential role of histidine 84 in elongation factor Tu for the chemical step of GTP hydrolysis on the ribosome. J Mol Biol 332: 689–699

Diaconu M, Kothe U, Schlunzen F, Fischer N, Harms JM, Tonevitsky AG, Stark H, Rodnina MV, Wahl MC (2005) Structural basis for the function of the ribosomal L7/12 stalk in factor binding and GTPase activation. Cell 121: 991–1004

Diedrich G, Spahn CM, Stelzl U, Schafer MA, Wooten T, Bochkariov DE, Cooperman BS, Traut RR, Nierhaus KH (2000) Ribosomal protein L2 is involved in the association of the ribosomal subunits, tRNA binding to A and P sites and peptidyl transfer. EMBO J 19: 5241–5150

Dorner S, Polacek N, Schulmeister U, Panuschka C, Barta A (2002) Molecular aspects of the ribosomal peptidyl transferase. Biochem Soc Trans 30: 1131–1136

Drummond DA, Wilke CO (2009) The evolutionary consequences of erroneous protein synthesis. Nat Rev Genet 10: 715–724

Effraim PR, Wang J, Englander MT, Avins J, Leyh TS, Gonzalez RL, Jr., Cornish VW (2009) Natural amino acids do not require their native tRNAs for efficient selection by the ribosome. Nat Chem Biol 5: 947–953

Erlacher MD, Lang K, Wotzel B, Rieder R, Micura R, Polacek N (2006) Efficient ribosomal peptidyl transfer critically relies on the presence of the ribose 2'-OH at A2451 of 23S rRNA. J Am Chem Soc 128: 4453–4459

Fahnestock S, Neumann H, Shashoua V, Rich A (1970) Ribosome-catalyzed ester formation. Biochemistry 9: 2477–2483

Gromadski KB, Daviter T, Rodnina MV (2006) A uniform response to mismatches in codon-anticodon complexes ensures ribosomal fidelity. Mol Cell 21: 369–377

Gromadski KB, Rodnina MV (2004a) Kinetic determinants of high-fidelity tRNA discrimination on the ribosome. Mol Cell 13: 191–200

Gromadski KB, Rodnina MV (2004b) Streptomycin interferes with conformational coupling between codon recognition and GTPase activation on the ribosome. Nat Struct Mol Biol 11: 316–322

Hesslein AE, Katunin VI, Beringer M, Kosek AB, Rodnina MV, Strobel SA (2004) Exploration of the conserved A+C wobble pair within the ribosomal peptidyl transferase center using affinity purified mutant ribosomes. Nucl Acids Res 32: 3760–3770

Ilag LL, Videler H, McKay AR, Sobott F, Fucini P, Nierhaus KH, Robinson CV (2005) Heptameric (L12)6/L10 rather than canonical pentameric complexes are found by tandem MS of intact ribosomes from thermophilic bacteria. Proc Natl Acad Sci USA 102: 8192–8197

Jenner L, Demeshkina N, Yusupova G, Yusupov M (2010) Structural rearrangements of the ribosome at the tRNA proofreading step. Nat Struct Mol Biol 17: 1072–1078

Johansson M, Bouakaz E, Lovmar M, Ehrenberg M (2008) The kinetics of ribosomal peptidyl transfer revisited. Mol Cell 30: 589–598

Katunin VI, Muth GW, Strobel SA, Wintermeyer W, Rodnina MV (2002) Important contribution to catalysis of peptide bond formation by a single ionizing group within the ribosome. Mol Cell 10: 339–346

Kim DF, Green R (1999) Base-pairing between 23S rRNA and tRNA in the ribosomal A site. Mol Cell 4: 859–864

Kingery DA, Pfund E, Voorhees RM, Okuda K, Wohlgemuth I, Kitchen DE, Rodnina MV, Strobel SA (2008) An uncharged amine in the transition state of the ribosomal peptidyl transfer reaction. Chem Biol 15: 493–500

Knudsen C, Wieden HJ, Rodnina MV (2001) The importance of structural transitions of the switch II region for the functions of elongation factor Tu on the ribosome. J Biol Chem 276: 22183–22190

Knudsen CR, Clark BF (1995) Site-directed mutagenesis of Arg58 and Asp86 of elongation factor Tu from *Escherichia coli*: effects on the GTPase reaction and aminoacyl-tRNA binding. Protein Eng 8: 1267–1273

Koch M, Huang Y, Sprinzl M (2008) Peptide-bond synthesis on the ribosome: no free vicinal hydroxy group required on the terminal ribose residue of peptidyl-tRNA. Angew Chem Int Ed Engl 47: 7242–7245

Konevega AL, Soboleva NG, Makhno VI, Semenkov YP, Wintermeyer W, Rodnina MV, Katunin VI (2004) Purine bases at position 37 of tRNA stabilize codon-anticodon interaction in the ribosomal A site by stacking and Mg²⁺-dependent interactions. RNA 10: 90–101

Kothe U, Rodnina MV (2006) Delayed release of inorganic phosphate from elongation factor Tu following GTP hydrolysis on the ribosome. Biochemistry 45: 12767–12774

Kothe U, Rodnina MV (2007) Codon reading by tRNA^Ala with modified uridine in the wobble position. Mol Cell 25: 167–174

Kothe U, Wieden HJ, Mohr D, Rodnina MV (2004) Interaction of helix D of elongation factor Tu with helices 4 and 5 of protein L7/12 on the ribosome. J Mol Biol 336: 1011–1021

Kramer EB, Farabaugh PJ (2007) The frequency of translational misreading errors in *E. coli* is largely determined by tRNA competition. RNA 13: 87–96

Krayevsky AA, Kukhanova MK (1979) The peptidyltransferase center of ribosomes. Prog Nucleic Acid Res Mol Biol 23: 1–51

Liang ST, Xu YC, Dennis P, Bremer H (2000) mRNA composition and control of bacterial gene expression. J Bacteriol 182: 3037–3044

Maden BE, Monro RE (1968) Ribosome-catalyzed peptidyl transfer. Effects of cations and pH value. Eur J Biochem 6: 309–316

Maguire BA, Beniaminov AD, Ramu H, Mankin AS, Zimmermann RA (2005) A protein component at the heart of an RNA machine: the importance of protein L27 for the function of the bacterial ribosome. Mol Cell 20: 427–435

Mansilla F, Knudsen CR, Laurberg M, Clark BF (1997) Mutational analysis of *Escherichia coli* elongation factor Tu in search of a role for the N-terminal region. Protein Eng 10: 927–934

Marshall RA, Aitken CE, Dorywalska M, Puglisi JD (2008) Translation at the single-molecule level. Annu Rev Biochem 77: 177–203

McClory SP, Leisring JM, Qin D, Fredrick K (2010) Missense suppressor mutations in 16S rRNA reveal the importance of helices h8 and h14 in aminoacyl-tRNA selection. RNA 16: 1925–1934

Miyoshi T, Nomura T, Uchiumi T (2009) Engineering and characterization of the ribosomal L10-L12 stalk complex. A structural element responsible for high turnover of the elongation factor G-dependent GTPase. J Biol Chem 284: 85–92

Mohr D, Wintermeyer W, Rodnina MV (2002) GTPase activation of elongation factors Tu and G on the ribosome. Biochemistry 41: 12520–12528

Netzer WJ, Hartl FU (1997) Recombination of protein domains facilitated by co-translational folding in eukaryotes. Nature 388: 343–349

Nissen P, Hansen J, Ban N, Moore PB, Steitz TA (2000) The structural basis of ribosome activity in peptide bond synthesis. Science 289: 920–930

Noller HF, Hoffarth V, Zimniak L (1992) Unusual resistance of peptidyl transferase to protein extraction procedures. Science 256: 1416–1419

O'Connor M, Dahlberg AE (1995) The involvement of two distinct regions of 23 S ribosomal RNA in tRNA selection. J Mol Biol 254: 838–847

Ogle JM, Brodersen DE, Clemons WM, Jr., Tarry MJ, Carter AP, Ramakrishnan V (2001) Recognition of cognate transfer RNA by the 30S ribosomal subunit. Science 292: 897–902

Ogle JM, Carter AP, Ramakrishnan V (2003) Insights into the decoding mechanism from recent ribosome structures. Trends Biochem Sci 28: 259–266

Ogle JM, Murphy FV, Tarry MJ, Ramakrishnan V (2002) Selection of tRNA by the ribosome requires a transition from an open to a closed form. Cell 111: 721–732

Ogle JM, Ramakrishnan V (2005) Structural insights into translational fidelity. Annu Rev Biochem 74: 129–177

Pan D, Zhang CM, Kirillov S, Hou YM, Cooperman BS (2008) Perturbation of the tRNA tertiary core differentially affects specific steps of the elongation cycle. J Biol Chem 283: 18431–18440

Pape T, Wintermeyer W, Rodnina MV (1998) Complete kinetic mechanism of elongation factor Tu-dependent binding of aminoacyl-tRNA to the A site of the *E. coli* ribosome. EMBO J 17: 7490–7497

Pape T, Wintermeyer W, Rodnina MV (1999) Induced fit in initial selection and proofreading of aminoacyl-tRNA on the ribosome. EMBO J 18: 3800–3807

Pape T, Wintermeyer W, Rodnina MV (2000) Conformational switch in the decoding region of 16S rRNA during aminoacyl-tRNA selection on the ribosome. Nat Struct Biol 7: 104–107

Parker J (1989) Errors and alternatives in reading the universal genetic code. Microbiol Rev 53: 273–298

Pavlov MY, Watts RE, Tan Z, Cornish VW, Ehrenberg M, Forster AC (2009) Slow peptide bond formation by proline and other N-alkylamino acids in translation. Proc Natl Acad Sci USA 106: 50–54

Pestka S (1972) Peptidyl-puromycin synthesis on polyribosomes from Escherichia coli. Proc Natl Acad Sci USA 69: 624–628

Piepenburg O, Pape T, Pleiss JA, Wintermeyer W, Uhlenbeck OC, Rodnina MV (2000) Intact aminoacyl-tRNA is required to trigger GTP hydrolysis by elongation factor Tu on the ribosome. Biochemistry 39: 1734–1738

Polacek N, Gaynor M, Yassin A, Mankin AS (2001) Ribosomal peptidyl transferase can withstand mutations at the putative catalytic nucleotide. Nature 411: 498–501

Proshkin S, Rahmouni AR, Mironov A, Nudler E (2010) Cooperation between translating ribosomes and RNA polymerase in transcription elongation. Science 328: 504–508

Radzicka A, Wolfenden R (1995) A proficient enzyme. Science 267: 90–93

Rattenborg T, Nautrup Pedersen G, Clark BF, Knudsen CR (1997) Contribution of Arg288 of Escherichia coli elongation factor Tu to translational functionality. Eur J Biochem 249: 408–414

Rodnina MV, Fricke R, Kuhn L, Wintermeyer W (1995) Codon-dependent conformational change of elongation factor Tu preceding GTP hydrolysis on the ribosome. EMBO J 14: 2613–2619

Rodnina MV, Fricke R, Wintermeyer W (1994) Transient conformational states of aminoacyl-tRNA during ribosome binding catalyzed by elongation factor Tu. Biochemistry 33: 12267–12275

Rodnina MV, Gromadski KB, Kothe U, Wieden HJ (2005) Recognition and selection of tRNA in translation. FEBS Lett 579: 938–942

Rodnina MV, Pape T, Fricke R, Kuhn L, Wintermeyer W (1996) Initial binding of the elongation factor Tu·GTP·aminoacyl-tRNA complex preceding codon recognition on the ribosome. J Biol Chem 271: 646–652

Rodnina MV, Wintermeyer W (2001) Fidelity of aminoacyl-tRNA selection on the ribosome: kinetic and structural mechanisms. Annu Rev Biochem 70: 415–435

Ruan B, Palioura S, Sabina J, Marvin-Guy L, Kochhar S, Larossa RA, Soll D (2008) Quality control despite mistranslation caused by an ambiguous genetic code. Proc Natl Acad Sci USA 105: 16502–16507

Ruusala T, Ehrenberg M, Kurland CG (1982) Is there proofreading during polypeptide synthesis? EMBO J 1: 741–745

Sanbonmatsu KY, Joseph S, Tung CS (2005) Simulating movement of tRNA into the ribosome during decoding. Proc Natl Acad Sci USA 102: 15854–15859

Schmeing TM, Huang KS, Kitchen DE, Strobel SA, Steitz TA (2005a) Structural insights into the roles of water and the 2' hydroxyl of the P site tRNA in the peptidyl transferase reaction. Mol Cell 20: 437–448

Schmeing TM, Huang KS, Strobel SA, Steitz TA (2005b) An induced-fit mechanism to promote peptide bond formation and exclude hydrolysis of peptidyl-tRNA. Nature 438: 520–524

Schmeing TM, Ramakrishnan V (2009) What recent ribosome structures have revealed about the mechanism of translation. Nature 461: 1234–1242

Schmeing TM, Seila AC, Hansen JL, Freeborn B, Soukup JK, Scaringe SA, Strobel SA, Moore PB, Steitz TA (2002) A pretranslocational intermediate in protein synthesis observed in crystals of enzymatically active 50S subunits. Nat Struct Biol 9: 225–230

Schmeing TM, Voorhees RM, Kelley AC, Gao YG, Murphy FVt, Weir JR, Ramakrishnan V (2009) The crystal structure of the ribosome bound to EF-Tu and aminoacyl-tRNA. Science 326: 688–694

Schuette JC, Murphy FVt, Kelley AC, Weir JR, Giesebrecht J, Connell SR, Loerke J, Mielke T, Zhang W, Penczek PA, Ramakrishnan V, Spahn CM (2009) GTPase activation of elongation factor EF-Tu by the ribosome during decoding. EMBO J 28: 755–765

Selmer M, Dunham CM, Murphy FV, Weixlbaumer A, Petry S, Kelley AC, Weir JR, Ramakrishnan V (2006) Structure of the 70S ribosome complexed with mRNA and tRNA. Science 313: 1935–1942

Sievers A, Beringer M, Rodnina MV, Wolfenden R (2004) The ribosome as an entropy trap. Proc Natl Acad Sci USA 101: 7897–7901

Sorensen MA, Pedersen S (1991) Absolute in vivo translation rates of individual codons in Escherichia coli. The two glutamic acid codons GAA and GAG are translated with a threefold difference in rate. J Mol Biol 222: 265–280

Stark H, Rodnina MV, Wieden H-J, Zemlin F, Wintermeyer W, van Heel M (2002) Ribosome interactions of aminoacyl-tRNA and elongation factor Tu in the codon recognition complex. Nat Struct Biol 9: 849–854

Thompson J, Kim DF, O'Connor M, Lieberman KR, Bayfield MA, Gregory ST, Green R, Noller HF, Dahlberg AE (2001) Analysis of mutations at residues A2451 and G2447 of 23S rRNA in the peptidyltransferase active site of the 50S ribosomal subunit. Proc Natl Acad Sci USA 98: 9002–9007

Thompson RC, Dix DB, Gerson RB, Karim AM (1981) Effect of Mg^{2+} concentration, polyamines, streptomycin, and mutations in ribosomal proteins on the accuracy of the two-step selection of aminoacyl-tRNAs in protein biosynthesis. J Biol Chem 256: 6676–6681

Thompson RC, Stone PJ (1977) Proofreading of the codon-anticodon interaction on ribosomes. Proc Natl Acad Sci USA 74: 198–202

Trobro S, Åqvist J (2005) Mechanism of peptide bond synthesis on the ribosome. Proc Natl Acad Sci U S A 102: 12395–12400

Trobro S, Åqvist J (2006) Analysis of predictions for the catalytic mechanism of ribosomal peptidyl transfer. Biochemistry 45: 7049–7056

Valle M, Zavialov A, Li W, Stagg SM, Sengupta J, Nielsen RC, Nissen P, Harvey SC, Ehrenberg M, Frank J (2003) Incorporation of aminoacyl-tRNA into the ribosome as seen by cryo-electron microscopy. Nat Struct Biol 10: 899–906

Vetter IR, Wittinghofer A (1999) Nucleoside triphosphate-binding proteins: different scaffolds to achieve phosphoryl transfer. Q Rev Biophys 32: 1–56

Villa E, Sengupta J, Trabuco LG, LeBarron J, Baxter WT, Shaikh TR, Grassucci RA, Nissen P, Ehrenberg M, Schulten K, Frank J (2009) Ribosome-induced changes in elongation factor Tu conformation control GTP hydrolysis. Proc Natl Acad Sci USA 106: 1063–1068

Voorhees RM, Schmeing TM, Kelley AC, Ramakrishnan V (2010) The mechanism for activation of GTP hydrolysis on the ribosome. Science 330: 835–838

Voorhees RM, Weixlbaumer A, Loakes D, Kelley AC, Ramakrishnan V (2009) Insights into substrate stabilization from snapshots of the peptidyl transferase center of the intact 70S ribosome. Nat Struct Mol Biol 16: 528–533

Weinger JS, Parnell KM, Dorner S, Green R, Strobel SA (2004) Substrate-assisted catalysis of peptide bond formation by the ribosome. Nat Struct Mol Biol 11: 1101–1106

Wiborg O, Andersen C, Knudsen CR, Clark BF, Nyborg J (1996) Mapping *Escherichia coli* elongation factor Tu residues involved in binding of aminoacyl-tRNA. J Biol Chem 271: 20406–20411

Wohlgemuth I, Beringer M, Rodnina MV (2006) Rapid peptide bond formation on isolated 50S ribosomal subunits. EMBO Rep 7: 669–703

Wohlgemuth I, Brenner S, Beringer M, Rodnina MV (2008) Modulation of the rate of peptidyl transfer on the ribosome by the nature of substrates. J Biol Chem 283: 32229–32235

Wohlgemuth I, Pohl C, M. V R (2010) Optimization of speed and accuracy of decoding in translation. EMBO J:in press

Wolfenden R (1963) The mechanism of hydrolysis of amino acyl RNA. Biochemistry 338: 1090–1092

Youngman EM, Brunelle JL, Kochaniak AB, Green R (2004) The active site of the ribosome is composed of two layers of conserved nucleotides with distinct roles in peptide bond formation and peptide release. Cell 117: 589–599

Zaher HS, Green R (2009) Quality control by the ribosome following peptide bond formation. Nature 457: 161–166

17

Sense and nonsense recognition by the ribosome

Rodrigo F. Ortiz-Meoz, Shan L. He, Hani S. Zaher, and Rachel Green

1. Introduction

Translation is the molecular process that deciphers the language of nucleic acid (nucleotides) into the language of proteins (amino acids). As for other core molecular processes, *in vitro* reconstituted systems have been used to define key cellular components involved in translation. During the past decade or so, there has been an explosion of ribosomal structural information defining multiple functional states throughout the translation process. These unprecedented views have allowed for the formulation of detailed hypotheses concerning the molecular mechanisms of translation, though their evaluation relies on methodologies that can observe its dynamic nature. Our laboratory has been interested in a central molecular recognition process on the ribosome – how are substrates selected that correspond to the codon poised in the A site? This question is an interesting one because this process involves two remote functional centers that span across the subunit interface of the ribosome, and as such, necessarily involves long-range signal transduction. Here we detail in broad strokes some of our contributions, mostly obtained from biochemical approaches, while placing them in the context of the field. We note that this chapter is not intended as a comprehensive overview of the field.

2. The site of decoding on the small subunit

During the elongation cycle of translation, the mRNA is threaded through the small ribosomal subunit where decoding takes place. During this process, the ribosome selects either the cognate tRNA from among the pool of all cellular tRNAs ("tRNA selection") or the appropriate release factor ("RF selection"), depending on the codon presented in the A site. Of particular interest here, this selection takes place with a remarkable fidelity leading to misincorporation or premature termination at a frequency of only 1 in $10^3 - 10^5$ events (Zaher and Green, 2009a).

2.1. How does the recognition of sense codons work at a molecular level?

During tRNA selection, the ribosome utilizes two distinct mechanisms, kinetic proofreading and kinetic discrimination, to bring about the documented remarkable levels of accuracy. Kinetic proofreading is made possible by the fact that the aminoacyl tRNA (aa-tRNA) is presented to the ribosome in a ternary complex with elongation factor Tu (EF-Tu) and GTP, thus allowing irreversible GTP hydrolysis by EF-Tu to separate the process into two distinct steps, initial selection and proofreading (Hopfield, 1974; Thompson and Stone, 1977). Fundamentally this is a thermodynamically driven process where the ribosome has two opportunities to inspect the tRNA:mRNA interaction and preferentially discard near-cognate aa-tRNAs. Kinetic discrimination, often referred to as "induced fit", is a kinetically driven process and is thought to be the result of conformational changes that are stabilized by cognate tRNA-mRNA interactions and ultimately lead to accelerated rates of GTPase activation and tRNA accommodation, the rate limiting steps of tRNA selection (Pape et al., 1999; Gromadski and Rodnina, 2004). These two general mechanisms are included in the kinetic and thermodynamic framework for tRNA selection as defined by the Rodnina group (Core steps shown in Figure 1a) (Pape et al., 1998). Understanding how these thermodynamic and kinetic constraints are

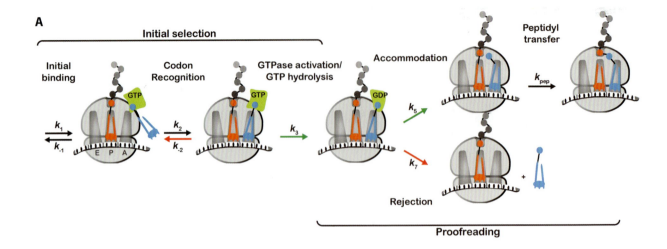

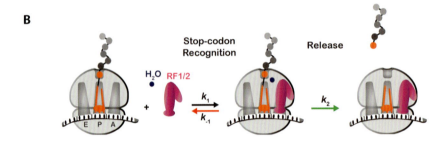

Fig. 1 tRNA- and RF-selection pathways. (A) Simplified scheme of tRNA-selection by the ribosome with only the core steps that contribute to discrimination shown. The process is divided into two phases, initial binding and proofreading separated by GTP hydrolysis. Forward rates that are accelerated for cognate tRNAs are depicted with green arrows while dissociation rates that are accelerated for near-cognate tRNAs are depicted with red arrows. (B) A minimal scheme for RF-selection by the ribosome consisting of two steps, a binding event followed by catalysis of peptide release. The rate of peptide release on stop codons is faster than on non-stop, and as a result the step is depicted with a green arrow. Moreover, the binding affinity of RF for stop codon ribosomal complexes is stronger relative to near-stop complexes, and as result the dissociation step is depicted with a red arrow.

brought to bear at a molecular level has been a focus of interest for the field for a number of years.

Available high-resolution crystal structures of 30S ribosomal subunits in the absence and presence of A-site-bound cognate and near-cognate tRNA anticodon stem loops (ASL) have taught us much about the decoding process (reviewed in Schmeing and Ramakrishnan, 2009). On a relatively local level (i. e. only considering interactions in the decoding center where the codon:anticodon interaction is monitored), universally conserved nucleotides A1492 and A1493 in the *apo* structure are found stacked within a bulge in helix 44 while nucleotide G530 adopts a *syn* conformation. Upon cognate ASL binding, A1492 and A1493 flip out from h44 to a position where they engage the minor groove of the tRNA:mRNA minihelix through A-minor interactions while G530 switches conformation from *syn* to *anti*. We think of these structural changes as constituting an "on" state for the decoding center (Figure 2a). This elaborate network of interactions stringently inspects the first and second positions of the codon:anticodon helix, permitting only Watson-Crick interactions. The 3rd position of the codon is monitored by other ribosome moieties including G530, C518 and portions of the ribosomal protein S12, ultimately permitting Watson-Crick and certain wobble interactions (Ogle et al., 2001; Murphy and Ramakrishnan, 2004). Other insights into structural features of decoding come from analysis of certain miscoding aminoglycosides bound to the small subunit. Paromomycin, for example, induces structural rearrangements in the decoding center that partially mimic those changes brought about on the binding of cognate tRNA, providing some immediate insight into how miscoding is stimulated by these molecules (Ogle et al., 2001).

These structural studies thus highlight potential key molecular contributors to tRNA selection, but do not

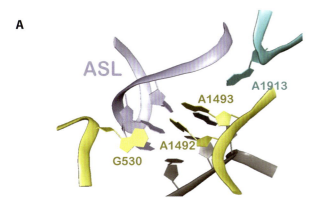

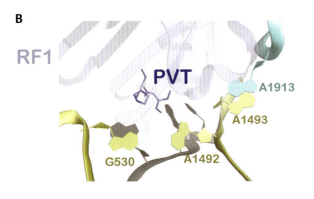

Fig. 2 Conformational changes in the decoding center of the ribosome. (A) Structural rearrangements in the decoding center upon tRNA anticodon stem-loop (ASL; light blue) binding. 16S rRNA residues G530, A1492 and A1493 are highlighted in yellow, while mRNA is depicted in grey. A1492 and A1493 inspect the geometry of the base pairing interactions of the first two positions of the codon:anticodon interaction. (B) Upon release factor 1 binding (RF1; light blue) distinct structural rearrangements are observed. G530 stacks with the 3rd position of the distorted mRNA (grey). While A1492 is again displaced from helix 44, A1493 remains within helix 44 and is stabilized there by stacking with residue A1913 in Helix 69 of the 23S rRNA (cyan). The PVT tripeptide motif of RF1 (partly responsible for recognizing the stop codon) is shown in dark blue.

on the forward and backward rate constants with the cognate aa-tRNAs closely match those observed for near-cognate aa-tRNAs on a wild type ribosome (Gromadski and Rodnina, 2004). Moreover, the effects of the decoding center mutations on the tRNA selection process can be rescued by the addition of paromomycin, suggesting that the base composition of A1492, A1493 and G530 is not key for the selection pathway *per se*. Instead, it seems that if the on state of the decoding center nucleotides is stabilized in some way, whether by strong docking interactions in the minor groove of the decoding helix or by paromomycin binding, tRNA selection is effectively promoted.

2.2. How does the recognition of stop codons work at a molecular level?

During RF selection, the ribosome selects the appropriate class-I release factor for decoding a particular stop codon (UAA or UAG by RF1 and UAA or UGA by RF2). As mentioned above, this process also occurs with remarkable fidelity wherein sense codons are recognized prematurely by release factors as seldom as 1 in 10^5 encounters (Jorgensen et al., 1993). Shown in Figure 1b is a minimal proposed scheme describing the interaction of class-I release factors with the ribosome, including an initial binding steps and subsequent catalysis; this process is less well characterized at a mechanistic level than tRNA selection. An understanding how this selection process is brought about at a molecular level is of considerable interest.

Early genetic and mutational studies identified the so called "tripeptide anticodon motifs" in RF1 (PxT) and RF2 (SPF) that appeared to be at least partially responsible for stop codon recognition (Ito et al., 2000). Recent structural studies of these two proteins bound to the ribosome have revealed additional structural elements in both the RFs and the 16S RNA that likely contribute to the accurate reading of stop codons (Korostelev et al., 2008; Laurberg et al., 2008; Weixlbaumer et al., 2008). Trivially, a different mechanism from that involved in tRNA selection must be at play since RF recognition of the codon involves protein-RNA interactions rather than the simple well-defined RNA-RNA interactions that determine tRNA selection. Indeed the structural studies verified the clear distinction between the two processes; while the stop codon – like in tRNA selection – is still inspected within the decoding center, the overall conformation of the three nucleotides in the codon is different, most notably with

define their actual contribution. As anticipated, mutation of the core decoding center nucleotides G530, A1492 and A1493 has substantial effects on tRNA selection. Aptamer-tagged *in vivo* assembled ribosomes carrying single nucleotide substitutions in these positions exhibit overall decreases in the rates for both GTPase activation and accommodation (as evaluated through the chemical steps of GTP hydrolysis and peptidyl transfer) with cognate tRNAs (Cochella et al., 2007). In contrast to wild-type ribosomes that almost never discard cognate aa-tRNAs at the proofreading step, A1492 and A1493 variant ribosomes were found to discard more than half of the cognate tRNAs at this stage. As such, the combined effects of the mutations the third-position nucleotide being unstacked from the second-position one. Similarly, the conformational status of the decoding center nucleotides in the presence of the bound RFs is quite distinct from that observed during tRNA selection. While A1492 rotates out of helix 44 (h44) to form part of the floor of the decoding pocket, A1493 remains within the helix, and G530, instead of rotating from the *syn* to the *anti* conformation, now stacks directly on the third nucleotide of the stop codon (Figure 2b). An interesting surprise was to see nucleotide 1913 in helix 69 (H69) of the large subunit reaching into the decoding center where it interacts with A1493 during RF selection (Laurberg et al., 2008). Indeed, helix 69 has been implicated in release factor function on the ribosome in previous biochemical experiments (Ali et al., 2006). As for the RF-specific interactions, a number of specific amino acids, including residues within the PxT and SPF motifs, are involved in "decoding" the three different stop codons. Residues shared by RF1 and RF2 are involved in the recognition of the U at position 1 common to all three stop codons, while the A and G nucleotides at positions 2 and 3 are recognized by distinct elements within these factors.

As for tRNA selection, these RF-bound ribosome structures are a jackpot for thinking about detailed molecular mechanisms. An obvious question is to what extent does the identity of the various molecular elements contribute to the efficiency of the process? Our laboratory has systematically explored the biochemical consequences of mutating nucleotides in the decoding region of the 16S RNA, assuming that their identities would be critical to release activity (since they are positioned right in the middle of the action and exhibit dynamic behavior). To our surprise, we found that mutations at G530, A1492 and A1493 had no discernible effect on the maximal rates for peptide release with cognate stop codons, though effects were substantial with so called "near-stop" codons (Youngman et al., 2007). Moreover, substitution of the codon with deoxyribonucleotides had dramatic effects on tRNA selection parameters and no effect on peptide release. Finally, while the aminoglycoside antibiotic paromomycin stimulates the tRNA selection process (leading to overall increases in miscoding), the same drug substantially inhibits RF-mediated stop codon recognition.

In terms of the most obvious conclusions, we now understand that paromomycin inhibits RF selection because it induces conformational rearrangements of A1492 and A1493 (so that they stick out of helix 44

into the decoding center) that are incompatible with RF binding. What remains puzzling is why, for example, mutations in the decoding center nucleotides have no consequence for release, since these nucleotides appear to be involved in significant interactions with the RFs and since we observe synthetic effects with the near-stop codons. While we propose that the release process is generally robust, making the actual identity of the decoding center nucleotides non-essential for supporting the "on state", we have no particular molecular rationale to explain such robustness.

3. The site of catalysis on the large subunit

After the aminoacyl-tRNAs and release factors are "selected" by the ribosome based on their compatibility with the codon in the A site, a catalytic event in the large subunit active site takes place. In the case of elongation, the α-amino group of the A-site aminoacyl-tRNA attacks the electron deficient carbonyl carbon of the peptidyl-tRNA to form a new peptide bond. In the termination reaction, the class-I RF works with the ribosome to orient an H_2O molecule for nucleophilic attack of the same carbonyl carbon, but here releasing the peptide from the peptidyl-tRNA. Despite the relatively long distance between the decoding and peptidyl transferase centers, there is good reason to believe that the activities in these sites are highly coordinated. The phrase "induced fit" is often used to describe the conformational state that is stabilized by the triggering events of codon recognition (for sense or stop codons) and leads to maximal long-range function. This section will focus on deciphering these functional connections.

3.1. How is peptidyl transferase activity impacted by tRNA-mediated sense codon recognition?

The obvious physical connectivity between the anticodon end of the tRNA, which interacts with the decoding center, and the CCA end, which interacts with the peptidyl transferase center, supports the notion that the small and large subunit functional centers communicate with one another; an idea that has been corroborated by several biochemical approaches. For example, during tRNA selection and associated peptidyl transfer reaction, pre-steady state kinetics indicate that cognate codon:anticodon interactions trigger both faster GTPase activation and accommodation (Pape et al., 1999; Gromadski and Rodnina, 2004). These two events (that precede actual catalysis) thus involve long-range conformational rearrangements in the ribosome that appear to be triggered by structural changes initiating in the decoding center.

We have learned much about how the peptidyl transferase center locally promotes the catalysis of peptide bond formation. An important initial event for this reaction is a rearrangement of the catalytic center (induced fit) on binding of the aminoacyl-tRNA as documented by biochemical (Youngman et al., 2004) and structural (Schmeing et al., 2005b) approaches. The A loop (and in particular G2553) likely plays a key role in triggering these conformational rearrangements in the large subunit (Brunelle et al., 2006). As for catalysis itself, there is little apparent reliance on nearby universally conserved nucleotides A2506, U2585 and U2602 for optimal catalysis (Polacek et al., 2003; Youngman et al., 2004), though we note that the Polacek group has argued for an important role for the 2'-OH of A2451 in this reaction (Erlacher et al., 2005; Erlacher et al., 2006). What appears to be most critical to catalysis *per se* in this active site is the 2'-OH of the peptidyl-tRNA that is poised to function in a proposed proton shuttle mechanism (Weinger et al., 2004; Schmeing et al., 2005a), though this view too has been debated (Koch et al., 2008).

3.2. How is release activity impacted by RF1-mediated stop codon recognition?

The decoding and peptidyl transfer centers are also physically connected during the termination reaction, but by the class-I release factor protein rather than the aminoacyl-tRNA. However, for the release reaction, there is no chaperone that guides the RF (as EF-Tu guides the aa-tRNA during tRNA selection), and there is no documented rate-limiting step that precedes the actual catalytic event (like accommodation). Interestingly, early kinetic studies showed that the rate of catalysis of peptide release depends on the codon recognition event – k_{cat} is faster on stop codons than on near-stop codons (Freistroffer et al., 2000). Our recent studies showed that paromomycin acts purely as a competitive inhibitor (exhibiting only $K_{1/2}$ effects) with authentic stop codons, but as a non-competitive inhibitor with near-stop codons with lingering effects on the k_{cat} of peptide release (Youngman et al., 2007).

These data together suggest that on recognition of authentic stop codons by the RF, special events take place that re-organize the decoding center, making it unable to bind paromomycin and thereby displacing it. These events thus conspire in triggering specific conformational changes that somehow affect the structure of active site moieties critical to chemistry, thereby speeding up catalysis.

From a chemical perspective, peptide release and peptidyl transfer are similar reactions, differing only in the identity of the attacking nucleophile (water vs. primary amino group, respectively). Despite these similarities, the peptide release reaction is inherently more difficult to perform because water is a relatively poorer nucleophile and it is not pre-ordered through attachment to an easily positioned macromolecular structure like the A-site tRNA. As such, peptide release may depend on more stringent mechanisms for controlling the orientations of potential substrates, and we therefore expect the catalytic site to be more sensitive to mutational perturbations. Fittingly, mutagenesis of conserved inner shell nucleotides in the peptidyl transferase center, including the ribose at A2451, has substantial effects on the rates of peptide release (Polacek et al., 2003; Youngman et al., 2004; Amort et al., 2007). Moreover, as for peptidyl transfer, the 2'-OH of the peptidyl-tRNA appears to be essential for catalysis of peptide release (Brunelle et al., 2008).

As for the release factors themselves, both RF1 and RF2 share a universally conserved GGQ tripeptide motif that is present at the site of catalysis. The glycine residues are seen to adopt a backbone conformation that would be sterically hindered by any other amino acids (Laurberg et al., 2008), and indeed, substitution of either glycine residue with an alanine results in dramatic losses in release activity (Zavialov et al., 2002; Shaw and Green, 2007). The conserved glutamine residue has been the focus of much attention. While RFs carrying mutations at this residue show modest defects in the rate of release (Seit-Nebi et al., 2001; Mora et al., 2003), nucleophile partitioning analysis has shown that this residue is critical for specifying water as the nucleophile in the reaction (Shaw and Green, 2007). Consistent with this, structural views place the glutamine backbone amide within hydrogen bonding distance of the leaving group of the reaction, the 3'-OH of the P-site tRNA, in a very central position in the active site (Laurberg et al., 2008). This hypothesis was further supported by subsequent mutational studies by the same group (Korostelev et al., 2008). We note, however, that another structural study found that the amide of the glutamine side chain is more centrally located in the active site, leading to the proposal that this group is more likely to play a significant role in catalysis (Weixl-baumer et al., 2008). Substitution of other conserved amino acids in this domain of the release factor failed to identify other functional groups essential to catalysis (Shaw and Green, 2007). As for peptidyl transfer, it seems likely that interaction of the RFs with the A loop is important for the release reaction, especially since peptide release chemistry is stimulated by binding of deacylated tRNA to the A site (Caskey et al., 1971).

4. What is the pathway for signal transduction between the functional centers of small and large subunits?

4.1. The pathway through the ribosome

X-ray crystal structures of the individual ribosomal subunits and of the various 70S complexes provide clues as to which molecular features might be important for such long-range communication. For example, in addition to the local structural changes in the small subunit decoding center described above, the 30S subunit undergoes more global structural changes on recognition of cognate, but not near-cognate, tRNAs (Ogle et al., 2002). The subunit undergoes a rotation of the head and shoulder domains, altering interactions between key proteins S4 and S5 and between S12 and RNA element h44, all in the small subunit. Predictably, paromomycin induces a similar sort of domain closure. Genetic experiments have identified mutations in both regions that appear to impact the overall fidelity of protein synthesis: mutations affecting the S12/h44 contact region on the interface side of the ribosome are generally *restrictive* (meaning that they make fewer mistakes) and are proposed to destabilize the closed conformation, while mutations at the S4/S5 contact region are generally *ram* (*ribosomal ambiguity*, meaning that they make more mistakes) and are proposed to stabilize the closed conformation (reviewed at length in (Zaher and Green, 2009a)). Our recent pre-steady state kinetic analysis of *restrictive* and *ram* ribosomes support these general conclusions, but surprisingly reveal that these variants are predominantly affected at different stages in the tRNA selection process (in the proofreading and initial selection phases, respectively) (Zaher and Green, 2010a). At present, structures of 70S termination complexes do not reveal observable evidence of such domain closure on stop codon recog-

nition, though we think it likely that similar rearrangements must be involved.

What other features in the ribosome itself might merit particular attention? As discussed earlier, during peptide release, nucleotide 1913 in the large subunit H69 reaches across the subunit interface, directly interacting with A1493, a key nucleotide in the decoding center of the small subunit. This interaction must be important for signal transduction for this reaction. Indeed, H69 itself has long been implicated in fidelity through genetic (O'Connor and Dahlberg, 1995; O'Connor, 2009) and biochemical (Ali et al., 2006) approaches. However, more often the contributions of this region to tRNA and RF selection have been difficult to distinguish largely due to the numerous ribosome components (both rRNA and rproteins) that are likely to be involved in discerning the identity and position of the tRNA (Jenner et al., 2010).

4.2. The pathway through the tRNA

In the end, the simplest route for signal transduction must involve the components that physically span the remote sites – the aminoacyl-tRNAs and the release factors. The aminoacyl-tRNAs bridge the subunit interface in multiple states – before GTPase activation (and hydrolysis), in the so-called A/T state, and after accommodation in the A/A state. Cryo-EM reconstructions initially provided some clues about the physical status of the bound tRNA that might be important for the signal relay (Stark et al., 2002; Valle et al., 2002). More recently, crystal structures became available that allow us to evaluate the tRNA and neighboring ribosome elements in both states at a higher resolution (Schmeing et al., 2009). The most striking difference between these two states is that the tRNA is distorted from its ground state when bound in the A/T state and still attached to EF-Tu, though interestingly, the conformational state of the decoding center in these two structures is indistinguishable. The distortion begins at the top of the anticodon helix, specifically disrupting interactions within the tRNA between positions 27:43, 25:45 and 26:44. Another notable distortion of the tRNA is seen in the D stem which is displaced from the T and acceptor stems by ~5 Å, altering several other interactions (15:48 and 16:59). An overlay of the A/T and A/A tRNAs is shown in Figure 3A.

Numerous studies have supported the notion that the tRNA itself might be directly involved in signal transduction. A number of years ago, Rodnina, Uh-

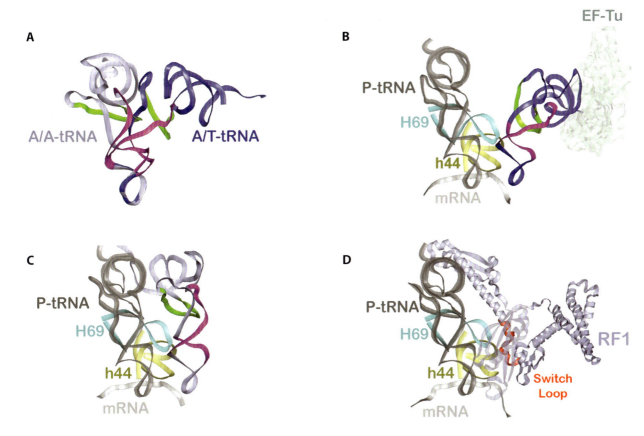

Fig. 3 Concerted conformational changes of the ribosome and its substrates. (A) Comparison of the structures of ribosome-bound tRNA in the A/A (light blue) and A/T (dark blue) states. The distortion of the A/T tRNA is centered around the top of the anticodon stem (residues 25–30 and 40–48; magenta) and the D-stem (residues 8 to 16; green). (B) The A/T tRNA is stabilized by interactions with EF-Tu (light green), H69 of the 23S rRNA (cyan), and h44 interactions in the decoding center (yellow). (C) The A/A tRNA shares similar contact with h44 but has a slightly altered contact with H69. (D) Release factor 1 (RF1; light blue) bound to the ribosomal A site. The key structural rearrangement of the switch loop (red), which allows domain 3 of the protein to reach the catalytic center, is facilitated by a pocket formed by H69 (cyan) upon stop-codon recognition.

lenbeck and colleagues cut the tRNA into two semi-functional pieces, an anticodon-D stem fragment and an acceptor-T stem fragment, and while each could bind and function on the ribosome, communication between the two functional centers was disrupted (Piepenburg et al., 2000). Subsequent biochemical studies in a number of laboratories established that even more subtle changes to the tRNA impact fidelity by constitutively activating tRNA selection independent of appropriate codon-anticodon interactions in the decoding center–the tRNA itself in these cases is triggering downstream events in the tRNA selection pathway (Cochella and Green, 2005; Pan et al., 2008; Ledoux et al., 2009). Moreover, a very recent *in-vitro* selection experiment in our laboratory identified several new mutations in tRNA^Trp that promote miscoding at high levels (Ortiz-Meoz and Green, 2010). Many of these mutations cluster in the region of the A/T-bound tRNA that shows the greatest deformation, suggesting that the miscoding phenotype is related to their ability to easily sample this configuration.

The structural distortion that defines the A/T tRNA appears to be stabilized by just a few crucial interactions with the ribosome (Figure 3B). At the anticodon end of the tRNA, the previously discussed decoding center nucleotides play an obvious role in stabilizing the cognate tRNA binding through A-minor and other interactions. Further up the molecule, the D stem of the tRNA contacts the tip of the universally conserved H69 from the 50S subunit, though in somewhat different configurations in the A/T and A/A states (Figure 3B, C). As mentioned above, the importance of H69 in overall fidelity has been documented at the genetic level (O'Connor and Dahlberg, 1995; O'Connor, 2009). And, while deletion of H69 had little to no impact on elongation rates (as assessed through poly(Phe) synthesis), there were subtle effects on fidelity in the same assay and more substantial effects on release (Ali et al., 2006). We have recently further defined a role for H69 in dictating the fidelity of tRNA selection, showing that the identity of A1913 is critical in allowing efficient selection of certain near-cognate or miscoding tRNA species (Ortiz-Meoz and Green, 2010).

Another critical feature of tRNA selection of course involves GTPase activation in the A/T ribosome complex. The details of this interaction are well characterized in the recent crystal structure of A/T bound tRNA (Schmeing et al., 2009), and its importance is supported by data showing that mutation of nucleotide 1067 of the 23S rRNA in the L11 binding region disrupts ternary complex binding and GTPase activation for cognate tRNA interactions (Saarma et al., 1997). These interactions are certainly related to the domain closure movements described earlier and its role in tRNA selection.

These studies together support the view that a distorted A/T tRNA structure is critical in the pathway of signaling events that originate in the decoding center and end in acceptance in the A/A state (and ultimate peptide bond formation). The universality of the mechanism is underscored by the very similar structures assumed by different aminoacyl-tRNAs in the A/T state (Li et al., 2008; Schmeing et al., 2009).

4.3. The pathway through the RF

The release factors appear to bridge the subunit interface in a single conformation, now well documented with several high-resolution structures (Korostelev et al., 2008; Laurberg et al., 2008; Weixlbaumer et al., 2008). There is no functional equivalent of EF-Tu during RF selection, and thus the semi-accommodated codon sampling state (A/T) of the tRNA has no equivalent during stop codon recognition. Accordingly, the class-I release factors appear to utilize a rather distinct signal transduction mechanism. As discussed above, optimal catalysis of peptide release (by domain 3 of the RF) depends on recognition of authentic stop codons (cognate) in the decoding center (by domain 2 of the RF) (Freistroffer et al., 2000; Youngman et al., 2007). In the published structure, these two domains are seen to pack against one another in what is referred to as the "closed" conformation (Vestergaard et al., 2001), though we know these factors adopt an "open" conformation in solution (Vestergaard et al., 2005) and on the ribosome.

How then does stop-codon recognition trigger conformational rearrangements that allow for optimal docking of the RF for catalysis? As it turns out, docking of an extended RF depends on substantial rearrangement of a "switch" loop (Laurberg et al., 2008) that connects domains 3 and 4 (Figure 3D). As we recall, during stop-codon recognition, residue 1913 of the 23S rRNA stacks with residue 1493 in the decoding center. This stacking interaction opens a pocket in the ribosome structure that accommodates the newly-rearranged switch loop, thus stabilizing the binding of the RF in its extended conformation.

We have used tethered structural probing (with probes located at three positions on the RF: in the decoding domain, on the switch loop, and in the catalytic center) to visualize conformational rearrangements in the ribosome that are triggered by RF-mediated stop codon recognition (He and Green, 2010). Docking of the RF on an authentic stop-codon complex (as compared to a near-stop complex) yields clear structural signatures: (i) The decoding center becomes largely inaccessible to cleavage with localized hydroxyl radicals, presumably because of solvent exclusion on snug recognition of UAA, (ii) the switch helix more closely approaches subunit interaction site B1, and (iii) the cleavages in the catalytic center are focused on regions near the A loop (instead of on H70 and H89). This analysis thus yields a detailed view of how these two recognition events (cognate and near-cognate) are distinct from one another.

5. Does retrospective editing rely on the same signaling pathways?

In our laboratory, we recently identified a surprising activity on the bacterial ribosome that appears to be important for quality control during translation (Zaher and Green, 2009b). At a biochemical level, we found that when a mismatched codon-anticodon helix is positioned in the P site of the ribosome, the level of selectivity during tRNA or RF selection in the A site is substantially reduced. When such mismatches are found in both the P and the E sites, the effects on fidelity are even more dramatic. The ultimate result of these changes in the specificity of the ribosome is the premature termination of protein synthesis, and a commensurate increase in the fidelity of translation (Figure 4). We propose that these ribosomal activities have evolved to minimize the levels of miscoding and possibly even more importantly, frameshifting, during translation.

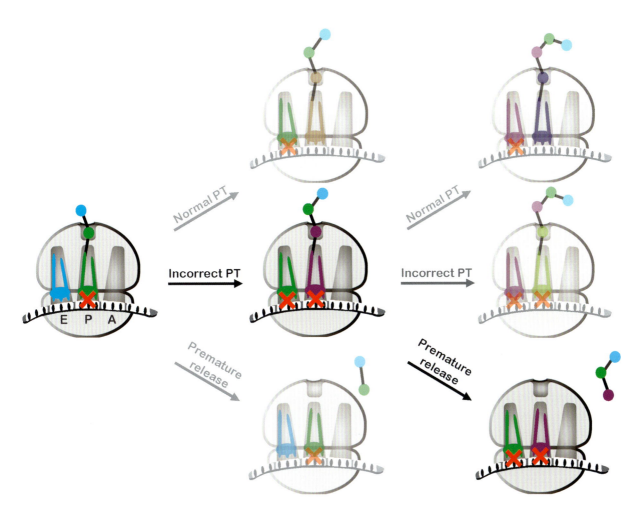

Fig. 4 Proposed model for post peptidyl-transfer quality control. The scheme shows the likely outcome of an initial P-site codon-anticodon mispairing. Following a missense error, the ribosome loses substrate selectivity and becomes more prone to react with non-cognate tRNAs and RFs, with the former being more likely at first. The resulting doubly mismatched complex appears to become even more error-prone, and eventually is recognized by a class-I release factor resulting in premature termination. The dominant path is shown in bold, while the less favorable partitioning pathways are shown as faded cartoons.

There remain many questions to resolve in order to fully understand this process of retrospective editing. To begin with, we would like to understand at a molecular level how subtle perturbations in the P (and possibly E) site decoding helix transmit their effects into the A site. Based on our transient kinetic analysis of this phenomenon, we have learned that the P-site mismatch results in decreases in the off-rates of near-cognate tRNAs from the ribosome (k_{-2} and k_7, according to the Rodnina scheme, (Gromadski and Rodnina, 2004)) and also in increases in the rates of GTPase activation and accommodation (k_3 and k_5, (Gromadski and Rodnina, 2004)) (Zaher and Green, 2010b). These effects closely mimic those of the aminoglycoside paromomycin on the tRNA selection process (Pape et al., 2000). So, in light of these biochemical results, we might argue that the miscoding phenotype of the P-site-perturbed ribosome reports on a conformational state of the ribosome that is somewhat understood. For example, we might predict that A1492 and A1493 are displaced from h44 of the decoding center, stabilizing the "on" state of the ribosome even in the absence of cognate aminoacyl-tRNA. On the other hand, such a conformation of the decoding center would be inhibitory for release factor interactions, and so this would be inconsistent with our data. Perhaps a rather distinct conformation of the A site is stabilized by the P-site disruption, thus explaining the losses in fidelity during both tRNA and RF selection. Such questions will be further explored through a combination of structural and biochemical approaches.

6. Conclusions

Here we have summarized some key findings from the past several years that focus on understanding the processes of tRNA and RF selection by the ribosome. This sophisticated ribonucleoprotein machine is able to select the substrate that corresponds to the tri-nucleotide codon poised in the A site with remarkable fidelity. Not surprisingly, there are similarities in these selection events which happen in common sites, but there are also differences that, at a minimum, reflect the fact that these substrates are physically composed of different building blocks (nucleic acid vs. protein). Understanding the molecular features that allow for this level of discrimination during these events is central to understanding both our evolution and biology.

Our contributions to this field have largely been biochemical, attempting to define the contributions of specific molecular features to the overall process through *in vitro* assays that recapitulate the individual steps. These efforts have been guided along the way by increasingly detailed high-resolution structures of the ribosome, now with multiple substrates and factors bound and trapped at various stages of translation. While our biochemical answers are sometimes narrow in focus, the results are often simpler to interpret than those performed in complex systems. Our goal is to connect what we learn through biochemistry with what has been learned from both genetic and structural approaches, filling in some of the dynamic details of the molecular mechanism that cannot be obtained otherwise. In the end, the molecular switches described throughout this review that define the substrate selection processes on the ribosome are likely central to all ribosome-based events, including its regulation. As we move forward, we hope to apply these same biochemical approaches to increasingly complex questions, revealing the molecular basis of ribosome function.

References

Ali IK, Lancaster L, Feinberg J, Joseph S, Noller HF (2006) Deletion of a conserved, central ribosomal intersubunit RNA bridge. Mol Cell 23: 865–874

Amort M, Wotzel B, Bakowska-Zywicka K, Erlacher MD, Micura R, Polacek N (2007) An intact ribose moiety at A2602 of 23S rRNA is key to trigger peptidyl-tRNA hydrolysis during translation termination. Nucleic Acids Res 35: 5130–5140

Brunelle JL, Shaw JJ, Youngman EM, Green R (2008) Peptide release on the ribosome depends critically on the 2' OH of the peptidyl-tRNA substrate. RNA 14: 1526–1531

Brunelle JL, Youngman EM, Sharma D, Green R (2006) The interaction between C75 of tRNA and the A loop of the ribosome stimulates peptidyl transferase activity. RNA 12: 33–39

Caskey CT, Beaudet AL, Scolnick EM, Rosman M (1971) Hydrolysis of fMet-tRNA by peptidyl transferase. Proc Natl Acad Sci USA 68: 3163–3167

Cochella L, Brunelle JL, Green R (2007) Mutational analysis reveals two independent molecular requirements during transfer RNA selection on the ribosome. Nat Struct Mol Biol 14: 30–36

Cochella L, Green R (2005) An active role for tRNA in decoding beyond codon:anticodon pairing. Science 308: 1178–1180

Erlacher MD, Lang K, Shankaran N, Wotzel B, Huttenhofer A, Micura R, Mankin AS, Polacek N (2005) Chemical engineering of the peptidyl transferase center reveals an important role of the 2'-hydroxyl group of A2451. Nucleic Acids Res 33: 1618–1627

Erlacher MD, Lang K, Wotzel B, Rieder R, Micura R, Polacek N (2006) Efficient ribosomal peptidyl transfer critically relies on the presence of the ribose 2'-OH at A2451 of 23S rRNA. J Am Chem Soc 128: 4453–4459

Freistroffer DV, Kwiatkowski M, Buckingham RH, Ehrenberg M (2000) The accuracy of codon recognition by polypeptide release factors. Proc Natl Acad Sci USA 97: 2046–2051

Gromadski KB, Rodnina MV (2004) Kinetic determinants of high-fidelity tRNA discrimination on the ribosome. Mol Cell 13: 191–200

He SL, Green R (2010) Visualization of codon-dependent conformational rearrangements during translation termination. Nat Struct Mol Biol 17: 465–470

Hopfield JJ (1974) Kinetic proofreading: a new mechanism for reducing errors in biosynthetic processes requiring high specificity. Proc Natl Acad Sci USA 71: 4135–4139

Ito K, Uno M, Nakamura Y (2000) A tripeptide 'anticodon' deciphers stop codons in messenger RNA. Nature 403: 680–684

Jenner L, Demeshkina N, Yusupova G, Yusupov M (2010) Structural rearrangements of the ribosome at the tRNA proofreading step. Nat Struct Mol Biol (Epub ahead of print)

Jorgensen F, Adamski FM, Tate WP, Kurland CG (1993) Release factor-dependent false stops are infrequent in *Escherichia coli*. J Mol Biol 230: 41–50

Koch M, Huang Y, Sprinzl M (2008) Peptide-bond synthesis on the ribosome: no free vicinal hydroxy group required on the terminal ribose residue of peptidyl-tRNA. Angew Chem Int ed 47: 7242–7245

Korostelev A, Asahara H, Lancaster L, Laurberg M, Hirschi A, Zhu J, Trakhanov S, Scott WG, Noller HF (2008) Crystal structure of a translation termination complex formed with release factor RF2. Proc Natl Acad Sci USA 105: 19684–19689

Laurberg M, Asahara H, Korostelev A, Zhu J, Trakhanov S, Noller HF (2008) Structural basis for translation termination on the 70S ribosome. Nature 454: 852–857

Ledoux S, Olejniczak M, Uhlenbeck OC (2009) A sequence element that tunes *Escherichia coli* tRNA(Ala)(GGC) to ensure accurate decoding. Nat Struct Mol Biol 16: 359–364

Li W, Agirrezabala X, Lei J, Bouakaz L, Brunelle JL, Ortiz-Meoz RF, Green R, Sanyal S, Ehrenberg M, Frank J (2008) Recognition of aminoacyl-tRNA: a common molecular mechanism revealed by cryo-EM. EMBO J 27: 3322–3331

Mora L, Zavialov A, Ehrenberg M, Buckingham RH (2003) Stop codon recognition and interactions with peptide release factor RF3 of truncated and chimeric RF1 and RF2from *Escherichia coli*. Mol Microbiol 50: 1467–1476

Murphy FV 4th, Ramakrishnan V (2004) Structure of a purine-purine wobble base pair in the decoding center of the ribosome. Nat Struct Mol Biol 11: 1251–1252

O'Connor M (2009) Helix 69 in 23S rRNA modulates decoding by wild type and suppressor tRNAs. Mol Genet Genomics 282: 371–380

O'Connor M, Dahlberg AE (1995) The involvement of two distinct regions of 23 S ribosomal RNA in tRNA selection. J Mol Biol 254: 838–847

Ogle JM, Brodersen DE, Clemons WM, Jr., Tarry MJ, Carter AP, Ramakrishnan V (2001) Recognition of cognate transfer RNA by the 30S ribosomal subunit. Science 292: 897–902

Ogle JM, Murphy FV, Tarry MJ, Ramakrishnan V (2002) Selection of tRNA by the ribosome requires a transition from an open to a closed form. Cell 111: 721–732

Ortiz-Meoz R, Green R (2010) Functional elucidation of a key contact between tRNA and the large ribosomal subunit rRNA during decoding. RNA 16: 2002–2013

Pan D, Zhang CM, Kirillov S, Hou YM, Cooperman BS (2008) Perturbation of the tRNA tertiary core differentially affects specific steps of the elongation cycle. J Biol Chem 283: 18 431–18440

Pape T, Wintermeyer W, Rodnina M (1999) Induced fit in initial selection and proofreading of aminoacyl-tRNA on the ribosome. EMBO J 18: 3800–3807

Pape T, Wintermeyer W, Rodnina MV (1998) Complete kinetic mechanism of elongation factor Tu-dependent binding of aminoacyl-tRNA to the A site of the *E. coli* ribosome. EMBO J 17: 7490–7497

Pape T, Wintermeyer W, Rodnina MV (2000) Conformational switch in the decoding region of 16S rRNA during aminoacyl-tRNA selection on the ribosome. Nat Struct Biol 7: 104–107

Piepenburg O, Pape T, Pleiss JA, Wintermeyer W, Uhlenbeck OC, Rodnina MV (2000) Intact aminoacyl-tRNA is required to trigger GTP hydrolysis by elongation factor Tu on the ribosome. Biochemistry 39: 1734–1738

Polacek N, Gomez MJ, Ito K, Xiong L, Nakamura Y, Mankin A (2003) The critical role of the universally conserved A2602 of 23S ribosomal RNA in the release of the nascent peptide during translation termination. Mol Cell 11: 103–112

Saarma U, Remme J, Ehrenberg M, Bilgin N (1997) An A to U transversion at position 1067 of 23 S rRNA from *Escherichia coli* impairs EF-Tu and EF-G function. J Mol Biol 272: 327–335

Schmeing TM, Huang KS, Kitchen DE, Strobel SA, Steitz TA (2005a) Structural insights into the roles of water and the 2' hydroxyl of the P site tRNA in the peptidyl transferase reaction. Mol Cell 20: 437–448

Schmeing TM, Huang KS, Strobel SA, Steitz TA (2005b) An induced-fit mechanism to promote peptide bond formation and exclude hydrolysis of peptidyl-tRNA. Nature 438: 520–524

Schmeing TM, Ramakrishnan V (2009) What recent ribosome structures have revealed about the mechanism of translation. Nature 461: 1234–1242

Schmeing TM, Voorhees RM, Kelley AC, Gao YG, Murphy FV 4th, Weir JR, Ramakrishnan V (2009) The crystal structure of the ribosome bound to EF-Tu and aminoacyl-tRNA. Science 326: 688–694

Seit-Nebi A, Frolova L, Justesen J, Kisselev L (2001) Class-1 translation termination factors: invariant GGQ minidomain is essential for release activity and ribosome binding but not for stop codon recognition. Nucleic Acids Res 29: 3982–3987

Shaw JJ, Green R (2007) Two distinct components of release factor function uncovered by nucleophile partitioning analysis. Mol Cell 28: 458–467

Stark H, Rodnina MV, Wieden HJ, Zemlin F, Wintermeyer W, van Heel M (2002) Ribosome interactions of aminoacyl-tRNA and elongation factor Tu in the codon-recognition complex. Nat Struct Biol 9: 849–854

Thompson RC, Stone PJ (1977) Proofreading of the codon-anticodon interaction on ribosomes. Proc Natl Acad Sci USA 74: 198–202

Valle M, Sengupta J, Swami NK, Grassucci RA, Burkhardt N, Nierhaus KH, Agrawal RK, Frank J (2002) Cryo-EM reveals an active role for aminoacyl-tRNA in the accommodation process. EMBO J 21: 3557–3567

Vestergaard B, Sanyal S, Roessle M, Mora L, Buckingham RH, Kastrup JS, Gajhede M, Svergun DI, Ehrenberg M (2005) The SAXS solution structure of RF1 differs from its crystal structure and is similar to its ribosome bound cryo-EM structure. Mol Cell 20: 929–938

Vestergaard B, Van LB, Andersen GR, Nyborg J, Buckingham RH, Kjeldgaard M (2001) Bacterial polypeptide release factor RF2 is structurally distinct from eukaryotic eRF1. Mol Cell8: 1375–1382

Weinger JS, Parnell KM, Dorner S, Green R, Strobel SA (2004) Substrate-assisted catalysis of peptide bond formation by the ribosome. Nat Struct Mol Biol 11: 1101–1106

Weixlbaumer A, Jin H, Neubauer C, Voorhees RM, Petry S, Kelley AC, Ramakrishnan V (2008) Insights into translational termination from the structure of RF2 bound to the ribosome. Science 322: 953–956

Youngman EM, Brunelle JL, Kochaniak AB, Green R (2004) The active site of the ribosome is composed of two layers of conserved nucleotides with distinct roles in peptide bond formation and peptide release. Cell 117: 589–599

Youngman EM, He SL, Nikstad LJ, Green R (2007) Stop codon recognition by release factors induces structural rearrangement of the ribosomal decoding center that is productive for peptide release. Mol Cell 28: 533–543

Zaher HS, Green R (2010a) Hyperaccurate and error-prone ribosomes exploit distinct mechanisms during tRNA selection. Mol Cell 39: 110–120

Zaher H, Green R (2010b) Kinetic basis for global loss of fidelity arising from mismatches in the P-site codon:anticodon, RNA 1980–1989

Zaher HS, Green R (2009a) Fidelity at the molecular level: lessons from protein synthesis. Cell 136: 746–762

Zaher HS, Green R (2009b) Quality control by the ribosome following peptide bond formation. Nature 457: 161–166

Zavialov AV, Mora L, Buckingham RH, Ehrenberg M (2002) Release of peptide promoted by the GGQ motif of class-I release factors regulates the GTPase activity of RF3. Mol Cell 10: 789–798

Rate and accuracy of messenger RNA translation on the ribosome

18

Magnus Johansson, Ka Weng Ieong, Johan Åqvist, Michael Y. Pavlov and Måns Ehrenberg

1. Selection pressure on size, rate and accuracy of ribosomes in growing bacteria

In the bacterial cell there is an ever evolving network of metabolic pathways, with well-defined flow-stoichiometries for maximal yield of biomass in different growth contexts (Ibarra et al., 2002; Feist et al., 2009). In line with this, any change in size or kinetic efficiency of an enzyme or macromolecular complex like the ribosome will affect the growth rate of the cell (Ehrenberg and Kurland, 1984; Kurland et al., 2003). When the intracellular control systems maintain optimal flow couplings for maximal growth rate, there is a simple relation between change in size or kinetic efficiency of an enzyme system and the bacterial growth rate. This can be used to assess the fitness loss or gain of mutations in enzyme systems and thus the probability of fixation of gain of function mutations (Kurland et al., 2003). With μ_0 defined as the wild type (exponential) growth rate, μ the altered growth rate in a mutant, $\delta\mu = \mu - \mu_0$ and s the fitness parameter, we have $\mu = \mu_0(1+s)$, where $s = \delta\mu/\mu_0$. The relative change, $\delta X/X$, in kinetic efficiency ($X = k$) or size ($X = N$) of an enzyme system will induce a relative change, $\delta\mu/\mu$, in the growth rate as determined by the logarithmic gain, $a_{\mu X}$:

$$s = \frac{\delta\mu}{\mu_0} = a_{\mu X}\frac{\delta X}{X} . \qquad (1)$$

The logarithmic gain $a_{\mu X}$ is given by (M. Ehrenberg, unpublished):

$$a_{\mu k} = +\frac{\rho_P}{\rho_0}; \; a_{\mu N} = -\frac{\rho_P}{\rho_0} . \qquad (2)$$

Here, ρ_P is the intracellular concentration of amino acid residues invested in the enzyme system and ρ_0 is the bulk concentration of all amino acid residues in the cell. The bacterial ribosome contains almost 10 000 amino acids and is present at high concentration in the bacterial cell (Bremer and Dennis, 2008), making the mass fraction ρ_{Ribo}/ρ_0 very large compared to the mass fractions of virtually any other enzymatic system. Accordingly, there is an exceptionally large selection pressure on the bacterial ribosome for rapid peptide elongation and small size.

The selection pressure for high accuracy in messenger RNA translation can, in principle, be assessed by taking into account the negative impact of amino acid substitution errors in proteins, the trade-off between accuracy and kinetic efficiency (Ehrenberg and Kurland, 1984; Kurland and Ehrenberg, 1984) as well as the inhibitory action of near- or non-cognate ternary complexes on protein elongation (Johansson et al., 2008b). The bacterial ribosome maintains high accuracy using two consecutive steps for selection of cognate aminoacyl-tRNA: initial selection (I) of ternary complex before hydrolysis of GTP on EF-Tu and proofreading selection (F) of aminoacyl-tRNA after GTP hydrolysis (Thompson and Stone, 1977; Ruusala et al., 1982). The (normalized) overall accuracy (A) is given by the ratio between the kinetic efficiency of cognate (k^c_{cat}/K^c_m) and near-cognate (k^{nc}_{cat}/K^{nc}_m) amino acid incorporation in a nascent peptide chain. These efficiency parameters are in general defined as (Fersht, 1999):

$$\frac{k^c_{cat}}{K^c_m} = k^c_l P^c_{prod} = k^c_l P^c_l \cdot P^c_F;$$

$$\frac{k^{nc}_{cat}}{K^{nc}_m} = k^{nc}_l P^{nc}_{prod} = k^{nc}_l P^{nc}_l \cdot P^{nc}_F , \qquad (3)$$

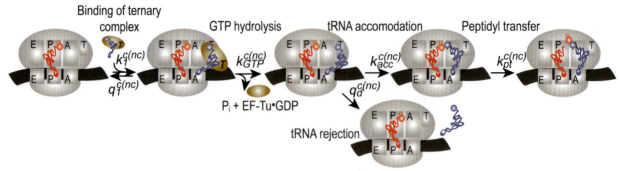

Scheme 1

so that:

$$A = \left(\frac{k_{cat}^c}{K_m^c}\right) / \left(\frac{k_{cat}^{nc}}{K_m^{nc}}\right) = \frac{k_I^c P_I^c}{k_I^{nc} P_I^{nc}} \cdot \frac{P_F^c}{P_F^{nc}} = I \cdot F \quad (4)$$

Here, k_I is the rate constant for ternary complex association to the ribosomal A site, P_I is the probability that binding of ternary complex leads to hydrolysis of GTP on EF-Tu, P_F is the probability that the reaction proceeds to peptidyl transfer after GTP hydrolysis on EF-Tu. Essential features of aa-tRNA selection by the mRNA programmed ribosome can be illustrated by the (simplistic) Scheme 1:

Ternary complex (aa-tRNA·EF-Tu·GTP) enters the ribosomal A site with rate constant k_I, from which it dissociates with rate constant q_I, or is subjected to GTP hydrolysis with rate constant k_{GTP}. After GTP hydrolysis on EF-Tu, aa-tRNA dissociates from the ribosome with rate constant k_q or is accommodated in the A site with rate constant k_{acc}. In the special case of Scheme 1, the probabilities in Eq. (3) become: $P_I^c = k_{GTP}^c / (k_{GTP}^c + q_I^c)$; $P_F^c = k_{acc}^c / (k_{acc}^c + q_d^c)$; $P_I^{nc} = k_{GTP}^{nc} / (k_{GTP}^{nc} + q_I^{nc})$ and $P_F^{nc} = k_{acc}^{nc} / (k_{acc}^{nc} + q_d^{nc})$, leading to the following expressions for the initial *(I)* and proofreading *(F)* selection:

$$I = \frac{k_I^c}{k_I^{nc}} \frac{1 + (q_I^{nc} / k_{GTP}^{nc})}{1 + (q_I^c / k_{GTP}^c)} = \frac{d_{II} + d_I \cdot a_I}{1 + a_I};$$

$$F = \frac{1 + (q_d^{nc} / k_{acc}^{nc})}{1 + (q_d^c / k_{acc}^c)} = \frac{1 + d_F \cdot a_F}{1 + a_F};$$

$$d_I = \frac{k_I^c q_I^{nc} k_{GTP}^c}{k_I^{nc} k_{GTP}^{nc} q_I^c} = d_{II} \cdot \frac{q_I^{nc} k_{GTP}^c}{q_I^c k_{GTP}^{nc}} = e^{-\Delta\Delta G_I^0 / RT};$$

$$d_F = \frac{q_d^{nc} / k_{acc}^{nc}}{q_d^c / k_{acc}^c} = e^{-\Delta\Delta G_F^0 / RT}. \qquad (5)$$

Here, I is determined by three parameters: (i) the discrimination parameter $d_{II} = k_I^c / k_I^{nc}$ given by the ratio

between the rate constants for association of cognate and near-cognate ternary complex to the A site; (ii) the discard parameter $a_I = q_I^c / k_{GTP}^c$, which determines the kinetic efficiency k_{cat}/K_m of the cognate reaction and (iii) the discrimination parameter d_I. The proofreading factor F is determined by the discrimination parameter d_F and the discard parameter $a_F = q_d^c / k_{acc}^c$ of the cognate aa-tRNA.

The fundamental parameter d_I is determined by the standard free energy difference between the near-cognate and cognate transition states for GTPase activation (Fersht, 1999), while d_F is determined by the free energy difference between near-cognate and cognate transition states for accommodation into the A site of the ribosome. Notably, I and F approach their maximal values d_I and d_F when the discard parameters a_I and a_F for the cognate reaction become very large. We note that the kinetic efficiency of the ribosome (k_{cat}/K_m) is essentially unaffected by steps following peptidyl transfer, which motivates the use of the simplified Scheme 1 in our discussion of ribosomal accuracy and kinetics.

The kinetic efficiency of cognate peptide bond formation (k_{cat}^c/K_m^c) is related to the accuracy of initial selection (I) and proofreading (F) through:

$$\frac{k_{cat}^c}{K_m^c} = k_I^c \frac{d_I - I}{d_I - d_{II}} \cdot \frac{d_F - F}{d_F - 1} \quad . \qquad (6)$$

When the rate constants for association of cognate and near cognate ternary complex to the A site are similar, (Gromadski and Rodnina, 2004), $d_{II} \approx 1$ and Eq. 6 takes the simple form used in (Johansson et al., 2008b). Relation 6 implies that the ribosome has its highest kinetic efficiency ($k_{cat}/K_m = k_I$) when there is little or no ($I = d_{II}$; $F = 1$; $A = d_{II}$) discrimination against near-cognate substrates. When the ribosome has its highest possible accuracy ($I = d_I$; $F = d_F$; $A = d_I \cdot d_F$), its kinetic efficiency is zero ($k_{cat}^c / K_m^c = 0$).

The average rate, v, of cognate peptide elongation in the living cell may be hampered not only by reduced kinetic efficiency at high accuracy as predicted by Eq. (6), but also by the rate-inhibitory action of near- and non-cognate tRNAs. Such inhibition is in general described by an expression of the type:

$$v = \frac{k_{cat}^c}{1 + \frac{K_m^c}{[T_3^c]} \cdot \left(1 + [T_3^{nc}]/K_m^{nc} + [T_3^{nonc}]/K_m^{nonc}\right)} \quad . \quad (7)$$

Here, $[T_3^c]$, $[T_3^{nc}]$ and $[T_3^{nonc}]$ are the *in vivo* concentrations of cognate, near- and non-cognate ternary complexes, respectively. Their corresponding K_m-values are K_m^c, K_m^{nc} and K_m^{nonc}, respectively. For simplicity, cognate, near-cognate and non-cognate ternary complexes have been contracted to three homogeneous groups. Importantly, Eq. (7) predicts that the extent of inhibition of the cognate reaction by near- and non-cognate ternary complexes will decrease with the decreasing ratios of cognate to near- or non-cognate K_m-values. Strong affinities of ternary complexes to the ribosomal A site are associated with low accuracy and, according to Eq. (7), with a strong inhibition of protein elongation by near- and non-cognate ternary complexes. Therefore, the rate of protein synthesis may first increase with increasing accuracy of tRNA selection due to reduced inhibition of the cognate reaction until the kinetic efficiency (k_{cat}^c/K_m^c) decreases according to the general rate-accuracy law in Eq. (6) (Johansson et al., 2008b), as will be further discussed below.

From these considerations we suggest that the bacterial cell has evolved to a physiological state where protein synthesis is unhampered by the inhibitory action of near- or non-cognate tRNAs, the error frequency of amino acid substitutions in nascent peptide chains is so small that synthesis of erroneous proteins does not significantly reduce the growth rate and the single step accuracies I and F are well below their limits d_I and d_F. In this parameter range with $d_{II} = 1$, Eq. (6) is approximated by:

$$\frac{k_{cat}^c}{K_m^c} = k_1^c (1 - \frac{I}{d_I} - \frac{F}{d_F}) \quad . \quad (8)$$

Reconstituted systems for mRNA translation in the test tube are not only very sensitive to the concentrations of the tRNA substrates and auxiliary protein translation factors, but also to buffer conditions and energy supply. Most such systems operate with peptide elongation rates and accuracy orders of magnitude smaller than those in the living cell. But does it really matter if the rate and accuracy of the biochemistry of protein synthesis greatly deviate from the evolutionarily optimized rate and accuracy of protein synthesis in the living cell? We suggest that it, indeed, does and that such deviations may be preventive for progress in ribosome research in a number of fundamental ways concerning (i) systems biology modeling of bacterial cells with input data from biochemistry; (ii) connecting population genetics and molecular evolution of bacteria in a quantitative bacterial physiology; (iii) determining genuine rate constants for the different steps of mRNA translation that would reveal the (authentic) physical constraints under which evolution works to attain the (here postulated) optimal kinetic performance of ribosomes in the living cell; (iv) understanding how essential features of ribosome structure have evolved under high selection pressure to minimize the deleterious effects of physical constraints on ribosome kinetics. In what follows, we will first describe early attempts to design an optimal system for protein synthesis in the test tube with *in vivo* like properties. Then we will consider the accuracy of tRNA selection by the mRNA programmed ribosome in the test tube and in the living cell and highlight the importance of calibrating biochemical with *in vivo* data. In a next section we discuss the different steps that lead to peptidyl transfer and, in particular, discuss current proposals of rate limiting steps and how they might be rationalized in terms of (authentic) physical constraints.

2. Experimental results

2.1. Early experiments on rate and accuracy

In order to increase the rate and accuracy of protein synthesis in the test tube and bring it *au pair* with the rate and accuracy in the living cell, Jelenc et al. (Jelenc and Kurland, 1979) designed a buffer system with "energy regeneration" for high speed and high accuracy. This "polymix" phosphate buffer contains, in addition to magnesium and calcium ions, the naturally occurring polyamines spermidine and putrescine along with ATP and GTP. It also contains phosphoenol pyruvate (PEP) and pyruvate kinase (PK) to regenerate ATP and GTP from ADP and GDP, respectively. It normally contains also myokinase (MK) to transform AMP, produced from ATP in aminoacylation of tRNAs, to ADP. The ionic composition of polymix was optimized for rapid poly(Phe) synthesis and high accuracy of Phe-tRNA^{Phe} selection on poly(U) programmed ribosomes in the presence of near cognate tRNAs, like Leu-tRNA$_2^{Leu}$ (tRNA$_{GAG}^{Leu}$) and Leu-tRNA$_4^{Leu}$ (tRNA$_{cmnmUmAA}^{Leu5}$). By synchronized initiation of poly(U) translation with NAc-[³H]Phe-tRNA^{Phe}, this system elongates poly(Phe)-chains at a rate of about 10 amino acids per second per ribosome and a missense error frequency around 10^{-4}, i. e. it has *in vivo* like properties (Wagner et al., 1982) with respect to the accuracy (Edelmann and Gallant, 1977; Stahl et al., 2004; Kramer and Farabaugh, 2007) and rate (Young and Bremer, 1976) of protein synthesis in the bacterial cell. Single turnover experiments with quench-flow techniques at 37 °C revealed a GTP hydrolysis time in ribosome-bound ternary complex of 10 ms and a 20 ms time for all steps following GTP hydrolysis leading to peptide bond formation (Bilgin et al., 1992). The system was used to demonstrate the existence of proofreading (Hopfield, 1974; Ninio, 1975) in a translocation competent poly(U) translation system with elongation factor G (EF-G) (Ruusala et al., 1982), generalizing the earlier finding of proofreading in a partial system lacking EF-G (Thompson and Stone, 1977). We note that misreading of UUU by Leu-tRNA$_2^{Leu}$ (cognate codons CUU, CUC) occurred with a normalized accuracy, $A = I \cdot F$, of 10^4 partitioned equally between initial selection, $I = 100$, and proofreading selection, $F = 100$ (Ruusala et al., 1982). The same system was also used to estimate the K_m-values for near-cognate reading of UUU codons by Leu-tRNA$_2^{Leu}$ and Leu-tRNA$_4^{Leu}$ as $>10^{-4}$ M and $>2 \cdot 10^{-5}$ M, respectively (Bilgin et al., 1988). It was also shown that a non-cognate ternary complex containing Val-tRNA^{Val} at concentrations up to $5 \cdot 10^{-4}$ M had no inhibitory effect on cognate UUU reading by Phe-tRNA^{Phe} or on UUU misreading by Leu-tRNA$_2^{Leu}$ or Leu-tRNA$_4^{Leu}$. At the same time, the K_m-value for the cognate reaction was estimated as $6.5 \cdot 10^{-7}$ M (Ruusala et al., 1984). Assuming these K_m-values to be relevant for protein elongation in the living *E. coli* cell, it follows from Eq. (7) above that there is virtually no inhibition of protein synthesis *in vivo* by negative interference from near- and non-cognate ternary complexes (Bilgin et al., 1988). This indicated to us that the optimized system for poly(U) translation was compatible with protein synthesis in the living cell (Bremer and Dennis, 2008), since it fulfilled the efficiency criteria expected from a system subjected to high selection pressure for rapid and accurate peptide bond formation.

2.2. More recent experiments on the accuracy of aminoacyl-tRNA selection

In 2004, a detailed scheme for the determinants of accuracy of aminoacyl-tRNA selection for peptidyl transfer was presented (Gromadski and Rodnina, 2004). 70S initiation complexes (ICs), containing a short open reading frame mRNA, had fMet-tRNA^{fMet} in the AUG programmed P site and the A site was programmed either with a UUU (Phe) or CUC (Leu) codon. Fast kinetics experiments with stopped-flow or quench-flow were used to derive rate constants for the different steps leading from the free ternary complex Phe-tRNA^{Phe}•EF-Tu•GTP reacting with a UUU (cognate reaction) or CUC (near cognate reaction) codons in the A site of 70S ICs all the way to peptide bond formation. These experiments were conducted in a polyamine containing buffer (Gromadski and Rodnina, 2004) that differed considerably from our polymix buffer (Jelenc and Kurland, 1979) in that it did not contain an *in situ* PEP-dependent energy regeneration system. An important result obtained in this study was the observation of a very large difference in the rate of GTPase activation between a cognate (k_{GTP}=260 s^{-1}) and a near cognate (k_{GTP}=0.4 s^{-1}) ternary complex (Gromadski and Rodnina, 2004). This observation validated a previous conclusion from the same group based on a model system for poly(U) translation that correct codon–anticodon interaction in the A site induces a conformational change in the ribosome / EF-Tu ("induced-fit" mechanism) that promotes GTP hydrolysis (Pape et al., 1999). The overall accuracy *(A)* measured by Gromadski et al was 940, partitioned in an initial selection *(I)* of 63 and a proofreading selection *(F)* of 16 (Gromadski and Rodnina, 2004). From the detailed kinetic scheme presented in that study (Gromadski and Rodnina, 2004) one can for the first time estimate the value of the intrinsic discrimination parameter d_I in initial selection of a cognate ternary complex. The estimated d_I-value of 226 000 is unexpectedly high in comparison to the very modest current initial selection (I) of 63 (Johansson et al., 2008b). The intrinsic accuracy of proofreading (d_F) is unknown, but a qualified guess, based on structural arguments (Ogle and Ramakrishnan, 2005), is that $d^F = d^I = d$. With $I = 63$, $F = 16$ and $d = 226 000$, the efficiency loss term $(I+F)/d$ in Eq. 8 above is $\sim 2 \cdot 10^{-4}$. If I and F were increased to 1000 each, so that $A = 10^6$, as in (Johansson et al., 2008a), the loss term would be about 1 % and thus very small, showing that with such a large d_I the accuracy could be adjusted to much higher values with almost no loss in the kinetic efficiency of the cognate reaction. It is, however, possible that the intrinsic selectivity d_l in (Gromadski and Rodnina, 2004) was overestimated. Its determination depends on the assumption that ternary complex with a GTPase deficient mutant of EF-Tu or with EF-Tu in complex with a non-cleavable GTP analogue dissociates from the ribosome with approximately the same rate as a native, GTP-containing ternary complex before GTPase activation. Further experiments will be required to settle this issue.

From the rate constants of the detailed kinetic scheme in (Gromadski and Rodnina, 2004) and assuming that peptide bond formation is rate limiting for protein elongation, K_m-values for cognate, near-cognate and non-cognate ternary complex interaction follow as, 0.07, 0.18 and 0.6 µM, respectively. Assuming these K_m-values to be relevant in the bacterial cell and assuming the *in vivo* concentrations of cognate, near cognate and non cognate ternary complexes to be 2, 15 and 83 µM, respectively (Gromadski et al., 2006), it follows from Eq. (7) that peptide elongation rate is severely reduced by near and, in particular, non-cognate ternary complex (Gromadski et al., 2006; Johansson et al., 2008 b). As described in Section 1 above, there would be a large selection pressure against this type of protein synthesis design in a population of growing bacteria (Johansson et al., 2008 b).

Our polymix-based system for cell free protein synthesis combines rapid peptidyl transfer (Ehrenberg et al., 1990; Bilgin et al., 1992; Johansson et al., 2008 a) with high accuracy of tRNA selection (Johansson et al., 2008 a) and small or negligible inhibition of the cognate protein synthesis by near- or non-cognate ternary complexes (Bilgin et al., 1988; Johansson et al., 2008 a). Since all components of the polymix buffer are present in E. coli (Jelenc and Kurland, 1979) it is, we propose, very likely that similar or even more advantageous ionic conditions have evolved in bacteria. A key component in the tuning of ribosomal accuracy is the free Mg^{2+} concentration. When $[Mg^{2+}]$ increases, the accuracy of ternary complex selection decreases drastically (Jelenc and Kurland, 1979; Pape et al., 1999; Gromadski et al., 2006). The reason is that Mg^{2+} ions stabilize the binding of all ternary complexes, cognate and near-cognate alike, to the A site without correspondingly reducing the rate constant for GTPase activation (Gromadski et al., 2006). In the polymix buffer the *free* concentration of Mg^{2+} is reduced from the total concentration of 5 mM by the presence of strongly chelating ATP (1 mM) and GTP (1 mM) molecules to about 3 mM. It is further reduced by the presence of

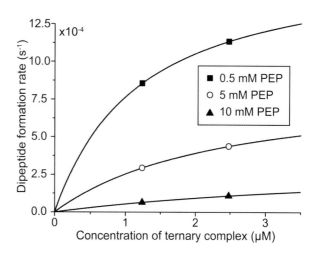

Fig. 1 Effect of PEP concentration on the kinetic efficiency of near-cognate codon misreading. The rate of the misreading of near cognate CUU by Phe-tRNA^{Phe} was measured at 37 °C at two different concentrations of ternary complex for three different concentrations of PEP in the polymix buffer indicated in the figure. The kinetic efficiency parameter, k_{cat}/K_m was estimated by fitting each curve to a Michaelis-Menten equation. The rates were measured from the time course of the misreading reaction upon mixing 70S initiation complexes mRNA-programmed with a CUU codon in the A site with ternary complex EF-Tu•GTP•Phe-tRNA^{Phe} at concentrations indicated in the figure. The 70S initiation complexes (0.5 µM final concentration) were prepared in polymix buffer containing 1 mM ATP, 1 mM GTP, initiation factors 1–3, MK (myokinase), PK (pyruvate kinase) and the indicated concentrations of PEP. Ternary complexes (the final concentration indicated in the figure), EF-Tu•GTP•Phe-tRNA^{Phe}, were prepared separately also in polymix buffer containing 1 mM ATP, 1 mM GTP, EF-Ts, PheRS, Phe, MK, PK and the indicated concentrations of PEP.

phosphoenol pyruvate (PEP), a weak chelator of Mg^{2+} ions, that serves as energy supply to regenerate GTP and ATP in the polymix buffer (Jelenc and Kurland, 1979). From preliminary experiments (Figure 1) it follows that the k_{cat}/K_m parameter for the reading of near-cognate codons by ternary complexes decreased roughly 30-fold when the PEP concentration increased from 0.5 to 10 mM. Using $K_d \approx 6$ mM for Mg^{2+} chelation by PEP (Wold and Ballou, 1957), we estimate that the free Mg^{2+} concentration decreased from 2.8 to 1.1 mM as the PEP concentration increased from its lowest to its highest value in this interval. As expected, the k_{cat}/K_m parameter for the reading of cognate codons by ternary complexes is much less sensitive to Mg^{2+} concentration (data not shown), meaning that the 30-fold reduction in k_{cat}/K_m for near-cognate codon reading would approximate the reduction in codon reading accuracy associated with reduction of the free Mg^{2+} concentration from about 2.8 to 1.1 mM. We note also that the 70S ribosomes retained their full activity at the

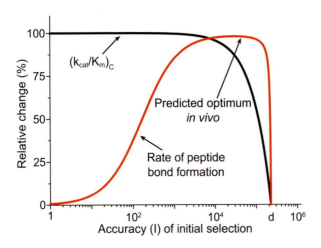

Fig. 2 The rate-accuracy trade-off. The black line shows how the kinetic efficiency (k_{cat}/K_m) of the ribosome in cognate peptide bond formation depends on the accuracy I of initial selection. The proofreading factor F is assumed here to be constant. The red line shows how the rate of peptide elongation with cognate substrates varies with accuracy I. At low accuracy, the ribosomes are inhibited by strong binding of near-and non-cognate ternary complexes to the A site, making the overall rate of peptide elongation very slow. For details of the calculations see (Johansson et al., 2008b).

largest PEP concentration of 10 mM during an incubation time of 60 min. Ribosome activity was checked by separate pre-incubation of 0.5 µM 70S initiation complex (IC) and 2 µM cognate ternary complex (T3) at 37 °C during different times from 10 to 60 min. The extent of dipeptide formation did not depend on the pre-incubation time (data not shown), demonstrating fully retained ribosome activity during 60 min, which was the longest incubation time for near-cognate dipeptide formation in the experiment in Figure 1. This is an important control, since Mg^{2+} ions are important also for the stability of the ribosome, and one would therefore expect a limit to how much the free Mg^{2+} concentration can be reduced without ribosome inactivation (Weiss et al., 1973; Weiss and Morris, 1973). An important role of the polyamines in the polymix buffer (Jelenc and Kurland, 1979) is, we propose, to stabilize the 70S ribosome complex at the low free Mg^{2+} concentrations necessary for high accuracy and low ribosome inhibition by near- and non-cognate ternary complexes.

A conservative estimate of the concentration range of free Mg^{2+} ions in the *E. coli* cell is 1 to 4 mM, but it is likely that the concentration is confined to the 1 to 2 mM range (Alatossava et al., 1985). Our preliminary results in Figure 1 demonstrate large variation in the accuracy of tRNA selection by the mRNA programmed ribosome in the 1 to 3 mM range of free Mg^{2+}. We sug-

gest that the accuracy in the living cell has been optimized for maximal growth rate, and careful calibration of rate and accuracy of our polymix buffer system to the accuracy and rate, of peptide elongation in the living cell is a project of high priority in the laboratory. This implies comparison of well defined data sets for the accuracy at different codons *in vitro* and *in vivo*. It also implies that the whole elongation cycle, including translocation, must be accounted for *in vitro*, as in our early poly(U)-based translation system (Wagner et al., 1982).

3. The path to ribosomal peptidyl transfer

3.1. Temperature dependence

The temperature dependence of the rate of ribosome catalyzed peptidyl transfer from initiator tRNA (fMet-tRNA^{fMet}) to the aminoacyl-tRNA model-compound puromycin was studied (Sievers et al., 2004) and compared to that of the non-catalyzed reaction (Sievers et al., 2004; Schroeder and Wolfenden, 2007). Interestingly, ribosome catalyzed peptide bond formation has higher activation enthalpy and much smaller activation entropy than the non-catalyzed reaction. This means that unlike most enzymatic reactions, ribosome catalyzed peptide bond formation is entropy driven (Sievers et al., 2004). It had been suggested that ribosomal catalysis of peptide bond formation was simply based on precise juxtaposition of the P-site peptidyl-tRNA and the A-site aminoacyl-tRNA in the peptidyl transfer center (Hansen et al., 2002), seemingly in line with the observed entropy driven catalysis of this reaction in a model system with puromycin (Sievers et al., 2004). As an alternative scenario, it was also suggested that the reduction in activation entropy could be caused by the ordering of water molecules in the peptidyl transfer center (PTC) (Sievers et al., 2004). Soon after, Åqvist and collaborators used molecular dynamics (MD) simulation methods, based on crystal structures of the 50S subunit from *Haloarcula marismortui* (Nissen et al., 2000; Hansen et al., 2002), to identify the transition state for ribosome-catalyzed peptide bond formation and estimate the activation enthalpy and entropy (Trobro and Åqvist, 2005; Trobro and Åqvist, 2006). They found high activation enthalpy and very small activation entropy for ribosome-catalyzed peptidyl transfer, in line with the experimental observations (Sievers et al., 2004). Trobro and Åqvist identified an ordered network of H-bonds involving water molecules and ribosomal RNA bases, which forms in the ground state and greatly reduces its entropy. The very same network remains along the reaction path giving the transition state the same small entropy as the ground state. Since the activation entropy is the difference in entropy between ground state and transition state, this explains the low activation entropy of ribosome catalyzed peptide bond formation. Trobro and Åqvist also concluded that ordering by juxtaposition of the reaction substrates cannot account for the experimentally observed entropy driven peptide bond formation on the ribosome (Trobro and Åqvist, 2005; Trobro and Åqvist, 2006).

The reason why the temperature dependence of the rate of ribosomal peptidyl transfer had been carried out with puromycin (Sievers et al., 2004), rather than with a native aa-tRNA, in the A site is that peptide bond formation had been considered to be kinetically masked by a preceding and rate limiting step, identified as the accommodation of aa-tRNA in the A site (Rodnina et al., 1994; Pape et al., 1998). We noted, however, that reported tRNA accommodation times at 37 °C measured using fluorescence-labeled tRNAs (Hesslein et al., 2004; Beringer et al., 2005; Bieling et al., 2006) are much longer than the average time for peptidyl transfer estimated in the optimized poly(U) system (Bilgin et al., 1992) or in our optimized system for translation of heteropolymeric mRNAs (Johansson et al., 2008a; Pavlov et al., 2009). That is, the overall times for peptidyl transfer at 37 °C after mixing of ternary complex with ribosomes containing a P-site peptidyl-tRNA varied between 120 and 500 ms (Hesslein et al., 2004; Beringer et al., 2005; Bieling et al., 2006), while we observed peptidyl transfer times of less than 20 ms between GTP hydrolysis on EF-Tu in the ternary complex and peptide bond formation (Bilgin et al., 1992; Johansson et al., 2008a; Pavlov et al., 2009). This suggested to us that a rate-limiting accommodation step preceding peptide bond formation could have been eliminated in our optimized system for protein synthesis, thereby making it possible to estimate the temperature dependence of the chemical rate of peptide bond formation with native tRNA substrates.

Accordingly, we used our buffer and energy regeneration system to design a series of experiments with initiator tRNA (fMet-tRNA^{fMet}) in the P site and a UUU codon for Phe in the A site (Johansson et al., 2008a). These 70S initiation complexes (ICs) were pre-incubated at different temperatures in the interval 10–37 °C and rapidly mixed with Phe-tRNA^{Phe}-containing ternary complex in a quench-flow instrument. The extents of GTP hydrolysis on EF-Tu and fMet-Phe

dipeptide formation at different quenching times were analyzed by thin layer chromatography and HPLC to determine the rate of peptide bond formation at each temperature. We found large activation enthalpy and insignificant activation entropy for the peptide bond formation reaction on the ribosome, very similar to the estimates obtained earlier with puromycin (Sievers et al., 2004) and in line with MD simulations (Trobro and Åqvist, 2005; Trobro and Åqvist, 2006). Although compatible with the idea that in our optimized system the rate limiting step for peptidyl transfer is the chemistry of peptide bond formation, one could not exclude the possibility that the similarity between our results and those by (Sievers et al., 2004) were coincidental. We note, however, that the MD simulations by the Åqvist group suggested a physical upper limit to the chemical rate of peptide bond formation, while we have no theoretical knowledge of a putative upper rate limit for the rate of tRNA accommodation. This is an interesting question in light of the high selection pressure for rapid protein synthesis in the living cell as discussed in Section 1 above. That is, if tRNA accommodation is rate limiting for protein synthesis *in vivo*, then we postulate the existence of a physical constraint of this step. One might think that such physical constraint could be the time for an aa-tRNA to freely rotate from its A/T to its accommodated A/A state. We note, however, that this cannot be the case, since free rotational diffusion of tRNA occurs in the ~100 ns range, corresponding to a rate constant of ~10^7 s^{-1}, many orders of magnitude larger than the compounded rate constants that we have estimated for peptidyl transfer (~100 s^{-1} at 37 °C).

3.2. pH dependence[1]

Peptide bond formation proceeds through a nucleophilic attack of the α-amino group of the amino acid on the A-site aa-tRNA on the ester carbonyl carbon of the peptide chain on the peptidyl-tRNA in the P site. The pK_a-values of the α-amino groups of free amino acid methyl or ethyl esters vary in the range 6.8 to 8.6 units (Hay and Porter, 1967; Hay and Morris, 1970) and these esters appear to have pK_a-values very similar to their corresponding aa-tRNAs (Wolfenden, 1963). One would therefore expect to observe amino acid-specific inhibition of ribosome-catalyzed peptide bond formation as the pH-value decreases from above to be-

---

1   Johansson et al. (2011)

low the pK$_a$-value of the α-amino group of the A-site bound aa-tRNA. A pH-dependent variation of the rate of ribosomal peptidyl transfer has previously been observed only for aminoacyl-tRNA analogues in the A site, like puromycin, C-puromycin and CC-puromycin (Katunin et al., 2002; Brunelle et al., 2006; Beringer and Rodnina, 2007) and others (Okuda et al., 2005). The pH-dependence of the rate of peptidyl transfer to these analogues is complex. The pH dependent rate change always reflects titration of *two* protons in the case of puromycin (Katunin et al., 2002; Brunelle et al., 2006). For C-puromycin the results were not conclusive: Beringer and Rodnina suggested the titration of two protons but also claimed that their data could equally well be explained by titration of one proton (Beringer and Rodnina, 2007). At the same time, Green and collaborators suggested the titration of only one proton in the case of C-puromycin (Brunelle et al., 2006). In the case of CC-puromycin, the pH dependence of peptide bond formation is very weak and qualitatively different from the pH dependence in the puromycin and C-puromycin cases (Beringer and Rodnina, 2007).

The reason why the amino acid-specific pH-dependence of peptide bond formation was studied with the aminoacyl-tRNA analogues, rather than with native aminoacyl-tRNAs in the A site, can be traced to the notion that the chemistry of peptide bond formation with native tRNAs is completely masked by rate-limiting accommodation of aminoacyl-tRNA in the A site (Rodnina et al., 1994; Pape et al., 1998). If true, this would mean that with native tRNAs one can only study the pH-dependence of the rate of aminoacyl-tRNA accommodation in the A site, about which we have no a priori knowledge. In line with this view, experiments from the Rodnina group revealed virtually no pH dependence for the rate of peptidyl transfer to Phe-tRNA^{Phe} in the A site from fMet-tRNA^{fMet} in the P site (Bieling et al., 2006). We note, however, that the rate of peptidyl transfer was small, less than 7 s^{-1} in those experiments, performed at 37 °C, and that the errors were large. We measured much larger peptidyl transfer rates, ~100 s^{-1}, for the same reaction at 37 °C (Johansson et al., 2008a; Pavlov et al., 2009). We conjectured therefore that the pH-dependence of the rate of the chemistry of peptide bond formation could, after all, be studied in a system where a rate limiting accommodation step had become much faster due to optimal buffer conditions, provided that the experimental precision was sufficiently high. We therefore selected a group of six aminoacyl-tRNAs in which the pK$_a$-values of the α-amino groups, calculated from the pK$_a$-values

of the corresponding amino acid methyl or ethyl esters (Wolfenden, 1963; Hay and Porter, 1967; Hay and Morris, 1970), varied from its smallest (pK$_a$=6.8 for Asn) to its largest (pK$_a$=8.6 for Pro) value among the twenty canonical amino acids. We ran quench-flow experiments with a concentration of 70S ICs (with fMet-tRNA^{fMet} in the P site) in large excess over the ternary complex concentration and simultaneously monitored the time evolution of GTP hydrolysis on EF-Tu and peptide bond formation for each of the six aminoacyl-tRNAs. We measured the average times for GTP hydrolysis (τ_{hyd}) and peptide bond formation (τ_{dip}) for each type of ternary complex at pH-values in the 6–8 interval, and estimated the average time for peptidyl transfer after GTP hydrolysis (τ_{pep}) from $\tau_{pep} = \tau_{dip} - \tau_{hyd}$ as illustrated in Figure 3. For all six aminoacyl-tRNAs T_{pep} increased with decreasing pH, as expected from protonation of their α-amino groups. We operationally defined a pK$_a$-value (pK_a^{obs}) for each one of these curves as the pH-value at which τ_{pep} had increased to twice its asymptotically smallest value (τ_{pep}^{min}) at high pH:

$$\frac{\tau_{pep}^{min}}{\tau_{pep}} = \frac{1}{1 + 10^{\left(pK_a^{obs} - pH\right)}} \quad . \tag{9}$$

Note that τ_{pep}^{min} is the average time of peptidyl transfer to a completely unprotonated amino group and thus corresponds to the largest possible rate of peptidyl transfer. In the cases of Gly and Pro aminoacyl-tRNAs the pH-dependence of τ_{pep}^{min} revealed the titration of a single proton in a model independent way, which we identified as the titration of the α-amino group. From this we inferred single proton titrations also in the cases of the other four tRNAs. We fitted our data accordingly and estimated pK_a^{obs} values varying from 5.9 (Asn) to 7.8 (Pro). The estimated pK_a^{obs}-values were downshifted to varying extents in relation to the pK_a-values of the corresponding amino acid esters and aminoacyl-tRNAs: the smallest shift of 0.4 units in the Gly and the largest shift of 1.7 units in the Ile case. In general, the pK_a^{obs}-value relates to the true pK$_a$-value, pK_a^{pt}, of the α-amino group of the A-site aminoacyl-tRNA through

$$pK_a^{obs} = pK_a^{pt} + \log_{10}\left(\tau_{pt}^{min} / \tau_{pep}^{min}\right) \quad . \tag{10}$$

Here, τ_{pt}^{min} is the average time for the chemistry of peptide bond formation when the A-site aa-tRNA is deprotonated. These downshifts have two extreme interpretations.

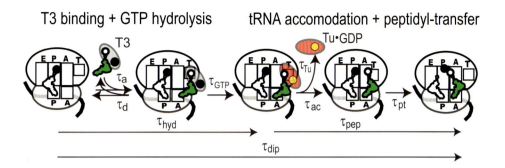

Fig. 3 Relation between average times of di-peptide formation, GTP hydrolysis and peptidyl transfer. The reaction of dipeptide formation is initiated by mixing ternary complexes T3 with 70S ICs containing fMet-tRNA in the P-site and a cognate codon for the ternary complex T3 in the A-site. The time of peptidyl-transfer (τ_{pep}) is calculated subtracting the experimentally measured time of GTP hydrolysis (τ_{GTP}) from the time of di-peptide formation (τ_{dip}) measured in the same experiment. The time of peptidyl transfer (τ_{pep}) is itself the sum of the average times of accommodation (τ_{ac}) and the chemical step of peptide bond formation (τ_{pt}).

In the first scenario, pK_a^{pt} is the same on and off the ribosome, and the downshifts depend solely on a pH-independent step, e. g. accommodation, preceding the chemistry of peptide bond formation. The different down shifts for the different tRNAs are in this scenario explained by different $\tau_{pt}^{min}/\tau_{pep}^{min}$ ratios with the largest ratio ($\tau_{pt}^{min}/\tau_{pep}^{min} \approx 0.4$) for Gly and the smallest ratio ($\tau_{pt}^{min}/\tau_{pep}^{min} \approx 0.02$) for Ile. Taking into account that $\tau_{pep} = \tau_{ac}+\tau_{pt}$ (see Figure 3), one finds that the time of Gly-tRNA accommodation, τ_{ac}, was only 50% longer than the minimal time, τ_{pt}^{min}, of the chemical step at high pH values. Further, it follows from Eq. (10) that in the Gly case the pH-dependent term, τ_{pt}, in τ_{pep} became larger than the pH-independent term (e. g. accommodation, τ_{ac}) at pH values below pK_a^{obs}. We concluded, therefore, that even in this extreme scenario the time, τ_{pt}, for the chemistry of peptide bond in the Gly ($pK_a^{obs} = 7.4$) case dominated the total time, $\tau_{pep} = \tau_{ac}+\tau_{pt}$ of peptidyl transfer already at physiological pH (7.4), implying that at this or lower pH-values the relative contribution to τ_{pep} of the step(s) preceding the chemistry of peptide bond formation is small. A similar conclusion holds in the Pro case ($pK_a^{obs} = 7.8$) for pH-values below 7.8. This means that the notion of tRNA accommodation as a rate limiting step for peptidyl transfer is not universally true, and that the chemistry of peptidyl transfer between native tRNA substrates is a significant determinant of protein elongation rate in the living cell, at least in the Gly and Pro cases. Our results also demonstrate that the chemistry of peptidyl transfer with native tRNA substrates can be studied in biochemical experiments at pH 7.5 in the Gly and Pro cases and at pH = 6 in the other four cases. These results show, in addition, that a slow Pro

incorporation in protein synthesis ($\tau_{pep} \approx 50$ ms for Pro versus $\tau_{pep} \approx 11$ ms for Phe) (Pavlov et al., 2009) can be partially explained by protonation of its amino-group at physiological pH.

In the second extreme scenario, the chemistry of peptide bond formation dominates the time for peptidyl transfer at all pH-values, i. e. $\tau_{pt}^{min}/\tau_{pep}^{min} \approx 1$. At the same time, the chemical environment of the peptidyl transfer center induces amino acid dependent downshifts in pK_a^{pt} in relation to their bulk solution values. The second scenario is supported by MD simulations, based on crystal structures of the peptidyl transfer center (Schmeing et al., 2005; Selmer et al., 2006) and the Linear Interaction Energy (LIE) method (Carlsson et al., 2008). These simulations predict that the pK_a-values of the α-amino groups of ribosome-bound aminoacyl-tRNAs are downshifted in relation to those of the free aminoacyl-tRNAs in an amino acid dependent way. The LIE predictions correlate very well with the *relative* extents to which the pK_a^{obs}-values are down shifted. Such a strong correlation follows directly from scenario two, but is not expected in scenario one suggesting that, in fact, the chemistry of peptide bond formation may under our experimental condition dominate the peptidyl transfer reaction time for all six aminoacyl-tRNAs, including Phe-tRNA[Phe]. However, the *absolute* magnitude of the LIE predicted downshifts depends on the dielectric constant of the surroundings of the peptidyl transfer center, which is an as yet unknown parameter. To obtain the absolute values of the LIE predicted downshifts, the calculations must be calibrated by an experiment in which the chemistry of peptide bond formation is directly measured without the complication of preceding steps (Ieong,

K., in progress). If successful, such a calibration can then be used to precisely estimate the time partitioning between the chemistry of peptide bond formation and preceding steps following GTP hydrolysis on EF-Tu for all six aminoacyl-tRNAs in this study and, eventually, for all naturally occurring aminoacyl-tRNAs.

Acknowledgements

The research grants from the Swedish Research Council and NIH/USA to M. E. are thankfully acknowledged.

References

Alatossava T, Jutte H, Kuhn A, Kellenberger E (1985) Manipulation of intracellular magnesium content in polymyxin B nonapeptide-sensitized *Escherichia coli* by ionophore A23 187. J Bacteriol 162: 413–419

Beringer M, Bruell C, Xiong L, Pfister P, Bieling P, Katunin VI, Mankin AS, Bottger EC, Rodnina MV (2005) Essential mechanisms in the catalysis of peptide bond formation on the ribosome. J Biol Chem 280: 36065–36072

Beringer M, Rodnina MV (2007) Importance of tRNA interactions with 23S rRNA for peptide bond formation on the ribosome: studies with substrate analogs. Biol Chem 388: 687–691

Bieling P, Beringer M, Adio S, Rodnina MV (2006) Peptide bond formation does not involve acid-base catalysis by ribosomal residues. Nat Struct Mol Biol 13: 423–428

Bilgin N, Claesens F, Pahverk H, Ehrenberg M (1992) Kinetic properties of *Escherichia coli* ribosomes with altered forms of S12. J Mol Biol 224: 1011–1027

Bilgin N, Ehrenberg M, Kurland C (1988) Is translation inhibited by noncognate ternary complexes? FEBS Lett 233: 95–99

Bremer H, Dennis PP (2008) Modulation of Chemical Composition and Other Parameters of the Cell at Different Exponential Growth Rates. In: Böck A, Curtiss III, R, Kaper JB, Karp PD, Neidhardt FC, Nyström T, Slauch JM, Squires CL, Ussery D, Schaechter E (eds.), EcoSal-*Escherichia coli* and Salmonella: Cellular and Molecular Biology. *http://www. ecosal. org*. ASM Press, Washington, DC

Brunelle JL, Youngman E M., Sharma D, Green R (2006) The interaction between C75 of tRNA and the A loop of the ribosome stimulates peptidyl transferase activity. RNA 12: 33–39

Carlsson J, Boukharta L, Åqvist J (2008) Combining docking, molecular dynamics and the linear interaction energy method to predict binding modes and affinities for non-nucleoside inhibitors to HIV-1 reverse transcriptase. J Med Chem 51: 2648–2656

Edelmann P, Gallant J (1977) Mistranslation in *E. coli*. Cell 10: 131–137

Ehrenberg M, Bilgin N, Kurland C (1990) Design and use of a fast and accurate in vitro translation system. In Spedding G (ed.), Ribosomes and protein synthesis. A practical approach. IRL Press at Oxford University Press, Oxford, UK, pp 101–128

Ehrenberg M, Kurland CG (1984) Costs of accuracy determined by a maximal growth rate constraint. Q Rev Biophys 17: 45–82

Feist A M, Herrgard MJ, Thiele I, Reed JL, Palsson BO (2009) Reconstruction of biochemical networks in microorganisms. Nat Rev Microbiol 7: 129–143

Fersht A (1999) Structure and mechanism in protein science: a guide to enzyme catalysis and protein folding. W. H. Freeman and Company, New York

Gromadski KB, Daviter T, Rodnina MV (2006) A uniform response to mismatches in codon-anticodon complexes ensures ribosomal fidelity. Mol Cell 21: 369–377

Gromadski KB, Rodnina MV (2004) Kinetic determinants of high-fidelity tRNA discrimination on the ribosome. Mol Cell 13: 191–200

Hansen JL, Schmeing TM, Moore PB, Steitz TA (2002) Structural insights into peptide bond formation. Proc Natl Acad Sci USA 99: 11 670–11 675

Hay RW, Morris PJ (1970) Proton ionisation constants and kinetics of base hydrolysis of some α-amino-acid esters in aqueous solution. Part II. J Chem Soc B, 1577–1582

Hay RW, Porter LJ (1967) Proton ionisation constants and kinetics of base hydrolysis of some α-amino-acid esters in aqueous solution. J Chem Soc B, 1261–1264

Hesslein AE, Katunin VI, Beringer M, Kosek AB, Rodnina MV, Strobel SA (2004) Exploration of the conserved A+C wobble pair within the ribosomal peptidyl transferase center using affinity purified mutant ribosomes. Nucleic Acids Res 32: 3760–3770

Hopfield JJ (1974) Kinetic proofreading: a new mechanism for reducing errors in biosynthetic processes requiring high specificity. Proc Natl Acad Sci USA 71: 4135–4139

Ibarra RU, Edwards JS, Palsson BO (2002) *Escherichia coli* K-12 undergoes adaptive evolution to achieve in silico predicted optimal growth. Nature 420: 186–189

Jelenc PC, Kurland CG (1979) Nucleoside triphosphate regeneration decreases the frequency of translation errors. Proc Natl Acad Sci USA 76: 3174–3178

Johansson M, Bouakaz E, Lovmar M, Ehrenberg M (2008a) The kinetics of ribosomal peptidyl transfer revisited. Mol Cell 30: 589–598

Johansson M, Ieong KW, Trobro S, Strazewski P, Aqvist J, Pavlov MY, and Ehrenberg M (2011) pH-sensitivity of the ribosomal peptidyl transfer reaction dependent on the identity of the A-site aminoacyl-tRNA. Proc Natl Acad Sci USA 108: 79–84

Johansson M, Lovmar M, Ehrenberg M (2008b) Rate and accuracy of bacterial protein synthesis revisited. Curr Opin Microbiol 11: 141–147

Katunin VI, Muth GW, Strobel SA, Wintermeyer W, Rodnina MV (2002) Important contribution to catalysis of peptide bond formation by a single ionizing group within the ribosome. Mol Cell 10: 339–346

Kramer EB, Farabaugh PJ (2007) The frequency of translational misreading errors in *E. coli* is largely determined by tRNA competition. RNA 13: 87–96

Kurland CG, Canback B, Berg OG (2003) Horizontal gene transfer: a critical view. Proc Natl Acad Sci USA 100: 9658–9662

Kurland CG, Ehrenberg M (1984) Optimization of translation accuracy. Prog Nucleic Acid Res Mol Biol 31: 191–219

Ninio J (1975) Kinetic amplification of enzyme discrimination. Biochimie 57: 587–595

Nissen P, Hansen J, Ban N, Moore PB, Steitz TA (2000) The structural basis of ribosome activity in peptide bond synthesis. Science 289: 920–930

Ogle JM, Ramakrishnan V (2005) Structural insights into translational fidelity. Annu Rev Biochem 74: 129–177

Okuda K, Seila AC, Strobel SA (2005) Uncovering the enzymatic pKa of the ribosomal peptidyl transferase reaction utilizing a fluorinated puromycin derivative. Biochemistry 44: 6675–6684

Pape T, Wintermeyer W, Rodnina M (1999) Induced fit in initial selection and proofreading of aminoacyl-tRNA on the ribosome. Embo J, 18, 3800–3807

Pape T, Wintermeyer W, Rodnina MV (1998) Complete kinetic mechanism of elongation factor Tu-dependent binding of aminoacyl-tRNA to the A site of the *E. coli* ribosome. Embo J, 17, 7490–7497

Pavlov MY, Watts RE, Tan Z, Cornish VW, Ehrenberg M, Forster AC (2009) Slow peptide bond formation by proline and other N-alkylamino acids in translation. Proc Natl Acad Sci USA, 106, 50–54

Rodnina MV, Fricke R, Wintermeyer W (1994) Transient conformational states of aminoacyl-tRNA during ribosome binding catalyzed by elongation factor Tu. Biochemistry 33: 12267–12275

Ruusala T, Andersson D, Ehrenberg M, Kurland CG (1984) Hyperaccurate ribosomes inhibit growth. Embo J 3: 2575–2580

Ruusala T, Ehrenberg M, Kurland CG (1982) Is there proofreading during polypeptide synthesis? Embo J, 1, 741–745

Schmeing TM, Huang KS, Strobel SA, Steitz TA (2005) An induced-fit mechanism to promote peptide bond formation and exclude hydrolysis of peptidyl-tRNA. Nature 438: 520–524

Schroeder GK, Wolfenden R (2007) The rate enhancement produced by the ribosome: an improved model. Biochemistry 46: 4037–4044

Selmer M, Dunham CM, Murphy FV 4th, Weixlbaumer A, Petry S, Kelley AC, Weir JR, Ramakrishnan V (2006) Structure of the 70S ribosome complexed with mRNA and tRNA. Science 313: 1935–1942

Sievers A, Beringer M, Rodnina MV, Wolfenden R (2004) The ribosome as an entropy trap. Proc Natl Acad Sci USA, 101, 7897–7901

Stahl G, Salem SN, Chen L, Zhao B, Farabaugh PJ (2004) Translational accuracy during exponential, postdiauxic, and stationary growth phases in Saccharomyces cerevisiae. Eukaryot Cell 3: 331–338

Thompson RC, Stone PJ (1977) Proofreading of the codon-anticodon interaction on ribosomes. Proc Natl Acad Sci USA, 74, 198–202

Trobro S and Åqvist J (2005) Mechanism of peptide bond synthesis on the ribosome. Proc Natl Acad Sci USA, 102, 12395–12400

Trobro S and Åqvist J (2006) Analysis of predictions for the catalytic mechanism of ribosomal peptidyl transfer. Biochemistry 45: 7049–7056

Wagner EG, Jelenc PC, Ehrenberg M, Kurland CG (1982) Rate of elongation of polyphenylalanine in vitro. Eur J Biochem 122: 193–197

Weiss RL, Kimes BW, Morris DR (1973) Cations and ribosome structure. 3. Effects on the 30S and 50S subunits of replacing bound Mg 2+ by inorganic cations. Biochemistry 12: 450–456

Weiss RL, Morris DR (1973) Cations and ribosome structure. I. Effects on the 30S subunit of substituting polyamines for magnesium ion. Biochemistry 12: 435–441

Wold F, Ballou CE (1957) Studies on the enzyme enolase. I. Equilibrium studies. J Biol Chem 227: 301–312

Wolfenden R (1963) The mechanism of hydrolysis of amino acyl RNA. Biochemistry 2: 1090–1092

Young R, Bremer H (1976) Polypeptide-chain-elongation rate in *Escherichia coli* B/r as a function of growth rate. Biochem J, 160, 185–194

Mutations in 16S rRNA that decrease the fidelity of translation

19

Sean P. McClory, Aishwarya Devaraj, Daoming Qin, Joshua M. Leisring, and Kurt Fredrick

1. Introduction

In the past decade, tremendous progress has been made in elucidating the structure and function of the ribosome (reviewed in Schmeing and Ramakrishnan, 2009). Numerous x-ray crystal structures and cryo-electron microscopic (cryo-EM) reconstructions of the ribosome with and without various substrates, factors, and antibiotics have been solved. At the same time, extensive biochemical studies have led to compelling kinetic models for the major steps of protein synthesis. While these studies give us a high-resolution picture of the ribosome and suggest a series of events involved in translation, the roles of specific ribosomal elements in particular events of the process remain unclear. Studies of mutations that confer altered function, particularly those in rRNA, will undoubtedly provide insight about these structure-function relationships.

Most model organisms carry multiple copies of ribosomal RNA (rRNA) genes; *Escherichia coli*, for example, has seven. This has generally hampered genetic studies of rRNA, because the effects of a mutation in one gene are often masked by the remaining wild-type copies in the cell. An indication of this technical problem is that most chromosomal mutations that affect the fidelity of translation have mapped to single-copy genes encoding ribosomal proteins or translation factors, rather than to rRNA genes (Haggerty and Lovett, 1997; Rosset and Gorini, 1969; Strigini and Gorini, 1970; Sussman et al., 1996).

In the 1980's, de Boer and coworkers pioneered a strategy to circumvent this problem for 16S rRNA (Hui and de Boer, 1987). They replaced the Shine-Dalgarno (SD) sequence (5'-GGAGG-3') of a reporter gene with either 5'-CCTCC-3' or 5'-GTGTG-3'. In both cases, the endogenous wild-type ribosomes no longer recognized the reporter mRNA efficiently. Then, compensatory substitutions in the anti-Shine-Dalgarno (ASD) sequence (to 5'-GGAGG-3' or 5'-CACAC-3', respectively) of a plasmid-borne 16S rRNA gene were made. When either of these altered 16S rRNA genes was expressed in the presence of the appropriate reporter construct, a subpopulation of ribosomes was generated in the cell that specifically translated the reporter. These engineered ribosomes with altered specificity in initiation were termed "specialized" ribosomes. In the 1990's, Cunningham and colleagues optimized this genetic system (Lee et al., 1996). They simultaneously randomized the SD region of the reporter (in this case, the chloramphenicol acetyl transferase gene) and the ASD region of the 16S rRNA gene (with a bias against the endogenous sequences) and selected for functional combinations. Among the numerous functional SD-ASD pairings recovered, a "winning" pair was identified (SD: 5'-AUCCC-3'; ASD: 5'-GGGGU-3') that increased both the efficiency and selectivity of the system. The use of such specialized ribosome strains provides two clear advantages for genetic studies of 16S rRNA: (i) Translation of the reporter gene is completely dependent on the single plasmid-encoded (specialized) 16S rRNA allele, hence phenotypes can be directly attributed to the mutant (specialized) ribosomes and are not masked by the endogenous (wild-type) ribosomes; (ii) mutations in the specialized 16S rRNA do not influence cell growth, hence any mutation can be analyzed without the concern of indirect physiological effects on reporter gene expression.

Based on the work of Cunningham, we constructed the indicator strain KLF2674, harboring the *lacZ* gene with the specialized SD sequence 5'-ATCCC-3' (SD*) in single-copy on a recombinant λ prophage (Table 1). The *lacZ* gene is preceded by a phage promoter (P$_{ant}$), which directs constitutive transcription of *SD*-lacZ* in the strain. We also made plasmid pKF207, which

Table 1 Indicator strains used in the genetic screens

Strain[a]	Mutation in SD*-lacZ reporter gene[b]
KLF2674	None (control)
KLF2672	Start codon changed from ATG to ATC
KLF4001	Codon 461 changed from GAA to GAT[c]
KLF2723	Codon 585 changed from TGG to TGA
KLF3406	Codons 250–252 replaced with GGGTTTTATC[d]
KLF3361	Codons 250–252 replaced with GGGTTTTAGC

[a] All strains have the genetic background *F- ara Δ(gpt-lac)5 Δ (recA-srl)306 srl301::Tn10*.

[b] The *lacZ* gene is on the chromosome (as part of a recombinant λ prophage) and has the specialized Shine-Dalgarno sequence 5'-ATCCC-3' (SD*).

[c] Codon 461 corresponds to an active-site glutamate residue of β-galactosidase. The substitution E461D essentially abolishes enzymatic activity (Cupples and Miller, 1988).

[d] This shifty sequence derives from *argI* and was shown previously to promote +1 frameshifting (Fu and Parker, 1994).

contains the 16S rRNA gene with the complementary ASD sequence 5'-GGGGT-3' (ASD*), under transcriptional control of the arabinose-inducible P$_{BAD}$ promoter (Abdi and Fredrick, 2005; Qin et al., 2007; Qin and Fredrick, 2009). Expression of the specialized 16S rRNA from pKF207 in KLF2674 cells results in a > 100-fold increase in β-galactosidase activity. This

system has a practical advantage over those described previously in that the reporter gene and the specialized 16S rRNA gene are unlinked.

In recent years, we have performed several screens for informational suppressor mutations in 16S rRNA using this genetic system (McClory et al., 2010; Qin and Fredrick, 2009). Each screen employed a different indicator strain carrying a particular mutation in *lacZ* (Table 1). These included a change of the start codon from ATG to ATC, a missense mutation at codon 461 (Cupples and Miller, 1988, 1989), a nonsense mutation at codon 585, and a "shifty sequence" that promotes +1 frameshifting (Fu and Parker, 1994) introduced in place of codons 250–252. All screens were performed in an analogous way. Plasmid pKF207 was mutagenized by propagation in the mutator strain XL1-Red (Stratagene) and then transformed into each of the various indicator strains. Transformants were then screened on plates containing X-gal and arabinose for those with increased levels of β-galactosidase activity. There was a bias for transition mutations due to the use of XL1-Red, which was made evident when site-directed transversions were subsequently engineered (see below). In this chapter, we summarize the mutations obtained and discuss their implications for ribosome function.

Table 2 Summary of 16S rRNA mutations that decrease the fidelity of initiation

Mutation	Location	Conservation[a]	No. isolates in screen (fold-increase in spurious initiation)[b]	Phenotype(s) reported previously[c]
U13G	h1	B, A, E	1 (2.5 ± 0.2)	
U789C	h24	B, A,	1 (5.2 ± 0.4)	
A790G	h24	B, A, E	1 (7.2 ± 1.0)	
G886A	h27	b, A, E	1 (2.5 ± 0.2)	Decreases nonsense suppression (in yeast)[d]
U1083C	h37	B, E	1 (1.8 ± 0.1)	
A1092G	h37	B, a	1 (1.7 ± 0.1)	
G1094A	h37	B, A	1 (2.5 ± 0.3)	
C1389U	h28	b, A	1 (1.7 ± 0.1)	
A1410G	h44		4 (3.4 ± 0.3)	
C1411U	h44		1 (2.0 ± 0.1)	
U1414C	h44		2 (2.2 ± 0.2)	
G1486A	h44	b	1 (1.3 ± 0.1)	

[a] Conservation of substituted nucleotide. Uppercase letters B, A, and E denote > 98 % conservation in the domains Bacteria, Archaea, and Eukarya, respectively; lowercase letters indicate > 90 % conservation; an omission of letters indicates < 90 % conservation.

[b] From Qin & Fredrick, 2009. Isolates were considered independent only if they originated from separate preparations of mutagenized pKF207. Values shown in parentheses correspond to the rate of

initiation from AUC relative to AUG for the mutant ribosomes, normalized to that for the control ribosomes. Data represent the normalized quotient of two means ± standard error for ≥ 3 independent experiments.

[c] Blank spaces indicate that no phenotypes have been reported, at least to our knowledge.

[d] Velichutina et al. (2000).

2. Ribosomes defective in start codon selection

Our first screen was for mutations that increase translation from the near-cognate start codon AUC (Qin and Fredrick, 2009). From 18 independent isolates, 13 different mutations in 16S rRNA were identified. All but one of these mutations increased initiation from AUC relative to AUG (i.e., decreased the fidelity of start-codon selection). The exception was a change in the ASD* sequence (5'-GGGGU-3' to 5'-GGGAU-3'). This mutation, which effectively converts the single U-G wobble pair in the SD*-ASD* helix to a Watson-Crick U-A pair, increased translation from both AUC and AUG by three-fold. The other mutations, which decreased initiation fidelity, clustered to three regions of the 30S subunit–the 790 loop, helix 44 (h44), and the neck region (Table 2, Figures 1–2).

The strongest phenotypes were conferred by mutations in the 790 loop (U789C and A790G) (Table 2,

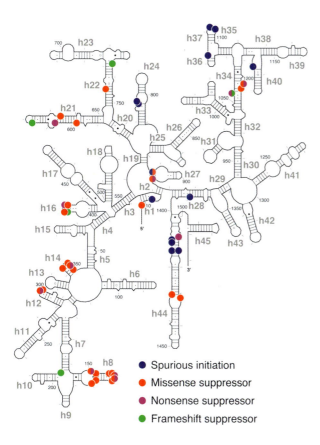

Fig. 1 Where the mutations map on the secondary structure of 16S rRNA. Colored circles show the positions of mutations identified in various screens for decreased translational fidelity (as indicated). Bi-colored circles indicate mutations that were recovered in two different screens.

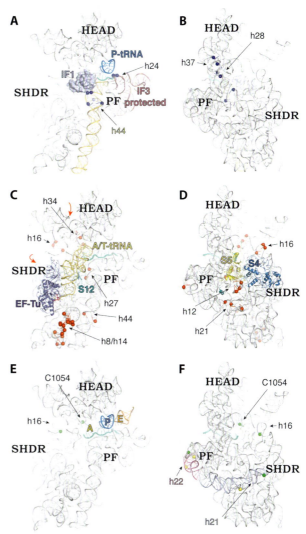

Fig. 2 Where the mutations map on the tertiary structure of 16S rRNA. (A-B) Mutations that decrease the fidelity of initiation (Qin and Fredrick, 2009) are depicted as blue spheres. Interface (A) and solvent (B) perspectives of 16S rRNA (PDB 1HR0; Carter et al., 2001) are shown, with bound IF1 in pale blue, and h44 highlighted in yellow. P-site tRNA (blue) and mRNA (green) from PDB 2J00 (Selmer et al., 2006) have been modeled in by superimposition of the corresponding 16S rRNA chain. Nucleotides protected by IF3 from hydroxyl radical cleavage (Dallas and Noller, 2001) are shown in rose. (C-D) Mutations that decrease the accuracy of decoding (McClory et al., 2010) are depicted as red spheres. Interface (C) and solvent (D) views of 16S rRNA (PDB 2WRN; Schmeing et al., 2009) are shown. This structure contains EF-Tu (indigo) and aminoacyl-tRNA (yellow) in the intermediate A/T state. Proteins S4, S5, and S12 are shown in blue, yellow, and teal, respectively. Red arrows (C) indicate rotations of the head and shoulder during domain closure (Ogle and Ramakrishnan, 2005). (E–F) Mutations isolated as +1frameshift suppressors in this work are shown as green spheres. Interface (E) and solvent (F) views of 16S rRNA (PDB 2J00; Selmer et al., 2006) are shown, with P-site tRNA (blue), E-site tRNA (orange), and mRNA (green). Previously identified mutations that promote -1frameshifting (Léger et al., 2007) are indicated with yellow spheres. Helices h21 and h22 are highlighted in pale blue and pink, respectively. 30S shoulder (SHDR), platform (PF), and head (HEAD) domains are indicated throughout.

Figure 2A). Because this loop was understood to contribute to the IF3 binding site (Dallas and Noller, 2001; Tapprich et al., 1989), we tested the effects of these mutations on IF3 binding. Indeed, both mutants decreased the affinity of IF3 for the 30S subunit substantially (~10-fold), suggesting that they act by destabilizing IF3 (Qin et al., 2007; Qin and Fredrick, 2009).

The second group of mutations mapped to h44, just "down" from the 30S A site (Figure 2A). Structural studies have indicated that binding of IF1 to the 30S subunit distorts this region of h44, locally displacing one strand and disrupting base pairs A1413-G1487 and U1414-G1486 (Carter et al., 2001). Mutation A1410G, which lies immediately adjacent to the factor, decreased the affinity of IF1 for the subunit, but mutations further down the helix (e. g., U1414C) did not (Qin and Fredrick, 2009). Two of the isolated mutations (U1414C and G1486A) each convert U1414-G1486 to a Watson-Crick pair. Conversion of the adjacent non-canonical A1413-G1487 pair to a Watson-Crick pair by site-directed mutagenesis also led to spurious initiation (Qin and Fredrick, 2009). Together these data suggest that the initiation process is controlled in part by the conformational state of h44, and thus particular structural perturbations of h44 cause defects in start-codon selection. Interestingly, h44 forms several contacts to the 50S subunit and these fidelity mutations lie right between inter-subunit bridges B2a and B3 (Selmer et al., 2006; Zhang et al., 2009). This raises the possibility that these mutations predominantly affect the 50S docking step of initiation and thereby confer their fidelity defects.

The third group of mutations mapped to the neck region of the subunit (Figure 2B). One mutation (C1389U) localized to the neck-helix (h28) itself, while the three others mapped to h37, which packs up against h28. These mutations presumably affect the dynamics of the head domain, and hence implicate head movement in the molecular mechanism of initiation. However, a better understanding of how these mutations act will require further characterization of the corresponding mutant subunits.

3. Ribosomes defective in aminoacyl-tRNA selection

We next screened for missense and nonsense suppressor mutations in 16S rRNA (McClory et al., 2010). To identify missense suppressors, we used indicator strain KLF4001 (Table 1), in which codon 461 of SD*-lacZ

was changed from GAA (Glu) to GAT (Asp). Glutamate 461 of LacZ contributes to the active site, and an aspartate at this position abolishes enzymatic activity (Cupples and Miller, 1988, 1989). Hence, generation of active β-galactosidase in this strain requires misreading of GAU by Glu-tRNA. To identify nonsense suppressors, we used strain KLF2723, in which codon 585 of SD*-lacZ was changed from TGG (Trp) to TGA (stop). In this strain, production of full-length (active) β-galactosidase requires readthrough of the UGA stop codon. An increase in UGA readthrough might result from a defect in decoding (e. g., misreading by Trp-tRNA) and/or a defect in RF2-dependent termination.

From the two screens, we obtained 132 independent isolates that revealed 34 different mutations (Table 3, Figures 1–2) (McClory et al., 2010). Most of the mutations recovered as nonsense suppressors were also recovered as missense suppressors, and the remainder clustered near missense suppressors. Subsequent characterization of all 34 mutations showed that all but one increased misreading of GAU in vivo. The exception was C1054U, which decreased misreading of GAU by two-fold. However, substitution of C1054 to A and G increased misreading of GAU by 20-fold and 40-fold, respectively (McClory et al., 2010). Hence all 34 nucleotides identified play a role in aminoacyl-tRNA selection.

Four mutations mapped in or near the A site (Figure 2C), consistent with previous genetic studies (Gregory and Dahlberg, 1995; Murgola et al., 1995; Pagel et al., 1997). Presumably, these mutations perturb the A site and thereby confer their phenotypes. Mutations C1200U and G1491A conferred stronger effects on UGA readthrough than on GAU misreading (Table 3; McClory et al., 2010), suggesting that these mutations inhibit RF2-dependent termination or differentially influence the fidelity of decoding, depending on the tRNA species involved.

Many mutations clustered to distinct regions near interfaces between the 30S shoulder domain and other parts of the ribosome (Figure 2C, D; Figure 3; McClory et al., 2010). The positions of these mutations implicate shoulder movement in the mechanism of decoding, lending functional support for the domain closure model of decoding proposed by Ramakrishnan and coworkers (Ogle and Ramakrishnan, 2005). One cluster mapped to h12 on the solvent side of the subunit, near the interface between ribosomal proteins S4 and S5 (Figure 2D). These h12 mutations may destabilize the open state of the 30S subunit, as has been proposed for mutations at the S4-S5 interface (Ogle and Rama-

Table 3 Summary of 16S rRNA mutations that decrease the fidelity of elongation

Mutation	Location	Conservation[a]	No. isolates in screen (fold-increase in suppression)[b]			Phenotype(s) reported previously[f]
			Missense[c]	Nonsense[d]	Frameshift[e]	
A7G	5' end		2 (1.6 ± 0.2)	0 (1.3 ± 0.1)	0	
A151G	h8	BAE	1 (2.9 ± 0.4)	4 (1.9 ± 0.3)	0	
C153U	h8	E	1 (2.8 ± 0.4)	0	0	
G158A	h8	Ae	1 (2.7 ± 0.2)	0	0	
G159A	h8	BA	6 (3.0 ± 0.2)	0	0	
A160G	h8	BAE	9 (7.6 ± 1.0)	3 (3.9 ± 1.4)	0	
A161G	h8	BA	8 (6.2 ± 0.9)	3 (5.9 ± 2.0)	0	Dominant negative for growth[g]
A162G	h8	BA	1 (2.0 ± 0.1)	0	0	
G168A	h8	aE	2 (3.8 ± 0.5)	0	0	
U170A	h8	BE	1 (2.2 ± 0.2)	0	0	
G299A	h12	Bae	15 (9.8 ± 1.3)	5 (9.5 ± 1.4)	0	Dominant negative for growth[g]
A300G	h12	BA	1 (5.0 ± 0.4)	0	0	
U343C	h14	Ba	1 (2.9 ± 0.4)	0	0	
ins345A[h]	h14	NA	0 (3.1 ± 0.2)	1 (3.5 ± 0.3)	0	
G346A	h14	BaE	3 (2.6 ± 0.3)	0	0	
G347A	h14	BaE	5 (8.3 ± 0.9)	3 (4.8 ± 0.4)	0	
G347U	h14	BaE	1 (15.9 ± 1.6)	0 (4.6 ± 0.9)	0	
ΔG348	h14	Ba	0 (3.5 ± 0.5)	1 (3.6 ± 0.5)	0	
A349G	h14	B	1 (2.5 ± 0.3)	0 (2.4 ± 0.3)	0	
C419U	h16		0	0	1 (1.9 ± 0.6)	
ΔU420	h16	b	1 (3.2 ± 0.4)	0	0	
G423A	h16	b	0 (2.1 ± 0.3)	1 (1.9 ± 0.2)	0	
G424A	h16	B	1 (1.8 ± 0.2)	0 (1.5 ± 0.2)	0	
U598C	h21	bA	1 (2.9 ± 0.3)	0	0	
G606A	h21		0 (2.1 ± 0.4)	1 (2.2 ± 0.3)	0	
G617A	h21	BA	0	0	1 (1.7 ± 0.6)	
C634U	h21	b	4 (2.5 ± 0.3)	6 (3.1 ± 0.4)	0	
G661A	h22	ae	1 (1.4 ± 0.1)	0	0	
C735U	h22		0	0	1 (2.1 ± 0.7)	
G886A	h27	bAE	1 (2.0 ± 0.1)	0	0	Decreases nonsense suppression (in yeast)[i]
U911C	h27	BAE	1 (3.0 ± 0.5)	1 (1.2 ± 0.1)	0	Increases nonsense suppression[j]
C1054U	h34	BAE	0 (0.6 ± 0.1)	4 (2.4 ± 0.3)	28 (20 ± 7)	Increases nonsense suppression[k,l]
C1200U	h34	BAE	1 (3.6 ± 0.6)	16 (28.0 ± 3.5)	0	Increases nonsense and frameshift suppression[k,m]
C1203U	h34	B	3 (2.1 ± 0.3)	0	0	
A1430G	h44		4 (4.0 ± 0.4)	0	0	
C1469U	h44		1 (3.5 ± 0.3)	0	0	Suppresses streptomycin dependence[n]
G1491A	h44		0 (2.0 ± 0.3)	4 (8.6 ± 0.8)	0	Increases nonsense and frameshift suppression[k]

ins, insertion; Δ, deletion; NA, not applicable.

[a] Conservation of substituted nucleotide. Uppercase letters B, A, and E denote >98% conservation in the domains *Bacteria*, *Archaea*, and *Eukarya*, respectively; lowercase letters indicate >90% conservation; Omission of letters indicates <90% conservation.

[b] Isolates were considered independent only if they originated from separate preparations of mutagenized pKF207. Values shown in parentheses correspond to the relative error rate, calculated as described (McClory et al., 2010). The absence of parentheses indicates that the data were not collected.

[c] From McClory et al., 2010. In this case, codon 461 of *lacZ* was changed from GAA to GAT. Production of active β-galactosidase requires misreading of GAU by Glu-tRNA.

[d] From McClory et al., 2010. In this case, codon 585 of *lacZ* was changed from TGG to TGA. Production of active β-galactosidase requires read-through of UGA.

[e] This work. In this case, codons 250–252 of *lacZ* were replaced with the sequence 5'-GGG TTT TAG C-3'. This sequence, based on a previously characterized frameshift site in *argI* (Fu and Parker, 1994), disrupts the *lacZ* reading frame. Production of full-length β-galactosidase requires +1 frameshifting by the translational machinery.

[f] Blank spaces indicate that no phenotypes have been reported, at least to our knowledge.

[g] Yassin et al. (2005).

[h] Insertion of A after C345.

[i] Velichutina et al. (2000).

[j] Lodmell and Dahlberg (1997).

[k] Gregory and Dahlberg (1995).

[l] Pagel et al. (1997).

[m] Moine and Dahlberg (1994).

[n] Allen and Noller (1991).

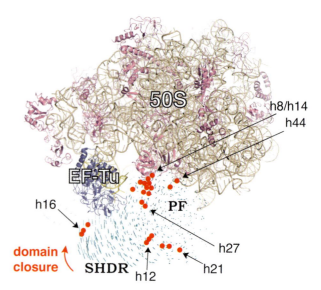

Fig. 3 Mutations that decrease the accuracy of decoding cluster near interfaces between the shoulder domain and other part of the ribosome. Positions of mutations that promote miscoding (red spheres; McClory et al., 2010) are shown in relation to rotation of the shoulder during domain closure. The "open" conformation of the 30S subunit (PDB 1J5E; Wimberly et al., 2000) has been overlaid on the structure of the 70S ribosome in the "closed" conformation bound to the ternary complex (PDB 2WRN, 2WRO; Schmeing et al., 2009) by superimposition of S5. Blue lines connect phosphorus atoms of 16S rRNA in the two conformational states and hence reflect their approximate movement during domain closure. The red arrow indicates the direction of this movement.

krishnan, 2005). A second group mapped nearby in h21, a long helix that spans the backside of the subunit and docks into the shoulder domain, forming contacts with h4, S4 and S16 (Figure 2D). Helix 21 appears to provide a structural support for the shoulder domain, and the mutations may compromise this putative role. A third set of mutations mapped to the terminus of h16 (Figure 2C, D), and each mutation is predicted to disrupt the tetraloop structure. This tetraloop forms a contact to S3 (a protein of the head domain) that is partially disrupted upon domain closure (Ogle et al., 2001). Hence the h16 mutations are predicted to destabilize the open state, which may explain their fidelity phenotypes.

Nearly half of the mutations mapped to h8 or h14, helices that interact with each other near the EF-Tu binding site (Figure 2C; McClory et al., 2010). Helices h8 and h14 also interact with the 50S subunit to form bridge B8. Recent cryo-EM studies revealed that the switch 1 motif of EF-Tu contacts h14 near the h8-h14 junction, which raised the possibility that h14 plays a direct role in activating the GTPase domain of EF-Tu during decoding (Villa et al., 2009). To investigate this possibility, we truncated h14 by deleting two base

pairs (McClory et al., 2010). This mutation (h14Δ2) is predicted to disrupt the putative contact to EF-Tu and the tertiary contact to h8. Ribosomes harboring h14Δ2 retained substantial activity *in vivo*, although they were error-prone, misreading GAU ten times more frequently than control ribosomes (McClory et al., 2010). *In vitro*, h14Δ2 stimulated the rate of single-turnover EF-Tu-dependent GTP hydrolysis. This effect was most pronounced for EF-Tu bearing a near-cognate aminoacyl-tRNA, consistent with the fidelity phenotypes observed *in vivo*. Very similar data were seen when h8 was truncated by three basepairs, suggesting that disruption of the h8-h14 interface suffices to confer these effects *in vivo* and *in vitro* (McClory et al., 2010). These data argue against a role for h14 in activating EF-Tu and instead suggest that h14 acts with h8 to negatively regulate GTP hydrolysis and thereby increase the stringency of decoding. We also engineered a two-base-pair insertion in h14, and this mutation resulted in complete loss of translation activity *in vivo* (McClory et al., 2010). Normally, the terminus of h14 docks against L19 and L14 to form bridge B8

(Selmer et al., 2006). Lengthening of h14 by two base pairs is predicted to push the subunits apart in this region, which appears to be intolerable for translation. Based on these data, we have proposed that bridge B8 normally acts to constrain shoulder movement, and hence mutations that comprise B8 decrease the energy barrier for inward shoulder rotation and thereby reduce the stringency of initial selection (McClory et al., 2010). This simple model explains the effects of mutations on both sides of bridge B8, those in h8/h14 of the 30S subunit and those in L19 of the 50S subunit (Maisnier-Patin et al., 2007).

4. Two screens for mutations that increase +1 frameshifting

We also performed two screens for 16S rRNA mutations that increase +1 frameshifting. We first employed indicator strain KLF3406 (Table 1), in which the shifty sequence from the *argI* gene (5'-GGG TTT TAT C-3') was inserted in place of codons 250–252 of *lacZ*. This shifty sequence appears to promote slippage of peptidyl-tRNA^{Phe} from the zero frame UUU to the overlapping +1 frame UUU (Fu and Parker, 1994). Codons 248–253 of *lacZ* correspond to a loop of β-galactosidase in which insertions can be made without loss of enzymatic activity (Feliu and Villaverde, 1998). Indeed, replacement of codons 250–252 of *lacZ* with the sequence GGG TTT TAT (which lacks the frameshift) resulted in only a small reduction (~10%) in β-galactosidase activity (data not shown). KLF3406 was transformed with various preparations of mutagenized pKF207, colonies were screened for increased β-galactosidase activity, and 20 independent isolates were obtained. Disappointingly, all isolates contained the identical mutation in the 3' end of the 16S rRNA gene which changed the ASD* sequence to 5'-GGGAT-3'. Subsequent analysis confirmed that this mutation increased the frequency of initiation rather than +1 frameshifting, since it increased expression of both *SD*-lacZ (control)* and *SD*-lacZ (+1fs)* to the same degree (data not shown). The reason why this screen failed was unclear. However, the basal level of β-galactosidase in KLF3406 was lower than we had expected based on previous studies of the *argI* shifty sequence, suggesting that decoding of the zero frame UAU codon was considerably faster in the context of our *lacZ* construct. This prompted us to perform another screen using KLF3361, which is identical to KLF3406 except that the shifty sequence (5'-GGG

TTT TAG C-3') encodes the stop codon UAG after the zero frame UUU (Table 1). We reasoned that the stop codon should increase ribosome pausing and thus the frequency of +1 frameshifting at this site. In this second screen, 31 independent mutations that mapped to 4 positions were identified (Table 3, Figures 2E-F). Mutation C1054U was isolated 28 times and stimulated +1 frameshifting (in KLF3361) to the largest degree (by 20-fold). Mutations C419U, G617A, and C735U were each isolated once and had considerably smaller effects (~2-fold). None of these mutations were able to suppress the +1 frameshift mutation of KLF3406.

Suppression of the +1 frameshift mutation in KLF3361 could result from an increased rate of peptidyl-tRNA^{Phe} slippage into the +1 frame and/or a decreased rate of RF1-dependent termination in response to the zero frame UAG. Previous studies have shown that mutation of A-site nucleotide C1054 can increase UAG readthrough and decrease the catalytic efficiency of RF1-dependent termination (Arkov et al., 1998; Arkov et al., 2000; Pagel et al., 1997), suggesting that the ability of C1054U to suppress the +1 frameshift in KLF3361 may stem primarily from an inhibition of RF1 activity. We also tested the purine subsitutions and found that C1054G increased suppression of this frameshift by eight-fold while C1054A had no effect (data not shown). Interestingly, this trend (U > G >> A) differs markedly from that seen for UGA nonsense suppression in KLF2723 (A > G > U) (McClory et al., 2010). Whether this is related to the distinct release factors involved in the two suppression events remains to be determined. Mutations C419U and G617A map to h16 and h21 (Figure 2E, F), respectively, clustering with mutations identified in the other elongation fidelity screens (Figure 1). These mutations modestly enhance frameshifting in KLF3361, perhaps by reducing the rate of RF1-dependent termination. Mutation C735U lies in h22 of the platform domain, near the h22/h23 junction (Figure 2F). Other mutations in h22 (at positions 666 and 739) that increase programmed -1 frameshifting were identified previously (Leger et al., 2007). All of these h22 mutations lie not far from the 30S E site, which has been shown to play a role in reading frame maintenance (Devaraj et al., 2009). It is possible that these h22 mutations indirectly destabilize codon-anticodon pairing in the E site and hence promote both +1 and -1 frameshifting.

5. Previous rRNA mutations

5.1. Mutations missed previously

In previous work, Gregory and Dahlberg (1995) isolated a number of nonsense (UGA) suppressor mutations in the rRNA genes. They mutagenized a strain containing an opal mutation within *trpA* [*trpA(UGA243)*] and a plasmid-borne *rrnB* operon, and selected for Trp+ prototrophs. They recovered three independent isolates of C1054U (16S), ten of C1200U (16S), one of G1491A (16S), and 11 of G1093A (23S). We obtained the same three 16S rRNA mutations (which localize in or near the A site) as well as mutations in h8, h12, h14, and h21, which were not recovered previously. Why the latter mutations were missed in the earlier study remains unclear, although several potential reasons are worth mentioning. One possibility is that the C1054U, C1200U, and G1491A alleles are more dominant in nature than the other alleles. The earlier study employed what we will refer to henceforth as the "conventional genetic system," in which both the plasmid-encoded mutant ribosomes and the chromosomally-encoded (wild-type) ribosomes translate the reporter gene (*trpA(UGA243)* in this case) (Gregory and Dahlberg, 1995). In this situation, certain mutant ribosomes (encoded by recessive alleles) may compete poorly for the reporter mRNA, thus masking their phenotypes. Another possibility is that certain alleles confer a growth defect in the conventional system and hence the corresponding strains were not recovered as Trp+ prototrophs. Consistent with this idea, A161G and G299A have been picked up in a screen for mutations that confer a dominant negative growth phenotype (Yassin et al., 2005). Finally, it may be that the only relevant difference involves the sensitivity of the two assays. Gregory and Dahlberg (1995) used a selection for Trp+, which presumably requires some threshold level of TrpA expression. Only a subset of mutations (e. g., C1054U, C1200U, G1491A) may have allowed sufficient readthrough of *trpA(UGA243)* to generate colonies on plates lacking tryptophan. In our system, C1200U and G1491A strongly increase UGA readthrough (by 30-fold and 9-fold, respectively) (Table 3; McClory et al., 2010). C1054U confers a weaker phenotype in KLF2723 but a very strong phenotype in KLF3361 (Table 3), suggesting that effects of C1054U may be highly context dependent.

5.2. Certain mutations reported previously to decrease fidelity cause loss of function in the specialized ribosome system

Several mutations in 16S rRNA were reported previously to decrease the fidelity of both initiation and elongation (O'Connor et al., 1997). In that study, a conventional system involving the reporter *lacZ* was used. Among the mutations associated with decreased fidelity were ΔC1400 and G529U. We did not recover any mutations near these positions and obtained only one mutation (G886A) that had an effect on both initiation and elongation fidelity (Figure 1). This prompted us to compare the effects of ΔC1400 and G529U in the two genetic systems (Table 4). In the specialized ribosome system, these mutations decreased translation from AUG by 60 to 70-fold, to near background levels, suggesting that ribosomes harboring ΔC1400 or G529U are largely inactive. These mutations also decreased translation from non-canonical start codons AUC and CUG by at least 40-fold, to the detection limit of the assay. In the *infC362* background, translation from AUC and CUG by ribosomes harboring ΔC1400 or G529U was detected, but reduced by >100-fold and ~60-fold, respectively, compared to the control strain. Neither mutation increased translation from AUC or CUG relative to AUG, based on the residual activities observed.

We considered the possibility that loss of activity conferred by the ΔC1400 and G529U mutations in our strains was due to reduced levels of mutant 30S subunits. To test this, we grew our strains in the presence of arabinose to induce expression of the mutant 16S rRNA, fractionated the corresponding lysates by sedimentation through sucrose gradients, and used primer extension to determine the relative amount of mutant 16S rRNA in the 30S, 70S, and polysome fractions (data not shown). For ΔC1400, mutant 16S rRNA was present in the 30S fraction at levels similar to the control strain (~20% of total) and was also detected in the 70S and polysome fractions, albeit at substantially lower levels. These data are inconsistent with the idea that reduced activity conferred by ΔC1400 results from decreased levels of 30S subunits. There was no evidence for partially assembled subunits from the A_{260} trace during gradient fractionation, further suggesting that ΔC1400 does not alter biogenesis of the 30S subunit. We tried to similarly analyze the distribution of 16S rRNA containing G529U in various ribosomal fractions. However, primer extension in this region resulted in a ladder of bands, regardless of the template, and thus the experiment could not be easily interpreted.

Table 4 Spurious initiation observed upon expression of 16S rRNA containing ΔC1400 and G529U depends on recognition of the reporter mRNA by endogeneous (wild type) ribosomes

Start Codon	16S allele	Specialized system[a]		Conventional system[b]
		infC+	*infC362*[c]	*infC+*
AUG	Vector only	11 ± 0.7	9.1 ± 2	NR
	Control (WT)	1500 ± 90	1300 ± 40	7300 ± 200
	ΔC1400	21 ± 0.4	18 ± 0.4	8700 ± 500
	G529U	24 ± 0.8	20 ± 0.6	6500 ± 600
AUC	Vector only	< 0.12[d]	< 0.12	NR
	Control (WT)	5.8 ± 0.3	30 ± 2	87 ± 5
	ΔC1400	< 0.12	0.26 ± 0.007	390 ± 20
	G529U	< 0.12	0.49 ± 0.03	480 ± 30
CUG	Vector only	< 0.12	< 0.12	NR
	Control (WT)	5.3 ± 0.2	30 ± 3	115 ± 13
	ΔC1400	< 0.12	0.19 ± 0.01	322 ± 4
	G529U	< 0.12	0.43 ± 0.05	326 ± 11

NR, not reported.

[a] This work. In these strains, the plasmid-encoded 16S rRNA contains an alternative (specialized) ASD sequence to allow specific translation of the *lacZ* reporter mRNA, which carries the complementary SD element. Values (mean ± SEM) are β-galactosidase units determined from the cleavage of CPRG, as described (Qin et al., 2007).

[b] From O'Connor et al., 1997. In these strains, the plasmid-encoded 16S rRNA contains a natural ASD, thus the corresponding ribosomes and the endogenous wild-type ribosomes both recognize the *lacZ* reporter mRNA. Values (mean ± SEM) are β-galactosidase Miller units.

[c] This mutation in the gene encoding IF3 (*infC*) has been shown previously to increase spurious initiation (Sussman et al., 1996).

[d] 0.12 represents the lower limit of detection.

These findings raise the possibility that increased spurious initiation, nonsense suppression, and frameshift suppression in the conventional system can be conferred indirectly by the mutant 30S subunits. This could, for example, be due to sequestration of translation factors or tRNAs by the mutant ribosomes, causing the endogenous wild-type ribosomes to translate with less fidelity. It will be of interest to further investigate whether mutant subunits can indirectly alter translation *in vivo*, as the results may hold relevance for numerous previous studies that employed the conventional system to characterize rRNA mutations.

6. Concluding remarks

In this chapter, we summarize our recent genetic studies of 16S rRNA. Using a specialized ribosome system with *lacZ* as the reporter, we screened for mutations that increase spurious initiation from AUC, increase miscoding by Glu-tRNA, increase readthrough of UGA, and increase +1 frameshifting at a site involving peptidyl-tRNA[Phe] slippage. Dozens of novel mutations were identified, implicating specific rRNA elements in different stages of translation. Mutations in h44 and the neck region decreased the fidelity of initiation, suggesting that conformational changes in h44 and the head domain are involved in start-codon selection (Qin and Fredrick, 2009). Mutations that stimulated miscoding clustered near interfaces between the shoulder domain and other parts of the ribosome, lending support to the domain closure model proposed by Ramakrishnan and coworkers. Subsequent analysis of ribosomes containing mutations in h8 and h14 showed that these elements negatively regulate GTP hydrolysis by EF-Tu to ensure accurate decoding (McClory et al., 2010). Further characterization of altered-function ribosomes identified in these screens should shed more light on the molecular mechanisms that govern protein synthesis.

Acknowledgements

This work was supported by grants from the National Institutes of Health (GM072 528) and the National Science Foundation (MCB0 840 996).

References

Abdi NM, Fredrick, K (2005) Contribution of 16S rRNA nucleotides forming the 30S subunit A and P sites to translation in *Escherichia coli*. RNA 11: 1624–1632

Allen PN, Noller HF (1991) A single base substitution in 16S ribosomal RNA suppresses streptomycin dependence and increases the frequency of translational errors. Cell 66: 141–148

Arkov AL, Freistroffer DV, Ehrenberg M, Murgola EJ (1998) Mutations in RNAs of both ribosomal subunits cause defects in translation termination. EMBO J 17: 1507–1514

Arkov AL, Freistroffer DV, Pavlov MY, Ehrenberg M, Murgola EJ (2000) Mutations in conserved regions of ribosomal RNAs decrease the productive association of peptide-chain release factors with the ribosome during translation termination. Biochimie 82: 671–682

Carter AP, Clemons WM, Jr., Brodersen DE, Morgan-Warren RJ, Hartsch T, Wimberly BT, Ramakrishnan, V (2001) Crystal structure of an initiation factor bound to the 30S ribosomal subunit. Science 291: 498–501

Cupples CG, Miller JH (1988) Effects of amino acid substitutions at the active site in *Escherichia coli* beta-galactosidase. Genetics 120: 637–644

Cupples CG, Miller JH (1989) A set of lacZ mutations in *Escherichia coli* that allow rapid detection of each of the six base substitutions. Proc Natl Acad Sci USA 86: 5345–5349

Dallas A, Noller HF (2001) Interaction of translation initiation factor 3 with the 30S ribosomal subunit. Mol Cell 8: 855–864

Devaraj A, Shoji S, Holbrook ED, Fredrick, K (2009) A role for the 30S subunit E site in maintenance of the translational reading frame. RNA 15: 255–265

Feliu JX, Villaverde, A (1998) Engineering of solvent-exposed loops in *Escherichia coli* beta-galactosidase. FEBS Lett 434: 23–27

Fu C, Parker, J (1994) A ribosomal frameshifting error during translation of the argI mRNA of *Escherichia coli*. Mol Gen Genet 243: 434–441

Gregory ST, Dahlberg AE (1995) Nonsense suppressor and antisuppressor mutations at the 1409–1491 base pair in the decoding region of *Escherichia coli* 16S rRNA. Nucleic Acids Res 23: 4234–4238

Haggerty TJ, Lovett ST (1997) IF3-mediated suppression of a GUA initiation codon mutation in the recJ gene of *Escherichia coli*. J Bacteriol 179: 6705–6713

Hui A, and de Boer HA (1987) Specialized ribosome system: preferential translation of a single mRNA species by a subpopulation of mutated ribosomes in *Escherichia coli*. Proc Natl Acad Sci USA 84: 4762–4766

Lee K, Holland-Staley CA, Cunningham PR (1996) Genetic analysis of the Shine-Dalgarno interaction: selection of alternative functional mRNA-rRNA combinations. RNA 2: 1270–1285

Leger M, Dulude D, Steinberg SV, Brakier-Gingras, L (2007) The three transfer RNAs occupying the A, P and E sites on the ribosome are involved in viral programmed -1 ribosomal frameshift. Nucleic Acids Res 35: 5581–5592

Lodmell JS, Dahlberg AE (1997) A conformational switch in *Escherichia coli* 16S ribosomal RNA during decoding of messenger RNA. Science 277: 1262–1267

Maisnier-Patin S, Paulander W, Pennhag A, Andersson DI (2007) Compensatory evolution reveals functional interactions between ribosomal proteins S12, L14 and L19. J Mol Biol 366: 207–215

McClory SP, Leisring JM, Qin D, Fredrick, K (2010) Missense suppressor mutations in 16S rRNA reveal the importance of helices h8 and h14 in aminoacyl-tRNA selection. RNA 16: 1925–1934

Moine H, Dahlberg AE (1994) Mutations in helix 34 of *Escherichia coli* 16 S ribosomal RNA have multiple effects on ribosome function and synthesis. J Mol Biol 243: 402–412

Murgola EJ, Pagel FT, Hijazi KA, Arkov AL, Xu W, Zhao SQ (1995) Variety of nonsense suppressor phenotypes associated with mutational changes at conserved sites in *Escherichia coli* ribosomal RNA. Biochem Cell Biol 73: 925–931

O' Connor M, Thomas CL, Zimmermann RA, Dahlberg AE (1997) Decoding fidelity at the ribosomal A and P sites: influence of mutations in three different regions of the decoding domain in 16S rRNA. Nucleic Acids Res 25: 1185–1193

Ogle JM, Brodersen DE, Clemons WM, Jr., Tarry MJ, Carter AP, Ramakrishnan, V (2001) Recognition of cognate transfer RNA by the 30S ribosomal subunit. Science 292: 897–902

Ogle JM, Ramakrishnan, V (2005) Structural insights into translational fidelity. Annu Rev Biochem 74: 129–177

Pagel FT, Zhao SQ, Hijazi KA, Murgola EJ (1997) Phenotypic heterogeneity of mutational changes at a conserved nucleotide in 16 S ribosomal RNA. J Mol Biol 267: 1113–1123

Qin D, Abdi NM, Fredrick, K (2007) Characterization of 16S rRNA mutations that decrease the fidelity of translation initiation. RNA 13: 2348–2355

Qin D, Fredrick, K (2009) Control of translation initiation involves a factor-induced rearrangement of helix 44 of 16S ribosomal RNA. Mol Microbiol 71: 1239–1249

Rosset R, Gorini, L (1969) A ribosomal ambiguity mutation. J Mol Biol 39: 95–112

Schmeing TM, Ramakrishnan, V (2009) What recent ribosome structures have revealed about the mechanism of translation. Nature 461: 1234–1242

Schmeing TM, Voorhees RM, Kelley AC, Gao YG, Murphy FV 4th, Weir JR, Ramakrishnan, V (2009) The crystal structure of the ribosome bound to EF-Tu and aminoacyl-tRNA. Science 326: 688–694

Selmer M, Dunham CM, Murphy FV 4th, Weixlbaumer A, Petry S, Kelley AC, Weir JR, Ramakrishnan, V (2006) Structure of the 70S ribosome complexed with mRNA and tRNA. Science 313: 1935–1942

Strigini P, Gorini, L (1970) Ribosomal mutations affecting efficiency of amber suppression. J Mol Biol 47: 517–530

Sussman JK, Simons EL, Simons RW (1996) *Escherichia coli* translation initiation factor 3 discriminates the initiation codon in vivo. Mol Microbiol 21: 347–360

Tapprich WE, Goss DJ, Dahlberg AE (1989) Mutation at position 791 in *Escherichia coli* 16S ribosomal RNA affects processes involved in the initiation of protein synthesis. Proc Natl Acad Sci USA 86: 4927–4931

Velichutina IV, Dresios J, Hong JY, Li C, Mankin A, Synetos D, Liebman SW (2000) Mutations in helix 27 of the yeast Saccharomyces cerevisiae 18S rRNA affect the function of the decoding center of the ribosome. RNA 6: 1174–1184

Villa E, Sengupta J, Trabuco LG, LeBarron J, Baxter WT, Shaikh TR, Grassucci RA, Nissen P, Ehrenberg M, Schulten K, Frank, J (2009) Ribosome-induced changes in elongation factor Tu conformation control GTP hydrolysis. Proc Natl Acad Sci USA 106: 1063–1068

Wimberly BT, Brodersen DE, Clemons WM, Jr., Morgan-Warren RJ, Carter AP, Vonrhein C, Hartsch T, Ramakrishnan, V (2000) Structure of the 30S ribosomal subunit. Nature 407: 327–339

Yassin A, Fredrick K, Mankin AS (2005) Deleterious mutations in small subunit ribosomal RNA identify functional sites and potential targets for antibiotics. Proc Natl Acad Sci USA 102: 16 620–16 625

Zhang W, Dunkle JA, Cate JH (2009) Structures of the ribosome in intermediate states of ratcheting. Science 325: 1014–1017

Decoding and deafness: Two sides of a coin

20

Rashid Akbergenov, Dmitry Shcherbakov, Tanja Matt, Stefan Duscha, Martin Meyer,
Déborah Perez Fernandez, Rashmi Pathak, Shinde Harish, Iwona Kudyba,
Srinivas R. Dubbaka, Sandrina Silva, Maria del Carmen Ruiz Ruiz, Sumantha Salian,
Andrea Vasella and Erik C. Böttger

1. The ribosome as drug target: the issue of selectivity

Antibiotics used in clinical medicine for the treatment of infectious diseases frequently target bacterial protein synthesis, as illustrated by macrolides, ketolides, lincosamides, oxazolidinones, aminoglycosides, and tetracyclines (Gale et al., 1981). In general, antibiotics target the ribosome at sites of functional relevance, e. g. the sites of decoding, translocation, and peptidyl transfer. The emergence of antibiotic resistance and the toxicity associated with some of the available agents ask for a further exploitation of the ribosome as a drug target.

The principle of antimicrobial chemotherapy dates back to Paul Ehrlich's concept of selective toxicity, i. e. specificity. The ribosome is a highly conserved structure present in all three kingdoms of life, archaea, bacteria, and eukarya. How can ribosomal inhibitors satisfy the principle of selective toxicity and discriminate between, e. g. bacterial and eukaryotic ribosomes? Compared to the bacterial ribosome, the eukaryotic ribosome comes in two flavors: the cytoplasmic and the mitochondrial ribosome. The components of the cytoribosome are encoded by chromosomal genes as are the mitoribosomal proteins, while the rRNA components of the mitoribosome are encoded by the mitochondrial genome.

An important means to understand a ribosomal inhibitor's basis for specificity or lack thereof is the analysis of bacterial resistance mutations (Edlind, 1989; Mathis et al., 2004; Mathis et al., 2005). Central here is the concept of "informative sequence positions" – i. e., the search for polymorphic residues (amino acid position of a ribosomal protein, nucleotide residue of ribosomal RNA) as a determinant of drug resistance in bacteria (Bottger et al., 2001; Bottger, 2007). The iden-

tification of a polymorphic residue as a determinant of ribosomal resistance provides information about the specificity of a ribosomal antibiotic, and particularly an answer to the question as to whether and why a drug affects the bacterial as opposed to the eukaryotic ribosome. The rationale for this hypothesis can best be illustrated by focusing on two examples which represent the very extremes: Macrolides and hygromycin B.

Macrolide antibiotics show substantial selectivity. Bacterial ribosomes are highly susceptible to macrolide antibiotics, while in eukaryotes both mitochondrial and cytoplasmic ribosomes are naturally resistant to these agents. As a result, macrolide antibiotics are characterized by the virtual absence of target-related toxicity. Different mutational alterations are associated with bacterial resistance to macrolides and these affect either nucleotides of 23S rRNA or ribosomal proteins L4 and L22 (Sigmund et al., 1984; Sander et al., 1997; Pfister et al., 2004; Poehlsgaard and Douthwaite, 2005). Particularly instructive is an A→G alteration in 23S rRNA which is associated with high-level bacterial resistance to macrolides (Sander et al., 1997; Pfister et al., 2004). *In-silico* sequence analyses revealed that both the eukaryotic cytoribosome and the eukaryotic mitoribosome carry a guanine at 23S rRNA position 2058 (Sor and Fukuhara, 1982; Bottger et al., 2001), readily explaining the natural non-susceptibility of these structures to macrolide compounds. It is a matter of ongoing debate as to whether additional, as yet unidentified ribosomal polymorphisms contribute to this effect (Bommakanti et al., 2008).

Hygromycin B is a universal inhibitor of translation which lacks selectivity, affecting ribosomes from archaea, bacteria and eukarya (Gale et al., 1981). Mutational alterations conferring bacterial drug resistance all map to the hygromycin-binding site within helix 44 (h44) of 16S rRNA. Significantly, drug-resistance-con-

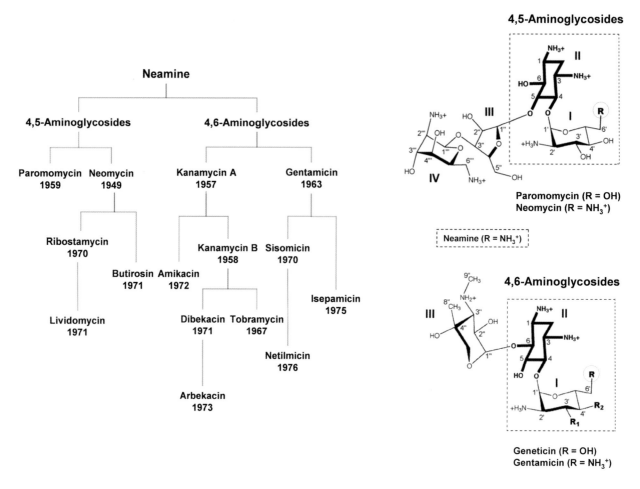

Fig. 1 The structural relationships of the 2-deoxystreptamine family of antibiotics.

ferring rRNA mutations are restricted to universally conserved nucleotides (Pfister et al., 2003b). The observation that ribosomal alterations mediating resistance to hygromycin B exclusively involve universally conserved nucleotides within rRNA explains the lack of specificity and general toxicity of this ribosomal inhibitor.

2. Aminoglycosides

2.1. Chemistry

Aminoglycosides form a large family of water-soluble, polycationic amino sugars that are used as broad-spectrum antibacterial agents. In the 60 years since their discovery, the aminoglycoside antibiotics have seen unprecedented use. Discovered in the 1940s (Schatz et al., 1944), these compounds were the long-sought remedy for tuberculosis and other serious bacterial

infections. Over the following decades, numerous new aminoglycosides were isolated and many new derivatives were synthesized (see Figure 1). The adverse side effects associated with all of them (Chambers, 1996), in particular renal toxicity (reversible) and ototoxicity (irreversible), led to a decline of their use in most countries in the 1970s and 1980s. Nevertheless, aminoglycosides are still among the most commonly used antibiotics worldwide due to their high efficacy, lack of drug-related allergy, and low cost.

Aminoglycosides are produced by different strains of soil actinomycetes. Aminoglycoside antibiotics are low-molecular-weight molecules of approximately 300–900 daltons. All natural and semi-synthetic aminoglycosides share a similar structure which is characterized by three to four rings. The most clinically relevant group of aminoglycosides, the 4,5-disubstituted and 4,6-disubstituted 2-deoxystreptamines, are characterized by a common neamine core, constituted by a glucopyranosyl moiety (ring I) glycosidically

A

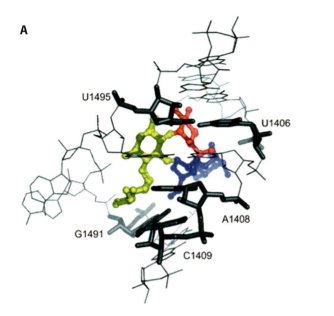

B

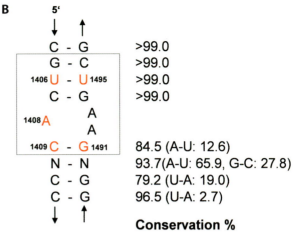

Fig. 2 (A) View of the three-dimensional structure of the A site in complex with 4,5- and 4,6-disubstituted 2-deoxystreptamines: the common neamine core is denoted in yellow; ring III of the 4,6-aminoglycosides (tobramycin) is denoted in red; rings III and IV of the 4,5-aminoglycosides (paromomycin) are denoted in blue. Note that A1492 and A1493 are shown in their flipped-out conformation. (B) Secondary structure of the decoding A site in bacteria (the degree of conservation is indicated). (Reprinted from Trends in Biotechnology 24: 145–147, 2006. © 2006 with permission from Elsevier)

linked to position 4 of a 2-deoxystreptamine (ring II). The core is further substituted by one or two glycosyl residues connected to position 5 or 6 of the deoxystreptamine moiety (see Figure 1). Several amino and hydroxyl groups are attached to the various rings of the structure, i. e., individual aminoglycosides differ from each other by the number, position and type of substituents.

2.2. Drug binding pocket

Accurate decoding is central in protein synthesis. It occurs at the aminoacyl-tRNA decoding site (A site) on the small ribosomal subunit where interactions between the aa-tRNA anticodon and the mRNA codon are recognized. Aminoglycosides bind to the decoding A site that corresponds to h44 in bacterial 16S rRNA. The structure of many of the aminoglycosides complexed to their rRNA target has been determined by X-ray crystal structure analysis, revealing the details of drug-target interactions at atomic resolution (Carter et al., 2000; Vicens and Westhof, 2001, 2003; Francois et al., 2005). A schematic view of the three-dimensional structure of the decoding site complexed with 4,5- and 4,6-aminoglycosides is shown in Figure 2A. The neamine core of these compounds binds in a similar fashion, i. e. ring I stacks upon G1491 and contacts A1408 (*Escherichia coli* numbering). The additional saccharide units, attached to either position 5 or 6 of the 2-deoxystreptamine, are oriented differently: (i) for 4,5-aminoglycosides, ring III reaches down the drug binding pocket towards nucleotide G1491, (ii) for 4,6-aminoglycosides, ring III reaches up the drug binding pocket towards residue U1406.

The common neamine core formed by rings I and II of the 2-deoxystreptamines is mainly responsible for proper drug binding. The binding pocket is closed at the upper site by base-pair G1405-C1496 and at the lower stem by base-pair C1409-G1491. Ring I intercalates into the internal loop formed by A1408, A1492 and A1493, and is crucial for the insertion into the decoding site. Despite the variations in chemical composition of the aminoglycosides, ring I binds always in the same orientation – it stacks upon G1491 and forms a pseudo base-pair interaction with the Watson-Crick edge of adenine 1408. Here, the ring oxygen of ring I accepts a hydrogen from the N6 of adenine, and the (protonated) amino- or hydroxyl group at position 6' donates a hydrogen to N1 of adenine, accounting for two direct hydrogen bonds between ring I and A1408 (Vakulenko and Mobashery, 2003; Francois et al., 2005). Additionally, hydroxyl groups at positions 3' and 4' of ring I can provide hydrogen bonds to the phosphate groups of A1492 and A1493, thereby further stabilizing the location of ring I. Ring II forms hydrogen bonds to G1494 and U1495 as well as to the phosphate groups of A1493 and G1494. Rings III and IV of the 4,5-aminoglycosides reach down the stem towards base-pair 1409–1491 and 1410–1490, enabling the hydroxyl group at position 5" of ring III to contact N7 of G1491. Ring III of the 4,6-aminoglycosides is in a different location and forms hydrogen bonds to the Hoogsteen side of G1405.

There is a large and diverse population of about 50 different aminoglycoside-modifying enzymes which transfer acetyl, phosphoryl, and adenyl groups in a cofactor-dependent manner to virtually every amino or hydroxyl group (for review, see (Vakulenko and Mobashery, 2003)). Most of these modifications will have two effects – disruption of a specific contact, e. g. hydrogen bonding, and introduction of a bulky side chain that interferes with drug conformation and positioning. The 2'- and 6'-NH_3^+ groups of ring I as well as the 3 position of ring II are acetylated by the action of acetyl coenzyme A-dependent N-acetyltransferases. The introduction of an acetyl group at 6'-NH_3^+ disrupts the hydrogen bond between the 6'-NH_3^+ moiety and A1408 and results in high-level drug resistance. Modification of the 2'-NH_3^+ group most likely influences the overall conformation by introducing a bulky side chain that interferes with proper positioning. The amino group at position 3 of ring II makes hydrogen-bonded contacts to N7 of G1494 and the phosphate groups of A1493 or G1494; acetylation of 3-NH_3^+ would prevent the formation of these interactions. Adenylation of 4'-OH of ring I as well as phosphorylation of ring I 3'-OH would prevent the insertion of ring I into the A site and H-bonding to the phosphate oxygens O-2 of A1493 and A1492, respectively.

2.3. Mechanisms of action

Aminoglycosides affect protein synthesis by inducing codon misreading and by inhibiting translocation of the tRNA-mRNA complex (Davies and Davis, 1968; Benveniste and Davies, 1973; Cabanas et al., 1978). The intimate connection between these two seemingly unrelated activities has been rationalized by X-ray crystallography and an elegant series of kinetic experiments. Upon binding of the 2-deoxystreptamines into the A-site loop, residues A1492 and A1493 are displaced; they are flipped out and positioned to interact with the shallow/minor groove of the codon-antico-

don helix (Carter et al., 2000). Such a conformational switch of A1492 and A1493 is normally observed only upon accommodation of a cognate tRNA-mRNA complex, thus providing a molecular explanation for the miscoding properties of these compounds (for review (Ogle and Ramakrishnan, 2005)). By doing so, aminoglycosides increase the affinity of the tRNA to the A site, i. e. they stabilize the pre-translocation state of the ribosome, and thereby increase the energy barrier of translocation (Peske et al., 2004; Feldman et al., 2010). In addition, the conformational change of the ribosome with A1492 and A1493 arrested in their flipped-out configuration presumably blocks conformational changes that are required for tRNA movement. Aminoglycosides do bind reversibly to the ribosome (Feldman et al., 2010) and, depending on the drug's off-rate, translocation may resume.

Besides miscoding, aminoglycosides also induce stop-codon read-through (Palmer et al., 1979; Singh et al., 1979). This feature has resulted in attempts to exploit these compounds for treatment of genetic diseases associated with premature stop codons (Howard et al., 1996; Hainrichson et al., 2008). The termination of protein synthesis occurs through the specific recognition of a stop codon in the A site of the ribosome by a release factor (RF), which then catalyzes the hydrolysis of the nascent protein chain from the P-site peptidyl-tRNA. The conformational changes required for release factor binding and peptide release are notably different from those of tRNA binding. In particular, the aminoglycoside-induced flipping out of A1492 and A1493 results in a steric clash with RF binding and prevents correct docking (Laurberg et al., 2008; Weixlbaumer et al., 2008). Consequently, aminoglycosides promote tRNA binding to result in miscoding, and abolish RF binding (Brown and Tate, 1994; Salas-Marco and Bedwell, 2005; Youngman et al., 2007), both consequences favoring stop codon read-through.

2.4. Toxicity

For more than 40 years, a major drawback to the use of aminoglycosides has been their ototoxicity, as aminoglycoside treatment may result in the destruction of sensory hair cells of the inner ear and a consequent permanent loss in hearing ability. Despite efforts to limit the antibiotic dose, hearing loss afflicts more than one fourth of patients receiving these drugs (Chambers, 1996). The outer hair cells of the mammalian cochlea are preferentially and irreversibly destroyed by

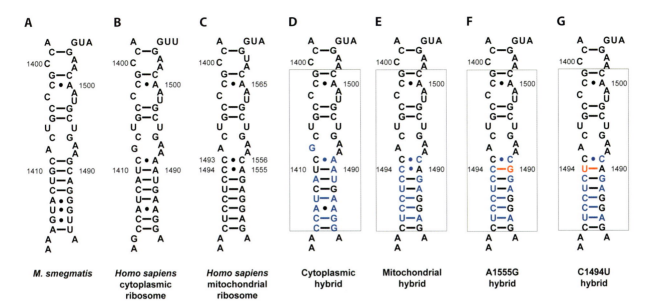

Fig. 3 Secondary-structure comparison of decoding-site rRNA sequences in the small ribosomal subunit. (A) Decoding region of 16S rRNA h44 in bacterial ribosomes of *M. smegmatis*; rRNA nucleotides are numbered according to the bacterial nomenclature, i. e., to homologous *E. coli* 16S rRNA positions. (B) Homologous 18S rRNA sequence in human cytoplasmic ribosomes; rRNA residues are numbered according to the bacterial nomenclature. (C) Homologous 12S rRNA sequence in human mitochondrial ribosomes; rRNA residues are numbered according to the mitochondrial nomenclature: mtDNA 1494 and 1555 correspond to *E. coli* 16S rRNA 1410 and 1490. (D-G) Decoding site rRNA of human-bacterial hybrid ribosomes. The transplanted helix is boxed, and nucleotide positions depicted in blue represent residues that are specific for human rRNA. Mutations that are associated with congenital deafness, mtDNA position 1555A to G (F), and 1494C to U (G) are highlighted in red. (Modified from Proc Natl Acad Sci USA 105: 3244–3249, 2008, with permission of the publisher © 2008 National Academy of Sciences, USA)

aminoglycosides, although, after prolonged treatment, the inner hair cells are also affected. Although the physiological and morphological aspects of aminoglycoside ototoxicity have been extensively studied in the past decades, research on the mechanisms involved has been frustrating, and the molecular mechanisms underlying toxicity are not well understood. In addition, neither drug uptake mechanisms nor intracellular targets have been elucidated. While it is in general agreed that aminoglycosides have to be taken up by hair cells in order to become cochleatoxic, there is no direct correlation between the concentration reached in the inner ear by different aminoglycosides and the magnitude of their ototoxic potential. These and other findings (reviewed in (Henley and Schacht, 1988)) make the early suggestion untenable that pharmacokinetic characteristics, i. e. selective tissue accumulation/penetration of aminoglycosides in the inner ear, are responsible for organ-specific damage.

The identification of the molecular target and the mechanisms by which aminoglycosides exert their toxic effects remained controversial. While studies have been plentiful, they mostly yielded anecdotal evidence of potential sites of drug action without proving

causal relations to cell destruction. Aminoglycosides reportedly exhibit a large number of documented effects on various sub-cellular structures and metabolic pathways, including effects on DNA, RNA, and protein synthesis, energy metabolism and ion transport, and synthesis or degradation of prostaglandins, gangliosides, mucopolysaccharides, and lipids (reviewed in (Schacht, 1986)). It is unclear which – if any – of these is involved in the compounds' ototoxicity, as several of these actions relate to the positive charge that aminoglycosides carry at physiological pH and that enable them to bind to a variety of negatively charged cell components, or to displace cations from their binding sites. In addition, such a multitude of potential target sites is unlikely to represent the cause of toxicity. It is, however, the primary damage-initiating event that needs to be known, if one wishes to ameliorate the adverse side effects of aminoglycosides. As it is, none of these observations establishes any firm causal relationship to the compounds' in-vivo ototoxicity.

Circumstantial evidence points to a possible role for the mitoribosome in aminoglycoside ototoxicity: (i) Mitochondrial ribosomes are more closely related to the bacterial ribosome than to the eukaryotic cytoplasmic ribosome; (ii) antibiotics which exhibit *in-vitro* activity on mitoribosomes are associated with toxicity *in vivo*, as it is the case for chloramphenicol and linezolid (Bottger 2007); (iii) attenuation of aminoglycoside ototoxicity by antioxidants relates to reactive oxygen species, oxidation and mitochondrial respiration (Garetz, Altschuler et al. 1994; Clerici, Hensley et al. 1996; Sha and Schacht 1999; Sha, Qiu et al. 2006; Chen, Huang et al. 2007). Ototoxicity of aminoglycoside antibiotics occurs in both a sporadic dose-dependent and an inherited fashion. The latter is associated with mutations in mitochondrial rRNA, i. e. A1555G and C1494U (Prezant et al., 1993; Zhao et al., 2004); in wild-type 12S rRNA, C1494forms a non-canonical base pair with A1555. Both mutations are found in the penultimate helix of the mitoribosomes' small subunit 12S rRNA, which is part of the A site, and replace a non-canonical base-pair with a Watson-Crick base-pair (see Figure 3). These mutations not only sensitize carriers to aminoglycosides but also predispose them to non-syndromic hearing loss (reviewed in (Fischel-Ghodsian, 1999)). It is most intriguing that the clinical phenotype of homoplasmic carriers with the corresponding mitochondrial rRNA mutation, which is present in all mitochondria throughout various tissues, is apparently limited to defects of the organ of Corti (cochlea). Based on this fortuitous genetic predisposition, one can ask the question whether there is a common link, or a common mechanism, involved in the pathogenesis of aminoglycoside-induced hearing loss, predisposition of the A1555G/C1494U mutations to non-syndromic deafness, and A1555G/C1494U associated hyper-susceptibility to aminoglycoside ototoxicity.

3. Establishment of a genetic model to study eukaryotic A-site function

The presence of several hundreds to thousands of mitochondrial organelles in a single eukaryotic cell, each organelle carrying multiple copies of its genome, made genetic manipulations of mitochondrial rRNA in higher eukaryotes impossible and frustrated any such attempts. Mainly because of the absence of suitable experimental models, the mechanistic link – if any – between mitochondrial rRNA mutation and disease has remained largely elusive.

For this reason and to establish an experimental and genetically tractable model for the study of human A-site function, we performed domain shuffling experiments (Hobbie et al., 2007; Hobbie et al., 2008b). The successful replacement of A-site residues in h44 of bacterial 16S rRNA with that of various eukaryotic homologs has demonstrated that the A-site rRNA behaves as an autonomous domain that can be exchanged between different species for study of function. Replacement of a 34-nucleotide part of h44 of bacterial 16S rRNA with the corresponding human homolog resulted in rRNA decoding sites corresponding to those in cytosolic and mitochondrial (wild-type and mutant) human ribosomes. Thus, we essentially exchanged the bacterial A site in h44 of 16S rRNA with various versions of its eukaryotic counterpart, resulting in bacterial hybrid ribosomes with a fully functional eukaryotic decoding site (Figure 3). These hybrid ribosomes provide a unique opportunity to study aminoglycoside susceptibility of the various eukaryotic ribosomes potentially targeted by aminoglycosides, i. e., cytoplasmic ribosomes, mitochondrial ribosomes, and mitochondrial C1494U and A1555G mutant ribosomes (Hobbie et al., 2008a).

To establish and validate the hybrid ribosome approach, we compared the drug susceptibility pattern of cytoplasmic-bacterial hybrid ribosomes, i. e. bacterial ribosomes with a human cytoplasmic A site, to that of rabbit reticulocyte ribosomes (the decoding site rRNA of human cytosolic ribosomes is identical to that of most vertebrates, including rabbit). Using various assays of translation activity, the susceptibility of the cytoplasmic-bacterial hybrid ribosomes towards 4,5- and 4,6-aminoglycosides was found to be virtually identical to that of rabbit reticulocyte ribosomes, i. e. native eukaryotic ribosomes (Table 1). Given that various components of the ribosome contribute to decoding, such as ribosomal proteins S4 and S12 (Ogle and Ramakrishnan, 2005), it is remarkable that bacterial ribosomes carrying only the central part of eukaryotic cytosolic h44 show a drug susceptibility pattern that is super-imposable on that of native eukaryotic ribosomes (Hobbie et al., 2007). These data indicate (i) that phylogenetically variable components of the ribosome besides h44 contribute little to aminoglycoside susceptibility, (ii) that the A site behaves as an autonomous domain which can be dissected from the remaining ribosome and shuffled between ribosomes of different phylogenetic origins for study of function, and (iii) that transplanting h44 of the eukaryotic cytoplasmic ribosome's decoding region is a valid approach for studying the specificity of aminoglycoside antibiotics. More recently, gene replacement techniques have been used to study the ribosomal RNA determinants

Table 1 Drug susceptibility of bacterial ribosomes, bacterial-cytoplasmic hybrid ribosomes and rabbit reticulocyte ribosomes. Drug susceptibility was assessed by aminoglycoside-induced inhibition of luciferase synthesis (IC50, μM). IC50 values represent the drug concentrations in μM that are required to inhibit synthesis of functional luciferase to 50%.

	Bacterial ribosomes	Cytoplasmic hybrid ribosomes	Rabbit reticulocyte ribosomes
Paromomycin	0.03	4.7	6.5
Neomycin	0.04	13	18
Netilmicin	0.05	64	58
Kanamycin	0.05	86	67
Geneticin	0.03	0.8	0.2

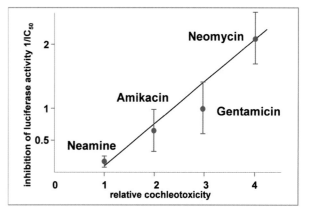

Minimal inhibitory concentrations (MIC, μg/ml)

	Mitochondrial hybrid ribosomes	Cytoplasmic hybrid ribosomes
Neomycin	16 - 32	> 1024
Amikacin	32 - 64	> 1024
Gentamicin	64 - 128	> 1024
Tobramycin	64 - 128	> 1024
Neamine	1024	> 1024

Fig. 4 Top: Drug susceptibility of bacterial-eukaryotic hybrid ribosomes carrying the mitochondrial versus cytoplasmic A site. Drug susceptibility was assessed by determination of minimal inhibitory drug concentrations towards recombinant bacterial cells with the indicated hybrid ribosome. Bottom: Aminoglycoside-mediated inhibition of luciferase expression in mitochondrial hybrid ribosomes correlates with cochleatoxicity.

of aminoglycoside resistance in the lower eukaryote *Saccharomyces cerevisiae*. Residues G1645 (homologous to *E. coli* A1408) and A1754 (homologous to *E. coli* G1491) of yeast cytoplasmic rRNA were identified as key nucleotides for drug binding (compare to the A site of the human cytoplasmic small ribosomal subunit in Figure 3). Alteration of G1645 and A1754 in yeast to their corresponding eubacterial residue, i. e., G1645 to A and A1754 to G, resulted in aminoglycoside susceptibility at a level similar to that of *E. coli* (Fan-Minogue and Bedwell, 2008). These results corroborate the conclusions from bacterial site-directed mutagenesis studies ((Pfister et al., 2003a; Hobbie et al., 2005; Pfister et al., 2005; Hobbie et al., 2006b), for review (Hobbie et al., 2006a)) and investigations on bacterial-eukaryotic hybrid ribosomes that residues 1408 and 1491 are key to selectivity.

Testing the mitochondrial hybrids for aminoglycoside susceptibility revealed that, in comparison to the natural resistance of cytoplasmic ribosomes, the mitochondrial ribosome is characterized by a more heterogeneous pattern of drug susceptibility (see Figure 4) that varies with the aminoglycoside studied. Importantly, the potency of a series of aminoglycosides in inhibiting mitochondrial hybrid ribosome function correlated with the relative cochleatoxicity of the respective compound in humans (Hobbie et al., 2008a). The correlation between these two measures is consistent with the hypothesis that aminoglycoside-induced cochleatoxicity relates to the drugs' activity against mitochondrial ribosomes.

Why do mitochondrial mutations A1555G and C1494U predispose to deafness and why are they associated with hyper-susceptibility to aminoglycoside-induced ototoxicity? Previous interpretations of the

pathogenic mechanisms involved (Guan, 2006) are hampered by a misalignment of rRNA residues (dating back to as early as 1993) postulating *E. coli* positions 1409-1491 as homologous to mitochondrial residues 1494-1555 (Hutchin et al., 1993). Refined alignments have unequivocally established that mitochondrial 12S rRNA positions C1494 and A1555 are not equivalent to bacterial 16S rRNA positions 1409 and 1491, but are instead homologous to bacterial 16S rRNA positions 1410 and 1490 (see Figure 3). Referring to studies of aminoglycoside susceptibility in bacterial ribosomes with alterations of the C1409-G1491 interaction is misleading when addressing the hyper-susceptibility of C1494U and A1555G mutant mitochondria to aminoglycosides (De Stasio et al., 1989; De Stasio and Dahlberg, 1990), since both mitochondrial 12S rRNA wild-type and C1494U or A1555G mutants have a C-C opposition at the corresponding *E. coli* positions

Table 2 Drug susceptibility of bacterial wild-type mitochondrial hybrid ribosomes (C1493-C1556), bacterial mitochondrial C1556G mutant hybrid ribosomes (C1493-1556G) and bacterial ribosomes (C1409-G1491). Drug susceptibility was assessed by MIC determinations (mg/L).

	Neomycin	Gentamicin	Tobramycin	Kanamycin
Bacterial-mitochondrial wt hybrids (C1493-C1556)	16.0–32.0	64.0–128.0	64.0–128.0	128.0
Bacterial-mitochondrial mutant hybrids C1556G (C1493-1556G)	1.0	1.0	1.0	1.0–2.0
Bacteria (C1409-G1491)	1.0	1.0	1.0	1.0–2.0

1409-1491. Thus, investigations on the C1409-G1491 interaction in bacteria can hardly serve as a model for the mitochondrial 12S rRNA alterations. In an effort to unambiguously assign the mitochondrial positions to the corresponding bacterial homologs, we constructed mitochondrial hybrid ribosomes with a C1556G replacement, resulting in a C1493-1556G Watson-Crick pair, identical to the bacterial C1409-G1491 homolog. The drug susceptibility pattern of the corresponding recombinant mutants was identical to that of bacterial ribosomes with a C1409-G1491 pair (see Table 2). These data unambiguously demonstrate that the mitochondrial 1493-1556 base pair is homologous to the bacterial 1409-1491 base pair and forms the base of the A-site loop.

Cell-free translation assays of the various bacterial-mitochondrial hybrid ribosomes revealed that the pathogenic mutations A1555G and C1494U affect translation fidelity. In comparison to bacterial hybrid ribosomes with a wild-type mitochondrial A site, the bacterial hybrid ribosomes with a deafness-associated mutant mitochondrial A site exhibited significantly reduced ribosomal accuracy (Hobbie et al., 2008b). In addition, and compared to the wild-type mitochondrial hybrids, the mitochondrial hybrids with an A1555G or a C1494U mutation are much more susceptible to aminoglycoside-induced inhibition of protein synthesis (by at least one order of magnitude) (see Table 3). Footprinting studies demonstrated that the mutants' increased susceptibility to drug action is associated with increased affinity for the aminoglycoside. Further studies revealed that the inherent infidelity of the A1555G and C1494U mutant mitochondrial hybrid ribosome in protein synthesis is excessively aggravated by aminoglycoside antibiotics (Hobbie et al., 2008a).

Taken together, these results converge on mitochondrial mistranslation as a key element in the pathogenesis of both mitochondrial rRNA polymorphism-associated and aminoglycoside-induced deafness. The results described above provide experimental support for aminoglycoside-induced dysfunction of the mitochondrial ribosome and experimental evidence for

Table 3 Mitochondrial A1555G and C1494U mutant hybrid ribosomes and aminoglycoside susceptibility. Drug susceptibility was assessed by whole cell assays and MIC determinations (top) and in cell-free translations assays by defining IC_{50} inhibitory values (bottom).

Minimal inhibitory concentrations (MIC, µg/ml)

	Mitochondrial hybrid	Mitochondrial A1555G mutant hybrid	Mitochondrial C1494U mutant hybrid
Gentamicin	64–128	16–32	16–32
Tobramycin	64–128	16	16
Amikacin	32–64	2–4	2–4

Inhibition of luciferase synthesis (IC_{50}, µM)

	Mitochondrial hybrid	Mitochondrial A1555G mutant hybrid	Mitochondrial C1494U mutant hybrid
Gentamicin	5.7	0.6	0.7
Tobramycin	7.8	0.8	0.9
Amikacin	7.0	0.4	0.5

a mechanistic link between mitochondrial A1555G and C1494U mutations and hyper-susceptibility to aminoglycosides. The exquisite tissue-specific action of aminoglycoside toxicity (that is, ototoxicity) is likely to involve additional factors, e. g. susceptibility to reactive oxygen species, drug uptake. Based on our experimental findings we suggest a unifying mechanism for both sporadic dose-dependent and inherited hyper-susceptibility to aminoglycoside-induced ototoxicity. This mechanism takes place at the intracellular level and involves dysfunction of mitochondrial protein synthesis. The hyper-susceptibility of mitochondrial A1555G and C1494U alterations to aminoglycosides is due to two mechanisms that act in concert: mutation-mediated increase in drug affinity and drug-induced exacerbation of the mutants' inherent defect in translational accuracy.

Disease pathogenesis due to mistranslation may involve two mechanistically different aspects of faulty protein synthesis: synthesis of dysfunctional proteins and the cellular response to misfolded proteins. The proteins encoded by human mitochondrial DNA and translated by mitochondrial ribosomes are subunits of the respiratory chain and oxidative phosphorylation pathway at the inner mitochondrial membrane. Lymphoblastoid cell lines with mutation A1555G or C1494U in mitochondrial 12S rRNA have been reported to show a decreased rate of oxygen consumption indicating impaired activity of the respiratory chain (Guan et al., 2001). Increased mistranslation of mitochondrial genes might also account for a misfolded protein response (Schroder and Kaufman, 2005), resulting in the cochlear alterations observed in symptomatic and asymptomatic carriers of the A1555G mutation (Bravo et al., 2006).

4. Aminoglycoside drug development

Mainly in the early 1970s, numerous new aminoglycosides were isolated and many derivatives were synthesized (see Figure 1). The known adverse effects of aminoglycosides, i. e. reversible nephrotoxicity and irreversible ototoxicity, are common to all of them (Chambers, 1996). Attempts by medicinal chemists to improve their antibacterial activity were invariably associated with increases in toxicity – thus, the more potent the compound was as an antibacterial, the more toxic it was. The chemical synthesis process was driven by trial and error, focusing on antibacterial potency, and there was little rationale that would allow the

separation of antibacterial activity from toxicity. Having identified mitochondrial mistranslation as a key element in aminoglycoside-induced deafness (Hobbie, Akshay et al. 2008), it constitutes a formidable challenge to build upon this insight and to develop aminoglycosides which are more selective and consequently may not be plagued by toxicity related to their mechanism of action.

Nucleotides 1408, 1409 and 1491 of 16S rRNA of the small ribosomal subunits' A site are critical for positioning ring I of the 2-deoxystreptamine aminoglycoside compounds (Pfister et al., 2003a; Hobbie et al., 2005; Pfister et al., 2005; Hobbie et al., 2006b, for review Hobbie et al., 2006a). The interspecies variability of these nucleotide residues provides the basis for the specificity and selectivity of aminoglycosides (Recht et al., 1999; Bottger et al., 2001; Bottger, 2007; Hobbie et al., 2008a). The drug binding site is distinctly different between bacterial ribosomes and those from the cytoplasm and mitochondria of eukaryotes (see Figure 3). The decoding A site in bacteria is characterized by A1408, the 1409–1491 base pair (84% C-G, 13% A-U) and a Watson-Crick pair between residues 1410–1490 (66% A-U, 28% G-C). In *Homo sapiens*, the cytoplasmic ribosome's A site is defined by G1408, a C1409-A1491 interaction and a U1410-A1490 pair; the mitochondrial A site is characterized by A1408, a C1409-C1491 (1493-1556) interaction and a C1410-A1490 (1494-1555) interaction (*E. coli* numbering with the corresponding mitochondrial numbering in brackets). In addition, the human mitochondrial A site is characterized by disease-associated polymorphisms, i. e. A1555G and C1494U (see above).

The link between deafness and malfunction of the mitochondrial ribosome allows for the hypothesis that aminoglycoside ototoxicity is related to drug mechanism of action and intimately associated with the compounds' antimitoribosomal activity. This hypothesis can in principle be tested, provided that it is possible to separate the antibacterial activity of aminoglycosides from the compounds' activity towards the eukaryotic ribosome. Is this a reasonable assumption, given the minute differences in the drug binding pocket? Is it possible to redirect binding at the level of target specificity, so as to develop aminoglycosides which are more selective? If possible, the corresponding compounds will allow for challenging the hypothesis experimentally and to provide proof that the drug-induced malfunction of the mitochondrial ribosome is indeed the key to aminoglycoside ototoxicity. Demonstrating the proof-of-principle would provide the required and

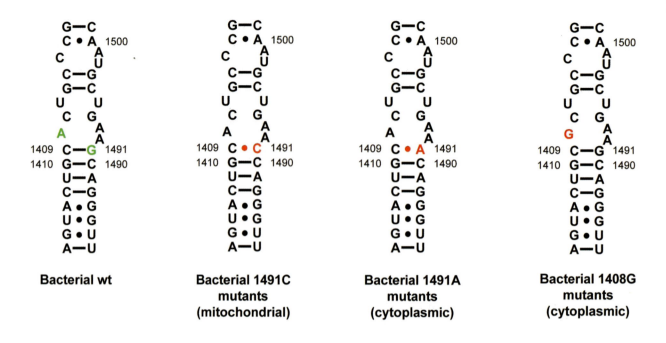

Fig. 5 Use of genetic mutants in drug development. The key residues involved in specificity, i.e. bacterial 1408A and 1491G are shown in green, the corresponding eukaryotic homologous positions in red.

long-sought rationale for a hypothesis-driven chemical synthesis approach – with the aim of synthesizing aminoglycoside compounds that have lost their antimitoribosomal activity but retained their activity towards bacterial ribosomes.

The available crystal structures of drug-ribosome complexes (Carter et al., 2000; Vicens and Westhof, 2001, 2003; Francois et al., 2005) provide an excellent starting point for compound development by rational design. However, exclusively structure-directed chemical transformations are unlikely to lead to success for various reasons. Most importantly, the knowledge available is simply not sufficient to rely on compound development by molecular design. The typical crystal structure resolution of around 3.0–3.2 Å leaves many uncertainties. In addition, crystallography provides a static picture, and does not capture internal motions of the ribosome during translation. Three-dimensional structures of drug-ribosome complexes at a useful crystallographic resolution are still not available for eukaryotic ribosomes, although hopefully they may become available in a not too distant future. For drug development it is essential to have an experimental system available which allows to readily assess and optimize a compound's target specificity. Ideally, such a model does not reflect a static snap-shot, but captures the whole complexity of ribosome motion.

In an effort to further understand and modify drug-target interactions we have synthesized a range of aminoglycosides derivatives, initially by focusing on substituents which at the structural level are located close to the polymorphic rRNA residues present in the drug binding pocket, namely substituents 6' and 4' of ring I (Pathak et al., 2005; Pathak et al., 2008). Based on the insight gathered, we have initiated a drug discovery program where we combine genetics and chemical synthesis with the view of modifying drug-target interactions to redirect drug binding. In the chosen strategy, we use a collection of mutant ribosomes to guide the synthesis of novel aminoglycoside derivatives. The point mutations in the bacterial drug binding pocket were chosen to reflect the polymorphic nucleotide residues present in the drug binding site, i. e., 16S rRNA residues 1408 and 1491 (see Figure 5).

A target-based screening approach is in part problematic, as it may result in compounds which show good selectivity at the target level, but no antibacterial activity at the whole cell level due to the lack of bacterial cell wall penetration. Given how little is known about the mechanisms by which antibiotics enter the bacterial cell, there is no rational basis that could di-

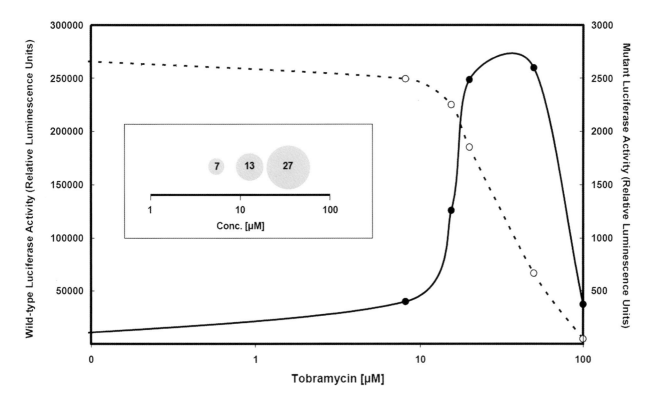

Fig. 6 Assessment of protein synthesis inhibition and misreading induction in cell-free translation assays of firefly luciferase (Fluc) activity on mitochondrial hybrid ribosomes. Flucwt (closed lines, closed circles) was used to assess drug-induced inhibition of protein synthesis. Flucmut (broken lines, open circles) was used to as-sess drug-induced misreading. Inset: Relative misreading at 25%, 50% and 75% inhibition of protein synthesis (IC25, IC50, IC75) in comparison to the untreated control (set as 1). The circle diameter indicates the relative misreading (Flucmut/Flucwt).

rect compound improvement in this area. Taking these problems into account, we screen the compounds by MIC assays that define the minimal concentration required to inhibit bacterial growth. Besides assessing drug specificity, this assay allows to address whether the compounds have antibacterial activity, as evidenced by the compounds' MIC towards wild-type bacteria. Compounds which pass this screening assay are then extensively characterized in cell-free translation assays using the set of recombinant ribosomes with single point mutations in the drug binding site described in Figure 5, complemented by hybrid ribosomes which carry the various eukaryotic drug binding pockets (cytoplasmic, mitochondrial wild-type, mitochondrial A1555G/C1494U mutant; see Figure 3). Susceptibility testing of the various mutant ribosomes provides a means to assess whether the chemical substituents introduced into the parental scaffold affect drug selectivity.

Given that we have linked aminoglycoside ototoxicity to drug-induced misreading, we have established an assay to study misreading in a sensitive gain-of-function assay. Mutation of amino acid at position 245 in the active site of firefly luciferase results in loss of enzymatic activity, with enzymatic function being partly restored by misreading (Grentzmann et al., 1998; Salas-Marco and Bedwell, 2005). Shown as an example in Figure 6 are mitohybrid ribosomes and the effect of tobramycin on protein synthesis as determined in cell-free translation assays. Inhibition of protein synthesis (reflecting translocation inhibition) is indicated by the compound's ability to inhibit synthesis of wild-type luciferase activity. Tobramycin-induced misreading is indicated by the compound's ability to induce firefly activity through amino acid misincorporation at position 245. Figure 6 also indicates the intimate mechanistic association between misreading and translocation inhibition.

Using this panoply of assays addressing the compounds' antibacterial and selectivity properties, iterative rounds of chemical synthesis and susceptibility testing are executed with the view of increasing compound selectivity to ultimately result in the synthesis of aminoglycoside derivatives which have lost their activity towards the eukaryotic ribosome but have retained their antibacterial activity.

Acknowledgements

Work in the authors' laboratory has been supported by the University of Zurich, the European Commission and the Swiss National Science Foundation. Erik C. Böttger would like to dedicate this paper to Rosmarie Welter-Enderlin; she will always be remembered. ECB would like to thank the former and current members of the laboratory for their contributions, in particular Sven N. Hobbie. Part of this paper was written when ECB was a Visiting Professor at the Institut Pasteur, Paris, where he thanks P. Courvalin for his hospitality and stimulating discussions. We apologize to all colleagues whose work could not be cited due to space limitations.

References

Benveniste R, Davies J (1973) Structure-activity relationships among the aminoglycoside antibiotics: role of hydroxyl and amino groups. Antimicrob Agents Chemother 4: 402–409

Bommakanti AS, Lindahl L, Zengel JM (2008) Mutation from guanine to adenine in 25S rRNA at the position equivalent to *E. coli* A2058 does not confer erythromycin sensitivity in Sacchromyces cerevisae. RNA 14: 460–464

Bottger EC (2007) Antimicrobial agents targeting the ribosome: the issue of selectivity and toxicity – lessons to be learned. Cell Mol Life Sci 64: 791–795

Bottger EC, Springer B, Prammananan T, Kidan Y, Sander P (2001) Structural basis for selectivity and toxicity of ribosomal antibiotics. EMBO Rep 2: 318–323

Bravo O, Ballana E, Estivill X (2006) Cochlear alterations in deaf and unaffected subjects carrying the deafness-associated A1555G mutation in the mitochondrial 12S rRNA gene. Biochem Biophys Res Commun 344: 511–516

Brown CM, Tate WP (1994) Direct recognition of mRNA stop signals by *Escherichia coli* polypeptide chain release factor two. J Biol Chem 269: 33164–33170

Cabanas MJ, Vazquez D, Modolell J (1978) Inhibition of ribosomal translocation by aminoglycoside antibiotics. Biochem Biophys Res Commun 83: 991–997

Carter AP, Clemons WM, Brodersen DE, Morgan-Warren RJ, Wimberly BT, Ramakrishnan V (2000) Functional insights from the structure of the 30S ribosomal subunit and its interactions with antibiotics. Nature 407: 340–348

Chambers HF (1996) Chemotherapy of microbial diseases. In: Hardman JG, Limbird LE (eds.) Goodmann & Gilman's the Pharmaceutical Basis of Therapeutics. McGraw-Hill, pp 1103–1121

Chen Y, Huang WG, Zha DJ, Qiu JH, Wang JL, Sha SH, Schacht J (2007) Aspirin attenuates gentamicin ototoxicity: from the laboratory to the clinic. Hear Res 226: 178–182

Clerici WJ, Hensley K, DiMartino DL, Butterfield DA (1996) Direct detection of ototoxicant-induced reactive oxygen species generation in cochlear explants. Hear Res 98: 116–124

Davies J, Davis BD (1968) Misreading of ribonucleic acid code words induced by aminoglycoside antibiotics. The effect of drug concentration. J Biol Chem 243: 3312–3316

De Stasio EA, Dahlberg AE (1990) Effects of mutagenesis of a conserved base-paired site near the decoding region of *Escherichia coli* 16 S ribosomal RNA. J Mol Biol 212: 127–133

De Stasio EA, Moazed D, Noller HF, Dahlberg AE (1989) Mutations in 16S ribosomal RNA disrupt antibiotic–RNA interactions. EMBO J 8: 1213–1216

Edlind TD (1989) Susceptibility of Giardia lamblia to aminoglycoside protein synthesis inhibitors: correlation with rRNA structure. Antimicrob Agents Chemother 33: 484–488

Fan-Minogue H, Bedwell DM (2008) Eukaryotic ribosomal RNA determinants of aminoglycoside resistance and their role in translational fidelity. RNA 14: 148–157

Feldman MB, Terry DS, Altman RB, Blanchard SC (2010) Aminoglycoside activity observed on single pre-translocation ribosome complexes. Nat Chem Biol 6: 244

Fischel-Ghodsian N (1999) Mitochondrial deafness mutations reviewed. Hum Mutat 13: 261–270

Francois B, Russell RJ, Murray JB, Aboul-ela F, Masquida B, Vicens Q, Westhof E (2005) Crystal structures of complexes between aminoglycosides and decoding A site oligonucleotides: role of the number of rings and positive charges in the specific binding leading to miscoding. Nucleic Acids Res 33: 5677–5690

Gale EF, Cundliffe E, Reynolds PE, Richmond MH, Waring MJ (1981) The molecular basis of antibiotic action. John Wiley & Sons Ltd, London UK

Garetz SL, Altschuler RA, Schacht J (1994) Attenuation of gentamicin ototoxicity by glutathione in the guinea pig in vivo. Hear Res 77: 81–87

Grentzmann G, Ingram JA, Kelly PJ, Gesteland RF, Atkins JF (1998) A dual-luciferase reporter system for studying recoding signals. RNA 4: 479–486

Guan MX (2006) Mitochondrial DNA mutations associated with aminoglycoside ototoxicity. J Audiolog Med 4: 170–178

Guan MX, Fischel-Ghodsian N, Attardi G (2001) Nuclear background determines biochemical phenotype in the deafness-associated mitochondrial 12S rRNA mutation. Hum Mol Genet 10: 573–580

Hainrichson M, Nudelman I, Baasov T (2008) Designer aminoglycosides: the race to develop improved antibiotics and compounds for the treatment of human genetic diseases. Org Biomol Chem 6: 227–239

Henley CM, 3rd, Schacht J (1988) Pharmacokinetics of aminoglycoside antibiotics in blood, inner-ear fluids and tissues and their relationship to ototoxicity. Audiology 27: 137–146

Hobbie SN, Akshay S, Kalapala SK, Bruell CM, Shcherbakov D, Bottger EC (2008a) Genetic analysis of interactions with eukaryotic rRNA identify the mitoribosome as target in aminoglycoside ototoxicity. Proc Natl Acad Sci USA 105: 20 888–20 893

Hobbie SN, Bruell C, Kalapala S, Akshay S, Schmidt S, Pfister P, Bottger EC (2006a) A genetic model to investigate drug-target interactions at the ribosomal decoding site. Biochimie 88: 1033–1043

Hobbie SN, Bruell CM, Akshay S, Kalapala SK, Shcherbakov D, Bottger EC (2008b) Mitochondrial deafness alleles confer misreading of the genetic code. Proc Natl Acad Sci USA 105: 3244–3249

Hobbie SN, Kalapala SK, Akshay S, Bruell C, Schmidt S, Dabow S, Vasella A, Sander P, Bottger EC (2007) Engineering the rRNA decoding site of eukaryotic cytosolic ribosomes in bacteria. Nucleic Acids Res 35: 6086–6093

Hobbie SN, Pfister P, Bruell C, Sander P, Francois B, Westhof E, Bottger EC (2006b) Binding of neomycin-class aminoglycoside antibiotics to mutant ribosomes with alterations in the A site of 16S rRNA. Antimicrob Agents Chemother 50: 1489–1496

Hobbie SN, Pfister P, Brull C, Westhof E, Bottger EC (2005) Analysis of the contribution of individual substituents in 4,6-aminoglycoside-ribosome interaction. Antimicrob Agents Chemother 49: 5112–5118

Howard M, Frizzell RA, Bedwell DM (1996) Aminoglycoside antibiotics restore CFTR function by overcoming premature stop mutations. Nat Med 2: 467–469

Hutchin T, Haworth I, Higashi K, Fischel-Ghodsian N, Stoneking M, Saha N, Arnos C, Cortopassi G (1993) A molecular basis for human hypersensitivity to aminoglycoside antibiotics. Nucleic Acids Res 21: 4174–4179

Laurberg M, Asahara H, Korostelev A, Zhu J, Trakhanov S, Noller HF (2008) Structural basis for translation termination on the 70S ribosome. Nature 454: 852–857

Mathis A, Wild P, Boettger EC, Kapel CM, Deplazes P (2005) Mitochondrial ribosome as the target for the macrolide antibiotic clarithromycin in the helminth Echinococcus multilocularis. Antimicrob Agents Chemother 49: 3251–3255

Mathis A, Wild P, Deplazes P, Boettger EC (2004) The mitochondrial ribosome of the protozoan Acanthamoeba castellanii is the target for macrolide antibiotics. Mol Biochem Parasitol 135: 225–229

Ogle JM, Ramakrishnan V (2005) Structural insights into translational fidelity. Annu Rev Biochem 74: 129–177

Palmer E, Wilhelm JM, Sherman F (1979) Phenotypic suppression of nonsense mutants in yeast by aminoglycoside antibiotics. Nature 277: 148–150

Pathak R, Böttger EC, Vasella A (2005) Design and synthesis of aminoglycoside antibiotics to selectively target 16S ribosomal RNA position 1408. Helvetica Chimica Acta 88: 2967–2985

Pathak R, Perez-Fernandez D, Nandurdikar R, Kalapala SK, Böttger EC, Vasella A (2008) Synthesis and evaluation of paromomycin derivatives modified at C(4') Helvetica Chimica Acta 91: 1533–1551

Peske F, Savelsbergh A, Katunin VI, Rodnina MV, Wintermeyer W (2004) Conformational changes of the small ribosomal subunit during elongation factor G-dependent tRNA-mRNA translocation. J Mol Biol 343: 1183–1194

Pfister P, Hobbie S, Brull C, Corti N, Vasella A, Westhof E, Bottger EC (2005) Mutagenesis of 16S rRNA C1409-G1491 base-pair differentiates between 6'OH and 6'NH3+ aminoglycosides. J Mol Biol 346: 467–475

Pfister P, Hobbie S, Vicens Q, Bottger EC, Westhof E (2003a) The molecular basis for A-site mutations conferring aminoglycoside resistance: relationship between ribosomal susceptibility and X-ray crystal structures. Chembiochem 4: 1078–1088

Pfister P, Jenni S, Poehlsgaard J, Thomas A, Douthwaite S, Ban N, Bottger EC (2004) The structural basis of macrolide-ribosome binding assessed using mutagenesis of 23S rRNA positions 2058 and 2059. J Mol Biol 342: 1569–1581

Pfister P, Risch M, Brodersen DE, Bottger EC (2003b) Role of 16S rRNA Helix 44 in Ribosomal Resistance to Hygromycin B. Antimicrob Agents Chemother 47: 1496–1502

Poehlsgaard J, Douthwaite S (2005) The bacterial ribosome as a target for antibiotics. Nat Rev Microbiol 3: 870–881

Prezant TR, Agapian JV, Bohlman MC, Bu X, Oztas S, Qiu WQ, Arnos KS, Cortopassi GA, Jaber L, Rotter JI, et al. (1993) Mitochondrial ribosomal RNA mutation associated with both antibiotic-induced and non-syndromic deafness. Nat Genet 4: 289–294

Recht MI, Douthwaite S, Puglisi JD (1999) Basis for prokaryotic specificity of action of aminoglycoside antibiotics. EMBO J 18: 3133–3138

Salas-Marco J, Bedwell DM (2005) Discrimination between defects in elongation fidelity and termination efficiency provides mechanistic insights into translational readthrough. J Mol Biol 348: 801–815

Sander P, Prammananan T, Meier A, Frischkorn K, Bottger EC (1997) The role of ribosomal RNAs in macrolide resistance. Mol Microbiol 26: 469–480

Schacht J (1986) Molecular mechanisms of drug-induced hearing loss. Hear Res 22: 297–304

Schatz A, Bugie E, Waksman SA (1944) Streptomycin, a substance exhibiting antibiotic activity against gram-positive and gram-negative bacteria. Proceedings of the Society for Experimental Biology and Medicine 55: 66–69

Schroder M, Kaufman RJ (2005) The mammalian unfolded protein response. Annu Rev Biochem 74: 739–789

Sha SH, Qiu JH, Schacht J (2006) Aspirin to prevent gentamicin-induced hearing loss. N Engl J Med 354: 1856–1857

Sha SH, Schacht J (1999) Stimulation of free radical formation by aminoglycoside antibiotics. Hear Res 128: 112–118

Sigmund CD, Ettayebi M, Morgan EA (1984) Antibiotic resistance mutations in 16S and 23S ribosomal RNA genes of Escherichia coli. Nucleic Acids Res 12: 4653–4663

Singh A, Ursic D, Davies J (1979) Phenotypic suppression and misreading Saccharomyces cerevisiae. Nature 277: 146–148

Sor F, Fukuhara H (1982) Identification of two erythromycin resistance mutations in the mitochondrial gene coding for the large ribosomal RNA in yeast. Nucleic Acids Res 10: 6571–6577

Vakulenko SB, Mobashery S (2003) Versatility of aminoglycosides and prospects for their future. Clin Microbiol Rev 16: 430–450

Vicens Q, Westhof E (2001) Crystal structure of paromomycin docked into the eubacterial ribosomal decoding A site. Structure 9: 647–658

Vicens Q, Westhof E (2003) Crystal structure of geneticin bound to a bacterial 16S ribosomal RNA A site oligonucleotide. J Mol Biol 326: 1175–1188

Weixlbaumer A, Jin H, Neubauer C, Voorhees RM, Petry S, Kelley AC, Ramakrishnan V (2008) Insights into translational termination from the structure of RF2 bound to the ribosome. Science 322: 953–956

Youngman EM, He SL, Nikstad LJ, Green R (2007) Stop codon recognition by release factors induces structural rearrangement of the ribosomal decoding center that is productive for peptide release. Mol Cell 28: 533–543

Zhao H, Li R, Wang Q, Yan Q, Deng JH, Han D, Bai Y, Young WY, Guan MX (2004) Maternally inherited aminoglycoside-induced and nonsyndromic deafness is associated with the novel C1494T mutation in the mitochondrial12S rRNA gene in a large Chinese family. Am J Hum Genet 74: 139–152.

Ribosomal protein S5, ribosome biogenesis and translational fidelity

21

Biswajoy Roy-Chaudhuri, Narayanaswamy Kirthi, Teresa Kelley and Gloria M. Culver

1. Overview

In an actively growing *Escherichia coli* cell, the rates of ribosome production and cell growth are directly correlated. It has been estimated that approximately 40% of the available energy during exponential growth of *E. coli* is devoted to ribosome biogenesis. Given these observations and the importance of accurate and functional ribosomes for gene expression, it is likely that the process of ribosome biogenesis is regulated and accurate.

The many structures of ribosomes and ribosomal subunits that have emerged over the last decade are an invaluable tool for understanding the process of ribosome biogenesis. These snapshots offer insight into a functional "end-state" conformation that could result from the assembly process. Additionally, a structural framework now exists for understanding how specific mutations linked to subunit assembly and function might act at the molecular level. Our work is focused on understanding the dynamic process of *E. coli* 30S ribosomal subunit biogenesis.

2. Biogenesis of bacterial 30S ribosomal subunit

Seven rDNA and nineteen ribosomal protein (r-protein)-containing operons are committed to production of ribosomal components in *E. coli*. Along with these core components, there are many additional factors that act during the production of ribosomal subunits. A variety of enzymes are involved in this process; in addition to RNA polymerase, nucleases, modification (rRNA and r-protein) factors, GTPases, and helicases all appear to play a role in this cascade [for reviews, see (Connolly and Culver, 2009; Kaczanowska and Ryden-Aulin, 2007; Wilson and Nierhaus, 2007)].

There are also many additional gene products that appear to play a role in ribosome biogenesis. However, the specific roles of these factors in the production of ribosomal subunits remain elusive.

The process of 30S subunit assembly has been studied more extensively than the analogous process for 50S subunits. For the *E. coli* 30S or small subunit, studies have been focused for many years on the association of r-proteins using extensive in vitro experiments [for review, see (Culver, 2003)]. Both 16S rRNA and the 21 r-proteins can be either isolated from cellular subunits or can be produced in recombinant form (proteins) or transcribed in vitro (rRNA) (Culver and Noller, 2000; Krzyzosiak et al., 1987). These purified components have been used to define the order of r-protein addition and the consequences of r-protein binding to 16S rRNA [see (Grondek and Culver, 2004) for example]. Many studies of 16S rRNA folding have also revealed some of the aspects of this process. These studies together with *in-vivo* work, suggest that kinetic traps for productive 16S rRNA folding exist and that these traps may explain the strong correlation between biogenesis mutants and cold sensitivity (Dammel and Noller, 1993; Woodson, 2008) as these traps would be significantly harder to overcome at low temperature. More recently, an increased interest in the role of extra-ribosomal assembly factors and in the temporal and kinetic biogenesis events in vivo has emerged. Thus, there are still many unanswered questions about the assembly of both bacterial ribosomal subunits.

2.1. Ribosomal protein S5

Ribosomal protein S5 has been implicated to function during both 30S subunit assembly and translational accuracy. For many years, the role of S5 in translational fidelity has been appreciated due to analyses of mutants with defects in this process. Ribosome ambiguity (*ram*) mutants have been isolated in both S4 and S5 (Figure 1) (Birge and Kurland, 1970; Deusser et al., 1970; Piepersberg et al., 1975a; Piepersberg et al., 1975b). Strains that carry these *ram* mutations demonstrate significantly higher levels of translational errors. These mutations have been identified in isolation and in combination with hyperaccurate (*restrictive*) mutations in r-protein S12 (Bjorkman et al., 1999; Ito and Wittmann, 1973). Using both genetic and structural information, a model for 30S subunit structural changes during decoding has been proposed (Ogle et al., 2003; Ogle et al., 2002). In the 30S subunit, S4 and S5 have a significant interaction surface, and mutations that lead to the *ram* phenotype can disrupt elements at this interaction surface (Figure 1). Thus, the model suggests that accuracy requires the appropriately timed opening of this interface with concomitant changes in the shoulder and head of the 30S subunit to form a closed structure. *Ram* mutations at this interface would therefore allow premature closing of the 30S subunit during decoding and thus increase incorporation errors during the process of decoding. S5 has been implicated in resistance to the antibiotic spectinomycin (spc) as well. Spectinomycin is an aminoglycoside that inhibits translation by sterically blocking the swiveling of the head domain of the 30S subunit, thereby disrupting the translocation of mRNA and tRNA (Borovinskaya et al., 2007). Mutations in S5 have also been isolated that yield strains with resistance to this antibiotic (De Wilde, 1973; Funatsu et al., 1972). The location of the spc-resistance and the *ram* mutations are distinct within the architecture of S5 [see (Kurland C. G., 1996)]. Thus S5 has at least two structural regions that are important for normal ribosome function. Additionally, mutations in S5 have been linked to defects in ribosome biogenesis *in vivo*. A screen for mutations that altered ribosome assembly allowed the identification of the S5 (*rpsE*) locus as important for this process (Guthrie et al., 1969; Nashimoto et al., 1971). These strains had altered monosome/polysome profiles and thus appeared to alter ribosome biogenesis *in vivo*. These strains were also resistant to spc and exhibited growth defects at normal growth temperatures and cold sensitivity. Further characterization of these strains or the specific changes in S5 that resulted in these phenotypes was not reported. Moreover, likely due to the rather severe growth phenotypes associated with these strains, they did not remain viable after many years of attempted

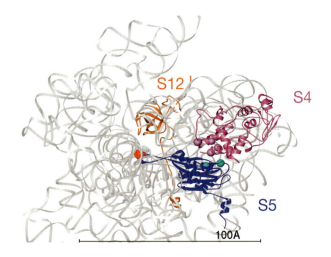

Fig. 1 Positions of ribosomal proteins S4, S5 and S12 that modulate translational fidelity in prokaryotes. Mutations in S4 and S5 (indicated by magenta and cyan spheres respectively) that confer the ram phenotype lie at the interface of r-proteins S4 (shown in magenta) and S5 (shown in blue). A portion of the 16S rRNA backbone is indicated in grey. The glycine residue at a universally conserved position (28 in *E. coli*) is located in loop 2 of S5 away from the interface of S4-S5 and is indicated by a red sphere; this glycine is substituted with aspartate in our study. Image based on 2AW7.pdb (Schuwirth et al., 2005).

storage (M. Nomura, personal communication). The mutations that lead to biogenesis defects were thus unexamined and it was thus not known if they would map to the same region of S5 as the previous mutations or if they would be unique.

Ribosomal protein S5 is very highly conserved across phylogeny with its counterpart in eukaryotic cells using the rps2 designation (Nakao et al., 2004). S5 is located in the vicinity of the functional center of the 30S subunit but is on the opposite face of the small subunit from the decoding site. The majority of the protein resides at the top of the body with a loop segment (loop 2) in a rather extended beta conformation that thus allows this protein to interact with the head as well (Wimberly et al., 2000) (Figure 2). S5 directly interacts with S4 and S8 along with RNA elements from all 4 major domains of 16S rRNA. Thus S5 is part of an extended network that maybe important for maintenance of the appropriate architecture or architectural changes critical for 30S subunit function.

2.2. Experimental rationale and summary

Our interest in 30S subunit assembly led to a re-evaluation of the role of S5 in this process. We isolated *E. coli* strains that were both spc-resistant and cold-sensitive

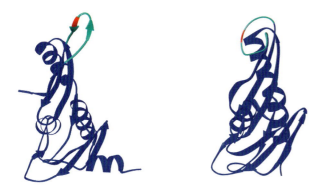

Fig. 2 Different conformations of loop 2 (indicated in cyan) in the structures of r-protein S5. Loop 2 houses Gly28 (indicated in red), which is mutated to aspartate in our study. The three-dimensional structure of S5from the *E. coli* 70S ribosome (2AW7.pdb (Schuwirth et al., 2005)) is shown on the left, the structure of unbound S5 from *Bacillus stearothermophilus* (1PKP. pdb; Ramakrishnan and White, 1992) on the right. As shown, loop 2forms an extended structure in the 70S ribosome but is folded back in the free S5 structure suggesting the dynamic nature of loop 2.

(Kirthi et al., 2006). We established that these mutations were in protein S5 and that mutation of Gly28 to Asp was crucial for these phenotypes. Gly28 is part of the loop 2 element of S5 and is proximal to, but unique from, other reported S5 spc-resistant mutants. We also demonstrated that strains carrying mutant S5(G28D) as the sole chromosomally encoded copy of this gene were defective in ribosome biogenesis, with the most prominent defects observed for the 30S subunit. It was established that these strains were also more error-prone than their isogenic parental counterparts, in spite of the location of the isolated mutation at a site distinct from the S4/S5 interface. Extragenic high-copy suppressors of the cold-sensitive phenotype established that RimJ could suppress the known phenotypes associated with S5(G28D) (Roy-Chaudhuri et al., 2008). Suppression of the fidelity defect was surprising as the ribosomes still contained the S5(G28D) protein and we demonstrated that RimJ did not associate with functional 70S ribosomes. Examination of 16S rRNA processing revealed that the S5(G28D) mutant strains were defective in end maturation and that RimJ could rescue this defect (Roy-Chaudhuri et al., 2010). The correlation between 16S rRNA maturation and fidelity defects was further examined. In a strain lacking an enzyme involved in generating the mature 5' end of 16S rRNA, ribosomes contained immature 16S rRNA and were more error-prone. Alternative structures at the 5' end of 16S rRNA in the presence and absence of the leader sequence had been proposed. We examined

both rRNA folding in leader containing ribosomes and also examined mutant forms of 16S rRNA that were unable to form some of these alternative structures. Our data support a model where basepairing between nucleotide −10 and −1 in the leader and elements of mature 16S rRNA normally found in helix 1 results in defects in fidelity. Models rationalizing the importance of these interactions will be discussed.

2.3. Isolation of mutations in protein S5

Upon isolation of strains that were simultaneously spc-resistant and cold- sensitive, the *rpsE* locus was amplified and sequenced (Kirthi et al., 2006). In all examined isolates, the Gly28 was changed to Asp [S5(G28D)]. In several of the isolates this was the only identified change, while a few isolates contained this mutation and others within the same region of S5. Given the severity of the effects of the single G28 to D substitution, this mutant form was selected for further analysis.

The position of this mutation was very intriguing. Gly28 of S5 resides in loop 2 of the structure and this loop has been observed in two different conformations using structural biological approaches (Davies et al., 1998; Schuwirth et al., 2005; Wimberly et al., 2000). Moreover, Gly28 is universally conserved. The extended beta sheet of loop 2 interacts with elements of the head of the 30S subunits and thus the presence of this specific amino acid (small and uncharged) may be critical for 30S subunit structure and function.

The causal relationship between this specific mutation and the observed phenotypes of cold sensitivity and spc-resistance was established by P1 transduction using lysate from the S5(G28D) strain into a wild-type background (Kirthi et al., 2006). Spc-resistant bacteria were isolated and observed to be cold-sensitive. Sequencing the *rpsE* gene confirmed that the G28D mutation was present in this newly generated strain; thus this amino acid substitution was responsible for the phenotypic changes.

Mutations in r-protein S5 can lead to defects in 30S subunit biogenesis (Nomura, 1974). To determine if this specific mutation in S5 could alter ribosome production *in vivo*, ribosome profiles from the parental and mutant strains were examined at both the permissive and non-permissive temperatures. At both temperatures, defects in SSU assembly and in the overall amount of 70S ribosomes were evident (see Table 1). At both temperatures, but particularly at low (non-permissive) temperature there are virtually no peaks

Table 1 Relative amounts of free 30S subunits or 70S ribosomes in wild-type and S5(G28D) *E. coli* strains

		30S	70S
Parent	permissive	0.25	0.75
	non-permissive	0.33	0.67
S5(G28D)	permissive	0.75	0.25
	non-permissive	1.0 (0.25*)	#

permissive = 37 °C
non-permissive = 24 °C
* estimate of precursor 30S particle relative to the free 30S subunit
# a reliable determination of the amount of 70S ribosome was not possible due to limiting material

Table 2 Miscoding in *E. coli* S5(G28D)

Miscoding event	parent	S5(G28D)
+1 frameshifting	1.0	3.2
−1 frameshifting	1.0	3.8
Premature UAA	1.0	7.8
Premature UAG	1.0	3.0
Premature UGA	1.0	1.5

corresponding to 70S ribosomes and this may account for the severe growth defect of the S5(G28D) strain. Additionally, at low temperature a peak that resembles a 30S subunit precursor/intermediate is observed in the mutant strain.

2.4. Processing of 16S rRNA is also defective in S5(G28D) strains

To determine if rRNA processing was also altered in the strain with the S5(G28D) mutation, 16S rRNA maturation was also examined (Roy-Chaudhuri et al., 2010). Primer extension analysis revealed that in the S5 mutant strain 16S rRNA maturation was impaired and it appears that RNase III cleavage occurs but that subsequent processing events are influenced. Surprisingly, substantial amounts of precursor 16S rRNA was found not only in free 30S subunit but also in the 70S ribosomal fractions from the mutant strain.

2.5. S5(G28D) alters translational fidelity

Although the mutation at position 28 of S5 is not proximal to the previously characterized ribosome ambiguity (*ram*) mutants (Figure 1), translational fidelity assays were carried out in this strain. Surprisingly, enhanced levels of +1 and -1 frameshifting and nonsense suppression were observed in the S5(G28D) strain when compared to its parental strain (Table 2).

2.6. Isolation of rimJ as a suppressor of cold sensitivity

To further analyze the role of S5 and the specific G28D mutation on biogenesis and fidelity, it would be useful to systematically dissect the two important functions. As the cold sensitive phenotype was likely correlated with the 30S subunit biogenesis defect, we posited that isolation of extragenic suppressors of the cold sensitivity might allow for separation of the biogenesis and fidelity functions. Using an *E. coli* genomic library, we isolated a set of plasmids that could allow growth of S5(G28D) at low temperature (Roy-Chaudhuri et al., 2008). All of the plasmids contained overlapping genomic fragments that encompassed the *rimJ* gene. This was particularly interesting as RimJ is known to be the N-terminal acetyltransferase that modifies S5 (Yoshikawa et al., 1987). Thus the likelihood of this being the genomic fragment responsible for suppression was further examined. Overexpression of the *rimJ* coding region decreased the doubling time of S5(G28D) to nearly that of the parental strain and supported growth at low temperatures. Moreover, overexpression of RimJ also returned the ribosome profiles and 16S rRNA processing to near wild-type levels (Roy-Chaudhuri et al., 2010). These findings suggested that RimJ suppressed the ribosome biogenesis defects associated with S5(G28D).

2.7. RimJ can alleviate the fidelity defect associated with S5(G28D)

To determine if RimJ overepxression in the S5(G28D) mutant strain had any effect of translational fidelity, in vivo miscoding assays were performed. Overexpression of RimJ decreased miscoding events in S5(G28D) significantly (Roy-Chaudhuri et al., 2008). This result was quite surprising considering the 70S ribosomes still contain S5(G28D). This finding indicates that the fidelity defects are not due to S5(G28D) directly but are secondary to another change that can be suppressed by overexpression of RimJ.

2.8. Strains with defects in 16S rRNA maturation have fidelity defects

One possible explanation for these data would be if 16S rRNA maturation status impacts translational fidelity. For this to be the case, immature 16S rRNA would need to accumulate in the 70S ribosomes, and indeed this was the case for the S5(G28D) strain (Roy-Chaudhuri et al., 2010). Additionally, RimJ overexpression alleviated the 16S rRNA maturation defect. We thus examined whether additional strains with 16S rRNA maturation defects accumulated precursor 16S rRNA in 70S ribosomes and if there was a correlation between this accumulation and fidelity. A variety of strains that have demonstrated defects in 30S subunit biogenesis have been shown to retard 16S rRNA maturation (Bubunenko et al., 2006; Bylund et al., 1998; Connolly et al., 2008; Inoue et al., 2003). We tested three such strains (deletion of r-protein S15: ΔrpsO; deletion of rRNA modification enzyme KsgA: ΔksgA; deletion of biogenesis factor RimM; ΔrimM). Although it had been demonstrated that precursor 16S rRNA accumulated in such strains (Bubunenko et al., 2006; Bylund et al., 1998; Connolly et al., 2008), we first demonstrated that indeed this precursor 16S rRNA was found in 70S ribosomes (Roy-Chaudhuri et al., 2010). Secondly, we observed that in each case the level of miscoding was enhanced in these strains compared to their isogenic parental strain. Thus, the correlation between defects in fidelity and 16S rRNA maturation was observed and suggested that these processes might be related.

One caveat to the above experiments is that the ribosomes themselves are altered in each case beyond the 16S rRNA maturation defect. To more directly assess the role of 16S rRNA maturation in maintaining fidelity, we examined accumulation of precursor 16S rRNA in 70S ribosomes and fidelity in a strain lacking RNase G. RNase G is an endoribonuclease that cleaves precursor 16S rRNA to form the mature 5' end (Figure 3). Previous work demonstrated that the gene encoding this protein could be deleted and that precursor 16S accumulated in the total RNA population (Li et al., 1999; Wachi et al., 1999). We observed that precursor 16S rRNA did assemble into the 70S ribosomal population and that translational fidelity was impaired

Table 3 Miscoding in the *E. coli rng* deletion strain

Miscoding event	parent	ΔRNaseG
+1 frameshifting	1.0	1.8
−1 frameshifting	1.0	1.3
Premature UAA	1.0	4.0
Premature UAG	1.0	2.8
Premature UGA	1.0	2.0

(Table 3) although not to the same extent as in some of the other strains tested (Roy-Chaudhuri et al., 2010). Thus, these data indicate that there is a direct correlation between the inclusion of precursor 16S rRNA in 70S ribosomes and translational accuracy.

2.9. The type of precursor sequence matters

During these experiments we observed that in different strains alternate forms of precursor 16S rRNA were found in the translating population. In the S5(G28D) mutant strains the majority of the precursor 16S rRNA was in the RNase III cleaved form (long form) but not further processed. Thus, either the leader or trailer sequences could have contributed to the changes in fidelity. However, in the RNase G deletion strain, the majority of precursor 16S rRNA had been processed by

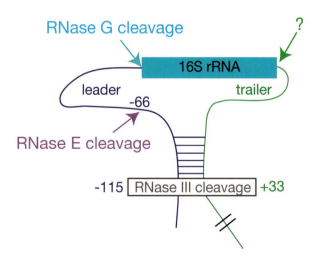

Fig. 3 Schematic of 16S rRNA processing in prokaryotes. RNase III endoribonuclease cleaves the primary transcript to generate precursor 16S rRNAs that are subsequently processed by endonucleases RNase E and RNase G to generate the 5' end of mature 16S rRNA and an yet unidentified enzyme to yield the mature 3' end. Arrows indicate the positions of cleavages by the various endoribonucleases (Figure is not drawn to scale).

Table 4 Relative changes in reactivity to chemical probing at A7 and A8 in 30S subunits and 70S ribosomes isolated from various *E. coli* strains

	30S	70S
parent	1.0	1.0
S5(G28D)	0.4	0.6
parent	1.0	1.0
Drng	0.6	0.8

RNase E and is fully mature at the 3' end (short form) (Li et al., 1999). Thus, changes in fidelity correlate with the presence of a short leader of 66 nucleotides. Previous work suggested that in the presence of leader sequences alternative structures could form that might preclude formation of mature helix 1 in 16S rRNA (Young & Steitz, 1978; Dammel & Noller, 1993; Dennis et al., 1997) (see Figure 4). Our data indicate that the shorter of the two precursor forms of 16S rRNA is sufficient to compromise accuracy (Roy-Chaudhuri et al., 2010); functional changes associated with the longer precursor form are yet to be fully examined but data suggest that decreased fidelity is also associated with inclusion of this precursor 16S rRNA in 70S ribosomes (Connolly et al., 2008; Roy-Chaudhuri et al., 2010). In the aforementioned model alternative structures form when nucleotides from −1 to −10 of the precursor 16S rRNA leader sequence base-pair with nucleotides 7–16 of mature 16S rRNA, thus disfavoring formation of helix 1 between mature sequence elements 9–12 and 22–25 (Figure 4). We used chemical probing to examine if helix 1 formation was disrupted in 70S ribosomes bearing either the long or short precursor forms. In all

cases it appears that helix 1 occupancy was decreased in the presence of 16S rRNA bearing leader sequences (Roy-Chaudhuri et al., 2010). Leader sequence correlates with both changes in fidelity and in disruption of mature 16S rRNA architecture. To further examine if the proposed base pairing between the leader and elements of helix 1 was responsible for the change in fidelity, nucleotides −10 to −1 in the leader were deleted and fidelity in stains dependent on this form of 16S rRNA for growth was examined in terms of translational miscoding. In the strain bearing the mutant precursor 16S rRNA, the short form of precursor 16S rRNA, which has lost the ability to form the proposed alternative structure (see Figure 4) does accumulate in 70S ribosomes, but these strains do not exhibit an increase in miscoding (Roy-Chaudhuri et al., 2010). Thus, miscoding is not correlated with simply having the leader sequence but with the specific sequence of the leader in this region. Moreover, these data suggest that the loss of nucleotides −1 to −10 negatively impacts 30S subunit biogenesis as there is a marked decrease in 16S rRNA maturation and a concomitant increase in precursor 16S rRNA inclusion in the 70S ribosome population. Thus, the interaction between the leader and mature sequences may be critical for appropriate 30S subunit formation *in vivo*.

2.10. The paradox of the leader

Our data suggest that the leader sequence is necessary for appropriate maturation of 16S rRNA and that its removal is necessary for functional ribosomes to have the appropriate level of accuracy to maintain life. Thus, the presence of the leader is paradoxical. It would ap-

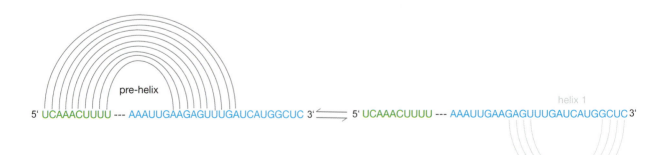

Fig. 4 Competing helices formed at the 5' end of 16S rRNA in presence of short leader sequences. Ten residues (green) immediately upstream of the 5' end of mature 16S rRNA (blue) in the leader are predicted to base-pair with mature regions to form a pre-helix which is shown to compete with helix 1 formed with mature regions. The arcs indicate canonical base pairing interactions.

pear that given the conserved nature of the leader, its importance to biogenesis in bacteria outweigh the potential negative impact on translational fidelity.

It has been posited that the leader of 16S rRNA in bacteria acts as a chaperone in a manner akin to the role of U3 in 18S rRNA maturation in eukaryotes (Hughes, 1996; Dennis et al., 1997). It remains to be determined if this is the case, but the distinction been cis-acting and trans-acting elements would alleviate the translational fidelity consequence in eukaryotes. Also the compartmental distinction between ribosome biogenesis and translation in eukaryotes, nucleolus *vs.* cytosol, might act to limit premature particles from entering the translational cycle. Thus, eukaryotes may have evolved a more "fail-safe" system to synthesize ribosomes with the appropriate level of maturation and therefore accuracy. In eukaryotic organisms, recent studies have suggested that there is cross-talk between ribosome biogenesis and cell cycle progression. In yeast, depletion of small subunit processosome proteins prevents re-entry into the cell cycle (Bernstein and Baserga, 2004) and in animal cells aberrant rRNA processing and ribosome assembly leads to cell-cycle arrest in a p53-dependent manner (Pestov et al., 2001). We propose that, similar to other cellular pathways, cells with aberrant rRNA processing and ribosome biogenesis either arrest the cell cycle or produce erroneous proteins.

Our results also have bearing on the evolutionary and functional significance of processing of the bacterial small subunit rRNA in ribosome biogenesis. It has been shown that *in-vitro* reconstitution of 30S ribosomes from precursor 16S rRNA is energetically more favorable than from mature 16S rRNA (Mangiarotti et al., 1975). Precursor 16S rRNA can assemble into 30S particles at low temperatures more efficiently than mature 16S rRNA. This might be facilitated by alternative conformational features in precursor 16S rRNA as compared to mature rRNA (Klein et al., 1985) and reflect co-transcriptional subunit assembly *in vivo.* Since 30S ribosomes with precursor RNA would be compromised for translational fidelity, as expected from this study, we hypothesize that cells have fine-tuned the biogenesis pathway so as to maintain quality control while attempting to optimize energetics. However, when cells are challenged with defects that alter 30S subunit biogenesis such as with the S5(G28D) mutation, they can still survive, as 30S subunits bearing precursor 16S rRNA can participate in translation, albeit with reduced accuracy. The ability to synthesize proteins is paramount to survival even at the ex-

pense of errors. This can be considered analogous to error-prone DNA replication and recombination. *In vivo* when DNA repair systems fail to correct damage, in some instances cells may recruit various damage tolerance strategies that allow continued replication thus sustaining life instead of arresting cell division, entering senescence or undergoing apoptosis. These strategies allow errors in the fidelity of replication but increase the overall chances of cell proliferation and survival.

References

Bernstein KA, Baserga SJ (2004) The small subunit processome is required for cell cycle progression at G1. Mol Biol Cell 15: 5038–5046

Birge EA, Kurland CG (1970) Reversion of a streptomycin-dependent strain of *Escherichia coli*. Mol Gen Genet 109: 356–369

Bjorkman J, Samuelsson P, Andersson DI, Hughes D (1999) Novel ribosomal mutations affecting translational accuracy, antibiotic resistance and virulence of Salmonella typhimurium. Mol Microbiol 31: 53–58

Borovinskaya MA, Shoji S, Holton JM, Fredrick K, Cate JH (2007) A steric block in translation caused by the antibiotic spectinomycin. ACS Chem Biol 2: 545–552

Bubunenko M, Korepanov A, Court DL, Jagannathan I, Dickinson D, Chaudhuri BR, Garber MB, Culver GM (2006) 30S ribosomal subunits can be assembled in vivo without primary binding ribosomal protein S15. RNA 12: 1229–1239

Bylund GO, Wipemo LC, Lundberg LA, Wikstrom PM (1998) RimM and RbfA are essential for efficient processing of 16S rRNA in *Escherichia coli*. J Bacteriol 180: 73–82

Connolly K, Culver G (2009) Deconstructing ribosome construction. Trends Biochem Sci 34: 256–263

Connolly K, Rife JP, Culver G (2008) Mechanistic insight into the ribosome biogenesis functions of the ancient protein KsgA. Mol Microbiol 70: 1062–1075

Culver GM (2003) Assembly of the 30S ribosomal subunit. Biopolymers 68: 234–249

Culver GM, Noller HF (2000) In vitro reconstitution of 30S ribosomal subunits using complete set of recombinant proteins. Meth Enzymol 318: 446–460

Dammel CS, Noller HF (1993) A cold-sensitive mutation in 16S rRNA provides evidence for helical switching in ribosome assembly. Genes Dev 7: 660–670

Davies C, Bussiere DE, Golden BL, Porter SJ, Ramakrishnan V, White SW (1998) Ribosomal proteins S5 and L6: high-resolution crystal structures and roles in protein synthesis and antibiotic resistance. J Mol Biol 279: 873–888

De Wilde M (1973) Identification of the substitution amino acid responsible for resistance to spectinomycin in a mutant of *Escherichia coli*. Arch Int Physiol Biochim 81: 369

Dennis PP, Russell AG, Moniz De Sa M (1997) Formation of the 5' end pseudoknot in small subunit ribosomal RNA: involvement of U3-like sequences. RNA 3: 337–343

Deusser E, Stoffler G, Wittmann HG (1970) Ribosomal proteins. XVI. Altered S4 proteins in *Escherichia coli* revertants from streptomycin dependence to independence. Mol Gen Genet 109: 298–302

Funatsu G, Schiltz E, Wittmann HG (1972) Ribosomal proteins. XXVII. Localization of the amino acid exchanges in protein S5from two *Escherichia coli* mutants resistant to spectinomycin. Mol Gen Genet 114: 106–111

Grondek JF, Culver GM (2004) Assembly of the 30S ribosomal subunit: positioning ribosomal protein S13 in the S7 assembly branch. RNA 10: 1861–1866

Guthrie C, Nashimoto H, Nomura M (1969) Studies on the assembly of ribosomes in vivo. Cold Spring Harb Symp Quant Biol 34: 69–75

Hughes JM (1996) Functional base-pairing interaction between highly conserved elements of U3 small nucleolar RNA and the small ribosomal subunit RNA. J Mol Biol 259: 645–654

Inoue K, Alsina J, Chen J, Inouye M (2003) Suppression of defective ribosome assembly in a rbfA deletion mutant by overexpression of Era, an essential GTPase in *Escherichia coli*. Mol Microbiol 48: 1005–1016

Ito T, Wittmann HG (1973) Amino acid replacements in proteins S5 and S12 of two *Escherichia coli* revertants from streptomycin dependence to independence. Mol Gen Genet 127: 19–32

Kaczanowska M, Ryden-Aulin M (2007) Ribosome biogenesis and the translation process in *Escherichia coli*. Microbiol Mol Biol Rev 71: 477–494

Kirthi N, Roy-Chaudhuri B, Kelley T, Culver GM (2006) A novel single amino acid change in small subunit ribosomal protein S5 has profound effects on translational fidelity. RNA 12: 2080–2091

Klein BK, Staden A, Schlessinger D (1985) Alternative conformations in *Escherichia coli* 16S ribosomal RNA. Proc Natl Acad Sci USA 82: 3539–3542

Krzyzosiak W, Denman R, Nurse K, Hellmann W, Boublik M, Gehrke CW, Agris PF, Ofengand J (1987) In vitro synthesis of 16S ribosomal RNA containing single base changes and assembly into a functional 30S ribosome. Biochemistry 26: 2353–2364

Kurland CG and Ehrenberg M (1996) Limitations of translational accuracy. In: Neidhardt FC, Curtiss R. III, Ingraham JL, Lin ECC, Low BK, Magasanik B, Reznikoff WS, Riley M, Schaechter M, Umbarger HE (eds) *Escherichia coli* and Salmonella: cellular and molecular biology. ASM Press, Washington, DC, Vol. 1, pp 979–1004

Li Z, Pandit S, Deutscher MP (1999) RNase G (CafA protein) and RNase E are both required for the 5' maturation of 16S ribosomal RNA. EMBO J 18: 2878–2885

Mangiarotti G, Turco E, Perlo C, Altruda F (1975) Role of precursor 16S RNA in assembly of *E. coli* 30S ribosomes. Nature 253: 569–571

Nakao A, Yoshihama M, Kenmochi N (2004) RPG: the Ribosomal Protein Gene database. Nucleic Acids Res 32:D168–170

Nashimoto H, Held W, Kaltschmidt E, Nomura M (1971) Structure and function of bacterial ribosomes. XII. Accumulation of 21S particles by some cold-sensitive mutants of *Escherichia coli*. J Mol Biol 62: 121–138

Nomura M, Held W (1974) Reconstitution of ribosomes: Studies of ribosome structure, function and assembly. In: Nomura

M, Tissières A, Lengyel P, (eds) Ribosomes. Cold Spring Harbor Laboratory Press, Cold Spring Harbor, New York pp 193–223

Ogle JM, Carter AP, Ramakrishnan V (2003) Insights into the decoding mechanism from recent ribosome structures. Trends Biochem Sci 28: 259–266

Ogle JM, Murphy FV, Tarry MJ, Ramakrishnan V (2002) Selection of tRNA by the ribosome requires a transition from an open to a closed form. Cell 111: 721–732

Pestov DG, Strezoska Z, Lau LF (2001) Evidence of p53-dependent cross-talk between ribosome biogenesis and the cell cycle: effects of nucleolar protein Bop1 on G(1)/S transition. Mol Cell Biol 21: 4246–4255

Piepersberg W, Bock A, Wittmann HG (1975a) Effect of different mutations in ribosomal protein S5 of *Escherichia coli* on translational fidelity. Mol Gen Genet 140: 91–100

Piepersberg W, Bock A, Yaguchi M, Wittmann HG (1975b) Genetic position and amino acid replacements of several mutations in ribosomal protein S5 from *Escherichia coli*. Mol Gen Genet 143: 43–52

Ramakrishnan, V, White, SW (1992) The structure of ribosomal protein S5 reveals sites of interaction with 16S rRNA. Nature 358: 768–771

Roy-Chaudhuri B, Kirthi N, Culver GM (2010) Appropriate maturation and folding of 16S rRNA during 30S subunit biogenesis are critical for translational fidelity. Proc Natl Acad Sci USA 107: 4567–4572

Roy-Chaudhuri B, Kirthi N, Kelley T, Culver GM (2008) Suppression of a cold-sensitive mutation in ribosomal protein S5 reveals a role for RimJ in ribosome biogenesis. Mol Microbiol 68: 1547–1559

Schuwirth BS, Borovinskaya MA, Hau CW, Zhang W, Vila-Sanjurjo A, Holton JM, Cate JH (2005) Structures of the bacterial ribosome at 3.5 A resolution. Science 310: 827–834

Wachi M, Umitsuki G, Shimizu M, Takada A, Nagai K (1999) *Escherichia coli* cafA gene encodes a novel RNase, designated as RNase G, involved in processing of the 5' end of 16S rRNA. Biochem Biophys Res Commun 259: 483–488

Wilson DN, Nierhaus KH (2007) The weird and wonderful world of bacterial ribosome regulation. Crit Rev Biochem Mol Biol 42: 187–219

Wimberly BT, Brodersen DE, Clemons WM, Jr., Morgan-Warren RJ, Carter AP, Vonrhein C, Hartsch T, Ramakrishnan V (2000) Structure of the 30S ribosomal subunit. Nature 407: 327–339

Woodson SA (2008) RNA folding and ribosome assembly. Curr Opin Chem Biol 12: 667–673

Yoshikawa A, Isono S, Sheback A, Isono K (1987) Cloning and nucleotide sequencing of the genes rimI and rimJ which encode enzymes acetylating ribosomal proteins S18 and S5 of *Escherichia coli* K12. Mol Gen Genet 209: 481–488

Young RA, Steitz JA (1978) Complementary sequences 1700 nucleotides apart form a ribonuclease III cleavage site in *Escherichia coli* ribosomal precursor RNA. Proc Natl Acad Sci USA 75: 3593–3597.

Section IV Elongation and ribosome dynamics

Exploring the structural dynamics of the translational machinery using single-molecule fluorescence resonance energy transfer

22

Daniel D. MacDougall and Ruben L. Gonzalez, Jr.

1. Introduction

The ribosome can be regarded as a molecular machine that converts chemical and thermal energy into productive mechanical work (Spirin, 2002; Frank and Gonzalez, 2010). This chemo- and thermomechanical view of ribosome function is fueling current efforts to identify the mobile components of the ribosomal machine, characterize the structural dynamics of these components, and develop an understanding of how these dynamics are regulated in order to direct mechanical processes during protein synthesis. It is within this context that single-molecule fluorescence resonance energy transfer (smFRET) (Ha, 2001) has emerged as a powerful tool for investigating the structural dynamics and mechanical properties of the translating ribosome. Because this chapter marks its first appearance in this volume, the first half of this chapter provides a brief introduction to smFRET that is specifically framed around its use as a tool for investigating the structural dynamics of the translational machinery. Our intent here is not to provide a comprehensive or detailed review of smFRET (for that we refer the reader to excellent reviews by Ha and co-workers (Ha, 2001; Roy et al., 2008)), but rather to provide a basic understanding of the technique and highlight the strengths and limitations that are most important for understanding and interpreting smFRET studies of protein synthesis.

In the second half of this chapter, we use one of the most dynamic steps in protein synthesis, the movement of the messenger RNA (mRNA)-transfer RNA (tRNA) complex through the ribosome during the translocation step of translation elongation, as an example with which to demonstrate the unique mechanistic information that can be obtained from smFRET studies of protein synthesis. Specifically, we describe how smFRET studies of ribosomal pre-translocation complexes have enabled the discovery and characterization of thermally activated structural fluctuations of the pre-translocation complex. We discuss in detail how modulations of these fluctuations are used to regulate and drive the translocation reaction. We close by briefly highlighting how similar results from smFRET studies of additional steps in protein synthesis are giving rise to new paradigms describing the mechanism and regulation of protein synthesis.

2. Single-molecule fluorescence resonance energy transfer

2.1. Physical principles underlying fluorescence resonance energy transfer

Fluorescence resonance energy transfer (FRET) (Förster, 1946) is a photophysical process involving two fluorophores, termed the donor and the acceptor. In a typical FRET experiment, the donor is directly illuminated by an excitation light source and, upon absorption of a photon, undergoes a transition to an excited electronic state. From its excited state, the donor can emit a photon and relax back to its ground electronic state, a process known as fluorescence. Alternatively, the excited donor can transfer energy to the acceptor *via* a non-radiative dipole-dipole coupling mechanism known as FRET, such that the acceptor now undergoes a transition to an excited electronic state. Subsequent relaxation of the acceptor back to its electronic ground state through the emission of a photon now results in fluorescence from the acceptor.

The efficiency of FRET (E_{FRET}) is given by $E_{FRET} = (1 + (R/R_0)^6)^{-1}$ where R is the distance between the donor and acceptor dipoles and R_0, known as the Förster distance, is the value of R at which $E_{FRET} = 0.50$ (Figure 1A). R_0 for a specific donor-acceptor pair is a constant that is given by $R_0 = 9.78 \times 10^3 \, (\Phi_D \bullet \kappa^2 \bullet n^{-4} \bullet J(\lambda))^{1/6}$ where Φ_D is the fluorescence quantum yield of the donor in the absence of the acceptor, κ^2 is a geometric factor that depends on the relative orientation of the donor and acceptor transition dipole moments, n is the refractive index of the medium in which the energy transfer occurs, and $J(\lambda)$ is the overlap between the fluorescence emission spectrum of the donor and the absorbance spectrum of the acceptor. Thus, while E_{FRET} scales with the inverse sixth power of R, precise determination of an absolute distance using an experimentally measured value of E_{FRET} requires either careful determination of R_0 (including Φ_D, κ^2, n, and $J(\lambda)$) or, more practically, careful experimental calibration of E_{FRET} *versus* R using a biomolecule of known conformation and thus known R. Depending on the specific donor-acceptor pair used, typical values of R_0 render E_{FRET} sensitive to distances in the range of $10–100$ Å. This exquisite sensitivity to distances on the Å length scale makes E_{FRET} an effective molecular ruler that has evolved into a powerful biophysical tool for investigating biomolecular structure and dynamics.

2.2. Single-molecule studies uncover unique mechanistic information

Ensemble FRET experiments have been in widespread use for over fifty years and have provided unprecedented insights into the structure and dynamics of biomolecules and biomolecular complexes (reviewed in (Wu and Brand, 1994; Clegg, 1995; Selvin, 1995, 2000; Jares-Erijman and Jovin, 2003; Hwang et al., 2009)). However, because ensemble FRET experiments report the mean value of E_{FRET} averaged over the trillions of individual biomolecules that collectively form the ensemble, any heterogeneity in the structure or dynamics of the biomolecules comprising the ensemble can generate a distorted mean value of E_{FRET} that is difficult or even impossible to interpret. Such heterogeneity can be generally divided into two categories: static heterogeneity, which originates from variations in the structure or dynamics of individual biomolecular subpopulations across the ensemble, and dynamic heterogeneity, which originates from the asynchronous transitioning of individual biomolecules between multiple states, each of which is structurally or dynamically distinct (Hwang et al., 2009). Single-molecule FRET (smFRET) experiments complement ensemble FRET

experiments by permitting the ensemble of biomolecules to be dissected into sub-populations (in the case of static heterogeneity) and/or states (in the case of dynamic heterogeneity), each of which exhibit characteristic structural and/or dynamic properties (Figure 1B and C). In so doing, smFRET experiments provide a powerful opportunity to investigate the structural, dynamic, and biochemical properties of individual subpopulations or states.

The ability to parse static and dynamic heterogeneity using smFRET can be of great mechanistic importance; this is because both types of heterogeneity can be exploited to regulate reaction mechanisms. As an example of mechanistically important dynamic heterogeneity, consider an enzyme that exists in a dynamic equilibrium between multiple conformational states, but where, in an extreme case, only one state can progress along the reaction pathway. In such a situation, precise control over the rate with which this state is sampled by the individual enzyme molecules within the ensemble and/or over the stability of this state relative to the other accessible states, provides an effective mechanism for regulating the enzymatic reaction. By permitting the identification and characterization of individual states, including functionally competent states, smFRET experiments provide an opportunity to collect mechanistically important data that are unique from, and complementary to, that obtained from ensemble FRET experiments.

2.3. Design of donor-acceptor labeling schemes

Perhaps the most significant challenges to smFRET studies of the translational machinery are the design of donor-acceptor labeling schemes and the technical challenges involved in the fluorescence labeling of translation components. To be mechanistically informative, donor-acceptor labeling schemes must be designed and implemented such that the structural rearrangement of interest yields a change in the distance between the donor and the acceptor that generates an experimentally detectable change in E_{FRET}. The fluorescence labeling of translation components that is required to meet this criterion must be: (i) efficient, such that a large population of the observed ribosomal complexes contain both a donor and an acceptor; (ii) specific, such that any heterogeneity detected within the ensemble of ribosomal complexes reflects static or dynamic heterogeneity of the complexes rather than heterogeneity in the positions of the donor or accep-

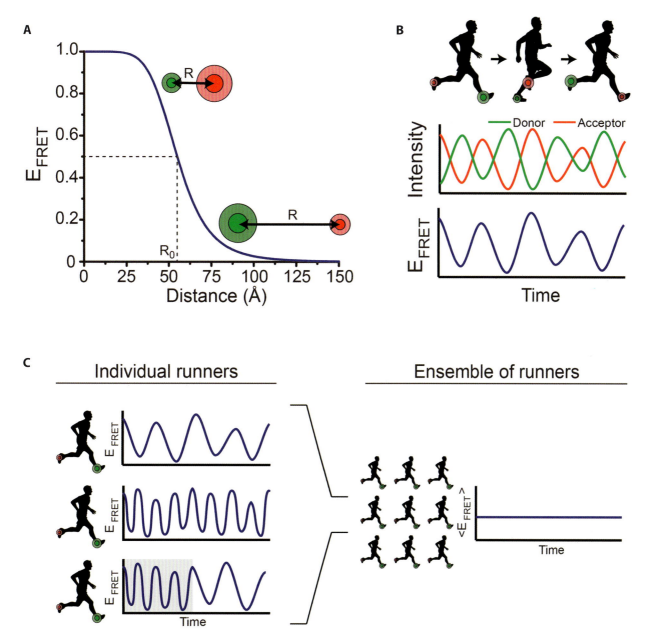

Fig. 1 FRET at the single-molecule and ensemble levels. (A) Plot of the FRET efficiency (E_{FRET}) as a function of the distance (R) between a donor fluorophore (green sphere) and an acceptor fluorophore (red sphere) with an R_0 of 55 Å. When $R < R_0$, $E_{FRET} > 0.50$, when $R = R_0$, $E_{FRET} = 0.50$, and when $R > R_0$, $E_{FRET} < 0.50$. (Adapted from (Roy et al., 2008) with permission from Macmillan Publishers Ltd). (B) A macroscopic example of a FRET experiment. Consider a single runner with a donor-acceptor pair attached to his sneakers (artwork based on an original illustration from iStockPhoto. com, (Laurence Dean). As the runner strides, the distance between the donor-acceptor pair periodically increases and decreases. Consequently, the donor and acceptor emission intensities (I_D and I_A, respectively) *versus* time trajectory yields periodic, anti-correlated increases and decreases in I_A and I_D that are characteristic of FRET. Likewise, the corresponding E_{FRET} *versus* time trajectory (where $E_{FRET} = I_A / (I_A + I_D)$) exhibits periodic increases and decreases in E_{FRET}. From this E_{FRET} *versus* time trajectory it is possible to determine the average stride length and rate of the runner, information that is critical to a full description of the mechanics of running. (C) Representative E_{FRET} *versus* time trajectories from three sub-populations of runners that can be distinguished by their different stride rates (left panel). Static heterogeneity arises from sub-populations of runners with either slow (top row) or fast (middle row) stride rates. Dynamic heterogeneity arises from a sub-population of runners (bottom row) who stochastically alternate between fast (grey shaded box) and slow stride rates. Despite this heterogeneity, analysis of hundreds of individual E_{FRET} *versus* time trajectories can provide the average stride rates of the slow and fast sub-populations of runners (or slow and fast phases of running), information that would be obscured in the ensemble-averaged E_{FRET} (<E_{FRET}>) *versus* time trajectory of the ensemble of runners (right panel).

tor; and (iii) minimally perturbative, such that the presence of the donor-acceptor pair does not significantly interfere with the biochemical activities of the relevant translation components. Over the past several years, we and others have developed highly purified *in vitro* translation systems which, in combination with a battery of previously developed standard biochemical assays, have allowed development and validation of numerous donor-acceptor labeling schemes for smFRET studies of protein synthesis (reviewed in (Frank and Gonzalez, 2010) and discussed below in Section 3.3). Rather than providing a detailed description of the design, implementation, and validation of donor-acceptor labeling schemes here, we instead refer the reader to a recently published chapter on this topic (Fei et al., 2010) and dedicate the remainder of this section to a discussion of the optical setup for smFRET experiments and analysis of the resulting data.

2.4. Total internal reflection fluorescence microscopy

The most important requirement for the detection of fluorescence emission from single molecules is a highly sensitive fluorescence microscope. A total internal reflection fluorescence microscope (TIRFM), which combines a totally internally reflected laser illumination source with wide-field optics and an electron-multiplying charge-coupled device (EMCCD) camera detector, offers such sensitivity (Axelrod et al., 1984; Axelrod, 2003; Joo and Ha, 2007) (Figure 2A). In a typical, prism-based TIRFM, the laser beam is aligned, collimated, and focused through a fused silica prism onto a quartz microfluidic flowcell. Upon encountering the interface between the quartz, having an index of refraction n_q, and the aqueous buffer containing the fluorescence-labeled ribosomal complexes, having an index of refraction $n_b < n_q$, the incident laser beam is totally internally reflected away from the quartz-buffer interface and back into the quartz at all angles greater than the "critical angle," θ_c, given by $\theta_c = \sin^{-1} n_b/n_q$. Regardless of the total internal reflection of the incident laser beam at the quartz-buffer interface, a weak evanescent electromagnetic field propagates into the medium of lesser index of refraction (i. e. the buffer) in the plane of incidence of the laser beam and in a direction that runs parallel along the quartz-buffer interface. The intensity of the weak evanescent field decays exponentially with increasing distance from the quartz-buffer interface, therefore selectively illuminating only

a thin layer of the buffer that is adjustable within a depth range of 70–300 nm. Because the excitation of molecules in the bulk buffer is limited by localization of the evanescent field to a thin layer of the buffer just beyond the quartz-buffer interface, the signal-to-noise ratio of a TIRFM is significantly greater than a conventional epi-fluorescence microscope, yielding very high sensitivity fluorescence detection.

Since the evanescent field generated by total internal reflection is confined to a thin layer of buffer just beyond the quartz-buffer interface, it is necessary to localize ribosomal complexes near the quartz surface of the microfluidic flowcell (Figure 2B). As a result, various approaches have been developed for tethering ribosomal complexes to the quartz surface in a manner that brings them within the evanescent field while preserving their biochemical activity (Blanchard et al., 2004b; Uemura et al., 2007; Wang et al., 2007). All such approaches combine a surface passivation method that renders the quartz surface relatively inert to non-specific binding of translation components with an affinity-based surface tethering method that allows specific tethering of ribosomal complexes (Rasnik et al., 2005). The most commonly used method involves passivating the quartz surface with a mixture of polyethylene glycol (PEG) and biotinylated PEG (Ha et al., 2002). Subsequent incubation of the PEG- and biotinylated PEG-derivatized microfluidic flowcell with streptavidin, followed by incubation with a ribosomal complex assembled onto a biotinylated mRNA (Blanchard et al., 2004b) or, alternatively, a directly biotinylated ribosomal subunit (Wang et al., 2007), then allows tethering of the ribosomal complex *via* a biotin-streptavidin-biotin bridge. In addition to confining ribosomal complexes within the evanescent field, surface tethering allows the fluorescence emission from individual, spatially localized donor-acceptor pairs to be observed for extended periods of time, limited only by the irreversible, oxygen-mediated photobleaching of the donor or acceptor (Hubner et al., 2001; Piwonski et al., 2005; Renn et al., 2006). It should be noted that observation times in smFRET experiments are typically extended through the use of enzymatic oxygen scavenging systems (Benesch and Benesch, 1953; Patil and Ballou, 2000; Ha, 2001; Aitken et al., 2008) and the photostabilities of the fluorophores are additionally enhanced through the addition of small-molecule triplet-state quenchers that suppress unwanted blinking of the fluorophores (Gonzalez Jr. et al., 2007; Aitken et al., 2008; Dave et al., 2009).

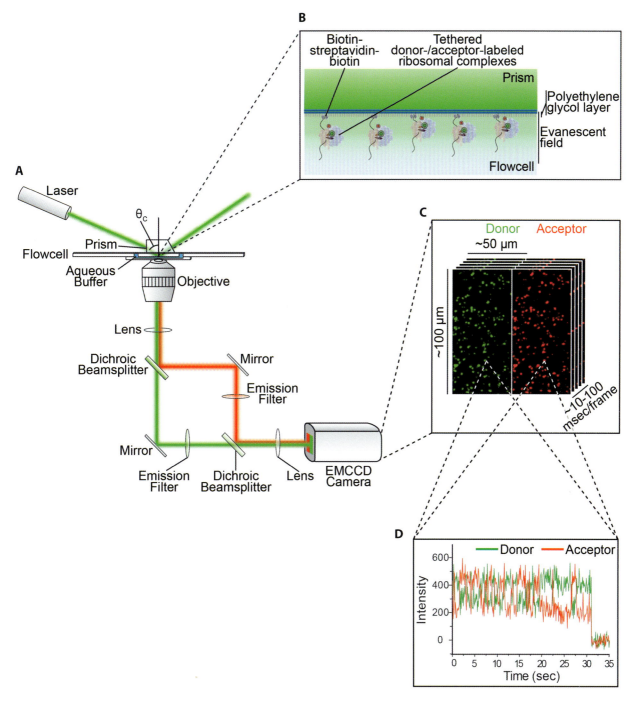

Fig. 2 Single-molecule fluorescence detection using a prism-based total internal reflection fluorescence microscope. (A) Principles of operation and typical optical setup of a total internal reflection fluorescence microscope. See Section 2.4 for a detailed description. (B) Inset showing an enlargement of the quartz/buffer interface and depicting the tethering of donor-/acceptor-labeled ribosomal complexes onto the polyethylene glycol (PEG)/biotinylated PEG-passivated quartz surface using a biotin-streptavidin-biotin bridge. See Section 2.4 for a detailed description. (C) Inset showing an enlargement of a single image recorded by the EMCCD. Typically, the donor and acceptor signals from 200–400 spatially resolved ribosomal complexes located within a 100 μm × 50 μm field-of-view are simultaneously imaged onto two separate halves of the capacitor array within the EMCCD camera. Individual images are recorded as a digital video with a typical frame rate in the tens of frames per sec (i.e., a time resolution in the tens of msec per frame). (D) Representative donor and acceptor emission intensities *versus* time trajectory derived from a single donor-/acceptor-labeled ribosomal complex within the field-of-view (Fei et al., 2009).

In addition to the high sensitivity arising from its total internal reflection illumination mode, the TIRFM is a wide-field instrument, therefore allowing simultaneous excitation and detection of fluorescence signals from several hundred individual ribosomal complexes. Upon selective donor excitation *via* total internal reflection illumination, fluorescence emissions from several hundred spatially localized donor-acceptor pairs, each arising from a single surface-tethered ribosomal complex, are simultaneously collected through a microscope objective. A system of lenses, dichroic beamsplitters, mirrors, and emission filters is then used to: (i) separate the emitted fluorescence into individual donor and acceptor channels; (ii) filter out any remaining traces of the total internally reflected illumination source from each channel; and (iii) direct the two channels to the EMCCD camera such that the donor and acceptor signals from several hundred spatially resolved ribosomal complexes are simultaneously imaged onto two separate sectors of the capacitor array within the EMCCD camera (Figure 2C). Individual images are recorded as a digital video with a typical frame rate in the tens of frames per sec (i. e. a time resolution in the tens of msec per frame). Despite the limited time resolution, it is the combination of the TIRFM's wide-field operation and high sensitivity that has thus far made it the instrument of choice for smFRET studies of protein synthesis.

2.5. Limitations of single-molecule fluorescence microscopies

Despite the current dominance of TIRFM, it is important to note a few limitations of this approach that significantly impact smFRET studies of protein synthesis. One limitation of TIRFM is the experimental time resolution, which is restricted by the rate with which the EMCCD camera reads out an individual frame. Using the typical settings on a current state-of-the-art EMCCD camera, the time resolution is limited to tens of msec per frame. Even using settings allowing for the maximal readout rate, albeit at the expense of significantly reduced signal-to-noise ratio, the time resolution of a current state-of-the-art EMCCD camera is limited to several msec per frame (Cornish and Ha, 2007). Therefore, structural rearrangements of ribosomal complexes occurring on timescales faster than ~10 msec can lead to time-averaged values of E_{FRET}, making it difficult or impossible to confidently identify and investigate these conformational changes.

Given that, *in vivo*, a single elongation cycle occurs on a timescale of ~50–200 msec at 37 °C (i. e. corresponding to a rate of ~5–20 amino acids per sec (Kennell and Riezman, 1977; Bremer and Dennis, 1987; Sorensen and Pedersen, 1991; Liang et al., 2000; Proshkin et al., 2010), with rate-limiting structural rearrangements of the elongating ribosomal complex expected to occur on timescales of a few msec to a couple hundred msec at 20–37 °C (Pape et al., 1998; Savelsbergh et al., 2003; Gromadski and Rodnina, 2004; Pan et al., 2007), it is perhaps inevitable that some subset of ribosomal complex dynamics will be poorly defined or even completely missed by TIRFM. Indeed, nearly all TIRFM-based smFRET studies of protein synthesis to date report at least one example of a conformational process that is missed due to time-averaged values of E_{FRET} (for a particularly clear example involving tRNA fluctuations within ribosomal complexes at low Mg^{2+} concentrations, see (Blanchard et al., 2004b)). In principle, the time resolution of smFRET experiments can be extended to a few μsec per timestep by replacing the TIRFM equipped with an EMCCD camera detector with a confocal fluorescence microscope equipped with an avalanche photodiode detector (Cornish and Ha, 2007). However, confocal microscopy is constrained in that it entails excitation and detection of a single ribosomal complex at a time, significantly increasing the amount of time required to collect datasets large enough to be statistically significant. Consequently, only a single example of a confocal microscopy-based smFRET experiment on ribosomal complexes has thus far been reported (Blanchard et al., 2004b).

An additional limitation of both TIRFM and confocal fluorescence microscopy is a restriction on the maximum concentration of fluorescence-labeled molecules that can be maintained within the microfluidic flowcell during smFRET experiments. As described in Section 2.4, the evanescent field produced in a TIRFM confines the excitation of fluorescence-labeled components to just a thin layer of buffer beyond the quartz-buffer interface. Using point-like illumination and detection, a confocal fluorescence microscope likewise confines the excitation of fluorescence-labeled components to a small (typically diffraction-limited) focal volume (Pawley, 2006). Despite this, concentrations of fluorescence-labeled components exceeding several tens of nM will substantially lower the signal-to-noise of both TIRFM and confocal fluorescence microscopy. Therefore, it remains difficult to perform smFRET experiments involving the delivery of physiologically relevant concentrations of fluorescence-labeled ribosomal subunits, tRNAs, or translation factors into the microfluidic flowcell. This limitation has been recently overcome through the development of so-called zero-mode waveguides (Levene et al., 2003; Moran-Mirabal and Craighead, 2008), which decrease the effective illumination volume by several orders of magnitude over that achieved by total internal reflection. This approach enables experiments to be conducted at near-physiological concentrations of fluorescence-labeled components (Uemura et al., 2010) (reviewed in the chapter by Uemura and Puglisi of this volume).

A third limitation of TIRFM, which also applies to confocal fluorescence microscopy and the use of zero-mode waveguides, is the potential ambiguity of smFRET data collected on systems with multiple donor-acceptor pairs (Hohng et al., 2004; Clamme and Deniz, 2005; Munro et al., 2010a). These experiments are challenging because the spectral overlap between the desired donor-acceptor pairs that is required to generate the desired FRET signals generally gives rise to unavoidable spectral overlap between alternative donor-acceptor pairs that can generate unwanted FRET signals. In order to avoid convoluting the desired FRET signals with unwanted FRET signals, care must therefore be taken in designing the labeling scheme such that the distances between unwanted donor-acceptor pairs remains large enough that their FRET efficiency is minimized or eliminated. Even if unwanted FRET signals can be minimized or eliminated, however, the significant spectral overlap between the desired donor-acceptor pairs that is required to generate the desired FRET signals opposes the spectral separation that is subsequently required to effectively separate the emitted fluorescence into the various individual donor and acceptor channels. Even careful optimization of the optical setup yields incompletely separated donor and acceptor fluorescence emissions with very low signal-to-noise ratios, invariably degrading the quality of the smFRET data to the point where it cannot be quantitatively analyzed without extensive and rigorous correction of the spectral bleedthrough among the various donor and acceptor channels.

2.6. Analysis of smFRET data

Technically detailed general procedures for the analysis of smFRET data have been recently described (Joo and Ha, 2007; Blanco and Walter, 2010) and a practical guide aimed at the non-expert is also available (Roy et al., 2008). Drawing from this framework, in this sec-

tion we provide a brief summary of the steps involved in the analysis of complex smFRET data such as that obtained from smFRET studies of protein synthesis and highlight those aspects of these procedures which have the most impact on the analysis and interpretation of the data.

The first step in the analysis of smFRET data is to extract E_{FRET} *versus* time trajectories, or smFRET trajectories, from the digital video output of the EMCCD camera. This is a fairly straightforward step that is usually done using a semi- or fully automated procedure that involves: (i) identification of single fluorophores in the donor and acceptor images recorded by the EMCCD camera; (ii) alignment of the donor and acceptor images to generate superimposed single donor-acceptor pairs; (iii) plotting of donor and acceptor emission intensities *versus* time trajectories for each donor-acceptor pair; (iv) spectral bleedthrough correction of donor emission into the acceptor channel and, if necessary, acceptor emission into the donor channel; (v) baseline correction of the donor and acceptor emission intensities such that the background from the donor and acceptor channels following photobleaching averages to zero intensity; and (vi) plotting of E_{FRET} *versus* time trajectories from the bleedthrough- and baseline-corrected donor and acceptor emission intensities *versus* time trajectories using the equation $E_{FRET} = I_A/(I_A + I_D)$, where I_A and I_D are the emission intensities of the acceptor and the donor, respectively.

At this point, only those trajectories that can be shown to arise from *bona fide* FRET between a single donor-acceptor pair are selected for further analysis. As this selection is typically based on visual inspection or a combination of semi-automated procedures and visual inspection, it is important to minimize user bias by defining a set of objective selection criteria. Because such criteria will be specific to each project and may vary across research groups, it is important that they be clearly reported in individual publications. Commonly applied criteria include requirements that: (i) the emission intensities of the donor and acceptor are within the intensity distributions expected for single donors and acceptors, respectively; (ii) the donor and acceptor undergo photobleaching in a single time step; and (iii) changes in donor and acceptor emission intensities are anti-correlated. For examples of selection criteria specific to smFRET studies of protein synthesis, see (Blanchard et al., 2004b; Munro et al., 2007; Wang et al., 2007; Cornish et al., 2008; Fei et al., 2008).

The detection of single fluorophores at the time resolutions common to smFRET experiments invariably leads to noisy raw smFRET trajectories. In order to avoid missing mechanistically important features obscured by the noisy nature of the data or, conversely, to avoid over- or misinterpreting noise as a mechanistically important feature of the data, statistically rigorous inference of the data should be performed. This is typically achieved by using a hidden Markov model to identify discrete conformational states within the noisy raw smFRET trajectories and to determine the most probable path (i. e., the idealized trajectory) through these conformational states (Qin et al., 1996; Andrec et al., 2003; McKinney et al., 2006; Talaga, 2007; Bronson et al., 2009; Liu et al., 2010). An important aspect of developing a hidden Markov model of a raw smFRET trajectory is determination of the model complexity (i. e., the number of conformational states that can be confidently inferred from the raw smFRET trajectory) and the model parameters (i. e., the distribution of E_{FRET} values and transition rates associated with each of the inferred conformational states). This is most commonly accomplished using maximum likelihood-based methods (Qin et al., 1996; McKinney et al., 2006; Liu et al., 2010) which seek to find the parameters that maximize the probability of the data given the model. These methods have a tendency to overestimate the model complexity, however, often leading to problems with overfitting of the data (Bronson et al., 2009; Liu et al., 2010). Thus, care should be taken in identifying additional (i. e. intermediate) conformational states using maximum likelihood-based hidden Markov modeling of smFRET trajectories; ideally, the authenticity of intermediate states identified in this way should be verified though structural and/or biochemical studies. More recently, an alternative to maximum likelihood, termed maximum evidence, has been suggested for the analysis of smFRET trajectories (Bronson et al., 2009). This method, which seeks to find the model that maximizes the probability of the data, naturally avoids overfitting and can have a significantly lower tendency to overestimate the number of conformational states that can be confidently identified in the raw smFRET trajectories.

Further analysis of the raw smFRET trajectories and/or idealized trajectories is highly dependent on the experimental question that is being addressed by the smFRET experiment (Roy et al., 2008). In general, however, equilibrium properties of the system can be derived from the population distribution of observed values of E_{FRET} over a large number of individual smFRET trajectories. The rates of transition between the various conformational states of a system that is in a dynamic conformational equilibrium can be obtained from exponential fits to the distribution of dwell times spent at each conformational state before transitioning to each of the other conformational states (Colquhoun and Hawkes, 1995) or by using the transition probability matrix that results directly from hidden Markov model analysis (McKinney et al., 2006). Non-equilibrium properties of the system can be assessed by monitoring the evolution of E_{FRET} as a function of time for a reaction in which a substrate or ligand is stopped-flow delivered to a surface-tethered biomolecule of interest (Blanchard et al., 2004a).

3. smFRET studies of the translocation step of translation elongation

3.1. Translocation of the mRNA-tRNA complex through the ribosome

During the elongation stage of protein synthesis, the ribosome sequentially adds amino acids to a growing polypeptide chain at a rate of ~5–20 amino acids sec^{-1} at 37 °C *in vivo* (Kennell and Riezman, 1977). With the addition of each amino acid, the ribosome repetitively cycles through three major steps: aminoacyl-tRNA (aa-tRNA) selection (Rodnina et al., 2005), peptide bond formation (Beringer and Rodnina, 2007), and translocation (Shoji et al., 2009). Upon accommodation of an aa-tRNA into the ribosomal A (aa-tRNA binding) site at the end of aa-tRNA selection, peptide bond formation results in the transfer of the nascent polypeptide from the peptidyl-tRNA at the ribosomal P (peptidyl-tRNA binding) site to the A-site aa-tRNA. The resulting ribosomal pre-translocation (PRE) complex is the substrate on which the GTPase elongation factor G (EF-G) will act to catalyze translocation by precisely one codon. During translocation, the newly deacylated tRNA at the P site moves into the ribosomal E (exit) site, the newly formed peptidyl-tRNA at the A site moves into the P site, and the next mRNA codon moves into the A site. The resulting ribosomal post-translocation (POST) complex is now ready to participate in aa-tRNA selection during the next round of the elongation cycle. How the ribosome accomplishes the rapid and precise movement of the mRNA-tRNA complex during translocation continues to be the subject of intense investigation using genetic, biochemical, structural, and, most recently, smFRET approaches (reviewed in (Shoji et al., 2009; Aitken et al., 2010; Dunkle and Cate, 2010; Frank and Gonzalez, 2010)).

3.2. Structural rearrangements of the translational machinery important for translocation

Based on the ribosome's universally conserved two-subunit architecture, Spirin (Spirin, 1968) and Bretscher (Bretscher, 1968) were the first to hypothesize that translocation proceeds *via* an intermediate configuration of the tRNAs within the PRE complex that is somehow coupled to a relative rearrangement of the small (30S in *Escherichia coli*) and large (50S in *E. coli*) ribosomal subunits. Experimental validation of this hypothesis first came from chemical probing studies (Moazed and Noller, 1989), subsequently confirmed by ensemble FRET experiments (Odom et al., 1990), demonstrating that peptide bond formation results in the spontaneous rearrangement of the ribosome-bound tRNAs from their "classical" P/P (denoting the 30S P/50S P sites) and A/A configurations, into intermediate "hybrid" P/E and A/P configurations. As the naming convention suggests, in their hybrid configurations the aminoacyl acceptor ends of the P- and A-site tRNAs have moved into the 50S subunit E and P sites, respectively, while their anticodon stem-loops remain bound at the 30S subunit P and A sites. EF-G likely promotes further rearrangements of the aminoacyl acceptor end of the A-site peptidyl-tRNA within the 50S subunit P site (Borowski et al., 1996; Pan et al., 2007) and subsequently catalyzes the movement of the tRNA anticodon stem-loops and the associated mRNA from the 30S subunit P and A sites into the 30S subunit E and P sites, respectively, thus completing the translocation reaction.

A decade after the chemical probing studies, cryogenic electron microscopic (cryo-EM) reconstructions of PRE complex analogs containing vacant A sites (PRE^{-A} complexes) and stabilized through the binding of EF-G in the presence of GDPNP, a non-hydrolyzable GTP analog, allowed visualization of the P/E tRNA configuration and of possibly associated large-scale conformational rearrangements of the PRE^{-A} complex (Frank and Agrawal, 2000; Valle et al., 2003). Comparison of cryo-EM reconstructions of PRE^{-A} complexes in the presence and absence of EF-G(GDPNP) revealed three major conformational changes. These were: (i) the aforementioned movement of the deacylated P-site tRNA from the P/P to

the P/E configuration; (ii) the ~20 Å movement of a universally conserved and highly dynamic domain of the 50S subunit E site, the L1 stalk, from an open to a closed conformation such that it establishes a direct interaction with the central fold, or elbow, domain of the P/E-configured tRNA; and (iii) the counter-clockwise, ratchet-like rotation of the 30S subunit with respect to the 50S subunit (when viewed from the solvent side of the 30S subunit) from a non-rotated to a rotated subunit orientation.

Hereafter we will refer to the two conformations of the PRE^{-A} complex observed in the cryo-EM studies described above as global state 1 (GS1), observed in the absence of EF-G(GDPNP) and encompassing non-rotated subunits, classically bound tRNAs, and an open L1 stalk, and global state 2 (GS2), observed in the presence of EF-G(GDPNP) and encompassing rotated subunits, hybrid-bound tRNAs, and a closed L1 stalk (Figure 3). We note that analogous terms have been introduced by Frank and co-workers (Macro State I and Macro State II; Frank et al., 2007), Noller and co-workers (non-rotated/classical and rotated/hybrid (Ermolenko et al., 2007b), and Cate and co-workers (R_0 and R_F; Zhang et al., 2009). Regardless of the differing terminologies, as originally hypothesized by Spirin and Bretscher, the conformational changes of the ribosome and the ribosome-bound tRNAs encompassed by the GS1-to-GS2 transition are expected to play a major role in facilitating the translocation reaction. Indeed, biochemical evidence lends support to the notion that GS2 represents an authentic on-pathway translocation intermediate (Dorner et al., 2006; Horan and Noller, 2007).

Using the available crystal and cryo-EM structures as guides, numerous donor-acceptor labeling schemes have been developed to investigate the dynamics of ribosome and tRNA conformational changes within PRE and PRE^{-A} complexes by smFRET, a subset of which will be discussed here. smFRET between A- and P-site tRNAs labeled within their elbow regions was initially shown to report on the occupancy of the classical and hybrid tRNA binding configurations (Blanchard et al., 2004b). Movement of the L1 stalk from an open to a closed conformation has been tracked through smFRET between donor and acceptor fluorophores attached to ribosomal proteins L1 and L9 (Fei et al., 2009) (an alternative L1-L33 smFRET signal has also been used for this purpose in an independent study; Cornish et al., 2009). In the closed conformation, the L1 stalk can form intermolecular contacts with the elbow domain of the P/E-configured tRNA;

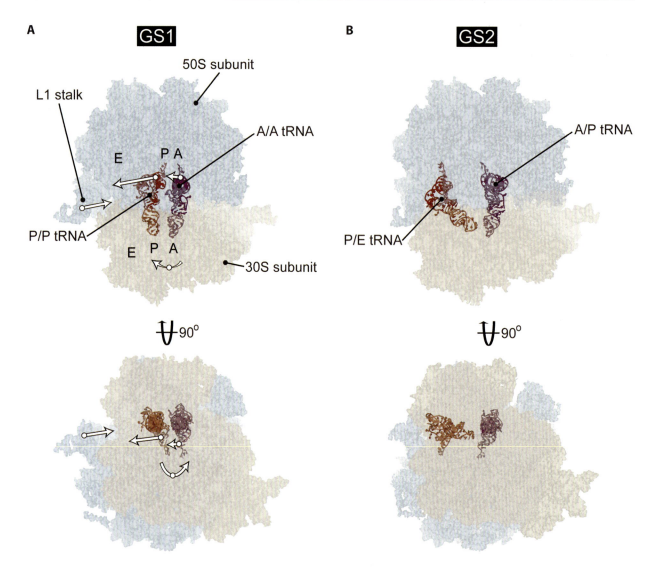

Fig. 3 Structural models of the GS1 and GS2 states of the PRE complex. Structural models were generated by flexible-fitting of atomic resolution structures into cryo-EM maps using molecular dynamics (Agirrezabala et al., 2008; Trabuco et al., 2008) kindly provided by Joachim Frank. The 30S and 50S ribosomal subunits were rendered in semi-transparent space-filling representations such that the P- and A-site tRNAs bound within the inter-subunit space and rendered in cartoon representations could be clearly visualized. Shown are perspectives from a side-view of the PRE complex illustrating the E-, P-, and A-tRNA binding sites (top row) as well as a view from the solvent-accessible surface of the 30S subunit (bottom row), obtained by 90° rotation of the side-view so that the 50S subunit is now behind the visual plane. (A) The GS1 state encompasses classically bound tRNAs, an open L1 stalk, no interaction between the L1 stalk and the P/P-configured tRNA, and a non-rotated subunit orientation. The white arrows mark the relative directions in which the tRNAs and the L1 stalk move and the 30S subunit rotates with respect to the 50S subunit during the transition from the GS1 state to the GS2 state. (B) The GS2 state encompasses hybrid-bound tRNAs, a closed L1 stalk, an interaction between the L1 stalk and the P/E-configured tRNA, and a rotated subunit orientation.

an smFRET signal between fluorophore-labeled L1 and P-site tRNA was developed to report on the formation and disruption of these contacts (Fei et al., 2008) (a similar L1-tRNA smFRET signal has also been used for this purpose in an independent study (Munro et al., 2010a)). Finally, inter-subunit rotation has been monitored through smFRET signals developed by reconstituting donor- and acceptor-labeled ribosomal proteins into 30S and 50S subunits; results obtained with an S6-L9 smFRET signal will be described below (Ermolenko et al., 2007a; Cornish et al., 2008). An additional inter-subunit FRET signal has been developed by hybridizing fluorescence-labeled oligonucleotides to helical extensions engineered into helix 44 of 16S ribosomal RNA (rRNA) within the 30S subunit and helix 101 of 23S rRNA within the 50S subunit (Dorywalska et al., 2005; Marshall et al., 2008). This latter signal, however, seems insensitive to the inter-subunit rotation observed by cryo-EM, instead reporting on an as yet undefined inter-subunit conformational switch that is uniquely triggered upon deacylation of the P-site tRNA *via* peptide bond formation and uniquely reset upon translocation of the A-site peptidyl-tRNA into the P site (Marshall et al., 2008; Aitken and Puglisi, 2010; Frank and Gonzalez, 2010).

Steady-state smFRET experiments on POST complexes prepared using the donor-acceptor pairs described above predominantly yield smFRET trajectories that stably sample a single FRET state with a distinct value of E_{FRET}, with the exception of the L1-L9 and the L1-L33 pairs. The specific dynamics of the L1-L9 and L1-L33 smFRET signals within a particular POST complex instead depend on the presence and identity of the E-site tRNA. Collectively, the data suggest that, prior to aa-tRNA accommodation into the A site and peptide bond formation, POST complexes primarily exist in a stable GS1-like structural state in which: (i) tRNAs exhibit a strong preference for their classical configurations; (ii) the L1 stalk primarily favors either the open conformation or a half-closed conformation that is unique to POST complexes carrying an E-site tRNA; (iii) interactions between the L1 stalk and the P-site tRNA are not made; and (iv) the majority of ribosomes are fixed in the non-rotated subunit orientation.

Deacylation of the P-site peptidyl-tRNA *via* peptidyl transfer to either aa-tRNA or the antibiotic puromycin at the A site yields PRE and PRE^{-A} complexes, respectively. Puromycin, which mimics the aminoacyl end of aa-tRNA, binds at the 50S A site and participates in peptide bond formation, ultimately dissociating from the 50S A site and leaving deacylated tRNA at the P site (Traut and Monro, 1964). In the absence of EF-G, the majority of steady-state smFRET experiments on PRE/PRE^{-A} complexes prepared using the donor-acceptor pairs described above yield smFRET trajectories that stochastically fluctuate between two FRET states with distinct values of E_{FRET} (Figure 4). In all cases, one of the observed FRET states could be assigned to GS1, the second to GS2 and Taken together, these experiments strongly suggest that, upon peptide bond formation and in the absence of EF-G, the entire PRE complex can stochastically fluctuate between GS1 and GS2, using the surrounding thermal bath as its sole energy source.

PRE/PRE^{-A} complexes therefore provide an excellent example of a dynamically heterogeneous system.

In an ensemble FRET experiment the asynchronous transitioning of individual PRE or PRE^{-A} complexes between GS1 and GS2 would be expected to yield a single, population-averaged value of E_{FRET}; indeed, such population-averaged values of E_{FRET} based on ensemble FRET experiments using several of the donor-acceptor pairs described above have been reported (Johnson et al., 1982; Paulsen et al., 1983; Odom et al., 1990; Ermolenko et al., 2007a; Ermolenko et al., 2007b). By eliminating this population averaging, the smFRET experiments: (i) reveal that PRE/PRE^{-A} complexes exist in a dynamic conformational equilibrium, fluctuating stochastically between GS1 and GS2; (ii) allow dissection of individual smFRET trajectories into time intervals spent in the GS1 or GS2 states, so identified by their characteristic values of E_{FRET}; (iii) enable detailed thermodynamic and kinetic characterization of the individual GS1 and GS2 states; and, as we shall see below, (iv) open the door to a still unfolding series of studies into the role that thermally activated structural fluctuations of the PRE complex may play in the mechanism and regulation of translocation.

Strong support for the interpretation of the smFRET data presented above has come from two ensemble FRET studies in which the S6-L9 FRET signal was used to monitor inter-subunit rotation as a function of experimental conditions favoring either the classical or hybrid tRNA configurations (Ermolenko et al., 2007a; Ermolenko et al., 2007b). These ensemble experiments showed that inter-subunit rotation in PRE/PRE^{-A} complexes can indeed occur in the absence of EF-G and demonstrated that PRE/PRE^{-A} ribosomes can be stabilized in the non-rotated or rotated subunit orientations by imposing experimental conditions that favor the classical or hybrid tRNA configurations, respectively. These data strongly suggest that the non-rotated subunit orientation is thermodynamically favored when the tRNAs are in the classical configuration (i. e. the GS1 state), while the rotated subunit orientation is thermodynamically favored when the tRNAs are in the hybrid configuration (i. e. the GS2 state). In complete agreement with this view, two recent cryo-EM studies in which particle classification methods were applied to a PRE complex revealed the existence of two classes of particles with structures corresponding to GS1 and GS2 (Agirrezabala et al., 2008; Julian et al., 2008).

The results of the smFRET and ensemble FRET experiments described above demonstrate that access to GS2 – and thus forward progression along the translocation reaction coordinate – can occur in the absence of EF-G and GTP hydrolysis. Indeed, full

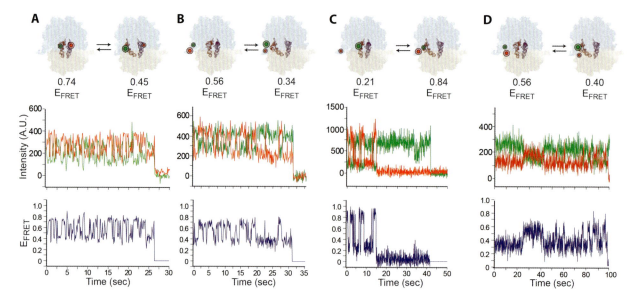

Fig. 4 E_{FRET} *versus* time trajectories derived from PRE complexes undergoing thermally activated fluctuations between GS1 and GS2. Structural models of GS1 and GS2 (top row) are displayed as in the top row of Figure 3. The approximate positions of the donor and acceptor fluorophores corresponding to each donor-acceptor labeling scheme are shown as green and red spheres, respectively. Representative donor and acceptor emission intensities *versus* time trajectories (middle row) are shown in green and red, respectively. The corresponding E_{FRET} *versus* time trajectories (bottom row), calculated using $E_{FRET} = I_A / (I_A + I_D)$, where I_A and I_D are the emission intensities of the acceptor and the donor, respectively, are shown in blue. (A) The tRNA-tRNA smFRET signal fluctuates between 0.74 (classical tRNA configuration, GS1) and 0.45 (hybrid tRNA configuration, GS2) values of E_{FRET} (Adapted from (Blanchard et al., 2004b) with permission from The National Academy of Sciences, USA). (B) The L1-L9 smFRET signal fluctuates between 0.56 (open L1 stalk conformation, GS1) and 0.34 (closed L1 stalk conformation, GS2) values of E_{FRET} (Reprinted from (Fei et al., 2009) with permission from The National Academy of Sciences, USA). (C) The L1-tRNA smFRET signal fluctuates between 0.21 (open L1 stalk not interacting with P/P-configured tRNA, GS1) and 0.84 (closed L1 stalk interacting with P/E-configured tRNA, GS2) values of E_{FRET} (Reprinted from (Fei et al., 2008) with permission from Elsevier). (D) The S6-L9 inter-subunit smFRET signal fluctuates between 0.56 (non-rotated subunit orientation, GS1) and 0.40 (rotated subunit orientation, GS2) values of E_{FRET}. (Adapted from Cornish et al., 2008, with permission from Elsevier)

rounds of spontaneous translation elongation have been observed *in vitro* in a factor-free environment, in which the ribosome moves slowly but directionally along the mRNA template to generate polypeptides of defined length (Pestka, 1969; Gavrilova and Spirin, 1971; Gavrilova et al., 1976). It seems, therefore, that many, if not all, of the conformational rearrangements required for translocation can be accessed with the input of thermal energy alone. The fluctuations of PRE/PRE^{-A} complexes observed by smFRET represent dynamic events that are likely important for promoting the movement of the mRNA-tRNA complex during translocation; these fluctuations may thus increase the probability that spontaneous translocation will occur.

3.3. Identification and characterization of intermediate states connecting GS1 and GS2

The tRNA and ribosome structural rearrangements that constitute transitions between GS1 and GS2 are undoubtedly complex, involving significant local and global reconfigurations of ribosome-ribosome and ribosome-tRNA interactions (Korostelev et al., 2008). Despite this complexity, however, the majority of smFRET studies to date report fluctuations between just two major FRET states corresponding to GS1 and GS2. Since individual transitions between GS1 and GS2 must necessarily occur *via* some pathway (or, more likely, *via* any one of numerous parallel pathways), the failure of the majority of smFRET studies to identify any intermediate states connecting GS1 and GS2 most likely arises from either (i) the limited time resolution (typically 25–100 msec per frame in studies of ribosome and tRNA dynamics) with which TIRFM-based smFRET studies can resolve energetically unstable, and thus transiently sampled, intermediate states or (ii) the limited sensitivity with which a specific donor-acceptor pair can be used to detect the distance change associated with the formation of a particular intermediate state. It may therefore be necessary to perform smFRET experiments using higher time resolution confocal microscopy (up to several µsec per frame) (Cornish and Ha, 2007) and/or alternative donor-acceptor labeling schemes in order to capture intermediate states that may exist on the pathway(s) connecting GS1 and GS2.

Despite the failure of most of the smFRET studies of PRE/PRE^{-A} complexes to detect any intermediate states, two studies have used maximum likelihood-based hidden Markov modeling of smFRET trajectories to identify an intermediate FRET state using a tRNA-tRNA smFRET signal (Munro et al., 2007) and two intermediate FRET states using an L1 stalk-tRNA smFRET signal (Munro et al., 2010a). Structurally, the tRNA-tRNA intermediate FRET state has been assigned to a PRE complex containing P/E- and A/A-configured tRNAs, an intermediate configuration which had been previously proposed on the basis of tRNA mutagenesis experiments (Pan et al., 2006) and ensemble kinetic experiments (Pan et al., 2007) based on a previously established kinetic scheme of translocation (Savelsbergh et al., 2003). Contrasting with the tRNA-tRNA intermediate FRET state, the two intermediate L1-tRNA FRET states remain to be structurally or biochemically characterized. A somewhat surprising feature of the tRNA-tRNA and L1-tRNA intermediate FRET states that have been thus far identified is that, in all cases, their equilibrium populations are greater than that of the FRET state assigned to GS2, implying that the thermodynamic stabilities of the intermediate states are actually greater than that of GS2. We anticipate that identification of these new intermediate states will continue to drive structural, biochemical, and smFRET studies aimed at elucidating the physical basis underlying transitions between GS1 and GS2.

3.4. Allosteric regulation of the GS1/GS2 dynamic equilibrium

Using the tRNA-tRNA, L1-L9 (or L1-L33), L1-tRNA, and S6-L9 inter-subunit smFRET signals, numerous studies have demonstrated that both the equilibrium distributions of FRET states corresponding to GS1 and GS2 as well as the rates of transitions between these FRET states are highly sensitive to experimental conditions. These include the concentration of Mg^{2+} ions, the absence, presence, identity, and acylation state of the tRNAs, the absence or presence of translation factors, and the absence or presence of antibiotics targeting the ribosome. Assuming that the conformational changes associated with the GS1-to-GS2 transition are a fundamental part of the translocation mechanism and that GS2 is an obligatory intermediate on the translocation pathway, these observations suggest that specific control over the dynamics of the PRE ribosome, through the acceleration/deceleration of conformational changes and the associated stabilization/destabilization of specific conformational states, could provide an effective means for regulating the translocation reaction. In this view, ribosomal ligands may function by rectifying the intrinsic conformational dynamics of the PRE complex in order to promote or inhibit the translocation reaction. Indeed, as described below, modulation of PRE complex dynamics through changes in experimental conditions can often be strongly correlated with the effect of those changes on the rate of translocation.

The dynamic exchange of tRNAs between classical and hybrid configurations necessarily requires the remodeling of multiple tRNA-rRNA and tRNA-ribosomal protein interactions; this suggests that the classical/hybrid tRNA equilibrium may be modulated by the concentration of Mg^{2+} ions in solution, since Mg^{2+} is known to play a crucial role in the folding and stabilization of RNA structures (Draper, 2004). An investigation of the Mg^{2+} dependence of a tRNA-tRNA smFRET signal over a range of 3.5 to 15 mM Mg^{2+} within a PRE complex carrying deacylated tRNA^{fMet} at the P site and a peptidyl-tRNA^{Phe} analog, N-acetyl-Phe-tRNA^{Phe}, at the A site revealed a Mg^{2+}-dependent shift in the equilibrium distribution of classical and hybrid configurations (Kim et al., 2007). Specifically, at low concentrations of Mg^{2+} (3.5 mM) the hybrid configuration is favored. However, the equilibrium fraction of the classical configuration increases with increasing Mg^{2+}, with the classical and hybrid configurations becoming equally populated at ~4 mM Mg^{2+}. Analysis of the transition rates between classical and hybrid configurations revealed that this occurs primarily through a Mg^{2+}-dependent stabilization of the classical configuration that decreases the rate of classical-to-hybrid tRNA transitions while leaving the rate of hybrid-to-classical tRNA transitions unaffected. In structural terms, this is interpreted to mean that classically bound tRNAs form a more extensive and compact network of Mg^{2+}-stabilized tRNA-rR-

NA and/or tRNA-ribosomal protein interactions. At higher Mg^{2+} concentrations (~7 mM and above), the classical configuration is almost exclusively favored on account of the decreased rate of classical-to-hybrid transitions. These results correlate strongly with – and offer a mechanistic explanation for – the known inhibitory and stimulatory effects, respectively, of high and low Mg^{2+} concentration on the rate of translocation. At high Mg^{2+} concentrations (~30 mM), translocation is blocked almost entirely, even in the presence of EF-G (Spirin, 1985), which can be rationalized by a Mg^{2+}-induced stalling of the classical-to-hybrid tRNA transition evidenced by smFRET. At the other extreme of low Mg^{2+} (~3 mM), spontaneous translocation can proceed rapidly (Spirin, 1985), an effect presumably linked to the accelerated rate of the classical-to-hybrid transition under low-Mg^{2+} conditions. smFRET evidence thus suggests that perturbations to the rate of the classical-to-hybrid tRNA transition can affect the rate of translocation, implying that manipulation of the GS1-to-GS2 transition may serve to regulate this critical step in protein synthesis.

Changes in the acylation state and identity of the P- and A-site tRNAs within ribosomal elongation complexes have similarly been found to influence the energetics of ribosome and tRNA conformational fluctuations. As discussed above, the presence of a peptide on the P-site tRNA largely suppresses ribosome and tRNA dynamics within a POST complex, whereas PRE/PRE^{-A} complexes bearing a deacylated P-site tRNA exhibit pronounced ribosome and tRNA dynamics. In addition, PRE/PRE^{-A} complex dynamics have been shown to be sensitive to the identity of the P-site tRNA. For example, a comparison of inter-subunit rotation dynamics within four different PRE^{-A} complexes differing only in the identity of the deacylated P-site tRNA (tRNA^{fMet}, tRNA^{Phe}, tRNA^{Tyr}, and tRNA^{Met} were used), revealed distinct thermodynamic and kinetic parameters underlying reversible inter-subunit rotation (Cornish et al., 2008). Likewise, PRE^{-A} complexes containing either tRNA^{fMet} or tRNA^{Phe} at the P site exhibit differences in the rate of L1 stalk and L1 stalk-P-site tRNA interaction dynamics that mirror those observed in the inter-subunit rotation experiments (Fei et al., 2008; Cornish et al., 2009; Fei et al., 2009; Munro et al., 2010a). Different P-site tRNA species, therefore, make sufficiently unique contacts with the ribosome such that they influence large-scale structural rearrangements of the PRE/PRE^{-A} ribosome in distinct ways. Similarly, the presence and acylation state of the A-site tRNA appears to influence the ther-

modynamic and kinetic behavior of conformational equilibria monitored by the individual smFRET signals. For example, the presence of A-site dipeptidyl-tRNA *versus* aa-tRNA increases the population of the hybrid tRNA configuration by increasing the rate of classical-to-hybrid tRNA transitions, as monitored by tRNA-tRNA smFRET (Blanchard et al., 2004b). Likewise, using the L1-tRNA smFRET signal, addition of aa-tRNA to PRE^{-A} complexes caused a slight increase in the rate with which the L1 stalk-P/E tRNA interaction is formed, with minimal effect on the rate with which this interaction is disrupted; occupancy of the A site by a tripeptidyl-tRNA increased the forward rate by an additional 6-fold, again with minimal effect on the reverse rate (Fei et al., 2008). Finally, the presence of a peptidyl-tRNA at the A site of PRE complexes shifts the equilibrium from the open to the closed L1 stalk conformation, as monitored by the L1-L9 smFRET signal, primarily by accelerating the rate of the open-to-closed L1 stalk transition (Fei et al., 2009).

From the data discussed above, a picture begins to emerge in which large-scale conformational rearrangements of the entire PRE complex can be rectified and allosterically controlled through even subtle and highly localized changes in interactions between the ribosome and its ligands (i. e., the presence of dipeptidyl- *versus* aa-tRNA at the A site). This feature of the PRE complex has apparently been exploited by antibiotics targeting the ribosome which often function by inhibiting the dynamics of the translational machinery. Indeed, smFRET studies have provided evidence that translocation inhibitors specifically interfere with the conformational dynamics of the PRE complex. For example, the potent translocation inhibitor viomycin, which binds at the interface between the 30S and 50S subunits (Yamada et al., 1978; Moazed and Noller, 1987; Johansen et al., 2006; Stanley et al., 2010), has been shown by smFRET to almost exclusively stabilize the rotated subunit orientation of the PRE complex, consistent with previous ensemble studies (Ermolenko et al., 2007b), and to almost completely suppress transitions between the rotated and non-rotated subunit orientations (Cornish et al., 2008). Interestingly, while chemical probing experiments suggest that viomycin stabilizes the P/E configuration of the P-site tRNA (Ermolenko et al., 2007b), tRNA-tRNA smFRET experiments suggest that viomycin instead decreases the rate of fluctuations between the classical and hybrid tRNA configurations, with conflicting results on whether the classical or hybrid configurations of the tRNAs are stabilized (Kim et al., 2007; Feldman et al., 2010).

smFRET investigations of PRE complexes have also been conducted in the presence of a collection of aminoglycoside antibiotics (Feldman et al., 2010), drugs that bind to helix 44 within 16S rRNA, stabilizing a conformation of the universally conserved 16S rRNA nucleotides A1492 and A1493 in which they are displaced from helix 44, adopting extra-helical positions that allow them to interact directly with the codon-anticodon minihelix at the 30S A site (Carter et al., 2000). All of the aminoglycosides tested were shown to suppress tRNA dynamics, specifically decreasing the rate of transitions out of the classical tRNA configuration and thus driving a net stabilization of the classical configuration. Although the observed effects are small (~1.5-fold rate decreases corresponding to ~1.7-fold increases in the stability of the classical tRNA configuration, on average), the magnitude of these effects elicited by each of the aminoglycosides tested correlates with the reduction in translocation rate observed in the presence of each drug (Peske et al., 2004; Feldman et al., 2010). Therefore, inhibition of transitions into the hybrid state and the resulting stabilization of the classical state represent a general mechanism for translocation inhibition by aminoglycosides, with subtle differences in the chemical structure of the antibiotic dictating the degree of inhibition. Taken together, the viomycin and aminoglycoside studies discussed above illustrate that inhibition of ribosome and/or tRNA dynamics within the PRE complex represents a general inhibition strategy leveraged by a variety of antibiotics targeting the ribosome.

3.5. EF-G-mediated control of the GS1/GS2 dynamic equilibrium during translocation

Perhaps the most dramatic effect on ribosome and tRNA dynamics within PRE/PRE^{-A} complexes is elicited by EF-G. smFRET studies have revealed that binding of EF-G(GDPNP) to PRE^{-A} complexes leads to stabilization of all structural features characterizing GS2: ribosomal subunits are stabilized in their rotated orientation, the L1 stalk is stabilized in the closed conformation, and the P-site tRNA is stabilized in its P/E configuration, where it forms a long-lived intermolecular interaction with the L1 stalk (Cornish et al., 2008; Fei et al., 2008; Cornish et al., 2009; Fei et al., 2009; Munro et al., 2010b) (Figure 5). Particularly remarkable is the stabilization of the closed state of the L1 stalk, which demonstrates that binding of EF-G(GDPNP) to the ribosome's GTPase factor binding

site can allosterically regulate L1 stalk dynamics ~175 Å away at the ribosomal E site. A major role of EF-G therefore appears to be its ability to bias intrinsic conformational fluctuations of the ribosome and tRNAs towards the on-pathway translocation intermediate GS2. In accord with the ability of the ribosome to translocate either in the forward (Pestka, 1969; Gavrilova and Spirin, 1971; Gavrilova et al., 1976; Bergemann and Nierhaus, 1983) or reverse (Shoji et al., 2006; Konevega et al., 2007) directions in the absence of EF-G, one of EF-G's main mechanistic functions may be to stabilize GS2, preventing reverse fluctuations along the translocation reaction coordinate and thus guiding the directionality of a process that the ribosome is inherently capable of coordinating on its own. This model finds strong support from biochemical experiments demonstrating that EF-G(GDPNP) stimulates the rate of translocation ~1,000-fold relative to uncatalyzed spontaneous translocation, and that GTP hydrolysis in the EF-G(GTP)-catalyzed reaction provides an additional rate enhancement of only ~50-fold (Rodnina et al., 1997; Katunin et al., 2002). GTP hydrolysis, which, based on fast kinetics measurements, precedes any further rearrangements of the aminoacyl acceptor end of the A-site peptidyl-tRNA within the large subunit P site (Borowski et al., 1996; Pan et al., 2007) as well as the movement of the mRNA-tRNA duplex on the small subunit, likely leads to conformational changes in EF-G as well as additional conformational changes of the PRE complex that promote these steps of the translocation reaction (Rodnina et al., 1997; Katunin et al., 2002; Savelsbergh et al., 2003).

Completion of the translocation reaction converts the PRE complex into a POST complex in which non-rotated subunits and classical tRNA configurations characteristic of GS1 prevail, and ribosome and tRNA dynamics are suppressed (Cornish et al., 2008; Fei et al., 2008). This effect has been observed in real time through pre-steady state smFRET experiments in which EF-G(GTP) was stopped-flow delivered to PRE complexes labeled with the inter-subunit S6-L9 smFRET pair (Cornish et al., 2008). Each PRE complex exhibits fluctuations between the non-rotated and rotated subunit orientations until the delivery of EF-G(GTP), at which point EF-G(GTP) binds to the PRE complex, catalyzes full translocation, and converts the PRE complex into a POST complex, thereby stabilizing the non-rotated subunit orientation. Similar pre-steady state smFRET experiments have been performed by stopped-flow delivery of an EF-Tu(GTP) aa-tRNA ternary complex and EF-G(GTP) to a POST

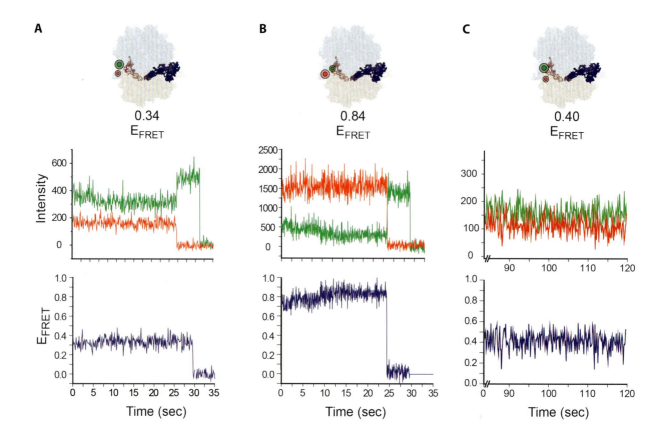

Fig. 5 E_{FRET} *versus* time trajectories derived from PRE^{-A} complexes stabilized in GS2 through their interactions with EF-G(GDPNP). A structural model of EF-G(GDPNP) bound to a PRE^{-A} complex (top row) generated by flexible-fitting of atomic resolution structures into a cryo-EM map using molecular dynamics was kindly provided by Joachim Frank. In this model, EF-G(GDPNP) was rendered in a space-filling representation and is shown in blue. The approximate positions of the donor and acceptor fluorophores corresponding to the three donor-acceptor labeling schemes are shown as green and red spheres, respectively. Donor and acceptor emission intensities *versus* time trajectories (middle row) and E_{FRET} *versus* time trajectories (bottom row) are shown as in Figure 4. (A) The L1-L9 smFRET signal is stabilized at the 0.34 (closed L1 stalk conformation, GS2) value of E_{FRET} (Reprinted from Fei et al., 2009, with permission from The National Academy of Sciences, USA). (B) The L1-tRNA smFRET signal is stabilized at the 0.84 (closed L1 stalk interacting with P/E-configured tRNA, GS2) value of E_{FRET} (Reprinted from (Fei et al., 2008) with permission from Elsevier). (C) The S6-L9 inter-subunit smFRET signal is stabilized at the 0.40 (rotated subunit orientation, GS2) value of E_{FRET} (Adapted from Cornish et al., 2008, with permission from Elsevier). Note that the analogous experiment using a tRNA-tRNA smFRET signal cannot be performed due to the opposing requirements for the presence of an A-site tRNA within a PRE complex in order to generate the smFRET signal and the absence of an A-site tRNA within a PRE^{-A} complex to establish stable binding of EF-G(GDPNP).

complex bearing fluorophore-labeled L1 and P-site peptidyl-tRNA (Fei et al., 2008). Stopped-flow delivery thus allows a full elongation cycle of aa-tRNA selection, peptidyl transfer, and EF-G(GTP)-catalyzed translocation to be monitored in real time using the L1-tRNA smFRET signal. The resulting smFRET trajectories exhibit a sharp transition from a low FRET state to a high FRET state upon peptidyl transfer (corresponding to formation of the intermolecular contacts between the L1 stalk and the P/E tRNA), followed by stable occupancy of the high FRET state until fluorophore photobleaching. This is in contrast to the analogous experiment performed in the absence of EF-G(GTP), where

the initial transition from the low FRET state to the high FRET state is followed by fluctuations between the two FRET states (corresponding to repetitive formation and disruption of L1-tRNA contacts). These results suggest that during EF-G(GTP)-catalyzed translocation, the intermolecular interactions formed between the L1 stalk and the P/E-configured tRNA are maintained during the movement of the deacylated tRNA from the hybrid P/E configuration into the classical E/E configuration. Formation and maintenance of these interactions provides a molecular rationale to help explain how the L1 stalk facilitates the translocation reaction (Subramanian and Dabbs, 1980).

3.6. Translation factors direct thermally activated conformational processes throughout all stages of protein synthesis

In the previous sections we have described how the conformational dynamics of the ribosome and its tRNA substrates are modulated during the translocation step of the elongation cycle, providing a regulatory mechanism that is exploited by EF-G to promote tRNA movements during translocation, as well as by antibiotics targeting the ribosome that impede this process. Beyond elongation, the initiation, termination, and ribosome recycling stages of protein synthesis all involve transitions between GS1 and GS2 (Agrawal et al., 2004; Klaholz et al., 2004; Allen et al., 2005; Gao et al., 2005; Myasnikov et al., 2005; Gao et al., 2007). Indeed, smFRET experiments using an inter-subunit FRET signal have demonstrated how initiation factors, particularly initiation factor 2, regulate functionally important inter-subunit dynamics during the assembly of an elongation competent ribosomal initiation complex during translation initiation (Marshall et al., 2009). Likewise, smFRET experiments have revealed how release factors and ribosome recycling factor rectify and thereby regulate the thermally activated GS1/GS2 dynamic equilibrium during the termination and ribosome recycling stages of protein synthesis (Sternberg et al., 2009). Therefore, modulation of the ribosome's global architecture through factor-dependent shifts in the translational machinery's conformational equilibria may serve as a general paradigm for translation regulation throughout all stages of protein synthesis.

4. Conclusions and future perspectives

In this chapter we have presented a basic overview of smFRET, including a discussion of the advantages and limitations of applying this biophysical technique to studies of the structural dynamics of protein synthesis. Using the translocation step of translation elongation as an example, we have described how smFRET experiments enable time-resolved observations of large-scale conformational rearrangements of the translational machinery, providing a unique opportunity to thermodynamically and kinetically characterize conformational processes that, while fundamental to the mechanism of protein synthesis, are generally obscured in ensemble studies. Beyond these smFRET studies of translocation, the donor-acceptor pairs described here as well as additional donor-acceptor pairs developed using fluorescence-labeled ribosomes, tRNAs, and translation factors are enabling a rapidly growing number of smFRET investigations of specific conformational processes encompassing every stage of protein synthesis (recently reviewed in (Aitken et al., 2010; Frank and Gonzalez, 2010)).

A major theme emerging from the collective body of smFRET studies of protein synthesis is the stochastic nature of individual steps within the mechanism of translation, in which thermal fluctuations of the ribosome and its tRNA substrates permit sampling of meta-stable conformational states on a complex multi-dimensional free energy landscape (Munro et al., 2009; Fischer et al., 2010; Frank and Gonzalez, 2010). An additional major theme is the ability of translation factors to regulate and direct the conformational equilibria of the ribosomal particle and its tRNA substrates throughout all stages of protein synthesis. By accelerating/decelerating particular conformational transitions and stabilizing/destabilizing particular conformational states, translation factors guide the directionality of conformational processes intrinsic to the ribosome-tRNA complex (Frank and Gonzalez, 2010). In an analogous way, smFRET characterization of the effect of antibiotics on the conformational dynamics of the translational machinery is revealing that these drugs often exert their inhibitory activities through the inhibition of the large-scale structural rearrangements that are required for protein synthesis to proceed rapidly.

Looking to the future, the dynamics of many mechanistically important, highly mobile ribosomal domains remain to be characterized using smFRET. Likewise, many functionally important conformational changes of the translational machinery have been suggested by structural work but have yet to be investigated using smFRET. A particularly exciting example is provided by the L7/L12 protein stalk of the 50S subunit's GTPase-associated center, which is thought to recruit translation factors to the ribosome and control biochemical steps such as GTP hydrolysis and inorganic phosphate release (Mohr et al., 2002; Savelsbergh et al., 2005). Characterizing the nature and timescale of L7/L12 stalk movements with respect to the ribosome, as well as the organization and timing of its interactions with translation factors throughout all stages of protein synthesis would greatly advance our mechanistic understanding of this universally conserved and essential structural element of the ribosome. Similarly, smFRET provides a means with which to characterize the thermodynamics and kinetics underlying putative movements of the head domain of the 30S subunit, which have been suggested to play important regulatory roles during translation initiation (Carter et al., 2001) as well as during the aa-tRNA selection (Ogle et al., 2002) and translocation (Spahn et al., 2004) steps of translation elongation.

Efforts to obtain a complete mechanistic understanding of the structural dynamics of the translating ribosome will benefit from the development of new technologies and experimental platforms. Recent advances, such as: (i) three-wavelength experiments using multiple donors and acceptors (Hohng et al., 2004) allowing simultaneous tracking of multiple conformational changes and investigation of the degree of conformational coupling within the translational machinery (Munro et al., 2010a; Munro et al., 2010b); (ii) new illumination strategies enabling single-molecule detection of surface-tethered, fluorescence-labeled biomolecules in a physiologically relevant, micromolar background concentration of freely diffusing, fluorescence-labeled ligands (Levene et al., 2003; Uemura et al., 2010); and (iii) new data analysis algorithms permitting increasingly unbiased analysis of smFRET trajectories (Bronson et al., 2009), will allow ever more complex mechanistic questions to be addressed. We envision that these advances will be particularly important in extending smFRET techniques from the studies of prokaryotic protein synthesis described here to studies of the significantly more complex and highly regulated translational machinery of eukaryotic organisms.

Acknowledgements

Work in the Gonzalez laboratory is supported by a Burroughs Wellcome Fund CABS Award (CABS 1 004 856), an NSF CAREER Award (MCB 0 644 262), an NIH-NIGMS grant (GM 084 288–01), and an American Cancer Society Research Scholar Grant (RSG GMC-117 152) to R. L. G. We thank Dr. Jingyi Fei, Ms. Margaret Elvekrog, and Prof. Dmitri Ermolenko for critically reading and commenting on the chapter and all members of the Gonzalez laboratory for many incisive discussions.

References

Agirrezabala X, Lei J, Brunelle JL, Ortiz-Meoz RF, Green R, Frank J (2008) Visualization of the hybrid state of tRNA binding promoted by spontaneous ratcheting of the ribosome. Mol Cell 32: 190–197

Agrawal RK, Sharma MR, Kiel MC, Hirokawa G, Booth TM, Spahn CM, Grassucci RA, Kaji A, Frank J (2004) Visualization of ribosome-recycling factor on the *Escherichia coli* 70S ribosome: functional implications. Proc Natl Acad Sci USA 101: 8900–8905

Aitken CE, Marshall RA, Puglisi JD (2008) An oxygen scavenging system for improvement of dye stability in single-molecule fluorescence experiments. Biophys J 94: 1826–1835

Aitken CE, Petrov A, Puglisi J (2010) Ribosome dynamics and translation. Annu Rev Biophys (in press)

Aitken CE, Puglisi JD (2010) Following the intersubunit conformation of the ribosome during translation in real time. Nat Struct Mol Biol 17: 793–800

Allen GS, Zavialov A, Gursky R, Ehrenberg M, Frank J (2005) The cryo-EM structure of a translation initiation complex from *Escherichia coli*. Cell 121: 703–712

Andrec M, Levy RM, Talaga DS (2003) Direct determination of kinetic rates from single-molecule photon arrival trajectories using hidden Markov models. J Phys Chem A 107: 7454–7464

Axelrod D (2003) Total internal reflection fluorescence microscopy in cell biology. Biophotonics, Pt B. San Diego: Academic Press, pp 1–33

Axelrod D, Burghardt TP, Thompson NL (1984) Total internal-reflection fluorescence. Annu Rev Biophys Bioeng 13: 247–268

Benesch RE, Benesch R (1953) Enzymatic removal of oxygen for polarography and related methods. Science 118: 447–448

Bergemann K, Nierhaus KH (1983) Spontaneous, elongation factor G independent translocation of *Escherichia coli* ribosomes. J Biol Chem 258: 15105–15113

Beringer M, Rodnina MV (2007) The ribosomal peptidyl transferase. Mol Cell 26: 311–321

Blanchard SC, Gonzalez RL, Kim HD, Chu S, Puglisi JD (2004a) tRNA selection and kinetic proofreading in translation. Nat Struct Mol Biol 11: 1008–1014

Blanchard SC, Kim HD, Gonzalez RL, Jr., Puglisi JD, Chu S (2004b) tRNA dynamics on the ribosome during translation. Proc Natl Acad Sci USA 101: 12 893–12 898

Blanco M, Walter NG (2010) Analysis of complex single-molecule FRET time trajectories. Meth Enzymol 472: 153–178

Borowski C, Rodnina MV, Wintermeyer W (1996) Truncated elongation factor G lacking the G domain promotes translocation of the 3' end but not of the anticodon domain of peptidyl-tRNA. Proc Nat Acad Sci USA 93: 4202–4206

Bremer H, Dennis PP (1987) Modulation of chemical composition and other parameters of the cell by growth rate. In: Neidhardt FC, ed. *Escherichia coli* and *Salmonella typhimurium*: cellular and molecular biology. Washington, DC: American Society for Microbiology, pp 1553–1569

Bretscher MS (1968) Translocation in protein synthesis: a hybrid structure model. Nature 218: 675–677

Bronson JE, Fei J, Hofman JM, Gonzalez Jr. RL, Wiggins CH (2009) Learning rates and states from biophysical time series: A Bayesian approach to model selection and single-molecule FRET data. Biophys J 97: 3196–3205

Carter AP, Clemons WM, Brodersen DE, Morgan-Warren RJ, Wimberly BT, Ramakrishnan V (2000) Functional insights from the structure of the 30S ribosomal subunit and its interactions with antibiotics. Nature 407: 340–348

Carter AP, Clemons WM, Jr., Brodersen DE, Morgan-Warren RJ, Hartsch T, Wimberly BT, Ramakrishnan V (2001) Crystal structure of an initiation factor bound to the 30S ribosomal subunit. Science 291: 498–501

Clamme JP, Deniz AA (2005) Three-color single-molecule fluorescence resonance energy transfer. Chemphyschem 6: 74–77

Clegg RM (1995) Fluorescence resonance energy transfer. Curr Opin Biotechnol 6: 103–110

Colquhoun D, Hawkes AG (1995) The principles of the stochastic interpretation of ion-channel mechanism. In: Sakmann B, Heher E, eds. Single Channel Recording. New York: Plenum Press, pp 397–482

Cornish PV, Ermolenko DN, Noller HF, Ha T (2008) Spontaneous intersubunit rotation in single ribosomes. Mol Cell 30: 578–588

Cornish PV, Ermolenko DN, Staple DW, Hoang L, Hickerson RP, Noller HF, Ha T (2009) Following movement of the L1 stalk between three functional states in single ribosomes. Proc Natl Acad Sci USA 106: 2571–2576

Cornish PV, Ha T (2007) A survey of single-molecule techniques in chemical biology. ACS Chem Biol 2: 53–61

Dave R, Terry DS, Munro JB, Blanchard SC (2009) Mitigating unwanted photophysical processes for improved single-molecule fluorescence imaging. Biophys J 96: 2371–2381

Dorner S, Brunelle JL, Sharma D, Green R (2006) The hybrid state of tRNA binding is an authentic translation elongation intermediate. Nat Struct Mol Biol 13: 234–241

Dorywalska M, Blanchard SC, Gonzalez Jr. RL, Kim HD, Chu S, Puglisi JD (2005) Site-specific labeling of the ribosome for single-molecule spectroscopy. Nucleic Acids Res 33: 182–189

Draper DE (2004) A guide to ions and RNA structure. RNA 10: 335–343

Dunkle JA, Cate JH (2010) Ribosome structure and dynamics during translation and termination. Annu Rev Biophys 39: 227–244

Ermolenko DN, Majumdar ZK, Hickerson RP, Spiegel PC, Clegg RM, Noller HF (2007a) Observation of intersubunit movement of the ribosome in solution using FRET. J Mol Biol 370: 530–540

Ermolenko DN, Spiegel PC, Majumdar ZK, Hickerson RP, Clegg RM, Noller HF (2007b) The antibiotic viomycin traps the ribosome in an intermediate state of translocation. Nat Struct Mol Biol 14: 493–497

Fei J, Bronson JE, Hofman JM, Srinivas RL, Wiggins CH, Gonzalez RLJ (2009) Allosteric collaboration between elongation factor G and the ribosomal L1 stalk directs tRNA movements during translation. Proc Natl Acad Sci USA 106: 15702–15707

Fei J, Kosuri P, MacDougall DD, Gonzalez RL, Jr (2008) Coupling of ribosomal L1 stalk and tRNA dynamics during translation elongation. Mol Cell 30: 348–359

Fei J, Wang J, Sternberg SH, MacDougall DD, Elvekrog MM, Pulukkunat DK, Englander MT, Gonzalez RL, Jr (2010) A highly purified, fluorescently labeled in vitro translation system for single-molecule studies of protein synthesis. Meth Enzymol 472: 221–259

Feldman MB, Terry DS, Altman RB, Blanchard SC (2010) Aminoglycoside activity observed on single pre-translocation ribosome complexes. Nat Chem Biol 6: 54–62

Fischer N, Konevega AL, Wintermeyer W, Rodnina MV, Stark H (2010) Ribosome dynamics and tRNA movement by time-resolved electron cryomicroscopy. Nature 466: 329–333

Förster T (1946) Energiewanderung und fluoreszenz. Naturwissenschaften 33: 166–175

Frank J, Agrawal RK (2000) A ratchet-like inter-subunit reorganization of the ribosome during translocation. Nature 406: 318–322

Frank J, Gao H, Sengupta J, Gao N, Taylor DJ (2007) The process of mRNA-tRNA translocation. Proc Natl Acad Sci USA 104: 19 671–19 678

Frank J, Gonzalez RL, Jr (2010) Structure and dynamics of a processive Brownian motor: the translating ribosome. Annu Rev Biochem 79: 381–412

Gao H, Zhou Z, Rawat U, Huang C, Bouakaz L, Wang C, Cheng Z, Liu Y, Zavialov A, Gursky R, Sanyal S, Ehrenberg M, Frank J, Song H (2007) RF3 induces ribosomal conformational changes responsible for dissociation of class I release factors. Cell 129: 929–941

Gao N, Zavialov AV, Li W, Sengupta J, Valle M, Gursky RP, Ehrenberg M, Frank J (2005) Mechanism for the disassembly of the posttermination complex inferred from cryo-EM studies. Mol Cell 18: 663–674

Gavrilova LP, Kostiashkina OE, Koteliansky VE, Rutkevitch NM, Spirin AS (1976) Factor-free ("non-enzymic") and factor-dependent systems of translation of polyuridylic acid by *Escherichia coli* ribosomes. J Mol Biol 101: 537–552

Gavrilova LP, Spirin AS (1971) Stimulation of "non-enzymic" translocation in ribosomes by p-chloromercuribenzoate. FEBS Lett 17: 324–326

Gonzalez Jr. RL, Chu S, Puglisi JD (2007) Thiostrepton inhibition of tRNA delivery to the ribosome. RNA 13: 2091–2097

Gromadski KB, Rodnina MV (2004) Kinetic determinants of high-fidelity tRNA discrimination on the ribosome. Mol Cell 13: 191–200

Ha T (2001) Single-Molecule Fluorescence Resonance Energy Transfer. Methods 25: 78–86

Ha T, Rasnik I, Cheng W, Babcock HP, Gauss GH, Lohman TM, Chu S (2002) Initiation and re-initiation of DNA unwinding by the *Escherichia coli* Rep helicase. Nature 419: 638–641

Hohng S, Joo C, Ha T (2004) Single-molecule three-color FRET. Biophys J 87: 1328–1337

Horan LH, Noller HF (2007) Intersubunit movement is required for ribosomal translocation. Proc Natl Acad Sci USA 104: 4881–4885

Hubner CG, Renn A, Renge I, Wild UP (2001) Direct observation of the triplet lifetime quenching of single dye molecules by molecular oxygen. J Chem Phys 115: 9619–9622

Hwang LC, Hohlbein J, Holden SJ, Kapanidis AN (2009) Single-molecule FRET: Methods and biological applications. In: Handbook of Single-Molecule Biophysics (Hinterdorfer P, van Oijen AM, eds.) New York: Springer, pp 130–132

Jares-Erijman EA, Jovin TM (2003) FRET imaging. Nat Biotechnol 21: 1387–1395

Johansen SK, Maus CE, Plikaytis BB, Douthwaite S (2006) Capreomycin binds across the ribosomal subunit interface using tlyA-encoded 2'-O-methylations in 16S and 23S rRNAs. Mol Cell 23: 173–182

Johnson AE, Adkins HJ, Matthews EA, Cantor CR (1982) Distance moved by transfer RNA during translocation from the A site to the P site on the ribosome. J Mol Biol 156: 113–140

Joo C, Ha T. 2007. Single-molecule FRET with total internal reflection microscopy. In: Single Molecule Techniques: a Laboratory Manual (Selvin PR, Ha T, eds.). Cold Spring Harbor, New York: Cold Spring Harbor Laboratory Press, pp 3–36

Julian P, Konevega AL, Scheres SH, Lazaro M, Gil D, Wintermeyer W, Rodnina MV, Valle M (2008) Structure of ratcheted ribosomes with tRNAs in hybrid states. Proc Natl Acad Sci USA 105: 16924–16927

Katunin VI, Savelsbergh A, Rodnina MV, Wintermeyer W (2002) Coupling of GTP hydrolysis by elongation factor G to translocation and factor recycling on the ribosome. Biochemistry 41: 12806–12812

Kennell D, Riezman H (1977) Transcription and translation initiation frequencies of the *Escherichia coli* lac operon. J Mol Biol 114: 1–21

Kim HD, Puglisi J, Chu S (2007) Fluctuations of transfer RNAs between classical and hybrid states. Biophys J 93: 3575–3582

Klaholz BP, Myasnikov AG, Van Heel M (2004) Visualization of release factor 3 on the ribosome during termination of protein synthesis. Nature 427: 862–865

Konevega AL, Fischer N, Semenkov YP, Stark H, Wintermeyer W, Rodnina MV (2007) Spontaneous reverse movement of mRNA-bound tRNA through the ribosome. Nat Struct Mol Biol 14: 318–324

Korostelev A, Ermolenko DN, Noller HF (2008) Structural dynamics of the ribosome. Curr Opin Chem Biol 12: 674–683

Levene MJ, Korlach J, Turner SW, Foquet M, Craighead HG, Webb WW (2003) Zero-mode waveguides for single-molecule analysis at high concentrations. Science 299: 682–686

Liang ST, Xu YC, Dennis P, Bremer H (2000) mRNA composition and control of bacterial gene expression. J Bacteriol 182: 3037–3044

Liu Y, Park J, Dahmen KA, Chemla YR, Ha T (2010) A comparative study of multivariate and univariate hidden Markov modelings in time-binned single-molecule FRET data analysis. J Phys Chem B 114: 5386–5403

Marshall RA, Aitken CE, Puglisi JD (2009) GTP hydrolysis by IF2 guides progression of the ribosome into elongation. Mol Cell 35: 37–47

Marshall RA, Dorywalska M, Puglisi JD (2008) Irreversible chemical steps control intersubunit dynamics during translation. Proc Natl Acad Sci USA 105: 15364–15369

McKinney SA, Joo C, Ha T (2006) Analysis of single-molecule FRET trajectories using hidden Markov modeling. Biophys J 91: 1941–1951

Moazed D, Noller HF (1987) Chloramphenicol, erythromycin, carbomycin and vernamycin B protect overlapping sites in the peptidyl transferase region of 23S ribosomal RNA. Biochimie 69: 879–884

Moazed D, Noller HF (1989) Intermediate states in the movement of transfer RNA in the ribosome. Nature 342: 142–148

Mohr D, Wintermeyer W, Rodnina MV (2002) GTPase activation of elongation factors Tu and G on the ribosome. Biochemistry 41: 12 520–12 528

Moran-Mirabal JM, Craighead HG (2008) Zero-mode waveguides: sub-wavelength nanostructures for single molecule studies at high concentrations. Methods 46: 11–17

Munro JB, Altman RB, O'Connor N, Blanchard SC (2007) Identification of two distinct hybrid state intermediates on the ribosome. Mol Cell 25: 505–517

Munro JB, Altman RB, Tung CS, Cate JH, Sanbonmatsu KY, Blanchard SC (2010a) Spontaneous formation of the unlocked state of the ribosome is a multistep process. Proc Natl Acad Sci USA 107: 709–714

Munro JB, Altman RB, Tung CS, Sanbonmatsu KY, Blanchard SC (2010b) A fast dynamic mode of the EF-G-bound ribosome. EMBO J 17: 770–781

Munro JB, Sanbonmatsu KY, Spahn CM, Blanchard SC (2009) Navigating the ribosome's metastable energy landscape. Trends Biochem Sci 34: 390–400

Myasnikov AG, Marzi S, Simonetti A, Giuliodori AM, Gualerzi CO, Yusupova G, Yusupov M, Klaholz BP (2005) Conformational transition of initiation factor 2 from the GTP- to GDP-bound state visualized on the ribosome. Nat Struct Mol Biol 12: 1145–1149

Odom OW, Picking WD, Hardesty B (1990) Movement of tRNA but not the nascent peptide during peptide bond formation on ribosomes. Biochemistry 29: 10734–10744

Ogle JM, Murphy FV, Tarry MJ, Ramakrishnan V (2002) Selection of tRNA by the ribosome requires a transition from an open to a closed form. Cell 111: 721–732

Pan D, Kirillov S, Zhang CM, Hou YM, Cooperman BS (2006) Rapid ribosomal translocation depends on the conserved 18–55 base pair in P-site transfer RNA. Nat Struct Mol Biol 13: 354–359

Pan D, Kirillov SV, Cooperman BS (2007) Kinetically competent intermediates in the translocation step of protein synthesis. Mol Cell 25: 519–529

Pape T, Wintermeyer W, Rodnina MV (1998) Complete kinetic mechanism of elongation factor Tu-dependent binding of aminoacyl-tRNA to the A site of the *E. coli* ribosome. EMBO (Eur Mol Biol Organ) J 17: 7490–7497

Patil PV, Ballou DP (2000) The use of protocatechuate dioxygenase for maintaining anaerobic conditions in biochemical experiments. Anal Biochem 286: 187–192

Paulsen H, Robertson JM, Wintermeyer W (1983) Topological arrangement of two transfer RNAs on the ribosome: fluorescence energy transfer measurements between A and P site-bound tRNA[Phe]. J Mol Biol 167: 411–426

Pawley JB (2006) Handbook of biological confocal microscopy. New York City: Springer

Peske F, Savelsbergh A, Katunin VI, Rodnina MV, Wintermeyer W (2004) Conformational changes of the small ribosomal subunit during elongation factor G-dependent tRNA-mRNA translocation. J Mol Biol 343: 1183–1194

Pestka S (1969) Studies on the formation of transfer ribonucleic acid-ribosome complexes. VI. Oligopeptide synthesis and translocation on ribosomes in the presence and absence of soluble transfer factors. J Biol Chem 244: 1533–1539

Piwonski H, Kolos R, Meixner A, Sepiol J (2005) Optimal oxygen concentration for the detection of single indocarbocyanine molecules in a polymeric matrix. Chem Phys Lett 405: 352–356

Proshkin S, Rahmouni AR, Mironov A, Nudler E (2010) Cooperation between translating ribosomes and RNA polymerase in transcription elongation. Science 328: 504–508

Qin F, Auerbach A, Sachs F (1996) Estimating single-channel kinetic parameters from idealized patch-clamp data containing missed events. Biophys J 70: 264–280

Rasnik I, McKinney SA, Ha T (2005) Surfaces and orientations: much to FRET about? Acc Chem Res 38: 542–548

Renn A, Seelig J, Sandoghdar V (2006) Oxygen-dependent photochemistry of fluorescent dyes studied at the single molecule level. Mol Phys 104: 409–414

Rodnina MV, Gromadski KB, Kothe U, Wieden HJ (2005) Recognition and selection of tRNA in translation. FEBS Lett 579: 938–942

Rodnina MV, Savelsbergh A, Katunin VI, Wintermeyer W (1997) Hydrolysis of GTP by elongation factor G drives tRNA movement on the ribosome. Nature 385: 37–41

Roy R, Hohng S, Ha T (2008) A practical guide to single-molecule FRET. Nat Methods 5: 507–516

Savelsbergh A, Katunin VI, Mohr D, Peske F, Rodnina MV, Wintermeyer W (2003) An elongation factor G-induced ribosome rearrangement precedes tRNA-mRNA translocation. Mol Cell 11: 1517–1523

Savelsbergh A, Mohr D, Kothe U, Wintermeyer W, Rodnina MV (2005) Control of phosphate release from elongation factor G by ribosomal protein L7/12. EMBO J 24: 4316–4323

Selvin PR (1995) Fluorescence resonance energy transfer. Meth Enzymol 246: 300–334

Selvin PR (2000) The renaissance of fluorescence resonance energy transfer. Nat Struct Biol 7: 730–734

Shoji S, Walker SE, Fredrick K (2006) Reverse translocation of tRNA in the ribosome. Mol Cell 24: 931–942

Shoji S, Walker SE, Fredrick K (2009) Ribosomal Translocation: One Step Closer to the Molecular Mechanism. ACS Chem Biol 27: 27

Sorensen MA, Pedersen S (1991) Absolute in vivo translation rates of individual codons in *Escherichia coli*. The two glutamic acid codons GAA and GAG are translated with a threefold difference in rate. J Mol Biol 222: 265–280

Spahn CM, Gomez-Lorenzo MG, Grassucci RA, Jorgensen R, Andersen GR, Beckmann R, Penczek PA, Ballesta JP, Frank J (2004) Domain movements of elongation factor eEF2 and the eukaryotic 80S ribosome facilitate tRNA translocation. EMBO J 23: 1008–1019

Spirin AS (1968) How does the ribosome work? A hypothesis based on the two subunit construction of the ribosome. Curr Mod Biol 2: 115–127

Spirin AS (1985) Ribosomal translocation: facts and models. Prog Nucleic Acid Res Mol Biol 32: 75–114

Spirin AS (2002) Ribosome as a molecular machine. FEBS Lett 514: 2–10

Stanley RE, Blaha G, Grodzicki RL, Strickler MD, Steitz TA (2010) The structures of the anti-tuberculosis antibiotics viomycin and capreomycin bound to the 70S ribosome. Nat Struct Mol Biol 17: 289–293

Sternberg SH, Fei J, Prywes N, McGrath KA, Gonzalez RL, Jr (2009) Translation factors direct intrinsic ribosome dynamics during translation termination and ribosome recycling. Nat Struct Mol Biol 16: 861–868

Subramanian AR, Dabbs ER (1980) Functional studies on ribosomes lacking protein L-1 from mutant escherichia coli. Eur J Biochem 112: 425–430

Talaga DS (2007) COCIS: Markov processes in single molecule fluorescence. Curr Opin Colloid Interface Sci 12: 285–296

Traut RR, Monro RE (1964) The Puromycin Reaction and its Relation to Protein Synthesis. J Mol Biol 10: 63–72

Uemura S, Aitken CE, Korlach J, Flusberg BA, Turner SW, Puglisi JD (2010) Real-time tRNA transit on single translating ribosomes at codon resolution. Nature 464: 1012–1017

Uemura S, Dorywalska M, Lee TH, Kim HD, Puglisi JD, Chu S (2007) Peptide bond formation destabilizes Shine-Dalgarno interaction on the ribosome. Nature 446: 454–457

Valle M, Zavialov AV, Sengupta J, Rawat U, Ehrenberg M, Frank J (2003) Locking and unlocking of ribosomal motions. Cell 114: 123–134

Wang Y, Qin H, Kudaravalli RD, Kirillov SV, Dempsey GT, Pan D, Cooperman BS, Goldman YE (2007) Single-molecule structural dynamics of EF-G–ribosome interaction during translocation. Biochemistry 46: 10 767–10 775

Wu P, Brand L (1994) Resonance energy transfer: methods and applications. Anal Biochem 218: 1–13

Yamada T, Mizugichi Y, Nierhaus KH, Wittmann HG (1978) Resistance to viomycin conferred by RNA of either ribosomal subunit. Nature 275: 460–461

Zhang W, Dunkle JA, Cate JH (2009) Structures of the ribosome in intermediate states of ratcheting. Science 325: 1014–1017

Real-time monitoring of single-molecule translation

23

Sotaro Uemura and Joseph D. Puglisi

1. Introduction

Translation is a complex process, involving repetitive dynamic interaction of RNA and protein substrates with the ribosome. Mechanistic investigations have delineated the global mechanism and timescale for individual steps in protein synthesis. Static structural views from crystallography and cryoelectron microscopy have revealed the detailed architecture of the ribosome, and how it interacts with tRNA and protein factor ligands. These results have provided snapshots for conformational states of the ribosome that may occur during translation. However, conformational and compositional dynamics that underlie translation remain poorly understood.

To explore the dynamics in translation, we have developed a range of single-molecule fluorescence approaches. Our initial efforts focused on single-molecule fluorescence resonance energy transfer (FRET) to uncover the pathway of tRNA selection on the ribosome. We then subsequently used FRET to understand conformational dynamics of the ribosomal particle during initiation and elongation (Aitken and Puglisi et al., 2010). Other groups have performed elegant single molecule analyses of ligand and ribosomal dynamics that are reviewed elsewhere (Marshall et al., 2009; Aitken et al., 2010).

Recently we have used zero-mode waveguides (ZMWs) and sophisticated detection instrumentation to allow real-time observation of translation at physiologically-relevant (μM) ligand concentrations (Uemura et al., 2010). We observe the transit of tRNAs – labeled with different fluorophores – on single translating ribosomes and have determined the number of tRNA molecules simultaneously bound to the ribosome at each codon of an mRNA (Uemura et al., 2010). The methods described in this chapter have broad ap-plications to the study the mechanism and regulation of translation.

2. Real-time translation in zero-mode waveguides

Single-molecule fluorescence experiments probe the dynamics of single translating ribosomes. Traditional single-molecule fluorescence applied to translation has used immobilization of translation complexes and Total Internal Reflection Fluorescence (TIRF) microscopy for excitation and detection of fluorescence (Figure 1). The evanescent wave caused by the interaction of the laser beam and glass surface leads to excitation of fluorophores within about 200 nm from the surface. This technique depends on distinguishing a localized and bound fluorophore from those freely diffusing in solution. However, this contrast is lost in TIRF at concentrations > 50 nM fluorescent ligand in solution. Translation *in vivo*, like many biological processes, requires ligand concentrations that are 50–1000-fold higher. New technologies are thus needed to detect and distinguish single-molecule fluorescence under biologically relevant concentrations. Zero-mode waveguides (Figure 1) are nanophotonic confinement structures consisting of a circular hole of 50–200 nm diameter in a metal cladding film on a solid transparent substrate (Levene et al., 2003). In conjunction with laser-excited fluorescence, they provide observation volumes on the order of zeptoliters (10^{-21} l), three to four orders of magnitude smaller than far-field excitation volumes, allowing fluorescence detection in the μM range. Arrays of thousands of ZMW can be simultaneously observed, providing single molecule data in high throughput.

The technical challenges of using ZMWs for single-molecule analysis of biomolecules are manifold. The

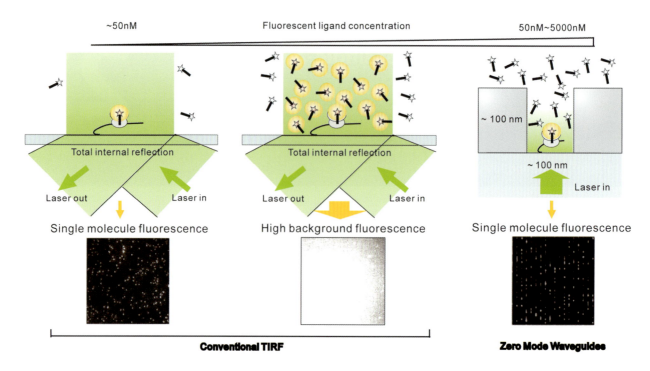

Fig. 1 Comparison between conventional TIRF and zero-mode waveguides. At fluorescent ligand concentrations lower than ~ 50 nM, each individual fluorescence spot can be visualized in con-

ventional TIRF (left). However, at higher than ~50 nM, spots are not distinguished due to high background fluorescence (middle). Zero-mode waveguides overcome this limitation (right).

ZMW arrays must be manufactured with precision and reproducibility, and the glass and aluminum surfaces must be passivated to avoid non-specific surface interaction of biomolecules. Sensitive optical instrumentation is required to illuminate and detect fluorescence from multiple wavelengths with millisecond time resolution. Recent advances in fabrication (Foquet et al., 2008), surface chemistry (Korlach et al., 2008), and detection instrumentation (Lundquist et al., 2008) have permitted direct monitoring of DNA polymerization in ZMWs (Eid et al., 2009). The binding of labeled ligands to an enzyme immobilized in a ZMW is detected as a pulse of fluorescent light (Figure 1).

The experiments using ZMWs to study translation require specific immobilization of ribosomes within the ZMW nanostructures, and the use of fluorescent ligands to follow translation. We normally immobilize ribosomal complexes on glass surfaces using biotin-streptavidin interactions; the surface contains biotinylated polyethylene glycol molecules, which bind to tetravalent streptavidin, and subsequently can bind a biotinylated biopolymer. We have used this approach in ZMWs. Ribosomal complexes containing tRNAs were immobilized using a biotinylated mRNA, as in our prior single-molecule investigations. tRNA binding to single ribosomal complexes was monitored us-

ing tRNAs that were specifically dye-labeled at their elbow positions without affecting their function (Blanchard et al., 2004a,b). Ribosomes were immobilized in ZMWs as 70S initiation complexes – containing fMet-tRNA^{fMet}(Cy3) – assembled on biotinylated mRNAs, which were tethered to the biotin-PEG-derivatized bottom of ZMWs through neutravidin-biotin linkages; mRNAs contained 5'-UTR and Shine-Dalgarno sequences, an initiation codon and a coding sequence of 12 codons, terminated by a stop (UAA) codon followed by four phenylalanine codons. Cy3 fluorescence from an immobilized complex confirmed the presence of initiator tRNA and marked a properly assembled and immobilized ribosome in a ZMW (Figure 2a).

Immobilization of the 20-nm 70S ribosome particle readily occurs in the *ca.* 100-nm ZMW. As shown in Figure 2b, immobilization of ribosomal complexes occurs in a concentration-dependent manner, following Poisson statistics for the occupancy of a ZMW hole. Blocking of ZMW-immobilized streptavidin by addition of free biotin eliminated ribosome immobilization (Figure 2b). These results showed that the ZMW surfaces are amenable to studies of large RNA-protein assemblies without non-specific surface interactions.

To probe initially the function of ZMW-immobilized ribosomes, we reproduced prior FRET measure-

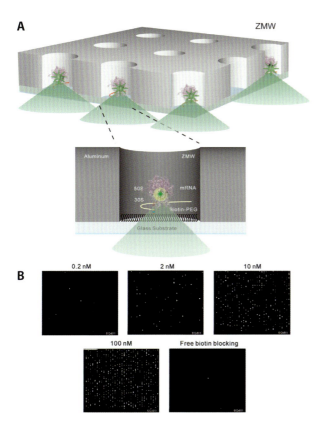

Fig. 2 Specific immobilization of ribosomal complexes in ZMW arrays. (A) Schematic illustration of ZMW construction. Ribosomal complexes are specifically immobilized in the bottom of derivatized ZMWs using biotinylated mRNAs. ZMW surface-immobilized, initial ribosome complex contains Cy3-labeled fMet-tRNA^{fMet}. (B) The population of fluorescence occupancy increased as the concentration of ribosome complex increased. Blocking neutravidin binding sites by pre-incubation with biotin prevents immobilization in the presence of 100 nM ribosomal complex.

ments between tRNAs. In these experiments, Cy3-labeled initiator tRNA^{fMet} was bound in the ribosomal P site and ternary complexes Phe-tRNA^{Phe}(Cy5)·EF-Tu·GTP were delivered to these ribosomal complexes in ZMWs. Binding and accommodation of tRNA^{Phe} in the A site leads to appearance and changes in FRET upon irradiation of only Cy3 with 532 nm illumination; we have extensively characterized the time course and FRET intensities for this process using single-molecule TIRF spectroscopy, and the results in ZMW were in quantitative agreement (Uemura et al., 2010). Thus, ZMWs do not perturb kinetics or conformational changes that underlie tRNA selection. This built confidence that we could observe translational processes in real time using ZMW-based detection.

3. Real-time monitoring of translation through sequential fluorescent-tRNA binding events

To observe translation in real time within ZMWs, we followed the sequential binding of the ternary complexes Phe-tRNA^{Phe}(Cy5)·EF-Tu·GTP and Lys-tRNA^{Lys}(Cy2)·EF-Tu·GTP to ribosomes programmed with mRNAs encoding 13 amino acids (M(FK)$_6$ and M(FKK)$_4$); ZMWs were illuminated simultaneously with 488, 532, and 642 nm excitation. Initiated ribosomes were identified by the presence of fMet-tRNA^{fMet}(Cy3); subsequent real-time arrival and occupation of tRNAs on translating ribosomes were detected as fluorescent pulses of appropriate color (Figure 3A). The sequence of the mRNA is readily distinguished from the pattern of fluorescent pulses (Figure 3B). At 200 nM ternary complex and 500 nM EF-G, ribosomes translate the entire mRNA (Figure 3B). The duration of most tRNA pulses is not limited by photobleaching at these high concentrations suggesting that the lifetime of each tRNA signal provides a signal for its transit time on the ribosome (see below). Addition of erythromycin, which binds to the exit tunnel of the ribosome (Schlunzen et al., 2001), blocks translation at 6–8 amino acids (Tenson et al., 2003), as expected (Figure 3C). These data strongly support the direct link between the pattern of tRNA pulses observed in the ZMW and translation.

The arrival of tRNAs at single ribosomes tracks the dynamic composition of the translational apparatus in real time. First tRNA arrival events are fast, as they do not depend on translocation. As predicted, the time between subsequent tRNA arrivals decreases with increasing EF-G concentrations between 30 and 500 nM (Uemura et al., 2010). For codons 2–12, the tRNA transit time is also strongly dependent on EF-G, as it represents at least two rounds of peptide bond formation and translocation (Uemura et al., 2010). Inhibition of EF-G by fusidic acid, which stabilizes EF-G·GDP on the ribosome post-translocation (Bodley et al., 1972), lengthens the transit time by 3.3-fold (Uemura et al., 2010). Arrival of the ribosome at the UGA stop codon after translation of 12 codons leads to a long pause from the remaining tRNA in the P site of the stalled ribosome. The dwell time for this last tRNA is 4.9-fold longer than for preceding pulses, underscoring that photobleaching of the P-site tRNA is not a significant problem using our approach at high factor concentrations: at 500 nM EF-G, the mean lifetime (4.1 s) of tRNAs bound to the ribosome is significantly shorter

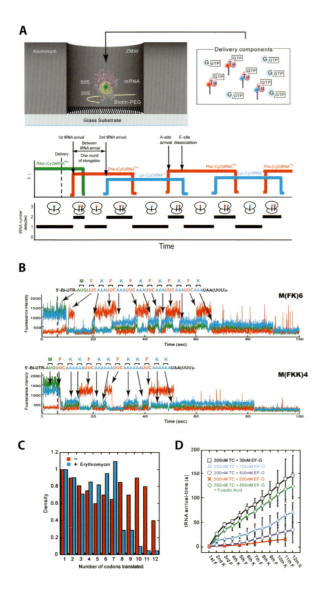

Fig. 3 Real-time translation at near-physiological concentrations. (A) Schematic of experimental setup (upper). Ternary complexes, Cy5-labeled Phe-tRNA^{Phe}-EF-Tu(GTP) and Cy2-labeled Lys-tRNA^{Lys}-EF-Tu(GTP), along with EF-G(GTP), are delivered to a ZMW surface-immobilized, initial ribosome complex containing Cy3-labeled fMet-tRNA^{fMet}. Fluorescence is excited by illumination at 488, 532, and 642 nm, and Cy2, Cy3, and Cy5 fluorescence are simultaneously detected. Expected signal sequence (lower panel): Initiation complexes are detected by fluorescence of fMet-(Cy3)tRNA^{fMet} bound at an initiation codon; fluorescent tRNAs are delivered as ternary complexes; arrival of Phe-(Cy5)tRNA^{Phe} or Lys-(Cy2)tRNA^{Lys} at the ribosomal A site is marked by red or blue fluorescent pulse. At high ternary complex concentrations, tRNA arrival times are fast (<< 1 s), and fluorescent pulses are overlapped, which indicates simultaneous occupancy by two tRNAs. The tRNA occupancy count is shown below the schematic trace. (B) Two heteropolymeric mRNAs encoding 13 amino acids each were used: (M(FK)$_6$ and M(FKK)$_4$. Translation was observed in the presence of 200 nM each of the ternary complexes of Phe-(Cy5) tRNA^{Phe} and Lys-(Cy2)tRNA^{Lys} as well as 500 nM EF-G as a series of fluorescent pulses that mirror the mRNA sequence. A long Cy2 pulse is observed upon arrival of the ribosome at the stop codon. (C) Event histograms for the translation of M(FK)$_6$ showing translation out to 12 elongation codons (red, n = 381). In the presence of 1 µM erythromycin, translation (blue, n = 201) is stalled at codon 8 of the mRNA. (D) Cumulative translation times for each codon in M(FK)$_6$ at 200 nM ternary complexes and 30, 100, or 200 nM EF-G; 500 nM ternary complexes and 500 nM EF-G, and 200 nM ternary complexes/500 nM EF-G in the presence of 1 µM fusidic acid. (Figure reproduced from Uemura et al. (2010). Reprinted by permission from Macmillan Publishers Ltd: Nature 464: 1012–1017, 2010)

4. Dynamic tRNA occupancy on the ribosome during translation

To define the mechanism linking tRNA arrival at the A site and release from the E site, we used these signals to measure the real-time tRNA occupancy of the ribosome during translation. This type of timing measurement is nearly impossible to perform at low concentrations of fluorescent ligand, since waiting times for bimolecular binding are too long. In our single-molecule traces, overlapping fluorescence pulses report on the number of tRNAs simultaneously bound to the ribosome, while appearance and departure of fluorescence indicates tRNA arrival and dissociation (Figure 3A). To determine the real-time occupancy of the ribosome at each codon, we post-synchronized 381 traces according to the arrival of aminoacyl-tRNA at each codon (Figure 5A). In this formulation, two-dimensional color plots reveal the time-dependent tRNA occupancy of hundreds of single ribosomes during each elongation cycle along the mRNA.

This analysis shows that EF-G driven translocation controls the number of tRNAs on the ribosome. At 30 nM EF-G, the two-tRNA state lasts ~6.3 s at each co-

than the photobleaching lifetime (17.3 s) observed in lower-concentration experiments. While paused on the stop codon, tRNA sampling events are observed with short lifetimes (~ 50 ms for Phe-tRNA^{Phe}(Cy5) or Lys-tRNA^{Lys}(Cy2),) (Figure 4A, B), which are clearly distinguishable from real tRNA transit events of > 1 s. These sampling events are consistent with a non-cognate ternary complex interaction (Rodnina et al., 2001) with the A site and their frequency is proportional to the concentration of ternary complex (Figure 4C). All trends discussed above were independent of the mRNA sequence.

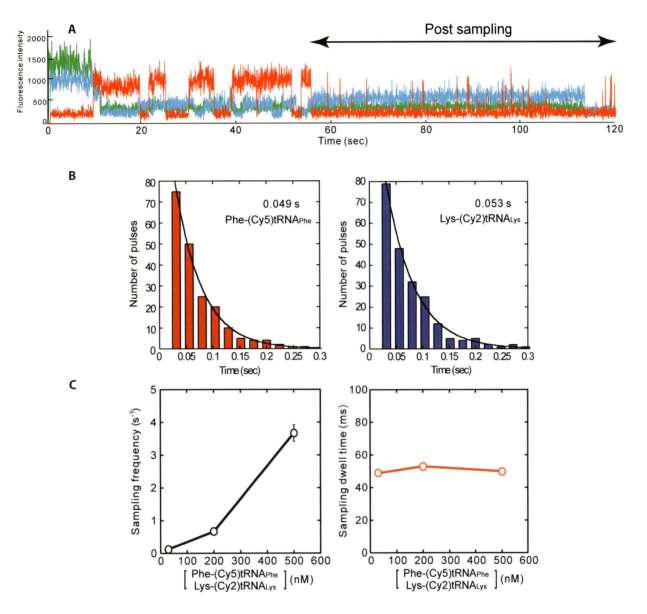

Fig. 4 A-site sampling on ribosomes stalled at the stop codon. (A) Fast sampling events at the stop codon position of the M(FK)6 template were observed in the presence of 200 nM ternary complex each of Phe-(Cy5)tRNA^{Phe} and Lys-(Cy2)tRNA^{Lys} and 500 nM EF-G. (B) Dwell time histograms individual sampling pulses of Phe-(Cy5)tRNA^{Phe} (left) and Lys-(Cy2)tRNA^{Lys} (right). Both histograms are well approximated by a single exponential fit. (C) The frequency of fast sampling increased linearly with TC concentration (left), while sampling dwell time did not depend on ternary complex concentration (right). (Figure reproduced from Uemura et al. (2010). Reprinted by permission from Macmillan Publishers Ltd: Nature 464: 1012–1017, 2010)

don, consistent with the estimated time for translocation. Increasing the concentration of EF-G to 500 nM shortens the lifetime of the two-tRNA state of the ribosome from 6.3 s to 1.5 s (Figure 5 A). These results suggest that EF-G binding and subsequent GTP hydrolysis drives the tRNA from the A/P and P/E hybrid states to the P and E sites, at which point the E-site tRNA rapidly dissociates (Figure 5 B). Consistent with this model, fusidic acid-stalled EF-G·GDP in the A site inhibits the arrival of the next tRNA and inhibits each round of

elongation, but does not affect the rate of tRNA dissociation from the E site.

5. Conclusion

The results presented here demonstrate that translation can be observed in real time using single ribosomes immobilized in ZMWs. The sequence of tRNA binding events reveals the encoding mRNA sequence. Full

A

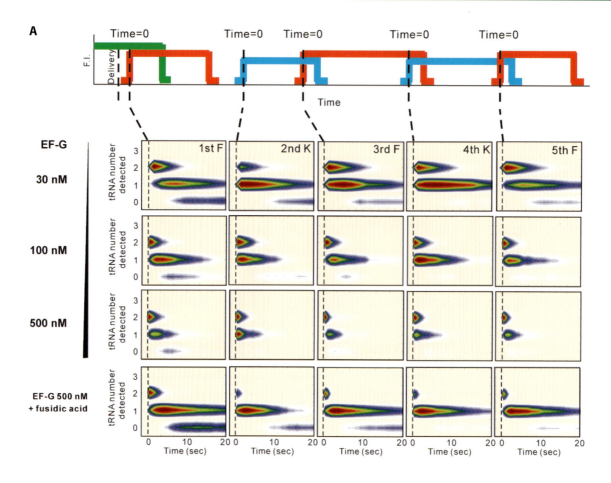

B

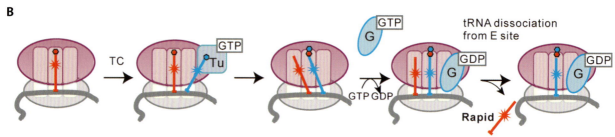

Fig. 5 Monitoring the dynamic tRNA occupancy of translating ribosomes. (A) Post-synchronization plots for time-resolved tRNA occupancy at codons 2–6 during translation of M(FK)$_6$. Two-dimensional histograms are post-synchronized in time with respect to each tRNA transit event (1st F~5th F) at 500, 100, and 30 nM EF-G, and at 500 nM EF-G in the presence of fusidic acid. (B) Model for tRNA transit through the ribosome during translation. Blue-labeled

tRNA arrives in the A-site, creating a 2-tRNA state with a red P-site-bound tRNA. This state forms an A/P hybrid state after peptide bond formation. Subsequent translocation of the two tRNA-mRNA complexes to the E and P sites is followed by rapid dissociation of red E-site tRNA. (Figure reproduced from Uemura et al. (2010). Reprinted by permission from Macmillan Publishers Ltd: Nature 464: 1012–17, 2010)

translation requires the presence of both EF-G and the appropriate ternary complexes. Ribosome-directed antibiotics interfere with translation as predicted by their mechanism of action: fusidic acid blocks the release of EF-G from the ribosome, slowing elongation, whereas erythromycin blocks elongation beyond 7 amino acids. The dynamics of tRNA binding events at each codon revealed slow initiation and long pauses upon encoun-

tering the stop codon; sampling of ternary complexes at the stop codon of stalled ribosomes is observed. Translation at micromolar concentrations of factors and ligands is efficient and rapid, avoiding limitations by dye photobleaching, and allows correlation of bimolecular binding events on single ribosomes.

The mechanism by which tRNAs transit through the ribosome during decoding, peptide bond formation and translocation was explored using our approach. Various models for interplay of the A and E site have been proposed. Recent dynamic and structural studies suggest that EF-G interaction within the A site may control the conformation of the E site (Gao et al., 2009). The ability to probe tRNA dynamics on the ribosome at high ternary complex and factors concentrations in ZMWs allowed us to determine the time-dependent composition of the ribosome at each codon during translation. These results show unambiguously that tRNA release from the E site is rapid once translocation has occurred and is not correlated to the arrival of the next tRNA (Figure 5B). This is consistent with a model of transient E site occupancy after translocation (Lill and Wintermeyer et al., 1987, Semenkov et al., 1996).

6. Future perspectives

The real-time system outlined here has broad application to the study of translation. As a bonus, the underlying mRNA sequence is revealed by the pattern of tRNA pulses observed during translation. Time-resolved analysis of compositional changes in the ribosome can be extended to initiation, elongation and release factor binding. We are currently investigating the timing of factor and subunit binding during initiation using dye-labeled components. In addition, the mechanism of action of antibiotics can be explored at each step and codon during translation; we are delineating the effects of aminoglycosides on elongation. Eukaryotic translational systems can be readily substituted to probe the dynamics of translational control and regulation (Petrov and Puglisi et al., 2010). The approach is ideally tuned to explore the timing of component binding and dissociation events during full translation while following the progress of a ribosome along an mRNA. Further technical advances will enhance the use of ZMWs for the study of translation. We have recently expanded our translation experiments to monitor four distinct dyes simultaneously. These improvements in methods and further optimization in instrumentation will permit observation of physiological translational events involved in the regulation of protein synthesis, such as frameshifting.

Acknowledgement

Supported by NIH GM51 266 (JDP). We thank A. Tsai and A. Petrov (Stanford) for encouragement and stimulating discussions. We thank J. Gray for performing ellipsometry experiments.

References

Aitken CE, Puglisi JD (2010) Following the intersubunit conformation of the ribosome during translation in real time. Nat Struct Mol Biol 17: 793–800
Aitken CE, Petrov A, Puglisi JD (2010) Single ribosome dynamics and the mechanism of translation. Annu Rev Biophys 39: 491–513
Blanchard SC, Gonzalez RL Jr, Kim HD, Chu S, Puglisi JD (2004a) tRNA selection and kinetic proofreading in translation. Nat Struct Mol Biol 11: 1008–1014
Blanchard SC, Kim HD, Gonzalez RL Jr, Puglisi JD, Chu S (2004b) tRNA dynamics on the ribosome during translation. Proc Natl Acad Sci USA 101: 12 893–12 898
Bodley JW, Godtfredsen WO (1972) Studies on translocation. XI. Structure-function relationships of the fusidane-type antibiotics. Biochem Biophys Res Commun 46: 871–877
Eid J, Fehr A, Gray J, Luong K, Lyle J, Otto G, Peluso P, Rank D, Baybayan P, Bettman B, Bibillo A, Bjornson K, Chaudhuri B, Christians F, Cicero R, Clark S, Dalal R, Dewinter A, Dixon J, Foquet M, Gaertner A, Hardenbol P, Heiner C, Hester K, Holden D, Kearns H, Kong X, Kuse R, Lacroix Y, Lin S, Lundquist P, Ma C, Marks P, Maxham M, Murphy D, Park I, Pham T, Phillips M, Roy J, Sebra R, Shen G, Sorenson J, Tomaney A, Travers K, Trulson M, Vieceli J, Wegener J, Wu D, Yang A, Zaccarin D, Zhao P, Zhong F, Korlach J, Turner S (2009) Real-time DNA sequencing from single polymerase molecules. Science 323: 133–138
Foquet M, Samiee KT, Kong X, Chauduri BP, Lundquist PM, Turner SW, Freudenthal J, Roitman DB (2008) Improved fabrication of zero-mode waveguides for single-molecule detection. J Appl Phys 103: 034301
Gao YG, Selmer M, Dunham CM, Weixlbaumer A, Kelly AC, Ramakrishnan V (2009) The structure of the ribosome with elongation factor G trapped in the posttranslational state. Science 326: 694–699
Korlach J, Marks PJ, Cicero RL, Gray JJ, Murphy DL, Roitman DB, Pham TT, Otto GA, Foquet M, Turner SW (2008) Selective aluminum passivation for targeted immobilization of single DNA polymerase molecules in zero-mode waveguide nanostructures. Proc Natl Acad Sci USA 105: 1176–1181
Levene MJ, Korlach J, Turner SW, Foquet M, Craighead HG, Webb WW (2003) Zero-mode waveguides for single-molecule analysis at high concentrations. Science 299: 682–686
Lill R, Wintermeyer W (1987) Destabilization of codon-anticodon interaction in the ribosomal exit site. J Mol Biol 196: 137–148
Lundquist PM, Zhong CF, Zhao P, Tomaney AB, Peluso PS, Dixon J, Bettman B, Lacroix Y, Kwo DP, McCullough E, Maxham M, Hester K, McNitt P, Grey DM, Henriquez C, Foquet M, Turner SW, Zaccarin D (2008) Parallel confocal detection of single molecules in real time. Opt Lett 33: 1026–1028
Marshall RA, Aitken CE, Puglisi JD (2009) GTP hydrolysis by IF2 guides progression of the ribosome into elongation. Mol Cell 35: 37–47

Petrov A, Puglisi JD (2010) Site-specific labeling of Saccharomyces cerevisiae ribosomes for single-molecule manipulations. Nucleic Acids Res. 38: e143

Rodnina MW, Wintermeyer W (2001) Fidelity of aminoacyl-tRNA selection on the ribosome: kinetic and structural mechanisms. Annu Rev Biochem 70, 415–435

Schlünzen F, Zarivach R, Harms J, Bashan A, Tocilj A, Albrecht R, Yonath A, Franceschi F (2001) Structural basis for the interaction of antibiotics with the peptidyl transferase centre in eubacteria. Nature 413: 814–821

Semenkov YP, Rodnina MV, Wintermeyer W (1996) The "allosteric three-site model" of elongation cannot be confirmed in a well-defined ribosome system from *Escherichia coli*. Proc Natl Acad Sci USA 93: 12 183–12 188

Tenson T, Lovmar M, Ehrenberg M (2003) The mechanism of action of macrolides, lincosamides and streptogramin B reveals the nascent peptide exit path in the ribosome. J Mol Biol 330: 1005–1014

Uemura S, Aitken CE, Korlach J, Flusberg BA, Turner SW, Puglisi JD (2010) Real-time tRNA transit on single translating ribosomes at codon resolution. Nature 464: 1012–1017

Underwood KA, Swartz JR, Puglisi JD (2005) Quantitative polysome analysis identifies limitations in bacterial cell-free protein synthesis. Biotechnol Bioeng 91: 425–435

Dynamic views of ribosome function: Energy landscapes and ensembles

24

P. C. Whitford, R. B. Altman, P. Geggier, D. S. Terry, J. B. Munro, J. N. Onuchic,
C. M. T. Spahn, K. Y. Sanbonmatsu, S. C. Blanchard

1. Introduction

Single-molecule fluorescence resonance energy transfer (smFRET) (reviewed in Munro et al., 2009) and cryo-electron microscopy (cryo-EM) investigations (Frank and Spahn, 2006; Spahn and Penczek, 2009; Fischer et al., 2010) of the translation apparatus reveal the ribosome's propensity to undergo large-scale fluctuations in conformation during function. Progress in these areas, building upon achievements in high-resolution structure determination of ribosomal subunits and functional complexes of the ribosome (Yusupov et al., 2001; Wekselman et al., 2009: Zhang et al., 2009; Gao et al., 2009; Demeshkina et al., 2010; Stanley et al., 2010), combined with an ever increasing breadth of computational modeling, simulation (Sanbonmatsu and Tung, 2007; Whitford et al., 2010a), and bioinformatics approaches (Roberts et al., 2008; Alexander et al., 2010), offers the potential to further broaden our understanding of the dynamic nature of the ribosome and translation components during protein synthesis. The large fluctuations observed by single-molecule studies, and the multitude of conformations reported by cryo-EM, make it clear that each "state" of the ribosome is in fact an ensemble of structurally similar configurations that are localized to a particular minimum on the free-energy landscape.

As single-molecule (Blanchard et al., 2004ab), cryo-EM (Frank and Spahn, 2006; Frank et al., 2007), rapid kinetics (Pape et al., 1999) and computational techniques (Sanbonmatsu et al., 2005) were first being extended to investigations of ribosome dynamics, the generally accepted goal was to identify specific "states" associated with biochemical steps in the translation process. This aim, driven by the canonical biochemical framework, held that structural and kinetic transitions in the ribosome were governed by factor- and GTP hydrolysis-driven transitions between a limited number of specific system configurations. However, the framing of the translation process in biochemical terms as a series of discrete "jumps" between distinct "states" mediated by specific translation factors often leads one to draw analogies between the ribosome and macroscopic machines which are inconsistent with both old and new observations in the field. First, the ribosome can synthesize protein in the absence of translation factors and nucleotide hydrolysis (Gavrilova et al., 1976), suggesting that translation is a function inherent to the ribosome. Second, as clearly demonstrated for other biopolymers (Frauenfelder et al., 1991), the ribosome is intrinsically dynamic. This aspect of ribosome function has now been highlighted by many recent smFRET measurements, wherein dynamics have been observed from a number of distinct structural perspectives. As evidence regarding the ribosome's intrinsically dynamic nature mounts (Munro et al., 2007, 2010 a,b; Cornish et al., 2008, 2009; Fei et al., 2008, 2009), it is increasingly apparent that the architecture of each translation component is designed to be inherently flexible and that further efforts must be made to describe function in this context.

An important consideration of dynamic systems is that even the lowest-energy "states" (e. g., the classical state), are in fact large ensembles of physically distinct configurations. "States" are better defined as a set of configurations with similar free energies that rapidly interconvert, fluctuating around a local free-energy minimum within a larger free-energy landscape. In this view, the energy landscape of ribosome function is characterized by distinct free energy basins, each with a relative free energy minimum and free-energy barriers separating it from other basins. A deeper understanding of the ribosome's transit between distinct basins on the free-energy landscape during function is central to delineating key features of both elemental translation processes as well as the cellular strategies for translational control of gene expression. The free-energy landscape is an intrinsic property of a given molecular system, independent of the experimental/theoretical method employed to study it and various methodological approaches have emerged as a means to probe ribosome function from this perspective. These efforts are highlighted by advances in the area of time-resolved cryo-EM investigations of the translocation reaction coordinate (Fischer et al., 2010) as well as single-molecule fluorescence and theoretical investigations.

Single-molecule FRET measurements, as well as simulations, provide a powerful means of probing the properties of the free-energy landscape by providing information about the time spent within a given energy minimum before transitioning to another. While the insights obtained through such investigations portend a shift in our understanding of the translation mechanism, an apparent lack of consensus has emerged regarding specific dynamic events, in particular, the rates and amplitudes of motion and the number of intermediate configurations observed during global rearrangements in ribosome structure (Blanchard, 2009). To the context of highlight early progress in monitoring ribosome activity at the single-molecule scale, this chapter attempts to define outstanding challenges facing the field and considerations that will be critical to delineating a quantitative framework for ribosome dynamics during function. Drawing from the protein folding community, a clear definition of the energy landscape of ribosome function is established, specific methods that can be used to measure features of the landscape are outlined, and the utility of the landscape view for understanding existing experimental data and protein synthesis regulation is explored. This template is provided to serve as a springboard towards next-generation experiments and a systems-level perspective on translation control.

2. Static states vs. ensembles of conformations

The first consideration to be made is that the term "state" implies the existence of a well-defined, deep, energetic minimum, where thermal fluctuations that lead to structurally-similar configurations are insignificant. In this view, an exogenous source of energy,

much greater than is available from the thermal bath, is necessary to mediate structural transitions in the system on a time scale necessary to support cellular translation. Recent smFRET and cryo-EM investigations of the pre-translocation complex provide evidence that distinct structural elements of the ribosome undergo spontaneous, transient transitions on a range of timescales (reviewed in Munro et al., 2009, 2010a,b). These data demonstrate the existence of metastable structural configurations of the ribosome stemming from that lead to reversible conformational events (subunit ratcheting/unratcheting; L1 stalk closure and opening; tRNA exchange between classical and hybrid configurations), whose relative populations depend on environmental variables and the ligands bound to the ribosome. Such observations suggest that the energetic barriers required to escape from each configuration of the system are small and the thermodynamic boundaries of each state are poorly defined. Correspondingly, a substantially more comprehensive picture of the energy landscape emerges where "states" are considered to be must be considered ensembles of structurally diverse configurations and thermal fluctuations are sufficient to promote substantial structural rearrangements. Moreover, if the energy landscape is not comprised of a single, large barrier between distinct configurations, a variety of possible transition routes between "states" in the system may exist. Here, the total number of routes between local minima can be collectively referred to as the transition state ensemble. As has been shown for protein structure-function relationships (Hyeon and Onuchic, 2007; Miyashita et al., 2003), the existence of structural heterogeneities within the ensemble suggests entropic contributions are critical to the architecture of the energy landscape, and correspondingly, ribosome function (Sievers et al., 2004; Whitford et al., 2010).

For the following discussions of energy landscapes, the differences between free energy, potential energy and enthalpy must be taken into consideration. The non-entropic component of the free energy can be referred to as enthalpy, effective enthalpy, or potential energy, depending on the context. A potential energy surface is distinct from the free-energy surface, as the latter also includes an entropic component.

A reduced description of the free-energy landscape of ribosome function is represented by the potential of mean force (*pmf*), the free energy of the system measured as a function of a particular coordinate. For instance, such a coordinate may describe movement of the tRNA through the ribosome, pro-

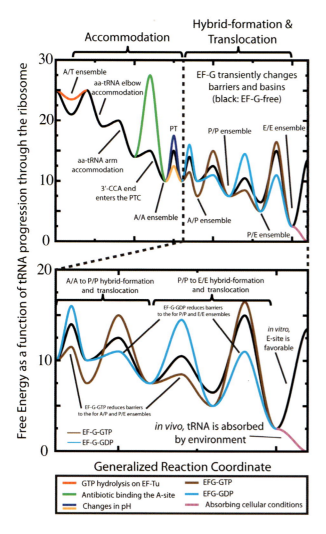

Fig. 1 Energy landscapes. Schematic representation of the potential of mean force (*pmf*) for a single tRNA molecule from the A/T configuration to when it leaves the E site. The black line represents the factor-free *pmf*. For simplicity, we discuss the *pmf* of a single tRNA molecule after averaging over the coordinates of the second tRNA. Perturbations to the landscape are schematized in color: (red) GTP hydrolysis by EF-Tu, and/or π releale, destabilizes the A/T ensemble; (green) antibiotic binding near the PTC could increase the barrier to aa-tRNA 3' CCA entry into the PTC since its displacement would be required for aa-tRNA accommodation; (blue/gold) changes in pH can alter the barrier/rate of peptidyl transfer (PT); (brown) EF-G-GTP can increase the rate of hybrid formation, and stability of the hybrid-state ensemble (i.e. lower barrier and lower minimum). (light blue) EF-G-GDP can increase the rate to, and stability of, the post-translocation ensemble. *In vivo* released E-site tRNAs are processed by cellular enzymes, making the chemical potential of spontaneous tRNA-binding energetically uphill (pink).

ceeding from the energy basin describing the A/T configuration of the ribosome to tRNA release from the E-site (Figure 1). Here, the potential of mean force is given by $pmf(\bar{X}) = -k_B T \ln(P(\bar{X}))$, where $P(\bar{X})$ the probability of configuration $\bar{X}$, and k_B is Boltzmann's

constant. In this schematic representation, the role of elongation factors can be depicted as transiently reducing the free-energy barriers associated with the transition state ensembles "ahead" of the tRNA molecule (in terms of movement of tRNA through the ribosome) and/or increasing the barriers associated with the transition state ensembles "behind" the tRNA. Similarly, the role of antibiotics can be represented as increasing the barriers associated with progression of the tRNA through the ribosome.

Quantitative descriptions of ribosome dynamics during translation are likely to emerge by applying the principles of energy landscape theory and the arsenal of quantitative tools largely developed in the protein and RNA folding fields (Bryngelson and Wolynes, 1989; Leopold et al., 1992; Onuchic and Wolynes, 1993; Onuchic et al., 2006; Wales, 2004; Onuchic et al., 2000; Thirumalai et al., 2010; Pincus, 2009; Thirumalai and Hyeon, 2005). This framework provides a uniform foundation for understanding the vast range of complex motions and energetic contributions associated with ribosome function. Understanding which features of the landscape are robust will be paramount in uncovering new mechanisms of control. If successful, this self-consistent, statistical physics-based framework may ultimately afford predictive control over ribosome function.

3. Choosing the energy landscape reaction coordinate

The first step in characterizing the energy landscape of any molecular process is to establish the proper reaction coordinate. A reaction coordinate is a measurable quantity one-dimensional measure of progress along a reaction process, and it is uniquely defined for a given conformation. Since molecular systems possess 3N configurational degrees of freedom, where N is the number of atoms, projecting a given motion onto a particular reaction coordinate reduces the dimensionality of the dynamic process under investigation. Generally, an optimal choice of reaction coordinate will capture accurately describe the barriers corresponding to the relevant kinetic steps of the system. This reaction coordinate should: (1) be continuous (i.e. local changes in structure are measured as local movements along the coordinate); (2) distinguish between starting and ending points of a transition and; (3) identify the transition state ensemble (TSE) (Nymeyer et al., 2000; Cho et al., 2006; Komatsuzaki et al., 2005; Das et al., 2006).

An ideal reaction coordinate identifies the ensemble of conformations (the TSE) that are equally likely to reach the starting ensemble and the ending ensemble, where the TSE corresponds to the maximum in the *pmf* along that reaction coordinate.

While the system's free energy landscape determines its kinetic and thermodynamic properties, the *pmf* depends on the experimental observable. The difference between the underlying barrier and the measured barrier is apparent by considering the statistical mechanical definition of the probability for a given configuration:

$$P(\bar{X}) = \frac{\Lambda \exp(-V(\bar{X})/k_B T)}{Z} ,$$

where $\bar{X}$ is a 3N-dimensional vector describing the coordinates of the system, $(\bar{X})/k$ is the potential energy at point $\bar{X}$, Z is the partition function of the system and Λ is the contribution to the partition function from the momenta integral. Here, the exponential in the numerator represents the velocity-independent, Boltzmann weight of a particular configuration. The denominator, Z, represents the unnormalized probability of all configurations. Thus, the pmf along any given reaction coordinate, Q, can be defined as

$$pmf(Q) = \frac{\Lambda \int_V \delta(Q(\bar{X}) - Q) \exp(-V(\bar{X})/k_B T) dR^3}{Z} ,$$

where the observed *pmf*, *pmf(Q)*, will depend on its projection from 3N-dimensional coordinates onto the reaction coordinate, $(\bar{X})-$. This discussion is particularly relevant to smFRET experiments where changes in FRET efficiency as a function of time are used to monitor the reaction coordinate but are only typically monitored from a single structural perspective. Any single fluorophore FRET pair can produce one slice of the multidimensional energy landscape. Thus, distinct dye locations used to monitor a given conformational event may therefore lead to distinct conclusions regarding the system's dynamics. Such considerations provide a plausible explanation for why different labeling strategies for monitoring L1 stalk dynamics on the ribosome yield a different number of states (Fei et al., 2008; Cornish et al., 2009; Munro et al., 2010b) and how spontaneous ratcheting dynamics may be observed using one set of labeling strategies (Cornish et al., 2008, 2009), while using others they are not observed (Marshall et al., 2008). That is, each local minimum may be composed of an ensemble of con-

figurations which is resolved by certain choices of dye locations and not by others. Specific reaction coordinates may mask or reveal more complicated features of the free-energy landscape (Garcia and Onuchic, 2003). The advent of technologies to monitor specific translation reactions from distinct structural perspectives (Cornish et al., 2008/9; Geggier et al., 2010) or multiple structural perspectives simultaneously (Munro et al., 2010a, b) can afford deeper insights into the relevant reaction coordinates of the process under observation. As additional technologies emerge to site-specifically position fluorophores in a greater number of positions within the translational apparatus, explicit control of the observed reaction coordinate can ultimately be obtained.

To provide a concrete example of how different reaction coordinates may lead to differential conclusions about the dynamics it is illustrative to examine a recently reported set of 312 independent all-atom simulations of tRNA accommodation (Whitford et al 2010). These data were used to calculate the probability distribution of the distance between the incoming aa-tRNA elbow and the P-site tRNA elbow (R_{Elbow}) as well as the distance between aminoacyl residues on A- and P-site tRNAs ($R_{3'}$) (Figure 2 A–D). The calculated probability distribution shows at least four highly populated ensembles accumulate during accommodation: the A/T ensemble, the elbow-accommodated ensemble, the elbow- and arm-accommodated ensemble, and the A/A ensemble. Along the $R_{3'}$ reaction coordinate (Figure 2 B), there are several peaks that closely reflect the full distribution of states transited by the system (Figure 2 A). In contrast, only two dominant peaks in the probability distribution are observed along the (R_{Elbow}) reaction coordinate (Figure 2 C). These data show that both individual structural perspectives are limited and that only when the distribution is measured as a function of both R_{Elbow} and $R_{3'}$ are the individual ensembles well separated.

Using a second reaction coordinate, the same simulations also revealed insights into the nature of L1 stalk motions. These data are of relevance to contextualizing recent smFRET studies of L1 protein motions in the pre-translocation complex (Munro et al., 2010a,b; Cornish et al., 2009; Fei et al., 2008). Here, the simulations showed large, thermally-accessible ($<6k_B T$) excursions in L1 stalk position as a function of time (Figure 2 E–H), where uncoupled twisting (± 20 degrees) and extension (± 20Å) motions of the L1 stalk occurred (Figure 2 F), which gave rise to a broadened ensemble of configurations (Figure 2 G). Such observations perhaps shed light

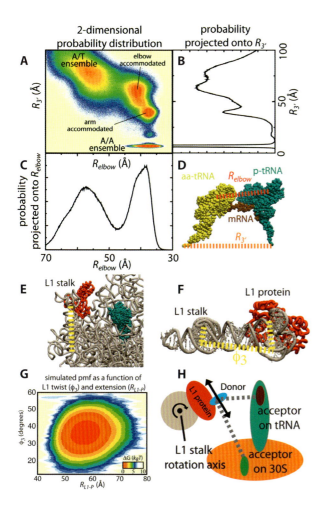

4. Further considerations in potential of mean force measurements

In addition to the aforementioned complexities associated with *pmf* measurement and calculation, it is also important to consider that the measured degrees of freedom in the smFRET measurement are in fact the distances between the transition dipoles of the fluorogenic centers, not the atoms to which they are attached. In most cases, the interdye distance can be approximated by their center of mass averaged over the imaging time interval. Anisotropic dye motions resulting from the steric contours of the molecule, or site to which it is attached, add a degree of uncertainty to this approximation. Rigorous comparison to the inter-atom and inter-dye distances also requires an estimation of how far the dyes extend away from the atoms to which they are attached.

As an example of the uncertainty in dye position in a smFRET experiment, fluctuations in inter-dye distance were calculated using all-atom simulations of the A/T ribosome configuration[1], where A- and P-site tRNAs were linked to Cy3 and Cy5 fluorophores, respectively, mirroring the labeling strategy used in smFRET imaging of aa-tRNA selection on the bacterial ribosome (Geggier et al., 2010) (Figure 3A). A naive estimate of the barrier height to aa-tRNA movement from an A/T to A/A position based on the experimentally-derived probability distribution[2] suggests a barrier-crossing free energy of ~0.6 k_BT. Based on such an analysis, the elbow-accommodation time would be estimated to be ~1–10 μs, substantially faster than anticipated from bulk kinetics (Rodnina and Wintermeyer, 2001; Johansson et al., 2008), previous smFRET experiments (Blanchard et al., 2004; Lee et al., 2007) as well as hidden-Markov modeling (Geggier et al., 2010). Here, a simple numerical exercise provides important insights into how functionally relevant barrier heights can be difficult to estimate from FRET probability distributions projected along the experimentally-obtained tRNA-elbow distance (Figure 3).

Fig. 2 Describing and using energy landscapes. (A) The probability of accommodation calculated from 312 independent all-atom simulations (Whitford et al 2010a) as a function of the distance between aa-tRNA and P-site tRNA elbows R_{Elbow} and amino acids $R_{3'}$. This distribution shows 4–6 peaks. (B) When the probability is calculated as a function of $R_{3'}$, not all peaks are clearly separated. (C) The probability distribution, as a function of R_{Elbow}, shows 2 dominant peaks. Some peaks in panel B are only present as shoulders in panel C. (D) Structural representation of R_{Elbow} and $R_{3'}$. (E) Structure of 23S rRNA (tan), P-site tRNA (cyan) and the L1 protein (red). (F) The L1 stalk "twist" coordinate, ϕ_3, is defined by the pseudo-dihedral angle formed by the C1' atoms of G2127, A2171, C2196 and G2093 on the 23S rRNA. (G) The *pmf* as a function of the distance between the P-site tRNA and protein L1, R_{L1-P}, and ϕ_3, calculated from all-atom simulations (Whitford et al 2010a), shows that large rearrangements in the L1 stalk (±20Å and ±20 degrees) occur with only a modest (few k_BT) energetic penalty. (H) Schematic that illustrates how rotation of the L1 stalk is more easily detectable by smFRET if the donor and acceptor fluorophores fall along the tangent (double arrow) of the rotation.

on why a different number of states, with distinct rates of motion, were observed by the different groups using varied labeling strategies to measure L1 motion (Figure 2H).

---

1 This near-A/T model was obtained by using a flexible-fitting algorithm (Orzechowski and Tama, 2008), in conjunction with a structure-based forcefield (Whitford et al 2009a, 2009b) and an unpublished cryo-EM map, as implemented elsewhere (Ratje et al., 2010).

2 This estimate was not reported in previous publications. Here, we simply used the logarithm of the probabilities obtained in the previous study to illustrate how the barrier height is likely underestimated by this approach.

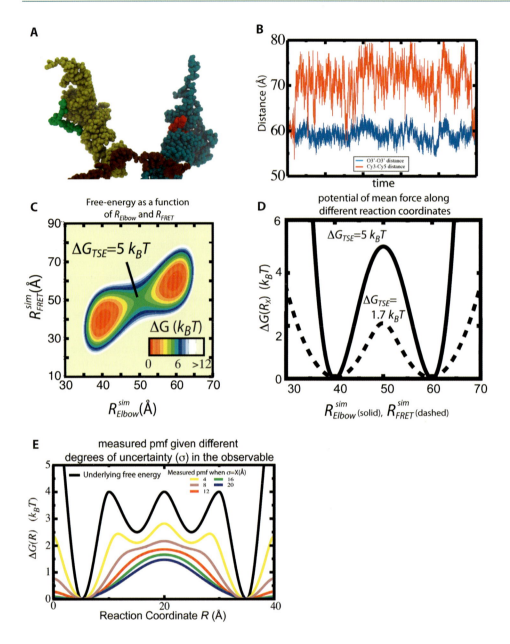

Fig. 3 Fluctuations in FRET dye position can alter the measured pmf. (A) Structural representation of aa-tRNA (left, yellow), P-site tRNA (right, cyan), mRNA (brown), Cy3 dye (green) and Cy5 dye (red). (B) Distance between the center of Cy3 and Cy5 (red) and R_{Elbow} as functions of time for an all-atom simulation of a near-A/T conformation. Fluctuations in R_{Elbow} are ~ 3 Å and fluctuations in dye distance is ~ 10 Å. (C) Hypothetical free-energy profile, with dispersion in dye distance of ~ 12 Å and a free-energy barrier of 5 k_BT. (D) Free energy in panel C projected onto the dye distance (dashed line) shows a barrier of ~ 1.7 kBT. When projected onto R_{Elbow} (solid line) the barrier is 5 k_BT. The pmf along the dye distance will likely underestimate the free-energy barrier. (E) Uncertainty in the projection onto a reaction coordinate can mask barriers in the landscape. A potential hypothetical free-energy profile is shown in (black). As done in panel C, this profile is described by a two-dimensional landscape, where the uncertainty in the observable (σ) is given a particular width and the "observed" pmf is then calculated. As sigma is increased, the intermediate minima are less pronounced.

When the free-energy barrier associated with a transition exceeds several k_BT, where barrier crossing events occur infrequently, then over reasonably long time intervals (ca. milliseconds) the time-averaged center of mass position of the Cy3 and Cy5 fluorophores provides a close approximation of the average tRNA position. In this regime, the Cy3 and Cy5 fluorophores tumble rapidly relative to the rates of large-scale tRNA motions and large changes in distance between the center of mass of the Cy3 and Cy5 fluorophores occur only during barrier-crossing events. However, when imaging time scales approach the rates of Brownian fluctuations in fluorophore position (e. g., microseconds), then the inter-dye distance fluctuates significantly even in the absence of appreciable changes in the A- and P-site tRNA inter-elbow distance. In this regime, dye mobility can contribute significantly to the experimentally measured *pmf*. Simulations suggest that dye positions may vary by as much as 12 Å on the nanosecond-microsecond time scale. To illustrate this point, the *pmf* of the A/T to A/A transition for both, R^{sim}_{Elbow} and R^{sim}_{FRET} probability distributions can be projected onto the reaction coordinates and analyzed according to statistical mechanics[3]:

$$\Delta G\left(R^{sim}_{FRET}\right) = -k_B T \ln\left(P\left(R^{sim}_{FRET}\right)\right) \quad .$$

This analysis shows that the estimated barrier height for motion based on R^{sim}_{FRET} (1.7 kBT) is substantially lower than for R^{sim}_{Elbow} (5 kBT), which is due to both experimental noise[4] and uncertainties in dye position (Figure 3D). As the rate of barrier-crossing is proportional to $e^{-\Delta G_{TSE}/k_B T}$, where ΔG_{TSE} is the barrier height and TSE is the transition state ensemble, such differences correspond to about a factor of 30 in timescale. Thus, broadening of the observed probability distribution due to the ratio of signal-to-noise and dye mobility considerations in smFRET measurements tend to provide underestimates of the physical barrier height and therefore overestimate the rates of motion. As the uncertainty in the observable increases, the likelihood of observing subtle features of the underlying energy landscape diminishes (Figure 3E). At the same time,

fast conformational events can be missed in smFRET measurements (Blanchard et al 2004a), having the opposite effect of providing estimates of the rates of motion that are slower than reflected by the true barrier heights. As the uncertainty in the projection onto a reaction coordinate changes, or if different reaction coordinates are employed, the measured *pmf* will be systematically altered.

Such considerations represent important challenges for all single-molecule fields. An important avenue by which to explore these issues is to simulate "noise-less" smFRET data according to explicit kinetic models of motion and then to simulate different types and amplitudes of noise to enable direct comparisons to the experimental data (Munro et al., 2007; Geggier et al., 2010). Progress may also be made by delineating methods for choosing the ideal reaction coordinate for a given transition. Such pursuits may be aided by (i) computational/theoretical efforts (Faradjian and Elber, 2004; Peng et al., 2010; Das et al., 2006) and (ii) more rigorous approaches for understanding the nature of fluorophore mobility in their biological contexts. While, in principle, it should be possible to choose an optimal reaction coordinate, performing experiments and simulations with probes in different positions may ultimately be necessary to provide the quantitative information necessary to determine robust features of the underlying energy landscape of a given conformational transition.

5. The role of diffusion in energy landscapes: connecting kinetics and thermodynamics

While no direct experimental measurements have been aimed at determining barrier-crossing attempt frequencies in the context of the translation apparatus, recent smFRET measurements of aa-tRNA selection (Geggier et al., 2010) and pre-translocation complex dynamics (Munro et al., 2010a, b) have observed what can be considered "non-productive" transition attempts to achieve key intermediate configurations during the process of aa-tRNA selection and translocation, respectively. During selection, aa-tRNA was described as sampling "GTPase-activated" and "accommodated" configurations prior to undergoing the chemical steps of GTP hydrolysis and peptide bond formation, respectively. The pre-translocation complex was shown to transiently achieve a FRET con-

---

3 Since the surface in Figure 2C is defined as a function of R^{sim}_{FRET} and R^{sim}_{Elbow}, "projecting" onto R^{sim}_{FRET} simply means calculating the appropriate sum over R^{sim}_{Elbow} for each value of R^{sim}_{FRET}.

4 In this discussion we are focusing on the contribution of dye uncertainty. Though, currently, additional sources of noise are likely comparable in scale.

figuration consistent with formation of the unlocked state, the putative intermediate for translocation. Such notions of non-productive conformations, which bear resemblance to the transition states of barrier crossing, are consistent with the stochastic nature of barrier-crossing processes. The following general relationship describing diffusive motions in molecular systems provides an important bridge between the free-energy landscape and the kinetic parameters of ribosome function (Bryngelson and Wolynes, 1989; Zwanzig, 1988):

$$\frac{1}{k} = \langle \tau \rangle = \int_{R_{start}}^{R_{end}} dQ \int_{\infty}^{R} dQ' \frac{\exp\left[(G(Q) - G(Q'))/k_B T\right]}{D}, \quad (1)$$

where k is the rate of barrier crossing, $\langle\tau\rangle$ is the mean-first passage time, Q is the reaction coordinate, D is the effective diffusion coefficient in reaction-coordinate space, and R_{start} and R_{end} are the endpoints of the transition. The diffusion coefficient D is often assumed constant and particle diffusion in three dimensions can be estimated according to the Stokes-Einstein relationship, $D = k_B T / 6\pi\eta\alpha$, where η is the viscosity of water and a is the hydrodynamic radius. Through analysis of kinetic models and experimental data, Fluitt et al. (2007) estimated the diffusion coefficient of ternary complex in solution to be ~2.6 μm^2/s. As the hydrodynamic radius of ternary complex is roughly two-fold larger than the tRNA molecule, tRNA diffusion in solution can be estimated to be ~5.2 μm^2/s. Scaling by an additional factor of one-third to account for the approximately one-dimensional diffusion of tRNA as it moves through the ribosome (i. e., along the R_{Elbow} vector) reduces this value to 1.7 $\mu m2$/s, in good agreement with explicit-solvent simulations of the 70S ribosome (205–301 ns each), estimating the diffusion in R_{Elbow} to be ~ 1 μm^2/s (Whitford et al., 2010b). Advances in experimental and computational methods (Nettels et al., 2008; Wong and Case, 2008) may allow for more precise estimates of the diffusive properties of tRNA inside the ribosome, though such studies have not been reported to date. Full characterization of the diffusive properties will allow for a quantitative relationship between kinetics and thermodynamics, as well as for direct comparisons to rates determined in bulk (Gromadski and Rodnina, 2004) and single-molecule methods. However, for sufficiently large energy barriers, the rate of barrier crossing (Eq. 1), can be simplified through the introduction of a prefactor, C, that approximates the diffusive prop-

erties, manifested as a barrier-crossing attempt frequency:

$$\frac{1}{k} = \langle \tau \rangle = \frac{1}{C} \exp(\Delta G_{TSE}/k_B T) . \quad (2)$$

One can think of the prefactor as describing how frequently the barrier crossing events are attempted, where the probability of successful crossing is determined by the barrier height. Investigations of the folding properties of small proteins suggest that C is ~ 0.1–20 μs^{-1} (Kubelka et al., 2004; Tang et al., 2009). Thirumalai and Hyeon (2005) similarly estimate C for RNA folding to be ~1 μs^{-1}. Explicit-solvent simulations of the 70S ribosome also suggest the prefactor associated with tRNA elbow-accommodation to be ~ 1 μs^{-1} (Whitford et al., 2010b). Here, a sharp distinction should be made between the prefactor associated with conformational rearrangements and the prefactor associated with the transition state for catalysis. Classical biochemical transition-state theory was developed in the context of enzyme catalysis, where the attempt frequency to the chemical transition state is governed by covalent bond stretching and bond angle bending. Since these vibrations occur on femtosecond to picosecond timescales, the attempt frequencies should occur in a similar time domain. However, in the case of conformational rearrangements where the attempt frequency is limited by multi-dimensional diffusion on the energy landscape, barrier-crossing attempts will be far more rare less frequent and dependent on the functionally-relevant degrees of freedom in the system. The observation of "non-productive" sampling events on path to productive aa-tRNA selection and translocation suggests that barrier-crossing attempts may be observed in both the microsecond and millisecond regimes. As the time-resolution of smFRET imaging experiments continues to increase, intermediates of this nature are likely to be more frequently detected and more critical to mechanistic discussions of ribosome function.

6. Energetic coupling between ribosomal compo nents

While the notion that the ribosome acts as a "thermal ratchet" (Spirin 2009) implies that the energetic barriers of large-scale ribosomal subunit motions are small relative to thermal fluctuations subunit motions are on the same order at $k_B T$, it has only recently been suggested that this may be the case throughout the translation cycle (Munro et al., 2009). Analogous to protein folding landscapes (Onuchic et al., 2000), broad energetic basins, combined with a low degree of energetic roughness, ensure physiologically-relevant timescales of function. Such a landscape is consistent with a scenario where many independent, thermally-accessible motions within the ribosome may be anticipated to occur, whose timescales are dependent on the diffusive properties of the system. Experimental observations of independent degrees of freedom in the pre-translocation complex support this notion (Munro et al., 2010a,b).

Formally, coupled degrees of freedom represent conformational events that are structurally distinct, but not energetically separable. A general expression of the total potential energy of a system, V, and the coupling between conformational events can be given by the equation:

$$V = W_U(V_1(R_1) + V_2(R_2)) + W_C V_C(R_1, R_2), \qquad (3)$$

where $V_i(R_i)$ is the separable contribution to the potential due to degree of freedom R_i, $V_C(R_1, R_2)$ is the inseparable contribution due to coupling between the R_1 and R_2, W_U is the scale of the uncoupled contribution and W_C is the scale of the coupling interaction. Assuming entropy is constant as a function of R_1 and R_2, the free energy of the system reduces to the potential energy. One can define the degree of coupling in the system, Γ, as the ratio of the coupled and uncoupled contributions, W_C/W_U. In the limit of uncoupled motions, $\Gamma = 0$. For weakly-coupled systems, $0 < \Gamma \ll 1$. When the contribution of the coupled and uncoupled terms are comparable, corresponding to a moderately, or loosely, coupled system, $\Gamma \sim 1$. When $\Gamma \gg 1$, the coupled interactions dominate the dynamics and the system should be considered strongly coupled. Figure 4 A–C illustrates how the energy landscape of a conformational process involving two structurally distinct degrees of freedom may change as Γ is increased. In the uncoupled case, each degree of freedom is independent, and there are two intermediate ensembles between the two endpoints (the upper-right and lower-left basins are considered the "endpoints"). At $\Gamma = 0.8$ the two intermediate ensembles become less populated, and it is more likely the system will move along both coordinates simultaneously. For $\Gamma = 6$, the two intermediates become highly destabilized and only two basins in the free energy landscape, the starting and ending configuration, are populated.

To relate such landscapes to experimental observables, trajectories of diffusive motion, similar which

are reminiscent, of to those that may be obtained from a single-molecule measurement may be generated via Brownian dynamics simulations, where in these simulations, the movement across the landscape is described by a "pseudo-particle" whose position evolves in time (Figure D–F). From such trajectories, the autocorrelation and cross-correlation functions of the variables can be calculated (Figure 4G–I). Here, both the distance trajectories (Figure 4D–F) and the correlation data (Figure 4G–I) reveal that movements along the two reaction coordinates are indeed independent for $\Gamma = 0$ and increasingly coupled as Γ increases. This illustrative example suggests that a reasonable experimental measure of the degree of coupling, Γ_{corr}, of a given process may be estimated by the ratio of the decay times for the cross-correlation function (τ_{cross}) and the autocorrelation function (τ_{auto}). When the degrees of freedom are uncoupled, τ_{cross} and Γ_{corr} are both zero. As coupling increases, τ_{cross} and τ_{auto} become comparable, and Γ_{corr} approaches 1. With continued advances in multi-color smFRET technologies, where multiple degrees of freedom can be probed simultaneously (Munro et al. 2010b), quantitative measures of the degree of coupling between any two degrees of freedom may be possible.

7. Energy landscapes and X-ray crystallography

At first glance, these dynamic views of ribosome function may appear at odds with the static structures obtained using x-ray crystallography and cryo-EM methods. However, if these two experimental approaches are considered in terms of the associated energy landscapes during data collection, the results of such studies are indeed consistent with ensemble descriptions.

For context, this discussion will focus on the process of cognate aa-tRNA accommodation, for which both structural (Villa et al., 2009; Schuette et al., 2009; Schmeing et al., 2009) and functional (Gromadski et al., 2004; Geggier et al., 2010; Whitford et al., 2010a) information have been recently obtained. Figure 5A shows a simplified schematic of a possible energy landscape for tRNA accommodation as it may occur in solution (black line).[5] In this context, there are many pos-

---

5 It should be noted that this schematic is not intended to be all-inclusive. That is, the landscape may possess substantially more detail (Geggier et al 2010 ; Whitford et al., 2010a), though for demonstrative purposes we used this overly-simplified representation.

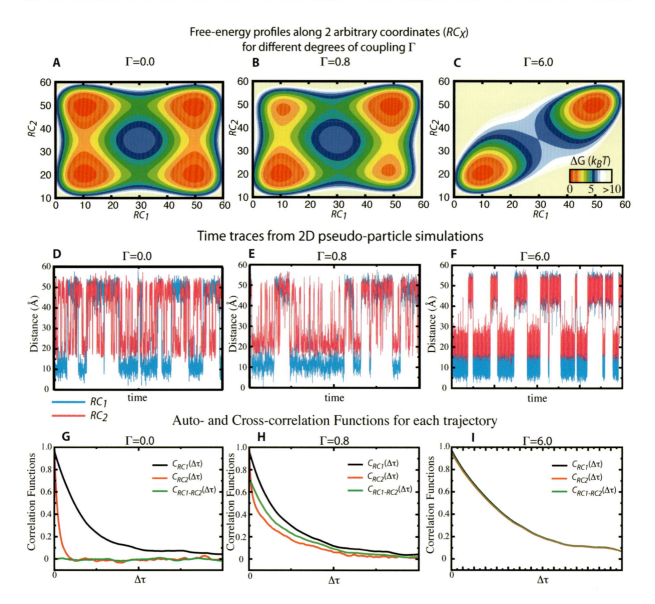

Fig. 4. Energy landscapes provide intuitive measures of coupling between coordinates. (A–C) Three energy landscapes with varied degrees of energetic coupling between the two reaction coordinates (Γ=0, 0.8 and 6). Each landscape is represented as the sum of two independent contributions to the landscape (one for each coordinate), each with two minima, plus an interaction term that harmonically couples the two coordinates (see main text). In (A), the coordinates are uncoupled (Γ=0), and the free-energy basins indicate it is equally likely to populate any combination of end states. That is, the free energy of the four basins is the same. Simulating diffusive motion of a "pseudo-particle" on this landscape (D) shows that movement along the two coordinates is independent, as verified by the nearzero cross-correlation function (G; labelled $C_{RC1\text{-}RC2}$).

When the energetic contribution of the coupling term is comparable (i.e., loosely coupled) to that of the uncoupled term (B) the intermediate basins are destabilized, (E) there is a visible correlation between the coordinates and (H) the timescale of decay of the cross correlation function approaches that of the autocorrelation functions for each coordinate (C_{RC1} and C_{RC2}). In the tightly-coupled case (C) there are no independent intermediate basins, the time traces of the coordinates (F) are visibly closely related and the cross- and auto-correlation functions are indistinguishable. While Γ provides a quantitative measure of the degree of coupling, in terms of energy, cross-correlation functions from multi-color smFRET can provide an experimental measure of coupling.

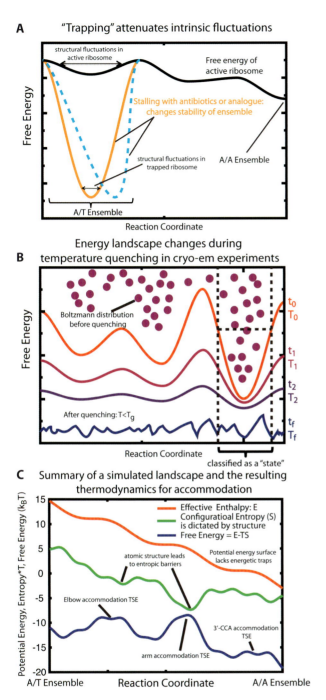

A "Trapping" attenuates intrinsic fluctuations

B Energy landscape changes during temperature quenching in cryo-em experiments

C Summary of a simulated landscape and the resulting thermodynamics for accommodation

Fig. 5. Energy landscapes in crystals, cryo-EM, and simulations. (A) When determining a crystal structure, it is necessary to "trap" the conformation. Energetically, this is represented as a significant decrease in the free energy of a particular ensemble (gold or dashed blue). Since the basin is not (infinitely deep and sharp, thermal fluctuations will lead to finite degrees of motion (arrows) inside the crystal. Certain fluctuations will be attenuated by lattice contacts between adjacent ribosomes in the crystal, and the energetic basin may shift (dashed blue). For example, because the L1 and L11 stalks are likely flexible (i. e. small energetic perturbations can significantly shift the distribution of configurations), the stalk ensembles will be affected by such interactions. (B) In a cryo-EM experiment, the pre-quenched ensemble follows a Boltzmann distribution. During quenching, if the glass transition temperature is reached $(T < T_g)$ faster than the reconfiguration time of the system, the final frozen ensemble will be a Boltzmann distribution at the original temperature $(T_X$ is the temperature at subsequent points in time t_X after quenching is initiated $t_X < t_{X+1})$. During data analysis, maximum-likelihood methods are used to identify "states". Future work should determine what a "classified state" is energetically, in energetic terms. (C) Functional forms of the energy landscape may be tested through simulation. When using an energetically "downhill" potential energy function for accommodation, steric barriers lead to highly-populated kinetic intermediates (Whitford et al., 2010a). Since these simulations displayed native-basin fluctuations that were consistent with anisotropic crystallographic analysis (Korostelev et al., 2008), explicit-solvent simulations, and normal mode calculations (Tama et al., 2003; Wang et al., 2004; Trylska et al., 2005), the coupling between the global fluctuations and the large-scale aa-tRNA movement into the ribosome could be characterized. Since these simulations exhibited similar tRNA-elbow movements as measured in smFRET (Geggier et al., 2010; Whitford et al., 2010a), reliable details of aa-tRNA acceptor arm accommodation and entry of the 3' CCA end into the PTC could be gleaned.

sible "effective perturbations" to the energy landscape upon dehydration and formation of crystal contacts in a lattice (depicted as gold and blue lines). Here, the effective perturbation may include the effects of agents, such as the antibiotics paromomycin and kirromycin, used to "trap" a particular conformation of the system. Collectively, these effects are depicted as increasing the depth of the A/T basin. While the depth of the basin may be increased by such perturbations, and possibly shifted (dashed blue line), the system is still subject

to energetic fluctuations arising from the surrounding environment. Even at the reduced temperatures employed during data acquisition, fluctuations of this nature can lead to structural fluctuations heterogeneities inside the crystal, which can be partially described by the B-factor. When the crystallizing conditions are sufficient to yield high-resolution scattering patterns, the energy basin is sharply defined and the distribution of the "trapped" ensemble becomes narrow enough for atomic resolution structure determination to be achieved. In such cases, the system is often referred to as a single "state". Nonetheless, the energy landscape framework necessarily applies and the system is simply a narrowly-defined ensemble, with reduced-amplitude motions.

While the ensemble nature of biological molecules has been rigorously demonstrated for proteins (Frauenfelder et al., 1979; 1991), where dynamics characterized as "forever kicking and screaming" have been noted (Weber 1975), this notion has only recently been exploited to study the motions inside ribosome crystals (Korostelev and Noller, 2007; Korostelev et al., 2008). Since each atom is subject to a unique set of interactions, the motions about the "trapped" basin are anisotropically distributed, both in direction and magnitude (Garcia, 1992; Garcia et al., 1997). Here, an analysis of the residual motions reveals that tRNA molecules fluctuate in the crystal lattice in directions consistent with the translocation reaction coordinate. These data support the notion that tRNAs remain in constant motion on the ribosome as suggested by single-molecule methods (Blanchard et al., 2004; Lee et al., 2007; Geggier et al., 2010; Whitford et al., 2010a) and that the effective perturbations of the crystal condition attenuate the overall magnitude of motions observed.

8. Energy landscapes and cryo-EM

Sample heterogeneity constitutes the most significant resolution-limiting factor when determining biomolecular structures, cryo-EM is also not immune to this difficulty (Spahn and Penczek, 2009). However, with the development of multiparticle methods of image processing it is now possible to directly visualize instrinsic conformational heterogeneity within ribosomal complexes (Klaholz et al., 2004; Schuette et al., 2009; Agirrezabala et al., 2008; Julian et al., 2008). By considering the relationship between temperature, the free-energy landscape and glass-like dynamics, one may understand the effects of "plunging" in terms of the ensemble descriptions presented here (Figure 5 B). Prior to freezing the sample in vitreous ice (quenching), the molecules under investigation are distributed among an ensemble of states according to the nature of the energy landscape. During quenching, the system responds to the perturbation and the energy landscape deforms in a time-dependent fashion. The rate and magnitude of the changes in the energy landscape depend on both the entropy of each configuration in the system (Wales, 2004) as well as the contribution of energetic interactions in each conformation. These energetic interactions can also be altered, such as the degree of screening of electrostatic interactions, which depends on the temperature-dependent dielectric constant of the solution (Landau et al., 1984). If quenching is sufficiently rapid, such that the glass transition temperature, T_g (Williams et al., 1955), is reached faster than the overall reconfiguration time of the system, then the ensemble of frozen molecules closely approximates the Boltzmann distribution of states in the system prior to freezing.

After freezing and imaging, multiparticle data processing methods enable sorting of the particle images and result in not one, but a set of reconstructions. However, while each reconstructions is often referred to as a "state", it represents the average of an ensemble of configurations, centered about a particular energetic minimum. High conformational flexibility of parts of a ribosomal complex can explain why certain parts of the density map may appear fragmented or may only be observed at low contour levels. For example, in the A/T state of the ribosome trapped during the accommodation process using the antibiotic kirromycin (Villa et al., 2009; Schuette et al., 2009), the authors found that the switch 1 helix of EF-Tu, a key structural element involved in GTPase activation, was in a disordered, dynamic state during the switch from the GTP form to the GDP form.

As recent landmark investigations of spontaneous reverse translocation have demonstrated (Fischer et al., 2010), the development of time-resolved cryo-EM techniques may be used to report on the *pmf* along a reaction coordinate. In the regime where over a million images can be obtained (Ratje et al., 2010), probability distributions can be calculated by projecting each image onto a distinct reaction coordinate. When one has thousands of counts per bin, the noise due to sampling should be on the order of $1/\sqrt{N} \approx 3\%$, assuming the uncertainty in the measured population scales as $\sqrt{N}$. Since the *pmf* is proportional to $\ln(P)$, an uncertainty in P by a factor of 1.03 corresponds to a difference in the *pmf* of $k_B T(\ln(1.03P) - \ln(P)) = 0.03\,k_B T$. The major caveat with such a venture is that, while the statistical noise due to sampling may be low, the uncertainty in the projection onto the coordinate and the noise present in each image can lead to much larger effects (see Figure 2). Similar to smFRET, characterizing the statistical properties of each of the reaction coordinate projections represents a substantial challenge towards using cryo-EM to reveal fine details of the energy landscape.

9. Exploring energy landscapes through simulation

As petaflop supercomputing (i. e., 10^{15} calculations per second) can now be achieved, molecular dynamics simulations of the 70S (and, in principle, 80S) ribosome functions are now accessible (Sanbonmatsu et al., 2005, Whitford et al., 2010a). With continued technical advances, as well as the development of new computational approaches, it is worthwhile to assess current computing practices and identify how each approach may shed light on the details of the energy landscape. Here, the general approaches, considerations and limitations in simulating ribosome functions are outlined in order for the reader to assess the field's directions and capabilities.

The first consideration in a simulation is to choose which force field (potential energy function) is appropriate for the questions of interest. In semi-empirical force fields, such as AMBER[6] (Pearlman et al., 1995), one uses a classical potential energy function, where interaction energies of the four nucleotides and twenty amino acids are calibrated to experimental quantities or to quantum mechanical calculations. These same energies are applied to any RNA or protein simulated and the system is immersed in a box of water molecules and excess ions. This approach often provides information about rapid (sub-microsecond) movement on the landscape. An alternative approach is to employ potential energy functions with pre-defined properties, such that the robust dynamical properties may be identified and tested against experiments. This second approach allows for much longer timescales (milliseconds) and can identify which properties of the ribosome's free energy landscape are most important in determining function.

The length of the simulation, the employed force field, and the process of interest must all be considered when determining what aspects of the landscape are being probed in each computational study. Here, we discuss recent simulations using semi-empirical force fields to illustrate how the timescale of each simulation allows for specific insights into the character of the landscape. Using enhanced sampled methods (Sanbonmatsu, 2006a; Vaiana and Sanbonmatsu, 2009; Garcia and Sanbonmatsu, 2000) thermodynamic convergence (i. e., calculated thermodynamic quantities do not change with additional sampling) of 16S decoding center dynamics with and without a ligand bound were achieved in 17,000 and 2,000 nanoseconds, respectively. These thermodynamic measurements showed that a stochastic gating mechanism of A1492/A1493 is likely a low-energy process, where there is a balance of motions of the A1492 and A1493 bases into and out of helix 44 during aminoglycoside recognition (Sanbonmatsu, 2006a). While simulations of the full 70S ribosome (~3 million atoms) are currently too expensive to measure

thermodynamics directly, the relaxation time of local fluctuations in tRNA position were measured to be on timescales of ~10 ns (Whitford et al., 2010b; 205–301 ns per simulation), which characterizes how tRNAs diffuse across the landscape. Simulations of three-way junctions (Besseová et al., 2010; 30–100 ns per simulation) demonstrated that the landscapes local to the crystal configurations are not deep energetic basins. Rather, rapid local reconfigurations on the landscape allow for flexibility, which is likely relevant during translation. Similar simulations with the CHARMM forcefield[7] (Brooks et al., 1983) of the nascent TnaC peptide inside the 23S tunnel (Trabuco et al., 2010; 120 ns per simulation) suggest that the interactions formed on short timescales (tens of nanoseconds) contribute to the arrested conformation being a deep minimum on the energy landscape.

The second approach to simulating ribosome landscapes is to employ effective potential energy functions that reflect principles suggested by ribosome experiments and physical arguments. Agreement between the resulting dynamics and experiments partially validates the proposed energy landscape. By varying the parameters of the simulation, one can also characterize features of the process that are robust and features that should be sensitive to the specific environmental conditions. Since the features of the potential energy surface are explicitly defined, effective perturbations may be introduced to understand how the ribosome responds to changes in the cellular environment.

Using effective potentials, the idea that accommodation is a downhill potential energy process was tested through simulations (Whitford et al., 2010a). The classical conformation was defined as the minimum of the potential energy function. By construction, the starting (A/T) configuration is higher in potential energy, which is consistent with the notion of tRNA molecules acting as molecular springs (Frank et al., 2005) that must be properly aligned to enter the A site (Sanbonmatsu, 2006b). Figure 5 C shows a schematic of the effective potential energy function used in that study. Since the configurational entropy is determined by the accessible phase space, and the potential energy was completely devoid of energetic roughness, kinetic intermediates were the result of changes in configurational entropy during accommodation. Since the kinetic proofreading mechanism (Hopfield, 1974)

6   It should be noted that, for nucleic acids, the AMBER forcefield is generally regarded (by the computational community) as the most well-developed potential for empirical simulations.

7   The vast majority of CHARMM development has focused on the parameterization of amino acids, and not nucleic acids.

requires barriers to be associated with the selection process, it was anticipated that entropic/steric barriers would exist and serve to reject non-cognate aa-tRNAs. While enthalpic interactions may drive the aa-tRNA into the ribosome, the simulation data suggested that the loss of configurational entropy resists accommodation. This entropic penalty resulted in reversible accommodation attempts and a multi-step accommodation process, much like what was observed through smFRET.

10. Future outlook

Understanding the mechanisms of translation control is expected to be a critical area of research in the coming decade. As the ribosome is the central component of the translation apparatus and an integration point for regulation, the implementation of quantitative tools to explore the molecular determinants of ribosome function in greater depth will be vital to this pursuit. The arsenal of structural, kinetic and computational methods already available for interrogating intrinsic conformational changes in the system and their relationship to function, have provided important new insights for the road ahead. The energy landscape underpinning ribosome conformational events has been shown to be sensitive to buffer conditions, tRNA ligands, translation factors and antibiotic inhibitors that bind directly to ribosomal RNA. These early insights speak to the inherently metastable nature of the ribosome's architecture, foreshadowing the existence of substantial distinctions in the energetics of ribosome function across domains of life and the potential to understand gene-specific regulatory events at the molecular scale. Such events include nascent chain stalling, stop codon readthrough, frameshifting as well as other dynamic recoding processes (Gesteland and Atkins, 1996).

The impact of single-molecule, cryo-EM, and theoretical/computational methods to advance our understanding of the mechanism of protein synthesis remains nascent. The implementation of multicolor FRET experiments, integrated platforms for simultaneous force and FRET measurements, high-throughput strategies, as well as sorting algorithms to further explore heterogeneities in the ensemble of particles are certain to provide access to numerous, untold discoveries. Continued advancement of the energy landscape framework of ribosome function through simulation will help provide a rigorous connection to an all-atom, molecular view of the system. As simulations are extending to experimentally attainable timescales (milliseconds), atomic simulations of translation processes will enable a direct integration of structural, computational and kinetic fields. Such efforts should aim to yield testable hypotheses about specific trajectories of motion. Experimentally validated trajectories may ultimately provide unforeseen insights into the basic translation mechanism as well as evolutionary distinctions that may exist between the translation apparatus of each organism. Such insights are expected to enrich our knowledge and understanding of cellular and therapeutic strategies for translation control.

References

Agirrezabala X, Lei J, Brunelle JL, Ortiz-Meoz RF, Green R, Frank J (2008) Visualization of the hybrid state of tRNA binding promoted by spontaneous ratcheting of the ribosome. Mol Cell 32: 190–197

Besseová I, Réblová K, Leontis NB, Sponer J (2010) Molecular dynamics simulations sugg est that RNA three-way junctions can act as flexible RNA structural elements in the ribosome. Nuc Acid Res 38:6247–6264; DOI:101093/nar/gkq414

Best RB, Hummer G (2010) Coordinate-dependent diffusion in protein folding. Proc Natl Acad Sci USA 19: 1088–1093

Blanchard SC, Gonzalez RL, Kim HD, Chu S, Puglisi JD (2004a) tRNA selection and kinetic proofreading in translation. Nat Struct Mol Bio 11: 1008–1014

Blanchard SC, Kim HD, Gonzalez RL Jr, Puglisi JD, Chu S (2004b) tRNA dynamics on the ribosome during translation. Proc Natl Acad Sci USA 101: 12893–12898

Blanchard SC (2009) Single-molecule observations of ribosome function. Curr Op Struct Bio 19: 103–109

Brooks BR, Bruccoleri RE, Olafson BD, States DJ, Swaminathan S, Karplus M (1983) CHARMM: A program for macromolecular energy, minimization, and dynamics calculations. J Comp Chem 4: 187–217

Bryngelson JD, Wolynes PG (1989) Intermediates and barrier crossing in a random energy model (with applications to protein folding). J Phys Chem 93: 6902–6915

Cho SS, Levy Y, Wolynes PG (2006) P versus Q: structural reaction coordinates capture protein folding on smooth landscapes. Proc Natl Acad Sci USA 103: 586: 591

Cornish PV, Ermolenko DN, Noller HF, Ha T (2008) Spontaneous intersubunit rotation in single ribosomes. Mol Cell 30: 578–588

Cornish P, Ermolenko D, Staple D, Hoang L, Hickerson R, Noller H, Ha T (2009) Following movement of the L1 stalk between three functional states in single ribosomes. Proc Natl Acad Sci USA106: 2571–2576

Das P, Moll M, Stamati H, Kavraki LE, Clementi C (2006) Low-dimensional, free-energy landscapes of protein-folding reactions by nonlinear dimensionality reduction. Proc Natl Acad Sci USA 103: 9885–9890

Demeshkina N, Jenner L, Yusupova G, Yusupov M (2010) Interactions of the ribosome with mRNA and tRNA. Curr Op Struct Biol 20: 325–332

Dorner S, Brunelle JL, Sharma D, Green R (2006) The hybrid state of tRNA binding is an authentic translation elongation intermediate. Nat Struct Mol Biol 13: 234–241

Dudko OK, Hummer G, Szabo A (2006) Intrinsic rates and activation free energies from single-molecule pulling experiments. Phys Rev Lett 96: 108101

Faradjian AK, Elber R (2004) Computing time scales from reaction coordinates by milestoning. J Chem Phys 120: 10880–10889

Fei J, Kosuri P, MacDougall DD, Gonzalez RL Jr. (2008) Coupling of ribosomal L1 stalk and tRNA dynamics during translation elongation. Mol Cell 30: 348–359

Fei J, Bronson JE, Hofman JM, Srinivas RL, Wiggins CH, Gonzalez RL (2009) Allosteric collaboration between elongation factor G and the ribosomal L1 stalk directs tRNA movement during translation. Proc Natl Acad Sci USA 106: 15702–15707

Fischer N, Konevega AL, Wintermeyer W, Rodnina MV, Stark H (2010) Ribosome dynamics and tRNA movement by time-resolved electron cryomicroscopy. Nature 466: 329–333

Fluitt A, Pienaar E, Viljoen H (2007) Ribosome kinetics and aatRNA competition determine rate and fidelity of peptide synthesis. Comp Bio Chem 31: 335–346

Frank J, Sengupta J, Gao H, Li W, Valle M, Zavialov A, Ehrenberg M (2005) The role of tRNA as a molecular spring in decoding, accommodation, and peptidyl transfer. FEBS Lett 579: 959–962

Frank J, Spahn CM (2006) The ribosome and the mechanism of protein synthesis. Rep Prog Phys 69: 1383–1417

Frank J, Gao H, Sengupta J, Gao N, Taylor DJ (2007) The process of mRNA-tRNA translocation. Proc Natl Acad Sci USA 104: 19671–19678

Frauenfelder H, Petsko GA, Tsernoglou D (1979) Temperature-dependent x-ray-diffraction as a probe of protein structural dynamics. Nature 280: 558–563

Frauenfelder H, Sligar SG, Wolynes PG (1991) The energy landscapes of motions of proteins. Science 254: 1598–1603

Gao Y-G, Selmer M, Dunham CM, Weixlbaumer A, Kelley AC, Ramakrishnan V (2009) The structure of the ribosome with elongation factor G trapped in the posttranslocational state. Science 326: 694–699

Garcia AE (1992) Large-amplitude nonlinear motions in proteins. Phys Rev Lett 68: 2696–2699

Garcia AE, Onuchic JN (2003) Folding a protein in a computer: An atomic description of the folding/unfolding of protein A. Proc Natl Acad Sci USA 13898–13903

Garcia AE, Krumhansl JA, Frauenfelder H (1997) Variations on a theme by Debye and Waller: From simple crystals to proteins. Prot Struct Func Gen 29: 153–160

Gavrilova LP, Kostiashkina OE, Koteliansky VE, Rutkevitch NM, Spirin AS (1976) Factor-free ("non-enzymic") and factor-dependent systems of translation of polyuridylic acid by *Escherichia coli* ribosomes. J Mol Biol 101: 537–552

Geggier P, Dave R, Feldman MB, Terry DS, Altman RB, Munro JB, Blanchard SC (2010) Conformational sampling of aminoacyl-tRNA during selection on the ribosome. J Mol Biol 399: 576–595; DOI:101016/j. jmb.2010.04038

Gesteland RF, Atkins JF (1996) Recoding: dynamic reprogramming of translation. Annu Rev Biochem 65: 741–68

Gromadski KB, Rodnina MV (2004) Kinetic determinants of high-fidelity tRNA discrimination on the ribosome. Mol Cell 13: 191–200

Hyeon C, Onuchic JN (2007) Mechanical control of the directional stepping dynamics of the kinesin motor. Proc Natl Acad Sci USA 104: 17382–17387

Hopfield JJ (1974) Kinetic proofreading: A new mechanism for reducing errors in biosynthetic processes requiring high specificity. Proc Natl Acad Sci USA 71: 4135–4139

Johansson M, Bouakaz E, Lovmar M, Ehrenberg M (2008) The kinetics of ribosomal peptidyl transfer revisited. Mol Cell 30: 589–598

Julian P, Konevega AL, Scheres SH, Lazaro M, Gil D, Wintermeyer W, Rodnina MV, Valle M (2008) Structure of ratcheted ribosomes with tRNAs in hybrid states. Proc Natl Acad Sci USA 105: 16924–16927

Klaholz BP, Myasnikov AG, van Heel M (2004) Visualization of release factor 3 on the ribosome during termination of protein synthesis. Nature 427: 862–865

Klepeis JL, Lindorff-Larsen K, Dror RO, Shaw DE (2009) Long-timescale molecular dynamics simulations of protein structure and function. Curr Opin Struct Biol 19: 120–127

Komatsuzuki T, Hoshino K, Matsunaga Y, Rylance GJ, Johnston RL, Wales DJ (2005) How many dimensions are required to approximate the potential energy landscape of a model protein? J Chem Phys 122: 084714

Korostelev A, Noller HF (2007) Analysis of structural dynamics in the ribosome byTLS crystallographic refinement. J Mol Biol 373: 1058–1070

Korostelev A, Asahara H, Lancaster L, Laurberg M, Hirschi A, Zhu J, Trakhanov S, Scott WG, Noller HF (2008) Crystal structure of a translation termination complex formed with release factor RF2. Proc Natl Acad Sci USA 105: 19684–19689

Landau LD, Lifshitz EM, Pitaevskii LP (1984) Electrodynamics of continuous media, 2nd ed. Reed Educational and Professional Publishing, Oxford

Lee T-H, Blanchard SC, Kim HD, Puglisi JD, Chu S (2007) The role of fluctuations in tRNA selection by the ribosome. Proc Natl Acad Sci USA 104: 13661–13665

Leopold PE, Montal M, Onuchic JN (1992) Protein folding funnels-A kinetic approach to the sequence structure relationship. Proc Natl Acad Sci USA 89: 8721–8725

Lu Q, Wang J (2009) Kinetics and statistical distributions of single-molecule conformational dynamics. J Phys Chem B 113: 1517–1521

Marshall RA, Dorywalska M, Puglisi JD (2008) Irreversible chemical steps control intersubunit dynamics during translation. Proc Natl Acad Sci USA 105: 15364–15369

Miyashita O, Onuchic JN, Wolynes PG (2003) Nonlinear elasticity, proteinquakes, and the energy landscapes of functional transitions in proteins. Proc Natl Acad Sci USA 100: 12570–12575

Munro JB, Altman RB, O'Connor, Blanchard SC (2007) Identification of two distinct hybrid-state intermediates on the ribosome. Mol Cell 25: 505–517

Munro JB, Sanbonmatsu KY, Spahn CMT, Blanchard SC (2009) Navigating the ribosome's metastable energy landscape. Trends Biochem Sci 34: 390–400

Munro JB, Altman RB, Tung C-S, Sanbonmatsu KY, Blanchard SC (2010a) A fast dynamic mode of EF-G-bound ribosome. EMBO J 29: 770–781

Munro JB, Altman RB, Tung C-S, Cate JDH, Sanbonmatsu KY, Blanchard SC (2010b) Spontaneous formation of the unlocked state of the ribosome is a multistep process. Proc Natl Acad Sci USA 107: 709–714

Nettels D, Gopich IV, Hoffman A, Schuler B (2007) Ultrafast dynamics of protein collapse from single-molecule photon statistics. Proc Natl Acad Sci USA 104: 2655–2660

Nettels D, Hoffmann A, Schuler B (2008) Unfolded protein and peptide dynamics investigated with single-molecule FRET and correlation spectroscopy from picoseconds to seconds. J Phys Chem B 112: 6137–6146

Nymeyer H, Socci ND, Onuchic JN (2000) Landscape approaches for determining the ensemble of folding transition states:

Success and failures hinge on the degree of frustration. Proc Natl Acad Sci USA 97: 634–639

Oliveira RJ, Whitford PC, Chahine J, Wang J, Onuchic JN, Leite VBP (2010) Exploring the origin of non-monotonic complex behavior and the effects of non-native interactions on the diffusive properties of protein folding. Biophys J (in press)

Onuchic JN, Wolynes PG (1993) Energy landscapes, glass transitions, and chemical reaction dynamics in biomolecular or solvent environment. J Chem Phys 98: 2218–2224

Onuchic JN, Nymeyer H, Garcia AE, Chahine J, Socci ND (2000) The energy landscape theory of protein folding: Insights into folding mechanisms and scenarios. Adv Protein Chem 53: 87–152

Onuchic JN, Kobayashi C, Miyashita O, Jennings P, Baldridge KK (2006) Exploring biomolecular machines: energy landscape control of biological reactions. Philos Trans Royal Soc 361: 1439–1443

Orzechowski M, Tama F (2008) Flexible fitting of high-resolution X-ray structures into cryo-electron microscopy maps using biased molecular dynamics simulations. Biophys J 95: 5692–5705

Pape T, Wintermeyer W, Rodnina M (1999) Induced fit in initial selection and proofreading of aminoacyl-tRNA on the ribosome. EMBO J 18: 3800–3807

Pearlman DA, Case DA, Caldwell JW, Ross WS, Cheatham TE III, DeBolt S, Ferguson D, Seibel G, Kollman P (1995) AMBER, a package of computer programs for applying molecular mechanics, normal mode analysis, molecular dynamics and free energy calculations to simulate the structural and energetic properties of molecules. Comp Phys Commun 91: 1–41

Penczek PA, Frank J, Spahn CMT (2006) A method of focused classification, based on the bootstrap 3D variance analysis, and its applications to EF-G-dependent translocation. 154: 184–194

Peng C, Zhang L, Head-Gordon T (2010) Instantaneous normal modes as an unforced reaction coordinate for protein conformational transitions. Biophys J 98: 2356–2364

Pérez A, Marchán I, Svozil D, Sponer J, Cheatham TE III, Laughton CA, Orozco M (2007) Refinement of the AMBER force field for nucleic acids: Improving the description of alpha/gamma conformers. Biophys J 92: 3817–3829

Pincus DL, Cho SS, Hyeon C, Thirumalai D (2009) Minimal models for proteins and RNA: From folding to function. In: Molecular biology of protein folding, Vol 84, pp 203–250. Elsevier Academic, San Diego, CA

Ratje AH, Loerke J, Mikolajka A, Brünner M, Hildebrand PW, Starosta A, Doenhoefer A, Connel SR, Fucini P, Mielke T, Whitford PC, Onuchic JN, Yu Y, Sanbonmatsu KY, Hartmann RK, Penczek PA, Wilson DN, Spahn CMT (2010) Head swivel on the ribosome facilitates translocation via intra-subunit tRNA hybrid sites. Nature 468: 713–716 (under review)

Roberts E, Sethi A, Montoya J, Woese CR, Luthey-Schulten Z (2008) Molecular signatures of ribosomal evolution. Proc Natl Acad Sci USA 105: 13953–13958

Roberts RW, Eargle J, Luthey-Schulten Z (2010) Experimental and computational determination of tRNA dynamics. FEBS Lett 584: 376–386

Rodnina MV, Wintermeyer W (2001) Fidelity of aminoacyl-tRNA selection on the ribosome: Kinetic and structural mechanisms. Annu Rev Biochem 70: 415–435

Sanbonmatsu KY (2006a) Energy landscape of the ribosomal decoding center. Biochimie 88: 1053–1059

Sanbonmatsu KY (2006b) Alignment/misalignment hypothesis for tRNA selection by the ribosome. Biochimie 88: 1075–1089

Sanbonmatsu KY, Tung C-S (2007) High performance computing in biology: Multimillion atom simulations of nanoscale systems. J Struct Bio 157: 470–480

Schmeing TM, Voorhees RM, Kelley AC, Gao Y-G, Murphy FV, Weir JR, Ramakrishnan V (2009) The crystal structure of the ribosome bound to EF-Tu and aminoacyl-tRNA. Science 326: 688–694

Schuette J-C, Murphy FC, Kelly AC, Weir JR, Geisebrecht J, Connell SR, Loerke J, Mielke T, Zhang W, Penczek PA, Ramakrishnan V, Spahn CMT (2009) GTPase activation of elongation factor EF-Tu by the ribosome during decoding. EMBO J 28: 1–11

Schuler B, Lipman EA, Eaton WA (2002) Probing the free-energy surface for protein folding with single-molecule fluorescence spectroscopy. Nature 419: 743–748

Sievers A, Beringer M, Rodnina MV, Wolfenden R (2004) The ribosome as an entropy trap. Proc Natl Acad Sci USA 101: 7897–7901

Spahn CMT, Penczek PA (2009) Exploring conformational modes of macromolecular assemblies by multiparticle cryo-EM. Curr Opin Struct Biol 19: 623–631

Spirin AS (2009) The ribosome as a conveying thermal ratchet machine. J Biol Chem 284: 21103–21119

Stanley RE, Blaha G, Grodzicki RL, Strickler MD, Steitz TA (2010) The structures of the anti-tuberculosis antibiotics viomycin and capreomycin bound to the 70S ribosome. Nat Struct Mol Biol 17: 289–293

Tama F, Valle M, Frank J, Brooks CL III (2003) Dynamic reorganization of the functionally active ribosome explored by normal mode analysis and cryo-electron microscopy. Proc Natl Acad Sci USA 100: 9319–9323

Tang J, Kang S-G, Saven, JG, Gai F (2009) Characterization of the cofactor-induced folding mechanism of a zinc-binding peptide using computationally designed mutants. J Mol Biol 389: 90–102

Trabuco LG, Harrison CB, Schreiner E, Schulten K (2010) Recognition of the regulatory nascent chain TnaC by the ribosome. Structure 18: 627–637

Thirumalai D, Hyeon C (2005) RNA and protein folding: Common themes and variations. Biochem 44: 4957–4970

Thirumalai D, O'Brien EP, Morrison G, Hyeon C (2010) Theoretical perspectives on protein folding. Annu Rev Biophys 39: 159–183

Trylska J, Tozzini V, McCammon JA (2005) Exploring global motions and correlations in the ribosome. Biophys J 89: 1455–1463

Vaiana AC, Sanbonmatsu KY (2009) Stochastic gating and drug-ribosome interactions. J Mol Biol 386: 648–661

Vanheel M, Frank J (1981) Use of multivariate statistics in analyzing the images of biological macromolecules. Ultramicroscopy 6: 187–194

Villa E, Sengupta J, Trabuco L, LeBarron J, Baxter WT, Shaikh TR, Grassucci RA, Nissen P, Ehrenberg M, Schulten K, Frank J (2009) Ribosome-induced changes in elongation factor Tu conformation control GTP hydrolysis. Proc Natl Acad Sci USA 106: 1063–1068

Wales DJ (1984) Energy Landscapes: Applications to clusters, biomolecules and glasses. Cambridge University Press, Cambridge

Wang J (2003) Statistics, pathways and dynamics of single molecule protein folding. J Chem Phys 118: 952–958

Wang Y, Rader AJ, Bahar I, Jernigan RL (2004) Global ribosome motions revealed with elastic network model. J Struct Biol 147: 302–314

Weber G (1975) Energetics of ligand binding to proteins. Adv Protein Chem 29: 1–83

Wekselmen I, Davidovich C, Agmon I, Zimmerman E, Rozenberg H, Bashan A, Birisio R, Yonath A (2009) Ribosome's mode of function: myths, facts and recent results. J Pept Sci 15: 122–130

Whitford PC, Miyashita O, Levy Y, Onuchic JN (2007) Conformational transitions of adenylate kinase: switching by cracking. J Mol Biol 366: 1661–1671

Whitford PC, Noel JK, Gosavi S, Schug A, Sanbonmatsu KY, Onuchic JN (2009a) An all-atom structure-based potential for proteins: Bridging minimal models with all-atom empirical forcefields. Prot Struc Func Bioinfo 75: 430–441

Whitford PC, Schug A, Saunders J, Hennelly SP, Onuchic JN, Sanbonmatsu KY (2009b) Nonlocal helix formation is key to understanding S-Adenosylmethionine-1 riboswitch function. Biophys J 96: L7-L9

Whitford PC, Geggier P, Altman RB, Blanchard SC, Onuchic JN, Sanbonmatsu KY (2010a) Accommodation of aminoacyl-tRNA into the ribosome involves reversible excursions along multiple pathways. RNA 16: 1196–1204; DOI: 101261/rna.2 035410

Whitford PC, Onuchic JN, Sanbonmatsu KY (2010b) Connecting energy landscapes with experimental rates for aminoacyl-tRNA accommodation in the ribosome J Amer Chem Soc 132: 13170–13171 (submitted)

Williams ML, Landel RF, Ferry JD (1955) Mechanical properties of substances of high molecular weight. 19. The temperature dependence of relaxation mechanisms in amorphous polymers and other glass-forming liquids. J Amer Chem Soc 77: 3701–3707

Wong V, Case DA (2008) Evaluating rotational diffusion from protein MD simulations. J Phys Chem B 112: 6013–6024.

Yang S, Roux B (2008) Src kinase conformational activation: thermodynamics, pathways, and mechanisms. PLOS Comp Biol 4: e1000047

Yusupov MM, Yusupova GZ, Baucom A, Leiberman K, Earnest TN, Cate JH, Noller HF (2001) Crystal structure of the ribosome at 5.5 Å. Science 292: 883–896

Zhang G, Feyunin I, Miekley O, Valleriani A, Moura A, Ignatova Z (2010) Global and local depletion of ternary complex limits translation elongation. Nuc Acid Res 38: 4778–4787

Zhang W, Dunkle A, Cate JHD (2009) Structures of the ribosome in intermediate states of ratcheting. Science 325: 1014–1017

Zwanzig R (1988) Diffusion in a rough potential. Proc Natl Acad Sci USA 85: 2029–2030

Ribosome dynamics: Progress in the characterization of mRNA-tRNA translocation by cryo-electron microscopy

25

Joachim Frank

1. Introduction

The ribosome is a highly complex molecular machine performing protein synthesis in all forms of life with an amazing degree of accuracy. Knowledge of its atomic structure, the result of pioneering work now honored by the award of the Nobel Prize, has prepared us for the next stage of inquiry, with the focus on the dynamics of the system. Key to understanding the mechanism of protein synthesis is provided by experimental data informing us about the ribosome's conformational changes and dynamic interactions with its functional ligands, mRNA, tRNA, EF-G and EF-Tu during the elongation cycle.

The composition of the ribosome from two subunits dedicated to the two primary functional activities, decoding and catalysis of peptidyl transfer, respectively, early on led to the speculation that the subunits move relative to each other during the elongation process (Bretscher, 1968; Spirin, 1969). Experimental evidence that the ribosome structure changes dynamically during this cyclic process goes back to Spirin and coworkers (1987), who observed changes in the particle's compactness. Another clue came from a study employing site-specific Pb^{2+} cleavage (Polacek et al., 2000). Three-dimensional reconstructions providing direct structural evidence of conformational changes were obtained by cryo-EM (Agrawal et al., 1998; 1999; Frank and Agrawal, 2000; Stark et al., 2000) prior to, or coincident with, the emergence of X-ray crystal structures (Ban et al., 2000; Harms et al., 2000; Wimberly et al., 2000 ;Yusupov et al., 2001). These observations, increasingly more detailed in the past decade (reviewed in Agirrezabala and Frank, 2009), indicated that intersubunit movement is only one of several dynamic changes affecting individual domains such as head, platform, and shoulder of the small subunit and L1

stalk, head domain, L7/L12 stalk, and L11 stalk of the large subunit. Each of these domain movements can be linked to subprocesses of the elongation cycle, such as factor binding and tRNA transport, and lead us to postulate that the different sizes of these domains, and the mechanical properties of associated hinges must be related to the differences in observed kinetic rates (see Savelsbergh et al., 2003).

The past decade of research into ribosomal dynamics has benefited from several developments: (i) availability of X-ray crystal structures for the bacterial ribosome (Cate et al., 2001) and its subunits (Ban et al., 2000; Harms et al., 2000; Wimberly et al., 2000); (ii) improvement of resolution of cryo-EM from the range of 12–15 Å (Gabashvili et al., 2000; Malhotra et al., 1998) to 5–7 Å (LeBarron et al., 2008; Seidelt et al., 2010; Villa et al., 2009); (iii) development and application of flexible fitting (Gao et al., 2003; Tama et al., 2004; Trabuco et al., 2008); (iv) development of powerful variance estimation and classification methods (Liao et al., 2010; Penczek et al. 2006; Scheres et al., 2007; Spahn and Penczek, 2009; Zhang et al., 2008); and (v) the development of the single-molecule FRET technique and its application to the ribosome (see Munro et al., 2009a; Frank and Gonzalez, 2010). While these developments represent a tremendous accumulation of experimental data and refinement of analytical tools, there has been a parallel effort to understand the ribosome as a complex molecular machine by molecular dynamics simulations either of its components (e. g. Li et al., 2006; Reblova et al., 2009) or, in the most ambitious feat, of the ribosome in its entirety (Sanbonmatsu, 2005).

As a consequence of these developments, the ribosome is now perceived as a thermally driven molecular machine with numerous degrees of freedom; specifically, it is considered a Brownian ratchet (Garai et al., 2009; Frank and Gonzalez, 2010; Moran et al., 2008;

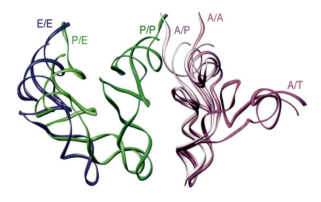

Fig. 1 The tRNA on its way through the ribosome. The different positions and conformations of the tRNA as it goes from the A/T to the E/E state. All classical positions (A/A, P/P, E/E) are from X-ray studies, and all hybrid positions (A/T, A/P, P/E) are obtained by flexible fitting of cryo-EM maps. Not shown are intermediate positions most recently discovered by cryo-EM and classification of pre-translocational complexes mentioned in the text. (Reproduced from Agirrezabala and Frank, 2009; © Cambridge University Press 2009)

Savelsbergh et al., 2003; Spirin, 2009), built in such a way that its normal motions are conducive to productive engagement of its parts and ligands (Frank et al., 2007; Tama et al., 2003). tRNA, one of the preeminent ligands, a molecule bestowed with extraordinary flexibility, goes through a large variety of conformations as it traverses the intersubunit space (Figure 1). Recent speculations seek to understand the actions of the postulated Brownian ratchet in greater detail, and center around the nature and action of the pawl (C. M. T. Spahn, personal communication; Zhang et al., 2010). The following is a brief account of these developments in the past three to four years, with particular focus on the mRNA-(tRNA)$_2$ translocation process.

2. Ribosomal dynamics – complementary aspects explored by smFRET and cryo-EM

Characterization of the dynamics of the actively translating ribosome during its work cycle has been from two primary sources: cryo-EM and single-molecule fluorescence resonance energy transfer (smFRET). Cryo-EM yields three-dimensional density maps representing ensemble averages that each characterizes the ribosome's conformation in a relatively stable, sufficiently populated state. Single-molecule FRET (smFRET), on the other hand, which came on the ribosome stage less than a decade ago with a pioneering study by Blanchard and coworkers (2004), provides informa-

tion on intramolecular distances, measured between strategically placed donor and acceptor fluorophores, and hence is able to report on those molecules that interconvert from one state to another – molecules, in other words, that are in transition between states and are poorly or not at all captured by cryo-EM. While the ribosome itself might be relatively stable and thus visible as a sturdy mass, ligands in motion and thus frozen in different conformations within an ensemble will be averaged into a blur. Thus the two techniques have complementary aspects which, when combined, give a more comprehensive view on the dynamics than either technique taken by itself.

One important development in the past few years has been that cryo-EM and smFRET studies have been increasingly engaged in a dialogue. Discoveries in cryo-EM such as the ratchet-like intersubunit rotation were confirmed and further analyzed as a stochastic process by bulk FRET (Ermolenko et al., 2007) or smFRET. There are examples for the opposite direction of the discovery process: Agirrezabala et al. (2008) and Julian et al. (2009), going by clues from smFRET (Kim et al., 2007; Cornish et al., 2008), discovered subpopulations in samples of a factor-free pre-translocation complex that were in two different states of rotation and tRNA binding positions. Two recent reviews make an effort to elaborate on the complementarity of the techniques alluded to above and attempt to provide a conceptual bridge (Frank and Gonzalez, 2010; Munro et al., 2009a). Schematic depictions of the free-energy landscape (Figure 2) help in this conceptualization and in the planning of future experiments. Actual free-energy estimates are available from lifetime data supplied by smFRET or, alternatively, from the observed ratios of subpopulations in cryo-EM datasets found by classification.

Given the sensitivity of translation to parameters such as temperature, concentration of Mg^{2+} and other ions, and presence/composition of polyamines (see Gromadski et al., 2006), the point has been made that relating observations from smFRET to those from cryo-EM requires the exact same samples to be analyzed (Frank and Gonzalez, 2010). Thus, interdisciplinary collaborations are necessary to derive a more meaningful, quantitative interpretation of the results coming from the two camps.

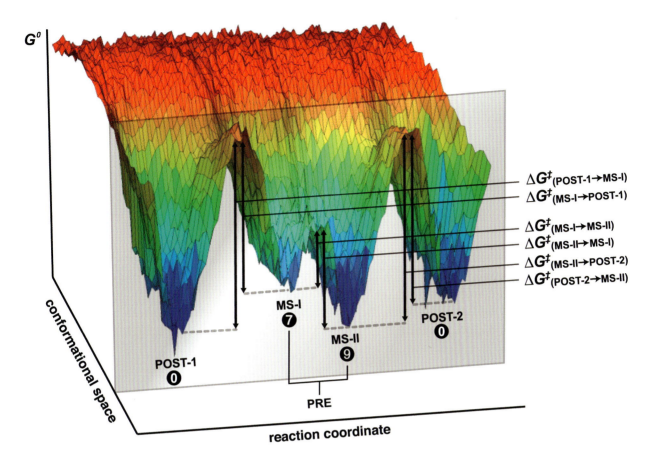

G^0

conformational space

reaction coordinate

$\Delta G^{\ddagger}$(POST-1→MS-I)
$\Delta G^{\ddagger}$(MS-I→POST-1)

$\Delta G^{\ddagger}$(MS-I→MS-II)
$\Delta G^{\ddagger}$(MS-II→MS-I)
$\Delta G^{\ddagger}$(MS-II→POST-2)
$\Delta G^{\ddagger}$(POST-2→MS-II)

POST-1 ❶
MS-I ❼
MS-II ❾
POST-2 ❶
PRE

Fig. 2 Schematic diagram of the free-energy landscape of the translating ribosome, deduced from life-time measurements in smFRET experiments. The diagram shows the free-energy profile along the reaction coordinate as we move from POST over PRE to the next POST state. Of note in the context of translocation is that the PRE complex exists in two states (valleys) MS-I and MS-II separated by a relatively low barrier. Thus, the PRE ribosome is capable of transitioning between MS-I and MS-II in the thermal environment, in the absence of EF-G. (Reproduced with permission from Frank and Gonzalez, 2010).

3. mRNA-tRNA translocation

In the translocation process, the mRNA, as well as the two tRNAs, must advance unidirectionally from the pre- to the post-translocation complex, such that the next codon is presented at the A site of the small subunit. Detailed cryo-EM studies have enabled us to put together (Taylor et al., 2007) a sequence of events, beginning with the pre-translocation complex, binding of EF-G, intersubunit rotation associated with the movement of the tRNAs from the A/A and P/P to the A/P and P/E configurations, GTP hydrolysis on EF-G followed by Pi release, conformational change of domain IV of EF-G, disengagement of G530, A1492 and A1493 from the mRNA-(tRNA)$_2$ moiety, small subunit head rotation, mRNA translocation, reverse intersubunit rotation, and reverse head rotation (see movie in Supplement to Taylor et al., 2007).

However, this sequential presentation is a vastly simplified version of the actual sequence of stochastic processes we can infer from smFRET studies, in which the Brownian ratchet character of the ribosome is dramatically evident. Stochastic intersubunit rotation in both directions, with rules of engagement altered by the intervention of some sort of pawl, is instrumental in this process. While previous work (e. g., Frank and Agrawal, 2000; Savelsbergh et al., 2003) used to depict the binding of EF-G in its GTP state as the driving force of the intersubunit rotation, several studies by a variety of techniques, including bulk FRET (Ermolenko et al., 2007), smFRET (Cornish et al., 2008), and cryo-EM (Agirrezabala et al., 2008; Julian et al., 2008), recently brought evidence for spontaneous intersubunit rotation in the absence of EF-G (Figure 3).

[Note: The phrase "ratchet-like rotation," introduced by Frank and Agrawal (2000), occasionally led

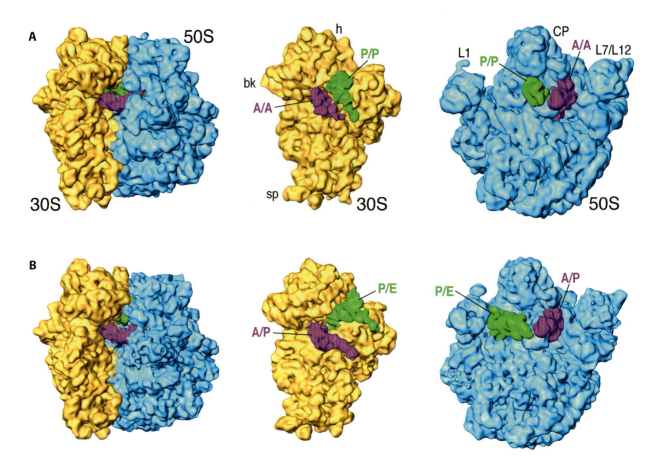

Fig. 3 Cryo-EM evidence for coexistence of PRE ribosomes in states MS-I (A) and MS-II (B). At 3.5 Mg2+ concentration, and in the absence of EF-G, a sample of the PRE complex consists of ribosomes in state MS-I (30S subunit unrotated, tRNAs in (A/A, P/P), L1 stalk outward) and those in state MS-II (30S subunit rotated, tRNAs in (A/P, P/E) in a ratio of 30:70. smFRET experiments (Kim et al., 2007; Cornish et al., 2008) showed earlier that the ribosomes fluctuate between the two states. (Reproduced from Agirrezabala et al., 2008; © Elsevier 2008).

to the use of "ratcheting" to refer to the intersubunit rotation. In view of the distinct meaning of a Brownian ratchet, important in the context of the discussion of ribosome function, we will henceforth avoid the term "ratcheting" in favor of "intersubunit rotation." Evidently, intersubunit rotation is only part of the ratchet mechanism.]

The role of EF-G in translocation is illuminated by the statement, backed by experimental evidence, that the ribosome in its intersubunit-rotated state is the true substrate of EF-G in its GTP form (Fei et al., 2009). In other words, very much in line with current models of enzyme kinetics, the conformation of the EF-G-bound ribosome already exists prior to EF-G binding, and is constantly sampled in the range of temperatures that support life, leading to the binding engagement of EF-G in its GTP form, followed by activation of GTP hydrolysis. The reason why spontaneous intersubunit rotation had not been seen before by

cryo-EM is two-fold: first, Mg^{2+} concentrations were usually in a range ($>> 3$ mM) where the unrotated form dominated, and second, the coexistence of two subpopulations with different conformations presents a complication that requires the application of sophisticated classification methods, which became available only recently.

Already the accumulation of evidence from several cryo-EM studies of factor binding strongly suggested that the ribosome performs its work by drawing from the energy of the thermal bath, as entirely unrelated processes of initiation, elongation, termination and recycling all go hand in hand with the intersubunit rotation (Frank et al., 2007). The terms "macrostate" (Frank et al., 2007) or "global state" (Fei et al., 2008) have been introduced to emphasize the fact that the ribosome undergoes a dramatic conformational change which alters its chemical properties in the intersubunit space. The extent of this change from macrostate I (MS-I) to

MS-II is evident from atomic representations obtained by flexible fitting (Gao et al., 2009 a; pdb MS-I: 3DG2, MS-II: 3DG0)

Intersubunit rotation is accompanied by movements of the tRNAs from the classical (A/A, P/P) to the hybrid (A/P, P/E) configurations, and by the movement of the L1 stalk into and out of the intersubunit space. These movements, according to recent smFRET data (Munro et al., 2009 b), appear to happen at different rates, but cryo-EM visualization of subpopulations in which the two states of intersubunit rotation go hand in hand with distinct configurations of the tRNAs and the L1 stalk (Agirrezabala et al., 2008; Julian et al., 2008) show that the movements are definitely in some way coupled.

In addition, there is recent accumulating evidence for the existence of intermediate states of the pre-translocational ribosome that are in equilibrium with the MS-I and MS-II states. SmFRET studies showed such states, which were, however, not highly populated (Munro et al., 2007; 2009 b; 2010). X-ray crystallography showed evidence for two intermediates with different degrees of intersubunit rotation, although the ribosome complex was not functionally relevant, with anticodon stem-loops substituting for tRNAs (Zhang et al., 2009). In cryo-EM, the separation of sub-populations co-existing in the projection dataset is a challenging problem that has only recently been addressed in a general way (Liao et al., 2010; Scheres et al., 2007; Spahn and Penczek, 2009). Using the maximum-likelihood algorithm of Scheres et al. (2007), we obtained reconstructions of a number of ribosome complexes with intermediate intersubunit rotations, as well as intermediate positions of the tRNAs and the L1 stalk. For wild-type ribosomes at $[Mg^{2+}]$ = 3.5 mM, we found a stable intermediate that is as highly populated as the classic state (X. Agirrezabala, H. Liao, J. Fu, J. L. Brunelle, R. F. Ortiz-Meoz, E. Schreiner, K. Schulten, R. Green, J. Frank, in preparation). In another study, employing a single-point mutation on the P loop of the 23S rRNA (Fu et al., 2009; 2011) at $[Mg^{2+}]$ = 15 mM, we found highly unusual positions of the tRNAs with contacts to the L1 stalk. As a plausible explanation, these intermediate configurations may be necessary because of the large distances to be bridged; thus they may act as "springboards" that ensure sufficiently high probability for completion of a particular productive forward pathway. In this context, the recent work by Fischer et al. (2010) on the back-translocation process is quite interesting, as it shows as many as 50 distinct conformations of the ribosome-(tRNA)$_2$ complex, some of which no doubt must reflect authentic intermediates in the forward translocation process.

4. Conclusion

mRNA-tRNA translocation, which requires large changes in the relative positions of all the players (subunits, tRNAs, mRNA), is a multi-step process whose individual steps have now become amenable to study both by cryo-EM and single-molecule FRET. Although the contours of the general mechanism can now be sketched out, we are still far away from a depiction of the exact sequence of events, and from a detailed mechanistic understanding of all the subprocesses. While cryo-EM, depicting highly populated states of relative stability, may create the illusion of seeing snapshots of a macroscopic machine in continuous, unidirectional motion, smFRET shows a drastically different picture, namely of a stochastic machine in constant jitter. Designing experiments with the goal to combine these complementary aspects presents a challenge for future studies.

As a final, more speculative note, the important part dynamics plays in translocation must have ramifications for the specific way the ribosome evolved from a precursor since evolutionary pressure will not work toward addition of useful peripheral components without ensuring that these components are compatible with the dynamics of the entire machine. Bokov and Steinberg's (2009) fascinating model for ribosome evolution, deduced from the hierarchy of tertiary contacts, might provide clues on how dynamical and structural features have co-evolved.

Acknowledgements

This work was supported by HHMI, NIH R37 GM29 169, and NIH R01 GM55 440.

References

Agirrezabala X, Lei J, Brunelle JL, Ortiz-Meoz RF, Green R, and Frank J (2008) Visualization of the hybrid state of tRNA binding promoted by spontaneous ratcheting of the ribosome. Mol Cell 32: 190–197

Agirrezabala X, Frank J (2009) Elongation in translation as a dynamic interaction among the ribosome, tRNA, and elongation factors EF-G and EF-Tu. Quart Rev Biophys 42: 159–200

Agrawal RK, Penczek P, Grassucci RA, Frank J (1998) Visualization of elongation factor G on the *Escherichia coli* 70S ribosome: The mechanism of translocation. Proc Natl Acad Sci USA. 95: 6134–6138

Agrawal RK, Heagle AB, Penczek P, Grassucci RA, Frank J (1999) EF-G-dependent GTP hydrolysis induces translocation accompanied by large conformational changes in the 70S ribosome, Nat Struct Biol 6: 643–647

Blanchard SC, Kim HD, Gonzalez Jr RL, Puglisi JD, Chu S (2004) tRNA dynamics on the ribosome during translation. Proc. Natl Acad. Sci. USA 101: 12 893–12 898

Bokov K, Steinberg SV (2009) A hierarchical model for evolution of 23S ribosomal RNA. Nature 457: 977–980

Bretscher MS (1968) Translocation in protein synthesis: a hybrid structure model. Nature 218: 675–677

Cornish PV, Ermolenko D. N, Noller HF, Ha T (2008) Spontaneous intersubunit rotation in single ribosomes. Mol Cell 30: 578–588

Ermolenko DN, Majumdar ZK, Hickerson RP, Spiegel PC, Clegg RM, Noller HF (2007) Observation of intersubunit movement of the ribosome in solution using FRET J Mol Biol 370: 530–540

Fei J, Kosuri P, MacDougall DD, Ruben L Gonzalez RL (2008) Coupling of Ribosomal L1 Stalk and tRNA Dynamics during Translation Elongation. Mol Cell 30: 348–359

Fei,J, Bronson JE, Hofman JM, Srinivas RL, Wiggins CH, Gonzalez RL Jr. (2009) Allosteric collaboration between elongation factor G and the ribosomal L1 stalk directs tRNA movements during translation. Proc Natl Acad Sci USA 106: 15 702–15 707

Fischer N, Konevega AL, Wintermeyer W, Rodnina MV, Stark H (2010) Ribosome dynamics and tRNA movement by time-resolved electron cryomicroscopy. Nature 466: 329–333

Frank J, Agrawal RK (2000) A ratchet-like inter-subunit reorganization of the ribosome during translocation. Nature 406: 318–322

Frank J, Zhu J, Penczek P, Li Y, Srivastava S, Verschoor A, Radermacher M, Grassucci R, Lata RK, Agrawal RK (1995) A model of protein synthesis based on cryo-electron microscopy of the *E. coli* ribosome. Nature 376: 441–444

Fu J, Kennedy D, Munro JB, Lei J, Blanchard SC, Frank J (2009) The P-site tRNA reaches the P/E position through intermediate positions. (Abstract) J. Biomol. Struct. Dyn. 26: 794–795

Fu J, Munro JB, Blanchard SC, Frank J (2011) Cryo-EM structures of the ribosome complex in intermediate states during tRNA translocation. Proc Natl Acad Sci USA (in press)

Gabashvili IS, Agrawal RK, Spahn CM, Grassucci RA, Svergun DI, Frank J, Penczek P (2000) Solution structure of the *E. coli* 70S ribosome at 11.5Å resolution. Cell 100: 537–549

Garai A, Chowdhury D, Ramakrishnan TV (2009) Stochastic kinetics of ribosomes: single motor properties and collective behavior. Phys Rev E 80: 011 908

Gao H, Sengupta J, Valle M, Korostelev A, Eswar N, Stagg SM et al. (2003) Study of the structural dynamics of the *E. coli* 70S ribosome using real-space refinement. Cell 113: 789–801

Gao H, LeBarron J, Frank J (2009a) Ribosomal dynamics: intrinsic instability of a molecular machine. In: Walter NG, Woodson SA, Batey RT (eds) Non-protein coding RNAs. Springer Berlin pp 303–316

Gao Y-G, Selmer M, Dunham CM, Weixlbaumer A, Kelley AC, Ramakrishnan V (2009b) The structure of the ribosome with elongation factor G trapped in the posttranslocational state. Science 326: 694–699

Gromadski KB, Daviter T, Rodnina MV (2006) A uniform response to mismatches in codon-anticodon complexes ensures ribosomal fidelity. Mol Cell 21: 369–377

Harms J, Tocilj A, Levin I, Agmon I, Holger Stark3, Ingo Kölln1, van Heel M, Cuff M, Schlünzen F, Bashan A, Franceschi F, Yonath A (2000) Elucidating the medium-resolution structure of ribosomal particles: an interplay between electron cryo-microscopy and X-ray crystallography. Structure 7: 931–941

Julián P, Konevega AL, Scheres SH W, Lázaro M, Gil D, Wintermeyer W, Rodnina MV, Valle M (2008) Structure of ratcheted ribosomes with tRNAs in hybrid states. Proc Natl Acad Sci USA 105: 16 924–16 927

Kim HD, Puglisi J, Chu S (2007) Fluctuations of transfer RNAs between classical and hybrid states. Biophys. J. 93: 3575–3582

Li W, Sengupta J, Rath BK, Frank J (2006) Functional conformations of the L11-ribosomal RNA complex revealed by correlative analysis of cryo-EM and molecular dynamics simulations. RNA 12: 1240–1253

Liao H, Frank J (2010) Classification by bootstrapping in single particle methods. Proc IEEE Int Symp on Biomedical Imaging: from nano to macro. IEEE Int Symp Biomedica (in press)

Malhotra A, Penczek P, Agrawal RK, Gabashvili IS, Grassucci RA, Junemann R, Burkhardt N, Nierhaus KH, Frank J (1998) *Escherichia coli* 70 S ribosome at 15Å resolution by cryo-electron microscopy: localization of fMet-tRNAfMet and fitting of L1 protein. J. Mol. Biol. 280: 103–116

Moazed D, Noller HF (1989) Intermediate states in the movement of transfer RNA in the ribosome. Nature 342: 142–148

Moran SJ, Flanagan IV, JF, Namy O, Stuart DI, Brierley I, Gilbert RJ C (2008) The mechanics of translocation: a molecular "Spring-and-Ratchet" system. Structure 16: 664–672

Munro J. B, Altman RB, O' Connor N, Blanchard SC (2007) Identification of two distinct hybrid state intermediates on the ribosome. Mol Cell 25: 505–517

Munro JB, Sanbonmatsu KY, Spahn CM, Blanchard SC (2009) Navigating the ribosome's metastable energy landscape. Trends Biochem. Sci. 34: 390–400

Munro JB, Altman RB, Tung C-S, Cate JH D, Sanbonmatsu KY, Blanchard SC (2010b) Spontaneous formation of the unlocked state of the ribosome is a multistep process. Proc Natl Acad USA 107: 709–714

Munro JB, Altman RB, Tung C-S, Sanbonmatsu KY, Blanchard SC (2009a) A fast dynamic mode of the EF-G-bound ribosome. EMBO J. 29: 770–781

Penczek PA, Frank J, Spahn CM T (2006) A method of focused classification, based on the bootstrap 3D variance analysis, and its application to EF-G-dependent translocation. J Struct Biol 154: 184–194

Polacek N, Patzke S, Nierhaus KH, Barta A (2000) Periodic conformational changes in rRNA: monitoring the dynamics of translating ribosomes. Mol Cell 6: 159–171

Reblova K, Razga F, Li W, Gao H, Frank J, Sponer J (2010) Dynamics of the base of ribosomal A-site finger revealed by molecular dynamics simulations and Cryo-EM. Nucl Acid Res 38: 1325–1340

Sanbonmatsu KY, Joseph S, Tung CS (2005) Simulating movement of tRNA into the ribosome during decoding. Proc Natl Acad Sci USA 102: 15 854–15 859

Savelsbergh A, Katunin VI, Mohr D, Peske F, Rodnina M. V, Wintermeyer W (2003) An elongation factor G-induced ribosome rearrangement precedes tRNA-mRNA translocation. Mol Cell 11: 1517–1523

Scheres SH, Gao H, Valle M, Herman GT, Eggermont PP, Frank J, Carazo JM (2007) Disentangling conformational states of macromolecules in 3D-EM through likelihood optimization. Nat Methods 4: 27–29

Schuwirth BS, Borovinskaya MA, Hau CW, Zhang W, Vila-Sanjurjo A, Holton JM, Cate JH (2005) Structures of the bacterial ribosome at 3.5 Å resolution. Science 310: 827–834

Seidelt B, Innis CA, Wilson DN, Gartmann M, Armache J-P, Villa E, Trabuco LG, Becker T, Mielke T, Schulten K, Steitz TA, Beckmann R (2010) Structural insight into nascent polypeptide chain–mediated translational stalling. Science 326: 1412–1415

Spirin AS (1969) A model of the functioning ribosome: locking and unlocking of the ribosome subparticles, Cold Spring Harbor Symp. Quant Biol 34: 197–207

Spirin AS, Baranov VI, Polubesov GS, Serdyuk IN, May RP (1987) Translocation makes the ribosome less compact. J Mol Biol194: 119–126

Spirin AS (2009) The ribosome as a conveying thermal ratchet machine. J Biol Chem 284: 21 103–21 119

Spahn CM T and Penczek PA (2009) Exploring conformational modes of macromolecular assemblies by multiparticle cryo-EM. Cur Opin Struct Biol 19: 623–631

Stark H, Mueller F, Orlova EV, Schatz M, Dube P, Erdemir T, Zemlin F, Brimacombe R, van Heel M (1995) The 70S *Escherichia coli* ribosome at 23Å resolution: fitting the ribosomal RNA. Structure 3: 815–821

Stark H, Rodnina MV, Wieden HJ, van Heel M, Wintermeyer W (2000) Large-scale movement of elongation factor G and ex- tensive conformational change of the ribosome during trans- location. Cell 100: 301–309

Tama F, Valle M, Frank J, Brooks CL, 3rd (2003) Dynamic reor- ganization of the functionally active ribosome explored by normal mode analysis and cryo-electron microscopy. Proc Natl Acad Sci USA 100: 9319–9323

Tama F, Miyashita O, Brooks CL (2004) Normal mode based flex- ible fitting of high-resolution structure into low-resolu- tion experimental data from cryo-EM. J Struct Biol 147: 315–326

Taylor DJ, Nilsson J, Merrill AR, Andersen GR, Nissen P, Frank J (2007) Structures of modified eEF2 80S ribosome complexes reveal the role of GTP hydrolysis in translocation. EMBO J. 26: 2421–2431

Valle M, Zavialov A, Sengupta J, Rawat U, Ehrenberg M, Frank J (2003) Locking and unlocking of ribosomal motions. Cell 114: 123–134

Wimberly BT, Brodersen DE, Clemons Jr WM, Morgan-Warren RJ, Carter A. P, Vonrhein C et al. (2000) Structure of the 30S ribosomal subunit. Nature 407: 327–339

Yusupov M, Yusupova GZ, Baucom A, Lieberman K, Earnest TN, Cate JH, Noller HF (2001), Crystal structure of the ribosome at 5.5 Å resolution. Science 292: 883–896

Zhang W, Kimmel M, Spahn CM, Penczek PA (2008) Heterogeneity of large macromolecular complexes revealed by 3D cryo-EM variance analysis. Structure 16: 1770–1776

Zhang W, Dunkle JA, Cate JH D (2009) Structures of the ribo- some in intermediate states of ratcheting. Science 325: 1014–1017

Functions of elongation factor G in translocation and ribosome recycling

26

Wolfgang Wintermeyer, Andreas Savelsbergh, Andrey L. Konevega, Frank Peske, Vladimir I. Katunin, Yuri P. Semenkov, Niels Fischer, Holger Stark, and Marina V. Rodnina

1. Introduction

Among the translation factors that assist the ribosome in synthesizing proteins, elongation factor G (EF-G) is the only one that functions in two different phases of protein synthesis, i. e. in the translocation step of the elongation phase and in ribosome disassembly following termination. During translocation two tRNAs move by large distances from one site of the ribosome to the next, adjacent site, with the coupled movement of mRNA by one codon. The process is promoted by EF-G and GTP hydrolysis to proceed at the velocity required for rapid protein synthesis in the cell. During ribosome recycling the ribosomal post-termination complex is dissociated into subunits; the reaction is brought about by EF-G together with the ribosome recycling factor (RRF) and requires GTP hydrolysis. Fundamental questions in understanding EF-G function are: (i) How does EF-G accelerate the movement of tRNAs together with the mRNA on the ribosome; (ii) how does EF-G cooperate with RRF to dissociate the ribosomes; and (iii) how are GTP hydrolysis and Pi release coupled to forward movement and ribosome disassembly? The aim of this review is to summarize the recent insights into the molecular mechanism of translocation and ribosome recycling and the role of EF-G in the two reactions. Detailed accounts focusing on different aspects of translocation can also be found in several recent reviews (Shoji et al., 2009; Dunkle and Cate, 2010; Frank and Gonzalez, 2010).

2. Translocation

2.1. Spontaneous movement of tRNAs

During translocation, peptidyl-tRNA moves from the A site of the ribosome to the P site and deacylated tRNA from the P site to the E site, from where it dissociates. The physiological, rapid reaction is promoted by EF-G and GTP hydrolysis. However, translocation can also proceed spontaneously, albeit slowly, in both forward and backward directions, implying that the propensity for movement is inherent to the ribosome-tRNA complex itself (Gavrilova et al., 1976; Fredrick and Noller, 2003; Shoji et al., 2006; Konevega et al., 2007). While they move from one ribosomal binding site to the other, the tRNAs assume intermediate configurations, and the ribosome undergoes dynamic structural changes. One example of such tRNA intermediates, the so-called tRNA hybrid states, was identified more than 20 years ago (Moazed and Noller, 1989). In the hybrid states, the anticodon stem-loops of the tRNAs reside in the A and P sites of the 30S subunit, while the respective acceptor ends are oriented towards the P and E sites of the large 50S subunit. The formation of the tRNA hybrid states correlates with the rotational movement (previously termed "ratchet" (Frank and Agrawal, 2000)) of the ribosomal subunits relative to each other (Agirrezabala et al., 2008; Julian et al., 2008). Formation of the hybrid/rotated state is an essential prerequisite of translocation, as blocking the ability of the ribosome to undergo rotational movement of the subunits abolishes translocation (Horan and Noller, 2007). The important role of hybrid-state formation for translocation has been shown for both the P/E state (Lill et al., 1989; Joseph and Noller, 1998; McGarry et al., 2005; Walker et al., 2008) and the A/P state (Semenkov et al., 2000; McGarry et al., 2005;

A

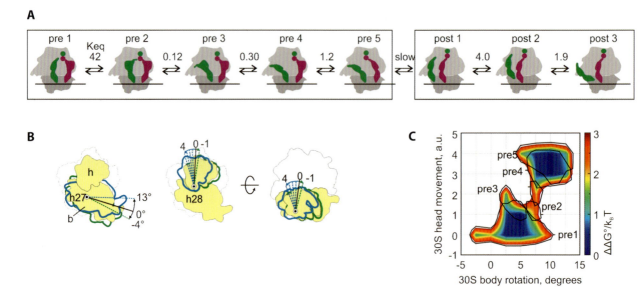

Fig. 1 Time-resolved cryo-EM analysis of spontaneous reverse translocation. (A) Trajectories of tRNA movement. The numbers above arrows represent equilibrium constants (K_{eq}), i.e. the ratio between the states n and n+1 that are in rapid equilibrium. The transition between the ensembles of post and pre states (indicated by square brackets), which entails the movement of the tRNA anticodon stem-loops on the 30S ribosomal subunit, is slow and rate-limiting. (B) Schematic of 30S body rotation (left) and 30S head movement (middle and right). The 30S body (b) rotates around a pivot point at helix 27 (h27) of 16S rRNA independent of the 30S head (h). In the ground state of the ribosome the rotation angle is 0°. The 30S head movement entails a tilt (middle) and a swiveling motion (right) around the neck region (h28), depicted in arbitrary units (a. u.) from −1 to 4 corresponding to an overall amplitude of 30 Å. Zero corresponds to the ground state of the ribosome. (C) Free-energy landscape of global ribosome conformation during hybrid-state formation. Black contour lines indicate the conformational space occupied by the different tRNA states (pre1 to pre5). The heat map indicates the free energy for the respective 30S sub-state. Adapted from (Fischer et al., 2010).

Dorner et al., 2006), suggesting that the hybrid/rotated state is an authentic intermediate of translocation. In addition to the hybrid/rotated state, some discrete intermediates have been identified by single-molecule FRET and cryo-EM techniques (reviewed in (Frank and Gonzalez, 2010)).

The full trajectory of spontaneous tRNA movement through the ribosome has been recently visualized by time-resolved cryo-EM (Fischer et al., 2010). From the analysis of 2,000,000 ribosome particles, 50 cryo-EM reconstructions of ribosome complexes with tRNAs at different stages of retro-translocation were obtained. The eight most distinct states were classified (Figure 1A), showing a quasi-continuous trajectory of tRNA movement through the ribosome. Each of these states comprises an ensemble of several (up to eleven) sub-states that share similar tRNA positions and tRNA-ribosome contacts, but vary in ribosome conformation, in keeping with single-molecule fluorescence data suggesting that motions of ribosome elements and tRNA movements are not strictly synchronized (Munro et al., 2010). tRNA movement entails the sequential step-by-step rupture and forma-

tion of ribosome-tRNA contacts (Fischer et al., 2010). Several ribosomal elements, such as helices 38 and 69 of 23S rRNA, protein L5, and the L1 stalk, move over long distances (between 8 Å and 40 Å) together with the tRNAs, thereby confining the space accessible for tRNA movement and reducing the number of contacts to be broken and formed at a time. In addition to the rotational movement of the 30S subunit body relative to the 50S subunit, a movement of the 30S subunit head relative to the body was observed, reminiscent of the "swivel" movement found in ribosome crystals (Dunkle and Cate, 2010) (Figure 1B). It is to be noted that the extent of 30S body rotation observed in this analysis is larger, between 14 and −6 degrees (Figure 1B), than anticipated based on previous work (Frank and Agrawal, 2000). This is mainly attributed to the higher temperature of specimen preparation, 18 °C vs. 4 °C, used in the cryo-EM analysis of Fischer et al. (2010). For each of the eight states of tRNA movement, there is a characteristic spectrum of nearly continuous 30S body rotations and 30S head movements (Fischer et al., 2010). On the other hand, it is also possible to observe a particular tRNA state over a wide range of different 30S conformations and, vice versa, to observe ribosomes in similar overall conformations that significantly differ in their tRNA positions. Thus, although 30S body and head movements correlate with tRNA movement, the coupling between them is loose.

Quantitative analysis of the distribution of particles between substates, as visualized by cryo-EM, suggests that the five states within the ensemble of pre-translocation states are in rapid equilibrium on the time scale of spontaneous tRNA movement (minutes); the same is true for the three post-translocation states (Figure 1A). Free-energy landscapes of ribosome conformations reveal that each tRNA state is characterized by an ensemble of sub-states representing global conformational states of the 30S subunit that fluctuate around a state with minimum free energy within the range of $2k_BT$ (Figure 1C). Changes in tRNA-ribosome interactions, in turn, shift the local energy minimum of the particular ribosome conformation. In this way, the tRNAs transiently rectify the structural fluctuations of the ribosome, thereby increasing the probability of further directional tRNA movement. The rapid equilibria between the various tRNA states indicate low kinetic barriers that allow the ribosome-tRNA complexes to rapidly sample all sub-states among pre or post states and to make use of multiple alternative pathways of movement. By coupling thermally activated large-scale conformational fluctuations of the ribosome, and in particular the 30S subunit, and of stochastic movements of the tRNAs, tRNA movement over long distances is promoted.

The direction of spontaneous translocation is determined by the thermodynamic gradient created by the different affinities of tRNA binding to the E, P, and A sites. Retro-translocation is favorable when the gain in affinity upon tRNA movement from the E to the P site compensates for the affinity loss resulting from the displacement of peptidyl-tRNA from the P to the A site (A. Konevega, unpublished data). For some tRNA combinations, spontaneous forward translocation is highly unfavorable (Shoji et al., 2006; Konevega et al., 2007). Likewise, the rate and preferential direction of hybrid/rotated state formation depends on the tRNAs in the A and P sites (Blanchard et al., 2004; Munro et al., 2007; Spiegel et al., 2007; Cornish et al., 2008). However, the rate of hybrid/rotated state formation for a model pre-translocation complex is about 2 s^{-1} (Cornish et al., 2008), which is orders of magnitude faster than the rate of spontaneous movement of the tRNA anticodon stem-loops on the 30S subunit, which is very low, in the range of 0.0001–0.01 s^{-1} (Shoji et al.,

2006; Konevega et al., 2007; Fischer et al., 2010). Thus, although the formation of the ratchet/hybrid state is an important intermediate of full spontaneous translocation, the translocation of the tRNA anticodons on the 30S subunit is the rate-limiting step of the reaction.

2.2. Kinetic mechanism of EF-G-dependent translocation

EF-G consists of five domains: domain 1 that binds GTP/GDP, domain 2 which is common among translational GTPases, and domains 3–5 which are specific to EF-G and form a unit that changes its orientation relative to domains 1 and 2. Kinetic studies provide a quantitative description of the elemental steps of translocation (Savelsbergh et al., 2003; Savelsbergh et al., 2005; Pan et al., 2007). EF-G is recruited to the ribosome through interactions with ribosomal protein L7/12 (Diaconu et al., 2005), of which the *E. coli* ribosome contains four copies (Figure 2). The interaction of EF-G with the ribosome induces a conformational change of EF-G that closes the nucleotide-binding pocket, resulting in a 30,000-fold stabilization of GTP binding (Wilden et al., 2006). The interaction is likely to involve the protein L7/12, the sarcin-ricin loop on the 50S subunit (Mohr et al., 2002; Diaconu et al., 2005; Gao et al., 2009) and the switch 1 region of EF-G (Ticu et al., 2009). At the same time GTP hydrolysis by EF-G, which is negligible on unbound factor, is accelerated to ~250 s^{-1} (37 °C) (Rodnina et al., 1997; Savelsbergh et al., 2003; Savelsbergh et al., 2005). The effect is independent of the occupancy of the tRNA binding sites of the ribosome (Table 1) and of hybrid state formation (Walker et al., 2008), suggesting that the binding of EF-G·GTP and subsequent GTP hydrolysis does not depend on the functional state of the ribosome. This is corroborated by steady-state kinetics, as the K_M values of turnover GTP hydrolysis are similar within a factor of five (0.2–1.1 μM) for ribosomes in various functional states (Table 1). As such, the influence of the presence or absence of a peptidyl residue on the P-site tRNA on EF-G affinity and catalysis is small, much smaller than proposed previously (Zavialov and Ehrenberg, 2003). As a consequence, a certain extent of "uncoupled" (Chinali and Parmeggiani, 1980) GTP hydrolysis by EF-G entering ribosomes that are not occupied by competing factors, such as EF-Tu or RF1/2, may take place.

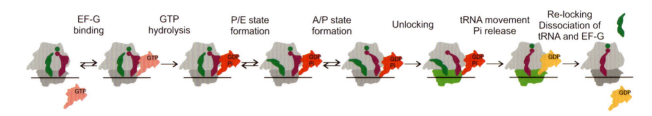

Fig. 2 Kinetic model of translocation. The 30S subunit is depicted in two conformations to indicate the locked (gray) and unlocked (green) state of the ribosome. EF-G undergoes conformational changes upon transition from the GTP-bound (pink) to the GDP·Pi-bound (red) and GDP-bound (yellow) forms. The sequence of steps is based on kinetic data (Savelsbergh et al., 2003; Savelsbergh et al., 2005; Pan et al., 2007; Walker et al., 2008).

EF-G binding stabilizes the ratchet/hybrid conformation of the ribosome, whereas the tRNA anticodons do not move at this stage (Valle et al., 2003). Because spontaneous formation of the rotated state (up to 2 s^{-1} at 25 °C; (Cornish et al., 2008)) is much slower than EF-G-catalyzed tRNA movement (10–15 s^{-1} at 25 °C;(Katunin et al., 2002; Feinberg and Joseph, 2006; Pan et al., 2007)), EF-G must accelerate the formation of the rotated state of the ribosome and the movement of the tRNAs into hybrid states. This can occur in a stepwise manner with the P/E state for the deacylated tRNA formed first, followed by the movement of the peptidyl-tRNA into the A/P state (Walker et al., 2008); the role of GTP hydrolysis in the acceleration of the hybrid state formation is unclear.

The tRNA-mRNA movement on the 30S subunit is preceded and rate-limited by a rearrangement ("unlocking") of the pre-translocation ribosome, which is driven by EF-G·GTP binding and GTP hydrolysis (Katunin et al., 2002; Savelsbergh et al., 2003). Unlocking presumably results in a change of the mobility of elements of the 30S subunit, thereby facilitating a swiveling movement of the 30S head (Schuwirth et

al., 2005). An involvement of the 30S head is also suggested by EF-G-induced protections of helix 34 of 16S rRNA (Matassova et al., 2001) and by the effects of mutations that affected the mobility of helix 34 (Kubarenko et al., 2006). Interestingly, a specific cleavage of 16S rRNA next to nucleotide 1493 in the decoding site by colicin E3 leads to an acceleration of tRNA movement in translocation (Lancaster et al., 2008), which can be rationalized in terms of facilitated structural fluctuations of the 30S subunit head as a result of cleavage. Translocation is inhibited by spectinomycin (Table 2) that binds to helix 34 (Carter et al., 2000; Borovinskaya et al., 2007) and, by interfering with the swiveling movement of the 30S head (Borovinskaya et al., 2007), appears to lock the 30S subunit in a conformation that is refractory to translocation (Peske et al., 2004). By contrast, streptomycin, by binding to the junction of helices 18, 27, and 44 of 16S rRNA (Carter et al., 2000), stabilizes a structure of the 30S subunit that is prone to translocation, an effect which almost completely compensates the 400-fold increase in affinity of the tRNA in the A site caused by the antibiotic (Table 2). Paromomycin, which binds to helix 44 at the decoding site, seems to inhibit translocation solely by stabilizing tRNA binding in the A site, whereas hygromycin B, which binds nearby, has a mixed effect, inhibiting translocation by both stabilizing the binding of A-site tRNA and interfering with a conformational change of the 30S subunit required for movement. Taken together, these results reveal the important role of movements within the small ribosomal subunit that follow unlocking for the movement of the tRNAs.

The movement of tRNAs and mRNA that takes place in the unlocked state of the ribosome appears to be rapid and spontaneous (Savelsbergh et al., 2003; Savelsbergh et al., 2005). As in the spontaneous reaction, during EF-G-catalyzed translocation the tRNAs are likely to move through a quasi-continuous landscape of intermediate states. It seems likely that the

Table 1 Kinetic parameters of EF-G-dependent GTP hydrolysis on various ribosome complexes

Ribosome complex	k_{cat} (s^{-1})	K_M (μM)	k_{app} (s^{-1})
Pre-translocation	6.9 ± 0.7	0.2 ± 0.1	80 ± 20
Post-translocation	7.4 ± 0.8	1.1 ± 0.3	90 ± 15
Post-termination	7.3 ± 0.3	0.5 ± 0.1	95 ± 20
Vacant	6.8 ± 0.6	0.2 ± 0.1	75 ± 20

Values of k_{cat} and K_M of multiple-turnover GTP hydrolysis were determined from the hyperbolic concentration dependencies of initial rates in titrations of EF-G (50 nM) with the respective ribosome complex (up to 2 μM). Values of k_{app} of single-round GTP hydrolysis were determined by quench-flow (0.1 μM ribosome complex, 0.8 μM EF-G) (Rodnina et al., 1997); the rate constant, as determined from the concentration dependence with vacant ribosomes, is k_{GTP} = 250 s^{-1} (Savelsbergh et al., 2003). All values for 37 °C.

Table 2 Inhibition of translocation by antibiotics binding to the 30S subunit

Antibiotic	Inhibition (-fold)	$\Delta\Delta G^{\neq}$	$\Delta\Delta G^{\circ}$	$\Delta\Delta G^{\circ}+\Delta\Delta G^{\neq}$ (kcal/mol)
None	1	-	-	-
Streptomycin	2	0.4	-2.3	-1.9
Spectinomycin	32	2.1	+0.8	2.9
Hygromycin B	320	3.5	-2.1	1.4
Paromomycin	160	3.1	-3.2	-0.1

$\Delta\Delta G^{\circ}$ and $\Delta\Delta G^{\neq}$ represent the differences of the free energy, ΔG°, of A-site binding and the free energy of activation, $\Delta G^{\neq}$, of translocation, respectively, of fMetPhe-tRNA^{Phe} determined in the presence and absence of antibiotic. A positive value of $\Delta\Delta G^{\circ}+\Delta\Delta G^{\neq}$ indicates an inhibition of translocation by interfering with conformational rearrangements of the 30S subunit that are involved in translocation. A negative value (as for streptomycin) indicates that the antibiotic stabilizes a conformation of the 30S subunit that is more prone to translocation than the conformation without antibiotic. Data taken from (Peske et al., 2004).

intermediate states of spontaneous translocation, both forward and backward, and EF-G-promoted translocation resemble each other structurally; however, their relative occupancy will probably be changed by the presence of EF-G and/or GTP hydrolysis. Future work applying time-resolved cryo-EM and single-molecule fluorescence is required to resolve these issues.

The full acceleration of translocation by EF-G and the turnover of the factor requires GTP hydrolysis. This raises the question to which extent Pi release is required for either step. We observe that tRNA-mRNA movement and Pi release are intrinsically rapid and independent of one another, as either reaction can be inhibited without affecting the other. For instance, tRNA-mRNA movement, but not Pi release, is blocked by antibiotics that bind in the 30S decoding region, such as paromomycin or hygromycin B (Peske et al., 2004) (Table 2), or by engineering a disulfide bridge into EF-G that restricts domain movement (Peske et al., 2000). The H583K mutation at the tip of domain 4 of EF-G has a similar effect, inhibiting translocation >100-fold without affecting Pi release (Savelsbergh et al., 2000). Conversely, several point mutations in the C-terminal domain of protein L7/12, which cause a strong inhibition of Pi release, have no effect on tRNA-mRNA movement, whereas the dissociation of EF-G from the ribosome is inhibited (Savelsbergh et al., 2005). In keeping with these results, movement is not affected by vanadate, an analog of Pi which strongly inhibits the dissociation of EF-G from the ribosome following GTP hydrolysis and translocation (Savelsbergh et al., 2009), explaining the well-known stalling of

EF-G on the ribosome when GTP is replaced with a non-hydrolyzable analog. The interaction of protein L7/12, presumably with the G' subdomain of EF-G, appears to be important for its binding to the ribosome and the conformational coupling between GTP hydrolysis, retention of Pi, and unlocking (Agrawal et al., 2001; Nechifor et al., 2007; Ticu et al., 2009). In summary, these results indicate that tRNA-mRNA movement takes place after the unlocking rearrangement, independent of and in parallel with Pi release and that Pi release is ultimately required for EF-G to assume a conformation that allows it to dissociate from the ribosome and the ribosome to return to the locked state.

The antibiotic fusidic acid blocks EF-G dissociation from the ribosome after translocation and Pi release (Savelsbergh et al., 2009 and references therein). The arrangement and conformation of EF-G stalled by fusidic acid on the post-translocation ribosome has been characterized by medium-resolution cryo-EM (Agrawal et al., 1998; Stark et al., 2000; Valle et al., 2003) and, at high resolution, by X-ray crystallography (Gao et al., 2009). When this conformation is compared with that of unbound EF-G, as determined crystallographically (Ævarsson et al., 1994; Czworkowski et al., 1994; Al-Karadaghi et al., 1996; Hansson et al., 2005), we see that in the ribosome-bound conformation domain 4 of EF-G is tilted relative to the body of the molecule formed of domains 1 and 2 and reaches into the A site of the 30S subunit. This effectively blocks backward tRNA movement into the A site while the ribosome is in the unlocked state, thus fixing the post-translocation state until the re-locked state has formed.

Given that the movement of tRNAs through the ribosome may occur spontaneously due to thermal fluctuations, how does EF-G accelerate translocation? Is EF-G a deterministic motor that couples the energy of GTP hydrolysis to the movement through direct mechano-chemical coupling, or does it act as a Brownian motor that biases thermal fluctuations towards forward movement by structural anisotropy produced by GTP cleavage?

According to the available data, EF-G accelerates translocation in three distinct ways. First, EF-G binding contributes to the formation of the hybrid/rotated state of the ribosome and, thereby, promotes the partial movement of the tRNAs from the classical towards the hybrid states. Second, EF-G accelerates the unlocking rearrangement of the ribosome and probably also the following conformational changes at the decoding region that are required for rapid tRNA movement. For full acceleration, GTP hydrolysis is required. It seems that EF-G effects unlocking in the GDP·Pi-bound form; the release of Pi is not required to accelerate unlocking. On the contrary, the retention of Pi may be necessary to prevent the premature collapse of EF-G into the unproductive GDP-bound conformation that eventually allows its dissociation from the ribosome. Third, domain 4 of EF-G appears to bias diffusion to produce forward movement by occupying the 30S subunit A site and blocking the backwards movement of peptidyl-tRNA. In conclusion, EF-G may function as a Brownian motor that accelerates tRNA-mRNA movement by (i) facilitating conformational rearrangements of the ribosome leading to movement and (ii) biasing forward movement by preventing backward movement, with domain 4 exerting the function of the pawl in a ratcheting device (Rodnina et al., 1997; Savelsbergh et al., 2009).

The latter way of energetic coupling seems to be particularly important in cases where forward movement as such would be thermodynamically unfavorable, e. g. due to tRNA preference for the pre-translocation state or to secondary structure in the mRNA. Force measurement on ribosomes translating an mRNA with a strong secondary structure suggested that translation occurs through successive translocation-and-pause cycles (Wen et al., 2008), pause lengths depending on the secondary structure of the mRNA. The applied force destabilized the secondary structure and decreased the duration of pauses, but did not affect translocation times. It is generally assumed that the ability to unwind secondary structure of mRNA is inherent in the ribosome, and it has been suggested that ribosomal proteins S3, S4, and S5 that reside at the mRNA entrance site are active in unwinding the mRNA (Takyar et al., 2005). An involvement of other helicases in mRNA unwinding is unlikely because the ribosome is able to move through mRNA hairpins

when only the two factors required for translation, EF-Tu and EF-G, and GTP are present. It is not known, however, whether the energy of GTP hydrolysis by EF-G serves to drive unwinding of the mRNA, thereby coupling forward movement and mRNA unwinding.

3. Ribosome recycling

3.1. The mechanism of ribosome recycling

Following termination, which entails the hydrolytic release of the completed nascent peptide from P site-bound peptidyl-tRNA and results in a ribosome with deacylated tRNA in the P site and an empty A site, EF-G, together with RRF (Hirashima and Kaji, 1973), brings about the rapid dissociation of the post-termination ribosome into subunits (Karimi et al., 1999; Ito et al., 2002; Fujiwara et al., 2004; Peske et al., 2005; Zavialov et al., 2005) (Figure 3). Subsequently, initiation factor 3 (IF3) binds to the 30S subunit, preparing the 30S subunit for initiation by promoting the release of tRNA and mRNA and preventing reassociation of the subunits (Karimi et al., 1999; Peske et al., 2005). RRF is an essential protein, indicating that rapid disassembly of post-termination ribosome complexes is essential for cell growth (Janosi et al., 1998).

The turnover rate of Pi release during ribosome recycling is about 1.5 s^{-1}, higher than the rate of ribosome disassembly, which is 0.3 s^{-1} (Peske et al., 2005; Savelsbergh et al., 2009). These data argue that, on average, EF-G can undergo several rounds of turnover GTP hydrolysis, before ribosome disassembly takes place. Thus, GTP hydrolysis and ribosome disassembly appear to be only loosely coupled, which may contribute to the quality control of recycling. While the evidence from chemical probing (Lancaster et al.,

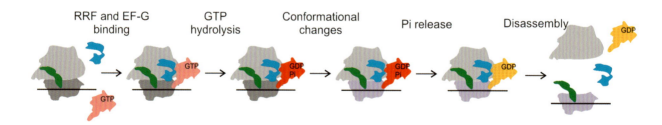

Fig. 3 Sequence of events in ribosome recycling. RRF (blue) and EF-G (pink) bind to the post-termination ribosome carrying deacylated tRNA (green) in the P site (P/E state). Following binding, GTP is hydrolyzed rapidly, and EF-G, in the GDP-Pi-bound form (red) effects conformational changes of the ribosome, in particular

of the 30S subunit (purple). Following Pi release, EF-G undergoes another conformational change (yellow), and ribosome disassembly takes place. Finally, RRF and EF-G are released, and the dissociation of tRNA and mRNA from the 30S subunit is promoted by IF3 (not shown).

2002), biochemical assays (Peske et al., 2005), and X-ray crystallography (Wilson et al., 2005; Weixlbaumer et al., 2007) suggests that RRF binding is precluded by peptidyl-tRNA in the P site (P/P state), in some cases spurious recycling of pre-termination ribosomes has been observed (Singh et al., 2008). Loose coupling between EF-G binding and ribosome disassembly may lower the risk of premature ribosome recycling in those cases where RRF binds to a post-translocation, rather than to a post-termination, ribosome.

3.2. The role of EF-G in ribosome recycling

The functions of EF-G in translocation and ribosome recycling resemble each other in several aspects, including effects of mutations in EF-G or the inhibition by antibiotics, but there are important differences in mechanistic detail. Unlike translocation, which takes place with both GTP and non-hydrolyzable GTP analogs, albeit at different rates, ribosome disassembly requires GTP hydrolysis (Karimi et al., 1999; Hirokawa et al., 2002; Peske et al., 2005; Zavialov et al., 2005). This suggests that ribosome disassembly involves a conformational change of EF-G that is induced only when Pi is released. In fact, the reaction is strongly inhibited by vanadate, an analog of Pi, while translocation is not (Savelsbergh et al., 2009). Another example is that ribosome disassembly is extremely sensitive to inhibition by fusidic acid at very low concentrations when compared with the inhibition of EF-G turnover that is observed only at much higher concentrations of fusidic acid (see below). The differences between the two reactions can probably be attributed to the fact that ribosome recycling is brought about by the heterodimeric assembly of EF-G with RRF on the ribosome.

The binding site of RRF on the ribosome has been mapped by directed hydroxyl radical footprinting (Lancaster et al., 2002), cryo-EM (Agrawal et al., 2004; Gao et al., 2005; Barat et al., 2007; Gao et al., 2007), and X-ray crystallography (Wilson et al., 2005; Borovinskaya et al., 2007; Weixlbaumer et al., 2007; Pai et al., 2008). RRF is bound at the interface between the subunits where it interacts mainly with the 50S subunit. RRF contacts helix 69 (H69) of 23S rRNA on the 50S subunit which connects to helix 44 (h44) of 16S rRNA on the 30S subunit to form inter-subunit bridge 2a. Bridge 2a is most important for holding the subunits together, as the deletion of H69 destabilizes the interaction between subunits considerably (Ali et al., 2006). A displacement of H69 by RRF alone has been observed with 50S subunits or 70S ribosomes from *E. coli* (Gao et al., 2005; Wilson et al., 2005; Borovinskaya et al., 2007; Gao et al., 2007), but not with ribosomes from *Thermus thermophilus* (Weixlbaumer et al., 2007). Single-molecule FRET data indicate that the binding of RRF to the post-termination complex influences the transitions between classic/non-rotated and hybrid/rotated states of the post-termination ribosome (Sternberg et al., 2009). The observations in the *E. coli* system suggest that EF-G, driven by GTP hydrolysis and Pi release, induces a movement of RRF that, in turn, displaces H69 and possibly other bridges, thereby destabilizing the interactions between the subunits such that disassembly is promoted. A movement of domain 2 of RRF, which is flexibly attached to domain 1, seems to play an important role in promoting ribosome disassembly (Gao et al., 2005; Gao et al., 2007).

The structure of the post-termination complex containing both RRF and EF-G is not known. There is evidence, however, indicating that the two factors interact on the ribosome. Interactions have been demonstrated by mutational analysis of EF-G and RRF (Ito et al., 2002), by monitoring factor binding to the ribosome (Seo et al., 2004), or by cryo-EM reconstructions of complexes of 50S ribosomal subunits with the two factors (Gao et al., 2005; Gao et al., 2007). According to these data, the interaction between the two factors involves residues in domain 3 of EF-G and the hinge region in RRF as well as in domain 4 of EF-G and domain 2 of RRF. The movement of domain 4 of EF-G induced by GTP hydrolysis and Pi release that is observed in the fusidic acid-stalled EF-G complex with the ribosome (Agrawal et al., 1998; Stark et al., 2000; Valle et al., 2003) may be relayed to RRF and drive the ribosome rearrangement that leads to the dissociation of the subunits, including a movement of H69. The latter scenario is consistent with a function in ribosome disassembly of EF-G as a molecular motor that brings about subunit dissociation at the expense of GTP hydrolysis and Pi release.

3.3. Ribosome recycling as *in-vivo* target of fusidic acid

The functions of EF-G on the ribosome are inhibited by a number of antibiotics. Generally, translocation and ribosome disassembly are inhibited in a similar way, typical examples being paromomycin, hygromycin B, viomycin, and thiostrepton (Table 3). However, there are exceptions, the most remarkable of which is fusidic acid. Fusidic acid does not inhibit single-round translocation, whereas it slows down EF-G/RRF-dependent ribosome disassembly about 100-fold. The inhibition becomes effective already at submicromolar concentrations of fusidic acid, about a thousand-fold less than required for the inhibition of EF-G turnover (Savelsbergh et al., 2009). This low concentration is close to inhibitory concentrations *in vivo*, suggesting that the inhibition of EF-G/RRF-dependent disassembly of the post-termination complex is the mechanism by which fusidic acid inhibits cell growth, rather than the inhibition of EF-G turnover in elongation, as is generally assumed. The IC50 value observed for EF-G/RRF-dependent ribosome disassembly *in vitro*, ~0.1 μM, is very close to the affinity of fusidic acid binding to EF-G on the ribosome, 0.2–0.4 μM, as determined by direct binding assays (Okura et al., 1970; Willie et al., 1975). Binding of fusidic acid to EF-G on the ribosome seems to block a conformational change of EF-G that follows GTP hydrolysis and Pi release and is required for EF-G to dissociate from the ribosome. Assuming that the same or a similar conformational change of EF-G, modulated by RRF, drives subunit dissociation, the inhibition by fusidic acid of ribosome disassembly may be explained by an inhibition of that transition.

4. Conclusions

The ribosome is a dynamic structure that undergoes spontaneous conformational changes driven by thermal energy. The function of EF-G in translocation is to rectify these changes towards directional movement of tRNAs and mRNA in translocation at the expense of GTP hydrolysis. We are beginning to understand how conformational coupling and uncoupling between ribosome and EF-G is orchestrated by precise timing of GTP hydrolysis and Pi release. The function of EF-G as the pawl in a Brownian ratcheting machine is understood quite well on the structural level. In contrast, the function of EF-G in promoting unlocking of the ribosome is less clear, as the complex of pretranslocation ribosomes with EF-G has not been characterized in detail. To determine that structure, its dynamics, and the role of GTP hydrolysis in inducing the structural changes that lead to translocation is one of the major challenges in the field. Another challenge is the structure of the complete ribosome·RRF·EF-G complex prior to ribosome disassembly, which needs to be determined and related to biochemical information. Complementing these structures with structural dynamics and kinetics, as obtained by bulk and single-molecule methods as well as time-resolved cryo-EM, will ultimately reveal the molecular mechanisms by which EF-G exerts its functions on the ribosome.

Table 3 Effect of antibiotics on translocation and ribosome disassembly

Antibiotic	Binding site	Translocation inhibition	Recycling inhibition
Tetracyclin	30S	+	-
Spectinomycin	30S	++	+
Streptomycin	30S	+	+
Paromomycin	30S	+++	++
Hygromycin B	30S	+++	+++
Viomycin	30S/50S	++++	++
Thiostrepton	50S	++++	++++
Sparsomycin	50S	-	+
Erythromycin	50S	+	-
Fusidic acid	EF-G	-	+++

Time courses of translocation or ribosome disassembly were measured by stopped-flow, monitoring fluorescence or light-scattering, respectively (Peske et al., 2004; Savelsbergh et al., 2009). –, no inhibition (translocation about 15 s^{-1}, recycling about 0.3 s^{-1}); +, two- to five-fold inhibition; ++, about 30-fold; +++, 100–300-fold; ++++, >10^4-fold.

Acknowledgement

We thank the past group members, Dagmar Mohr and Berthold Wilden, who have contributed to the work on EF-G.

References

Ævarsson A, Brazhnikov E, Garber M, Zheltonosova J, Chirgadze, al-Karadaghi S, Svensson LA, Liljas A (1994) Three-dimensional structure of the ribosomal translocase: elongation factor G from *Thermus thermophilus*. EMBO J 13: 3669–3677

Agirrezabala X, Lei J, Brunelle JL, Ortiz-Meoz RF, Green R, Frank J (2008) Visualization of the hybrid state of tRNA binding promoted by spontaneous ratcheting of the ribosome. Mol Cell 32: 190–197

Agrawal RK, Linde J, Sengupta J, Nierhaus KH, Frank J (2001) Localization of L11 protein on the ribosome and elucidation of its involvement in EF-G-dependent translation. J Mol Biol 311: 777–787

Agrawal RK, Penczek P, Grassucci RA, Frank J (1998) Visualization of elongation factor G on the *Escherichia coli* 70S ribosome: The mechanism of translocation. Proc Natl Acad Sci USA 95: 6134–6138

Agrawal RK, Sharma MR, Kiel MC, Hirokawa G, Booth TM, Spahn CM, Grassucci RA, Kaji A, Frank J (2004) Visualization of ribosome-recycling factor on the *Escherichia coli* 70S ribosome: functional implications. Proc Natl Acad Sci USA 101: 8900–8905

Al-Karadaghi S, Ævarsson A, Garber M, Zheltonosova J, Liljas A (1996) The structure of elongation factor G in complex with GDP: conformational flexibility and nucleotide exchange. Structure 4: 555–565

Ali IK, Lancaster L, Feinberg J, Joseph S, Noller HF (2006) Deletion of a conserved, central ribosomal intersubunit RNA bridge. Mol Cell 23: 865–874

Barat C, Datta PP, Raj VS, Sharma MR, Kaji H, Kaji A, Agrawal RK (2007) Progression of the ribosome recycling factor through the ribosome dissociates the two ribosomal subunits. Mol Cell 27: 250–261

Blanchard SC, Kim HD, Gonzalez RL, Jr., Puglisi JD, Chu S (2004) tRNA dynamics on the ribosome during translation. Proc Natl Acad Sci USA 101: 12 893–12 898

Borovinskaya MA, Pai RD, Zhang W, Schuwirth BS, Holton JM, Hirokawa G, Kaji H, Kaji A, Cate JH (2007) Structural basis for aminoglycoside inhibition of bacterial ribosome recycling. Nature Struct Mol Biol 14: 727–732

Carter AP, Clemons WM, Brodersen DE, Morgan-Warren RJ, Wimberly BT, Ramakrishnan V (2000) Functional insights from the structure of the 30S ribosomal subunit and its interactions with antibiotics. Nature 407: 340–348

Chinali G, Parmeggiani A (1980) The coupling with polypeptide synthesis of the GTPase activity dependent on elongation factor G. J Biol Chem 255: 7455–7459

Cornish PV, Ermolenko DN, Noller HF, Ha T (2008) Spontaneous intersubunit rotation in single ribosomes. Mol Cell 30: 578–588

Czworkowski J, Wang J, Steitz TA, Moore PB (1994) The crystal structure of elongation factor G complexed with GDP, at 2.7 Å resolution. EMBO J 13: 3661–3668

Diaconu M, Kothe U, Schlunzen F, Fischer N, Harms JM, Tonevitsky AG, Stark H, Rodnina MV, Wahl MC (2005) Structural basis for the function of the ribosomal L7/12 stalk in factor binding and GTPase activation. Cell 121: 991–1004

Dorner S, Brunelle JL, Sharma D, Green R (2006) The hybrid state of tRNA binding is an authentic translation elongation intermediate. Nature Struct Mol Biol 13: 234–241

Dunkle JA, Cate JH (2010) Ribosome structure and dynamics during translocation and termination. Annu Rev Biophys 39: 227–244

Feinberg JS, Joseph S (2006) Ribose 2'-hydroxyl groups in the 5' strand of the acceptor arm of P-site tRNA are not essential for EF-G catalyzed translocation. RNA 12: 580–588

Fischer N, Konevega AL, Wintermeyer W, Rodnina MV, Stark H (2010) Ribosome dynamics and tRNA movement by time-resolved electron cryomicroscopy. Nature 466: 329–333

Frank J, Agrawal RK (2000) A ratchet-like inter-subunit reorganization of the ribosome during translocation. Nature 406: 318–322

Frank J, Gonzalez RL (2010) Structure and dynamics of a processive Brownian motor: the translating ribosome. Annu Rev Biochem 79: 381–412

Fredrick K, Noller HF (2003) Catalysis of ribosomal translocation by sparsomycin. Science 300: 1159–1162

Fujiwara T, Ito K, Yamami T, Nakamura Y (2004) Ribosome recycling factor disassembles the post-termination ribosomal complex independent of the ribosomal translocase activity of elongation factor G. Mol Microbiol 53: 517–528

Gao N, Zavialov AV, Ehrenberg M, Frank J (2007) Specific interaction between EF-G and RRF and its implication for GTP-dependent ribosome splitting into subunits. J Mol Biol 374: 1345–1358

Gao N, Zavialov AV, Li W, Sengupta J, Valle M, Gursky RP, Ehrenberg M, Frank J (2005) Mechanism for the disassembly of the posttermination complex inferred from cryo-EM studies. Mol Cell 18: 663–674

Gao YG, Selmer M, Dunham CM, Weixlbaumer A, Kelley AC, Ramakrishnan V (2009) The structure of the ribosome with elongation factor G trapped in the posttranslational state. Science 326: 694–699

Gavrilova LP, Kostiashkina OE, Koteliansky VE, Rutkevitch NM, Spirin AS (1976) Factor-free ("non-enzymic") and factor-dependent systems of translation of polyuridylic acid by *Escherichia coli* ribosomes. J Mol Biol 101: 537–552

Hansson S, Singh R, Gudkov AT, Liljas A, Logan DT (2005) Crystal structure of a mutant elongation factor G trapped with a GTP analogue. FEBS Lett 579: 4492–4497

Hirashima A, Kaji A (1973) Role of elongation factor G and a protein factor on the release of ribosomes from messenger ribonucleic acid. J Biol Chem 248: 7580–7587

Hirokawa G, Kiel MC, Muto A, Selmer M, Raj VS, Liljas A, Igarashi K, Kaji H, Kaji A (2002) Post-termination complex disassembly by ribosome recycling factor, a functional tRNA mimic. EMBO J 21: 2272–2281

Horan LH, Noller HF (2007) Intersubunit movement is required for ribosomal translocation. Proc Natl Acad Sci USA 104: 4881–4885

Ito K, Fujiwara T, Toyoda T, Nakamura Y (2002) Elongation factor G participates in ribosome disassembly by interacting with ribosome recycling factor at their tRNA-mimicry domains. Mol Cell 9: 1263–1272

Janosi L, Mottagui-Tabar S, Isaksson LA, Sekine Y, Ohtsubo E, Zhang S, Goon S, Nelken S, Shuda M, Kaji A (1998) Evidence for *in vivo* ribosome recycling, the fourth step in protein biosynthesis. EMBO J 17: 1141–1151

Joseph S, Noller HF (1998) EF-G-catalyzed translocation of anticodon stem-loop analogs of transfer RNA in the ribosome. EMBO J 17: 3478–3483

Julian P, Konevega AL, Scheres SH, Lazaro M, Gil D, Wintermeyer W, Rodnina MV, Valle M (2008) Structure of ratcheted ribosomes with tRNAs in hybrid states. Proc Natl Acad Sci USA 105: 16 924–16 927

Karimi R, Pavlov MY, Buckingham RH, Ehrenberg M (1999) Novel roles for classical factors at the interface between translation termination and initiation. Mol Cell 3: 601–609

Katunin VI, Savelsbergh A, Rodnina MV, Wintermeyer W (2002) Coupling of GTP hydrolysis by elongation factor G to translocation and factor recycling on the ribosome. Biochemistry 41: 12 806–12 812

Konevega AL, Fischer N, Semenkov YP, Stark H, Wintermeyer W, Rodnina MV (2007) Spontaneous reverse movement of mRNA-bound tRNA through the ribosome. Nature Struct Mol Biol 14: 318–324

Kubarenko A, Sergiev P, Wintermeyer W, Dontsova O, Rodnina MV (2006) Involvement of helix 34 of 16 S rRNA in decoding and translocation on the ribosome. J Biol Chem 281: 35 235–35 244

Lancaster L, Kiel MC, Kaji A, Noller HF (2002) Orientation of ribosome recycling factor in the ribosome from directed hydroxyl radical probing. Cell 111: 129–140

Lancaster LE, Savelsbergh A, Kleanthous C, Wintermeyer W, Rodnina MV (2008) Colicin E3 cleavage of 16S rRNA impairs decoding and accelerates tRNA translocation on *Escherichia coli* ribosomes. Mol Microbiol 69: 390–401

Lill R, Robertson JM, Wintermeyer W (1989) Binding of the 3' terminus of tRNA to 23S rRNA in the ribosomal exit site actively promotes translocation. EMBO J 8: 3933–3398

Matassova AB, Rodnina MV, Wintermeyer W (2001) Elongation factor G-induced structural change in helix 34 of 16S rRNA related to translocation on the ribosome. RNA 7: 1879–1885

McGarry KG, Walker SE, Wang H, Fredrick K (2005) Destabilization of the P site codon-anticodon helix results from movement of tRNA into the P/E hybrid state within the ribosome. Mol Cell 20: 613–622

Moazed D, Noller HF (1989) Intermediate states in the movement of transfer RNA in the ribosome. Nature 342: 142–148

Mohr D, Wintermeyer W, Rodnina MV (2002) GTPase activation of elongation factors Tu and G on the ribosome. Biochemistry 41: 12 520–12 528

Munro JB, Altman RB, O'Connor N, Blanchard SC (2007) Identification of two distinct hybrid state intermediates on the ribosome. Mol Cell 25: 505–517

Munro JB, Altman RB, Tung CS, Cate JH, Sanbonmatsu KY, Blanchard SC (2010) Spontaneous formation of the unlocked state of the ribosome is a multistep process. Proc Natl Acad Sci USA 107: 709–714

Nechifor R, Murataliev M, Wilson KS (2007) Functional interactions between the G' subdomain of bacterial translation factor EF-G and ribosomal protein L7/L12. J Biol Chem 282: 36 998–37 005

Okura A, Kinoshita T, Tanaka N (1970) Complex formation of fusidic acid with G factor, ribosome and guanosine nucleotide. Biochem Biophys Res Commun 41: 1545–1550

Pai RD, Zhang W, Schuwirth BS, Hirokawa G, Kaji H, Kaji A, Cate JH (2008) Structural Insights into ribosome recycling factor interactions with the 70S ribosome. J Mol Biol 376: 1334–1347

Pan D, Kirillov SV, Cooperman BS (2007) Kinetically competent intermediates in the translocation step of protein synthesis. Mol Cell 25: 519–529

Peske F, Matassova NB, Savelsbergh A, Rodnina MV, Wintermeyer W (2000) Conformationally restricted elongation factor G retains GTPase activity but is inactive in translocation on the ribosome. Mol Cell 6: 501–505

Peske F, Rodnina MV, Wintermeyer W (2005) Sequence of steps in ribosome recycling as defined by kinetic analysis. Mol Cell 18: 403–412

Peske F, Savelsbergh A, Katunin VI, Rodnina MV, Wintermeyer W (2004) Conformational changes of the small ribosomal subunit during elongation factor G-dependent tRNA-mRNA translocation. J Mol Biol 343: 1183–1194

Rodnina MV, Savelsbergh A, Katunin VI, Wintermeyer W (1997) Hydrolysis of GTP by elongation factor G drives tRNA movement on the ribosome. Nature 385: 37–41

Savelsbergh A, Katunin VI, Mohr D, Peske F, Rodnina MV, Wintermeyer W (2003) An elongation factor G-induced ribosome rearrangement precedes tRNA-mRNA translocation. Mol Cell 11: 1517–1523

Savelsbergh A, Matassova NB, Rodnina MV, Wintermeyer W (2000) Role of domains 4 and 5 in elongation factor G functions on the ribosome. J Mol Biol 300: 951–961

Savelsbergh A, Mohr D, Kothe U, Wintermeyer W, Rodnina MV (2005) Control of phosphate release from elongation factor G by ribosomal protein L7/12. EMBO J 24: 4316–4323

Savelsbergh A, Rodnina MV, Wintermeyer W (2009) Distinct functions of elongation factor G in ribosome recycling and translocation. RNA 15: 772–780

Schuwirth BS, Borovinskaya MA, Hau CW, Zhang W, Vila-Sanjurjo A, Holton JM, Cate JH (2005) Structures of the bacterial ribosome at 3.5 Å resolution. Science 310: 827–834

Semenkov YP, Rodnina MV, Wintermeyer W (2000) Energetic contribution of tRNA hybrid state formation to translocation catalysis on the ribosome. Nat Struct Biol 7: 1027–1031

Seo HS, Kiel M, Pan D, Raj VS, Kaji A, Cooperman BS (2004) Kinetics and thermodynamics of RRF, EF-G, and thiostrepton interaction on the Escherichia coli ribosome. Biochemistry 43: 12 728–12 740

Shoji S, Walker SE, Fredrick K (2006) Reverse translocation of tRNA in the ribosome. Mol Cell 24: 931–942

Shoji S, Walker SE, Fredrick K (2009) Ribosomal translocation: one step closer to the molecular mechanism. ACS Chem Biol 4: 93–107

Singh NS, Ahmad R, Sangeetha R, Varshney U (2008) Recycling of ribosomal complexes stalled at the step of elongation in Escherichia coli. J Mol Biol 380: 451–464

Spiegel PC, Ermolenko DN, Noller HF (2007) Elongation factor G stabilizes the hybrid-state conformation of the 70S ribosome. RNA 13: 1473–1482

Stark H, Rodnina MV, Wieden H-J, van Heel M, Wintermeyer W (2000) Large-scale movement of elongation factor G and extensive conformational change of the ribosome during translocation. Cell 100: 301–309

Sternberg SH, Fei J, Prywes N, McGrath KA, Gonzalez RL, Jr (2009) Translation factors direct intrinsic ribosome dynamics during translation termination and ribosome recycling. Nature Struct Mol Biol 16: 861–868

Takyar S, Hickerson RP, Noller HF (2005) mRNA helicase activity of the ribosome. Cell 120: 49–58

Ticu C, Nechifor R, Nguyen B, Desrosiers M, Wilson KS (2009) Conformational changes in switch I of EF-G drive its directional cycling on and off the ribosome. EMBO J 28: 2053–2065

Valle M, Zavialov A, Sengupta J, Rawat U, Ehrenberg M, Frank J (2003) Locking and unlocking of ribosomal motions. Cell 114: 123–134

Walker SE, Shoji S, Pan D, Cooperman BS, Fredrick K (2008) Role of hybrid tRNA-binding states in ribosomal translocation. Proc Natl Acad Sci USA 105: 9192–9197

Weixlbaumer A, Petry S, Dunham CM, Selmer M, Kelley AC, Ramakrishnan V (2007) Crystal structure of the ribosome recycling factor bound to the ribosome. Nat Struct Mol Biol 14: 733–737

Wen JD, Lancaster L, Hodges C, Zeri AC, Yoshimura SH, Noller HF, Bustamante C, Tinoco I (2008) Following translation by single ribosomes one codon at a time. Nature 452: 598–603

Wilden B, Savelsbergh A, Rodnina MV, Wintermeyer W (2006) Role and timing of GTP binding and hydrolysis during EF-G-dependent tRNA translocation on the ribosome. Proc Natl Acad Sci USA 103: 13 670–13 675

Willie GR, Richman N, Godtfredsen WP, Bodley JW (1975) Some characteristics of and structural requirements for the interaction of 24,25-dihydrofusidic acid with ribosome – elongation factor G complexes. Biochemistry 14: 1713–1718

Wilson DN, Schluenzen F, Harms JM, Yoshida T, Ohkubo T, Albrecht R, Buerger J, Kobayashi Y, Fucini P (2005) X-ray crystallography study on ribosome recycling: the mechanism of binding and action of RRF on the 50S ribosomal subunit. EMBO J 24: 251–260

Zavialov AV, Ehrenberg M (2003) Peptidyl-tRNA regulates the GTPase activity of translation factors. Cell 114: 113–122

Zavialov AV, Hauryliuk VV, Ehrenberg M (2005) Splitting of the posttermination ribosome into subunits by the concerted action of RRF and EF-G. Mol Cell 18: 675–686

Mechanism and dynamics of the elongation cycle

27

Barry S. Cooperman, Yale E. Goldman, Chunlai Chen, Ian Farrell, Jaskarin Kaur, Hanqing Liu, Wei Liu, Gabriel Rosenblum, Zeev Smilansky, Benjamin Stevens, and Haibo Zhang

1. Introduction

Continued dramatic progress in the elucidation of the structures of the bacterial ribosome and its functional complexes has led to proposals for the detailed mechanisms of ribosome-catalyzed protein synthesis (Schmeing and Ramakrishnan, 2009; Agirrezabala and Frank, 2009). Ensemble rapid reaction kinetics (Antoun et al., 2006; Daviter et al., 2006; Dorner et al., 2006; Grigoriadou et al., 2007; Hetrick et al., 2009; Pan et al., 2007, 2008; Pape et al., 1998; Phelps and Joseph 2006; Rodnina et al., 1997; Savelsbergh et al., 2003; Walker et al., 2008; Wintermeyer et al., 2004; Zaher and Green, 2009; Zavialov and Ehrenberg, 2003) and single-molecule (Blanchard et al., 2004a,b; Cornish et al., 2008, 2009; Fei et al., 2008, 2009; Marshall et al., 2008, 2009; Munro et al., 2007, 2010a,b; Uemura et al., 2010; Wang et al., 2007) studies of the translational machinery in the past several years have resulted in increased understanding of many aspects of the initiation, elongation, and termination phases of protein synthesis, but many essential points remain to be elucidated.

A simplified version of the elongation cycle, based on known ribosomal structures (Schmeing and Ramakrishnan, 2009), is shown in Figure 1. The process begins with an incoming aminoacyl-tRNA (aa-tRNA), in the form of a ternary complex (TC) with EF-Tu and GTP, binding to the A/T site of the ribosome, either prior to (step 1a) or following (step 1b) deacylated tRNA dissociation from the ribosome. We denote these routes the 2:3 and 2:1 pathways, respectively, corresponding to the ribosome oscillating between having 2 and 3 or 2 and 1 bound tRNAs, respectively. The aa-tRNA is next accommodated into the A site (step 2a or 2b), with concomitant GTP hydrolysis and dissociation of EF-Tu·GDP and P_i. This is followed

rapidly by peptide bond formation (step 3), transferring the nascent peptide to the tRNA in the A site, and resulting in formation of the pre-translocation (PRE) complex. In the PRE complex, tRNAs fluctuate between classical (A/A and P/P sites, 30S/50S binding positions) and hybrid (A/P and P/E sites) states (step 4 and its reversal). EF-G·GTP binding to the hybrid PRE complex (step 5) induces translocation of peptidyl-tRNA and deacylated-tRNA into the P- and E-sites, respectively, with movement of the mRNA by one codon and with GTP hydrolysis (step 6). Dissociation of EF-G·GDP and P_i (step 7), leading to formation of the post-translocation (POST) complex, completes the cycle. Structural studies have demonstrated large scale motions of the ribosome during elongation that

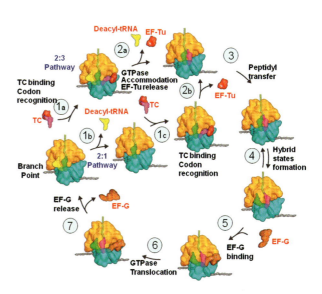

Fig. 1 The elongation cycle. Adapted from Schmeing and Ramakrishnan (2009). Note that ternary complex (TC) binding may precede (2:3 pathway) or follow (2:1 pathway) deacylated tRNA dissociation. Also noteworthy is the assumption that hybrid state formation precedes EF-G binding.

include a rotation (termed ratcheting) of the 30S and 50S subunits relative to each other (Frank and Agarwal, 2000), and a swiveling of the head domain of the 30S subunit (Zhang et al., 2009).

Here we present recent results from our laboratories that utilize ensemble stopped-flow fluorescence and quenched-flow ensemble kinetics as well as single-molecule fluorescence resonance energy transfer (smFRET) studies to investigate several aspects of the elongation cycle, focusing on translocation, the function of EF4 (LepA), A site/E site allosteric interaction, and ternary complex binding to the ribosome. An overview of these experiments is presented in Figure 2.

Most of our studies employ fluorescence-labeled components of the protein synthesis system, including tRNAs, mRNAs, protein factors, and ribosomal proteins. For tRNAs we introduce labels by exploiting the highly specific reduction chemistry of dihydrouridine. This approach, pioneered by Zachau and his co-workers (Wintermeyer et al., 1979) has been expanded by us to include the efficient introduction of

dyes suitable for single-molecule studies (Betteridge et al., 2007; Pan et al., 2009). Derivatizing positions in the D loop of tRNAs presents two very significant advantages. tRNAs derivatized in this manner generally retain substantial functionality in protein synthesis assays and almost all tRNAs (> 90 % in E. coli) contain dihydrouridine residues. For mRNAs, we use the 3' terminus labeling procedure (Peske et al., 2005; Studer et al., 2003). For both protein factors and ribosomal proteins we typically remove Cys residues present in the native structures and then introduce Cys residues at what are predicted to be sterically strategic sites, useful for FRET studies, based on known structures of ribosomal complexes.

Labeled ribosomal proteins are combined with ribosomes lacking such proteins to form ribosomes labeled at a unique site (Chen et al., 2011; Qin et al., 2009; Seo et al., 2006; Wang et al., 2007). In some of the studies reported below we utilize ribosomes labeled in protein L11 or L1, each of which falls within a mobile domain of the large (50S) ribosomal subunit that is strongly involved in the binding and movement of tRNA during the elongation cycle. L11 is part of the GTPase-associated center (GAC) (Valle et al., 2003a) which contributes part of the A/T and A sites. L1 is part of the L1 stalk that interacts with tRNA in the P, E and P/E sites.

2. Translocation

2.1. Ensemble studies

Earlier ensemble studies of translocation kinetics (Pan et al., 2007; Rodnina et al., 1997; Savelsbergh et al., 2003; Walker et al., 2008) led to a model of translocation in which EF-G·GTP binding to a PRE complex is followed by a very rapid GTP hydrolysis, conformational changes including subunit rotation ("ratcheting"), specific motions within the GAC (Diaconu et al., 2005; Frank and Agrawal, 2000; Seo et al., 2006) and conversion of the PRE complex to a hybrid PRE complex, denoted the INT complex (Pan et al., 2007). During formation of the INT complex, the 3' terminus of peptidyl-tRNA moves from a position in which it has negligible puromycin reactivity ($<10^{-3}$ relative to a POST complex) to a position in which it has appreciable puromycin reactivity (~5% of that of a POST complex) (Pan et al., 2007). Translocation is completed by further movement of the tRNAs into the P/P and E/E positions to form the POST complex, with dissoci-

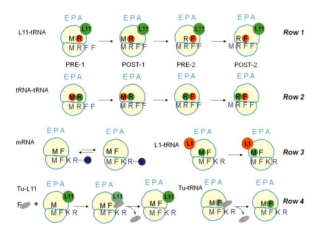

Fig. 2 Schematic of the complexes and reactions studied in the ensemble and smFRET studies of this work. The blue colored letters M, R, F and K indicate mRNA codons for fMet, Arg, Phe and Lys. The black letters M, R, F and K refer to tRNA^{fMet}, tRNA^{Arg}, tRNA^{Phe} and tRNA^{Lys}. Red dots indicate Cy5-labeling of L1 or tRNAs. Green dots indicate Cy3-labeling of L11 or tRNA. The blue dot indicates fluorescein-labeling of the 3' end of mRNA and the grey oval represents QSY-labeled EF-Tu. Rows 1 and 2: L11-tRNA (L11-t) and tRNA-tRNA (tt) FRET complexes, respectively, for elongation cycles 1 and 2. Row 3 (left): Fluorescein-labeled mRNA complexes for measuring EF4 interaction with PRE-1 and POST-1 complexes. Row 3 (right): L1-tRNA^{fMet} FRET complexes for measuring PRE-1 to POST-1 conversion. Similar complexes with Cy3-labeled tRNA^{Phe} measure PRE-2 to POST-2 conversion (not shown). Row 4: Reactions utilizing ternary complex containing QSY-labeled EF-Tu. (Left): Reaction with 70SIC complex containing Cy3-labeled L11. (Right): Release of EF-Tu during accommodation of Cy3-labeled Phe-tRNA^{Phe} into the A site.

ation of EF-G·GDP and P_i and reversion of the ribosome to the unratcheted conformation.

2.2. smFRET studies

More recently we have examined the first two elongation cycles of protein synthesis using ribosomes programmed with an mRNA coding for the initial tripeptide fMetArgPhe (Chen et al., 2011). In these experiments we use smFRET to monitor the distances, both in PRE complexes and during the translocation step resulting in POST complex formation, between position 87 in protein L11 and either tRNA^{Arg} (1st cycle) or tRNA^{Phe} (2nd cycle) (L11-t complexes – Figure 2, row 1) and between the D loops of either tRNA^{fMet} and tRNA^{Arg} (1st cycle) or tRNA^{Arg} and tRNA^{Phe} (2nd cycle) (tt complexes – Figure 2, row 2).

2.3. Dynamics of PRE complexes

All four PRE complexes, PRE-1(L11-t), PRE-1(tt), PRE-2-(L11-t) and PRE-2-(tt) show two FRET states with some molecules fluctuating between them and the others staying stably in one or the other of the FRET states. Two fluctuating tRNA-tRNA FRET states were reported in an earlier smFRET study of a PRE-I complex (Blanchard et al., 2004a,b), which identified the state having higher tRNA-tRNA FRET efficiency with the classical orientation of the tRNAs in the A/A and P/P positions, and the lower FRET state with tRNAs in the hybrid A/P and P/E positions. Cryo-electron microscopy (cryo-EM) studies (Agirrezabala et al., 2008) show that the hybrid positions of the tRNAs relative to the 30S subunit are similar to their classical positions, whereas their locations in the 50S subunit are displaced in the direction of translocation. These structural results support the assignment of the higher and lower FRET states in all four PRE complexes studied by us to the classical and hybrid states, respectively. The similarities between the L11-t and tt complexes suggest that the reversible classical to hybrid transition is monitored by changes in both L11-tRNA and tRNA-tRNA distances, indicating that the tRNAs and the L11 region move in concert.

2.4. Translocation of PRE complexes

Translocation was monitored in real time following stopped-flow injection of EF-G·GTP into a reaction vessel containing PRE-1(L11-t), PRE-2(L11-t), or PRE-2(tt) complexes. EF-G·GTP addition to each PRE complex results in a FRET change to a new state whose FRET efficiency agrees well with the value predicted for a POST complex, based on structural studies. Monitoring smFRET between either L11 and tRNA or tRNA and tRNA thus provides an unambiguous demonstration of the translocation process. Detailed analysis on smFRET translocation traces of both the L11-t and tt complexes demonstrates several significant points (Chen et al., 2011). (1) EF-G directly binds to both the classical and hybrid PRE complexes. This contrasts with a previous hypothesis (Figure 1, Fei et al., 2008) that A/A-bound peptidyl tRNA has to move to the A/P position prior to EF-G binding. (2) EF-G binding halts the fluctuations prior to catalyzing the translocation step. (3) Translocation from the classic state proceeds via formation of a transient intermediate, having a FRET indistinguishable from that of the hybrid state of the PRE complex, consistent with the stopped-flow results of Pan et al., (2007) described above. (4) Following translocation, ribosome conformational lability within the highly mobile L11 region of the ribosome (Schuwirth et al., 2005), is largely suppressed by the binding of the next cognate EF-Tu-tRNA·GTP ternary complex at the A site, perhaps because of contacts between the ternary complex and this region (Schmeing et al., 2009).

3. Enigmatic EF4(LepA)

There has been a recent surge of interest in EF4, stemming largely from the recent demonstration of its ability to catalyze back translocation (Qin et al., 2006), a process which occurs spontaneously, albeit quite slowly, following EF-G dissociation from the ribosome (Konevega et al., 2007; Shoji et al., 2006). EF4 is a ribosome-dependent GTPase with strong structural similarity to the translocation factor EF-G. Four out of the five EF-G domains – I, II, III, and V, but not IV nor the G' insertion in domain I – are also found in EF4. Domain IV of EF-G reaches into the A site in the decoding center and is thought to prevent back translocation (Connell et al., 2007; Nierhaus 1996; Rodnina et al., 1997). EF4 also has a unique C-terminal domain with a novel fold (Evans et al., 2008). The structure of the PRE complex resulting from prolonged incubation of the POST complex with EF4, has also been determined (Connell et al., 2008). In this complex, denoted PRE(L), the 3' end of peptidyl-tRNA occupies a puromycin-unreactive position, denoted the A/L position, which is intermediate between those found in the A/A and A/T sites. Elucidation of the *in-vivo* role of EF4 is also being pursued actively. Although DEF4 cells grown in rich medium have long been known to have no phenotype (Dibb and Wolfe, 1986), Nierhaus and co-workers (Pech et al., 2011) have recently demonstrated that certain stress conditions, including high salt, cause a ΔEF4 strain to be overgrown by wild-type bacterial cells, and causes EF4 to move from the membrane to the cytoplasm. In addition, Shoji et al., (2010) have shown a ΔEF4 *E. coli* strain to be hypersensitive to potassium tellurite and to penicillin.

3.1. EF4 catalysis of partial back translocation

In recently published work (Liu et al., 2010) we used a coordinated set of ensemble kinetic measurements, including changes in the puromycin reactivity of peptidyl tRNA and in the fluorescence of labeled tRNAs and mRNA, to elucidate the kinetic mechanism of EF4-catalyzed back translocation. This resulted in the formulation of the quantitative kinetic mechanism shown in Figure 3A. Here I_1, I_2, and I_3, formed via steps 1–3, respectively, are intermediates in the overall conversion of the POST complex to the PRE(L) complex, which is formed via step 4. PRE(L), just like the PRE complex formed in the absence of EF4, is virtually unreactive toward puromycin. All three intermediates in Figure 3A have ~12-fold less reactivity toward puromycin than the POST complex. Overall formation of I_3 from the POST complex, though relatively slow by the standards of the normal reactions of elongation cycle (k = 0.012 s^{-1}), nevertheless proceeds considerably

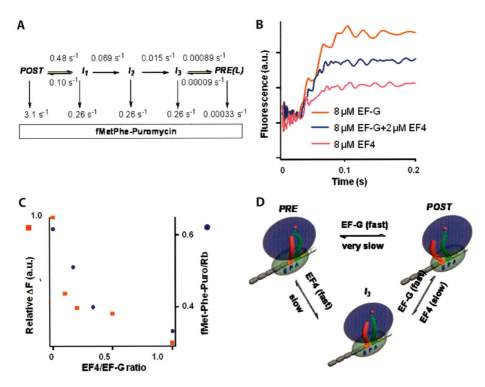

Fig. 3 EF4 reaction with a PRE-1 complex. (A) Kinetic scheme of EF4-catalyzed back translocation (see text for details). (B) EF4-dependent mRNA movement measured by change in the fluorescence of Flu-mRNA014 in a stopped-flow fluorometer. Fluorescence was monitored using a 495-nm long-pass filter on excitation at 460 nm. PRE complexes (0.1 μM) were rapidly mixed with 8 μM EF-G·GTP (red trace) or 8 μM EF4·GTP (pink trace) or 8 μM EF-G·GTP + 2 μM EF4·GTP (blue trace). (C) EF4 competition with EF-G for binding to PRE complex. (■) EF4-dependent mRNA movement measured as in panel B. PRE complexes (0.1 μM) were rapidly mixed with EF4·GTP (5 μM) or EF-G·GTP (5 μM) + EF4·GTP at the indicated ratios. Shown is the relative change in fluorescence, DF, equal to 1.0 in the absence of EF4, after 1 s. (●) fMetPhe reactivity toward puromycin. PRE complexes (0.1 μM) were rapidly mixed with 3 mM puromycin and EF-G·GTP (3 μM) or EF-G·GTP (3 μM) + EF4·GTP at the indicated ratios and quenched with acid at 3.5 sec. (D) Tentative model for the EF4 mechanism of action *in vivo*.

more rapidly than the formation of PRE(L) from I_3. In fact, the rate constant for the latter reaction, 9×10^{-4} s^{-1}, is only slightly higher than the overall rate constant for PRE formation measured in the absence of EF4, 5.7×10^{-4} s^{-1}. This comparison suggests that EF4 performs an interrupted catalysis, stopping at the formation of I_3 rather than catalyzing the complete process of back translocation culminating in PRE complex formation. During overall formation of I_3, movements of the tRNA core regions and of mRNA are closely coupled to one another, but are sometimes decoupled from movement of the 3' end of peptidyl-tRNA (the acceptor arm). For example, step 1 is accompanied by a change in the position of the acceptor arm, with little or no movement of mRNA or either of the tRNA cores, while movement of mRNA and the two tRNA cores in step 3 is not accompanied by movement of the acceptor arm.

Peptidyl-tRNA in the I_3 complex is likely to occupy a position falling in between the A and P sites, based on its substantial reactivity toward puromycin. This position is almost certainly different from the A/L position referred to above. Our results also suggest that tRNA^{fMet} might occupy a position within the 30S subunit that is intermediate between the P and E sites, possibly similar to the position that fMet-tRNA^{fMet} is thought to adopt during 70S initiation complex formation.

3.2. EF4 competes with EF-G for binding to PRE complex

The relative slowness of the partial EF4-catalyzed back translocation process, mentioned above, raises the question of whether the POST complex is the principal substrate for EF4. In more recent work, we have examined the interaction of EF4 with a PRE-1 complex containing tRNA^{fMet} and fMetPhe-tRNA-^{Phe}. Two experimental measures, the fluorescence of fluorescein-mRNA (Figure 2, row 3, left) and reactivity of fMetPhe toward puromycin, provide compelling evidence that EF4 reacts with PRE complex as rapidly as does EF-G, and in a competitive fashion with EF-G. This competition is not surprising considering the extensive structural homology between these two G proteins.

Prior work has established that, following rapid mixing of the PRE complex with EF-G•GTP, a rapid lag phase is followed by an increase in both fluorescence intensity of fluorescein-labeled mRNA and puromycin reactivity of peptidyl-tRNA (Pan et al., 2007; Qin et al., 2010; Rodnina et al., 1997; Savelsbergh et al., 2003).

Similar kinetics are seen on EF4•GTP addition to PRE complex, although the complex formed at saturating EF4 contains a fluorescein-mRNA with lower fluorescence intensity (Figure 3B) and lower puromycin reactivity than that seen at saturating EF-G. Evidence of the competition between EF-G and EF4 for PRE complex comes from results presented in Figure 3B, C, with the levels of fluorescein-mRNA fluorescence and of reactivity of peptidyl-tRNA toward puromycin dropping monotonically to levels seen in the presence of saturating EF4 alone when PRE complex is mixed with a fixed amount of EF-G and increasing levels of added EF4. These results suggest an apparent affinity of EF4 for the PRE complex that is higher than that of EF-G.

Using smFRET analysis, as described above, we also have shown that adding EF4 to the PRE complex shifts the distribution of states in favor of the hybrid (lower FRET) state, for either PRE-1(L11-t) or PRE-1(tt) complexes. However, this state is altered from a true PRE or true POST complex, given the results presented in Figure 3B, C.

3.3. A tentative model for the EF4 mechanism of action *in vivo*

The complex that is formed on EF4 addition to PRE complex has properties similar to those of I_3 formed on addition of EF4 to POST complex, and our current working hypothesis is that they are indeed the same. Based on the rate data presented above, it is likely that I_3, which we have shown can rapidly react with EF-G to form the POST complex, is formed predominately via EF4 interaction with PRE complex. These points are summarized in Figure 3D.

What biological purpose might I_3 formation serve? Qin et al., (2006) have shown that added EF4 significantly increases the fraction of active protein made in a cell-free coupled transcription-translation system. EF4 capture of the PRE complex, by competing with EF-G, will at least transiently interrupt normal polypeptide elongation and could, by facilitating co-translational protein folding (Zhang et al., 2009), increase the fraction of active protein produced, as suggested by Shoji et al., (2010). Despite our results showing that EF4 significantly outcompetes EF-G in binding to a PRE complex, such competition should be rather infrequent (<1 in 10 elongation cycles) under normal growth conditions, when EF4 is present in bacterial cytoplasm at approximately 0.02 copies/ribosome, some 50-fold less than EF-G (assuming that the EF-G concentration equals that of the ribosome; Gordon, 1970). However, under conditions of growth stress the EF4 concentration in the bacterial cytoplasm rises by almost an order of magnitude (K. Nierhaus, private communication), significantly raising the probability of EF4 capture. It is also possible that specific ribosome conformations, which could become prominent when specific sequences of mRNA are being translated, also influence the competition between EF-G and EF4. This is because slowing the elongation cycle would only be expected to have salutary effects on the activities of a limited set of proteins. Clearly further experiments will be needed to examine this point.

4. Release of E-site tRNA

A point of ambiguity in our understanding of the elongation cycle concerns the relative timing of deacylated-tRNA release from the E site of a POST complex as compared with the binding of cognate aminoacyl-tRNA to the A site (Semenkov et al., 1996; Dinos et al., 2005; Wilson and Nierhaus, 2006). This point is important to resolve, because E-site binding has been postulated to modulate a variety of ribosomal functions, including maintenance of reading frame (Marquez et al., 2004) and the fidelity of A-site tRNA selection (Geigenmuller and Nierhaus 1990). We use both ensemble FRET and smFRET studies to address this question, which can be posed as whether TC binding follows the 2:3 or 2:1 pathway, as defined in Figure 1.

4.1. Ensemble stopped-flow experiments.

Following Fei et al., (2008, 2009) we have prepared ribosomes containing a Cy5-labeled T202C-L1 variant (Fei et al., (2008); *E. coli* wild-type L1 lacks Cys) and used them, in combination with tRNAs labeled with either Cy3 or rhodamine 110 (rhd) (Betteridge et al., 2009; Pan et al., 2009) in the D loop, in ensemble stopped-flow FRET experiments measuring rates of L1-tRNA and tRNA-tRNA distance changes following POST complex formation that parallel recent single molecule experiments (Fei et al., 2008; Munro et al., 2010a). In these experiments (Figure 4A–C) PRE complexes containing tRNA^{fMet} and either fMetPhe-tRNA^{Phe} or Phe-tRNA^{Phe} in the A-site are rapidly mixed with EF-G. GTP and changes in the FRET efficiency between L1^{Cy5} and tRNA^{Met}(Cy3) (Lt traces) or between tRNA^{Met}(Cy3) and either fMetPhe-tRNA^{Phe}(rhd) or Phe-tRNA^{Phe}(rhd) (tt

traces) are monitored. As expected from smFRET results (Blanchard et al., 2004a,b; Chen et al., 2011; Fei et al., 2008), neither Lt nor tt FRET efficiency changes much during PRE to POST conversion, and the large FRET decreases are only seen following POST formation, corresponding presumably to tRNA^{fMet} dissociation from the ribosome. In each case, the rates of tt

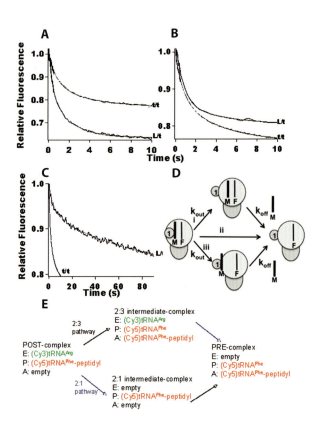

Fig. 4 FRET experiments. (A)–(D) Ensemble stopped-flow FRET. PRE complexes programmed with mRNA initially coding for fMetPheLys were rapidly mixed with EF-G·GTP (final concentration 1 μM). L1 position is indicated by circle. Decreases in Cy5 fluorescence (black traces) or Cy3 fluorescence (gray traces) were measured on excitation of Cy3 (540 nm) or rhodamine110 (480 nm), respectively. (A) Buffer I and (B) buffer J. Black or gray traces are for samples containing Cy5-labeled T202C-L1, tRNA^{fMet}(Cy3), and fMetPhe-tRNA^{Phe} or tRNA^{fMet}(Cy3) and fMetPhe-tRNA^{Phe}(rhd), respectively. (C) Black or gray traces are for samples containing Cy5-labeled T202C-L1, tRNA^{fMet}(Cy3), and Phe-tRNA^{Phe} or tRNA^{fMet}(Cy3) and Phe-tRNA^{Phe}(rhd), respectively. Buffer I (50 mM Tris-HCl [pH 7.5], 70 mM NH₄Cl, 30 mM KCl, 7 mM MgCl₂, and 1 mM DTT). Buffer J (20 mM HEPES-KOH [pH 7.5 at 0 °C], 150 mM NH₄Ac, 4.5 mM MgAc₂, 4 mM 2-mercaptoethanol, 0.05 mM spermine, and 2 mM spermidine). (D) Scheme showing three possible reaction pathways for loss of tRNA^{fMet} from POST complexes. Pathways i, ii and iii correspond to results presented in panels A, B and C, respectively. (E) Schematic of a single-molecule FRET experiment to distinguish 2:3 from 2:1 pathways in the conversion of a POST-2 complex to a PRE-3 complex. Ribosomes are programmed with an mRNA coding for the initial sequence fMetArgPhePhe. The black arrows represent binding of TC to the A site, and the blue arrows represent dissociation of deacyl tRNA from the E site (Chen et al., in preparation).

FRET loss are very similar (1.5 ± 0.3 s^{-1}). They differ in that the rate of Lt FRET decrease is three to four-fold faster than the rate of tt FRET decrease in Figure 4A, proceeds at about the same rate in Figure 4B, and is much slower in Figure 4C. These results lead to the scheme shown in Figure 4D, in which either L1 moves to an open conformation (Agrawal et al., 1999; Cornish et al., 2009; Fischer et al., 2010) without tRNA^{fMet} attached prior to tRNA^{fMet} dissociation from the ribosome (path i), or the movement of L1 to an open conformation coincides with tRNA^{fMet} dissociation (path ii), or L1 moves to an open conformation with tRNA^{fMet} attached, and tRNA^{fMet} then dissociates only very slowly (path iii). The results shown in Figure 4A and 4B were obtained for identical complexes containing tRNA^{fMet} and fMetPhe-tRNA^{Phe}, but in different buffers (7 mM Mg^{2+} and no polyamines, buffer I, Figure 4A; 4.5 mM Mg^{2+} and polyamines, buffer J), corresponding to those employed in previous studies of translocation. The experiments shown in Figure 4C were performed in buffer J, but with Phe-tRNA^{Phe} in place of fMetPhe-tRNA^{Phe}, and demonstrate the very large retarding effect that this replacement has on tRNA^{fMet} release.

4.2. smFRET experiments

The experiments shown in Figure 4A-D provide an indication of the potential complexities involved in measuring E-site tRNA dissociation in real-time, and have prompted us to conduct smFRET experiments to further explore this issue. For these experiments (Figure 4E) we use Cy3- and Cy5-labeled tRNAs (Figure 2, row 2) and a total internal reflection fluorescence (TIRF) microscope in combination with alternating-laser excitation (Kapanidis et al., 2004). The technique enables observation of both donor (Cy3) fluorescence and donor-sensitized acceptor (Cy5) fluorescence (i. e., FRET) in 532 nm laser-illuminated frames alternating with Cy5 fluorescence in 640 nm laser-illuminated frames. These capabilities allow us to distinguish quite readily between the 2:1 and 2:3 pathways (Figure 1) in individual ribosomes by the amplitude of fluorescence and by FRET between adjacent tRNAs in either the A and P sites or the P and E sites (Figure 4E).

A series of studies, employing several mRNAs and examining up to four elongation cycles, shows that partitioning between the 2:3 and 2:1 pathways is modulated by several factors. For example, in agreement with the ensemble results quoted above, the fraction of POST complexes proceeding by the 2:3 pathway increases in buffer J as compared to buffer I and for peptides lacking N-terminal fMet as compared with those that contain fMet. This latter point may reflect the more general result that the fraction of 2:3 pathway decreases dramatically as the length of nascent peptide chain increases, a possible reflection of the ribosome changing from an initiation mode to an elongation mode of translation (Tenson and Hauryliuk 2009).

5. EF-Tu interactions on the ribosome

During the binding of cognate TC to the ribosome in the A/T state, EF-Tu interacts directly with protein L11 within the GAC while maintaining its interaction with aminoacyl-tRNA (Schmeing et al., 2009). As a result of GTP hydrolysis and aminoacyl-tRNA accomodation into the A site, EF-Tu·GDP dissociates from the ribosome. Despite extensive ensemble kinetic studies on TC complex interaction with the ribosome (early work by the Rodnina-Wintermeyer group is reviewed in Daviter et al., 2006; more recent work includes Cochella and Green, 2005; Cochella et al., 2007; Kothe and Rodnina, 2006, 2007; Pan et al., 2008) and detailed computer modeling of the process (Whitford et al., 2010), the rates of EF-Tu release from L11 and from aminoacyl-tRNA during the process of its dissociation from the ribosome are not known.

To investigate this process further, we have labeled an *E. coli* EF-Tu mutant (generously provided by Joanna Perła-Kaján and Wlodeck Mandecki) at position 348 with a fluorescence quencher (QSY9) (denoted EF-Tu^{QSY}) and used it to monitor the EF-Tu interaction with both L11 labeled at position 87 (Figure 2, row 4, left side) and aminoacyl-tRNA labeled in the D loop (Figure 2, row 4, right side). As can be seen in Figure 5A, these three positions form an almost equilateral triangle within the ribosome, at distances that are quite appropriate for sensitive monitoring by FRET between the donor-acceptor pairs chosen.

EF-Tu^{QSY} forms a fully functional TC with Phe-tRNA^{Phe}. Rapid mixing of this TC with a 70SIC programmed with an mRNA coding for the initial sequence fMetPhe and containing Cy3-labeled L11 (denoted 70SIC^{Cy3}) leads to the biphasic change in Cy3 fluorescence shown in Figure 5B, in which an initial quenching of Cy3 fluorescence on TC binding is followed by a regain in fluorescence intensity as EF-Tu^{QSY}·GDP is released from the ribosome. EF-Tu^{QSY} also forms a fully functional TC with Phe-tRNA^{Phe}(Cy3), in

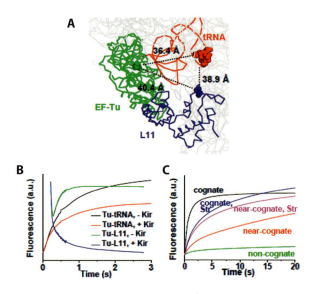

Fig. 5 (A) Distances between residue numbers 348 (EF-Tu), 87 (L11) and 16 (aminoacyl-tRNA) in the kirromycin-stabilized complex with aminoacyl-tRNA bound in the A/T state. Protein Data Bank accession codes 2WRN, 2WRO, 2WRQ, and 2WRR. (B and C) Ensemble stopped-flow FRET measurements on rapid mixing of Phe-TC containing EF-Tu^{QSY} with 70SIC in the presence and absence of kirromycin. (B) All traces are for 70SIC rapidly mixed with TC containing EF-Tu^{QSY}. Kirromycin (100 μM) was premixed with 70SIC. Traces marked Tu-tRNA or Tu-L11 contained Phe-tRNA^{Phe}(Cy3) (0.1 μM) and 70SIC (0.4 μM) or 70SIC^{Cy3} (0.1 μM) and TC (0.4 μM), respectively. (C) Miscoding and streptomycin (100 μM) effects, as measured by EF-Tu^{QSY} effects on Phe-tRNA^{Phe}(Cy3) fluorescence. The codons at the A site were: cognate UUC; near-cognate CUC; non-cognate CGU. Concentrations are after mixing.

Figure 5C demonstrate the ability of the assay based on the use of EF-Tu^{QSY} and Phe-tRNA^{Phe}(Cy3) to distinguish between cognate, near-cognate, and noncognate interactions, and illustrate how the differences between cognate and near-cognate tRNAs are almost entirely suppressed in the presence of streptomycin, in accord with previous results. (Gromadski and Rodnina, 2004).

6. Future perspectives

The increasing availability of detailed structures of functional ribosomal complexes permits significant improvements in the design of fluorescent derivatives of components of the translation apparatus for FRET experiments that probe dynamic changes accompanying transformation of the ribosome during the elongation cycle, and, more generally, during all of translation. Such FRET experiments can be performed in ensemble or single molecule mode, as described above. Single molecule FRET experiments offer two important advantages over ensemble FRET experiments for studying individual steps of translation (Ha, 2001). First, when there is a rapidly equilibrating distribution of populations characterized by different FRET efficiencies within a given state of the ribosome, for instance within the PRE complex (Blanchard et al., 2004a,b), single molecule FRET allows direct determination of the distribution, whereas an ensemble experiment will only give a single FRET efficiency, corresponding to a weighted average of the individual FRET values. Second, smFRET can identify mechanistically important intermediates that may be too short-lived and/or weakly populated to be detected by ensemble measurements. On the other hand, ensemble measurements, by determining average values, provide an important constraint for smFRET experiments, testing whether the surface immobilization procedures that are often used in smFRET experiments yield ribosomes that are as functionally active as those in solution. In addition, ensemble experiments allow direct comparison of rates of chemical reactions (peptidyl transfer, GTP hydrolysis) with rates of fluorescence change. Such comparisons are invaluable for formulating detailed kinetic mechanisms, and are generally not accessible by single molecule approaches. Clearly these two approaches are highly complementary, and their joint application to problems of mechanism should play an increasing role in efforts to understand ribosome function.

which the Cy3 fluorescence is substantially quenched. Rapid mixing of this doubly-labeled TC with unlabeled 70SIC programmed as above shows no change in Cy3 fluorescence on initial binding to the ribosome, but does lead to an increase in Cy3 fluorescence (Figure 5B) as Phe-tRNA^{Phe}(Cy3) is accomodated into the A site (demonstrated as well by parallel rate measurements of fMetPhe formation, not shown) and EF-Tu^{QSY}·GDP is released from the ribosome.

The results presented in Figure 5B demonstrate the utility of using labeled EF-Tu to investigate the details of EF-Tu interaction with the ribosome in reactions leading to PRE complex formation. In one application of these assays, we have found that the presence of kirromycin completely prevents any movement of EF-Tu away from L11, while allowing an apparently partial dissociation of EF-Tu from aminoacyl-tRNA (Figure 5B). The assays will also be useful for screening of the ability of antibiotics to induce misreading. For example, the results presented in

References

Agirrezabala X, Lei J, Brunelle JL, Ortiz-Meoz RF, Green R, Frank J (2008) Visualization of the hybrid state of tRNA binding promoted by spontaneous ratcheting of the ribosome Mol Cell 32: 190–197

Agirrezabala X, Frank J (2009) Elongation in translation as a dynamic interaction among the ribosome, tRNA, and elongation factors EF-G and EF-Tu. Q Rev Biophys 42: 159–200

Agrawal RK, Heagle AB, Penczek P, Grassucci RA, Frank J (1999) EF-G-dependent GTP hydrolysis induces translocation accompanied by large conformational changes in the 70S ribosome. Nat Struct Biol, 6: 643–647

Antoun A, Pavlov MY, Lovmar M, Ehrenberg M (2006) How initiation factors tune the rate of initiation of protein synthesis in bacteria. EMBO J 25: 2539–2550

Betteridge T, Liu H, Gamper H, Kirillov S, Cooperman BS, Hou YM (2007) Fluorescent labeling of tRNAs for dynamics experiments. RNA 13: 1594–1601

Blanchard SC, Gonzalez RL, Kim HD, Chu S, Puglisi JD (2004a) tRNA selection and kinetic proofreading in translation. Nat Struct Mol Biol 11: 1008–1014

Blanchard SC, Kim HD, Gonzalez RL Jr, Puglisi JD, Chu S (2004b) tRNA dynamics on the ribosome during translation. Proc Natl Acad Sci USA 101: 12893–12898

Chen C, Stevens B, Kaur J, Liu H Wang, Y, Zhang H, Rosenblum G, Smilansky Z, Goldman YE, Cooperman BS (2011) Single-molecule fluorescence measurements of ribosomal translocation dynamics Mol Cell (in press)

Cochella L, Brunelle JL, Green R (2007) Mutational analysis reveals two independent molecular requirements during transfer RNA selection on the ribosome. Nat Struct Mol Biol 14: 30–36

Cochella L, Green R (2005) An active role for tRNA in decoding beyond codon:anticodon pairing Science 308: 1178–1180

Connell SR, Takemoto C, Wilson DN, Wang H, Murayama, K, Terada, T, Shirouzu, M, Rost, M, Schuler, M, Giesebrecht J, Dabrowski M, Mielke T, Fucini P, Yokoyama S, Spahn CM (2007) Structural basis for interaction of the ribosome with the switch regions of GTP-bound elongation factors. Mol Cell 25: 751–764

Cornish PV, Ermolenko DN, Staple DW, Hoang L, Hickerson RP, Noller HF, Ha T (2009) Following movement of the L1 stalk between three functional states in single ribosomes. Proc Natl Acad Sci USA 106: 2571–2576

Cornish PV, Ermolenko DN, Noller, HF, Ha T (2008) Spontaneous intersubunit rotation in single ribosomes. Mol Cell 30: 578–588

Daviter T, Gromadski KB, Rodnina MV (2006) The ribosome's response to codon-anticodon mismatches. Biochimie 88: 1001–1011

Diaconu M, Kothe U, Schlünzen F, Fischer N, Harms JM, Tonevitsky AG, Stark H, Rodnina MV, Wahl MC (2005) Structural basis for the function of the ribosomal L7/12 stalk in factor binding and GTPase activation. Cell 121: 991–1004

Dinos G, Kalpaxis DL, Wilson DN, Nierhaus KH (2005) Deacylated tRNA is released from the E site upon A site occupation but before GTP is hydrolyzed by EF-Tu. Nucleic Acids Res 33: 5291–5296

Dorner S, Brunelle JL, Sharma D, Green R (2006) The hybrid state of tRNA binding is an authentic translation elongation intermediate. Nat Struct Mol Biol 13: 234–241

Evans RN, Blaha G, Bailey S, Steitz TA (2008) The structure of LepA, the ribosomal back translocase. Proc Natl Acad Sci USA 105: 4673–4678

Fei J, Kosuri P, MacDougall DD, Gonzalez RL Jr. (2008) Coupling of ribosomal L1 stalk and tRNA dynamics during translation elongation. Mol Cell 30: 348–359

Fei J, Bronson JE, Hofman JM, Srinivas RL, Wiggins CH, Gonzalez RL Jr. (2009) Allosteric collaboration between elongation factor G and the ribosomal L1 stalk directs tRNA movements during translation. Proc Natl Acad Sci USA 106: 15702–15707

Fischer N, Konevega AL, Wintermeyer W, Rodnina MV, Stark H (2010) Ribosome dynamics and tRNA movement by time-resolved electron cryomicroscopy. Nature 466: 329–333

Frank J, Agrawal RK (2000) A ratchet-like inter-subunit reorganization of the ribosome during translocation. Nature 406: 318–322

Geigenmuller U, Nierhaus KH (1990) Significance of the third tRNA binding site, the E site, on E. coli ribosomes for the accuracy of translation: an occupied E site prevents the binding of non-cognate aminoacyl-tRNA to the A site. EMBO J 9: 4527–4533

Gordon J (1970) Regulation of the in vivo synthesis of the polypeptide chain elongation factors in Escherichia coli. Biochemistry 9: 912–917

Grigoriadou, C, Marzi, S, Kirillov, S, Gualerzi, CO Cooperman, BS (2007) A quantitative kinetic scheme for 70 S translation initiation complex formation. J Mol Biol 373: 562–572

Gromadski KB, Rodnina MV (2004) Streptomycin interferes with conformational coupling between codon recognition and GTPase activation on the ribosome. Nat Struct Mol Biol. 11: 316–322

Ha T (2001) Single-molecule fluorescence resonance energy transfer. Methods 25: 78–86

Hetrick B, Lee K, Joseph S (2009) Kinetics of stop codon recognition by release factor 1. Biochemistry 48: 11178–11184

Kapanidis AN, Lee NK, Laurence TA, Doose S, Margeat E, Weiss S (2004) Fluorescence-aided molecule sorting: Analysis of structure and interactions by alternating-laser excitation of single-molecules. Proc Natl Acad Sci USA 101: 8936–8941

Konevega AL, Fischer N, Semenkov Y P, Stark H, Wintermeyer W, Rodnina, MV (2007) Spontaneous reverse movement of mRNA-bound tRNA through the ribosome. Nat Struct Mol Biol 14: 318–324

Kothe U, Rodnina MV (2007) Codon reading by tRNA[Ala] with modified uridine in the wobble position. Mol Cell 25: 167–174

Kothe U, Rodnina MV (2006) Delayed release of inorganic phosphate from elongation factor Tu following GTP hydrolysis on the ribosome. Biochemistry 45: 12767–12774

Liu H, Pan D, Pech M, Cooperman BS (2010) Interrupted catalysis: the EF4 (LepA) effect on back-translocation. J Mol Biol 396: 1043–1052

Márquez V, Wilson DN, Tate WP, Triana-Alonso F, Nierhaus KH (2004) Maintaining the ribosomal reading frame: The influence of E site during translational regulation of release factor 2. Cell 118: 45–55

Marshall RA, Aitken CE, Puglisi, JD (2009) GTP hydrolysis by IF2 guides progression of the ribosome into elongation. Mol Cell 35: 37–47

Marshall, R A, Dorywalska, M Puglisi J D (2008) Irreversible chemical steps control intersubunit dynamics during translation. Proc Natl Acad Sci USA 105: 15364–15369

Munro JB, Altman RB, O'Connor N, Blanchard SC (2007) Identification of two distinct hybrid state intermediates on the ribosome. Mol Cell 25: 505–517

Munro JB, Altman RB, Tung CS, Cate JH, Sanbonmatsu KY, Blanchard SC (2010a) Spontaneous formation of the unlocked state of the ribosome is a multistep process. Proc Natl Acad Sci USA 107: 709–714

Munro JB, Altman RB, Tung CS, Sanbonmatsu KY, Blanchard SC (2010b) A fast dynamic mode of the EF-G-bound ribosome. EMBO J 29: 770–781

Nierhaus KH (1996) An elongation factor turn-on. Nature 379: 491–492

Pan D, Kirillov SV, Cooperman BS (2007) Kinetically competent intermediates in the translocation step of protein synthesis. Mol Cell 25: 519–529

Pan D, Qin H, Cooperman BS (2009) Synthesis and functional activity of tRNAs labeled with fluorescent hydrazides in the D-loop. RNA 15: 346–354

Pan D, Zhang CM, Kirillov S, Hou YM, Cooperman BS (2008) Perturbation of the tRNA tertiary core differentially affects specific steps of the elongation cycle. J Biol Chem 283: 18431–18440

Pape T, Wintermeyer W, Rodnina M V (1998) Complete kinetic mechanism of elongation factor Tu-dependent binding of aminoacyl-tRNA to the A site of the E coli ribosome. EMBO J 17: 7490–7497

Pech M, Karim Z, Yamamoto H, Kitakawa M, Qin Y, Nierhaus KH (2011) Elongation factor 4 (EF4/LepA) accelerates protein synthesis at increased Mg^{2+} concentrations. Proc Natl Acad Sci USA 108: 3199–3203

Peske F, Rodnina MV, Wintermeyer W (2005) Sequence of steps in ribosome recycling as defined by kinetic analysis. Mol Cell 18: 403–412

Phelps SS, Joseph S (2005) Non-bridging phosphate oxygen atoms within the tRNA anticodon stem-loop are essential for ribosomal A site binding and translocation. J Mol Biol 349: 288–301

Qin Y, Polacek N, Vesper O, Staub E, Einfeldt E, Wilson DN, Nierhaus, KH (2006) The highly conserved LepA is a ribosomal elongation factor that back-translocates the ribosome. Cell 127: 721–733

Rodnina MV, Savelsbergh A, Katunin VI, Wintermeyer W (1997) Hydrolysis of GTP by elongation factor G drives tRNA movement on the ribosome. Nature 385: 37–41

Rheinberger HJ, Nierhaus KH (1987) The ribosomal E site at low Mg2+: coordinate inactivation of ribosomal functions at Mg2+ concentrations below 10 mM and its prevention by polyamines. J Biomol Struct Dyn 5: 435–546

Savelsbergh A, Katunin VI, Mohr D, Peske F, Rodnina MV, Wintermeyer W (2003) An elongation factor G-induced ribosome rearrangement precedes tRNA-mRNA translocation. Mol Cell 11: 1517–1523

Schmeing TM, Voorhees RM, Kelley AC, Gao YG, Murphy FV 4 th, Weir JR, Ramakrishnan V (2009) The crystal structure of the ribosome bound to EF-Tu and aminoacyl-tRNA. Science 326: 688–994

Schmeing, TM, Ramakrishnan, V (2009) What recent ribosome structures have revealed about the mechanism of translation. Nature 461: 1234–1242

Semenkov YP, Rodnina MV, Wintermeyer W (1996) The "allosteric three-site model" of elongation cannot be confirmed in a well-defined ribosome system from *Escherichia coli*. Proc Natl Acad Sci USA 93: 12 183–12 188

Seo HS, Kiel M, Pan D, Raj VS, Kaji A, Cooperman BS (2006) EF-G-dependent GTPase on the ribosome. Conformational change and fusidic acid inhibition. Biochemistry 45: 2504–2514

Shoji S, Janssen BD, Hayes CS, Fredrick K (2010) Translation factor LepA contributes to tellurite resistance in *Escherichia coli* but plays no apparent role in the fidelity of protein synthesis. Biochimie 92: 157–163

Shoji, S, Walker, S E, Fredrick, K (2006) Back translocation of tRNA in the ribosome. Mol Cell 24: 931–942

Studer SM, Feinberg JS, Joseph S (2003) Rapid kinetic analysis of EF-G dependent mRNA translocation in the ribosome. J Mol Biol 327: 369–381

Tenson T, Hauryliuk V (2009) Does the ribosome have initiation and elongation modes of translation? Mol Microbiol 72: 1310–1315

Uemura S, Aitken CE, Korlach J, Flusberg BA, Turner SW, Puglisi JD (2010) Real-time tRNA transit on single translating ribosomes at codon resolution. Nature 464: 1012–1017

Valle M, Zavialov A, Li W, Stagg SM, Sengupta J, Nielsen RC, Nissen P, Harvey SC, Ehrenberg M, Frank J (2003a) Incorporation of aminoacyl-tRNA into the ribosome as seen by cryo-electron microscopy. Nat Struct Biol 10: 899–906

Valle M, Zavialov A, Sengupta J, Rawat U, Ehrenberg M, Frank J (2003b) Locking and unlocking of ribosomal motions. Cell 114: 123–134

Walker SE, Shoji S, Pan D, Cooperman BS, Fredrick K (2008) Role of hybrid tRNA-binding states in ribosomal translocation. Proc Natl Acad Sci USA 105: 9192–9197

Wang Y, Qin H, Kudaravalli RD, Kirillov SV, Dempsey GT, Pan D, Cooperman BS, Goldman YE (2007) Single-molecule structural dynamics of EF-G-ribosome interaction during translocation. Biochemistry 46: 10767–10775

Whitford PC, Geggier P, Altman RB, Blanchard SC, Onuchic JN, Sanbonmatsu KY (2010) Accommodation of aminoacyl-tRNA into the ribosome involves reversible excursions along multiple pathways RNA 16: 1196–1204

Wilson DN, Nierhaus KH (2006) The E site story: the importance of maintaining two tRNAs on the ribosome during protein synthesis. Cell Mol Life Sci 63: 2725–2737

Wintermeyer W, Schleich HG, Zachau HG (1979) Incorporation of amines or hydrazines into tRNA replacing wybutine or dihydrouracil. Meth Enzymol 59: 110–121

Wintermeyer W, Peske F, Beringer M, Gromadski KB, Savelsbergh A, Rodnina MV (2004) Mechanisms of elongation on the ribosome: dynamics of a macromolecular machine. Biochem Soc Trans 32: 733–737

Zaher HS, Green R (2009) Fidelity at the molecular level: lessons from protein synthesis. Cell 136: 746–762

Zavialov AV, Ehrenberg M (2003) Peptidyl-tRNA regulates the GTPase activity of translation factors. Cell 114: 113–122

Zhang G, Hubalewska M, Ignatova, Z (2009) Transient ribosomal attenuation coordinates protein synthesis and co-translational folding. Nat Struct Mol Biol 16: 274–280

Zhang W, Dunkle JA, Cate J H (2009) Structures of the ribosome in intermediate states of ratcheting. Science 325: 1014–1017

Studies on the mechanisms of translocation and termination

28

Harry F. Noller, Dmitri N. Ermolenko, Andrei Korostelev, Martin Laurberg, Jianyu Zhu, Haruichi Asahara, Laura Lancaster, Lucas Horan, Alexander Hirschi, John Paul Donohue, Sergei Trakhanov, Clint Spiegel, Robyn Hickerson, Peter Cornish, Taekjip Ha

1. Introduction

This chapter addresses two long-standing questions concerning ribosome structure and function: (i) How are the mRNA and tRNAs moved through the ribosome following formation of each peptide bond? and (ii) how does recognition of a stop codon result in hydrolysis of peptidyl-tRNA? Not surprisingly, results from structural biology have played an important part in formulating mechanistic models for both of these processes. Although structural information is essential for understanding the detailed molecular mechanisms of such processes, it is in itself insufficient for establishing whether or not they are correct. There are already sufficient published examples of false mechanistic inferences based on ribosome structures to remind us that such models need to be tested experimentally, preferably by diverse approaches. Key aspects of the standard models for translocation and termination have emerged from structural observations – cryoEM reconstructions and x-ray crystallography, respectively. Both models have been subjected to experimental tests of various kinds, a process that continues in many laboratories. In the first part of this chapter, we describe the results of experiments using both single-molecule and bulk fluorescence methods to examine the relationship between intersubunit movement, hybrid-states binding of tRNA and translocation. In the second part, we discuss a model for the mechanism of translation termination based on the x-ray crystal structures of the translation termination complexes, and some experimental tests of the model.

2. Studies on the Mechanism of Translocation

Of all the fundamental molecular mechanisms in biology, the coordinated, coupled translocation of mRNA and tRNA through the ribosome must be among the most complex and daunting to understand. This process requires rapid and precise movement of two 25kD tRNAs along with their associated mRNA, over distances of 20–70Å, while maintaining the integrity of their respective codon-anticodon base pairs. There is considerable evidence that translocation involves internal movements of the ribosome, including intersubunit rotation and movement of the head of the 30S subunit, the L1 stalk and other features. Productive translocation also requires the participation of elongation factor EF-G, which catalyzes the process; its role continues to be poorly understood.

Early chemical probing studies showed that the tRNAs are translocated through the ribosome in two steps, first on the 50S subunit, then on the 30S subunit (Moazed and Noller, 1989). During its movement through the ribosome (Figure 1), a tRNA is delivered by EF-Tu·GTP to the ribosome in the A/T state in which the anticodon is bound to the codon in the 30S A site and the aminoacyl end of the tRNA is bound to EF-Tu (Moazed et al, 1988). Hydrolysis of GTP results in the release of EF-Tu·GDP and the accommodation of the aminoacyl end of the tRNA into the A site of the 50S subunit (the classical A/A state). Following peptide bond formation, the newly formed peptidyl-tRNA moves into the A/P hybrid state, the first step of translocation. This first step is promoted by binding of EF-G·GTP, although it can occur in the absence of GTP (Spiegel et al., 2007; Moazed and Noller, 1989). EF-G then catalyzes translocation on the 30S subunit, moving the peptidyl-tRNA from the A/P state to the P/P

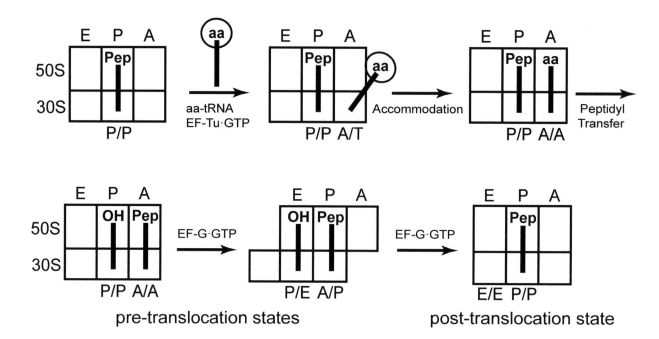

Fig. 1 Schematic illustration of the hybrid-states mechanism for tRNA movement through the ribosome during one round of elongation. The 70S ribosome is indicated by a rectangle, divided into its 30S and 50S subunits and partitioned into the A, P, E tRNA binding sites. The binding states of the tRNAs (shown as bold vertical lines) as previously defined (Moazed and Noller, 1989) are indicated at the bottom of each complex. The aminoacyl, peptidyl and deacylated states of the tRNAs are indicated by aa, Pep and OH, respectively. EF-Tu is represented by a circle.

classical state. The peptidyl-tRNA becomes deacylated when the next peptide bond is formed; the affinity of the 50S E site for the acceptor end of deacylated tRNA stabilizes movement of the tRNA from the P/P state to the hybrid P/E state. The deacylated tRNA is then moved from the P/E to the E/E state, catalyzed by EF-G·GTP, and released from the ribosome by a GTP-dependent process (Spiegel et al., 2007).

Separate movement of the tRNAs on the two ribosomal subunits suggested the possibility that these processes are coupled to intersubunit movement (Moazed and Noller, 1989). Cryo-EM reconstructions by Frank and Agrawal (2000) of ribosome complexes containing bound EF-G showed a noticeable counterclockwise rotation of the 30S subunit around an axis perpendicular to the subunit interface. They proposed that translocation is driven by a ratcheting mechanism involving intersubunit rotation (Frank and Agrawal, 2000; Frank et al., 2007). Intersubunit movement in solution has been demonstrated using changes in the FRET signal between probes attached to the 30S and 50S subunits (Ermolenko et al., 2007a). A combination of FRET and chemical probing experiments showed that the binding states of tRNAs in ribosomes bound to rotated ribosomes were indistinguishable from hybrid states (Ermolenko et al., 2007b; Spiegel et al., 2007). The requirement for intersubunit movement was tested by creating a reversible intersubunit crosslink with an engineered disulfide bond between proteins S6 and L2 (Horan and Noller, 2007). Formation of the crosslink stalled protein synthesis by specifically blocking translocation; reversal of the crosslink by reducing the disulfide immediately restored translocation.

Single-molecule intersubunit FRET experiments helped to address a central question: What supplies the energy required for intersubunit rotation? FRET was measured between fluorescent probes attached to protein S6 or S11 in the 30S subunit and L9 in the 50S subunit in single ribosomes (Cornish et al., 2008). When a deacylated tRNA was bound to the P site, either with a vacant A site or a peptidyl-tRNA bound to the A site, spontaneous intersubunit rotation was observed (Figure 2A, B). Transition density plots show only two significant peaks (Figure 2C), indicating only two intersubunit rotational states. Spontaneous rotation occurred in the complete absence of EF-G or GTP, showing that intersubunit rotation can be driven by thermal energy alone, and so has a very low free energy of activation. When a peptidyl-tRNA occupied

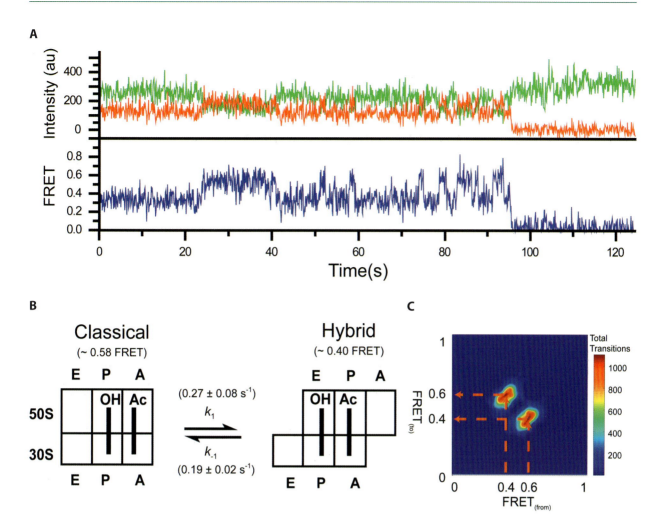

Fig. 2 Single-molecule FRET traces of ribosomes undergoing spontaneous intersubunit rotation between the classical and hybrid states. The 30S subunit was labeled with a donor fluor on protein S6 and the 50S subunit with an acceptor fluor on protein L9. (A) The green and red traces show fluorescence intensities of the donor and acceptor, respectively, and the blue trace shows the calculated FRET efficiency. (B) Illustration of the intersubunit movement, schematized as in Figure 1. Forward and reverse rates of intersubunit rotation (k_1 and k_{-1}) are indicated. (C) Transition density plot based on data from many independent smFRET measurements. (Figure reproduced from Cornish et al. (2008) Mol Cell 30: 578–588 © 2008 with permission from Elsevier)

the P site, no spontaneous rotation was observed; this can be explained by the strong specificity of the E site for tRNA with a deacylated acceptor end, preventing movement of peptidyl-tRNA from the P/P state into the P/E state. These experiments also showed that the occupancies of the hybrid (rotated) and classical (non-rotated) states were similar, indicating that the equilibrium between the rotated and non-rotated states is near one. Thus, intersubunit rotation is a process whose energetics are poised on a knife edge, minimizing the expenditure of energy to drive it forward or backward. However, the spontaneous rotation measured in these experiments is non-productive, resulting in no net translocation. Productive translocation requires binding of EF-G, either with GTP or its non-hydrolyzable

analogue GDPNP. Non-productive intersubunit rotation can be likened to slipping of a ratchet without intervention of a pawl to rectify the rotation into unidirectional movement. This means that the pawl of the translocational ratchet is created by the interaction of EF-G·GTP with the ribosome. The exact nature of the pawl is unknown, although it has been implicated in models suggesting that EF-G in the post state blocks the A site against reverse movement of tRNA as long as the unlocked state persists.

A second untested inference regarding the translocation mechanism is the timing of mRNA translocation. This was tested in stopped-flow experiments (Ermolenko and Noller, 2010) using FRET to measure intersubunit rotation and fluorescence quenching us-

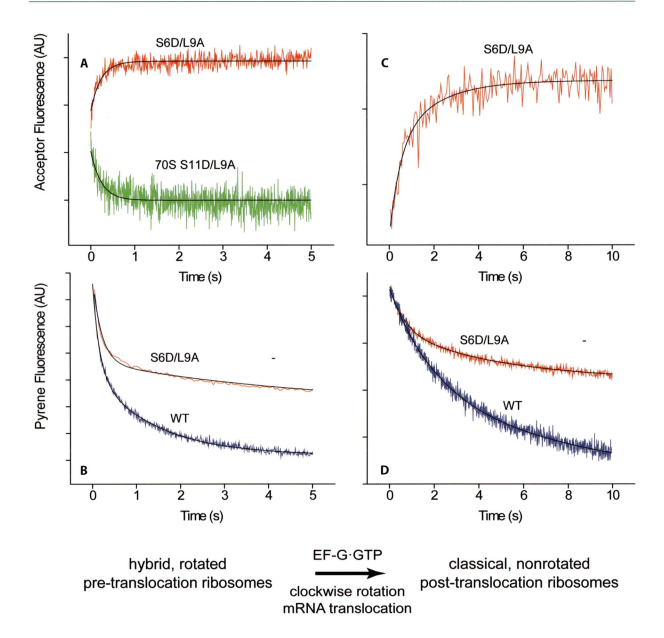

Fig. 3 Stopped-flow fluorescence measurements of intersubunit rotation and mRNA translocation. (A, C) Intersubunit rotation was measured by FRET using ribosomes labeled with a donor fluor on protein S6 (red traces) or S11 (green traces) and an acceptor on protein L9. An increase in S6-L9 FRET or a decrease in S11-L9 FRET is indicative of clockwise rotation of the 30S subunit relative to the 50S subunit. (B, D) mRNA translocation was measured using the fluorescence quenching assay of Studer et al. (2003), using either unlabeled wild-type ribosomes (WT) or ribosomes labeled with fluorescent probes on proteins S6 and L9 (S6D/L9A). Pretranslocation complexes were formed with N-Ac-Phe-tRNA in the A site and (A, B), deacylated elongator methionine tRNA or (C, D) initiator methionine tRNA in the P site and rapidly mixed with EF-G·GTP. Ribosomes were programmed with synthetic mRNAs containing a fluorescein probe attached to position +9. (Figure reproduced from Ermolenko and Noller 2010)

ing the method developed by Joseph and co-workers (Studer et al., 2003) to follow mRNA movement. Pre-translocation complexes containing a deacylated elongator tRNA^{Met} in the P site and N-Ac-Phe-tRNA in the A site were rapidly mixed with EF-G·GTP in a stopped-flow fluorimeter (Figure 3). An increase in S6-L9 FRET (or rapid decrease in S11-L9 FRET) was ob-

served (Figure 3A). These FRET changes are indicative of clockwise rotation of the 30S subunit, corresponding to the transition from the hybrid state to classical state. Quenching of mRNA fluorescence, i. e., mRNA movement, occurred at a similar rate (Figure 3B). When translocation was slowed by substituting deacylated initiator tRNA^{fMet}, both the rate of intersubunit

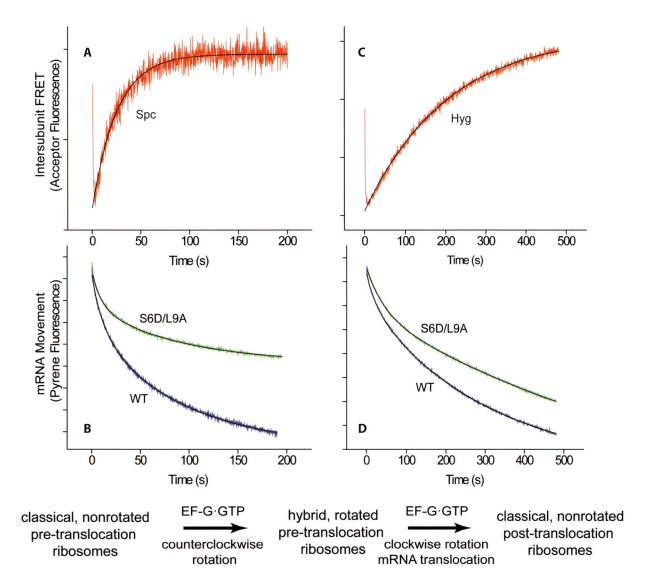

classical, nonrotated
pre-translocation
ribosomes
→ EF-G·GTP
counterclockwise
rotation
→ hybrid, rotated
pre-translocation
ribosomes
→ EF-G·GTP
clockwise rotation
mRNA translocation
→ classical, nonrotated
post-translocation
ribosomes

Fig. 4 Resolution of clockwise and counterclockwise intersubunit rotation. Pretranslocation complexes were formed as in Figure 1A, B and rapidly mixed with EF-G·GTP in the presence of the translocation inhibitors spectinomycin (A,B) or hygromycin B (C, D). Intersubunit rotation was measured by changes in FRET between fluors attached to proteins S6 and L9 (red traces). Movement of mRNA was followed by quenching of fluorescein-labeled mRNA in wild-type ribosomes (blue traces) or doubly-labeled ribosomes (green traces). A rapid counterclockwise rotation is observed at the very beginning of the traces in A, C, followed by a slower, clockwise rotation. Translocation of mRNA clearly occurs during the second, clockwise rotational step. (Figure reproduced from Ermolenko and Noller 2010)

rotation and the rate of mRNA movement slowed to a similar degree (Figure 3C, D). To resolve the two steps of intersubunit rotation, the translocation inhibitors spectinomycin and hygromycin B were used to further slow the rate of translocation. The results of this experiment (Figure 4A, C) reveal a rapid initial decrease in S6-L9 FRET, indicative of counterclockwise rotation, followed by a slower clockwise rotation; these events correspond to movement from the classical state to the hybrid state and reversal back to the classical state. The rate of movement of mRNA (Figure 4B, D) is clearly similar to the second (clockwise) step of intersubunit rotation in each case, and much slower than the first (counterclockwise) step. We conclude that translocation of mRNA occurs during the second step of intersubunit movement.

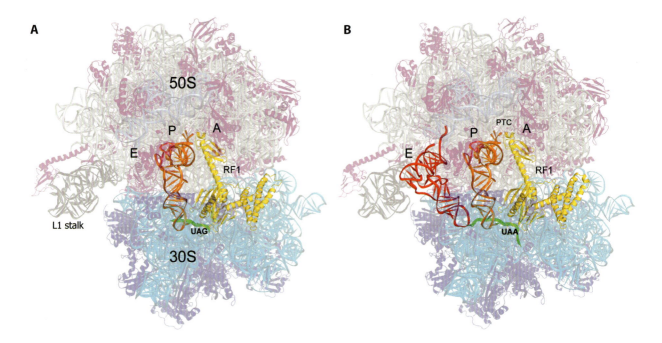

Fig. 5 X-ray crystal structures showing the position of release factor RF1 in translation termination complexes. RF1 was bound to complexes containing deacylated initiator tRNA in the P site and an mRNA containing either (A) a UAG (Korostelev et al., 2010) or (B) a UAA stop codon (Laurberg et al., 2008). Deacylated tRNA is also bound to the E site in (b). (Figure reproduced from Korostelev et al. 2010)

3. Studies on the Mechanism of Termination

Following the discovery of nonsense codons and release factors, these three fundamental questions concerning the mechanism of translation termination remained unanswered for more than four decades: (1) Since there are no tRNAs that recognize stop codons, how are they recognized – directly by the release factors, or indirectly by structural elements of the ribosome? (2) How is peptidyl-tRNA hydrolysis catalyzed – directly by the release factors, or (as suggested from inhibition by peptidyl-transferase inhibitors) by the ribosome itself? And, (3) how is peptidyl-tRNA hydrolysis so closely coupled to stop codon recognition that the frequency of premature termination is of the order of 10^{-5} in the absence of any GTP-consuming proof-reading step (Freistroffer et al., 2000)? Recently, the x-ray crystal structures of all four possible termination complexes have been solved (Laurberg et al., 2008; Korostelev et al., 2008; Weixlbaumer et al., 2008; Korostelev et al., 2010), providing the structural framework for addressing these questions in detail. Some aspects of the proposed mechanisms have already been subjected to experimental test.

In bacteria, recognition of the stop codons UAG, UAA is mediated by release factor RF1, and UGA, UAA by RF1. Nakamura and co-workers showed that the codon specificities of RF1 and RF2 could be swapped genetically by transplanting conserved elements of their respective amino acid sequences. Stop codon recognition was proposed to involve "tripeptide anticodons" – PVT for RF1 and SPF for RF2 (Ito et al., 2000). This proposal was supported by low-resolution cryo-EM and crystal structures, which placed these tripeptide elements in the decoding site of the 30S subunit (Klaholz et al., 2003; Rawat et al., 2003; Petry et al., 2005). This placement was confirmed by the higher-resolution structures (Figure 5), although the recognition elements have turned out to be much more complicated than a simple tripeptide (Figure 6). Indeed, the Thr residue of the PVT, Ser of the SPF motifs are critically involved in codon recognition; Thr186 helps to recognize U1 and A2 (Figure 6A and B), while Ser206 interacts exclusively with U1 (Figures 6C and D). Interactions with backbone amide groups of the N-terminal end of helix 5 restrict recognition to U in the first position in all four complexes. The first two codon bases are stacked on each other, but stacking of His193 on the second base separates the first two bases from

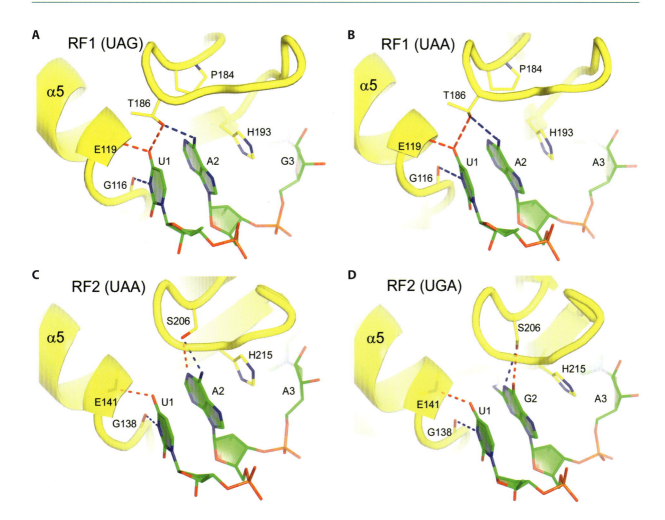

Fig. 6 Recognition of the first two bases of the three stop codons by the two release factors. Recognition of (A) UAG by RF1 (Korostelev et al., 2010); (B) UAA by RF1 (Laurberg et al., 2008); (C) UAA by RF2 (Korostelev et al., 2008) and (D) UGA by RF2 (Weixlbaumer et al., 2008). The release factors are shown in yellow and the stop codon in green. The first two bases of the stop codon are stacked on each other; His193 (or 215) stacks on the second base, separating it from the third base of the codon. (A, B) Thr186 from the PVT motif of RF1 and (C, D) Ser206 from the SPT motif of RF2 are involved in recognition of the second base of the codon. Interactions with the backbone of helix a5 restrict the first position to U in all four complexes. (Figure reproduced from Korostelev et al. 2010)

the third base, which lies orthogonal to the first two in a separate recognition pocket (Figure 7). For both factors, the N7 position of the third base is H-bonded to Thr194. In the case of RF1, the side-chain of Gln181 can recognize either A or G in the third position depending on its rotational orientation; in RF2, Gln181 is replaced by Val, restricting RF2 to recognition of A in the third position. Although the backbone of 16S rRNA contacts the backbone of the stop codons and the third base stacks on G530 of 16S rRNA, the bases are recognized completely by the release factors.

Prior to any structural studies, the conserved GGQ motif of the release factors was already implicated in catalysis of peptidyl-tRNA hydrolysis (Frolova et al., 1999; Seit-Nebi et al., 2001; Mora et al., 2003). Again,

cryo-EM and low-resolution x-ray crystal structures showed that the region of the release factors containing the GGQ was indeed located in the peptidyl-transferase site of the ribosome (Klaholz et al., 2003; Rawat et al., 2003; Petry et al., 2005). Although it was speculated that the side-chain of the conserved Gln residue would be involved in catalysis, the higher-resolution crystal structures show that its backbone amide nitrogen, rather than its side-chain, is positioned to stabilize both the product of the reaction and its transition state (Figure 8). The role of the side-chain of Gln230 was studied by Shaw and Green (2007), who showed that substitution with a variety of other amino acids had only modest effects on catalysis, but affected the ability of the factor to discriminate water from other

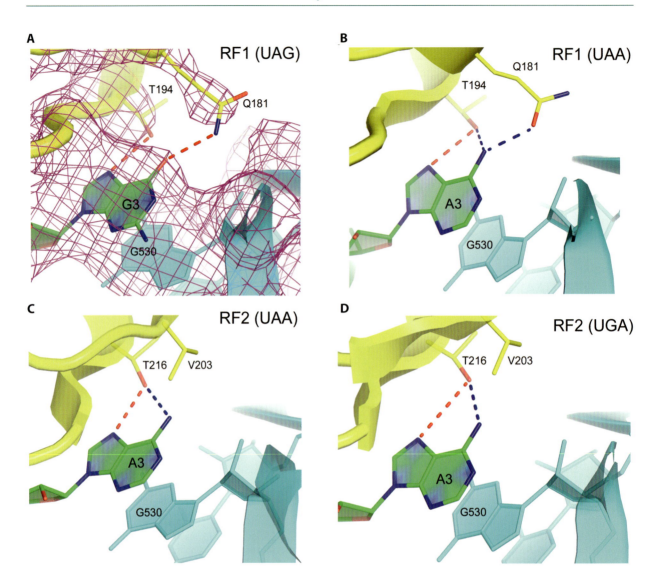

Fig. 7 Recognition of the third base of the stop codons occurs in a separate pocket in the two release factors. (A, B) RF1 (Laurberg et al., 2008; Korostelev et al., 2010) is able to recognize either G or A in the third position by rotation of the side-chain of Gln181. (C, D) In RF2, the Gln is replaced by Val203, allowing only recognition of A in the third position. (Figure reproduced from Korostelev et al. 2010)

potential attacking nucleophiles. We tested the importance of its backbone amide by substituting G230 with proline whose α-amide nitrogen is unable to donate a hydrogen bond (Korostelev et al., 2008). The result was to completely abolish peptidyl-tRNA hydrolysis (Figure 9). Although it can be argued that the effect of proline substitution could be caused by distorting the local conformation, we note that the backbone torsion angles of Gln230 are fully within the normal range observed for proline in high-resolution crystal structures. The proposed catalytic mechanism (Korostelev et al., 2010) is shown in Figure 10. The attacking nucleophilic water molecule is activated by the 2'-OH of ribose 76 of the P-site tRNA (Brunelle et al., 2008). The

α-amide nitrogen of Gln230 stabilizes the transition state by H-bonding to the developing oxyanion of the tetrahedral intermediate, and stabilizes the deacylated tRNA product by H-bonding to the 3'-OH of A76.

The crystal structures of the isolated release factors (Vestergaard et al., 2001; Shin et al., 2004) are very different from their structures bound to the ribosome. However, in solution, their conformations are in equilibrium between an extended form and a compact one that likely resembles that of the crystal structure of the isolated factor (Vestergaard et al., 2005; Zoldak et al., 2007). We have proposed that the high accuracy of translation termination is due to a mechanism based on interconversion between these two conformational

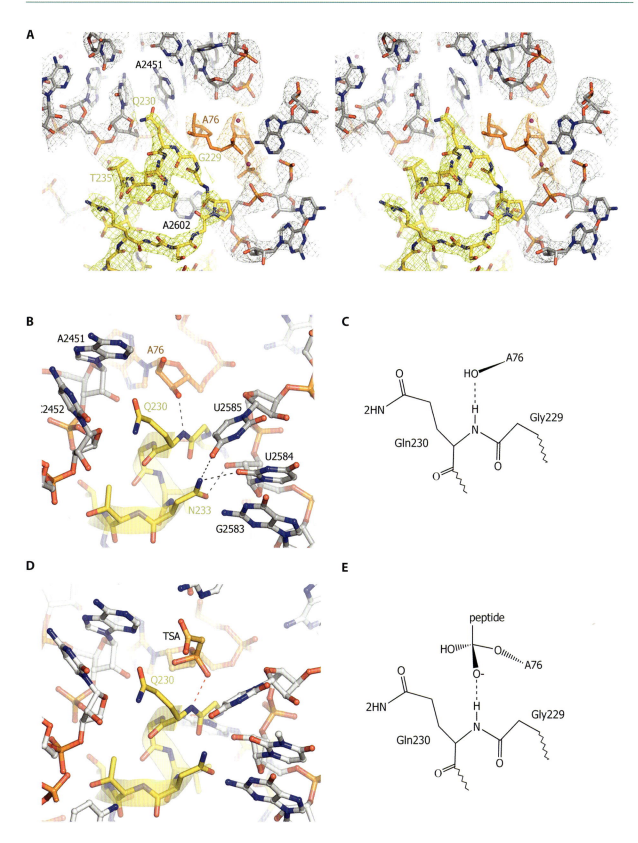

Fig. 8 Catalysis of peptidyl-tRNA hydrolysis by the backbone of the RF. (A) 3Fo-2Fc electron density map showing that the side-chain of Gln230 in the conserved GGQ motif of RF1 points away from the site of catalysis. Its backbone amide nitrogen is positioned to form H bonds that could act in catalyzing the hydrolysis reaction by (B) product stabilization or (C) transition-state stabilization. In (C) the structure of the transition-state analog determined by Schmeing et al. (2005) was superimposed on the structure of the termination complex. (Figure reproduced from Laurberg et al. 2008)

A

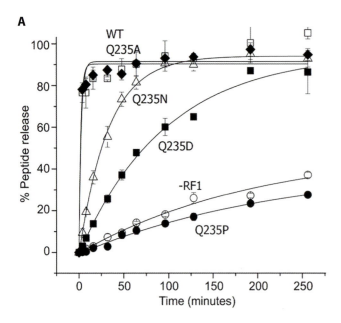

B

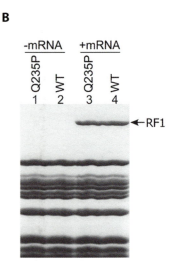

Fig. 9 Substitution of Gln230 with proline abolishes peptidyl-tRNA hydrolysis activity. (A) Peptide release was measure for a termination complex formed with (●) a mutant Q230P version of RF1 and other mutant versions of RF1. (B) The mRNA-dependent affinity of Q230P RF1 is similar to that of wild-type RF1. (Figure reproduced from Korostelev et al. 2008)

states (Laurberg et al., 2008). According to this model, the factor initially recognizes the stop codon in a compact form. Successful recognition results in formation of a binding pocket for the rearranged conformation of an extended loop (which we have termed the switch loop) that links domains 2 and 3 of the factor (Figure 11A). In the bound form, the loop rearranges to an α-helical conformation that creates an extension of α-helix 7 of the factor; when bound in the pocket, α7 directs the GGQ domain into the catalytic site. Critical elements of 16S and 23S rRNA contribute to the binding pocket, and can do so only when the reading head of the factor docks successfully on a stop codon. One of these elements is A1493, which, together with A1492, flips out from its position in helix 44 to bind to the minor groove of the codon-anticodon helix during sense-codon recognition (Ogle et al., 2001). In stop-codon recognition, A1493 remains in its ground state in helix 44; if it were to flip out, it would clash sterically with the domain 2 of the factor, preventing

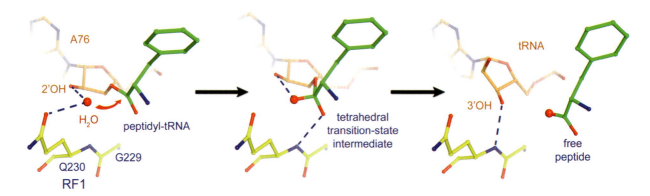

Fig. 10 Proposed mechanism of peptidyl-tRNA hydrolysis. *(left)* The attacking nucleophilic water molecule is activated by the 2'-OH of ribose 76 of the peptidyl-tRNA. *(middle)* The backbone α-amide nitrogen stabilizes the transition state by H-bonding to the developing oxyanion. *(right)* The backbone amide nitrogen also contributes to catalysis by product stabilization through H-bonding to the 3'-OH of the deacylated tRNA product. The side-chain of Gln230 discriminates water from other potential attacking nucleophiles (Brunelle et al. 2008). (Figure reproduced from Korostelev et al. 2010)

A

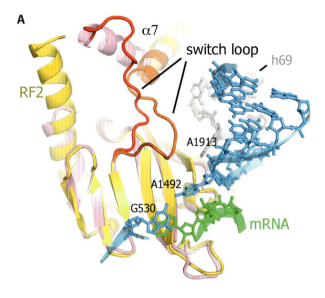

B

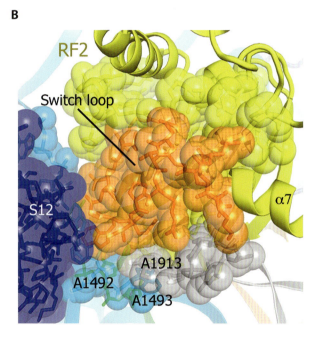

Fig. 11 Rearrangement of the switch loop of RF2. (A) The structures of the free and bound forms of the release factors differ dramatically. In the bound form, the switch loop rearranges to create an extension of α-helix 7. (B) The switch-loop extension of α-helix 7 binds in a pocket formed by features of 16S and 23S rRNA, protein S12, and domain 2 of the release factor. (Figure reproduced from Laurberg et al. 2008)

codon recognition. This allows A1913 at the tip of the loop of helix 69 of 23S rRNA to stack on A1493, where it forms one side of the binding pocket for the rearranged switch loop helix (Figure 11B). This model is supported by the observation that paromomycin, which stimulates sense codon-anticodon interaction, strongly inhibits translation termination (Youngman

et al., 2007). This can be explained by the fact that paromomycin stabilizes the flipped-out conformation of A1492 and A1493, because it binds to the site in helix 44 normally occupied by A1493 (Ogle et al., 2001); the flipped-out conformation of A1493 would thus clash with domain 2 of the factor, preventing stop codon recognition and failing to provide a surface for stacking of A1913. A second finding is that deletion of helix 69 of 23S rRNA combined with a two-amino acid deletion in the switch loop, which have only 10-fold or 3-fold effects, respectively, by themselves, has a large synthetic-negative effect, creating more than a thousand-fold decrease in the rate of peptidyl-tRNA hydrolysis by the release factors (Korostelev et al., 2010). This result indicates that the switch loop and helix 69 interact (directly or indirectly) in a common pathway, consistent with the proposed model.

Acknowledgements

We thank present and former members of the Noller laboratory for many helpful discussions, advice and reagents. This work was supported by grants from the NIH, NSF (to H. F. N.) and an NSF Career Award to T. H.

References

Brunelle JL, Shaw JJ, Youngman EM, Green R (2008) Peptide release on the ribosome depends critically on the 2' OH of the peptidyl-tRNA substrate. RNA 14: 1526–1531

Cornish PV, Ermolenko DN, Noller HF and Ha T (2008) Spontaneous intersubunit rotation in single ribosomes. Mol Cell 30: 578–588

Ermolenko DN, Noller HF (2010) mRNA translocation occurs during the second step of ribosomal intersubunit rotation. Nat Struct Mol Biol 18: 457–462

Ermolenko DN et al. (2007a) Observation of intersubunit movement of the ribosome in solution using FRET. J Mol Biol 370: 530–540

Ermolenko DN et al. (2007b) The antibiotic viomycin traps the ribosome in an intermediate state of translocation. Nat Struct Mol Biol 14: 493–497

Frank J, Agrawal RK (2000) A ratchet-like inter-subunit reorganization of the ribosome during translocation. Nature 406: 318–322

Frank J, Gao H, Sengupta J, Gao N, Taylor DJ (2007) The process of mRNA-tRNA translocation. Proc Natl Acad Sci USA 104: 19 671–678

Freistroffer DV, Kwiatkowski M, Buckingham RH, Ehrenberg M (2000) The accuracy of codon recognition by polypeptide release factors. Proc Natl Acad Sci USA 97: 2046–2051

Frolova LY et al. (1999) Mutations in the highly conserved GGQ motif of class peptidyl-1 polypeptide release factors abolish ability of human eRF1 to trigger peptidyl-tRNA hydrolysis. RNA 5: 1014–1020

Horan LH, Noller HF (2007) Intersubunit movement is required for ribosomal translocation. Proc Natl Acad Sci USA 104: 4881–4885

Ito K, Uno M, Nakamura Y (2000) A tripeptide 'anticodon' deciphers stop codons in messenger RNA. Nature 403: 680–684

Klaholz BP et al. (2003) Structure of the *Escherichia coli* ribosomal termination complex with release factor 2. Nature 421: 90–94

Korostelev A, Asahara H, Lancaster L, Laurberg M, Hirschi A, Zhu J, Trakhanov S, Scott WG, Noller HF (2008) Crystal structure of a translation termination complex formed with release factor RF2. Proc Natl Acad Sci USA, 105: 19 684–19 689

Korostelev A, Zhu J, Asahara H† and Noller HF (2010) Recognition of the amber UAG codon by release factor RF1. EMBO J 29: 2577–2585

Laurberg M, Asahara H, Korostelev A, Zhu J, Trakhanov S and Noller HF (2008) Structural basis for translation termination on the 70S ribosome. Nature, 454: 852–857

Moazed D, Robertson JM, Noller HF (1988) Interaction of elongation factors EF-G, EF-Tu with a conserved loop in 23S RNA. Nature 334, 362–364

Moazed D, Noller HF (1989) Intermediate states in the movement of transfer RNA in the ribosome. Nature 342: 142–148

Mora L, Heurgue-Hamard V, Champ S, Ehrenberg M, Kisselev LL, Buckingham RH (2003) The essential role of the invariant GGQ motif in the function and stability in vivo of bacterial release factors RF1 and RF2. Mol Microbiol 47: 267–275

Ogle JM et al. (2001) Recognition of cognate transfer RNA by the 30S ribosomal subunit. Science 292: 897–902

Petry S et al. (2005) Crystal structures of the ribosome in complex with release factors RF1 and RF2 bound to a cognate stop codon. Cell 123: 1255–1266

Rawat UB et al. (2003) A cryo-electron microscopic study of ribosome-bound termination factor RF2. Nature 421: 87–90

Schmeing TM, Huang KS, Strobel SA, Steitz TA (2005) An induced-fit mechanism to promote peptide bond formation and exclude hydrolysis of peptidyl-tRNA. Nature, 438: 520–524

Seit-Nebi A, Frolova L, Justesen J, Kisselev L (2001) Class-1 translation termination factors: invariant GGQ minidomain is essential for release activity and ribosome binding but not for stop codon recognition. Nucleic Acids Res 29: 3982–3987

Shaw JJ, Green R (2007) Two distinct components of release factor function uncovered by nucleophile partitioning analysis. Mol Cell 28: 458–467

Shin DH, Brandsen J, Jancarik J, Yokota H, Kim R, Kim SH (2004) Structural analyses of peptide release factor 1 from Thermotoga maritima reveal domain flexibility required for its interaction with the ribosome. J Mol Biol 341: 227–239

Spiegel PC, Ermolenko DN, Noller HF (2007) Elongation factor G stabilizes the hybrid-state conformation of the 70S ribosome. RNA 13: 1473–1482

Studer SM, Feinberg JS and Joseph S (2007) Rapid kinetic analysis of EF-G-dependent mRNA translocation in the ribosome. J Mol Biol 327: 369–381

Vestergaard B et al. (2001) Bacterial polypeptide release factor RF2 is structurally distinct from eukaryotic eRF1. Mol Cell 8: 1375–1382

Vestergaard B et al. (2005) The SAXS solution structure of RF1 differs from its crystal structure and is similar to its ribosome bound cryo-EM structure. Mol Cell 20: 929–938

Weixlbaumer A, Jin H, Neubauer C, Voorhees RM, Petry S, Kelley AC, Ramakrishnan V (2008) Insights into translational termination from the structure of RF2 bound to the ribosome. Science 322: 953–956

Youngman EM, He SL, Nikstad LJ, Green R (2007) Stop codon recognition by release factors induces structural rearrangement of the ribosomal decoding center that is productive for peptide release. Mol Cell 28: 533–543

Zoldak G, Redecke L, Svergun DI, Konarev PV, Voertler CS, Dobbek H, Sedlak E, Sprinzl M (2007) Release factors 2 from *Escherichia coli* and *Thermus thermophilus*: structural, spectroscopic and microcalorimetric studies. Nucleic Acids Res 35: 1343–1353

The mechanism by which tmRNA rescues stalled ribosomes

29

David Healey, Mickey Miller, Christopher Woolstenhulme, Allen Buskirk

1. Introduction

Not all translation reactions end in the synthesis of a full-length protein. In bacteria, ribosomes stall at the 3' end of mRNA transcripts lacking stop codons, as they cannot efficiently employ release factors for termination and recycling. Some non-stop mRNAs arise from defects in transcription. RNA polymerase occasionally terminates transcription prematurely; this can occur either as a result of pausing at specific sequences or encountering a tightly-bound protein on the DNA (Abo et al., 2000). Another likely source is the regular process of mRNA degradation. mRNAs are turned over quickly in bacteria, with an average half-life of about six or seven minutes (Bernstein et al., 2002; Selinger et al., 2003). Bacterial mRNAs are degraded by endonucleases and by processive 3' to 5' exonucleases (Condon, 2007). An exonuclease that collides with a translating ribosome leaves it stalled on the truncated transcript. Ribosome stalling constitutes a serious threat to the integrity of bacterial cells: roughly 1 in 250 of all translation reactions result in an irreversible arrest (Moore and Sauer, 2005). If these arrested ribosomes were not released, the majority of ribosomes would become inoperative within a single generation.

A translational quality control system in bacteria rescues stalled ribosomes with a small stable RNA known as tmRNA. This remarkable molecule possesses both transfer and messenger RNA activity: aminoacylated with alanine, tmRNA enters stalled ribosomes and adds Ala to the nascent peptide chain (Figure 1). Leaving the broken mRNA, the ribosome resumes translation on the tmRNA template, adding a short tag to the growing polypeptide and terminating translation at a stop codon. The stalled ribosome is recycled and the 11 amino acid tag marks the aborted nascent polypeptide for destruction by cellular proteases (Kei-

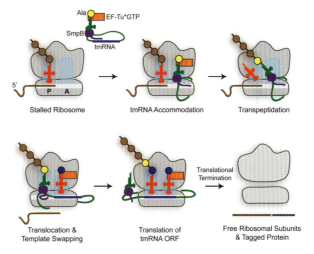

Fig. 1 Model of trans-translation. Alanyl-tmRNA (green) and its protein partner SmpB (purple) are delivered to stalled ribosomes by EF-Tu (orange). The nascent polypeptide is transferred to tmRNA. Translocation of tmRNA·SmpB to the P site releases the truncated mRNA and positions the tmRNA ORF (dark blue) in the ribosomal A site. Translation resumes on the tmRNA ORF, directing the addition of an additional ten-residue tag to the nascent polypeptide, after which termination occurs at a stop codon. This process recycles stalled ribosomes, allowing the subunits to dissociate, and tags the nascent peptide for degradation by proteases.

ler et al., 1996). Because the ribosome switches templates during protein synthesis, this process is called trans-translation (Figure 1).

tmRNA and its protein partner, small protein B (SmpB), are found in all fully-sequenced eubacterial genomes (Moore and Sauer, 2007). tmRNA is essential for viability or pathogenicity in some species of bacteria (Huang et al., 2000; Hutchison et al., 1999), and loss of the tmRNA gene causes sensitivity to various stresses in *E. coli* and *B. subtilis* (Muto et al., 2000; Oh and Apirion, 1991). The recycling of stalled ribosomes by tmRNA appears to be its primary function; tagging and destruction of aborted polypeptides appears to be secondary.

Here we review progress in elucidating the mechanism by which tmRNA rescues stalled ribosomes. We examine the molecules involved and the models that have arisen to explain how tmRNA recognizes and enters stalled ribosomes, releases them from truncated transcripts, and tags the proteins for destruction. We also identify and address some of the important mechanistic questions that remain to be answered. Readers interested in the degradation of tagged proteins (Moore and Sauer, 2007) or additional biological roles of tmRNA in various bacteria (Keiler, 2008) are referred to other recent reviews.

2. The structure and function of tmRNA

In *E. coli*, tmRNA is the product of a single gene (*ssrA*) that makes a primary transcript 457 nucleotides in length. While in other bacteria tmRNA levels are regulated by stress responses (Muto et al., 2000), in *E. coli* tmRNA is expressed from a constitutive Σ^{70}-like promoter (Komine et al., 1994; Oh et al., 1990). tmRNA is processed at the 5' terminus by the endonuclease RNase P (Komine et al., 1994). The 3' terminus is first processed by endonucleases RNase III and/or RNase E, followed by trimming by exonucleases RNase T and RNase PH (Li et al., 1998; Lin-Chao et al., 1999; Makarov and Apirion, 1992; Srivastava et al., 1990). These events are very similar to typical tRNA processing (Li and Deutscher, 2002; McClain et al., 1987). Recently tRNAse Z was found to be the primary 3' endonuclease for normal tRNAs, but it is yet unknown whether tRNAse Z also aids in 3' processing of tmRNA (Hartmann et al., 2009). The final tmRNA product is 363 nucleotides long and contains a tRNA-like domain (TLD), an mRNA-like domain with an open reading frame (ORF), several helices, and multiple pseudoknot structures (Figure 2) (Chauhan and Apirion, 1989; Felden et al., 1997; Komine et al., 1994; Williams and Bartel, 1996). tmRNA is resistant to nuclease degradation with a half-life of approximately 60 minutes. It is stabilized by its binding to SmpB; when SmpB is absent, the half-life of tmRNA suffers a four-fold reduction (Hanawa-Suetsugu et al., 2002; Moore and Sauer, 2005).

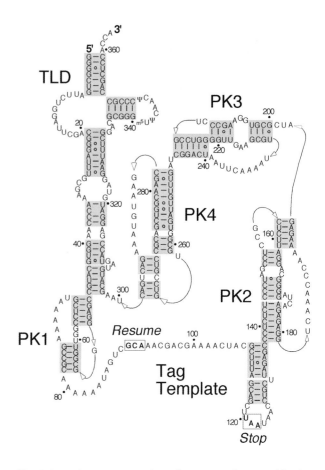

Fig. 2 Secondary structure of *E. coli* tmRNA. The tRNA-like domain (TLD), pseudoknots 1–4, and the open reading frame (including the resume and stop codons) are labeled.

2.1. tRNA-like domain

The tRNA-like domain (TLD) is formed through interactions between the mature 5' and 3' ends of tmRNA (Figure 2) (Komine et al., 1994). Like tRNA, the TLD contains an acceptor stem ending in 5'-CCA-3' (Felden et al., 1997; Williams, 2000; Williams and Bartel, 1996; Zwieb and Wower, 2000). This acceptor stem contains a G:U wobble pair that is recognized by alanyl-tRNA synthetase (AlaRS) (Komine et al., 1994; Nameki et al., 1999c). AlaRS is a particularly appropriate synthetase for tmRNA, because tmRNA lacks an anticodon stem, and unlike other synthetases, AlaRS does not need to bind the anticodon region to perform its function (Hou and Schimmel, 1988; McClain and Foss, 1988).

The tmRNA D loop varies from normal tRNAs in that it lacks dihydrouridine residues (Felden et al., 1998; Hanawa-Suetsugu et al., 2001). It is also much shorter than the traditional D arm and lacks a helical structure. It contains multiple conserved residues that constitute a binding site for SmpB as described below (Barends et al., 2001; Gutmann et al., 2003). The T arm more closely matches its tRNA counterpart (Williams, 2000; Zwieb and Wower, 2000); it even contains the same modified nucleotides (two pseudouridines and one 5-methyluridine) (Felden et al., 1998). Portions of the D loop interact with the T arm and SmpB to form a central core similar to that found in normal tRNA (Bessho et al., 2007). The T arm and acceptor stem also act as binding sites for elongation factor Tu (Barends et al., 2001; Gutmann et al., 2003; Valle et al., 2003).

2.2. Pseudoknots

Multiple pseudoknot structures exist in tmRNA from all bacterial species (Felden et al., 1997; Williams, 2000; Williams and Bartel, 1996; Zwieb and Wower, 2000). In *E. coli* tmRNA, the four pseudoknots are arranged such that one (PK1) is located upstream of the ORF while PK2, PK3, and PK4 are downstream (Figure 2). Early studies seemed to indicate that PK1 was essential to tmRNA tagging while the other three were dispensable. This was based on the observation that replacing PK1 with a single-stranded motif resulted in severely impaired aminoacylation and tagging, while replacing the other three only marginally reduced tmRNA function (Nameki et al., 1999b; Nameki et al., 2000). It was proposed that PK1 was essential for binding ribosomes in order to position the ORF properly (Nameki et al., 1999a; Valle et al., 2003).

However, more recent studies have shown that substitution of PK1 with a small, stable hairpin is able to support robust tmRNA tagging ability *in vivo* (Tanner et al., 2006; Wower et al., 2009). These results suggest that the role of PK1 in trans-translation is not ribosome binding, but rather stabilizing the structure of the region between the TLD and the ORF and preventing global misfolding of tmRNA (Tanner et al., 2006; Wower et al., 2009). Pseudoknots 2–4, though certainly less critical than PK1, also play a role in tmRNA function. Pseudoknot 2, 3, or 4 deletion mutants are unable to produce the same levels of mature tmRNA as wild-type (Wower et al., 2004). This finding suggests that PK2, PK3 and especially PK4 play a role in tmRNA maturation, folding, or stability.

2.3. Open reading frame

The open reading frame of tmRNA encodes a ten-amino-acid tag (ANDENYALAA) that is added to the C terminus of stalled peptides (Tu et al., 1995). The 5' end of the tag template is unstructured, providing a site for the ribosome to resume translation, while the 3' end of this sequence forms part of a conserved helix. A specific tag sequence is not required for the release of stalled ribosomes by tmRNA. The first codon (GCA) can be changed to nearly any other codon without affecting tmRNA function (O'Connor, 2007; Williams et al., 1999). The substitution of six histidine residues at the C terminus of the tag (ANDEHHHHHH) only slightly reduces tmRNA activity (Roche and Sauer, 2001). Mutation of the tag sequence, however, can inhibit the proteolytic degradation of the tagged protein (Roche and Sauer, 1999; Williams et al., 1999).

In *E. coli*, there are five proteases that degrade tagged proteins: ClpXP, ClpAP, Lon, FtsH and Tsp (Choy et al., 2007; Flynn et al., 2001; Herman et al., 1998; Keiler et al., 1996). The most robust of these, ClpXP, binds to the C terminus (residues LAA) of the peptide tag (Farrell et al., 2005; Gottesman et al., 1998; Levchenko et al., 2000; Lies and Maurizi, 2008). This process is enhanced by an adaptor protein, SspB, that tethers the protease to the tagged peptide by binding both the N-terminal region of the peptide tag (residues AAND) and the ClpX machinery (Flynn et al., 2001; Levchenko et al., 2000). As a result, bacterial cells are able to efficiently recognize and degrade peptide products from rescued ribosomes.

3. SmpB structure and function

The small, basic protein SmpB is essential to trans-translation; deletion of SmpB conveys all of the same phenotypes characteristic of tmRNA knockouts (Dulebohn et al., 2007; Karzai et al., 1999). SmpB binds tmRNA, enhances its aminoacylation, and prevents its degradation by RNase R (Hanawa-Suetsugu et al., 2002; Hong et al., 2005; Shimizu and Ueda, 2002). SmpB also binds to ribosomes and recruits tmRNA; cosedimentation experiments indicate that tmRNA does not bind to ribosomes in the absence of SmpB (Hanawa-Suetsugu et al., 2002; Karzai et al., 1999). There are currently no known functions of SmpB outside of the trans-translation process (Dulebohn et al., 2007).

3.1. Structure

The three-dimensional structure of SmpB was solved in solution by NMR (Dong et al., 2002; Nameki et al., 2005; Someya et al., 2003), and the SmpB-tmRNA complex was solved by x-ray crystallography (Figure 3) (Bessho et al., 2007; Gutmann et al., 2003). The core is an oligonucleotide binding (OB) fold: six antiparallel β-strands form a closed β-barrel, exposing two highly-conserved RNA-binding sites on opposite sides of the barrel (Dong et al., 2002; Gutmann et al., 2003; Someya et al., 2003). Similar OB-folds have been identified on several other RNA-binding proteins involved in translation, including the initiation factor IF1 (Dong et al., 2002; Murzin, 1993). SmpB's β-barrel is enclosed on one side by a long α-helix (Gutmann et al., 2003). Of the 160 amino acids in *E. coli* SmpB, the C-terminal 30 residues comprise a tail that, while unstructured in solution and not observable in NMR or crystal structures, performs an essential function in trans-translation. Deleting the tail abolishes tagging entirely (Sundermeier et al., 2005).

3.2. SmpB-tmRNA interactions

SmpB has two separate clusters of highly-conserved amino acids that each function as RNA-binding sites. One of these is a tmRNA-binding site, including E31, L91, N92, and K124. Mutations in these residues dramatically reduce SmpB-tmRNA interaction (Hanawa-Suetsugu et al., 2002; Nameki et al., 2005). SmpB binds tmRNA on the D loop of the TLD region (Gutmann et al., 2003). The binding is specific and has high affinity, with measures of K_d in the low nanomolar range (Hallier et al., 2006; Jacob et al., 2005; Karzai et al., 1999; Metzinger et al., 2008; Sundermeier et al., 2005).

SmpB helps tmRNA mimic the structure and function of alanine-specific tRNA during aminoacylation and entry to the ribosome (Figure 3). As mentioned above, tmRNA's TLD lacks the stem structure of the D stem-loop. SmpB compensates by stabilizing the D loop: residues R45, W118, and V41 interact with tmRNA nucleotides A8 and C48 to form the consecutive stacking structure that is normally formed by C13-G22 base pair in the tRNA D stem (Bessho et al., 2007). SmpB also associates with other conserved tmRNA nucleotides in this region. U17, C18, and A20 in the D loop, as well as U328 and U329 in the T stem, are protected from chemical modification by

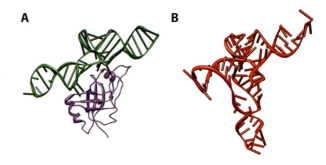

Fig. 3 SmpB mimics the anticodon stem-loop of canonical tRNAs. (A) Structure of the tmRNA tRNA-like domain (green) in complex with SmpB (purple) as reported by Bessho et al. (2007) (PDB 2CZJ). (B) *T. thermophilus* tRNA^{Ser} is a class II tRNA with an extended variable arm (PDB 1SER). Structures were rendered with Chimera.

SmpB (Nameki et al., 2005). Interestingly, the structure of the SmpB-tmRNA complex reveals that SmpB compensates for tmRNA's lack of an anticodon stemloop. SmpB structurally mimics the anticodon arm of a canonical tRNA (Figure 3) (Bessho et al., 2007), which has important implications for how tmRNA and SmpB enter stalled ribosomes (see below).

How many SmpBs bind to a single tmRNA? The stoichiometry of the tmRNA-SmpB complex has been the subject of some controversy. Optical biosensor and melting curve analysis (Nameki et al., 2005), as well as hydroxyl radical probing (Ivanova et al., 2007), assert that only a single SmpB binds tmRNA. In contrast, enzyme probing, UV crosslinking, footprinting, affinity labeling, and filter-binding assays predict up to three separate SmpB binding sites on tmRNA (Metzinger et al., 2005; Wower et al., 2002). Furthermore, surface plasmon resonance (SPR) indicates that the highest affinity binding site for SmpB is not the TLD but rather a site just upstream of tmRNA's ORF (Metzinger et al., 2008). This second binding site suggests a separate role for SmpB in helping set the reading frame on tmRNA, as discussed below.

3.3. SmpB-ribosome binding

Apart from the conserved tmRNA-binding site of SmpB, the protein also has a second site that is likely involved in binding the ribosome during trans-translation. This cluster of highly conserved residues (N17, K18, Y24, Y55, K131, K133, K134, and R139) is located on the opposite side of the β-barrel from the tmRNA binding domain. Mutation of these residues is detrimental to trans-translation activity, but has no effect on tmRNA binding (Dulebohn et al., 2007; Nonin-Lecomte et al., 2009).

Biochemical and structural studies have attempted to localize SmpB in the various steps as it moves through the ribosome, but its binding sites are still ill-defined. SmpB can interact with both the 30S and 50S subunits of the ribosome (Hallier et al., 2006; Ivanova et al., 2007; Kurita et al., 2007). SmpB footprints are located on the 50S subunit below the L7/L12-stalk (near the GTPase associated center) and on the 30S subunit in the vicinity of the P site. The higher-affinity binding partner appears to be the 30S subunit (Hallier et al., 2006). Hydroxyl radical probing suggests that SmpB helices α1 and α3 contact 16S rRNA in the P site (Kurita et al., 2007); the α1 helix contains some of the conserved residues discussed above.

Two separate pre-accommodation binding sites for SmpB were visualized in cryo-EM structures (Kaur et al., 2006); one SmpB is bound to the 30S A site in the decoding center, the other bound to the GTPase center of the 50S subunit. The 30S-bound SmpB also binds the D loop of tmRNA and has the geometry predicted by modeling the tmRNA-SmpB co-crystal data into the A site of the ribosome complexes, which suggests that this SmpB is functionally relevant (Bessho et al., 2007; Gutmann et al., 2003). While the 50S-binding site is consistent with the probing experiments, it does not match the observed co-crystal structure geometry. The conflicting results regarding the number of SmpB binding sites on tmRNA and on the ribosome have led to various models that include more than one SmpB molecule in certain steps of the tagging process.

3.4. One SmpB per tmRNA

The controversy regarding the number of SmpB molecules involved in trans-translation has been resolved in favor of a model in which each tmRNA binds a single SmpB both in solution and during its passage through the ribosome. SmpB and tmRNA exist in a 1:1 molar ratio in the cell (Sundermeier and Karzai, 2007). Since both get degraded unless they are complexed with the other (Moore and Sauer, 2005), this means that they must be bound together in a 1:1 complex. Accordingly, analyses of complexes isolated from the resumption of translation on tmRNA through termination have found that SmpB and tmRNA are in a 1:1 ratio (Bugaeva et al., 2008; Shpanchenko et al., 2005). Because SmpB replaces the anticodon arm in moving through the ribosome, the predicted binding sites for SmpB in

the 30S A and P sites are at their expected locations. The 50S-binding site is likely a biochemical artifact due to SmpB's high basicity. Structural studies now support this model as well; a new cryo-EM analysis of the post-accommodated state has only one SmpB bound (Cheng et al., 2010).

4. Recognition of stalled ribosomes

How are stalled ribosomes recognized by SmpB and tmRNA? It is clear that tmRNA and SmpB do not compete effectively with aminoacyl-tRNAs for binding to elongating ribosomes, as even a 20-fold overexpression of tmRNA and SmpB *in vivo* does not increase the level of tagged proteins (Moore and Sauer, 2005). tmRNA is blocked by the presence of downstream mRNA in elongating ribosomes. It was recognized early on that tmRNA only targets ribosomes with truncated mRNA templates (Keiler et al., 1996)

4.1. Empty A sites

In vitro experiments with purified components confirm that ribosomes transfer their nascent polypeptides to tmRNA with highest efficiency when there are six or fewer nucleotides in the A site (Ivanova et al., 2004). The rates of peptidyl transfer to tmRNA in this situation compare roughly to the termination reaction catalyzed by RF1 (Ivanova et al., 2004). Ribosomes bound to longer mRNAs react with tmRNA slightly more slowly, and the efficiency drops 20-fold when the mRNA reaches 15 nucleotides or more in length. Structural work revealed that 12–13 mRNA nucleotides downstream of the A site form electrostatic interactions with the highly basic S3, S4, and S5 proteins (Yusupova et al., 2001). This binding may anchor longer mRNAs in the A site and sterically prevent tmRNA and SmpB from binding, while shorter, unanchored mRNAs can loop out into the intersubunit space. In this model, steric occlusion allows tmRNA to distinguish between elongating ribosomes and stalled ones.

The S3, S4, and S5 proteins form a channel between the head and shoulder of the 30S subunit through which mRNA passes as it enters the ribosome (Yusupova et al., 2001). The positive charges that line this channel are expected to create electrostatic repulsion that could open the channel in the absence of mRNA. This open conformation has been seen in some crystal structures (Schluenzen et al., 2000). Possibly, the open channel conformation, which would not normally occur with elongating ribosomes, may serve as another recognition element for tmRNA binding (Moore and Sauer, 2007). Upon entering the ribosome, the template region of tmRNA must be positioned in this channel for tmRNA to be translated. The channel must open because the 5' and 3' ends of tmRNA are paired together–tmRNA is effectively a circular template that cannot be threaded through otherwise. It has not been resolved, however, whether placing the tmRNA template in this channel is involved with recognition of stalled ribosomes or occurs after tmRNA recruitment.

4.2. Removal of downstream mRNA

Truncated mRNAs are not the only source of translational stalling; ribosomes can stall in the middle of an mRNA as well. Strings of rare codons in overexpressed transcripts, for example, can stall ribosomes (Roche and Sauer, 1999). Similarly, certain nascent peptide sequences can cause the termination or peptidyl transfer reactions to be very inefficient (Collier et al., 2004; Hayes et al., 2002a, b). If these stalling events persist long enough for the downstream mRNA downstream to be degraded, the ribosomes become irreversibly arrested and have to be rescued by tmRNA (Hayes and Sauer, 2003; Li et al., 2006; Sunohara et al., 2004a; Sunohara et al., 2004b). When ribosomes stall in the middle of a transcript, downstream nucleotides must be removed in order for the ribosome to be recycled by tmRNA. The rate of degradation of downstream mRNA is enhanced when the RNA is not protected by translating ribosomes.

The mRNA downstream of the stalling site can be degraded in different ways. Often the mRNA is truncated at the 3' boundary of the ribosome, about 15 nt downstream from the P-site codon, beyond which 3' to 5' exonucleases are sterically blocked (Li et al., 2006; Sunohara et al., 2004a). In some cases, stalling leads to cleavage of the mRNA at the upstream ribosome boundary (Bjornsson and Isaksson, 1996; Loomis et al., 2001; Yao et al., 2008), presumably by initial endonucleolytic cleavage.

4.3. A-site cleavage

Alternatively, in a small number of specific cases, the mRNA is truncated at the A-site codon itself (Hayes and Sauer, 2003). This A-site cleavage is probably the result of endonucleases, though this is not formally proven, as no 3' product of the cleavage event has been detected. The one known A-site endonuclease is the RelE toxin, which cleaves mRNA and inhibits global translation in response to amino acid starvation (Pedersen et al., 2003). It cleaves with some sequence specificity, preferring CAG and UAG codons. Interestingly, it only cleaves RNA within the context of the ribosomal A site; RelE does not cleave RNA by itself (Pedersen et al., 2003). The mechanism of RelE cleavage and its ribosome dependence were clarified recently when the structure of RelE inside 70S ribosomes was solved (Neubauer et al., 2009).

A second A-site endonuclease is postulated to cleave ribosomes stalled during termination after proline codons (Garza-Sanchez et al., 2008). No known nuclease is responsible for this second activity, though it does require the RNase II exonuclease to degrade the downstream RNA to within 21 nucleotides of the P-site codon prior to A-site cleavage (Garza-Sanchez et al., 2009). It has been proposed that the reaction is catalyzed by the ribosome itself (Hayes and Sauer, 2003), presumably in a regulated manner, though no such cleavage has been observed in assays using pure components.

Because the *in vitro* studies show that mRNA in the A site inhibits peptidyl transfer to tmRNA, one might expect that A-site cleavage would be essential for tagging to occur. This does not appear to be the case *in vivo*, where the situation is rather more complex. In one example, stalling was reported at a protein ending in Lys-Lys-Arg-Arg sequences (with rare Arg codons). The tmRNA tag was added immediately after the second Lys codon (Garza-Sanchez et al., 2008). The mRNA was degraded to the 3' boundary, about 18 nucleotides from the P site, but only very small amounts of A-site cleavage were visible. The authors conclude that 3'-boundary cleavage is sufficient for recruiting tmRNA (Garza-Sanchez et al., 2008). Similar results were found in a separate study on an mRNA with five consecutive rare Arg codons: only boundary cleavage was detected but robust tagging by tmRNA was observed (Li et al., 2006). These data suggest that A-site cleavage may not be essential for tagging to occur *in vivo*, provided that the downstream RNA is processed back to the boundary. Additional experiments will be required to further investigate these discrepancies.

If the ribosome stalls with aminoacyl-tRNA or release factors trapped in the A site, tagging cannot occur; their presence in the A site blocks tmRNA and SmpB binding. This is the case with SecM, a leader peptide that regulates the downstream *secA* gene in response to changing levels of activity in the secretory machinery. When secretory capability is high, the machinery binds the signal peptide in SecM and pulls it out of the ribosome; when the secretory machinery is less active, ribosomes stall at the FxxxxWIxxxxGIRxGP sequence in SecM, changing the mRNA structure and increasing expression of SecA (Nakatogawa and Ito, 2002). Inhibition of tagging of stalled SecM is essential to maintaining the logic of the genetic switch. When the ribosome stalls at SecM, Pro-tRNA^{Pro} is bound in the A site as an important part of the stalling mechanism; it also blocks tmRNA-mediated tagging (Garza-Sanchez et al., 2006). Overexpression of SecM or other stalling peptide sequences can lead to tagging as the aminoacyl-tRNAs trapped in the A site are depleted and stalling occurs with no tRNA or release factor blocking the A site.

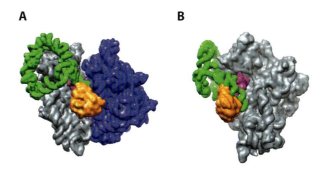

Fig. 4 Cryo-EM structure of the pre-accommodation tmRNA SmpB·EF-Tu complex bound to the 70S ribosome. (A) tmRNA (green) bound to SmpB (purple) (PDB 2OB7) and EF-Tu (orange) (PDB 1OB2) was fitted to the 70S complex using coordinates from Gillet et al. (2007). tmRNA is seen wrapped around the beak of the 30S subunit (grey) with the tmRNA TLD and SmpB bound in the A site. (B) Interface view of the 30S subunit shows SmpB bound near the decoding center. Structures were fitted and rendered using Chimera.

5. Entering the A site of stalled ribosomes

Because of its tRNA-like nature, tmRNA is able to interact with elongation factor Tu (EF-Tu) much in the same way as a canonical tRNA does. Reconstructions from cryo-electron microscopy (cryo-EM) show that, on the ribosome, EF-Tu binds the acceptor arm and the T arm of alanyl-tmRNA in a manner virtually identical to that of EF-Tu in complex with aminoacyl-tRNA (Valle et al., 2003). As with aminoacyl-tRNA, EF-Tu protects the alanyl-tmRNA ester bond from hydrolysis (Barends et al., 2000; Rudinger-Thirion et al., 1999). EF-Tu is likewise essential for the addition of alanine to a stalled peptide by alanyl-tmRNA; peptidyl transfer occurs only at a very slow rate in its absence (Hallier et al., 2004; Shimizu and Ueda, 2006).

The hybrid nature of tmRNA, however, raises problems as it enters the ribosome. During normal translation, the signal to accommodate the appropriate tRNA in the A site depends on correct codon-anticodon base pairing between mRNA and cognate tRNA. The ribosome recognizes the geometry of the codon-anticodon base pairs (Ogle et al., 2001; Ogle and Ramakrishnan, 2005). Conserved 16S nucleotides A1492 and A1493 flip out of a loop in helix 44 to bind the minor groove of the codon-anticodon duplex (Ogle et al., 2001). G530 also undergoes a *syn* to an *anti* conformational change to interact with the second and third base pairs. These local movements lead to global conformational changes, specifically a rotation of the head and shoulder of the 30S subunit toward the intersub-

unit space, effectively closing the 30S subunit over the codon-anticodon helix (Ogle et al., 2002). These global conformational changes are communicated to EF-Tu through its interactions with both of the ribosomal subunits and through distortion of the tRNA structure (Schmeing et al., 2009).

Since tmRNA lacks an anticodon and enters ribosomes that have no mRNA in their A sites, how does it trigger GTPase activity by EF-Tu? Structural and biochemical data suggest that, during ribosome rescue, the decoding center is engaged not by an RNA duplex but by the SmpB protein. In particular, the x-ray crystal structure of the tmRNA-SmpB complex (Figure 3) suggests an interaction between the SmpB tail and the decoding center. The C-terminal tail of SmpB, roughly 30 residues long, is located roughly where the anticodon loop of a normal tRNA would be. Placing the tmRNA-SmpB co-crystal structure into a tRNA-like orientation in the ribosome points the C-terminal tail toward the decoding center. Cryo-EM studies of 70S ribosomes bound to tmRNA, SmpB, EF-Tu in a pre-accommodation complex also orient the C-terminal tail toward the decoding center (Figure 4) (Kaur et al., 2006; Weis et al., 2010). The unstructured tail appears to take on a helical structure within the ribosome, as shown by analysis of the conserved sequence of the tail and the hydroxyl radical probing of 16S rRNA with SmpB tail residues ligated to Fe-BABE (Jacob et al., 2005; Kurita et al., 2010; Kurita et al., 2007).

The C-terminal SmpB tail is essential to tmRNA's ability to accept the nascent polypeptide. Mutations of conserved residues in the tail, particularly D(137) KR, abolish tagging by tmRNA *in vivo* and drastically reduce the rate of peptidyl transfer to tmRNA in vitro (Sundermeier et al., 2005). Hydroxyl radical and chemical probing experiments have shown that the SmpB tail binds nucleotides in the 30S A site, from the decoding center to the downstream mRNA channel (Kurita et al., 2007). Indeed, SmpB binding to the decoding center protects nucleotides A1492, A1493, and G530 from reacting with chemical probes and causes a shift of these nucleotides in NMR spectra of a small A-site mimic (Nonin-Lecomte et al., 2009). Taken together, the structural and biochemical data suggest a model in which SmpB binding to the decoding center triggers the conformational changes associated with canonical decoding, leading to EF-Tu activation and accommodation of tmRNA. Further work must be performed to determine exactly how the C-terminal tail might trigger those changes and license tmRNA entry into the ribosome.

6. Template swapping

What is the fate of a truncated mRNA once tmRNA enters the stalled ribosome? The mRNA's binding to the ribosome is stabilized initially by its interaction with the peptidyl-tRNA. After the stalled polypeptide is transferred to tmRNA, the defective mRNA template is ejected with the deacylated tRNA when translocation occurs (Ivanova et al., 2005). *In vivo*, the release of the stalled mRNA may be even faster than observed *in vitro*, as upstream ribosomes may facilitate mRNA release by pulling the loosely bound mRNA free of the leading ribosome (Ivanova et al., 2005). Once the truncated mRNA is released from the stalled ribosomes, it is targeted for decay while the ribosomes resume translation on the tmRNA ORF.

Several studies have shown that tmRNA facilitates the degradation of non-stop mRNAs (Mehta et al., 2006; Richards et al., 2006; Yamamoto et al., 2003). Non-stop mRNA decay in bacteria is dependent on SmpB, suggesting that degradation requires the tmRNA-SmpB complex to actively engage stalled ribosomes (Richards et al., 2006). How does tmRNA facilitate the degradation of truncated messages? Once the non-stop mRNA is released from the stalled ribosome, it is no longer protected from exonucleases that efficiently attack the 3' end of any mRNA lacking secondary structure (Yamamoto et al., 2003). Indeed, the half-life of non-stop mRNAs increases significantly in the absence of tmRNA (Mehta et al., 2006; Richards et al., 2006; Yamamoto et al., 2003). Another possible explanation is that tmRNA may recruit RNase R, a 3' to 5' exonuclease, to non-stop mRNAs. RNase R copurifies with the tmRNA-SmpB complex (Karzai and Sauer, 2001), and mutations in the 3' end of the ORF reduce non-stop mRNA degradation without affecting the tRNA or mRNA-like functions of tmRNA (Richards et al., 2006).

7. Selecting the reading frame on tmRNA

As the ribosome switches RNA templates, how is the appropriate codon selected for translation to resume on tmRNA? The transfer of ribosomes to tmRNA resembles a normal round of elongation and not a reinitiation event. No specialized initiator tRNA or protein factors are required, nor does tmRNA base pair with 16S rRNA like the Shine-Dalgarno sequence on mRNA does. It also appears that conserved secondary structural elements in tmRNA do not bind sites on the ribosome to position the first codon in the tmRNA ORF (the resume codon) properly. As discussed above, the four pseudoknots that dominate the tmRNA structure can be replaced with unrelated sequences with little or no loss of tmRNA activity (Nameki et al., 2000; Tanner et al., 2006; Wower et al., 2009). Frame selection does not result from base or structure-specific interactions of tmRNA with the ribosome directly.

7.1. tmRNA determinants of frame selection

The tmRNA nucleotides critical for frame selection lie upstream of the resume codon (Lee et al., 2001; Miller et al., 2008; Williams et al., 1999). These two key nucleotides, U85 and A86 in *E. coli*, are conserved in natural tmRNA sequences; A86 was also conserved in random mutagenesis and selection experiments (Williams et al., 1999). Mutations in either base lead to loss of tmRNA function and errors in frame selection *in vitro* and *in vivo*. The U85A mutation, for example, partially shifts translation to the –1 frame (Lee et al., 2001; Miller et al., 2008). Mutation of the universally conserved A86 leads to severe loss of function (Lee et al., 2001; Williams et al., 1999); the A86C mutation shifts translation entirely to the +1 frame (Miller et al., 2008). In contrast, the resume codon itself and the three nucleotides before it can be changed with little or no effect on tmRNA activity (Lee et al., 2001; O'Connor, 2007).

U85 and A86 appear to act as markers that cause translation to resume at a given distance downstream. The distance from PK1 to U85 is not critical for tagging, but insertions or deletions between A86 and the resume codon (G90) cause misreading of the resume codon (Lee et al., 2001; Miller et al., 2008). Taken together, these mutagenesis data support a model in which U85 and/or A86 bind to a ligand that draws the tmRNA template sequence into the ribosomal A site, placing the nucleotide four bases downstream as the first in the resume codon.

7.2. Protein determinants of frame selection

What is the ligand that binds upstream of the resume codon? One candidate that has been proposed is ribosomal protein S1, which was shown to crosslink to U85 (Wower et al., 2000). Cryo-EM structures of tmRNA bound inside 70S ribosomes reveal that S1 affects the structure of the tmRNA template sequence (Gillet et al., 2007). Though S1 cannot interact directly with tmRNA on the ribosome, it has been proposed that free S1 binds tmRNA and stabilizes a functional, open complex that is then passed to stalled ribosomes (Gillet et al., 2007). In support of this model, one study presents evidence that S1 is required for tmRNA to serve as a template *in vitro* (Saguy et al., 2007). In contrast, two studies using reconstituted translation systems (Qi et al., 2007; Takada et al., 2007) demonstrated that S1 is non-essential. The only genetic data available likewise argue against a role for S1 in trans-translation (McGinness and Sauer, 2004).

A more likely candidate is the SmpB protein. In addition to its well characterized binding site in the TLD, SmpB binds tmRNA upstream of the resume codon, reducing the accessibility of the upstream sequence to nucleases in probing assays (Metzinger et al., 2005). It has been reported that this interaction has a high affinity, comparable to SmpB's binding to the TLD (Metzinger et al., 2008). Intriguingly, mutations in tmRNA that alter frame selection also alter SmpB's interaction with U85 (Konno et al., 2007). While only one SmpB accompanies tmRNA through the ribosome, it seems that SmpB contacts tmRNA at different sites at each step.

Genetic evidence supports the idea that SmpB binding to the upstream region plays a role in frame selection. The A86C mutation in tmRNA leads to the total loss of tagging in the 0 frame and high levels of tagging in the +1 frame. SmpB mutants were identified

that suppress both of these defects, restoring activity and proper frame selection on A86C tmRNA (Watts et al., 2009). Intriguingly, the SmpB residues that were mutated (Tyr24, Val129, and Ala130) cluster together in a hydrophobic pocket far away from the TLD binding site. These results demonstrate that SmpB plays a biologically relevant role in setting the frame on tmRNA.

These data are consistent with the following hypothetical structural model. As described above, SmpB acts as an anticodon stem mimic in the SmpB-TLD complex; modeling this structure into the P site of the 70S ribosome shows that residues Tyr24, Val129 and Ala130 would be found on the A-site face of SmpB, not far from the 16S rRNA and the decoding center. An interaction between this region of SmpB and the tmRNA nucleotides U85 or A86 could draw the tmRNA ORF into the A site. With the first codon (GCA) lying in the mRNA channel in the decoding center, translation would begin with tmRNA as a template. Additional structural and biochemical studies need to be done to rigorously prove this model.

7.3. Following tmRNA through the ribosome

Cryo-EM structures of tmRNA and SmpB within the 70S ribosome indicate that the ORF alone lies along the mRNA channel and that pseudoknots 2–4 are organized into a large spiral encircling the beak of the 30S subunit (Kaur et al., 2006; Valle et al., 2003). The ribosome intersubunit bridges must melt to accommodate the passage of tmRNA through the ribosome. SmpB and tmRNA are translocated together, moving from the A site to the P site and then out through the E site. Although helix H5 unwinds for the 3' end of the ORF to be decoded, the pseudoknots do not melt during the process (Ivanov et al., 2002). Termination occurs with either release factor binding to the UAA stop codon, after which tmRNA is presumably recycled along with the ribosome subunits.

8. Alternative rescue mechanisms

In *E. coli*, tmRNA is not essential for cell viability, suggesting that there must be an alternative ribosome rescue pathway. Indeed, proteins synthesized from non-stop mRNAs accumulate in *E. coli* strains lacking tmRNA. Likewise, cells carrying tmRNA variants that encode alternate tag sequences yield a mixture of tagged and untagged proteins (Moore and Sauer, 2005; Roche and Sauer, 1999, 2001).

How are stalled ribosomes rescued independently of tmRNA? One pathway that has been suggested as an alternative mechanism for ribosome rescue is peptidyl-tRNA drop-off. Genetic analyses show that RF3, RRF, EF-G increase recycling in the absence of tmRNA (Singh et al., 2008; Singh and Varshney, 2004). The proposal is that these proteins could dissociate the subunits and peptidyl hydrolase (Pth) could hydrolyze the peptidyl-tRNAs released. In support of this model, it was found that growth defects arising from temperature sensitive Pth alleles are suppressed by overexpression of tmRNA. The mechanism is not well understood, but it appears that drop-off is only possible with nascent peptides less than 40 amino acid residues. A peptide longer than this may be partially folded outside of the peptide tunnel preventing passage backward through the tunnel (Janssen and Hayes, 2009; Kuroha et al., 2009).

Recent kinetic analyses of ribosome recycling in which peptidyl-tRNA turnover was measured directly cast doubt on the peptidyl-tRNA drop-off hypothesis. These studies argue that overexpression of RF3, RRF, Pth does not accelerate peptidyl-tRNA turnover on stalled ribosomes (Janssen and Hayes, 2009). The authors propose an alternative ribosome recycling pathway, based on evidence that ribosomes from *E. coli* can recycle from non-stop mRNAs in the absence of exogenous factors (Kuroha et al., 2009; Szaflarski et al., 2008). Perhaps the ribosome's hold on non-stop mRNA and peptidyl-tRNA is reduced due to the lack of mRNA interactions with the ribosome along the mRNA path (Kuroha et al., 2009).

9. Future prospects

In the last five years, structural studies have provided new insights into the interaction of the trans-translation machinery with itself and with the ribosome. Together, genetic, structural, and biochemical studies have resolved contradictions in the literature, yielding models that are well-supported and explain much about how ribosome rescue occurs. They have discovered and emphasized the critical role that SmpB plays in every step of ribosome rescue: stabilizing tmRNA, licensing tmRNA entry into ribosomes, setting the reading frame and moving with tmRNA through the ribosome. However, there are still many unresolved questions: how exactly does SmpB bypass the decod-

ing center to allow accommodation of tmRNA? What signal is transmitted to EF-Tu to hydrolyze GTP? How do SmpB and tmRNA interact to set the reading frame? What is the structure of complexes later in the trans-translation process, after SmpB and tmRNA have moved out of the A site? What alternate mechanisms rescue stalled ribosomes when tmRNA is not available? Additional studies need to be done to nail down the answers to these questions.

References

Abo T, Inada T, Ogawa K, Aiba, H (2000) SsrA-mediated tagging and proteolysis of LacI and its role in the regulation of lac operon. EMBO J 19: 3762–3769

Barends S, Karzai AW, Sauer RT, Wower J, Kraal, B (2001) Simultaneous and functional binding of SmpB and EF-Tu-TP to the alanyl acceptor arm of tmRNA. J Mol Biol 314: 9–21

Barends S, Wower J, Kraal, B (2000) Kinetic parameters for tmRNA binding to alanyl-tRNA synthetase and elongation factor Tu from *Escherichia coli*. Biochemistry-Us 39: 2652–2658

Bernstein JA, Khodursky AB, Lin PH, Lin-Chao S, Cohen SN (2002) Global analysis of mRNA decay and abundance in *Escherichia coli* at single-gene resolution using two-color fluorescent DNA microarrays. Proc Natl Acad Sci USA 99: 9697–9702

Bessho Y, Shibata R, Sekine S, Murayama K, Higashijima K, Hori-Takemoto C, Shirouzu M, Kuramitsu S, Yokoyama S (2007) Structural basis for functional mimicry of long-variable-arm tRNA by transfer-messenger RNA. Proc Natl Acad Sci USA 104: 8293–8298

Bjornsson A, Isaksson LA (1996) Accumulation of a mRNA decay intermediate by ribosomal pausing at a stop codon. Nucleic Acids Res 24: 1753–1757

Bugaeva EY, Shpanchenko OV, Felden B, Isaksson LA, Dontsova OA (2008) One SmpB molecule accompanies tmRNA during its passage through the ribosomes. FEBS Lett 582: 1532–1536

Chauhan AK, Apirion, D (1989) The gene for a small stable RNA (10Sa RNA) of *Escherichia coli*. Mol Microbiol 3: 1481–1485

Cheng K, Ivanova N, Scheres SH, Pavlov MY, Carazo JM, Hebert H, Ehrenberg M, Lindahl, M (2010) tmRNA. SmpB complex mimics native aminoacyl-tRNAs in the A site of stalled ribosomes. J Struct Biol 169: 342–348

Choy JS, Aung LL, Karzai AW (2007) Lon protease degrades transfer-messenger RNA-tagged proteins. J Bacteriol 189: 6564–6571

Collier J, Bohn C, Bouloc, P (2004) SsrA tagging of *Escherichia coli* SecM at its translation arrest sequence. J Biol Chem 279: 54 193–54 201

Condon, C (2007) Maturation and degradation of RNA in bacteria. Curr Opin Microbiol 10: 271–278

Dong G, Nowakowski J, Hoffman DW (2002) Structure of small protein B: the protein component of the tmRNA-SmpB system for ribosome rescue. EMBO J 21: 1845–1854

Dulebohn D, Choy J, Sundermeier T, Okan N, Karzai AW (2007) Trans-translation: The tmRNA-mediated surveillance mechanism for ribosome rescue, directed protein degradation, and nonstop mRNA decay. Biochemistry-Us 46: 4681–4693

Farrell CM, Grossman AD, Sauer RT (2005) Cytoplasmic degradation of ssrA-tagged proteins. Mol Microbiol 57: 1750–1761

Felden B, Hanawa K, Atkins JF, Himeno H, Muto A, Gesteland RF, McCloskey JA, Crain PF (1998) Presence and location of modified nucleotides in *Escherichia coli* tmRNA: structural mimicry with tRNA acceptor branches. EMBO J 17: 3188–3196

Felden B, Himeno H, Muto A, McCutcheon JP, Atkins JF, Gesteland RF (1997) Probing the structure of the E*scherichia coli* 10Sa RNA (tmRNA) RNA 3: 89–103

Flynn JM, Levchenko I, Seidel M, Wickner SH, Sauer RT, Baker TA (2001) Overlapping recognition determinants within the ssrA degradation tag allow modulation of proteolysis. Proc Natl Acad Sci USA 98: 10 584–10 589

Garza-Sanchez F, Gin JG, Hayes CS (2008) Amino acid starvation and colicin D treatment induce A-site mRNA cleavage in *Escherichia coli*. J Mol Biol 378: 505–519

Garza-Sanchez F, Janssen BD, Hayes CS (2006) Prolyl-tRNA(Pro) in the A-site of SecM-arrested ribosomes inhibits the recruitment of transfer-messenger RNA. J Biol Chem 281: 34 258–34 268

Garza-Sanchez F, Shoji S, Fredrick K, Hayes CS (2009) RNase II is important for A-site mRNA cleavage during ribosome pausing. Mol Microbiol 73: 882–897

Gillet R, Kaur S, Li W, Hallier M, Felden B, Frank, J (2007) Scaffolding as an organizing principle in trans-translation. The roles of small protein B and ribosomal protein S1. J Biol Chem 282: 6356–6363

Gottesman S, Roche E, Zhou Y, Sauer RT (1998) The ClpXP and ClpAP proteases degrade proteins with carboxy-terminal peptide tails added by the SsrA-tagging system. Genes Dev 12: 1338–1347

Gutmann S, Haebel PW, Metzinger L, Sutter M, Felden B, Ban, N (2003) Crystal structure of the transfer-RNA domain of transfer-messenger RNA in complex with SmpB. Nature 424: 699–703

Hallier M, Desreac J, Felden, B (2006) Small protein B interacts with the large and the small subunits of a stalled ribosome during trans-translation. Nucleic Acids Res 34: 1935–1943

Hallier M, Ivanova N, Rametti A, Pavlov M, Ehrenberg M, Felden, B (2004) Pre-binding of small protein B to a stalled ribosome triggers trans-translation. J Biol Chem 279: 25 978–25 985

Hanawa-Suetsugu K, Bordeau V, Himeno H, Muto A, Felden, B (2001) Importance of the conserved nucleotides around the tRNA-like structure of *Escherichia coli* transfer-messenger RNA for protein tagging. Nucleic Acids Res 29: 4663–4673

Hanawa-Suetsugu K, Takagi M, Inokuchi H, Himeno H, Muto, A (2002) SmpB functions in various steps of trans-translation. Nucleic Acids Res 30: 1620–1629

Hartmann RK, Gossringer M, Spath B, Fischer S, Marchfelder, A (2009) The making of tRNAs and more – RNase P and tRNase Z. Prog Mol Biol Transl Sci 85: 319–368

Hayes CS, Bose B, Sauer RT (2002 a) Proline residues at the C terminus of nascent chains induce SsrA tagging during translation termination. J Biol Chem 277: 33 825–33 832

Hayes CS, Bose B, Sauer RT (2002 b) Stop codons preceded by rare arginine codons are efficient determinants of SsrA tagging in *Escherichia coli*. Proc Natl Acad Sci USA 99: 3440–3445

Hayes CS, Sauer RT (2003) Cleavage of the A site mRNA codon during ribosome pausing provides a mechanism for translational quality control. Mol Cell 12: 903–911

Herman C, Thevenet D, Bouloc P, Walker GC, D'Ari, R (1998) Degradation of carboxy-terminal-tagged cytoplasmic proteins by the E*scherichia coli* protease HflB (FtsH) Genes Dev 12: 1348–1355

Hong SJ, Tran QA, Keiler KC (2005) Cell cycle-regulated degradation of tmRNA is controlled by RNase R and SmpB. Mol Microbiol 57: 565–575

Hou YM, Schimmel, P (1988) A simple structural feature is a major determinant of the identity of a transfer RNA. Nature 333: 140–145

Huang C, Wolfgang MC, Withey J, Koomey M, Friedman DI (2000) Charged tmRNA but not tmRNA-mediated proteolysis is essential for Neisseria gonorrhoeae viability. EMBO J 19: 1098–1107

Hutchison CA, Peterson SN, Gill SR, Cline RT, White O, Fraser CM, Smith HO, Venter JC (1999) Global transposon mutagenesis and a minimal Mycoplasma genome. Science 286: 2165–2169

Ivanov PV, Zvereva MI, Shpanchenko OV, Dontsova OA, Bogdanov AA, Aglyamova GV, Lim VI, Teraoka Y, Nierhaus KH (2002) How does tmRNA move through the ribosome? FEBS Lett 514: 55–59

Ivanova N, Lindell M, Pavlov M, Holmberg Schiavone L, Wagner EG, Ehrenberg, M (2007) Structure probing of tmRNA in distinct stages of trans-translation. RNA 13: 713–722

Ivanova N, Pavlov MY, Ehrenberg, M (2005) tmRNA-induced release of messenger RNA from stalled ribosomes. J Mol Biol 350: 897–905

Ivanova N, Pavlov MY, Felden B, Ehrenberg, M (2004) Ribosome rescue by tmRNA requires truncated mRNAs. J Mol Biol 338: 33–41

Jacob Y, Sharkady SM, Bhardwaj K, Sanda A, Williams KP (2005) Function of the SmpB tail in transfer-messenger RNA translation revealed by a nucleus-encoded form. J Biol Chem 280: 5503–5509

Janssen BD, Hayes CS (2009) Kinetics of paused ribosome recycling in *Escherichia coli*. J Mol Biol 394: 251–267

Karzai AW, Sauer RT (2001) Protein factors associated with the SsrA. SmpB tagging and ribosome rescue complex. Proc Natl Acad Sci USA 98: 3040–3044

Karzai AW, Susskind MM, Sauer RT (1999) SmpB, a unique RNA-binding protein essential for the peptide-tagging activity of SsrA (tmRNA) EMBO J 18: 3793–3799

Kaur S, Gillet R, Li W, Gursky R, Frank, J (2006) Cryo-EM visualization of transfer messenger RNA with two SmpBs in a stalled ribosome. Proc Natl Acad Sci USA 103: 16 484–16 489

Keiler KC (2008) Biology of trans-translation. Annu Rev Microbiol 62: 133–151

Keiler KC, Waller PR, Sauer RT (1996) Role of a peptide tagging system in degradation of proteins synthesized from damaged messenger RNA. Science 271: 990–993

Komine Y, Kitabatake M, Yokogawa T, Nishikawa K, Inokuchi, H (1994) A tRNA-like structure is present in 10Sa RNA, a small stable RNA from *Escherichia coli*. Proc Natl Acad Sci USA 91: 9223–9227

Konno T, Kurita D, Takada K, Muto A, Himeno, H (2007) A functional interaction of SmpB with tmRNA for determination of the resuming point of trans-translation. RNA 13: 1723–1731

Kurita D, Muto A, Himeno, H (2010) Role of the C-terminal tail of SmpB in the early stage of trans-translation. RNA 16: 980–990

Kurita D, Sasaki R, Muto A, Himeno, H (2007) Interaction of SmpB with ribosome from directed hydroxyl radical probing. Nucleic Acids Res 35: 7248–7255

Kuroha K, Horiguchi N, Aiba H, Inada, T (2009) Analysis of nonstop mRNA translation in the absence of tmRNA in *Escherichia coli*. Genes Cells 14: 739–749

Lee S, Ishii M, Tadaki T, Muto A, Himeno, H (2001) Determinants on tmRNA for initiating efficient and precise trans-translation: some mutations upstream of the tag-encoding sequence of *Escherichia coli* tmRNA shift the initiation point of trans-translation in vitro. RNA 7: 999–1012

Levchenko I, Seidel M, Sauer RT, Baker TA (2000) A specificity-enhancing factor for the ClpXP degradation machine. Science 289: 2354–2356

Li X, Hirano R, Tagami H, Aiba, H (2006) Protein tagging at rare codons is caused by tmRNA action at the 3' end of nonstop mRNA generated in response to ribosome stalling. RNA 12: 248–255

Li Z, Deutscher MP (2002) RNase E plays an essential role in the maturation of *Escherichia coli* tRNA precursors. RNA 8: 97–109

Li Z, Pandit S, Deutscher MP (1998) 3' exoribonucleolytic trimming is a common feature of the maturation of small, stable RNAs in *Escherichia coli*. Proc Natl Acad Sci USA 95: 2856–2861

Lies M, Maurizi MR (2008) Turnover of endogenous SsrA-tagged proteins mediated by ATP-dependent proteases in *Escherichia coli*. J Biol Chem 283: 22 918–22 929

Lin-Chao S, Wei CL, Lin YT (1999) RNase E is required for the maturation of ssrA RNA and normal ssrA RNA peptide-tagging activity. Proc Natl Acad Sci USA 96: 12 406–12 411

Loomis WP, Koo JT, Cheung TP, Moseley SL (2001) A tripeptide sequence within the nascent DaaP protein is required for mRNA processing of a fimbrial operon in *Escherichia coli*. Mol Microbiol 39: 693–707

Makarov EM, Apirion, D (1992) 10Sa RNA: processing by and inhibition of RNase III. Biochem Int 26: 1115–1124

McClain WH, Foss, K (1988) Changing the identity of a tRNA by introducing a G-U wobble pair near the 3' acceptor end. Science 240: 793–796

McClain WH, Guerrier-Takada C, Altman S (1987) Model substrates for an RNA enzyme. Science 238: 527–530

McGinness KE, Sauer RT (2004) Ribosomal protein S1 binds mRNA and tmRNA similarly but plays distinct roles in translation of these molecules. Proc Natl Acad Sci USA 101: 13 454–13 459

Mehta P, Richards J, Karzai AW (2006) tmRNA determinants required for facilitating nonstop mRNA decay. RNA 12: 2187–2198

Metzinger L, Hallier M, Felden, B (2005) Independent binding sites of small protein B onto transfer-messenger RNA during trans-translation. Nucleic Acids Res 33: 2384–2394

Metzinger L, Hallier M, Felden, B (2008) The highest affinity binding site of small protein B on transfer messenger RNA is outside the tRNA domain. RNA 14: 1761–1772

Miller MR, Healey DW, Robison SG, Dewey JD, Buskirk AR (2008) The role of upstream sequences in selecting the reading frame on tmRNA. BMC Biology 6: 29

Moore SD, Sauer RT (2005) Ribosome rescue: tmRNA tagging activity and capacity in *Escherichia coli*. Mol Microbiol 58: 456–466

Moore SD, Sauer RT (2007) The tmRNA system for translational surveillance and ribosome rescue. Annu Rev Biochem 76: 101–124

Murzin AG (1993) OB(oligonucleotide/oligosaccharide binding)-fold: common structural and functional solution for non-homologous sequences. EMBO J 12: 861–867

Muto A, Fujihara A, Ito KI, Matsuno J, Ushida C, Himeno, H (2000) Requirement of transfer-messenger RNA for the growth of Bacillus subtilis under stresses. Genes Cells 5: 627–635

Nakatogawa H, Ito, K (2002) The ribosomal exit tunnel functions as a discriminating gate. Cell 108: 629–636

Nameki N, Chattopadhyay P, Himeno H, Muto A, Kawai, G (1999a) An NMR and mutational analysis of an RNA pseudoknot of *Escherichia coli* tmRNA involved in trans-translation. Nucleic Acids Res 27: 3667–3675

Nameki N, Felden B, Atkins JF, Gesteland RF, Himeno H, Muto, A (1999b) Functional and structural analysis of a pseudoknot

upstream of the tag-encoded sequence in *E. coli* tmRNA. J Mol Biol 286: 733–744

Nameki N, Someya T, Okano S, Suemasa R, Kimoto M, Hanawa-Suetsugu K, Terada T, Shirouzu M, Hirao I, Takaku H, et al. (2005) Interaction analysis between tmRNA and SmpB from *Thermus thermophilus*. J Biochem 138: 729–739

Nameki N, Tadaki T, Himeno H, Muto, A (2000) Three of four pseudoknots in tmRNA are interchangeable and are substitutable with single-stranded RNAs. FEBS Lett 470: 345–349

Nameki N, Tadaki T, Muto A, Himeno, H (1999c) Amino acid acceptor identity switch of *Escherichia coli* tmRNA from alanine to histidine in vitro. J Mol Biol 289: 1–7

Neubauer C, Gao YG, Andersen KR, Dunham CM, Kelley AC, Hentschel J, Gerdes K, Ramakrishnan V, Brodersen DE (2009) The structural basis for mRNA recognition and cleavage by the ribosome-dependent endonuclease RelE. Cell 139: 1084–1095

Nonin-Lecomte S, Germain-Amiot N, Gillet R, Hallier M, Ponchon L, Dardel F, Felden, B (2009) Ribosome hijacking: a role for small protein B during trans-translation. EMBO Reports 10: 160–165

O'Connor, M (2007) Minimal translation of the tmRNA tag-coding region is required for ribosome release. Biochem Biophys Res Commun 357: 276–281

Ogle JM, Brodersen DE, Clemons WM, Jr., Tarry MJ, Carter AP, Ramakrishnan, V (2001) Recognition of cognate transfer RNA by the 30S ribosomal subunit. Science 292: 897–902

Ogle JM, Murphy FV, Tarry MJ, Ramakrishnan, V (2002) Selection of tRNA by the ribosome requires a transition from an open to a closed form. Cell 111: 721–732

Ogle JM, Ramakrishnan, V (2005) Structural insights into translational fidelity. Annu Rev Biochem 74: 129–177

Oh BK, Apirion, D (1991) 10Sa RNA, a small stable RNA of *Escherichia coli*, is functional. Mol Gen Genet 229: 52–56

Oh BK, Chauhan AK, Isono K, Apirion, D (1990) Location of a gene (ssrA) for a small, stable RNA (10Sa RNA) in the *Escherichia coli* chromosome. J Bacteriol 172: 4708–4709

Pedersen K, Zavialov AV, Pavlov MY, Elf J, Gerdes K, Ehrenberg, M (2003) The bacterial toxin RelE displays codon-specific cleavage of mRNAs in the ribosomal A site. Cell 112: 131–140

Qi H, Shimizu Y, Ueda, T (2007) Ribosomal protein S1 is not essential for the trans-translation machinery. J Mol Biol 368: 845–852

Richards J, Mehta P, Karzai AW (2006) RNase R degrades non-stop mRNAs selectively in an SmpB-tmRNA-dependent manner. Mol Microbiol 62: 1700–1712

Roche ED, Sauer RT (1999) SsrA-mediated peptide tagging caused by rare codons and tRNA scarcity. EMBO J 18: 4579–4589

Roche ED, Sauer RT (2001) Identification of endogenous SsrA-tagged proteins reveals tagging at positions corresponding to stop codons. J Biol Chem 276: 28 509–28 515

Rudinger-Thirion J, Giege R, Felden, B (1999) Aminoacylated tmRNA from *Escherichia coli* interacts with prokaryotic elongation factor Tu. RNA 5: 989–992

Saguy M, Gillet R, Skorski P, Hermann-Le Denmat S, Felden, B (2007) Ribosomal protein S1 influences trans-translation *in vitro* and *in vivo*. Nucleic Acids Res 35: 2368–2376

Schluenzen F, Tocilj A, Zarivach R, Harms J, Gluehmann M, Janell D, Bashan A, Bartels H, Agmon I, Franceschi F, et al. (2000) Structure of functionally activated small ribosomal subunit at 3.3 angstroms resolution. Cell 102: 615–623

Schmeing TM, Voorhees RM, Kelley AC, Gao YG, Murphy FV 4th, Weir JR, Ramakrishnan, V (2009) The crystal structure of the ribosome bound to EF-Tu and aminoacyl-tRNA. Science 326: 688–694

Selinger DW, Saxena RM, Cheung KJ, Church GM, Rosenow, C (2003) Global RNA half-life analysis in *Escherichia coli* reveals positional patterns of transcript degradation. Genome Res 13: 216–223

Shimizu Y, Ueda, T (2002) The role of SmpB protein in trans-translation. FEBS Lett 514: 74–77

Shimizu Y, Ueda, T (2006) SmpB triggers GTP hydrolysis of elongation factor Tu on ribosomes by compensating for the lack of codon-anticodon interaction during trans-translation initiation. J Biol Chem 281: 15 987–15 996

Shpanchenko OV, Zvereva MI, Ivanov PV, Bugaeva EY, Rozov AS, Bogdanov AA, Kalkum M, Isaksson LA, Nierhaus KH, Dontsova OA (2005) Stepping transfer messenger RNA through the ribosome. J Biol Chem 280: 18 368–18 374

Singh NS, Ahmad R, Sangeetha R, Varshney, U (2008) Recycling of ribosomal complexes stalled at the step of elongation in *Escherichia coli*. J Mol Biol 380: 451–464

Singh NS, Varshney, U (2004) A physiological connection between tmRNA and peptidyl-tRNA hydrolase functions in *Escherichia coli*. Nucleic Acids Res 32: 6028–6037

Someya T, Nameki N, Hosoi H, Suzuki S, Hatanaka H, Fujii M, Terada T, Shirouzu M, Inoue Y, Shibata T, et al. (2003) Solution structure of a tmRNA-binding protein, SmpB, from Thermus thermophilus. FEBS Lett 535: 94–100

Srivastava RK, Miczak A, Apirion, D (1990) Maturation of precursor 10Sa RNA in *Escherichia coli* is a two-step process: the first reaction is catalyzed by RNase III in presence of Mn2+. Biochimie 72: 791–802

Sundermeier TR, Dulebohn DP, Cho HJ, Karzai AW (2005) A previously uncharacterized role for small protein B (SmpB) in transfer messenger RNA-mediated trans-translation. Proc Natl Acad Sci USA 102: 2316–2321

Sundermeier TR, Karzai AW (2007) Functional SmpB-ribosome interactions require tmRNA. J Biol Chem 282: 34 779–34 786

Sunohara T, Jojima K, Tagami H, Inada T, Aiba, H (2004a) Ribosome stalling during translation elongation induces cleavage of mRNA being translated in *Escherichia coli*. J Biol Chem 279: 15 368–15 375

Sunohara T, Jojima K, Yamamoto Y, Inada T, Aiba, H (2004b) Nascent-peptide-mediated ribosome stalling at a stop codon induces mRNA cleavage resulting in nonstop mRNA that is recognized by tmRNA. RNA 10: 378–386

Szaflarski W, Vesper O, Teraoka Y, Plitta B, Wilson DN, Nierhaus KH (2008) New features of the ribosome and ribosomal inhibitors: non-enzymatic recycling, misreading and back-translocation. J Mol Biol 380: 193–205

Takada K, Takemoto C, Kawazoe M, Konno T, Hanawa-Suetsugu K, Lee S, Shirouzu M, Yokoyama S, Muto A, Himeno, H (2007)

In vitro trans-translation of Thermus thermophilus: ribosomal protein S1 is not required for the early stage of trans-translation. RNA 13: 503–510

Tanner DR, Dewey JD, Miller MR, Buskirk AR (2006) Genetic analysis of the structure and function of transfer messenger RNA pseudoknot 1. J Biol Chem 281: 10 561–10 566

Tu GF, Reid GE, Zhang JG, Moritz RL, Simpson RJ (1995) C-terminal extension of truncated recombinant proteins in *Escherichia coli* with a 10Sa RNA decapeptide. J Biol Chem 270: 9322–9326

Valle M, Gillet R, Kaur S, Henne A, Ramakrishnan V, Frank, J (2003) Visualizing tmRNA entry into a stalled ribosome. Science 300: 127–130

Watts T, Cazier D, Healey D, Buskirk, A (2009) SmpB contributes to reading frame selection in the translation of transfer-messenger RNA. J Mol Biol 391: 275–281

Weis F, Bron P, Rolland JP, Thomas D, Felden B, Gillet, R (2010) Accommodation of tmRNA-SmpB into stalled ribosomes: a cryo-EM study. RNA 16: 299–306

Williams KP (2000) The tmRNA website. Nucleic Acids Res 28: 168

Williams KP, Bartel DP (1996) Phylogenetic analysis of tmRNA secondary structure. RNA 2: 1306–1310

Williams KP, Martindale KA, Bartel DP (1999) Resuming translation on tmRNA: a unique mode of determining a reading frame. EMBO J 18: 5423–5433

Wower IK, Zwieb C, Wower, J (2004) Contributions of pseudoknots and protein SmpB to the structure and function of tmRNA in trans-translation. J Biol Chem 279: 54 202–54 209

Wower IK, Zwieb C, Wower, J (2009) *Escherichia coli* tmRNA lacking pseudoknot 1 tags truncated proteins *in vivo* and *in vitro*. RNA 15: 128–137

Wower IK, Zwieb CW, Guven SA, Wower, J (2000) Binding and cross-linking of tmRNA to ribosomal protein S1, on and off the *Escherichia coli* ribosome. EMBO J 19: 6612–6621

Wower J, Zwieb CW, Hoffman DW, Wower IK (2002) SmpB: a protein that binds to double-stranded segments in tmRNA and tRNA. Biochemistry 41: 8826–8836

Yamamoto Y, Sunohara T, Jojima K, Inada T, Aiba, H (2003) SsrA-mediated trans-translation plays a role in mRNA quality control by facilitating degradation of truncated mRNAs. RNA 9: 408–418

Yao S, Blaustein JB, Bechhofer DH (2008) Erythromycin-induced ribosome stalling and RNase J1-mediated mRNA processing in Bacillus subtilis. Mol Microbiol 69: 1439–1449

Yusupova GZ, Yusupov MM, Cate JH, Noller HF (2001) The path of messenger RNA through the ribosome. Cell 106: 233–241

Zwieb C, Wower, J (2000) tmRDB (tmRNA database) Nucleic Acids Res 28: 169–170

Section V Nascent peptide and tunnel interactions

Nascent peptide-mediated ribosome stalling promoted by antibiotics

30

Nora Vázquez-Laslop, Haripriya Ramu, Alexander Mankin

1. Introduction

Many mechanisms that regulate protein expression operate through mRNA. The sequence of the Shine-Dalgarno region and its accessibility influence the efficiency of initiation of translation; the choice of codons and mRNA secondary structure affect progression of the ribosome along mRNA at the elongation stage; and the mRNA context and identity of the stop codon determine how efficiently protein is released at the termination step. However, there is another important mechanism that controls expression of a number of genes at a principally different level. In this mechanism, the ribosome checks the structure of the polypeptide it is assembling; in response to certain nascent peptide sequences and, often, specific cellular cues, it modulates its activity. One of the most dramatic types of ribosomal response to the regulatory nascent peptide sequences is stalling. In the best-characterized cases, the nascent peptide–controlled stalling occurs at a dedicated regulatory open reading frame (ORF) that precedes the regulated gene or operon whose expression is transcriptionally or translationally attenuated. Ribosome stalling relieves the attenuation by either altering the mRNA secondary structure or interfering with binding of the transcription termination factors.

Two most impressive examples of gene regulation operating through this mechanism have been described in a series of influential papers from the laboratories of Ito and Yanofsky (Cruz-Vera et al., 2006; Cruz-Vera et al., 2005; Gong and Yanofsky, 2002; Muto et al., 2006; Nakatogawa and Ito, 2001; Nakatogawa and Ito, 2002). Expression of a key component of the cell secretion apparatus, SecA, is controlled by programmed translation arrest at codon 165 of a 170-codon *secM* ORF that precedes the *secA* gene. Ribosome stalling requires the presence of the nascent peptide "stalling" sequence at

the C-terminus of the nascent peptide and a prolyl-tRNA in the A site of the ribosome (Muto et al., 2006; Nakatogawa and Ito, 2002). The stability of the stalled complex depends on the activity of the cellular secretion apparatus: the stalled ribosome complex (SRC) persists for a long time when cellular secretion is tuned down. Expression of the *E. coli* tryptophanase operon, *tna*, is regulated by a 24-codon ORF *tnaC*. When tryptophan in the cell is abundant, the ribosome stalls at the last sense codon of *tnaC*, leading to the activation of transcription of the *tna* operon. The identity of amino acids at positions 12, 16, and 24 of the TnaC nascent peptide and the binding of tryptophan at a yet-unidentified site near the peptidyl transferase center (PTC) are required for the ribosome stalling. The spectrum of bacterial genes controlled by programmed translation arrest, which depends on interaction of the ribosome with the nascent peptide, likely extends beyond the *sec* and *tna* operons (Chiba et al., 2009; Tanner et al., 2009; Yap and Bernstein, 2009).

These recent studies of ribosome stalling at the *secM* and *tnaC* ORFs emphatically revealed an amazing ability of the ribosome to monitor the structure of the nascent peptide in the exit tunnel. However, the first description of nascent peptide-controlled programmed translation arrest dates back to the early 1980s when the laboratories of Weisblum and Dubnau demonstrated that the expression of an inducible macrolide resistance gene, *ermC*, is activated by stalling of the ribosome at an upstream 19-codon leader ORF (*ermCL*) (Gryczan et al., 1980; Horinouchi and Weisblum, 1980). The stalling occurs in the presence of an inducing antibiotic (e. g., erythromycin) that, as we now know, binds in the nascent peptide exit tunnel (NPET) (Figure 1). Most unexpectedly, however, the induction of *ermC* expression and thus, by inference, ribosome stalling critically depends on the se-

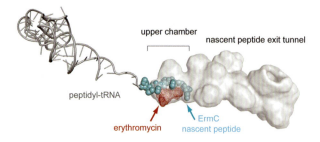

Fig. 1 Macrolide antibiotic and nascent peptide in the exit tunnel of the ribosome. The tunnel surface was extracted from the structure of *Thermus thermophilus* large ribosomal subunit (Voorhees et al., 2009) using the approach described in (Voss et al., 2006). The antibiotic (erythromycin, red) was docked on the basis of its placement in the *E. coli* ribosome (Dunkle et al., 2010); a 9-amino acid ErmCL nascent peptide (cyan) was modeled according to (Tu et al., 2005).

quence of the peptide encoded in the *ermCL* ORF. As we understand now, the requirement for a specific sequence reflected the operation of a fundamental mechanism of the nascent peptide–controlled translation arrest, assisted, in this case, by an antibiotic bound in the NPET (Weisblum, 1995). Subsequent studies of other antibiotic resistance genes showed that several of them are likely regulated through a similar mechanism that involves ribosome stalling within the first few codons of regulatory leader ORFs (Gryczan et al., 1984; Hue and Bechhofer, 1992; Kwak et al., 1991; Kwon et al., 2006; Murphy, 1985; Sandler and Weisblum, 1989).

Nascent peptide–dependent ribosome stalling represents a response of the ribosome to specific nascent peptide structures that was discovered because the SRC was long-lived. Other, more subtle types of nascent peptide–ribosome interactions likely lead to short-lived, paused ribosome complexes that may affect many important aspects of protein synthesis: protein folding, recoding, secretion, among others. Yet, despite the obvious importance of the ability of the ribosome to monitor and respond to the nascent peptide, we know very little, even at the very basic level, about the principles of nascent peptide recognition in the exit tunnel and the molecular mechanisms of the ribosomal response. In the next few sections, we will summarize what we have learned about nascent peptide-ribosome functional interactions from studies of drug- and nascent peptide-dependent translation arrest.

2. Characteristics of stalling peptides

2.1. How diverse are the peptides that direct drug-dependent ribosome stalling?

Inspection of the known macrolide resistance genes shows that the majority of them are preceded by characteristic short ORFs (Ramu et al. 2009; Roberts 2008; Subramanian et al. 2011). Many of these ORFs are furnished with well-defined ribosome binding

Table 1 Examples of peptides encoded in regulatory ORFs of macrolide resistance genes

Leader peptide[a]	Sequence
IAVV peptides	
ErmAL1	MCTC<u>IAVV</u>DITLSHL
ErmAL1[b]	MCTS<u>IAVV</u>EITLSHS
Erm36L	MGSP<u>SIAVT</u>RFRRF
IFVI peptides	
ErmAL2	MGMFS<u>IFVI</u>ERFHYQPNQK
ErmCL	MGIFS<u>IFVI</u>STVHYQPNKK
ErmGL2	MGLYS<u>IFVI</u>ETVHYQPNEK
ErmYL	MGNCS<u>LFVI</u>NTVHYQPNEK
RLR peptides	
EreAL	MLRSRAVALKQSYAL
Erm34L	MHF<u>IRLR</u>FLVLNK
ErmDL	MTHSM<u>RLR</u>FPTLNQ
MsrAL	MTASM<u>RLK</u>
MsrCL	MTASM<u>KLR</u>FELLNNN
Erm39L	MSVTY<u>IRLR</u>IT
ErmXL	MLISGTAFL<u>RLR</u>TNRKAFPTP
ErmQL	MIMNGGIAS<u>IRLR</u>R
EreAL	MTPNNSFKPT<u>PLR</u>GAA
ErmFL	MKTPTGLSGSISQ<u>RVR</u>TLVK
ErmWL	MGFSFTGSAF<u>IRLR</u>TA
Miscellaneous peptides	
ErmBL	MLVFQMRNVDKTSTVLKQT KNSDYADK
MefBL	MYLIFM
ErmGL1	MRIDDYCS
MphCL	MYQIKNGN
EreAL	MSLVIGEAKV
Erm37L	MRTAPEPWGW
MphBL	MAKEALEVQGS
ErmEL	MNKYSKRDAIN
ErmFL	MRVSVRVAACARC

[a] Sequence accession numbers and original references can be found in (Ramu et al. 2009; Roberts 2008; Subramanian et al. 2011)
[b] different sequence versions of the ErmAL1 leader peptides from the Genbank accession numbers AF002 716 and X03 216.

A

Leader peptide	Full peptide sequence and P-site codon in the stalled complex	toeprint signal
ErmAL1	MCTSI**AVV**EITLSHS	Strong
Erm36L	MGSPSI**AVT**RFRRF	Strong
ErmAL2	MGTFS**IFVI**NKVRYQPNQN	Strong
ErmCL	MGIFS**IFVI**STVHYQPNKK	Strong
ErmDL	MTHSM**RLR**FPTLNQ	Strong
Erm34L	MHFI**RLR**FLVLNK	Strong
ErmXL	MLISGTAFL**RLR**TNRKAFPTP	Strong
EreAL	**MLR**SRAVALKQSYAL	Strong
MsrCL	MTASM**KLR**FELLNNN	Strong
MsrSAL	MTASM**RLK**	Strong
MsrDL	MYLIFM	Strong
ErmBL	MLVFQMRNVDKTSTILKQTKNS-DYVDKYVRLIPTSD	Strong
ErmSL	MSMGIAARPPRAALLPPPSVPRSR	Weak
Erm38L	MSITSMAAPVAAFIRPRTA	Weak
EreAL	MTPNNSFKPTPLRGAA	Weak

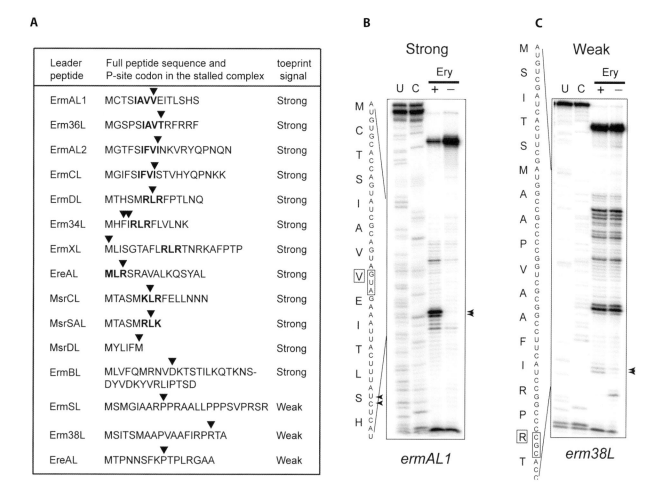

B Strong — *ermAL1*

C Weak — *erm38L*

Fig. 2 Nascent peptides in stalled ribosome complexes formed at the regulatory ORFs of macrolide resistance genes. (A) Sequences of the peptides encoded in ORFs where erythromycin-dependent ribosome stalling was characterized by toeprinting analysis (Vázquez-Laslop et al., 2010; Vázquez-Laslop et al., 2008; Ramu et al., 2011). The amino acid residue specified by the P-site codon in the stalled ribosome complex is indicated by a triangle. The consensus sequences of peptides belonging to the IAVV, IFVI, and RLR classes are shown in bold. Examples of strong and weak erythromycin-dependent toeprint signals are presented in panels B and C, respectively. In a typical toeprinting experiment, a DNA template corresponding to the leader peptide ORF equipped with the T7 promoter and a unified ribosome-binding site was translated in the *E. coli* cell-free ("PURE") system (Shimizu et al., 2001) in the absence or presence of 50μM erythromycin; the location of the stalled ribosome was determined by reverse transcriptase–mediated extension of a DNA primer annealed to the 3' end of the RNA transcript. The location of the reverse transcriptase stop ("toeprint signal," arrowheads) within the ORF sequence (U and C sequencing lanes are shown) helped to identify the codon in the P-site of the stalled ribosome (15–17 nucleotides "upstream" of the toeprint signal; boxed in the gene and peptide sequences).

sites, arguing in favor of their expression (Subramanian et al., 2011). Some of the peptides associated with the resistance genes show a significant degree of conservation, a strong indication of their functional importance. Thus, the peptides encoded in the leader ORFs of several functionally diverse genes, ranging from rRNA methylases to drug transporters, are characterized by the presence of a conserved sequence, IFVI, that is critical for SRC formation at the *ermCL* ORF (Mayford and Weisblum, 1989; Vázquez-Laslop et al., 2008) (Table 1). Somewhat related to them are peptides encoded

in the first leader ORF (*ermAL1*) of several variants of *ermA* genes and in the *erm36L* leader; these peptides comprise the sequence IAVV/T, which is also critical for stalling (see below). A subset of the leader peptides is characterized by the presence of an RLR sequence, often preceded by a hydrophobic amino acid (Met, Ile, or Leu). The precise location of this sequence within the peptide varies somewhat, although in most cases it is positioned within the 8–14 N-terminal residues. Besides these three major classes, a significant fraction of the leader peptides encoded in the leader ORFs show little similarity to each other (the miscellaneous category in Table 1). Taken together, the structures of regulatory peptides of macrolide resistance genes show an amazing variety. If only a fraction of these peptides were capable of directing drug-dependent SRC formation, similar to ErmCL, it would appear that the problem of stalling the ribosome at a defined site in mRNA has many different solutions.

In fact, experimental testing shows that SRCs are formed upon *in-vitro* translation of a variety of regulatory ORFs of macrolide resistance genes in the presence of erythromycin (Figure 2). Such experiments are carried out in an *E. coli* cell-free translation system composed of the purified components (Shimizu et al., 2001), where formation of a stable SRC can be detected by toeprinting (primer extension inhibition) analysis (Hartz et al., 1988). Several of the tested leader ORFs, including *ermAL1*, *ermAL2*, *ermBL*, *ermCL*, *ermDL*, *erm34L*, *erm36L*, *ermXL*, *ereAL*, and *msrSAL*, could direct formation of a stable SRC as could be judged by the high intensity of the toeprinting signal (Figure 2A, B). The diversity of peptide sequences encoded in these ORFs attests to the fuzzy nature of the signal that ribosome recognizes in such peptides.

A weaker toeprint was observed at the *ermSL* and *erm38L* ORFs and at the second site in the *ereAL* ORF, likely indicating transient ribosome pausing (Figure 2A, C). One should bear in mind, however, that leader ORFs that failed to yield stable SRCs in the presence of erythromycin might do so in the presence of other antibiotics. As we show later in this chapter, the ability of a peptide to cause ribosome stalling critically depends on the chemical structure of the inducing drug. It would not be surprising if leader peptide sequences of some resistance genes were evolutionary optimized to induce translation arrest with the assistance of macrolides other than erythromycin.

Although only a few known classes of drugs bind in the NPET and can directly assist in nascent peptide recognition, a number of genes that confer resistance to other antibiotics are also preceded by short ORFs with putative regulatory functions (reviewed in Lovett and Rogers, 1996; Subramanian et al., 2011). At least some of these ORFs may direct nascent peptide-dependent programmed translation arrest (Lodato et al., 2006; Morita et al., 2009; Rogers et al., 1990; Stasinopoulos et al., 1998). Programmed translation arrest induced by antibiotics that do not bind in the tunnel may engage molecular mechanisms principally different from those that function at the leader ORFs of macrolide-resistance genes (Lovett, 1996).

2.2. What is the length requirement of the nascent peptides that direct erythromycin-dependent ribosome stalling?

Macrolide antibiotics bind in the NPET at a short distance from the PTC and significantly narrow the tunnel aperture. When the growing peptide reaches the site of drug binding, steric hindrance leads to destabilization of peptidyl-tRNA association with the ribosome. As a result, erythromycin and similar macrolides cause peptidyl-tRNA drop-off during early rounds of translation when the nascent peptide is six to ten amino acids long (Menninger and Otto, 1982; Tenson et al., 2003). Therefore, if the site of erythromycin-induced programmed translation arrest were located at a significant distance from the 5' end of the leader ORF, the ribosome would have little chance of reaching it because of premature loss of peptidyl-tRNA. Indeed, at the leader ORFs that promote formation of the stable erythromycin-dependent SRCs (indicated by a strong toeprint signal), stalling takes place when the nascent peptide is not longer than ten amino acids (Figure 2A). Because erythromycin starts inciting peptidyl-tRNA drop-off when the nascent peptides are as short as six amino acids, not every ribosome that initiates translation of the regulatory ORF can reach the stalling site (Weisblum, 1998). For example, only about 10% of the ribosomes that initiate translation of *ermCL* can apparently reach the site of programmed stalling at its ninth codon (T. Tenson, personal communication). Such seemingly "inefficient" placement of the translation arrest site probably reflects the requirement for the nascent peptide to reach a certain length in order to direct formation of a sufficiently stable stalled complex. In agreement with this notion, removing one, two, and three codons from the 5' end of the *ermCL* gene progressively reduces the stability of the SRC (Vázquez-Laslop et al., 2008). One possible explanation is that efficient complex formation depends on interactions of the N-terminus of the nascent peptide with the tunnel elements located sufficiently far away from the PTC, which shorter nascent peptides simply would not reach. Such interactions may involve contacts with the extended loop of protein L22 exposed in the tunnel approximately 30 Å away from the PTC. Protein L22 appears to play an important role in drug-independent ribosome stalling at *secM* and *tnaC* ORFs (Cruz-Vera et al., 2005; Lawrence et al., 2008; Nakatogawa and Ito, 2002) as well as in erythromycin-dependent SRC formation during *ermCL* translation (Vázquez-Laslop *et al., 2008*). Nascent peptides protruding from the peptidyl-tRNA bound in the ribosomal P site need to be at least seven or eight amino acids long to come into contact with L22. Furthermore, peptidyl-tRNA esterified with very short peptides is prone to spontaneous rapid dissociation from the ribosome (Heurgué-Hamard et al., 2000). Therefore, even if the ribosome stops at early codons of the regulatory ORF, it might be difficult to sustain its stable association with peptidyl-tRNA and thus retain the SRC on the mRNA. On the other hand, if in only a fraction of the ribosomes the nascent peptide succeeds in "slithering" through the narrow hole left in the macrolide-occupied tunnel without prior dissociation of peptidyl-tRNA, then the SRC could be sufficiently stable.

Despite these considerations, at several of the regulatory ORFs (erm34L, ereAL), erythromycin-induced translation arrest takes place close to the initiator codon. The extreme case is arrest of the ribosome at the initiator codon of the ermXL cistron (Figure 2A). All these peptides belong to the RLR class. Yet, the ribosome stalls before the conserved sequence is synthesized. By comparison with the other peptides of this class (Figure 2A), one would think that the true, physiologically relevant arrest of translation should occur at the conserved RLR sequence. The aberrant premature stalling observed experimentally may depend on mRNA conformation rather than the nascent peptide structure. It may result from artificial conditions of the cell-free transcription-translation system, which relies on T7 RNA polymerase and lacks cellular helicases, that might help the ribosome in reaching the proper stalling site. Importantly, erythromycin-dependent arrest of the ribosome at the initiator codon of ermXL argues that binding of erythromycin in the NPET can affect one of the steps of translation initiation, none of which directly involve the exit tunnel. Therefore, the ribosome might be able to recognize the presence of the antibiotic in the tunnel and relay this information to one of the functional centers responsible for initiation of translation.

2.3. What are the nascent peptide sequence requirements for ribosome stalling?

Macrolide-dependent programmed translation arrest requires the presence of a nascent peptide with a specific amino acid sequence in the ribosome tunnel. Early genetic experiments that used ermC-based reporters suggested that the C-terminal segment of the nascent peptide in the ermCL SRC is important for inducible

expression of the resistance gene and identified several mutations that abolish its induction (Mayford and Weisblum, 1989). However, the genetic approach can be misleading, because the mutations may activate gene expression due to changes in mRNA secondary structure. In contrast, the in-vitro toeprinting analysis with its ability to directly detect SRC on mRNA can reliably identify the contribution of individual amino acids in the nascent peptide to SRC formation (Muto et al., 2006; Vazquez-Laslop et al., 2008).

Toeprinting studies combined with scanning mutagenesis showed that translation arrest at the ermCL and ermAL1 regulatory ORFs is controlled by the sequence of the four C-terminal amino acids in the corresponding nascent peptides (IFVI and IAVV, respectively). Mutations of any of these residues to alanine (or of the native alanine at position 6 of ermAL1 to glycine) practically abolished erythromycin-dependent ribosome stalling (Table 2). The IFVI sequence is conserved in a number of peptides encoded in the leaders of macrolide resistance genes, additionally underscoring its functional significance (Table 1; see (Subramanian et al., 2011) for a more extended list of leader peptide sequences). Mutations at the amino acid positions proximal to the N-termini of ErmCL or ErmAL1 peptides had little effect on stalling, revealing the lack of sequence-specific interactions of the N-terminal peptide segment with the tunnel elements. One should be cautious, however, in extrapolating this conclusion to the first N-terminal amino acid residue: because changing the nature of formyl-methionine is experimentally challenging, it is unclear whether the very N-terminus of the nascent peptide is engaged in any defined chemical interactions with the tunnel. In this regard, it is noteworthy that in the drug-dependent SRCs the location of the N-terminal formyl-methionine in the tunnel is close to the placement of important recognition elements of the longer peptides that direct drug-independent stalling (Table 2).

In both ErmCL and ErmAL1, the critical C-terminal segment of the nascent peptide on peptidyl-tRNA in the SRC is decisively hydrophobic (IFVI in ErmCL and IAVV in ErmAL1). It is obvious, however, that erythromycin-dependent translation arrest at ermCL or ermAL1 requires more than just a hydrophobic patch at the peptide's C-terminus. Not only alanine mutations, which preserve the general hydrophobic nature of the sequence, but even more subtle alterations, for example, changing Ile9 of ErmCL to Leu, may dramatically decrease the stalling properties of the peptide (Mayford and Weisblum, 1989; Vázquez-Laslop et al.,

Table 2 Positions in the stalling nascent peptides critical for programmed translation arrest. Critical amino acids are highlighted in green. Residues whose mutations have moderately negative effect on stalling are highlighted in pale-green

LIGAND	PEPTIDE	POSITION OF AMINO ACID RESIDUE RELATIVE TO THE C-TERMINUS																			P site	A site
		-19	-18	-17	-16	-15	-14	-13	-12	-11	-10	-9	-8	-7	-6	-5	-4	-3	-2	-1	P site	A site
Antibiotic	ErmAL1 [a]													M	C	T	S	I	A	V	V_8	E
	ErmCL												M	G	I	F	S	I	F	V	I_9	S
	ErmBL										M	L	V	F	Q	M	R	N	V		D_{10}	K
	ErmDL													M	T	H	S	M	R		L_7	R
None	SecM [a)]	P	Q	A	K	F	S	T	P	V	W	I	S	Q	A	Q	G	I	R	A	G_{165}	P
	SecM$_{Ms}$ [b]	P	G	S	P	H	F	C	L	S	P	S	Y	F	H	A	P	I	R	G	S_{165}	P
	FxxYxIWPP [c]						S	L	Q	K	R	L	F	Q	K	Y	G	I	W	P	P_{15}	P
	MifM [d]	R	I	T	T	W	I	R	K	V	F	R	M	N	S	P	V	N	D	E	E_{88}	D
Tryptophan	TnaC [a]	L	H	I	V	T	S	K	W	F	N	I	D	N	K	I	V	D	H	R	P_{24}	term

[a] Data for the Erm leader peptides are from (Vázquez-Laslop et al., 2010; Vázquez-Laslop et al., 2008; Ramu, Vázquez-Laslop, Mankin, in preparation), for SecM and TnaC are from (Nakatogawa and Ito, 2002; Gong and Yanofsky, 2002; Ito et al., 2010)

[b] Engineered mutant versions of SecM (Yap and Bernstein, 2009).

[c] Selected from a random library (Tanner et al., 2009).

[d] Programmed translation arrest at *mifM* in *Bacillus subtilis* regulates expression of *yidC2* that encodes a component of the membrane protein insertion pathway (Chiba et al., 2009).

2008). Thus, it appears that the mechanism of peptide recognition is fairly sophisticated and is characterized by fine resolution and high fidelity.

The general trend – recognition of the C-terminal region of the nascent peptide – that emerges from the analyses of ErmCL and ErmAL1 peptides holds true also for the peptide encoded in the *ermBL* ORF where the ribosome stalls at codon 10 of the regulatory ORF. Alanine scanning mutagenesis showed that the identities of Arg7, Val9 and Asp10 are critical for stalling (Table 2). Additionally, mutation of Val9 to Gly prevents the formation of the stalled complex. Impressively, the essential C-terminal sequence of the ErmBL nascent peptide (NVD) has no resemblance to the hydrophobic C-terminal sequences of ErmCL and ErmAL1 peptides, thereby exposing a considerable variability in the composition of the peptide's stalling sequences.

The idiosyncrasy in the sequence requirement becomes even more dramatic as far as the ErmDL peptide is concerned. Here, the C-terminal rule appears to be broken. In the *ermDL* regulatory ORF, where the ribosome stalls at codon 7, alanine mutations of Thr2, Met5, and Leu7, but not of the other peptide residues, prevented SRC formation (Table 2). In addition, mutation of Ser4 to Ile prevented SRC formation (Kwon et al., 2006; Ramu and Mankin, unpublished). ErmDL belongs to the RLR class of regulatory peptides (Table 1) which is characterized by considerable sequence diversity. It would be interesting to explore the distribu-

tion of amino acids critical for stalling in other RLR peptides: comparison of the location and nature of such residues may help to pinpoint their specific interactions with ribosomal sensors.

Very little is known about folding of the nascent peptide in the tunnel, which makes it difficult to decipher principles of recognition of the critical sequences. The ribosome may sense the overall fold of the essential segment of the stalling peptide or, alternatively, amino acids essential for stalling may be presented in the extended peptide chain for specific interactions with the tunnel sensors. Cryo-electron microscopy studies of the TnaC peptide in the stalled ribosome showed the nascent protein chain in an extended conformation, arguing in favor of recognition of individual amino acids (Seidelt et al., 2009). On the other hand, the reported contortion of the SecM nascent peptide structure upon completion of assembling its critical sequence seems to favor the fold-recognition model (Woolhead et al., 2006). The currently continuing effort in obtaining SRC complexes amenable to high-resolution structural analysis will definitely yield key insights into the folding state of the regulatory peptides within the tunnel.

Even this incomplete picture of distribution of critical amino acids in peptides that confer ribosome stalling already reveals a great variety of sequences able to direct temporal arrest of translation. The heterogeneity of nascent peptides that can control ribosome function suggests that nascent peptide-mediated signaling might be a fairly common mechanism of regulation of gene expression. The variety of possible functional outcomes of nascent peptide-mediated signaling – from pausing, to stalling, to targeting (Bornemann et al., 2008; Nakatogawa and Ito, 2002; Zhang et al., 2009) – calls for a high malleability of the ribosomal response elements that have to recognize the nascent peptide's instructions.

3. Recognition of stalling peptides by the ribosome

3.1. What are the ribosomal sensors of the nascent peptide?

In SRCs formed at the *erm* leader ORFs, the critical amino acid residues are clustered toward the C-termini of the nascent peptides and are thus located at the entrance of the exit tunnel. This is also the place where macrolide antibiotics, critical for SRC formation, bind. Accordingly, the ribosomal sensors that recognize the stalling signal should be confined to the upper chamber of the tunnel – the segment between the entrance at the PTC and the constriction formed by proteins L4 and L22, some 30 Å away. Up to the constriction site, the walls of the tunnel are formed exclusively of rRNA. Some of the 23S rRNA residues in this segment of the NPET likely play the role of sentinels that recognize the critical sequence of the short stalling leader peptides of macrolide resistance genes.

Mutational analysis of rRNA in conjunction with the *in-vitro* toeprinting assay identified several nucleotides essential for erythromycin-dependent SRC formation at *ermCL* and *ermAL1* leader ORFs (Table 3) (Vázquez-Laslop et al., 2010; Vázquez-Laslop et al., 2008; Ramu and Mankin, unpublished). While the viable mutations of A2062 (to U or C), A2503 (to G), or U1782 (to C) had little effect upon protein synthesis *in vitro*, they essentially abolished ribosome stalling at either *ermCL* or *ermAL1* cistrons. A2058 and A2059 are key components as well because they constitute the binding site of the macrolide antibiotics required for stalling. Experimental testing confirmed that mutation of A2058 to G prevented SRC formation at *ermCL* and *ermBL* ORFs (Vázquez-Laslop and Mankin, unpublished); by inference, it is assumed that the same mutation would abolish stalling at other *erm* leader cistrons. The same is likely true for A2059, whose substitutions prevent macrolide binding (Vester and Douthwaite, 2001), although mutations at this posi-

tion have not been experimentally tested. Mutations at three other 23S rRNA positions, U2586, U2587, and U2609, had moderate effects upon drug-dependent stalling at *ermCL* and *ermAL1* ORFs, whereas mutations at U790, G2583, and U2584 or an addition of an extra adenine in the run of four As (pos. 749–753) in the loop of helix 35 did not affect SRC formation.

The nucleotide residues that are important for the ribosomal response to the nascent peptide form two patches on the walls of the NPET. Four adenines (A2058, A2059, A2503, and A2062) constitute the "A-patch." On the opposite wall of the tunnel, the "U-patch" is built of U2609, U1782, and U2586 and also includes A2587 (Figure 3).

Three of the adenines of the A-patch, A2058, A2059, and A2503, form a ladder that leads from the macrolide binding site up toward the PTC A site; the upper rung of the ladder, A2503, is buttressed by A2062. The location of A2062 allows it to come into direct contact with the nascent peptide. The base of A2062 is one of the most mobile in the tunnel (Fulle and Gohlke, 2009; Starosta et al., 2010) and its orientation is very likely to be influenced by the presence of the nascent peptide. Therefore, this nucleotide is best suited to be the direct sensor of ErmCL or ErmAL1 peptides. A2058 and A2059 interact with the macrolide antibiotic bound in the tunnel. It could be argued that the role of these two residues in SRC formation is only passive because their orientation does not change when the drug binds to the vacant ribosome (Schluenzen et al., 2001; Tu et al., 2005). However, there is no information about the structural status of these residues when both the drug and the nascent peptide are simultaneously present in the tunnel. Furthermore, mutations of A2058 negatively affect translation arrest at the *secM* ORF, which does not require the presence of antibiotic (Nakatogawa and Ito, 2002). This argues that A2058 as well as its neighbor A2059 may play a more active role in sensing the presence of the stalling cues in the tunnel. The fourth of the critical adenines in the A-patch, A2503, appears to take part in a signal relay mechanism. A2503 is shielded from direct access to the nascent peptide by the bound macrolide molecule. However, the A2503G mutation has little influence upon erythromycin binding, ruling out its involvement as a direct antibiotic sensor (Vázquez-Laslop et al., 2010). Nonetheless, A2503 must be very sensitive to the exact placement of the nucleotide sensors of the drug and the nascent peptide. Being stacked upon A2059 and A2058, A2503 should respond to even subtle adjustments of these two residues. It also

Table 3 Effects of mutations in 23S rRNA on nascent peptide-dependent ribosome stalling

nucleotide [a]	erythromycin-dependent				drug-independent	
	ermAL	*ermBL*	*ermCL*	*ermDL*	*secM* [b]	*tnaC* [c]
A750 (+A)	none	none	none [d]	none	strong	strong
U790 (G)	none	none	nd [g]	none	nd	nd
U1782 (C)	strong	none	nd	none	nd	nd
A2058 (G)	strong [e]	strong [e]	strong [e]	strong [e]	strong	none [f]
A2059	strong [e]	strong [e]	strong [e]	strong [e]	nd	nd
A2062 (U)	strong	none	strong	none	strong	none
A2503 (G)	strong	none	strong	none	strong	none
G2583 (A)	none	none	nd	none	nd	strong
U2584 (C)	none	none	nd	none	nd	strong
U2586 (C)	moderate	none	none	none	nd	nd
A2587 (G)	moderate	none	nd	none	nd	nd
U2609 (C)	moderate	none	moderate	none	moderate	strong

[a] nucleotide residues of *E. coli* 23S rRNA; tested mutations are shown in parentheses.
[b] stalling is prolonged when secretion is sluggish.
[c] stalling requires binding of tryptophan to an unidentified site in the ribosome.
[d] gray shading indicates testing exclusively with the *in vivo* reporter, not verified by toeprinting.
[e] residue, whose mutations confer high level of erythromycin resistance and thus, expected to be essential for erythromycin binding (not tested experimentally in the stalling experiments).
[f] in the presence of high concentration of tryptophan.
[g] not determined.

comes into close contact with A2062; in one of the observed conformations of A2062, the two bases can form an N7 symmetric pair (Ban et al., 2000), suggesting that A2503 can perceive the orientation of A2062 in the tunnel. Furthermore, of the four residues in the A-patch, A2503 is the one closest to the PTC active site. This structural connection along with its sensitivity to the conformation of the other rRNA residues involved in drug-dependent SRC formation make A2503 an excellent candidate for relaying the stalling signal from the NPET to the PTC. In this regard, it is worth noting that in the majority of available crystallographic structures, A2503 has been modeled in the *syn* conformation (Ban et al., 2000; Schuwirth et al., 2005; Selmer et al., 2006). However, in some complexes the base was rotated into trans orientation (Harms et al., 2001; Jenner et al., 2007; Petry et al., 2005), indicating that both configurations are allowed in the context of the ribosome structure. Such *syn-anti* transition can be hypothetically exploited for propagating the stalling signal from the tunnel to the PTC.

In addition to its strategic position near the PTC, A2503 is one of the few 23S rRNA residues that are posttranscriptionally modified. C2 methylation of A2503 (by the RlmN methyltransferase (Toh et al.,

2008)) could favor the *syn-trans* re-orientation of the base and thus may facilitate the relay of the stalling signal through the ribosome. In support of this possibility, the induction of the *ermC* gene, controlled by ribosome stalling at *ermCL*, becomes more sluggish in the cells that lack C2 methylation at A2503 (Vázquez-Laslop et al., 2010).

The four nucleotides of the U-patch (U2609, U1782, U2586, and A2587) form the second sensor element contributing to nascent peptide recognition. Similar to the A-patch components, these nucleotides are structurally interconnected and, starting from U2609, lead up the course toward the P-site of the PTC. Mutation of U1782 to C dramatically reduced nascent peptide–dependent ribosome stalling at *ermAL1* (Ramu and Mankin, unpublished). In contrast, changes at the three other U-patch nucleotides had only moderate effects upon SRC formation at *ermCL* and *ermAL1*, suggesting that this sensor element plays an important but secondary role compared with the A-patch (Vázquez-Laslop et al., 2010; Ramu and Mankin, unpublished).

As mentioned in previous sections of this chapter, the amino acid sequences and the distribution of critical amino acid residues in the peptides encoded in

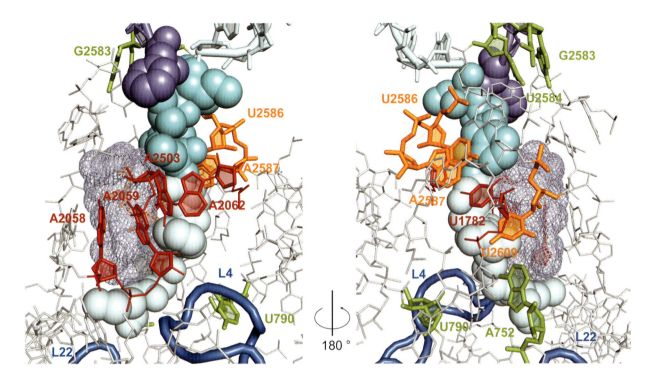

Fig. 3 Nascent peptide sensors in the ribosome exit tunnel. 23S rRNA residues whose mutations prevent ribosome stalling at *ermCL* and *ermAL1* regulatory ORFs are shown in red; nucleotides whose mutations moderately affect stalling are shown in orange; and the residues whose mutations had no influence upon stalling are shown in green. The rest of the rRNA residues of the tunnel wall are shown in gray. The loops of proteins L4 and L22 are shown in aquamarine. The 3' end of aminoacyl-tRNA bound in the A site is shown in purple, with the amino acid (Phe) atoms represented by spheres. Peptidyl-tRNA esterified with a nine-amino acid ErmCL nascent peptide is shown in pale cyan. Four C-terminal amino acids of ErmCL nascent peptide critical for stalling are shown in cyan. The erythromycin molecule is depicted as a mesh (silver-blue).

the *ermBL* and *ermDL* ORFs differ significantly from those in the ErmCL and ErmAL1 peptides (Tables 1 and 2). Could the same ribosome sensors recognize such different stalling cues? Experimental testing showed that most of the rRNA residues that play crucial roles in ribosome stalling at *ermCL* and *ermAL1* ORFs (A2503, A2062, U1782, U2609) appear to have little influence on the recognition of ErmBL or ErmDL peptides. None of the mutations that affected translation arrest at *ermCL* or *ermAL1* had any effect upon SRC formation at *ermBL* or *ermDL* (Vázquez-Laslop et al., 2010; Ramu and Mankin, unpublished). Of all the positions listed in Table 3, only the mutations of A2058 and A2059 that abolish drug binding are expected to prevent stalling at these regulatory cistrons. It is safe to conclude that the ribosome uses different mechanisms or, at the very least, different modules of the same mechanism to recognize diverse classes of regulatory nascent peptides.

The conclusion holds true even when the nascent peptides that direct drug-independent stalling, SecM and TnaC, are included. SecM converges with ErmCL and ErmAL1: mutations at A2062, A2503, and A2058

have similarly strong effects in all three systems (Na-katogawa and Ito, 2002; Vázquez-Laslop et al., 2010). At the same time, none of these mutations influence SRC formation at the *tnaC* ORF. Conversely, mutations that strongly affect TnaC-dependent reporter induction (e. g., at position G2583 or U2584) had no influence upon ribosome stalling at *ermAL1*, *ermCL*, or *secM* ORFs (Cruz-Vera et al., 2005; Vázquez-Laslop et al., 2010; Yang et al., 2009). Cryo-electron microscopic analysis of the ribosome stalled at the *tnaC* ORF showed proximity of the nascent peptide to several nucleotide residues in the NPET, and it suggested several scenarios of how the TnaC nascent peptide can be recognized and how the signal can be relayed to the PTC (Seidelt et al., 2009). Our mutational analysis indicates that the ribosome may use a variety of sensory/relay pathways to diagnose the presence of different stalling peptides.

Although rRNA appears to play the major role in nascent peptide–mediated translation arrest, ribosomal proteins also contribute to recognition of the regulatory signals. Indeed, deletion of three amino acids in the tunnel-exposed β-hairpin of L22 reduces efficiency of ribosome stalling at the *ermCL* ORF (Vázquez-Laslop et al., 2008). Mutations in L22 also negatively affect SRC formation at *secM* (Lawrence et al., 2008; Nakatogawa and Ito, 2002) and *tnaC* (Cruz-Vera et al., 2005) ORFs, suggesting an important role of this protein in the ribosomal response to functional nascent peptides (Berisio et al., 2003). Two other ribosomal proteins, which have access to the tunnel, L4 and L23, may modulate the ribosomal response to other types of regulatory nascent peptides (Bornemann et al., 2008; Lawrence et al., 2008).

3.2. What is broken in the stalled ribosome?

Although our knowledge about molecular details of drug- and nascent peptide–controlled translation arrest is rudimentary, the PTC emerges as the focal point of the stalling response mechanism. Because in the ribosome stalled at *ermCL* and *ermAL1* ORFs the nascent peptide is associated with tRNA bound in the ribosomal P site, it is clear that the dysfunctional ribosome is unable to catalyze peptide bond formation with the incoming aminoacyl-tRNA (Vázquez-Laslop et al., 2008; Ramu et al., 2011). This conclusion is corroborated by the slow release by puromycin of the ErmCL nascent peptide from the SRC (Vázquez-Laslop et al., 2008). Puromycin release of the nascent peptide is also inefficient in SRCs formed at *secM* or *tnaC* ORFs (Gong and Yanofsky, 2002; Muto et al., 2006), suggesting that disruption of catalytic properties of the PTC is the common mechanism of nascent peptide–controlled translation arrest.

Catalysis of the peptidyl transfer reaction requires proper positioning of peptidyl-tRNA in the P site, aminoacyl-tRNA in the A site, adjustment of the PTC structure, and possible chemical involvement of the 2' hydroxyl of A2451 (Erlacher et al., 2005; Rodnina et al., 2007; Schmeing et al., 2005). Which of the components of the catalytic mechanism is disrupted in the stalled ribosome? Given the association of the stalling nascent peptide with the P site, this site of the PTC seemed like a reasonable focal point of the stalling mechanism. Distorted placement of the stalling peptide in the tunnel could displace the carbonyl carbon atom of the peptidyl-tRNA ester bond from its optimal position for the nucleophilic attack of the a-amino group of aminoacyl-tRNA, thereby preventing peptide bond formation. However, although this scenario has not been ruled out, the recent data strongly point to the importance of the PTC A site in

establishing the translation arrest state of the stalled ribosome.

Mutational analysis showed that stalling at the *ermAL1* ORF was abolished with certain A-site codon mutations (Table 2) (Ramu et al., 2011). Codons specifying charged (Glu, Asp, Lys, Arg, His) and a few other types (Leu, Ile, Trp, Tyr, Pro) of amino acids as well as a stop codon were conducive to stalling, but SRC formation was significantly reduced or even completely prevented when Ala, Ser, Cys, Met, or Phe codons replaced the Glu9 codon of *ermAL1*. Subsequent biochemical analysis showed that it was the nature of the amino acid rather than the structure of tRNA that was pivotal for the ribosome's ability to catalyze transfer of the eight-amino acid ErmAL1 peptide to the A site-bound aminoacyl-tRNA. Thus, the PTC A site, which in a normally functional ribosome can accept any natural amino acid, becomes selective in the ribosome stalled at the *ermAL1* ORF. In contrast, translation arrest at the *ermCL* regulatory ORF shows little sensitivity to the nature of the codon in the A site of the stalled ribosome (Mayford and Weisblum, 1989; Vázquez-Laslop et al., 2008), showing that the A site in the *ermCL* stalled complex exists in such a restrictive state that none of the amino acid residues can be properly positioned for the nucleophilic attack on peptidyl-tRNA. Such a difference in the functionality of the PTC A site in different stalled complexes suggests that the structure of the nascent peptide in the NPET has a direct influence upon the properties of the PTC acceptor site.

The critical sequences of ErmCL and ErmAL1 stalling peptides (IFVI and IAVV, respectively) differ in two amino acid residues. However, a change of a single amino acid in the ErmCL nascent peptide, Phe7 to Ala (position -2 relative to the C terminus, Table 2), is sufficient to alter the properties of the A site from restrictive (as in wild-type ErmCL-stalled complex) to selective (as in the ErmAL1 complex) (Ramu et al., 2011). Changing the same Phe7 residue to Gly prevented SRC formation irrespective of the nature of the A-site amino acid, showing that in this case the A site remains in its "normal" versatile state. The picture that emerges from these results is that the structure of the nascent peptide is sensed in the NPET and the information is communicated to the A site of the PTC. Depending on the severity of the structural alteration, the properties of the A site can be changed either less dramatically, rendering it selective to certain types of acceptor amino acids or, more drastically, converting the A site into a restric-

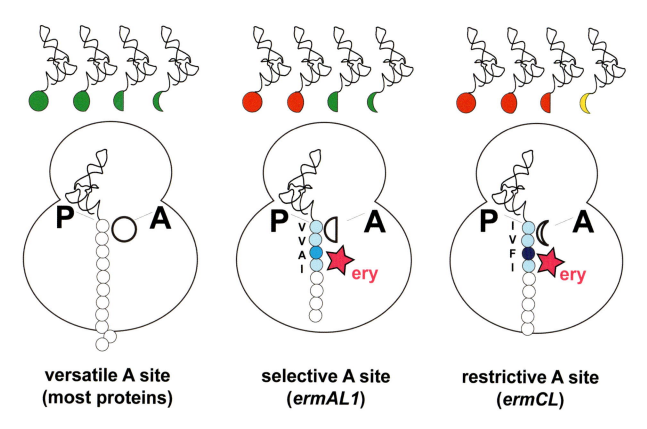

**versatile A site
(most proteins)**

**selective A site
(*ermAL1*)**

**restrictive A site
(*ermCL*)**

Fig. 4 The nascent peptide defines properties of the PTC A site. The peptide bond can be formed with any amino acid (green) placed by aminoacyl-tRNA in the versatile A site of the normally translating ribosome. Only some amino acids can be accommodated in the A site of the ribosome stalled at the *ermAL1* cistron; amino acids (red) which cannot be properly placed in the selective A site are conducive to the formation of stalled ribosome complex. The restrictive A site of the ribosome stalled at the *ermCL* open reading frame rejects most (red) and only poorly accommodates some (yellow) amino acids. The ribosome stalls irrespective of the A-site codon identity. The critical C-terminal sequences of ErmAL1 and ErmCL nascent peptides are colored pale cyan. The amino acid residue at position -2, whose identity directly affects properties of the A site, is shown in cyan (ErmAL1) or blue (ErmCL). The erythromycin molecule (ery) bound in the ribosome exit tunnel is shown as a magenta star.

tive state in which it is incapable of properly accommodating most amino acids (Figure 4).

A similar mechanism may operate in drug-independent translation arrest. Stalling at the *secM* ORF requires the presence of a proline codon in the A site, although proline is not incorporated into the nascent peptide in a significant fraction of the stalled complexes (Muto et al., 2006). The A-site proline is also critical for stalling sequences identified through genetic selection (Tanner et al., 2009). The ribosome stalled at the *tnaC* ORF is unable to catalyze peptide bond formation with Trp-tRNA or puromycin but, interestingly, is able to transfer the nascent peptide to A site-bound prolyl-tRNA (Gong and Yanofsky, 2002). Thus, in different stalled complexes, the A site appears to display a different degree of selectivity toward the acceptor substrate of the peptidyl transfer reaction.

4. What is the function of the antibiotic in drug-dependent ribosome stalling?

A distinctive feature of ribosome stalling at the regulatory ORFs of macrolide resistance genes is the requirement for the presence of an antibiotic in the NPET. What is the function of the drug ligand in the nascent peptide-controlled translation arrest? The aperture of the unobstructed NPET leaves the nascent peptide a significant degree of freedom in the tunnel. Binding of a macrolide molecule significantly narrows the tunnel (Dunkle et al., 2010; Schlunzen et al., 2001; Tu et al., 2005) but leaves an opening sufficiently wide for the nascent peptide to squeeze by (Tu et al., 2005). The current models of drug-dependent translation arrest presume that the stalling nascent peptide slithers through the narrow hole of the drug-bound tunnel rather than coiling up in the limited space between the PTC and the bound antibiotic (Figure 1). In this

A

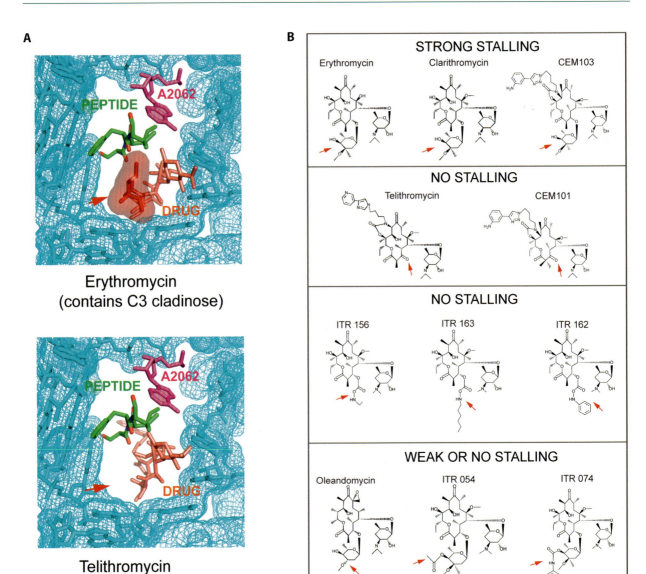

Erythromycin
(contains C3 cladinose)

Telithromycin
(does not contain C3 cladinose)

B

STRONG STALLING

Erythromycin   Clarithromycin   CEM103

NO STALLING

Telithromycin   CEM101

NO STALLING

ITR 156   ITR 163   ITR 162

WEAK OR NO STALLING

Oleandomycin   ITR 054   ITR 074

Fig. 5 The role of antibiotics in ribosome stalling. (A) The presence of C3 cladinose in the macrolide molecule narrows the tunnel and restricts the freedom of the nascent peptide placement (view from the PTC down the tunnel). (B) Macrolide drugs with C3 cladinose induce ribosome stalling at the *ermCL* cistron. Drugs that lack cladinose (ketolides), drugs with other bulky C3 substitutions, or antibiotics with a chemically modified C3 sugar cannot induce the formation of a stable SRC at *ermCL*.

scenario, the presence of an antibiotic would ensure tight contacts of the critical sequence of the nascent peptide with the A- and U-sensor patches on the tunnel walls.

This model is consistent with the observation that decreasing bulkiness of the drug molecule at the site of its presumed contact with the critical sequences of ErmCL or ErmAL1 peptides cripples the drug's ability to function as an accessory for ribosome stalling. The cladinose sugar, linked at the C3 position to the lactone ring of erythromycin, protrudes toward the PTC and comes into close contact with the critical C-terminal

segment of ErmCL or ErmAL1 stalling nascent peptides (Figure 5). Removing cladinose does not abolish binding of macrolides in the tunnel but completely prevents them from acting as inducers of ribosome stalling (Vázquez-Laslop et al., 2008). This observation has important clinical relevance: the inability of ketolides, a group of drugs lacking the C3 cladinose, to promote ribosome stalling makes them poor inducers of some of the macrolide resistance genes and thus increases the spectrum of their activity (Bonnefoy et al., 1997).

5. What could be the molecular mechanism of drug- and nascent peptide–induced translation arrest?

Let us summarize what we have learned so far about drug- and nascent peptide–dependent ribosome stalling at the regulatory ORFs of erm resistance genes.

ermCL and ermAL1

1. The ribosome stalls when the nascent peptide reaches the length of eight to nine amino acids.
2. The nature of the four C-terminal residues of the nascent peptide is important for stalling. The critical sequence has a well-defined hydrophobic character.
3. The chemical structure of the antibiotic ligand is critical for stalling. The ribosome recognizes a composite structure of the nascent peptide and the antibiotic.
4. Specific rRNA residues in the upper chamber of the NPET participate in sensing and/or communicating the stalling signal to the PTC.
5. The stalled ribosome is unable to catalyze peptide bond formation. Peptidyl-tRNA is bound in the classic P/P-state. The A site is altered from versatile to selective (*ermAL1*) or restrictive (*ermCL*) states.

ermBL and ermDL

1. SRC forms at the tenth codon of *ermBL* or the seventh codon of the *ermDL* ORFs.
2. ErmBL residues critical for stalling are clustered at the C-terminus of the nascent peptide, but in contrast to ErmAL1 or ErmCL, they do not have a well-defined hydrophobic character. In ErmDL peptides, essential residues are scattered through the length of the nascent peptide.
3. The lack of erythromycin's C3 cladinose or its chemical modification in the inducing macrolide antibiotic has no effect upon stalling.
4. None of the rRNA mutations in the NPET tested so far prevent stalling.
5. ErmBL: the nature of A-site amino acid is important; ErmDL: the Ala mutation of the A-site codon did not prevent SRC formation. It is not known whether the A-site amino acid is incorporated into the nascent peptide for either of the complexes.

Clearly, substantially more is known about translation arrest condoned by ErmCL and ErmAL1 peptides, compared with ErmBL and ErmDL peptides. Therefore, in the next paragraphs we will try to formulate a model of the drug- and nascent peptide–controlled translation arrest that accounts for the available experimental data pertaining to ErmAL1/ErmCL peptides, but will leave unmapped the *terra incognita* of ErmBL- and ErmDL-mediated stalling.

Binding of a cladinose-containing macrolide drug to the ribosome narrows the tunnel and delimits the possible passageway for the nascent peptide. The drug-bound ribosome can comfortably polymerize the first few amino acids of the nascent peptide, but when the N-terminus of the peptide reaches the antibiotic in the tunnel, its further progression is impeded. The obstruction for peptide progression leads to peptidyl-tRNA drop-off from a fraction of ribosomes, but in some of the translating ribosomes the peptide manages to thread through the opening left open in the tunnel. Once the critical amino acid sequence (IFVI in ErmCL or IAVV in ErmAL1) has been synthesized, it establishes a set of specific interactions with the sensor residues on the wall of the tunnel, A2062 in the A-patch and possibly U1782 and U2609 in the U-patch. The interactions of the nascent peptide with the tunnel sensors, especially those involving the amino acid residue at position -2 of the peptide (Phe7 in ErmCL and Ala6 in ErmAL1), are promoted by the cladinose sugar of the macrolide antibiotic. The proper nascent peptide-macrolide composite structure is additionally recognized from the antibiotic side by residues A2058, A2059, and possibly C2610 and/or G2505. The shift in the orientation of the tunnel sensors triggers a conformational rearrangement of the nucleotides of the relay pathway, including the posttranscriptionally modified A2503. The conformational switch is propagated to the A site of the PTC. A shift in the position of the A-site nucleotides, which, for example, changes the geometry of the A2451/C2452 cleft where side chains of some amino acids bind, renders the A site either selective or restrictive. The resulting inability of the PTC to catalyze peptidyl transfer stalls the ribosome. Threading the nascent peptide through the tunnel partially obstructed by the bound antibiotic and possibly engagement of protein L22 by the peptide's N terminus prevent peptidyl-tRNA drop-off and thus prolong the life of the stalled complex.

Many details of this mechanism are clearly speculative and may change as more experimental data become available. However, we believe they already provide a reasonable framework for subsequent testing of mechanistic and structural aspects of the nascent peptide-controlled translation arrest.

While molecular details of ribosome stalling at *ermAL1/ermCL* ORFs are starting to emerge, the information that has been gained about SRC formation at *ermBL* and *ermDL* ORFs is very scarce. The most intriguing fact is that none of the specific ribosomal sensors operating with ErmBL or ErmDL nascent peptides have been identified so far. Could it be that the stalling signal is propagated to the PTC through the peptide itself, without engaging tunnel residues?

What about the stalling mechanism at other leader ORFs of macrolide resistance genes? Do they fall into one or two well-defined categories, or does each peptide activate different sensors in the NPET? How do antibiotics that do not bind in the tunnel facilitate nascent peptide-dependent stalling? How do molecular mechanisms of drug-dependent stalling relate to the general principles of control of translation by the nascent peptide? What are the cellular genes that are regulated by nascent peptide-ribosome interactions? How has the mechanism of nascent peptide sensing evolved? Answering these and other related questions should illuminate the poorly understood but fundamentally important mechanisms that operate in the dark of the mysterious ribosomal tunnel.

Acknowledgements

We thank our colleagues, Dorota Klepacki, Sai Lakshmi Subramanian, Krishna Kannan, Jacqueline LaMarre, Pulkit Gupta, Anna Ochabowicz, and Dipti Panchal for help with some experiments mentioned in this review and for fruitful discussions. The work on nascent peptide-dependent ribosome stalling is supported by grant MCB-0 824 739 from the National Science Foundation, USA.

References

Ban N, Nissen P, Hansen J, Moore PB, Steitz TA (2000) The complete atomic structure of the large ribosomal subunit at 2.4 A resolution. Science 289: 905–920

Bemer-Melchior P, Juvin ME, Tassin S, Bryskier A, Schito GC, Drugeon HB (2000) *In vitro* activity of the new ketolide telithromycin compared with those of macrolides against Streptococcus pyogenes: Influences of resistance mechanisms and methodological factors. Antimicrob Agents Chemother 44: 2999–3002

Berisio R, Schluenzen F, Harms J, Bashan A, Auerbach T, Baram D, Yonath A (2003) Structural insight into the role of the ribosomal tunnel in cellular regulation. Nat Struct Biol 10: 366–370

Bonnefoy A, Girard AM, Agouridas C, Chantot JF (1997) Ketolides lack inducibility properties of MLS(B) resistance phenotype. J Antimicrob Chemother 40: 85–90

Bornemann T, Jockel J, Rodnina MV, Wintermeyer W (2008) Signal sequence-independent membrane targeting of ribosomes containing short nascent peptides within the exit tunnel. Nat Struct Molec Biol 15: 494–499

Chiba S, Lamsa A, Pogliano K (2009) A ribosome-nascent chain sensor of membrane protein biogenesis in *Bacillus subtilis*. EMBO J 28: 3461–3475

Cruz-Vera LR, Rajagopal S, Squires C, Yanofsky C (2005) Features of ribosome-peptidyl-tRNA interactions essential for tryptophan induction of *tna* operon expression. Mol Cell 19: 333–343

Cruz-Vera LR, Gong M, Yanofsky C (2006) Changes produced by bound tryptophan in the ribosome peptidyl transferase center in response to TnaC, a nascent leader peptide. Proc Natl Acad Sci USA 103: 3598–3603

Dunkle JA, Xiong L, Mankin AS, Cate JHD (2010) Structures of the *E. coli* ribosome with antibiotics bound near the peptidyl transferase center explain spectra of drug action. Proc Natl Acad Sci USA 107: 17 152–17 157

Erlacher MD, Lang K, Shankaran N, Wotzel B, Huttenhofer A, Micura R, Mankin AS, Polacek N (2005) Chemical engineering of the peptidyl transferase center reveals an important role of the 2'-hydroxyl group of A2451. Nucl Acids Res 33: 1618–1627

Fulle S, Gohlke H (2009) Statics of the ribosomal exit tunnel: implications for cotranslational peptide folding, elongation regulation, and antibiotics binding. J Mol Biol 387: 502–517

Gong F, Yanofsky C (2002) Instruction of translating ribosome by nascent peptide. Science 297: 1864–1867

Gryczan T, Israeli-Reches M, Del Bue M, Dubnau D (1984) DNA sequence and regulation of ermD, a macrolide-lincosamide-streptogramin B resistance element from *Bacillus licheniformis*. Mol Gen Genet 194: 349–356

Gryczan TJ, Grandi G, Hahn J, Grandi R, Dubnau D (1980) Conformational alteration of mRNA structure and the posttranscriptional regulation of erythromycin-induced drug resistance. Nucl Acids Res 8: 6081–6097

Harms J, Schluenzen F, Zarivach R, Bashan A, Gat S, Agmon I, Bartels H, Franceschi F, Yonath A (2001) High resolution structure of the large ribosomal subunit from a mesophilic eubacterium. Cell 107: 679–688

Hartz D, McPheeters DS, Traut R, Gold L (1988) Extension inhibition analysis of translation initiation complexes. Meth Enzymol 164: 419–425

Heurgue-Hamard V, Dincbas V, Buckingham RH, Ehrenberg M (2000) Origins of minigene-dependent growth inhibition in bacterial cells. EMBO J 19: 2701–2709

Horinouchi S, Weisblum B (1980) Posttranscriptional modification of mRNA conformation: mechanism that regulates erythromycin-induced resistance. Proc Natl Acad Sci USA 77: 7079–7083

Hue KK, Bechhofer DH (1992) Regulation of the macrolide-lincosamide-streptogramin B resistance gene ermD. J Bacteriol 174: 5860–5868

Ito K, Chiba S, Pogliano K (2010) Divergent stalling sequences sense and control cellular physiology. Bioch Biophys Res Commun 393: 1–5

Jenner L, Rees B, Yusupov M, Yusupova G (2007) Messenger RNA conformations in the ribosomal E site revealed by X-ray crystallography. EMBO Rep 8: 846–850

Kwak JH, Choi EC, Weisblum B (1991) Transcriptional attenuation control of ermK, a macrolide-lincosamide-streptogramin B resistance determinant from Bacillus licheniformis. J Bacteriol 173: 4725–4735

Kwon AR, Min YH, Yoon EJ, Kim JA, Shim MJ, Choi EC (2006) ErmK leader peptide: amino acid sequence critical for induction by erythromycin. Arch Pharm Res 29: 1154–1157

Lawrence M, Lindahl L, Zengel JM (2008) Effects on translation pausing of alterations in protein and RNA components of the ribosome exit tunnel. J Bacteriol 190: 5862–5869

Lodato PB, Rogers EJ, Lovett PS (2006) A variation of the translation attenuation model can explain the inducible regulation of the pBC16 tetracycline resistance gene in *Bacillus subtilis*. J Bacteriol 188: 4749–4758

Lovett PS (1996) Translation attenuation regulation of chloramphenicol resistance in bacteria – A review. Gene 179: 157–162

Lovett PS, Rogers EJ (1996) Ribosome regulation by the nascent peptide. Microbiol Rev 60: 366–385

Mayford M, Weisblum B (1989) ermC leader peptide. Amino acid sequence critical for induction by translational attenuation. J Mol Biol 206: 69–79

Menninger JR, Otto DP (1982) Erythromycin, carbomycin, and spiramycin inhibit protein synthesis by stimulating the dissociation of peptidyl-tRNA from ribosomes. Antimicrob Agents Chemother 21: 810–818

Morita Y, Gilmour C, Metcalf D, Poole K (2009) Translational control of the antibiotic inducibility of the PA5471 gene required for mexXY multidrug efflux gene expression in Pseudomonas aeruginosa. J Bacteriol 191: 4966–4975

Murphy E (1985) Nucleotide sequence of ermA, a macrolide-lincosamide-streptogramin B determinant in *Staphylococcus aureus*. J Bacteriol 162: 633–640

Muto H, Nakatogawa H, Ito K (2006) Genetically encoded but non-polypeptide prolyl-tRNA functions in the A site for SecM-mediated ribosomal stall. Mol Cell 22: 545–552

Nakatogawa H, Ito K (2001) Secretion monitor, SecM, undergoes self-translation arrest in the cytosol. Mol Cell 7: 185–192

Nakatogawa H, Ito K (2002) The ribosomal exit tunnel functions as a discriminating gate. Cell 108: 629–636

Petry S, Brodersen DE, Murphy FVt, Dunham CM, Selmer M, Tarry MJ, Kelley AC, Ramakrishnan V (2005) Crystal structures of the ribosome in complex with release factors RF1 and RF2 bound to a cognate stop codon. Cell 123: 1255–1266

Ramu H, Mankin A, Vazquez-Laslop N (2009) Programmed drug-dependent ribosome stalling. Mol Microbiol 71: 811–824

Ramu, H, Vázquez-Laslop, N, Klepacki, D, Dai, Q, Piccirilli, J, Micura, R, Mankin, A S (2011) Nascent peptide in the ribosome exit tunnel affects functional properties of the A-site of the peptidyl transferase center. Mol Cell 41: 321–330

Roberts MC (2008) Update on macrolide-lincosamide-streptogramin, ketolide, and oxazolidinone resistance genes. FEMS Microbiol Lett 282: 147–159

Rodnina MV, Beringer M, Wintermeyer W (2007) How ribosomes make peptide bonds. Trends Biochem Sci 32: 20–26

Rogers EJ, Kim UJ, Ambulos NP, Jr., Lovett PS (1990) Four codons in the cat-86 leader define a chloramphenicol-sensitive ribosome stall sequence. J Bacteriol 172: 110–115

Sandler P, Weisblum B (1989) Erythromycin-induced ribosome stall in the *ermA* leader: a barricade to 5'-to-3' nucleolytic cleavage of the ermA transcript. J Bacteriol 171: 6680–6688

Schluenzen F, Zarivach R, Harms J, Bashan A, Tocilj A, Albrecht R, Yonath A, Franceschi F (2001) Structural basis for the interaction of antibiotics with the peptidyl transferase centre in eubacteria. Nature 413: 814–821

Schmeing TM, Huang KS, Strobel SA, Steitz TA (2005) An induced-fit mechanism to promote peptide bond formation and exclude hydrolysis of peptidyl-tRNA. Nature 438: 520–524

Schuwirth BS, Borovinskaya MA, Hau CW, Zhang W, Vila-Sanjurjo A, Holton JM, Cate JH (2005) Structures of the bacterial ribosome at 3.5 A resolution. Science 310: 827–834

Seidelt B, Innis CA, Wilson DN, Gartmann M, Armache JP, Villa E, Trabuco LG, Becker T, Mielke T, Schulten K, Steitz TA, Beckmann R (2009) Structural insight into nascent polypeptide chain-mediated translational stalling. Science 326: 1412–1415

Selmer M, Dunham CM, Murphy FVt, Weixlbaumer A, Petry S, Kelley AC, Weir JR, Ramakrishnan V (2006) Structure of the 70S ribosome complexed with mRNA and tRNA. Science 313: 1935–1942

Shimizu Y, Inoue A, Tomari Y, Suzuki T, Yokogawa T, Nishikawa K, Ueda T (2001) Cell-free translation reconstituted with purified components. Nat Biotechnol 19: 751–755

Starosta AL, Karpenko VV, Shishkina AV, Mikolajka A, Sumbatyan NV, Schluenzen F, Korshunova GA, Bogdanov AA, Wilson DN (2010) Interplay between the ribosomal tunnel, nascent chain, and macrolides influences drug inhibition. Chem Biol 17: 504–514

Stasinopoulos SJ, Farr GA, Bechhofer DH (1998) *Bacillus subtilis* tetA(L) gene expression: evidence for regulation by translational reinitiation. Mol Microbiol 30: 923–932

Subramanian SL, Ramu, H., Mankin, A. S (2011) Inducible resistance to macrolide antibiotics. In: Antibiotic drug discovery and development. Dougherty TJ, Pucci, M. J. (eds.). New York, NY: Springer Publishing Company (in press)

Tanner DR, Cariello DA, Woolstenhulme CJ, Broadbent MA, Buskirk AR (2009) Genetic identification of nascent peptides that induce ribosome stalling. J Biol Chem 284: 34 809–34 818

Tenson T, Lovmar M, Ehrenberg M (2003) The mechanism of action of macrolides, lincosamides and streptogramin B reveals the nascent peptide exit path in the ribosome. J Mol Biol 330: 1005–1014

Toh SM, Xiong L, Bae T, Mankin AS (2008) The methyltransferase YfgB/RlmN is responsible for modification of adenosine 2503 in 23S rRNA. RNA 14: 98–106

Tu D, Blaha G, Moore PB, Steitz TA (2005) Structures of MLSBK antibiotics bound to mutated large ribosomal subunits provide a structural explanation for resistance. Cell 121: 257–270

Vázquez-Laslop N, Thum C, Mankin AS (2008) Molecular mechanism of drug-dependent ribosome stalling. Mol Cell 30: 190–202

Vázquez-Laslop N, Ramu, H., Klepacki, D., Mankin, A. S (2010) The key role of a conserved and modified rRNA residue in the ribosomal response to the nascent peptide. EMBO J 29: 3108–3117

Vester B, Douthwaite S (2001) Macrolide resistance conferred by base substitutions in 23S rRNA. Antimicrob Agents Chemother 45: 1–12

Voorhees RM, Weixlbaumer A, Loakes D, Kelley AC, Ramakrishnan V (2009) Insights into substrate stabilization from snapshots of the peptidyl transferase center of the intact 70S ribosome. Nat Struct Molec Biol 16: 528–533

Voss NR, Gerstein M, Steitz TA, Moore PB (2006) The geometry of the ribosomal polypeptide exit tunnel. J Mol Biol 360: 893–906

Weisblum B (1995) Insights into erythromycin action from studies of its activity as inducer of resistance. Antimicrob Agents Chemother 39: 797–805

Weisblum B (1998) Macrolide resistance. Drug Resist Updat 1: 29–41

Woolhead CA, Johnson AE, Bernstein HD (2006) Translation arrest requires two-way communication between a nascent polypeptide and the ribosome. Mol Cell 22: 587–598

Yang R, Cruz-Vera LR, Yanofsky C (2009) 23S rRNA nucleotides in the peptidyl transferase center are essential for tryptophanase operon induction. J Bacteriol 191: 3445–3450

Yap MN, Bernstein HD (2009) The plasticity of a translation arrest motif yields insights into nascent polypeptide recognition inside the ribosome tunnel. Mol Cell 34: 201–211

Zhang G, Hubalewska M, Ignatova Z (2009) Transient ribosomal attenuation coordinates protein synthesis and co-translational folding. Nat Struct Molec Biol 16: 274–280

Zhong P, Cao Z, Hammond R, Chen Y, Beyer J, Shortridge VD, Phan LY, Pratt S, Capobianco J, Reich KA, Flamm RK, Or YS, Katz L (1999) Induction of ribosome methylation in MLS-resistant *Streptococcus pneumoniae* by macrolides and ketolides. Microb Drug Resist 5: 183–188

Nascent polypeptide chains within the ribosomal tunnel analyzed by cryo-EM

31

Daniel N. Wilson, Shashi Bhushan, Thomas Becker and Roland Beckmann

1. An active role for the ribosomal tunnel during translation

The ribosome is a large macromolecular particle that synthesizes polypeptide chains from the substituent amino acid building blocks. The active site for peptide bond formation, the so-called peptidyl transferase center (PTC), is located in a cleft on the intersubunit side of the large ribosomal subunit (reviewed by (Polacek and Mankin, 2005; Simonovic and Steitz, 2009)). As the nascent polypeptide chain is being synthesized, it passes through a tunnel within the large subunit and emerges at the solvent side where protein folding occurs. The first hints for the presence of a ribosomal tunnel in the large subunit came from proteolysis protection and immuno-electron microscopy (EM) studies: Using IgG antibodies raised against β-galactosidase or the rubisco small subunit, Lake and coworkers could show that polypeptide chains emerge on the back of large subunit of the bacterial (*Escherichia coli*) 70S and eukaryotic (plant) 80S ribosome, respectively – some 75 Å from the intersubunit interface (Bernabeu and Lake, 1982; Bernabeu et al., 1983). This distance was consistent with the earlier findings that 30–40 C-terminal amino acids of nascent polypeptide chains are protected by eukaryotic and bacterial ribosomes from proteolysis (Malkin and Rich, 1967; Blobel and Sabatini, 1970; Smith et al., 1978).

Visualization of the tunnel within the large subunit was first seen from 3D image reconstructions of 2D arrays of chick embryo ribosomes (Milligan and Unwin, 1986) and then subsequently in *Bacillus stereothermophilus* large 50S subunits (Yonath et al., 1987). Cryo-EM reconstructions and X-ray crystallography structures presented the ribosomal tunnel with increasing resolution (Frank et al., 1995; Beckmann et al., 1997; Ban et al., 2000; Morgan et al., 2000; Becker et al.,

2009), revealing an 80–100 Å long conduit that varies in width between 10–20 Å. Progressive crosslinking of nascent polypeptide chains of increasing length with domains V, II, III, and I of the 23S rRNA is consistent with the path through the tunnel of the large ribosomal subunit (Choi and Brimacombe, 1998). Moreover, in cryo-EM structures of the yeast 80S ribosome bound to the Sec61 complex, the protein-conducting channel for protein transport across the endoplasmic reticulum, the ribosomal tunnel was aligned with the pore of the Sec61 complex, suggesting that nascent polypeptide chains can pass directly from the tunnel into the translocon (Beckmann et al., 1997). Recently, nascent polypeptide chains have been directly observed within the ribosomal tunnel extending from the PTC to the exit site on the back of the large subunit (Becker et al., 2009; Seidelt et al., 2009; Bhushan et al., 2010), as originally predicted by Lake and coworkers in the 1980's (Bernabeu and Lake, 1982; Bernabeu et al., 1983). The X-ray crystal structures of bacterial and archaeal ribosomes have revealed that the ribosomal tunnel is predominantly composed of ribosomal RNA (rRNA) (Ban et al., 2000; Nissen et al., 2000; Harms et al., 2001; Schuwirth et al., 2005; Selmer et al., 2006), consistent with an overall electronegative potential (Lu et al., 2007; Lu and Deutsch, 2008). In addition to rRNA, the extensions of the ribosomal proteins L4 and L22 (L17 in eukaryotes) contribute to formation of the tunnel wall, and form a so-called "constriction" where the tunnel narrows (Ban et al., 2000; Nissen et al., 2000). Near the tunnel exit, the ribosomal protein L39e is present in eukaryotic and archaeal ribosomes, whereas the extension of L23 (L25 in eukaryotes) occupies a similar position in bacteria (Harms et al., 2001; Schuwirth et al., 2005; Selmer et al., 2006).

Despite its universality, a functional role for the ribosomal tunnel is only beginning to emerge. For many years, the ribosomal tunnel was thought of only as a passive conduit for the nascent polypeptide chain, however, accumulating evidence indicates that, for some nascent chains, the tunnel plays a more active role (reviewed by Deutsch, 2003). In particular, a number of leader peptides induce translational stalling in response to the presence or absence of an effector molecule, and in doing so regulate translation of a downstream gene (reviewed by Lovett and Rogers, 1996; Tenson and Ehrenberg, 2002). Well-characterized examples include the bacterial SecM, ErmC and TnaC as well as eukaryotic AAP and CMV leader peptides, for which mutations in the leader peptide sequence, or in the ribosomal tunnel components themselves, can relieve the translational arrest (Morris and Geballe, 2000; Gong and Yanofsky, 2002; Nakatogawa and Ito, 2002; Vazquez-Laslop et al., 2008). Collectively, these data imply a direct interaction between specific residues of the leader peptide with distinct locations of the ribosomal tunnel.

2. TnaC-mediated translational stalling

One of the best characterized small molecule-dependent stalling mechanisms is that of TnaC, a leader peptide involved in the regulation of the *tryptophanase* (*tna*) operon of *Escherichia coli* (Gong and Yanofsky, 2002). In the *tna* operon, the *tnaC* regulatory leader is located upstream of two structural genes, *tnaA* and *tnaB*, encoding the enzyme tryptophanase and a tryptophan-specific permease, respectively (Gong and Yanofsky, 2002). The spacer region between the *tnaC* and *tnaA* genes contains several potential Rho-dependent transcription-termination sites, such that when free tryptophan levels are low in the cell, the TnaC leader peptide is translated and the ribosomes are released from the mRNA, allowing Rho to access and terminate transcription before the RNA polymerase reaches the *tnaA/B* genes. In the presence of free tryptophan, however, the TnaC peptide is translated, but termination and release of the TnaC nascent chain from the ribosome is prevented. The stalled TnaC•70S complex blocks the Rho-dependent transcription-termination sites and thus transcription of the downstream *tnaA/B* genes ensues (Gong and Yanofsky, 2002), ultimately leading to the removal of free tryptophan from the cytoplasm until non-inducing levels are restored. Site-directed mutagenesis studies have identified Trp12, Asp16 and Pro24 of the 24-residue TnaC peptide as being crucial for stalling (Gong and

Yanofsky, 2002; Cruz-Vera et al., 2005; Cruz-Vera and Yanofsky, 2008). In the stalled complex, TnaC•tRNA^Pro (Pro24) is located within the P site of the ribosome (Gong et al., 2001), indicating that Asp16 and Trp12 are retained within the exit tunnel. Moreover, mutations in ribosomal tunnel components also alleviate stalling (Cruz-Vera et al., 2005), suggesting that interaction between the TnaC nascent chain and the ribosomal tunnel is an essential feature of the stalling mechanism.

2.1. Cryo-EM of a stalled TnaC•70S complex

In order to structurally investigate TnaC-mediated translational stalling, it was necessary to generate a homogeneous TnaC•70S complex. This was achieved using an *E. coli* S30-based transcription-translation system (Jewett and Swartz, 2004; Liu et al., 2005), where a modified tnaC leader template containing a linker to an N-terminally located calmodulin-binding protein tag was translated in the presence of high concentrations (2 mM) of free tryptophan (Seidelt et al., 2009). The stalled TnaC•70S complex was purified from non-ribosomal factors using sucrose gradient density centrifugation, and then separated from non-translating ribosomes using a calmodulin sepharose matrix (Seidelt et al., 2009). Subsequently, cryo-EM and single-particle analysis was used to reconstruct the *Escherichia coli* TnaC•70S complex at 5.8 Å resolution (Figure 1A) (Seidelt et al., 2009). The structure of the TnaC•70S complex revealed additional density for the mRNA (red in Figure 1A) and a single peptidyl-tRNA within the intersubunit space of the TnaC•70S complex (green in Figure 1A). As expected, the tRNA was positioned at the P site of the ribosome, whereas density for the mRNA spanned the A, P, and E sites. Most strikingly, however, was the presence of additional density within the exit tunnel that could be attributed to the TnaC leader nascent chain (Figure 1A).

2.2. Interaction of the TnaC leader peptide with the ribosomal tunnel

Careful inspection of the ribosomal exit tunnel revealed that the density for the TnaC nascent chain fuses with the tunnel wall at a multitude of sites. These contact sites are distributed along the entire length of the tunnel and vary depending upon the threshold level (Figure 1). At the PTC, additional density

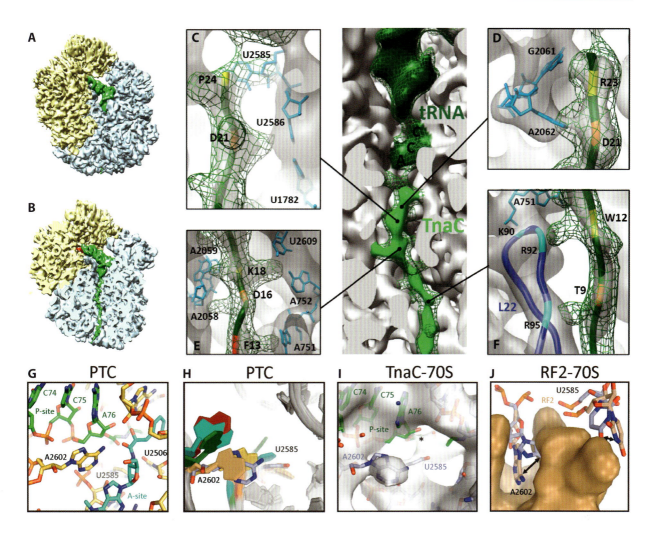

Fig. 1 Visualization of the stalled TnaC•70S ribosome complex. (A) Overview of TnaC•70S ribosome complex, with P-tRNA (green), 30S (yellow) and 50S (blue) indicated. (B) Transverse section of (A) to show ribosomal tunnel, TnaC-tRNA (green), and mRNA (red). (C)–(F) Contacts of the TnaC nascent chain (green mesh) with components of the ribosomal tunnel. (G) Relative location of the CCA-ends of A- (cyan) and P-tRNA (green) at the PTC (PDB1VQN) (Schmeing et al., 2005b). (H) Comparison of position of A2602 in various X-ray structures of ribosomal particles. (I) Distinct positions of A2602 and U2585 at the PTC of the TnaC•70S complex. (J) Comparison of the positions of A2602 and U2585 (light blue) from (I) with the positions of A2602 and U2585 (gold) from an RF2•70S complex (Weixlbaumer et al., 2008). RF2 is shown as gold surface representation. (Figure adapted from Seidelt et al., 2009)

is observed connecting Pro24 of TnaC and U2585 of the 23S rRNA (Figure 1C), whereas the neighboring U2586, together with U1782, appear to form a connection in the region where Asp21 is likely to be located (Figure 1C). Mutations in the U2585 region have been shown to reduce the maximum level of TnaC induction (Yang et al., 2009). The highly conserved Pro24 is essential for TnaC stalling, since Pro24Ala mutations abolish the Trp-dependent inhibition of TnaC-tRNA cleavage at the PTC (Cruz-Vera and Yanofsky, 2008). Very strong density links G2061 and A2062 to the region near residues Arg23 and Asp21, respectively, of TnaC (Figure 1D). Although A2062 has not been ana-

lyzed for its effects on TnaC stalling, mutations at this position have nevertheless been shown to relieve the translational arrest mediated by the ErmC leader peptide (Vazquez-Laslop et al., 2008).

Deeper in the tunnel, two connections are visible linking A2058 and A2059 with the nascent chain in the proximity of Asp16 and Lys18 (Figure 1E), which may explain the protection of these nucleotides from sparsomycin-enhanced chemical modification seen during tryptophan induced TnaC-stalling (Cruz-Vera et al., 2007). Asp16 is highly conserved within the TnaC leader peptide and Asp16Ala mutations abolish the Trp-dependent inactivation of the PTC (Cruz-

Vera and Yanofsky, 2008). Ribosomes with A2058G mutations are slightly more responsive to Trp-induced stalling in a *rrn* Δ6 strains (Cruz-Vera et al., 2005), whereas this mutation strongly alleviates secM-mediated translational stalling (Nakatogawa and Ito, 2002). Strong density that extends out from the TnaC nascent chain at the putative location of Lys18 fuses with the ribosomal tunnel where U2609 and A752 are located, whereas the adjacent nucleotide A751 appears to contact TnaC in the vicinity of Phe[13] (Figure 1E). Consistently, mutations at U2609 as well as an insertion at A751 have been reported to eliminate the induction by tryptophan (Cruz-Vera et al., 2005).

The TnaC nascent chain makes two major contacts with the β-hairpin of ribosomal protein L22 (Figure 1F): One connects Arg95 of L22 with the nascent chain near Thr9, whereas the other is found at the tip of the loop, where Lys90 and Arg92 are located, and fuses with TnaC in proximity of the highly conserved Trp12 residue (Figure 1F). This latter contact should be important for TnaC-stalling since (i) the spacing between Trp12 and Pro24 is critical for efficient stalling (Gong and Yanofsky, 2002; Cruz-Vera et al., 2005; Cruz-Vera and Yanofsky, 2008) and (ii) mutations of Trp12 in TnaC as well as Lys90 in L22 also eliminate tryptophan induction (Cruz-Vera et al., 2005). Additional evidence for the close proximity of Trp12 to the tip of L22 comes from crosslinks between the neighboring Lys11 with 23S rRNA nucleotides in the vicinity of A751 (Cruz-Vera et al., 2005), which also makes contact with the tip of the β-hairpin of L22 (Figure 1F).

2.3. TnaC-mediated inactivation of the PTC

The PTC of the ribosome is the site of peptide-bond formation and peptidyl-tRNA hydrolysis (reviewed by (Polacek and Mankin, 2005; Simonovic and Steitz, 2009)). The correct positioning of the substrates at the PTC, i. e. the CCA-ends of A- and P-tRNAs during peptide bond formation (Figure 1G) or the P-tRNA and the GGQ motif of the release factors (RFs) during termination, is critical to ensuring efficient catalysis. Specific conformational changes of highly conserved nucleotides of the 23S rRNA within the PTC have been associated with binding of different ligands to this active site, for example, the nucleotides A2602 and U2585 have been observed to adopt dramatically different conformations in ribosome structures depending upon the functional state (Figure 1H), for examples, see (Bashan et al., 2003; Schmeing et al.,

2005b; Wilson et al., 2005)). Because the mechanism of TnaC-mediated translational stalling results in the inactivation of the PTC (Gong and Yanofsky, 2002), it is interesting to examine the conformation of this region in the 70S•TnaC complex (Figure 1I). Density accounting for the P-tRNA and surrounding 23S rRNA nucleotides at the PTC is clearly observed, with A2602 appearing to adopt a very defined conformation in the 70S•TnaC complex that resembles the position of A2602 observed when the translation inhibitor sparsomycin is bound at the PTC (Schmeing et al., 2005a). In addition to A2602, continuous density between the nascent chain and the location of U2585, suggests that this flexible base shifts to interact with the Pro24 of TnaC (Figure 1I).

Inactivation of the PTC in the 70S•TnaC complex requires free tryptophan, the binding site of which has been proposed to overlap with that of the antibiotic sparsomycin (Cruz-Vera et al., 2006; Cruz-Vera et al., 2007; Cruz-Vera and Yanofsky, 2008). This proposal is attractive given the sparsomycin-like conformation of A2602 observed in the PTC of 70S•TnaC complex, but also because the possible stacking of the free tryptophan between the peptidyl-tRNA and A2602, in a manner analogous to sparsomycin, would explain the fixed conformation of A2602. Furthermore, binding of Trp-tRNA at the A site can also induce TnaC stalling in the absence of free tryptophan, leading to the suggestion that the free tryptophan molecule binds where the aminoacyl moiety of an A-tRNA is located at the PTC (Gong and Yanofsky, 2002), which also overlaps with the sparsomycin binding site. However, no additional density that could be attributed to the free tryptophan molecule is observed within the sparsomycin-binding site, nor in the A site of the PTC (Figure 1I), despite the purification and cryo-EM analysis of the 70S•TnaC complex in the presence of 2 mM tryptophan (Seidelt et al., 2009).

Nevertheless, the conformations of A2602 and U2585 observed in the 70S•TnaC complex are incompatible with simultaneous co-habitation of termination release factors (RFs) (Figure 1J) (Korostelev et al., 2008; Laurberg et al., 2008; Weixlbaumer et al., 2008; Jin et al., 2010). This suggests that even if RFs can still bind to the stalled 70S•TnaC complexes (Cruz-Vera et al., 2005), the fixed conformation of A2602 and U2585 would prevent correct positioning of the GGQ motif of the RF within the PTC that is necessary for efficient hydrolysis and release of the nascent chain from the P-tRNA (Korostelev et al., 2008; Laurberg et al., 2008; Weixlbaumer et al., 2008; Jin et al., 2010). Indeed, mutations at A2602 lead to defects in peptidyl-tRNA hydrolysis (Polacek et al., 2003; Youngman et al., 2004) indicating the importance of this residue for termination activity.

2.4. TnaC leader peptide: Extended versus compacted

The interpretation of the TnaC peptide with an extended conformation is in agreement with the density observed for the nascent chain displaying properties similar to the extensions of ribosomal proteins (Figure 2). At low thresholds, there is continuous density for the TnaC peptide throughout the entirety of the exit tunnel, however with increasing thresholds, the density for the TnaC peptide becomes fragmented and then disappears completely (Figure 2A). This characteristic behavior is similar to that observed for the density of the N-terminal extension of ribosomal protein L27 (Figure 2B). In contrast, α-helical ribosomal proteins, for example L29, maintain rod-like density for the helical regions at increasing thresholds, whereas density for the linking regions is lost (Figure 2C).

Since the resolution of the TnaC•70S maps is limited to ~6 Å, it is not possible to model side chains and therefore the interpretation of the contacts can only be considered as an approximation. Nevertheless, given that X-ray crystal structures are available for the *E. coli* 70S ribosome (Schuwirth et al., 2005) and the relative location of the P-tRNA on the ribosome (Yusupov et al., 2001; Selmer et al., 2006), these provide an excellent constraint to build a model for the TnaC nascent chain. With the CCA-end of the P-tRNA fixed, it was possible to fit a molecular model of all 24 amino acids of the TnaC leader peptide into the additional density within the exit tunnel using Rapper for the initial models (de Bakker et al., 2006) followed by a subsequent molecular dynamics flexible fitting (MDFF) procedure (Trabuco et al., 2008) (Figure 2D). The fitting resulted in an ensemble of models, all of which displayed an extended conformation of the nascent chain with root-mean-square fluctuations (RMSFs) for the Cα atoms smaller than 2 Å (Figure 2D). The excellent agreement between selected models of the TnaC peptide and the experimental density is striking.

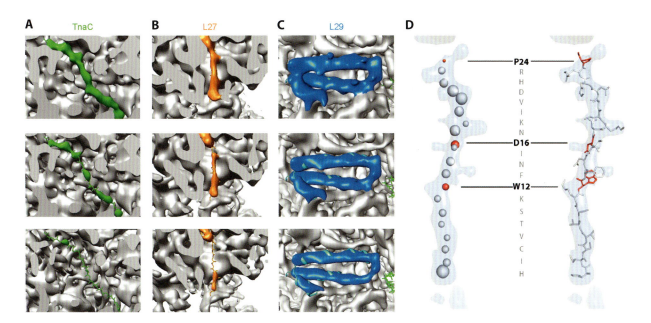

Fig. 2 Comparison of the cryo-EM density for the TnaC nascent chain with ribosomal proteins L27 and L29. (A) Series of three panels with increasing threshold (top to bottom) showing the same section through the TnaC•70S complex to reveal the nascent chain (green) within the exit tunnel. (B) Density for the extended N-terminal region of ribosomal protein L27 (orange), with each respective panel at the same threshold as for TnaC in panel A. (C) Density for the highly α-helical ribosomal protein L29 (blue) shown at identical thresholds as for TnaC nascent chain in panel A and L27 in panel B. In (A)–(C), the respective models for TnaC (green), L27 (orange) and L29 (cyan) are shown for reference. (Figure adapted from Seidelt et al., 2009). (D) Rapper and MDFF based fit of the TnaC model to the isolated TnaC nascent chain density, with the radii of the C(atoms correspond to the RMSF for 10 different TnaC conformers (left) and an all-atom representation of one of the ten TnaC conformers (right).

3. Protein folding within the confinement of the tunnel

The dimensions of the tunnel preclude the folding of domains as large as an IgG domain (~ 17 kDa), whereas α-helix formation is more feasible (Kramer et al., 2001; Voss et al., 2006). Comparisons of X-ray crystal structures of large subunits from various bacteria (*E. coli* (Schuwirth et al., 2005), *Deinococcus radiodurans* (Harms et al., 2001) and *Thermus thermophilus* (Yusupov et al., 2001; Selmer et al., 2006)) and archaea (*Haloarcula marismortui*) (Ban et al., 2000) with subnanometer cryo-EM reconstructions of 70S and 80S ribosomes from diverse origins (Halic et al., 2006a; Halic et al., 2006b; Chandramouli et al., 2008; Becker et al., 2009; Seidelt et al., 2009; Taylor et al., 2009), suggests that there is only limited conformational flexibility within the ribosomal tunnel and provide no evidence for large scale conformational rearrangement that could provide sufficient space to allow tertiary folding of nascent chains within the tunnel. Indeed, early biochemical studies indicated that fluorophores, such as pyrene and cascade yellow with a smallest dimension of ~ 4 Å were efficiently incorporated into nascent polypeptide chains, whereas larger fluorophores such as eosin, with a minimum dimension of ~ 11 Å, were much less efficiently incorporated (Ramachandiran et al., 2000).

A stereochemical analysis of the peptidyl transferase reaction led to the suggestion that the ribosome generates an α-helical conformation for all nascent polypeptide chains as they are being synthesized (Lim and Spirin, 1986). However, differences in proteolysis and antibody detection patterns of distinct synthetic and natural nascent polypeptide chains when bound to the ribosome suggested that in fact distinct nascent chains adopt different secondary structure conformations within the ribosomal tunnel (Picking et al., 1992; Tsalkova et al., 1998, reviewed by Kramer et al., 2001). More recently, fluorescence resonance energy transfer (FRET) studies have indicated that a transmembrane signal anchor sequence is compacted in a manner consistent with α-helix formation in the tunnel as it travels through the ribosome (Woolhead et al., 2004). Interestingly, in the same study the compaction of the transmembrane nascent chain was lost upon exiting the tunnel, suggesting that the tunnel plays a pivotal role in stabilizing the proposed helical conformation. Independent biochemical analyses also support the potential of the nascent chain to adopt compacted or helical conformations in the tunnel and have even identified specific regions of the tunnel that promote compaction (Kosolapov et al., 2004; Lu and Deutsch, 2005a, b; Kosolapov and Deutsch, 2009).

3.1. Cryo-EM of translating ribosomes with nascent chains of high helical propensity

Recently, cryo-EM has also been used to structurally investigate the potential of helix formation within the ribosomal tunnel (Bhushan et al., 2010). This study used a wheat germ *in-vitro* translation system using DNA templates which were truncated to remove the stop codon, thus trapping the translating ribosomes at the last codon (Bhushan et al., 2010). Two templates were used where different regions of the dipeptidylaminopeptidease B (DPAP-B) sequence were replaced with a short sequence encoding a peptide that has a strong propensity to form a hydrophilic α-helix in solution (Marqusee and Baldwin, 1987; Arai et al., 2001). The peptide contains five EAAAK repeats and adopts a standard [i + 4 (i) α-helix, in which every backbone N-H group donates a hydrogen bond to the backbone C=O group of the amino acid four residues earlier. In addition, each repeat of the helix is stabilized by a Glu$^-$Lys$^+$ salt bridge, leading to >80% helicity in aqueous solvent as determined by circular dichromism (CD) studies (Marqusee and Baldwin, 1987; Arai et al., 2001). When translation reaches the 3' end of the truncated mRNA, 115 amino acids have been translated and Asp-tRNA is located at the P site of the ribosome. In the Helix 1 construct, the helix-forming sequence is positioned at amino acids 72–96, whereas in Helix 2 it is located at amino acids 83–108, i. e. −19 and −7 from the Asp of the P-tRNA, respectively. Since the ribosomal tunnel can enclose 30–40 amino acids, both helix-forming sequences are predicted to be contained within the exit tunnel. Cryo-EM structures using these templates generate the 80S•Helix 1-RNC and 80S•Helix 2-RNCs, both of which showed strong density for a single peptidyl-tRNA within the intersubunit space as well as the presence of additional density within the exit tunnel that can be attributed to the nascent chain (Figure 3A, B) (Bhushan et al., 2010). In the tunnel of both the 80S•Helix-RNCs, the strongest region of density is observed for the N-terminal region of the nascent chain near the tunnel exit, however fragmented density is also observed within the upper and mid regions (Figure 3A, B).

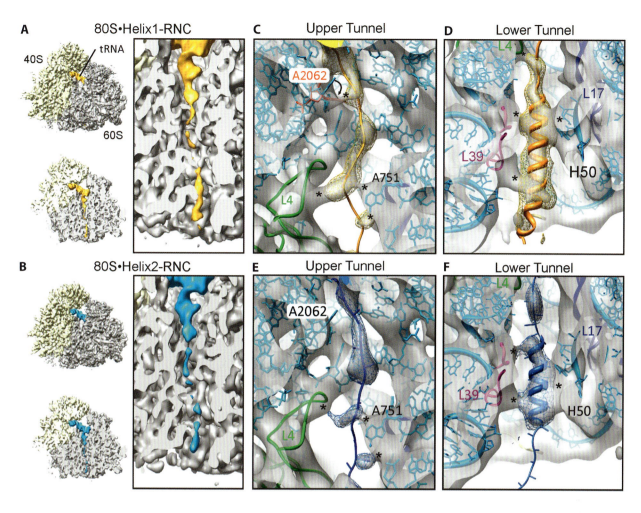

Fig. 3 Visualization of 80S•Helix-RNCs. Overview (top left) and transverse sections (bottom left and right panel) of the (A) 80S•Helix 1-RNC and (B) 80S•Helix 2-RNC, with peptidyl-tRNA in gold and blue, respectively. Small 40S (yellow) and large 60S (blue) ribosomal subunits as indicated. Contacts of the (C, D) Helix 1 (gold) and (E, F) Helix 2 (blue) nascent chains (mesh) with components of the (C, E) upper and (D, F) lower ribosomal tunnel. (Figure adapted from Bhushan et al., 2010)

3.2. Folding of helical peptides within distinct regions of the tunnel

The density for the nascent chain in the cryo-EM structure of the 80S•Helix 1-RNC appears to be extended in the upper tunnel region (Figure 3C), whereas density consistent with helix formation or compaction is evident in the lower tunnel region (Figure 3D). A similar trend is also observed in the maps of the 80S•Helix 2-RNC with the nascent chain in the upper tunnel region appearing to be predominantly in an extended conformation (Figure 3E), whereas the strongest density for the nascent chain is observed in the lower tunnel region (Figure 3F). If all five repeats of the EAAAK sequence had adopted a helical conformation in the 80S•Helix 2-RNC, one would also expect strong density to be present in the upper region of the tunnel, near

to the constriction between ribosomal proteins L4 and L17. Therefore, it was suggested that the proximal portion of the region with helical propensity is unable to adopt an α-helical conformation, but instead acquires an extended conformation (Bhushan et al., 2010). Although it is possible that compaction is present in the middle region of the tunnel but simply not observed due to flexibility, this is incompatible with the excellent agreement between the density and the model for the distal portion of the remaining helical stretch (Figure 2F). Furthermore, the relative confinement of the constriction region limits the degrees of conformational freedom when an a-helix forms within this region. The ability to form a helix near to the tunnel exit site, but not in the middle region of the tunnel, is also consistent with the zones of secondary structure formation identified by Deutsch and coworkers (Lu and Deutsch,

2005a). Curiously, more density for the nascent chain is observed in the upper region of the 80S•Helix 2-RNC compared to the 80S•Helix 1-RNC, suggesting an influence of the Helix 2 sequence on the conformation of adjacent region of the nascent chain. The influence of neighboring residues on the conformation of adjacent regions of the nascent chain has been observed for the folding in the tunnel of transmembrane regions of the Kv1.3 voltage-gated potassium channel (Tu and Deutsch, 2010). Similarly, the placement of the critical Arg163 during SecM-mediated translation stalling is dependent on the properties of neighboring residues within the nascent polypeptide chain (Yap and Bernstein, 2009). Although there appears to be slightly more compaction within this region of the 80S•Helix 2-RNC, the density is not consistent with α-helix.

3.3. Characteristic behavior of helical peptides within the ribosomal tunnel

Given that the nascent chain within the lower tunnel of the 80S•Helix 1-RNC forms a compacted or helical secondary structure, it is interesting to analyze the properties of the density for this region, relative to the presumably more extended region of the nascent chain in the upper tunnel (Figure 4). Filtering the 80S•Helix 1-RNC map at different resolutions, ranging from 6–7 Å to 10–11 Å reveals that electron density for the extended region of the nascent chain is only observable at 6–7 Å, whereas at lower resolutions 8–9 Å and 10–11 Å, very little density is observable (Figure 4 A). In contrast, electron density for the nascent chain is still observable in the lower region of the tunnel, even at 10–11 Å, consistent with the interpretation that this region adopts a compacted or helical structure (Figure 4 A). Similarly, contouring the 80S•Helix 1-RNC at increasing thresholds (Threshold 1 in Figure 4 A and increasing thresholds 2–4 of Figure 4 B) leads to a loss in electron density for the nascent chain in the upper region of the tunnel whereas density persists in the lower region, again supporting the extended and compacted conformations of the nascent chain in the upper and lower tunnel, respectively.

Analysis of the 80S•Helix 1-RNC and 80S•Helix 2-RNC suggests that the lower tunnel region supports, and maybe even promotes, the ability of nascent polypeptide chains to adopt compacted or helical conformations, whereas no helix formation was observed in the middle region of the tunnel (Figure 4 A, B) (Bhushan et al., 2010). In contrast, FRET studies have

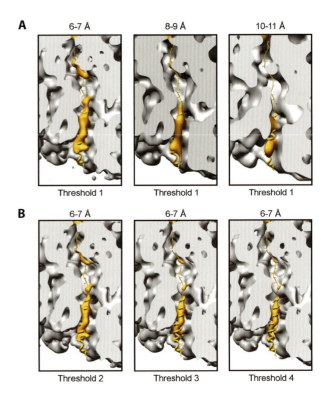

Fig. 4 The 80S•Helix 1-RNC nascent chain at different resolutions and thresholds. (A–B) Transverse sections of the 80S•Helix 1-RNC map (gray), showing the tunnel with nascent chain density (gold). (A) The 80S•Helix 1-RNC map filtered at (panels, left to right) 6–7 Å, 8–9 Å and (10–11 Å. (B) The 80S•Helix 1-RNC map shown with increasing thresholds (from left to right) relative to Threshold 1 shown in panel A. (Figure adapted from Bhushan et al., 2010)

suggested that only transmembrane helices maintain a helical conformation during its passage throughout the entire ribosomal tunnel (Woolhead et al., 2004). These diverse findings may result from the use of a very stable hydrophilic helix was utilized for the cryo-EM study, instead of a hydrophobic transmembrane helix as used in the FRET study. They reinforce the idea that nascent chains may exhibit different behavioral patterns within the tunnel. This is likely to result from a complex interplay between the environment of distinct regions of the tunnel and the nature of the nascent polypeptide chain. It will be interesting to structurally investigate the folding behavior of different types of helical sequences, such as transmembrane and signal anchor sequences with varying degrees of hydrophobicity, within distinct regions of the tunnel.

4. Implications of interaction of nascent polypeptide chains within the ribosomal tunnel

Because of the high structural conservation of the ribosomal tunnel between eukaryotes and prokaryotes, it is possible to make a direct comparison between the interaction pattern of a non-stalling nascent chain such as that present in the 80S•Helix 1-RNC (Bhushan et al., 2010), with a stalling sequence, such as the TnaC leader peptide present in the 70S•TnaC-RNC (Seidelt et al., 2009) (Figure 5). In the cryo-EM structure of the 70S•TnaC-RNC, multiple contacts are observed from the C-terminal region of the peptide to a distinct set of ribosomal components, consistent with the conservation and importance of this region for inducing translational stalling. Interestingly, in the upper region of the tunnel of the 80S•Helix 1-RNC, interac-

tion between the nascent chains and a subset of these components is also observed (Figure 3), for example, with the regions in the vicinity of A2062 and A751 (*E. coli* numbering) of the large subunit rRNA as well as the loops of ribosomal proteins L4 and L17 (L22 in bacteria) located at the constriction. The fact that the contacts observed here for non-stalling sequences are similar in location to those predicted for some of the known stalling leader peptides may indicate that these regions of the tunnel represent functional hotspots for tunnel-nascent chain interaction.

In addition to inducing translational stalling, the interaction between nascent polypeptide chains and the ribosomal tunnel can regulate the rate of translation (Lu and Deutsch, 2008) and has been suggested to act as a signal to recruit chaperones and translocation machinery at the tunnel exit site (reviewed by Kramer et al., 2009; Cabrita et al., 2010). Moreover, allowing, or

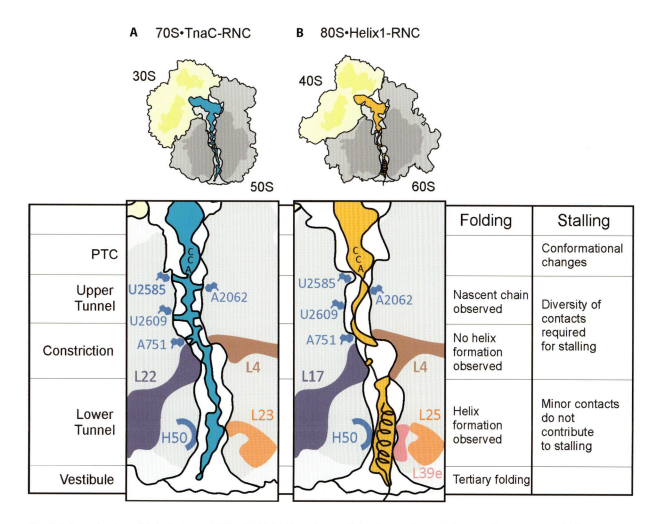

Fig. 5 Schematic view of (A) the bacterial TnaC•70S-RNC, with the eukaryotic (B) 80S•Helix 1-RNC, indicating the specific regions of the ribosomal tunnel that contribute to protein folding or translational stalling.

even promoting, α-helix formation (Lu and Deutsch, 2005a; Tu and Deutsch, 2010), when β-sheet formation is not yet possible, may have an impact on protein folding. Protein folding might occur using a hierarchy of secondary structure elements, with α-helix formation occurring first wherever possible. Such a scenario would considerably reduce the complexity of the theoretical conformational space that is necessary to be sampled before the correct fold is adopted. Additionally, it would also change the appearance of nascent peptides as substrates for chaperones acting co-translationally. Tertiary structure formation, such as α-helices and β-hairpins, has already been observed to occur near the tunnel exit (> 80 Å from tunnel start) where the tunnel widens significantly to form a vestibule (Evans et al., 2008; Lu and Deutsch, 2008; Kosolapov and Deutsch, 2009). Additionally, α-helix formation in the tunnel may be important for proteins containing α-helical domains destined for membrane insertion (Liao et al., 1997; Woolhead et al., 2004). Cotranslational targeting by the signal recognition particle (SRP), for example, may be promoted since the presence of a signal-anchor sequence within the tunnel promotes binding of SRP to the ribosome (Berndt et al., 2009), and α-helicity of the signal sequence is important for its recognition by SRP (Mingarro et al., 2000). Indeed, compaction of transmembrane domains in the ribosomal tunnel has been reported (Woolhead et al., 2004) and, a compacted conformation for the signal anchor sequence has been observed by cryo-EM to bind in the vestibule at the end of the ribosomal tunnel on *E. coli* ribosomes (Halic et al., 2006a). While these hypotheses need to be examined, the conservation of the dimensions of the ribosomal tunnel is consistent with its significance in providing nascent proteins with a defined first environment.

References

Arai R, Ueda H, Kitayama A, Kamiya N, Nagamune T (2001) Design of the linkers which effectively separate domains of a bifunctional fusion protein. Protein Eng 14: 529–532

Ban N, Nissen P, Hansen J, Moore PB, Steitz TA (2000) The complete atomic structure of the large ribosomal subunit at 2.4 Å resolution. Science 289: 905–920

Bashan A, Agmon I, Zarivach R, Schluenzen F, Harms J, Berisio R, Bartels H, Franceschi F, Auerbach T, Hansen HA, Kossoy E, Kessler M, Yonath A (2003) Structural basis of the ribosomal machinery for peptide bond formation, translocation, and nascent chain progression. Mol Cell 11: 91–102

Becker T, Bhushan S, Jarasch A, Armache JP, Funes S, Jossinet F, Gumbart J, Mielke T, Berninghausen O, Schulten K, Westhof E, Gilmore R, Mandon EC, Beckmann R (2009) Structure of monomeric yeast and mammalian Sec61 complexes interacting with the translating ribosome. Science 326: 1369–1373

Beckmann R, Bubeck D, Grassucci R, Penczek P, Verschoor A, Blobel G, Frank J (1997) Alignment of conduits for the nascent polypeptide chain in the ribosome- Sec61 complex. Science 278: 2123–2126

Bernabeu C, Lake JA (1982) Nascent polypeptide chains emerge from the exit domain of the large ribosomal subunit: immune mapping of the nascent chain. Proc Natl Acad Sci USA 79: 3111–3115

Bernabeu C, Tobin EM, Fowler A, Zabin I, Lake JA (1983) Nascent polypeptide chains exit the ribosome in the same relative position in both eukaryotes and prokaryotes. J Cell Biol 96: 1471–1474

Berndt U, Oellerer S, Zhang Y, Johnson AE, Rospert S (2009) A signal-anchor sequence stimulates signal recognition particle binding to ribosomes from inside the exit tunnel. Proc Natl Acad Sci USA 106: 1398–1403

Bhushan S, Gartmann M, Halic M, Armache JP, Jarasch A, Mielke T, Berninghausen O, Wilson DN, Beckmann R (2010) alpha-Helical nascent polypeptide chains visualized within distinct regions of the ribosomal exit tunnel. Nat Struct Mol Biol 17: 313–317

Blobel G, Sabatini DD (1970) Controlled proteolysis of nascent polypeptides in rat liver cell fractions. I. Location of the polypeptides within ribosomes. J Cell Biol 45: 130–145

Cabrita LD, Dobson CM, Christodoulou J (2010) Protein folding on the ribosome. Curr Opin Struct Biol 20: 33–45

Chandramouli P, Topf M, Menetret JF, Eswar N, Cannone JJ, Gutell RR, Sali A, Akey CW (2008) Structure of the mammalian 80S ribosome at 8.7 A resolution. Structure 16: 535–548

Choi K, Brimacombe R (1998) The path of the growing peptide chain through the 23S rRNA in the 50S ribosomal subunit; a comparative cross-linking study with three different peptide families. Nucleic Acids Res 26: 887–895

Cruz-Vera L, Rajagopal S, Squires C, Yanofsky C (2005) Features of ribosome-peptidyl-tRNA interactions essential for tryptophan induction of tna operon expression. Mol Cell 19: 333–343

Cruz-Vera LR, Gong M, Yanofsky C (2006) Changes produced by bound tryptophan in the ribosome peptidyl transferase center in response to TnaC, a nascent leader peptide. Proc Natl Acad Sci USA 103: 3598–3603

Cruz-Vera LR, New A, Squires C, Yanofsky C (2007) Ribosomal features essential for tna operon induction: tryptophan binding at the peptidyl transferase center. J Bacteriol 189: 3140–3146

Cruz-Vera LR, Yanofsky C (2008) Conserved residues Asp16 and Pro24 of TnaC-tRNAPro participate in tryptophan induction of Tna operon expression. J Bacteriol 190: 4791–4797

de Bakker PI, Furnham N, Blundell TL, DePristo MA (2006) Conformer generation under restraints. Curr Opin Struct Biol 16: 160–165

Deutsch C (2003) The birth of a channel. Neuron 40: 265–276

Evans MS, Sander IM, Clark PL (2008) Cotranslational folding promotes beta-helix formation and avoids aggregation in vivo. J Mol Biol 383: 683–692

Frank J, Zhu J, Penczek P, Li YH, Srivastava S, Verschoor A, Radermacher M, Grassucci R, Lata RK, Agrawal RK (1995) A model of protein synthesis based on cryo-electron microscopy of the *E. coli* ribosome. Nature 376: 441–444

Gong F, Ito K, Nakamura Y, Yanofsky C (2001) The mechanism of tryptophan induction of tryptophanase operon expression: tryptophan inhibits release factor-mediated cleavage of TnaC-peptidyl-tRNA(Pro). Proc Natl Acad Sci USA 98: 8997–9001

Gong F, Yanofsky C (2002) Instruction of translating ribosome by nascent peptide. Science 297: 1864–1867

Halic M, Blau M, Becker T, Mielke T, Pool MR, Wild K, Sinning I, Beckmann R (2006a) Following the signal sequence from ribosomal tunnel exit to signal recognition particle. Nature 444: 507–511

Halic M, Gartmann M, Schlenker O, Mielke T, Pool MR, Sinning I, Beckmann R (2006b) Signal recognition particle receptor exposes the ribosomal translocon binding site. Science 312: 745–747

Harms J, Schluenzen F, Zarivach R, Bashan A, Gat S, Agmon I, Bartels H, Franceschi F, Yonath A (2001) High resolution structure of the large ribosomal subunit from a mesophilic eubacterium. Cell 107: 679–688

Jewett MC, Swartz JR (2004) Mimicking the *Escherichia coli* cytoplasmic environment activates long-lived and efficient cell-free protein synthesis. Biotechnol Bioeng 86: 19–26

Jin H, Kelley AC, Loakes D, Ramakrishnan V (2010) Structure of the 70S ribosome bound to release factor 2 and a substrate analog provides insights into catalysis of peptide release. Proc Natl Acad Sci USA 107: 8593–8598

Korostelev A, Asahara H, Lancaster L, Laurberg M, Hirschi A, Zhu J, Trakhanov S, Scott WG, Noller HF (2008) Crystal structure of a translation termination complex formed with release factor RF2. Proc Natl Acad Sci USA 105: 19 684–19 689

Kosolapov A, Deutsch C (2009) Tertiary interactions within the ribosomal exit tunnel. Nat Struct Mol Biol 16: 405–411

Kosolapov A, Tu L, Wang J, Deutsch C (2004) Structure acquisition of the T1 domain of Kv1.3 during biogenesis. Neuron 44: 295–307

Kramer G, Boehringer D, Ban N, Bukau B (2009) The ribosome as a platform for co-translational processing, folding and targeting of newly synthesized proteins. Nat Struct Mol Biol 16: 589–597

Kramer G, Ramachandiran V, Hardesty B (2001) Cotranslational folding–omnia mea mecum porto? Int J Biochem Cell Biol 33: 541–553

Laurberg M, Asahara H, Korostelev A, Zhu J, Trakhanov S, Noller HF (2008) Structural basis for translation termination on the 70S ribosome. Nature 454: 852–857

Liao SR, Lin JL, Do H, Johnson AE (1997) Both lumenal and cytosolic gating of the aqueous ER translocon pore are regulated from inside the ribosome during membrane protein integration. Cell 90: 31–41

Lim VI, Spirin AS (1986) Stereochemical analysis of ribosomal transpeptidation. Conformation of nascent peptide. J Mol Biol 188: 565–574

Liu DV, Zawada JF, Swartz JR (2005) Streamlining *Escherichia coli* S30 extract preparation for economical cell-free protein synthesis. Biotechnol Prog 21: 460–465

Lovett PS, Rogers EJ (1996) Ribosome regulation by the nascent peptide. Microbiol Rev 60: 366–385

Lu J, Deutsch C (2005a) Folding zones inside the ribosomal exit tunnel. Nat Struct Mol Biol 12: 1123–1129

Lu J, Deutsch C (2005b) Secondary structure formation of a transmembrane segment in Kv channels. Biochemistry 44: 8230–8243

Lu J, Deutsch C (2008) Electrostatics in the ribosomal tunnel modulate chain elongation rates. J Mol Biol 384: 73–86

Lu J, Kobertz WR, Deutsch C (2007) Mapping the electrostatic potential within the ribosomal exit tunnel. J Mol Biol 371: 1378–1391

Malkin LI, Rich A (1967) Partial resistance of nascent polypeptide chains to proteolytic digestion due to ribosomal shielding. J Mol Biol 26: 329–346

Marqusee S, Baldwin RL (1987) Helix stabilization by Glu-...Lys+ salt bridges in short peptides of de novo design. Proc Natl Acad Sci USA 84: 8898–8902

Milligan RA, Unwin PNT (1986) Location of exit channel for nascent protein in 80S ribosome. Nature 319: 693–695

Mingarro I, Nilsson I, Whitley P, von Heijne G (2000) Different conformations of nascent polypeptides during translocation across the ER membrane. BMC Cell Biol 1: 3

Morgan DG, Menetret JF, Radermacher M, Neuhof A, Akey IV, Rapoport TA, Akey CW (2000) A comparison of the yeast and rabbit 80 S ribosome reveals the topology of the nascent chain exit tunnel, inter-subunit bridges and mammalian rRNA expansion segments. J Mol Biol 301: 301–321

Morris DR, Geballe AP (2000) Upstream open reading frames as regulators of mRNA translation. Mol Cell Biol 20: 8635–8642

Nakatogawa H, Ito K (2002) The ribosomal exit tunnel functions as a discriminating gate. Cell 108: 629–636

Nissen P, Hansen J, Ban N, Moore PB, Steitz TA (2000) The structural basis of ribosome activity in peptide bond synthesis. Science 289: 920–930

Picking WD, Picking WL, Odom OW, Hardesty B (1992) Fluorescence characterization of the environment encountered by nascent polyalanine and polyserine as they exit *Escherichia coli* ribosomes during translation. Biochemistry 31: 2368–2375

Polacek N, Gomez MJ, Ito K, Xiong L, Nakamura Y, Mankin A (2003) The critical role of the universally conserved A2602 of 23S ribosomal RNA in the release of the nascent peptide during translation termination. Mol Cell 11: 103–112

Polacek N, Mankin AS (2005) The ribosomal peptidyl transferase center: structure, function, evolution, inhibition. Crit Rev Biochem Mol Biol 40: 285–311

Ramachandiran V, Willms C, Kramer G, Hardesty B (2000) Fluorophores at the N terminus of nascent chloramphenicol acetyltransferase peptides affect translation and movement through the ribosome. J Biol Chem 275: 1781–1786

Schmeing TM, Huang KS, Kitchen DE, Strobel SA, Steitz TA (2005a) Structural insights into the roles of water and the 2' hydroxyl of the P site tRNA in the peptidyl transferase reaction. Mol Cell 20: 437–448

Schmeing TM, Huang KS, Strobel SA, Steitz TA (2005b) An induced-fit mechanism to promote peptide bond formation and exclude hydrolysis of peptidyl-tRNA. Nature 438: 520–524

Schuwirth B, Borovinskaya M, Hau C, Zhang W, Vila-Sanjurjo A, Holton J, Cate J (2005) Structures of the bacterial ribosome at 3.5 Å resolution. Science 310: 827–834

Seidelt B, Innis CA, Wilson DN, Gartmann M, Armache JP, Villa E, Trabuco LG, Becker T, Mielke T, Schulten K, Steitz TA, Beckmann R (2009) Structural insight into nascent polypeptide chain-mediated translational stalling. Science 326: 1412–1415

Selmer M, Dunham C, Murphy Ft, Weixlbaumer A, Petry S, Kelley A, Weir J, Ramakrishnan V (2006) Structure of the 70S ribosome complexed with mRNA and tRNA. Science 313: 1935–1942

Simonovic M, Steitz TA (2009) A structural view on the mechanism of the ribosome-catalyzed peptide bond formation. Biochim Biophys Acta 1789: 612–623

Smith WP, Tai PC, Davis BD (1978) Interaction of secreted nascent chains with surrounding membrane in *Bacillus subtilis*. Proc Natl Acad Sci USA 75: 5922–5925

Taylor DJ, Devkota B, Huang AD, Topf M, Narayanan E, Sali A, Harvey SC, Frank J (2009) Comprehensive molecular structure of the eukaryotic ribosome. Structure 17: 1591–1604

Tenson T, Ehrenberg M (2002) Regulatory nascent peptides in the ribosomal tunnel. Cell 108: 591–594

Trabuco LG, Villa E, Mitra K, Frank J, Schulten K (2008) Flexible fitting of atomic structures into electron microscopy maps using molecular dynamics. Structure 16: 673–683

Tsalkova T, Odom OW, Kramer G, Hardesty B (1998) Different conformations of nascent peptides on ribosomes. J Mol Biol 278: 713–723

Tu LW, Deutsch C (2010) A folding zone in the ribosomal exit tunnel for Kv1.3 helix formation. J Mol Biol 396: 1346–1360

Vazquez-Laslop N, Thum C, Mankin AS (2008) Molecular mechanism of drug-dependent ribosome stalling. Mol Cell 30: 190–202

Voss NR, Gerstein M, Steitz TA, Moore PB (2006) The geometry of the ribosomal polypeptide exit tunnel. J Mol Biol 360: 893–906

Weixlbaumer A, Jin H, Neubauer C, Voorhees R, Petry S, Kelley A, Ramakrishnan V (2008) Insights into translational termination from the structure of RF2 bound to the ribosome. Science 322: 953–956

Wilson DN, Schluenzen F, Harms JM, Yoshida T, Ohkubo T, Albrecht R, Buerger J, Kobayashi Y, Fucini P (2005) X-ray crystallography study on ribosome recycling: the mechanism of binding and action of RRF on the 50S ribosomal subunit. EMBO J 24: 251–260

Woolhead CA, McCormick PJ, Johnson AE (2004) Nascent membrane and secretory proteins differ in FRET-detected folding far inside the ribosome and in their exposure to ribosomal proteins. Cell 116: 725–736

Yang R, Cruz-Vera LR, Yanofsky C (2009) 23S rRNA nucleotides in the peptidyl transferase center are essential for tryptophanase operon induction. J Bacteriol 191: 3445–3450

Yap MN, Bernstein HD (2009) The plasticity of a translation arrest motif yields insights into nascent polypeptide recognition inside the ribosome tunnel. Mol Cell 34: 201–211

Yonath A, Leonard KR, Wittmann HG (1987) A tunnel in the large ribosomal subunit revealed by three-dimensional image reconstruction. Science 236: 813–816

Youngman EM, Brunelle JL, Kochaniak AB, Green R (2004) The active site of the ribosome is composed of two layers of conserved nucleotides with distinct roles in peptide bond formation and peptide release. Cell 117: 589–599

Yusupov MM, Yusupova GZ, Baucom A, Lieberman K, Earnest TN, Cate JH, Noller HF (2001) Crystal structure of the ribosome at 5.5 A resolution. Science 292: 883–896

Mechanistic insight into co-translational protein processing, folding, targeting, and membrane insertion

32

Daniel Boehringer, Basil Greber, Nenad Ban

1. Introduction

In the cell newly synthesized polypeptides are subjected to enzymatic processing, chaperone-assisted folding, and targeting to translocation pores at membranes concurrently with their synthesis by the ribosome (Figure 1). The major players in these events are, (i) ribosome-associated chaperones, (ii) nascent-chain-processing enzymes, (iii) the signal recognition particle – a complex that recognizes ribosomes that are translating membrane and some secretory proteins and targets them to the membrane – and (iv) the membrane-protein-insertion machinery – a large multi-subunit trans-membrane complex responsible for protein insertion into or translocation across membranes. The ribosome plays a major role in governing the interplay between the various factors involved. Using electron microscopy, crystallography and biochemical approaches, we investigated the structural and mechanistic aspects of the interaction between these factors and the ribosome.

2. The ribosomal tunnel

Proteins being synthesized by the ribosome first have to diffuse through the ribosomal nascent polypeptide exit tunnel before reaching the cytoplasm. The exit tunnel is not a passive environment for the traversing nascent polypeptides. Interactions with the tunnel may promote the initiation of secondary structures in nascent chains, but can also lead to translational arrest or pausing.

The tunnel resembles an unbranched tube of approximately 100 Å that accommodates a peptide stretch of approximately 30 amino acids in extended conformation or, provided that the formation of sec-

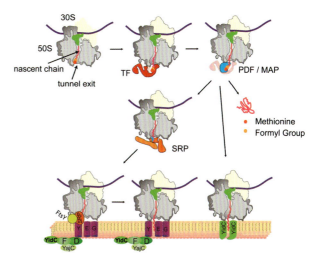

Fig. 1 Overview of protein folding and targeting in bacteria. Nascent chains first interact with the ribosome-associated chaperone trigger factor. The N-terminus of the nascent chain is processed by peptide deformylase and methionine aminopeptidase. Depending on their future function, proteins either fold inside the cytosol or are exported through or inserted into the inner membrane. Integral membrane proteins carrying a hydrophobic N-terminal signal anchor sequence are inserted into the inner membrane. This pathway strictly depends on the early co-translational recognition of the signal anchor sequence by SRP and subsequent targeting by the SRP receptor homologue FtsY to the translocon. Upon productive docking of the ribosome onto the translocon and release of SRP and FtsY, proteins are co-translationally inserted into the inner membrane. Alternatively, in bacteria, some nascent chains are inserted into the membrane by the YidC membrane protein insertase.

ondary structure is possible, up to 60 amino acids in an α-helical conformation (Picking et al., 1992; Tsalkova et al., 1998; Voss et al., 2006) (Figure 2A). In bacterial ribosomes, the tunnel wall is comprised predominantly of 23S ribosomal RNA (rRNA) and looped-out segments of the ribosomal proteins L4, L22 and L23 (Ban et al., 2000). The tunnel diameter varies in the range of 10–20 Å and includes a constriction located about

30 Å from the peptidyl transferase center formed by extended loops of L22 and L4. At its distal end, the ribosomal tunnel opens up in a funnel like shape (Ban et al., 2000; Voss et al., 2006). The rim of the tunnel exit is composed of rRNA, a ring of four ubiquitously conserved ribosomal proteins (L22, L23, L24, L29) and additional bacteria-specific proteins L32 and L17 (Nissen et al., 2000) (Figure 2B). These proteins, along with 23S rRNA, provide important interaction sites for the processing, folding and targeting factors.

Several studies suggest that ribosomes promote secondary structure formation of the nascent chain inside the ribosomal tunnel. Fluorescence resonance energy transfer (FRET) pairs placed at opposite ends of a nascent transmembrane segment were used to measure the degree of folding in the tunnel of eukaryotic ribosomes (Woolhead et al., 2004). Strikingly, inside the tunnel the distance between the two labels was as short as when the transmembrane domain was inserted into the membrane as an α-helix, suggesting that the ribosome was promoting helical secondary structure formation. Another study used pegylation of cysteines placed at defined positions in polypeptides to measure the relative compaction of arrested nascent chains (Lu and Deutsch, 2005). Secondary structure formation was observed within the tunnel, predominantly in regions close to the peptidyl transferase center and near the tunnel exit. Similar analyses suggest that co-translational acquisition of secondary structure, and even some tertiary structure, might occur at the tunnel exit where the tunnel widens towards the cytosol (Kosolapov and Deutsch, 2009). Consistent with these biochemical results, recent cryo-electron microscopy (cryo-EM) structures of stalled translating ribosomes showed evidence for secondary structure formation inside the ribosomal tunnel (Bhushan et al., 2010).

Nascent chains with certain sequences lead to translational pausing or arrest. Pausing might coordinate the binding of ribosome-associated factors or facilitate co-translational folding of nascent peptides (Marin, 2008). Another form of regulation involves short peptide stretches that cause transient translational arrest by sequence-specific interactions with the exit tunnel. One example is the 170 amino acid long secreted SecM peptide that regulates the translation of the downstream *secA* gene in response to the secretion status of the cell (Nakatogawa and Ito, 2001). During synthesis, the nascent SecM polypeptide interacts with the ribosomal exit tunnel to transiently arrest translation. Arresting the ribosome on the SecM message prevents the formation of a stable secondary

RNA structure in the region of the ribosome binding site and leads to the initiation of translation of SecA (Murakami et al., 2004). The SecM induced arrest of translation is based on interactions between 23S rRNA (bases A2058 and A749–753) and projections of L22 and L4 with parts of the C-terminal stalling sequence (F^{150}xxxxWIxxxxGIRAGP^{166}) in the nascent chain of

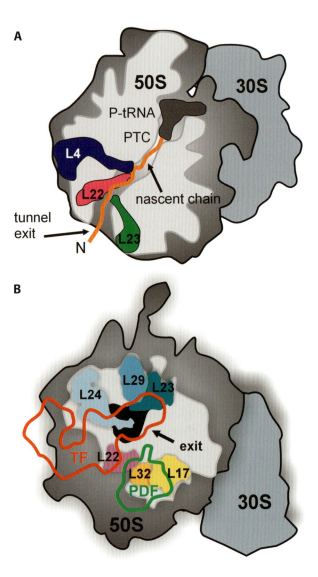

Fig. 2 The path of the nascent chain through the ribosome. (A) Schematic drawing of the ribosome (gray) sliced along the tunnel, showing the path of the nascent chain (orange) from the peptidyl transferase center (PTC) to the exit site. Proteins that interact with the nascent chain are color-coded (L4, blue; L22, magenta; L23, green). (B) The tunnel exit (black) of the large ribosomal subunit, showing the ribosomal proteins surrounding it and associated enzymes. The surface of the conserved proteins L23, L29, L24 (blue) and L22 (magenta) is indicated. Bacteria-specific proteins L17 and L32 are colored yellow. The projections of PDF (green) and the trigger factor (TF, red) on the bacterial ribosomal surface are shown as outlines.

SecM (Nakatogawa and Ito, 2002; Woolhead et al., 2006). In addition, the codon in the ribosomal A site is essential. Pro166-tRNA remains at the A site without forming a polypeptide bond with the P site-located peptidyl-tRNA^{Gly} (Muto et al., 2006). Another type of translational arrest occurs when the TnaC protein is synthesized in the regulation of the *Escherichia coli tna* operon of enzymes involved in tryptophan metabolism (Gong et al., 2001). Stalling on the *tnaC* transcript, which occurs in the presence of tryptophan, occludes the Rho dependent transcription terminator and activates the expression of the *tna* operon. Recent cryo-EM reconstructions of ribosomes stalled by the TnaC peptide sequence indicate that the conformation of the peptidyl transferase center is altered in order to interfere with the binding of release factor 2 and translation termination (Seidelt et al., 2009). Conformational changes in the active site may also cause translational arrest in the case of *SecM*. Interfering with translation termination allows the purification of stalled ribosomes *in vitro* and *in vivo*. Ribosomes stalled with the SecM stalling sequences, in particular, proved to be valuable for studying the interactions of various proteins with translating ribosomes in biochemical and structural studies (Schaffitzel and Ban, 2007).

The interactions between the nascent polypeptide and the ribosomal tunnel may even lead to conformational changes that are propagated towards the tunnel exit to signal the translational status of the ribosome. Recent results suggest that the presence of a nascent chain alters the interaction of the tunnel exit with targeting factors. In *E. coli*, nascent chains that are too short to reach the ribosomal surface still increase the affinity of SRP for ribosomes by ~100-fold (Bornemann et al., 2008). Signal transfer from the inside of the tunnel to the ribosomal surface occurs via a loop in L23 that reaches into the exit tunnel. This active recruitment of SRP to the ribosome is independent of a signal sequence in the nascent chain. Similar results were obtained with the eukaryotic SRP. In yeast, SRP also shows higher affinity for actively translating ribosomes harboring nascent chains buried in the exit tunnel than for vacant ribosomes (Berndt et al., 2009). Together these findings show that ribosomes can transmit information on the presence of a nascent chain from their interior to the surface to control the interaction with SRP.

3. Co-translational protein processing, folding and targeting factors

3.1. Trigger factor

The bacterial ribosome associated chaperone, trigger factor, can be cross-linked to nascent chains emerging from the ribosomal exit tunnel and, considering its abundance and affinity for the ribosome, is likely to be the first factor encountered by a newly synthesized protein (Figure 1) (Maier et al., 2005). Trigger factor is the only ribosome-associated chaperone in bacteria. In *E. coli*, trigger factor is a constitutively expressed protein, and under physiological conditions almost all ribosomes are likely to be in complex with a trigger factor molecule (Lill et al., 1988). Trigger factor was shown to prevent aggregation and promote refolding of substrate proteins *in vitro* (Huang et al., 2000; Maier et al., 2001). In the absence of nascent chains, trigger factor cycles on and off the ribosome with a mean residence time of 11–15 seconds (Kaiser et al., 2006; Maier et al., 2003). The presence of a nascent chain on the ribosome increases the affinity of trigger factor by 9- to 30-fold, depending on the length of the nascent chain and its folding status (Kaiser et al., 2006; Rutkowska et al., 2008). Some nascent chains remain associated with trigger factor even after their dissociation from the ribosome (Kaiser et al., 2006). Trigger factor is not essential, but the loss of trigger factor in cells lacking the DnaK chaperone, which has an overlapping set of substrate proteins, results in synthetic lethality at temperatures > 30 °C and causes misfolding and aggregation of several hundred different newly synthesized proteins (Deuerling et al., 1999; Teter et al., 1999). The chaperone function of trigger factor under these conditions was shown to depend on ribosome binding (Kramer et al., 2002).

The crystal structure of trigger factor provided important insights into its function (Ferbitz et al., 2004) (Figure 3A). Trigger factor folds into an elongated molecule with a dragon-like shape with protruding regions that can be considered as the tail, arms and head. The N-terminal ribosome binding domain harbors a conserved loop sequence involved in ribosome binding mainly through contacts with L23 close to the tunnel exit (Ferbitz et al., 2004; Kramer et al., 2002). This domain is connected to the peptidyl-prolyl isomerase (PPIase) domain located at the other end of the elongated molecule. The PPIase domain is dispensable for the general chaperone function of trigger factor (Kramer et al., 2004). The C-terminal domain lies between the head and the tail and forms a cradle with two protruding arms. This domain was shown to be important for the general chaperone function *in vitro* and *in vivo* (Merz et al., 2006; Zeng et al., 2006). In addition to the crystal structure of the full-length *E. coli* trigger factor, a fragment comprising its N-terminal ribosome-binding domain was co-crystallized in complex with the archaeal *Haloarcula marismortui* 50S ribosomal subunit (Ferbitz et al., 2004). By superimposing full-length trigger factor onto its ribosome-bound N-terminal domain, it was proposed that trigger factor forms a hydrophobic cradle directly under the exit of the nascent polypeptide tunnel of the 50S subunit. This cradle is open on both sides and appears to be large enough to accommodate small globular domains.

The cryo-electron microscopic reconstruction of *E. coli* trigger factor in complex with a translating ribosome (Figure 3B) revealed the conformation of the chaperone in its active state (Merz et al., 2008). For this study, *E. coli* ribosomes were stalled with the SecM stalling sequence fused to an α-spectrin SH3 domain. By crosslinking trigger factor to the nascent chain, a stable trigger factor-ribosome complex was obtained. The EM reconstruction of this complex shows that trigger factor forms an arch over the tunnel exit when it interacts with a nascent chain (Ferbitz, 2004) (Figure 3B, C). This confirmed the general binding mode of trigger factor that results in the formation of a protected folding space between the factor and the tunnel exit as suggested by crystallographic experiments (Ferbitz et al., 2004) (Figure 3C). Recently, the interaction of a substrate protein with free trigger factor has been investigated by determining the crystal structure of ribosomal protein S7 in complex with trigger factor (Martinez-Hackert and Hendrickson, 2009). The structure shows that protein S7 is bound between the C-terminal arms and the N-terminal domain similar as suggested for the binding of trigger factor to newly synthesized proteins on the ribosome.

The existence of a protected folding space for the nascent chain between trigger factor and the ribosome is supported by protease-protection experiments. The size of the substrates of trigger factor varies, and nascent chains of more than 100 amino acids have been shown to be protected from proteolytic degradation by trigger factor (Hoffmann et al., 2006; Tomic et al., 2006). Trigger factor also protects folded domains from protease degradation, consistent with the existence of a protected folding space (Hoffmann et al., 2006; Merz et al., 2006). However, using peptide libraries it was shown that trigger factor preferentially binds to hydro-

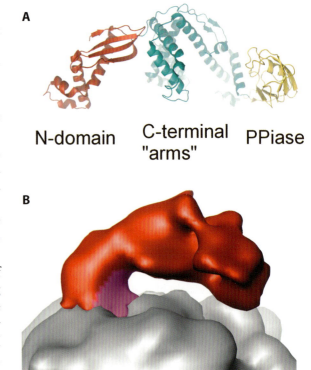

A N-domain   C-terminal "arms"   PPiase

B

C cradle

L23

Fig. 3 Structure of trigger factor on the ribosome. (A) Crystal structure of unbound *E. coli* trigger factor. The protein has an elongated shape with an N-terminal ribosome-binding domain (red), a peptidyl-prolyl cis/trans isomerase domain (yellow) and a C-terminal domain with two arms inserted in between (blue). (B) EM reconstruction of *E. coli* trigger factor bound to a ribosome-nascent chain complex (TF, red; 50S subunit, grey; nascent chain, purple). (C) Trigger factor (red) bound on the ribosome (rRNA grey, protein green) creates a protected folding space for the nascent chain. Surface representation of the crystal structure of trigger factor and the *E. coli* 50S ribosomal subunit (pdb: 2AW4) fitted into the EM density (sliced along the ribosomal tunnel).

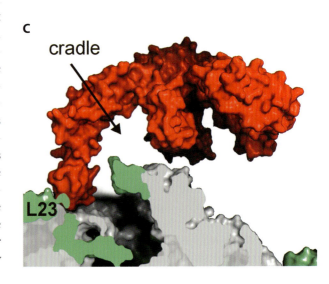

phobic peptide stretches (Patzelt et al., 2001). Furthermore, the interaction time of trigger factor is prolonged for more hydrophobic sequences (Kaiser et al., 2006).

The interactions of trigger factor with nascent chains of different lengths might be facilitated by the flexibility of the factor. Indeed, comparing the structures of free and substrate-bound trigger factor from *Thermotoga maritima* reveals a significant degree of flexibility, in that the N- and C-terminal domains are rotated by approximately 10° relative to each other, although the overall structural organization of trigger factor is unaffected (Hoffmann et al., 2006). The conformational flexibility of *E. coli* trigger factor upon substrate binding on the ribosome was also investigated using FRET (Kaiser et al., 2006). The change of the FRET intensities observed in this experiment indicates a conformational change that is in agreement with the conformational differences between the crystal structures of the free and substrate-bound *T. maritima* trigger factor (Martinez-Hackert and Hendrickson, 2009). Furthermore, the comparison of crystal structures of N-terminal fragments of trigger factor bound to 50S ribosomal subunits provides additional insights into its conformational flexibility. Two crystal structures of the N-terminal domain of trigger factor from *Deinococcus radiodurans* bound to large ribosomal subunits have been solved (Baram et al., 2005; Schlunzen et al., 2005). These structures show that, although trigger factor always contacts the ribosome by its conserved ribosome-binding loop, the remainder of the N-terminal domain can adopt a range of different orientations.

The interactions between the growing nascent chain and trigger factor were investigated using crosslinking (Lakshmipathy et al., 2007; Merz et al., 2008). These experiments indicate that the nascent chain is initially present in an extended conformation within the cradle formed by trigger factor (Lakshmipathy et al., 2007; Merz et al., 2008). Using a photoactivatable zero-length crosslinker at specific locations on trigger factor, it was possible to follow the interactions between the nascent polypeptide and trigger factor as a function of the chain length (Merz et al., 2008). These studies suggest that the largely unfolded nascent polypeptide extends along the hydrophobic cradle of trigger factor towards the PPIase domain before folding in the space between the two "arms" and the ribosome binding "tail" of the molecule.

The cellular role of trigger factor in protein folding is still a matter of ongoing research. It was shown that trigger factor prolongs the time for a nascent chain to

fold, leading to an increase in the yield of folded protein (Agashe et al., 2004). Some substrates are transferred from trigger factor to other chaperones that act in subsequent folding steps (Hartl and Hayer-Hartl, 2002). How the hand-over of nascent proteins from trigger factor to cytosolic chaperones such as DnaK is achieved is an interesting question in the field. Furthermore, trigger factor might also have a chaperone function even in its free state. The tight complex between free trigger factor and ribosomal protein S7 as observed in the *T. maritima* crystal structure suggests that trigger factor might be involved in ribosome biogenesis, stabilizing unstable proteins in the cytosol (Martinez-Hackert and Hendrickson, 2009). Furthermore, considering that trigger factor is likely the first protein to interact with nascent polypeptides on the ribosome raises questions regarding its role in their transfer to other factors and enzymes that co-translationally interact with newly synthesized proteins.

3.2. Methionine aminopeptidase and peptide deformylase

Bacteria, mitochondria, and plastids initiate translation by the binding of a specialized initiator tRNA charged with formylmethionine (Giglione et al., 2004). N-formylation is assumed to block the reactive amino group to prevent unfavourable side-reactions and to enhance the efficiency of translation initiation. The formyl group is removed by peptide deformylase (PDF) (Figure 1). The N-terminal methionine is subsequently removed by methionine aminopeptidase (MAP).

Bacterial PDF is an essential protein that co-translationally removes the formyl group of virtually all nascent proteins, a prerequisite for the subsequent action of MAP to cleave off the N-terminal methionine (Solbiati et al., 1999). Early work indicated that PDF co-purifies with *Bacillus subtilis* ribosomes (Takeda and Webster, 1968). Recent biochemical and crystallographic studies revealed that PDF from *E. coli* associates with the large subunit of ribosomes through a C-terminal helical extension that binds to a groove between ribosomal proteins L22 and L32 located next to the ribosomal exit tunnel (Bingel-Erlenmeyer et al., 2008) (Figure 4). *E. coli* PDF binding orients its active site optimally for an interaction with the emerging nascent polypeptide. Ribosome binding is functionally important *in vivo* since C-terminally truncated *E. coli* PDF that does not bind ribosomes reduces the

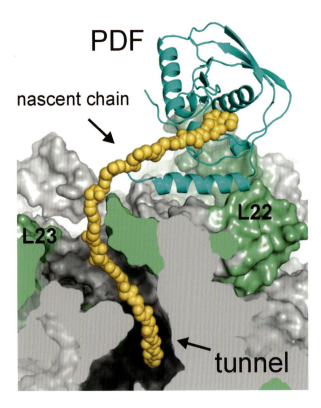

Fig. 4 Ribosome-associated peptide deformylase (PDF). PDF (cyan ribbon) is bound to ribosomal proteins L22 and L32 next to the ribosomal tunnel exit. The nascent chain interacting with PDF is modeled (yellow spheres).

viability and growth rate of cells under PDF-limiting conditions (Bingel-Erlenmeyer et al., 2008). In some Gram-positive bacteria a second type of PDF, PDF II, exists that differs in the C terminus and has not been characterized with respect to its possible interactions with the ribosome (Giglione et al., 2004).

MAPs catalyze the removal of N-terminal methionine from nascent peptide chains and are essential in all kingdoms of life (Giglione et al., 2004). Several lines of evidence suggest that some MAP family members interact with ribosomes. In *E. coli*, N-terminal methionine residues are co-translationally removed when the nascent chains reach a minimal length of just 40 amino acids (Ball and Kaesberg, 1973). Furthermore, yeast MAPs were identified as ribosome-associated proteins (Raue et al., 2007; Vetro and Chang, 2002). However, the ribosome-binding site of bacterial MAP has not been determined yet.

3.3. Signal recognition particle

A large proportion of proteins, 20–30% in *E. coli*, needs to be incorporated into membranes (Luirink et al., 2005). In bacteria, most inner membrane proteins (IMPs) are recognized co-translationally by the ubiquitous signal recognition particle (SRP) (Keenan et al., 2001) and targeted to SecYEG, which forms the protein-conducting channel in the plasma membrane (Rapoport, 2007). In contrast, most secreted proteins in bacteria are post-translationally translocated by SecYEG, mediated by the chaperone SecB, and the SecYEG-bound ATPase SecA. To enter either membrane insertion or translocation pathways, similar hydrophobic signal sequences at the N-terminus of the proteins are required.

The interaction of SRP with ribosome-nascent chain complexes (RNCs) displaying a signal sequence ensures that proteins are inserted into the membrane co-translationally (Doudna and Batey, 2004; Keenan et al., 2001) (Figure 1). The signal sequence in the nascent chain is bound by SRP as it emerges from the exit tunnel. Then, the RNC-SRP complex interacts with the membrane-bound SRP receptor in a GTP-controlled fashion and is targeted to the SecYEG translocon in the membrane (Angelini et al., 2005). The conformational rearrangements of SRP in this process coordinate the release of the RNC onto the translocon, transferring the nascent chain from SRP to the protein-translocating pore.

Bacterial SRP consists of the Ffh protein and 4.5S RNA. The Ffh protein comprises three domains (Keenan et al., 2001). The GTPase domain (G domain) and the ribosome binding N domain are connected via a long linker to the M domain, which tightly associates with the SRP RNA and binds the signal sequence. The crystal structure of a complex between the M domain of the Ffh protein and a fragment of the 4.5S RNA revealed the details of their interaction (Batey et al., 2000). The M domain uses a hydrophobic cleft on its surface to bind the hydrophic signal sequence in alpha-helical conformation (Janda et al., 2010). The binding is not specific for particular sequences but depends on the hydrophobicity and secondary structure-forming propensity (Adams et al., 2002; Lee and Bernstein, 2001).

SRP binds to ribosomes via its N domain even in the absence of the signal sequence (Gu et al., 2003) (Figure 5a). Specifically, the contact point on the ribosome is at protein L23 as demonstrated by crosslinking experiments (Gu et al., 2003). However, SRP appears to be bound flexibly or in different conformations, as cryo-EM studies show only partial density for SRP bound to ribosomes at protein L23 (Schaffitzel et al., 2006) (Figure 5). In this initial binding stage, SRP samples the nascent chain for the presence of a signal sequence. Binding of the signal sequence by the M domain results in the stabilization of SRP in one conformation (Schaffitzel et al., 2006). A more flexible mode of SRP binding to vacant ribosomes, as compared to its binding to translating ribosomes with a signal sequence exposed, is also indicated by the observation that 4.5S RNA was crosslinked to 23S rRNA only when a signal sequence was present (Rinke-Appel et al., 2002).

The complex between an *E. coli* RNC with a signal sequence and SRP was visualized by electron microscopy (Halic et al., 2006; Schaffitzel et al., 2006) showing SRP stabilized in one particular conformation. The structures show the M domain bound to the signal sequence at the tunnel exit. The SRP RNA bound by the M domain stretches across the tunnel, contacting the ribosome near protein L32 and L22. The NG domains of Ffh contact the ribosome at protein L23. In this conformation, the NG domain is not in contact with the 4.5S RNA, consistent with FRET measurements (Buskiewicz et al., 2009).

The bacterial SRP receptor, FtsY, contains N and G domains that are highly homologous to the NG domains of Ffh (Montoya et al., 1997). Both G domains bind GTP, and GTPase activation upon Ffh-FtsY complex formation leads to nascent chain release onto the translocon and complex disassembly. Crystal structures of the GTPase-activated complex show an extensive network of interactions between FtsY and Ffh (Egea et al., 2004; Focia et al., 2004). However, the formation of this complex is inefficient in solution (Zhang et al., 2008). The presence of an RNC displaying a signal sequence accelerates SRP-FtsY complex formation 100–400 fold (Zhang et al., 2009). This acceleration depends on the presence of 4.5S RNA (Zhang et al., 2008). Mutational analysis revealed that the 4.5S RNA tetraloop interacts with FtsY (Jagath et al., 2001) and specific interactions with the FtsY G domain are required for efficient complex formation (Shen and Shan, 2010). Correspondingly, mutating a single lysine residue, Lys399, to alanine in the FtsY G domain abrogates the stabilizing effect of RNCs on Ffh-FtsY complex formation almost completely (Shen and Shan, 2010). Therefore, the positioning of the 4.5S RNA in the SRP-RNC complex likely triggers the binding of FtsY and subsequent conformational rearrangements.

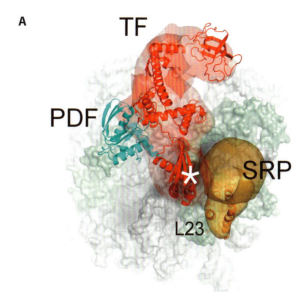

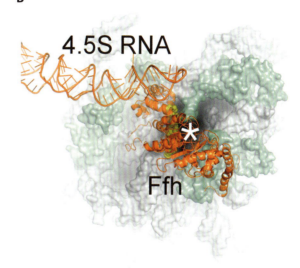

Fig. 5 Spatial arrangement of nascent chain–processing factors around the tunnel exit (star). (A) For short nascent chains (yellow spheres), trigger factor (TF, red ribbon, pdb 2VRH) might function as a passive router, facilitating nascent-chain processing by peptidyl deformylase (PDF, blue ribbon, pdb 2VHM) and sampling for signal sequences by the signal recognition particle (SRP). The flexibly bound SRP is represented by the EM density observed for SRP (orange surface) bound to the non-translating ribosome in a partially disordered state. The N domain of SRP as observed in the cryo-EM reconstruction (pdb 2J28) is modeled (orange ribbon). (B) Longer nascent chains displaying the signal sequence for protein translocation are bound by SRP (orange ribbon, pdb 2J28), which stably associates with the ribosome. In this conformation, SRP would overlap with the trigger factor bound to a translating ribosome, as observed by EM (EM density shown in transparent red).

3.4. The interplay of the factors during co-translational folding, processing and targeting

Considering the cellular concentrations of molecules that interact with nascent peptide chains, trigger factor is likely to be present at the tunnel exit of all ribosomes (Buskiewicz et al., 2004; Lill et al., 1988). Therefore, the binding of other processing and targeting factors has to be considered in this context. Modeling PDF bound to the ribosome (Bingel-Erlenmeyer et al., 2008) onto the structure of trigger factor bound to a translating ribosome (Merz et al., 2008) shows that the two factors can bind simultaneously (Figure 5A). PDF is present in sub-stoichiometric amounts relative to the ribosome, so it might use its rapid binding and dissociation kinetics to sample ribosomes for the emerging N-termini of nascent chains (Bingel-Erlenmeyer et al., 2008). In contrast, trigger factor dissociates slowly (Kaiser et al., 2006). In the model of simultaneous trigger factor and PDF binding, PDF can access the nascent chain through the lateral openings of the trigger factor cradle (Bingel-Erlenmeyer et al., 2008). Similarly, while trigger factor is bound to the ribosome, MAP might also bind and access the nascent chain for the subsequent step of nascent chain processing.

Trigger factor and SRP may simultaneously bind to the ribosome and compete for interactions with nascent chains, as demonstrated using crosslinking experiments (Buskiewicz et al., 2004; Raine et al., 2004). Although both factors share L23 as a docking site, their interaction sites on protein L23 appear to be non-overlappping (Figure 5A, B). The binding site of trigger factor is located on the side of L23 facing the tunnel exit (Ferbitz et al., 2004), whereas Ffh of SRP appears to be bound to another patch of L23, located further away from the tunnel exit (Halic et al., 2006) (Figure 5A). Thus, during sampling for the signal sequence, trigger factor and SRP might simultaneously bind at the tunnel exit. When the signal sequence is recognized by the M domain of SRP, SRP latches onto the ribosomal surface (Schaffitzel et al., 2006). Comparing SRP in this signal sequence-bound conformation with the structure of trigger factor bound to a translating ribosome reveals a clash of 4.5S RNA and trigger factor (Halic et al., 2006; Merz et al., 2006; Schaffitzel et al., 2006) (Figure 5B). This indicates that SRP will probably displace trigger factor when bound to a signal sequence of nascent chains destined for co-translational membrane insertion.

Trigger factor might aid the discrimination between membrane proteins, i.e. SRP substrates, and post-translationally secreted proteins (Luirink and Sinning, 2004). It was observed that the extent to which trigger factor competes with SRP for nascent-chain binding depends on the hydrophobicity of the nascent chain (Eisner et al., 2006; Ullers et al., 2006). SRP has a higher affinity for more hydrophobic signal sequences of co-translationally inserted proteins, compared to the less hydrophobic sequences of post-translationally secreted proteins (Lee and Bernstein, 2001; Valent et al., 1997). Consequently, in trigger factor-deletion strains, an increased number of ribosomes is found at the membrane (Lee and Bernstein, 2002; Ullers et al., 2007). Based on crosslinking data, it was suggested that the function of trigger factor is to shield less hydrophobic nascent chains of post-translationally secreted proteins from interaction with SRP (Beck et al., 2000). Thereby, trigger factor might increase the specificity of SRP for its substrates. The interplay of trigger factor and SRP probably facilitates the sorting of co-translationally inserted nascent chains into the SRP pathway and post-translationally inserted proteins into the SecB pathway.

As discussed above, the affinity of SRP for RNCs increases in response to the presence of nascent chains in the ribosomal tunnel. This signaling involves L23 and might also influence binding of trigger factor to RNCs. Therefore the binding surface on the ribosome might change in response to the translational status of the ribosome. It is likely that the affinity and the binding kinetics of all the involved factors are fine-tuned to regulate their exchange during co-translational nascent chain folding, processing and targeting. Detailed kinetics of the interaction of the nascent chain-interacting factors with nascent chains of different length and hydrophobicity will give more insight into these processes.

4. Membrane protein insertion

The interaction of SRP with its receptor transfers the nascent chain of transmembrane proteins onto the membrane-insertion machinery. The co-translational mode of membrane protein insertion ensures that the membrane proteins will be inserted with correct topology and prevents aggregation of the hydrophobic transmembrane segments. A dramatic conformational rearrangement of SRP is required to allow the insertion factors to access the tunnel exit. This process requires GTP hydrolysis on SRP and its receptor, followed by the handover of the nascent chain to the insertion factors (Xie and Dalbey, 2008). Two distinct pathways are responsible for co-translational protein insertion in bacteria (Figure 1). The SecYEG and YidC protein insertion pores insert the nascent membrane proteins while maintaining the membrane permeability barrier. Furthermore, the nascent transmembrane segments have to be released into the lipid bilayer, requiring a lateral opening of the insertion machinery.

4.1. Architecture of the SecYEG protein insertion pore

The Sec pathway, responsible for the bulk of protein insertion, depends on the Sec-translocon. The SecYEG complex forms the core of this insertion machinery, and a number of additional factors, among them SecD, SecF and YajC (Duong and Wickner, 1997) and YidC (Scotti et al., 2000), are part of the Sec holocomplex and participate in protein insertion and protein folding. In conjunction with the motor protein SecA, the SecYEG translocon can also act post-translationally in protein secretion (Natale et al., 2008). In this pathway, YidC may function to release its substrate nascent chains from SecYEG or as a chaperone to assist in membrane protein folding, complex assembly, and quality control (Beck et al., 2001; Kiefer and Kuhn, 2007; Nagamori et al., 2004; van Bloois et al., 2008). YidC interacts with transmembrane helices of membrane proteins as they are released from the SecYEG translocon (Urbanus et al., 2001; van der Laan et al., 2001). The interaction of YidC with the SecYEG translocon is thought to be mediated by the accessory complex SecDFYajC (Nouwen and Driessen, 2002). In the membrane, YidC is much more abundant than SecYEG (Driessen, 1994; Urbanus et al., 2002). Therefore, it is probably present both in a complex with SecYEG-SecD/F/YajC and in unbound form.

The cryo-EM structure of SecYEG bound to the translating *E. coli* ribosome (Mitra et al., 2005) shows SecYEG bound to the large ribosomal subunit above the polypeptide exit tunnel (Figure 6A). SecYEG is connected to the ribosome at three major contact sites. Two of those contact sites are mediated by contacts to 23S rRNA helices 59 and 24, respectively. Ribosomal protein L24 may contribute to the latter contact site. The third major contact point is mediated by ribosomal proteins L23 and L29 (Mitra et al., 2005). The ribosomal contacts in the SecYEG-ribosome-nascent chain complex are mostly in agreement with a more recent structure of the SecY complex bound to a non-

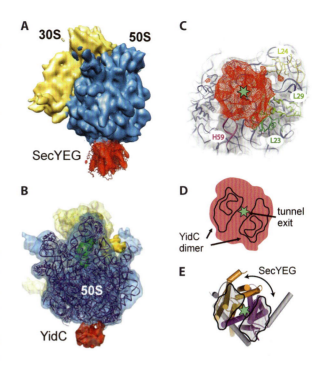

Fig. 6 Protein insertion by SecYEG and YidC. (A) Cryo-EM density of the *E. coli* RNC-SecYEG complex. SecYEG is bound at the tunnel exit of the RNC. The small subunit is shown in yellow, the large subunit in blue, and the SecYEG density in red. (B) Cryo-EM density of the *E. coli* RNC-YidC complex. YidC is bound at the tunnel exit of the RNC. The small subunit is shown in yellow, the large subunit in blue, with a crystal structure of the *E. coli* large ribosomal subunit fitted into the density, and the YidC density in red. (C) YidC bound at the tunnel exit of the large ribosomal subunit (gray mesh). The binding sites of YidC are highlighted: ribosomal proteins L23, L29, and L24, as well as helix 59 of the 23S rRNA. (D) Superposition of the YidC projection structure (Lotz et al., 2008) onto an outline of the YidC density as shown in panel C. (E) YidC and SecYEG might share a common overall architecture. Superposition of the YidC dimer projection structure (Lotz et al., 2008) with the SecYEβ translocon crystal structure (van den Berg et al., 2004). The claw-like opening movement of SecYEβ is indicated by an arrow.

translating ribosome (Menetret et al., 2007). These interaction sites overlap with the contacts of SRP on the ribosome. Consequently, SRP has to undergo a dramatic conformational remodeling to enable the access of the translocon to the tunnel exit during nascent chain handoff from the SRP M domain to the protein insertion pore. This process has to be tightly regulated to maintain the association of the ribosome with the membrane and efficiently transfer the nascent chain.

The oligomeric state of the active ribosome-bound Sec translocon in eukaryotic and bacterial cells has not yet been fully resolved. Early cryo-EM studies indicated oligomers of the Sec complex bound to non-translating (Beckmann et al., 1997; Menetret et al., 2005; Morgan et al., 2002) and translating (Beckmann et al., 2001;

Mitra et al., 2005) ribosomes. The most recent cryo-EM structures of translating eukaryotic ribosomes show a single Sec61 monomer embedded in a detergent micelle bound to the ribosome (Becker et al., 2009). Cryo-EM and mass spectrometry experiments also indicate that a monomer of SecYEG is bound to the non-translating bacterial ribosome (Menetret et al., 2007). These recent results suggested that, as indicated by biochemical studies (Osborne and Rapoport, 2007; Zimmer et al., 2008), a single copy of SecY forms the protein insertion pore. However, in the native membrane, SecYEG might exist in higher oligomeric states associated with the accessory factors SecDF, YajC and YidC (Bessonneau et al., 2002; Gold et al., 2010; Osborne and Rapoport, 2007; Zimmer et al., 2008). The *in-vitro* systems studied in the structural experiments using detergents will only show tightly bound SecYEG molecules and will not capture more transient interactions between components of the translocase machinery.

Detailed insight into how the Sec system inserts proteins was provided by the high-resolution crystal structure of the SecYEβ translocon (Van den Berg et al., 2004). The structure suggested that a single SecYEβ heterotrimer forms the translocation pore. The hourglass-shaped and mostly hydrophilic channel is enclosed by the SecY subunit and can be sealed by a short plug helix from the outside. The channel walls are mainly formed by four transmembrane helices in the center of SecY. The plug helix is displaced upon binding of a protein substrate to the translocon (Van den Berg et al., 2004; Zimmer et al., 2008). The SecY subunit is divided into two inverted pseudo-symmetric halves comprising transmembrane helices (TMHs) 1–5 and 6–10, respectively. This suggests that the release of the polypeptide into the lipid bilayer occurs through a lateral gate formed by a claw-like opening of the two SecY halves (Van den Berg et al., 2004). The lateral gate is flanked by TMHs 2b and 7, which had been implicated in recognition of the signal sequence in post-translational protein export previously (Plath et al., 1998; Van den Berg et al., 2004). Therefore, signal sequence recognition probably involves intercalation of the signal sequence into the lateral gate. More recent structures of bacterial protein-conducting channels in complex with SecA (Zimmer et al., 2008) or an antibody fragment (Tsukazaki et al., 2008) as well as biochemical experiments (du Plessis et al., 2009) provided additional evidence for some of the functional features of the Sec-translocon mentioned above, in particular regarding lateral gate opening and movements of the plug helix. In addition, these studies provided compel-

ling evidence that, indeed, only one channel is active in protein translocation (Zimmer et al., 2008), although a Sec translocon dimer may be present in proteoliposomes, and consequently, in vivo (Osborne and Rapoport, 2007).

4.2. Insertion of membrane proteins by YidC

The second co-translational pathway for membrane protein insertion is Sec-independent and depends critically on the YidC protein. It is therefore termed YidC-only pathway (van Bloois et al., 2005). The Sec-independent function of YidC is essential for cell viability (van Bloois et al., 2005). Substrates such as the c-subunit of the F_0F_1-ATPase (F_0c) (van der Laan et al., 2004; Yi et al., 2004; Yi et al., 2003), M13 procoat protein (Samuelson et al., 2000) and the Pf3 phage coat protein (Chen et al., 2002; Serek et al., 2004) can be inserted into proteoliposomes by purified YidC in a catalytic and cotranslational fashion with correct topology. Interestingly, F_0c mutants that are unable to oligomerize still depend on YidC for membrane insertion, indicating that YidC acts on these substrates at the insertion step, and not merely during complex assembly (Kol et al., 2006). YidC biology has been reviewed in some detail recently (Kol et al., 2008). YidC belongs to the Oxa1/YidC/Alb3 protein family that is characterized structurally by a conserved core of five transmembrane helices and functionally by the participation of its members in membrane protein biogenesis in bacteria, mitochondria and chloroplasts. In YidC, this conserved core is N-terminally extended by a periplasmic domain, which is anchored by an additional transmembrane helix at its N terminus (Yi and Dalbey, 2005).

The cryo-EM structure of YidC bound to the translating *E. coli* ribosome-nascent chain complex provided first insight into the mechanism of co-translational protein insertion by this insertase system (Kohler et al., 2009) (Figure 6A, B). The structure shows a YidC dimer bound above the polypeptide tunnel exit on the large ribosomal subunit (Figure 6C), most likely representing the form of YidC active in protein insertion in the YidC-only pathway. The EM density of the ribosome-bound YidC corresponds in size to a dimer of YidC as observed in a 10 Å projection map of YidC two-dimensional crystals (Lotz et al., 2008) (Figure 6D). The contact sites of YidC on the ribosome have been mapped to helix 59 of the 23S rRNA, ribosomal proteins L23 and L29, and ribosomal protein L24 and helix 24 of the 23S rRNA (Figure 6B). Strikingly, these contact points are shared with the non-homologous SecYEG translocon (Menetret et al., 2007; Mitra et al., 2005). Currently, high-resolution structural information on YidC is limited to structures of the periplasmic domain (Oliver and Paetzel, 2008; Ravaud et al., 2008), which do not provide insight into the protein insertion mechanism. However, the similar mode of ribosome binding of SecYEG and YidC indicates that both insertases are also similar on a functional level. Comparing the projection structure of the YidC dimer (Lotz et al., 2008) with the crystal structure of the monomeric archaeal SecYEβ complex (Van den Berg et al., 2004) reveals that the two have remarkably similar dimensions (Figure 6E), each consisting of 12 TMHs. Furthermore, four transmembrane helices that are important for protein insertion were identified by crosslinking to be located in the center of the YidC dimer, and consequently could form a protein insertion pore (Kohler et al., 2009). It was proposed that the two YidC molecules in the dimer could function in a similar way as the two domains of SecY, as mentioned above, with a channel in the center that opens by a separation of the component halves (Kohler et al., 2009). In the Sec-independent pathway, YidC could directly bind to nascent-chain substrates and act analogously to SecYEG. This suggests that a common molecular architecture for protein insertion pores might exist.

References

Adams H, Scotti PA, De Cock H, Luirink J, Tommassen, J (2002) The presence of a helix breaker in the hydrophobic core of signal sequences of secretory proteins prevents recognition by the signal-recognition particle in *Escherichia coli*. Eur J Biochem 269: 5564–5571

Agashe VR, Guha S, Chang HC, Genevaux P, Hayer-Hartl M, Stemp M, Georgopoulos C, Hartl FU, Barral JM (2004) Function of trigger factor and DnaK in multidomain protein folding: increase in yield at the expense of folding speed. Cell 117: 199–209

Angelini S, Deitermann S, Koch HG (2005) FtsY, the bacterial signal-recognition particle receptor, interacts functionally and physically with the SecYEG translocon. EMBO Rep 6: 476–481

Ball LA, Kaesberg P (1973) Cleavage of the N-terminal formylmethionine residue from a bacteriophage coat protein in vitro. J Mol Biol 79: 531–537

Ban N, Nissen P, Hansen J, Moore PB, Steitz TA (2000) The complete atomic structure of the large ribosomal subunit at 2.4 A resolution. Science 289: 905–920

Baram D, Pyetan E, Sittner A, Auerbach-Nevo T, Bashan A, Yonath, A (2005) Structure of trigger factor binding domain in biologically homologous complex with eubacterial ribo-

some reveals its chaperone action. Proc Natl Acad Sci USA 102: 12 017–12 022

Batey RT, Rambo RP, Lucast L, Rha B, Doudna JA (2000) Crystal structure of the ribonucleoprotein core of the signal recognition particle. Science 287: 1232–1239

Beck K, Eisner G, Trescher D, Dalbey RE, Brunner J, Muller, M (2001) YidC, an assembly site for polytopic *Escherichia coli* membrane proteins located in immediate proximity to the SecYE translocon and lipids. EMBO Rep 2: 709–714

Beck K, Wu LF, Brunner J, Muller, M (2000) Discrimination between SRP- and SecA/SecB-dependent substrates involves selective recognition of nascent chains by SRP and trigger factor. EMBO J 19: 134–143

Becker T, Bhushan S, Jarasch A, Armache JP, Funes S, Jossinet F, Gumbart J, Mielke T, Berninghausen O, Schulten K, et al. (2009) Structure of monomeric yeast and mammalian Sec61 complexes interacting with the translating ribosome. Science 326: 1369–1373

Beckmann R, Bubeck D, Grassucci R, Penczek P, Verschoor A, Blobel G, Frank, J (1997) Alignment of conduits for the nascent polypeptide chain in the ribosome-Sec61 complex. Science 278: 2123–2126

Beckmann R, Spahn CM, Eswar N, Helmers J, Penczek PA, Sali A, Frank J, Blobel, G (2001) Architecture of the protein-conducting channel associated with the translating 80S ribosome. Cell 107: 361–372

Berndt U, Oellerer S, Zhang Y, Johnson AE, Rospert S (2009) A signal-anchor sequence stimulates signal recognition particle binding to ribosomes from inside the exit tunnel. Proc Natl Acad Sci USA 106: 1398–1403

Bessonneau P, Besson V, Collinson I, Duong, F (2002) The SecYEG preprotein translocation channel is a conformationally dynamic and dimeric structure. EMBO J 21: 995–1003

Bhushan S, Gartmann M, Halic M, Armache JP, Jarasch A, Mielke T, Berninghausen O, Wilson DN, Beckmann, R (2010) alpha-Helical nascent polypeptide chains visualized within distinct regions of the ribosomal exit tunnel. Nat Struct Mol Biol 17: 313–317

Bingel-Erlenmeyer R, Kohler R, Kramer G, Sandikci A, Antolic S, Maier T, Schaffitzel C, Wiedmann B, Bukau B, Ban, N (2008) A peptide deformylase-ribosome complex reveals mechanism of nascent chain processing. Nature 452: 108–111

Bornemann T, Jockel J, Rodnina MV, Wintermeyer, W (2008) Signal sequence-independent membrane targeting of ribosomes containing short nascent peptides within the exit tunnel. Nat Struct Mol Biol 15: 494–499

Buskiewicz I, Deuerling E, Gu SQ, Jockel J, Rodnina MV, Bukau B, Wintermeyer, W (2004) Trigger factor binds to ribosome-signal-recognition particle (SRP) complexes and is excluded by binding of the SRP receptor. Proc Natl Acad Sci USA 101: 7902–7906

Buskiewicz IA, Jockel J, Rodnina MV, Wintermeyer, W (2009) Conformation of the signal recognition particle in ribosomal targeting complexes. RNA 15: 44–54

Chen M, Samuelson JC, Jiang F, Muller M, Kuhn A, Dalbey RE (2002) Direct interaction of YidC with the Sec-independent Pf3 coat protein during its membrane protein insertion. J Biol Chem 277: 7670–7675

Deuerling E, Schulze-Specking A, Tomoyasu T, Mogk A, Bukau, B (1999) Trigger factor and DnaK cooperate in folding of newly synthesized proteins. Nature 400: 693–696

Doudna JA, Batey RT (2004) Structural insights into the signal recognition particle. Annu Rev Biochem 73: 539–557

Driessen AJ (1994) How proteins cross the bacterial cytoplasmic membrane. J Membr Biol 142: 145–159 du Plessis DJ, Berrelkamp G, Nouwen N, Driessen AJ (2009) The lateral gate of SecYEG opens during protein translocation. J Biol Chem 284: 15 805–15 814

Duong F, Wickner, W (1997) Distinct catalytic roles of the SecYE, SecG and SecDFyajC subunits of preprotein translocase holoenzyme. EMBO J 16: 2756–2768

Egea PF, Shan SO, Napetschnig J, Savage DF, Walter P, Stroud RM (2004) Substrate twinning activates the signal recognition particle and its receptor. Nature 427: 215–221

Eisner G, Moser M, Schafer U, Beck K, Muller, M (2006) Alternate recruitment of signal recognition particle and trigger factor to the signal sequence of a growing nascent polypeptide. J Biol Chem 281: 7172–7179

Ferbitz L, Maier T, Patzelt H, Bukau B, Deuerling E, Ban, N (2004) Trigger factor in complex with the ribosome forms a molecular cradle for nascent proteins. Nature 431: 590–596

Focia PJ, Shepotinovskaya IV, Seidler JA, Freymann DM (2004) Heterodimeric GTPase core of the SRP targeting complex. Science 303: 373–377

Giglione C, Boularot A, Meinnel, T (2004) Protein N-terminal methionine excision. Cell Mol Life Sci 61: 1455–1474

Gold VA, Robson A, Bao H, Romantsov T, Duong F, Collinson, I (2010) The action of cardiolipin on the bacterial translocon. Proc Natl Acad Sci USA 107: 10 044–10 049

Gong F, Ito K, Nakamura Y, Yanofsky, C (2001) The mechanism of tryptophan induction of tryptophanase operon expression: tryptophan inhibits release factor-mediated cleavage of TnaC-peptidyl-tRNA(Pro) Proc Natl Acad Sci USA 98: 8997–9001

Gu SQ, Peske F, Wieden HJ, Rodnina MV, Wintermeyer, W (2003) The signal recognition particle binds to protein L23 at the peptide exit of the *Escherichia coli* ribosome. RNA 9: 566–573

Halic M, Blau M, Becker T, Mielke T, Pool MR, Wild K, Sinning I, Beckmann, R (2006) Following the signal sequence from ribosomal tunnel exit to signal recognition particle. Nature 444: 507–511

Hartl FU, Hayer-Hartl, M (2002) Molecular chaperones in the cytosol: from nascent chain to folded protein. Science 295: 1852–1858

Hoffmann A, Merz F, Rutkowska A, Zachmann-Brand B, Deuerling E, Bukau, B (2006) Trigger factor forms a protective shield for nascent polypeptides at the ribosome. J Biol Chem 281: 6539–6545

Huang GC, Li ZY, Zhou JM, Fischer, G (2000) Assisted folding of D-glyceraldehyde-3-phosphate dehydrogenase by trigger factor. Protein Sci 9: 1254–1261

Jagath JR, Matassova NB, de Leeuw E, Warnecke JM, Lentzen G, Rodnina MV, Luirink J, Wintermeyer, W (2001) Important role of the tetraloop region of 4.5S RNA in SRP binding to its receptor FtsY. RNA 7: 293–301

Janda CY, Li J, Oubridge C, Hernandez H, Robinson CV, Nagai, K (2010) Recognition of a signal peptide by the signal recognition particle. Nature 465: 507–510

Kaiser CM, Chang HC, Agashe VR, Lakshmipathy SK, Etchells SA, Hayer-Hartl M, Hartl FU, Barral JM (2006) Real-time observation of trigger factor function on translating ribosomes. Nature 444: 455–460

Keenan RJ, Freymann DM, Stroud RM, Walter, P (2001) The signal recognition particle. Annu Rev Biochem 70: 755–775

Kiefer D, Kuhn, A (2007) YidC as an essential and multifunctional component in membrane protein assembly. Int Rev Cytol 259: 113–138

Kohler R, Boehringer D, Greber B, Bingel-Erlenmeyer R, Collinson I, Schaffitzel C, Ban, N (2009) YidC and Oxa1form dimeric insertion pores on the translating ribosome. Mol Cell 34: 344–353

Kol S, Nouwen N, Driessen AJ (2008) Mechanisms of YidC-mediated insertion and assembly of multimeric membrane protein complexes. J Biol Chem 283: 31 269–31 273

Kol S, Turrell BR, de Keyzer J, van der Laan M, Nouwen N, Driessen AJ (2006) YidC-mediated membrane insertion of assembly mutants of subunit c of the F1F0 ATPase. J Biol Chem 281: 29 762–29 768

Kosolapov A, Deutsch, C (2009) Tertiary interactions within the ribosomal exit tunnel. Nat Struct Mol Biol 16: 405–411

Kramer G, Rauch T, Rist W, Vorderwulbecke S, Patzelt H, Schulze-Specking A, Ban N, Deuerling E, Bukan, B, (2002) L23 protein functions as a chaperone docking site on the ribosome. Nature 419: 171–174

Kramer G, Ramachandiran V, Horowitz PM, Hardesty, B (2002) The molecular chaperone DnaK is not recruited to translating ribosomes that lack trigger factor. Arch Biochem Biophys 403: 63–70

Lakshmipathy SK, Tomic S, Kaiser CM, Chang HC, Genevaux P, Georgopoulos C, Barral JM, Johnson AE, Hartl FU, Etchells SA (2007) Identification of nascent chain interaction sites on trigger factor. J Biol Chem 282: 12 186–12 193

Lee HC, Bernstein HD (2001) The targeting pathway of *Escherichia coli* presecretory and integral membrane proteins is specified by the hydrophobicity of the targeting signal. Proc Natl Acad Sci USA 98: 3471–3476

Lee HC, Bernstein HD (2002) Trigger factor retards protein export in *Escherichia coli*. J Biol Chem 277: 43 527–43 535

Lill R, Crooke E, Guthrie B, Wickner, W (1988) The "trigger factor cycle" includes ribosomes, presecretory proteins, and the plasma membrane. Cell 54: 1013–1018

Lotz M, Haase W, Kuhlbrandt W, Collinson, I (2008) Projection structure of yidC: a conserved mediator of membrane protein assembly. J Mol Biol 375: 901–907

Lu J, Deutsch, C (2005) Folding zones inside the ribosomal exit tunnel. Nat Struct Mol Biol 12: 1123–1129

Luirink J, Sinning, I (2004) SRP-mediated protein targeting: structure and function revisited. Biochim Biophys Acta 1694: 17–35

Luirink J, von Heijne G, Houben E, and de Gier JW (2005) Biogenesis of inner membrane proteins in *Escherichia coli*. Annu Rev Microbiol 59: 329–355

Maier R, Eckert B, Scholz C, Lilie H, Schmid FX (2003) Interaction of trigger factor with the ribosome. J Mol Biol 326: 585–592

Maier R, Scholz C, Schmid FX (2001) Dynamic association of trigger factor with protein substrates. J Mol Biol 314: 1181–1190

Maier T, Ferbitz L, Deuerling E, Ban, N (2005) A cradle for new proteins: trigger factor at the ribosome. Curr Opin Struct Biol 15: 204–212

Marin, M (2008) Folding at the rhythm of the rare codon beat. Biotechnol J 3: 1047–1057

Martinez-Hackert E, Hendrickson WA (2009) Promiscuous substrate recognition in folding and assembly activities of the trigger factor chaperone. Cell 138: 923–934

Menetret JF, Hegde RS, Heinrich SU, Chandramouli P, Ludtke SJ, Rapoport TA, Akey CW (2005) Architecture of the ribosome-channel complex derived from native membranes. J Mol Biol 348: 445–457

Menetret JF, Schaletzky J, Clemons WM, Jr., Osborne AR, Skanland SS, Denison C, Gygi SP, Kirkpatrick DS, Park E, Ludtke SJ, et al. (2007) Ribosome binding of a single copy of the SecY complex: implications for protein translocation. Mol Cell 28: 1083–1092

Merz F, Boehringer D, Schaffitzel C, Preissler S, Hoffmann A, Maier T, Rutkowska A, Lozza J, Ban N, Bukau B, et al. (2008) Molecular mechanism and structure of trigger factor bound to the translating ribosome. EMBO J 27: 1622–1632

Merz F, Hoffmann A, Rutkowska A, Zachmann-Brand B, Bukau B, Deuerling, E (2006) The C-terminal domain of *Escherichia coli* trigger factor represents the central module of its chaperone activity. J Biol Chem 281: 31 963–31 971

Mitra K, Schaffitzel C, Shaikh T, Tama F, Jenni S, Brooks CL, 3rd, Ban N, Frank, J (2005) Structure of the *E. coli* protein-conducting channel bound to a translating ribosome. Nature 438: 318–324

Montoya G, Svensson C, Luirink J, Sinning, I (1997) Crystal structure of the NG domain from the signal-recognition particle receptor FtsY. Nature 385: 365–368

Morgan DG, Menetret JF, Neuhof A, Rapoport TA, Akey CW (2002) Structure of the mammalian ribosome-channel complex at 17A resolution. J Mol Biol 324: 871–886

Murakami A, Nakatogawa H, Ito, K (2004) Translation arrest of SecM is essential for the basal and regulated expression of SecA. Proc Natl Acad Sci USA 101: 12 330–12 335

Muto H, Nakatogawa H, Ito, K (2006) Genetically encoded but non-polypeptide prolyl-tRNA functions in the A site for SecM-mediated ribosomal stall. Mol Cell 22: 545–552

Nagamori S, Smirnova IN, Kaback HR (2004) Role of YidC in folding of polytopic membrane proteins. J Cell Biol 165: 53–62

Nakatogawa H, Ito, K (2001) Secretion monitor, SecM, undergoes self-translation arrest in the cytosol. Mol Cell 7: 185–192

Nakatogawa H, Ito, K (2002) The ribosomal exit tunnel functions as a discriminating gate. Cell 108: 629–636

Natale P, Bruser T, Driessen AJ (2008) Sec- and Tat-mediated protein secretion across the bacterial cytoplasmic membrane–distinct translocases and mechanisms. Biochim Biophys Acta 1778: 1735–1756

Nissen P, Hansen J, Ban N, Moore PB, Steitz TA (2000) The structural basis of ribosome activity in peptide bond synthesis. Science 289: 920–930

Nouwen N, Driessen AJ (2002) SecDFyajC forms a heterotetrameric complex with YidC. Mol Microbiol 44: 1397–1405

Oliver DC, Paetzel, M (2008) Crystal structure of the major periplasmic domain of the bacterial membrane protein assembly facilitator YidC. J Biol Chem 283: 5208–5216

Osborne AR, Rapoport TA (2007) Protein translocation is mediated by oligomers of the SecY complex with one SecY copy forming the channel. Cell 129: 97–110

Patzelt H, Rudiger S, Brehmer D, Kramer G, Vorderwulbecke S, Schaffitzel E, Waitz A, Hesterkamp T, Dong L, Schneider-Mergener J, et al. (2001) Binding specificity of *Escherichia coli* trigger factor. Proc Natl Acad Sci USA 98: 14 244–14 249

Picking WD, Picking WL, Odom OW, Hardesty, B (1992) Fluorescence characterization of the environment encountered by nascent polyalanine and polyserine as they exit *Escherichia coli* ribosomes during translation. Biochemistry 31: 2368–2375

Plath K, Mothes W, Wilkinson BM, Stirling CJ, Rapoport TA (1998) Signal sequence recognition in posttranslational protein transport across the yeast ER membrane. Cell 94: 795–807

Raine A, Ivanova N, Wikberg JE, Ehrenberg, M (2004) Simultaneous binding of trigger factor and signal recognition particle to the *E. coli* ribosome. Biochimie 86: 495–500

Rapoport TA (2007) Protein translocation across the eukaryotic endoplasmic reticulum and bacterial plasma membranes. Nature 450: 663–669

Raue U, Oellerer S, Rospert S (2007) Association of protein biogenesis factors at the yeast ribosomal tunnel exit is affected by the translational status and nascent polypeptide sequence. J Biol Chem 282: 7809–7816

Ravaud S, Stjepanovic G, Wild K, Sinning I (2008) The crystal structure of the periplasmic domain of the *Escherichia coli* membrane protein insertase YidC contains a substrate binding cleft. J Biol Chem 283: 9350–9358

Rinke-Appel J, Osswald M, von Knoblauch K, Mueller F, Brimacombe R, Sergiev P, Avdeeva O, Bogdanov A, Dontsova, O (2002) Crosslinking of 4.5S RNA to the *Escherichia coli* ribosome in the presence or absence of the protein Ffh. RNA 8: 612–625

Rutkowska A, Mayer MP, Hoffmann A, Merz F, Zachmann-Brand B, Schaffitzel C, Ban N, Deuerling E, Bukau, B (2008) Dynamics of trigger factor interaction with translating ribosomes. J Biol Chem 283: 4124–4132

Samuelson JC, Chen M, Jiang F, Moller I, Wiedmann M, Kuhn A, Phillips GJ, Dalbey RE (2000) YidC mediates membrane protein insertion in bacteria. Nature 406: 637–641

Schaffitzel C, Ban, N (2007) Generation of ribosome nascent chain complexes for structural and functional studies. J Struct Biol 158: 463–471

Schaffitzel C, Oswald M, Berger I, Ishikawa T, Abrahams JP, Koerten HK, Koning RI, Ban, N (2006) Structure of the *E. coli* signal recognition particle bound to a translating ribosome. Nature 444: 503–506

Schlunzen F, Wilson DN, Tian P, Harms JM, McInnes SJ, Hansen HA, Albrecht R, Buerger J, Wilbanks SM, Fucini, P (2005) The binding mode of the trigger factor on the ribosome: implications for protein folding and SRP interaction. Structure 13: 1685–1694

Scotti PA, Urbanus ML, Brunner J, de Gier JW, von Heijne G, van der Does C, Driessen AJ, Oudega B, Luirink, J (2000) YidC, the *Escherichia coli* homologue of mitochondrial Oxa1 p, is a component of the Sec translocase. EMBO J 19: 542–549

Seidelt B, Innis CA, Wilson DN, Gartmann M, Armache JP, Villa E, Trabuco LG, Becker T, Mielke T, Schulten K, et al. (2009) Structural insight into nascent polypeptide chain-mediated translational stalling. Science 326: 1412–1415

Serek J, Bauer-Manz G, Struhalla G, van den Berg L, Kiefer D, Dalbey R, Kuhn, A (2004) *Escherichia coli* YidC is a membrane insertase for Sec-independent proteins. EMBO J 23: 294–301

Shen K, Shan SO (2010) Transient tether between the SRP RNA and SRP receptor ensures efficient cargo delivery during cotranslational protein targeting. Proc Natl Acad Sci USA 107: 7698–7703

Solbiati J, Chapman-Smith A, Miller JL, Miller CG, Cronan JE, Jr (1999) Processing of the N termini of nascent polypeptide chains requires deformylation prior to methionine removal. J Mol Biol 290: 607–614

Takeda M, Webster RE (1968) Protein chain initiation and deformylation in B. subtilis homogenates. Proc Natl Acad Sci USA 60: 1487–1494

Teter SA, Houry WA, Ang D, Tradler T, Rockabrand D, Fischer G, Blum P, Georgopoulos C, Hartl FU (1999) Polypeptide flux through bacterial Hsp70: DnaK cooperates with trigger factor in chaperoning nascent chains. Cell 97: 755–765

Tomic S, Johnson AE, Hartl FU, Etchells SA (2006) Exploring the capacity of trigger factor to function as a shield for ribosome bound polypeptide chains. FEBS Lett 580: 72–76

Tsalkova T, Odom OW, Kramer G, Hardesty, B (1998) Different conformations of nascent peptides on ribosomes. J Mol Biol 278: 713–723

Tsukazaki T, Mori H, Fukai S, Ishitani R, Mori T, Dohmae N, Perederina A, Sugita Y, Vassylyev DG, Ito K, et al. (2008) Conformational transition of Sec machinery inferred from bacterial SecYE structures. Nature 455: 988–991

Ullers RS, Ang D, Schwager F, Georgopoulos C, Genevaux, P (2007) Trigger Factor can antagonize both SecB and DnaK/DnaJ chaperone functions in *Escherichia coli*. Proc Natl Acad Sci USA 104: 3101–3106

Ullers RS, Houben EN, Brunner J, Oudega B, Harms N, Luirink, J (2006) Sequence-specific interactions of nascent *Escherichia coli* polypeptides with trigger factor and signal recognition particle. J Biol Chem 281: 13 999–14 005

Urbanus ML, Froderberg L, Drew D, Bjork P, de Gier JW, Brunner J, Oudega B, Luirink, J (2002) Targeting, insertion, and localization of *Escherichia coli* YidC. J Biol Chem 277: 12 718–12 723

Urbanus ML, Scotti PA, Froderberg L, Saaf A, de Gier JW, Brunner J, Samuelson JC, Dalbey RE, Oudega B, Luirink, J (2001) Sec-dependent membrane protein insertion: sequential interaction of nascent FtsQ with SecY and YidC. EMBO Rep 2: 524–529

Valent QA, de Gier JW, von Heijne G, Kendall DA, ten Hagen-Jongman CM, Oudega B, Luirink, J (1997) Nascent membrane and presecretory proteins synthesized in *Escherichia coli* associate with signal recognition particle and trigger factor. Mol Microbiol 25: 53–64

van Bloois E, Dekker HL, Froderberg L, Houben EN, Urbanus ML, de Koster CG, de Gier JW, Luirink, J (2008) Detection of cross-links between FtsH, YidC, HflK/C suggests a linked role for these proteins in quality control upon insertion of bacterial inner membrane proteins. FEBS Lett 582: 1419–1424

van Bloois E, Nagamori S, Koningstein G, Ullers RS, Preuss M, Oudega B, Harms N, Kaback HR, Herrmann JM, Luirink, J (2005) The Sec-independent function of *Escherichia coli* YidC is evolutionary-conserved and essential. J Biol Chem 280: 12 996–13 003

van den Berg B, Clemons WM, Jr., Collinson I, Modis Y, Hartmann E, Harrison SC, Rapoport TA (2004) X-ray structure of a protein-conducting channel. Nature 427: 36–44

van der Laan M, Bechtluft P, Kol S, Nouwen N, Driessen AJ (2004) F1F0 ATP synthase subunit c is a substrate of the novel YidC pathway for membrane protein biogenesis. J Cell Biol 165: 213–222

van der Laan M, Houben EN, Nouwen N, Luirink J, Driessen AJ (2001) Reconstitution of Sec-dependent membrane protein insertion: nascent FtsQ interacts with YidC in a SecYEG-dependent manner. EMBO Rep 2: 519–523

Vetro JA, Chang YH (2002) Yeast methionine aminopeptidase type 1 is ribosome-associated and requires its N-terminal zinc finger domain for normal function *in vivo*. J Cell Biochem 85: 678–688

Voss NR, Gerstein M, Steitz TA, Moore PB (2006) The geometry of the ribosomal polypeptide exit tunnel. J Mol Biol 360: 893–906

Woolhead CA, Johnson AE, Bernstein HD (2006) Translation arrest requires two-way communication between a nascent polypeptide and the ribosome. Mol Cell 22: 587–598

Woolhead CA, McCormick PJ, Johnson AE (2004) Nascent membrane and secretory proteins differ in FRET-detected folding far inside the ribosome and in their exposure to ribosomal proteins. Cell 116: 725–736

Xie K, Dalbey RE (2008) Inserting proteins into the bacterial cytoplasmic membrane using the Sec and YidC translocases. Nat Rev Microbiol 6: 234–244

Yi L, Celebi N, Chen M, Dalbey RE (2004) Sec/SRP requirements and energetics of membrane insertion of subunits a, b, and c of the *Escherichia coli* F1F0 ATP synthase. J Biol Chem 279: 39 260–39 267

Yi L, Dalbey RE (2005) Oxa1/Alb3/YidC system for insertion of membrane proteins in mitochondria, chloroplasts and bacteria (review) Mol Membr Biol 22: 101–111

Yi L, Jiang F, Chen M, Cain B, Bolhuis A, Dalbey RE (2003) YidC is strictly required for membrane insertion of subunits a and c of the F(1)F(0)ATP synthase and SecE of the SecYEG translocase. Biochemistry 42: 10 537–10 544

Zeng LL, Yu L, Li ZY, Perrett S, Zhou JM (2006) Effect of C-terminal truncation on the molecular chaperone function and dimerization of *Escherichia coli* trigger factor. Biochimie 88: 613–619

Zhang X, Kung S, Shan SO (2008) Demonstration of a multistep mechanism for assembly of the SRP x SRP receptor complex: implications for the catalytic role of SRP RNA. J Mol Biol 381: 581–593

Zhang X, Schaffitzel C, Ban N, Shan SO (2009) Multiple conformational switches in a GTPase complex control co-translational protein targeting. Proc Natl Acad Sci USA 106: 1754–1759

Zimmer J, Nam Y, Rapoport TA (2008) Structure of a complex of the ATPase SecA and the protein-translocation channel. Nature 455: 936–943

Section VI Evolution

33

Molecular palaeontology as a new tool to study the evolution of ribosomal RNA

Sergey V. Steinberg and Konstantin Bokov

1. Introduction

The ribosome is a large RNA-protein complex that performs the synthesis of proteins in all living organisms. The emergence of the ribosome has been a pivotal step in the evolution of life on earth. It is generally accepted that the ribosome emerged almost four billion years ago from the RNA world, in which the primordial chemical reactions of life were catalyzed by RNA (Crick, 1968; Gilbert, 1986). Correspondingly, the ancient ribosome represented an RNA body, while proteins were added to its structure later, when the ribosome became effective enough to synthesize them. The original ribosomal RNA (rRNA) is believed to have been a rather small molecule, which gradually expanded to the modern size through addition of new elements (Noller, 2004; Hury et al., 2006; Smith et al., 2008). In order to understand details of this evolutionary process, one cannot use the standard approach of aligning available nucleotide sequences of ribosomal RNA and constructing phylogenetic trees. Due to the nature of that approach, its ability to elucidate evolutionary events in the past is limited by the moment when all branches of the phylogenetic tree come together, which corresponds to the so-called Last Universal Common Ancestor (LUCA). On the other hand, because in all presently living organisms the ribosome core has essentially the same structure (Gutell et al., 1994; Doudna and Rath, 2002), it should have formed before the split of the tree of life in three major domains, i. e. before LUCA. This discrepancy makes the standard approach inapplicable to the problem of early ribosome evolution and necessitates the development of alternative approaches.

The determination of the tertiary structure of the two ribosomal subunits (Ban et al., 2000; Wimberly et al., 2000) and of the whole ribosome (Schuwirth et al.,

2005; Selmer et al., 2006) opened new possibilities for understanding how the ribosome emerged in evolution. Several attempts have been made to approach the problem using the available structural and biochemical data (Hury et al., 2006; Smith et al., 2008; Wuyts, 2001). In particular, based on the fact that the peptidyl transferase center, which is the active site of the ribosome, is positioned in the middle of the tertiary structure of the 23S rRNA (Nissen et al., 2000; Polacek and Mankin, 2005), it was considered as the most ancient part of the ribosome (Hury et al., 2006). Correspondingly, domain V of 23S rRNA, which contains the peptidyl transferase center and forms contacts with most of the other domains, was also suggested to be the most ancient domain of the molecule. Other parts of the 23S rRNA appeared later and thus were positioned farther from the center and closer to the surface of the modern ribosome. A more detailed description of the evolution of 23S rRNA was hardly possible without taking into consideration the real conformation of the ribosomal RNA, although how exactly the knowledge of the conformation could be used was not at all clear.

In this situation, we suggested that if indeed different parts of the ribosome structure emerged at different moments, it could be possible to distinguish between more recent and more ancient elements based on the way they interact with each other. More recent elements emerged after more ancient elements and are thus expected to structurally adapt for the proper interaction with the latter. Correspondingly, if we were able to detect signs of such adaptation, it would help us to determine the relative age of different elements and thus to figure out the order in which different elements joined the ribosome structure as it evolved. The idea that the evolution of rRNA could be followed not only based on the proximity of different elements to the peptidyl-transferase center, i. e. based on the global position of elements in the whole ribosome structure, but also locally, based on the interaction between neighboring elements, was eventually developed into a new approach that we call here molecular palaeontology. We tested this approach on 23S rRNA from *Escherichia coli* and surprisingly found that the tertiary structure of this molecule contains a lot of information on the early evolution of the ribosome, which has been practically undisturbed since four billion years. The analysis of this information has allowed us to develop a concerted, modular scheme of the evolution of 23S rRNA which is described in the present article.

2. Restrained evolution of the ribosome structure

We start the description of our approach by making some general suggestions concerning the situation in which the primordial ribosome was evolving. We divide the whole process of the ribosome evolution in two major periods, before and after the emergence of the first RNA molecule able to perform a ribosome-related function. Henceforth, we will call that RNA molecule the proto-ribosome. We may not know what the exact function of the proto-ribosome was, but this function should have been essential for the organism in which it appeared. We think that the proto-ribosome was relatively small, which allowed it to emerge as the result of a single mutagenic act. Prior to the appearance of the proto-ribosome, ribosome evolution could be described as random reshuffling of RNA chains that was limited only by the capacity of the RNA-synthesizing machinery of the organism. However, when the proto-ribosome emerged and spread in the population, the maintenance of the level of ribosome function achieved so far became an essential aspect for the competitiveness and survival of organisms.

The further expansion of the size of the ribosome is expected to have proceeded through a series of insertions in different regions of the polynucleotide chain. Although the location and the nucleotide sequence of each insertion presumably were completely random, only those insertions that made the ribosome more effective would have had a chance to spread in the population and to be passed down to the following generations. This aspect allows us to formulate general constraints that an insertion should comply with to be propagated in evolution. These constraints pertain to the location of the insertion in the tertiary structure of the ribosome, to the conformation of the inserted frag-

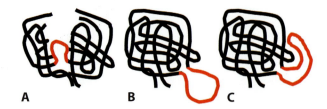

Fig. 1 Different types of accommodation of a newly emerged insertion to an existing proto-ribosome. (A) If the new insertion (red) occurred in a crowded area of the existing molecule (black), it could result in the breakage of the whole tertiary structure. (B, C) If the insertion occurred at the outskirts of the ribosome structure, it would not disturbe the tertiary structure. (B) If the insertion has a loose conformation, it would create problems for folding and integrity of the ribosome structure. (C) If the insertion has a compact structure and can be fixed on the surface of the ribosome, it would stabilize the whole structure and will have a chance to propagate in evolution.

ment, and to the way it interacts with the existing parts of the ribosome that emerged previously.

First, the necessity to preserve the level of the ribosome function that was already achieved can be formulated on the structural level as a requirement that the insertion must not disturb the positions of nucleotides that were already present in the ribosome. This can only be possible if in the tertiary structure the 5' and 3' termini of the inserted fragment are positioned close to each other. Also, the inserted fragment should not interfere with the existing parts of the tertiary structure. These constraints would disfavour insertions in crowded areas (Figure 1A) and would favour those insertions that occur at the outskirts of the existing structure (Figures 1B and C). The inserted fragment is also expected to be fixed on the surface of the growing ribosome structure (Figure 1C), because only then the addition of the new structural element would not compromise the stability of the whole ribosome.

The final set of constraints imposes limits on the types of interactions that the newly emerged fragment can form with the already existing parts of the ribosome. In the tertiary structure of the ribosomal RNA, one can find many cases when one region of the polynucleotide chain can acquire a particular conformation only in the presence of another region. The most typical cases of this kind represent Watson-Crick double helices and A-minor motifs. In a double helix, the conformation of each of the two strands is fixed through base pairing with the opposite strand. In the A-minor motifs, which consist of a stack of unpaired nucleotides, predominantly adenosines, that pack with a double helix (Nissen et al. 2001, Doherty et al. 2001), the adenosine stack acquires a particular conformation

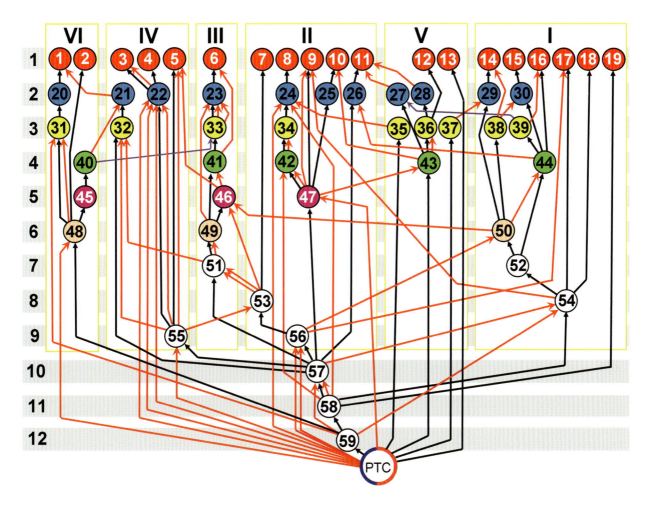

Fig. 2 Dismantling the tertiary structure of 23S rRNA. Removed elements are shown by numbered circles. The position of each element in the 23S rRNA secondary structure is shown in Figure 3A. Elements belonging to the same g eneration of acquired elements are shown with the same color. Generations are numbered from 1 to 12. Roman numerals indicate secondary structure domains. PTC stands for the symmetric structure in domain V containing the peptidyl-transferase center (the proto-ribosome, see also Fig- ure 3B). An arrow between two elements, e. g. A → B, indicates that the removal of A before B would compromise the integrity of the remaining ribosome. Black arrows denote covalent connections between the respective elements, red arrows A-minor interactions; violet arrows denote a double helix which would correspond to a non-local pseudoknot (Modified from Bokov and Steinberg, Nature 457: 977–980, 2009)

only upon the interaction, with the double helix. For all such cases, we suggest a general principle according to which structural integrity of more ancient elements cannot be dependent on the presence of more recently acquired elements. For double helical regions, the application of this principle means that both strands of a helix should emerge simultaneously as parts of the same fragment. For A-minor motifs, if means that the adenosine stack cannot be a more ancient acquisition of the ribosome than the corresponding double helix. The latter constraint represents a type of adaption of a newly emerged element to the existing structural context and provides the asymmetry between more ancient an more recent elements of the ribosome structure.

3. Dismantling the ribosome structure

Based on the constraints described above, we developed a procedure of systematic dismantling the structure of ribosomal RNA through elimination of those elements that are qualified as latest acquisitions of the ribosome. As an element we considered an individual double helix or a domain of stacked nucleotides that would form a stable compact arrangement upon addition to the ribosome structure. The 5' and 3' termini of an element qualified as a latest acquisition must be structurally close to each other. A qualified element must contain both strands of the same double helix. Finally, if a qualified element forms an A-minor motif with the remaining ribosome, it must contain the stack

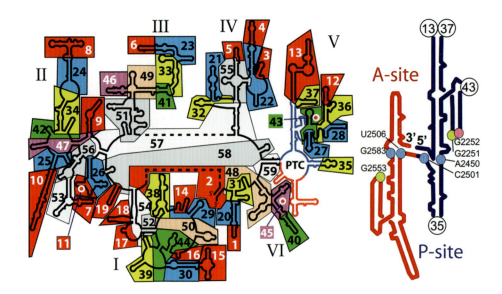

Fig. 3 (A) Positions of stepwise removed elements in the secondary structure of 23S rRNA. Numbers and colors of the elements correspond to those shown in Figure 2. The blue and red structure in domain V (PTC) stands for the proto-ribosome. The red and blue parts of the proto-ribosome correspond to the A and P sites, respectively. (B) Symmetry in the proto-ribosome. The three numbered circles stand for the elements of domain V immediately attached to the proto-ribosome. Colored circles show the positions of the nucleotides in the two parts of the proto-ribosome that coordinate nucleotides A76 (blue), C75 (yellow) and C74 (magenta) of tRNAs. The inclination of the lower double helix of the A-site part of the proto-ribosome indicates that in the ribosome structure its position is not completely symmetrical to the corresponding double helix in the P-site part (Modified from Bokov and Steinberg, Nature 457: 977–980, 2009)

of unpaired nucleotides that form this interaction, and not the double helix. All elements qualified as latest acquisitions of the ribosome thus constitute the last generation of acquired elements. Their removal would allow the identification of the second-last generation of acquired elements. The iterative procedure would be repeated until it reaches the proto-ribosome or until no more elements could be removed. Because this approach studies the process of development of a molecule of life through analysis of more and more ancient fossils of this molecule, it shares some features with palaeontology. Therefore, we name this approach the molecular palaeontology.

The application of the described dismantling algorithm to the tertiary structure of 23S rRNA from *E. coli* (Schuwirth et al., 2005) identified 59 elements which are distributed over 12 generations (Bokov and Steinberg, 2009). In Figure 2, these elements are depicted as numbered circles, and their positions in the 23S rRNA secondary structure are shown in Figure 3A. In Figure 2, the circles are connected by arrows, all of which go from a lower to a higher level. Each arrow A → B between elements A and B indicates that B is a more recent acquisition of the ribosome than A and that the emergence of B was dependent on the presence of A. Most arrows are either black or red. Black arrows indicate that elements A and B are

covalently connected in the way that the appearance of B before A would have compromised the integrity of the polynucleotide chain of 23S rRNA. Red arrows stand for the A-minor interactions between the double helix of element A and the nucleotide stack of element B. The appearance of B before A thus would have compromised the conformational integrity of element B.

There are also two magenta arrows connecting elements 40 → 33 and 39 → 27. These two arrows correspond to the so-called non-local pseudoknots and indicate a special type of dependency that we faced in our analysis. By definition, pseudoknots constitute secondary structure arrangements in which the loop of a stem-loop structure forms a double helix with a region outside this stem-loop. The requirement that in each helix both strands should be parts of the same element may create problems if in one of the two helices constituting a pseudoknot the strands belong to regions distant from each other in the secondary structure. 23S rRNA from *E. coli* contains eight pseudoknots, six of which are local in the sense that both helices constituting the pseudoknot are proximal to each other in the secondary structure of 23S rRNA and can thus belong to the same element. Only two pseudoknots are not local, those in which a double helix is formed between elements 40 and 33 and between

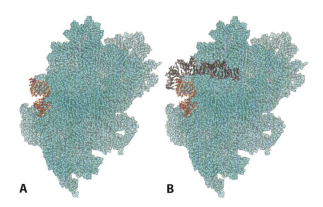

Fig. 4 The formation of the non-local pseudoknot 40→33. The blue area depicts the presumed structure of 23S rRNA just before the emergence of element 33. Element 40 (red + yellow) represents a stem capped by a tetraloop. (A) The integrity of element 40 does not require the presence of element 33. (B) The position of the newly emerged element 33 (black + magenta) can be stabilized by the formation of a double helix with element 40, which constitutes the pseudoknot.

elements 39 and 27 (Figure 2). A careful analysis of the latter two pseudoknots showed that in both of them one strand of the double helix is able to keep its structural integrity without forming base pairs with the opposite strand. For example, element 40 consists of a stem capped by a tetraloop, and it is the nucleotides of this tetra-loop that form the pseudoknot double helix with element 33 (Figure 4). Given that tetraloops represent compact stable arrangements, the integrity of element 40 does not need the presence of element 33. This aspect has allowed us to eliminate element 33 without jeopardising the integrity of element 40, even though the two elements form together a double helix. A similar situation pertains to elements 39 and 27, which together also form a double helix. Here, like in the previous case, the integrity of element 39 was judged not to be dependent on the presence of element 27. These two examples show that, on rare occasions, the requirement that two strands of the same helix should have emerged simultaneously can be violated. This can happen, however, only if, due to particular circumstances, the integrity of one of the two strands can be maintained without the presence of the opposite strand.

4. The origin of 23S rRNA

The removal of the 59 elements identified by the analysis described above eliminated 93% of the original 23S rRNA. The remaining part, which consists of 220

nucleotides, is located in domain V (Figure 3). The central region of this 220-nucleotide fragment forms the peptidyl-transferase center. Recently, it was observed that this fragment has a symmetric structure (Agmon et al., 2005). The symmetry is clearly seen on the levels of both secondary and tertiary structure (blue and red regions, Figure 3B). One half of this symmetric structure corresponds to the P site (blue), the other half to the A site (red). Moreover, there is a close correspondence between the positions of the nucleotides of the two halves that are involved in the fixation of the equivalent elements of the tRNAs in A and P sites (Samaha et al., 1995; Nissen et al., 2000; Kim and Green, 1999; Hansen et al., 2002). In the polynucleotide chain of the remaining part, the P-site half precedes the A-site half. The similarity between the two halves is so high that it is logical to suggest that they originated by a duplication of the same RNA fragment (Agmon et al., 2005). From this point of view, the evolution of 23S rRNA started with an initial fragment of about 110 nucleotides, which presumably was able to bind the CCA terminus of what would later become transfer RNA. The duplication of this fragment allowed the resulting molecule to bind two CCA termini simultaneously. Within this arrangement, the two CCA termini asso-

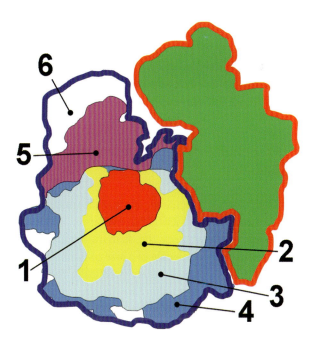

Fig. 5 The expansion of the 23S rRNA structure as it evolved. 1: the proto-ribosome, 2–5: the size of the 23S rRNA after 8 (2), 20 (3), 50 (4) and all 59 (5) elements have been added to the proto-ribosome. 6: the area of the 50S subunit that does not include ribosomal RNA. Green: 30S subunit (Modified from Bokov and Steinberg, Nature 457: 977–980, 2009)

ciated with the two halves are juxtaposed in space to allow for the transpeptidation reaction. Most probably, this dimer was already able to synthesize oligopeptides with random amino acid sequences, hence the designation proto-ribosome. This view is supported by the fact that in-vitro-selected small RNA molecules resembling the peptidyl-transferase center were able to perform transpeptidation (Zhang and Cech, 1997; 1998), thus demonstrating that this reaction does not require any other element of the ribosome structure.

All other elements of 23S rRNA were gradually added to the proto-ribosome, one element at a time, in essentially the same way. Each element could appear only when all elements that were required for its proper positioning had already been placed, as defined in Figure 2. New elements were added as insertions containing all necessary details to dock with the surface of the evolving ribosome without disturbing already existing parts. The most common way for a new element to be fixed on the ribosome surface would be through the formation of an A-minor interaction with an already existing double helix. As the number of added elements grew, the proto-ribosome became larger (Bokov and Steinberg, 2009). As a result, in the structure of the modern 23S rRNA, the age of different elements would generally correlate with their distance from the peptidyl transferase center (Figure 5). Even though this result may be considered as expected, the fact that it was not assumed at the outset of our analysis and has been obtained based solely on the consideration of local interactions between closely packed elements adds to the validity of the whole approach.

5. Probabilistic quantification of the model

In this section we provide a quantitative evaluation of the proposed model of 23S rRNA evolution. The question to address is whether the dismantling algorithm applied to an RNA molecule of a size of the 23S rRNA will always be able to eliminate practically the whole polynucleotide chain or, alternatively, for some structures it can be arrested before it reaches the end of the molecule. The question is equivalent to whether it is always possible to arrange all elements of an arbitrary RNA structure in the way shown in Figure 2, i.e. with all dependencies going from a lower to a higher level.

According to our analysis, the dismantling algorithm will always be able to reach the end of the molecule, as long as the dependencies between different

elements existing in this molecule do not form cycles. For a given secondary structure, the presence or absence of such cycles would depend on the particular scheme of A-minor interactions. We can analyze, for example, what would happen if dependence $59 \rightarrow 54$ (Figure 2) had the opposite direction. This would mean that contrary to what happens in real 23S rRNA, the adenosine stack of this A-minor interaction now belongs to element 59, while the double helix is located in element 54. After such reorientation of dependence $59 \rightarrow 54$, the dependencies between three elements 54, 57 and 59 would form the cycle $59 \rightarrow 57 \rightarrow 54 \rightarrow 59$. The existence of this cycle will make each of the three elements 54, 57 and 59 dependent on the presence of the other two. Due to such inter-dependence, the dismantling procedure will be automatically arrested when it reaches any element of this cycle. We thus can conclude that the elimination of all 59 elements of 23S rRNA has been possible due to the fact that the tertiary structure of the molecule does not contain cycles of dependency.

Interestingly, the reorientation of an A-minor interaction does not always lead to the formation of a cycle. For example, the reorientation of A-minor interaction $21 \rightarrow 1$ does not create a cycle and can be accommodated to the pattern of Figure 2 through the displacement of several elements: elements 1, 20 and 31 should move down one layer, while element 21 should move up one layer. After these accommodations, all dependencies will again be oriented from a lower to a higher level.

If the evolution of 23S rRNA proceeded according to the scenario proposed here, i.e. through consecutive incorporation of local insertions and formation of A-minor motifs between adenosine stacks of more recently emerged elements and double helices of more ancient elements, the automatic outcome will be that the tertiary structure of 23S rRNA is free of cycles of dependency. An alternative explanation would be that such cycles are absent by chance, while the directions of the A-minor interactions did not play any specific role in the evolution of 23S rRNA. To discriminate between the two hypotheses, we analyzed the scheme presented in Figure 2 and showed that, if the orientations of all A-minor motifs in 23S rRNA were not essential for evolution and were chosen randomly, the probability that such structure did not contain cycles of dependency would have been $P < 10^{-9}$. Such a low probability allows us to conclude that the absence of cycles is a fundamental property of 23S rRNA that is directly related to the particular trajectory of its emergence.

6. The simplicity of the ribosome structure

23S rRNA is a large molecule with a rather complex tertiary structure. However, the analysis we have performed indicates that a molecule of the size of 23S rRNA could have a much more complex tertiary structure. The strict constraints under which 23S rRNA has been evolving, namely the requirement for systematic incremental increase of ribosome efficiency, do not provide too much freedom for the appearance of exceedingly complex structures. In the following, the indicators of the relative simplicity of the tertiary structure of 23S rRNA are summarized:

(A.) The molecule is built based on the same simple concept applied to the structure over and over again. As a result, the tertiary structure of the whole molecule is characterized by global order, as illustrated by the common orientation of all A-minor motifs (Figure 2).

(B.) The requirement for the ribosome to have a stable structure after the addition of each new element limits the size of elements to be added. Indeed, for a large element that emerged spontaneously, the probability that it properly adapted to a rather heterogenic surface of the existing ribosome and became stably integrated would have been rather low. As a result, the average length of inserted elements was only 45 nucleotides and the length exceeded a hundred nucleotides only for two elements.

(C.) The structure of 23S rRNA is packed with A-minor interactions. The universal presence of the A-minor motif in the molecule can be explained by the relatively small size of newly emerged elements and the necessity to fix their positions on the surface of the ribosome. The fixation of an element with the help of an A-minor interaction does not require any special arrangement in more ancient elements except for the existence of a short double helical region. Given that double helices are common elements of RNA structure, such regions will always have been present in the vicinity of a new insertion. Also, a stack of unpaired adenosines can be arranged in almost any place of a newly emerged element, regardless of its structure. The A-minor motif thus seems to be the easiest way for the integration of a new element into the existing ribosome structure.

(D.) The necessity for the ribosome to maintain the stability of its structure after addition of each new element precludes the existence of long unpaired regions. This aspect explains the high percentage of nucleotides of 23S rRNA that are involved in double helices.

(E.) The prohibition of long unpaired regions eliminates the possibility for the emergence of non-local pseudoknots in which fragments belonging to distant parts of the secondary structure form a long double helix. In the two non-local pseudoknots present in the 23S rRNA, this helix contains only three or four base pairs. Moreover, in both pseudoknots the conformation of one strand of this helix can be stabilized by the particular structural context. The presence of such context seems to be essential for the ability of this strand to emerge before the appearance of the opposite strand.

7. The deteriorating evolution of mitochondrial ribosomes

The suggested model of the evolution of 23S rRNA is mainly based on the assumption of a systematic incremental increase of ribosome efficiency, which does not allow mutations rendering the ribosome less effective to propagate in evolution. This principle prohibits the emergence of elements whose structural integrity would require the emergence of additional elements in the future. It also prohibits deletions, thus providing for the incremental growth of the size of the ribosomal RNA. However, if at some moment the priorities changed and the incremental increase of the ribosome efficiency became no longer essential for the competitiveness of the organism, the evolutionary path could have taken a dramatic turn. A case of this type seems to have happened with mitochondrial ribosomes from protozoa, which, compared to bacterial ribosomes, have lost essential parts of their ribosomal RNA (O'Brien, 2002; Mears et al., 2002). We associate this phenomenon with the fact that the function of mitochondrial ribosomes in these organisms has become limited to the synthesis of only a dozen mitochondrial membrane proteins encoded in the mitochondrial genome. Given that membrane proteins have a substantially longer lifespan than most soluble proteins and thus are not needed to be renewed every few hours, the requirements for the efficiency of the mitochondrial ribosomes are less stringent. As a result, the length of the RNA of these ribosomes can be reduced, and still the organism will remain perfectly functional.

One may suggest that the deterioration of the ribosomal RNA in these mitochondria repeats in the reverse order the same steps that took place during the primary emergence of the ribosome. We do not support this suggestion and think that the two processes have very little, if anything, to do with one another.

They have also taken place in essentially different conditions: the deleted parts of RNA in mitochondrial ribosomes are usually replaced by proteins. This replacement would compensate, at least partly, for the loss of the RNA part, which results in a twice as high protein content in mitochondrial ribosomes compared to cytosolic ribosomes (O'Brien, 2002).

8. Tertiary structure versus nucleotide sequence

The results of our analysis strongly suggest that most elements of the modern 23S rRNA have practically the same tertiary structure as they had at the moment of their emergence. This suggestion correlates with the directionality of the A-minor interaction in the modern ribosome and with the fact that the ribosome core is conserved among all organisms. However, the conservation of the secondary and tertiary structure of ribosomal RNA still allowed some drift of the nucleotide sequence which is responsible for the existing variety of nucleotide sequences of ribosomal RNA. Interestingly, the analysis of ribosomal RNA sequences showed that the most variable parts correspond to double-helical regions, while unpaired regions exhibit much stronger conservation (Smit et al., 2007). This aspect further supports the idea that the preservation of tertiary arrangements, which are mostly modulated by unpaired regions, was more important for ribosome assembly and function than particular nucleotide sequences of double-helical regions.

Interestingly, while most sequence modifications of ribosomal RNA can be considered to result from neutral drift, some changes seem to have been essential. Thus, when the original duplication of the 110-nucleotide fragment created the proto-ribosome, the coexistence of two identical fragments in the same molecule should have created a problem for proper folding. In particular, those parts of the polynucleotide chain that in the original 110-nucleotide fragment formed double helices now got a possibility to form helices both within each half and between the two halves. To avoid such folding problems, there should have been a strong evolutionary pressure to make the nucleotide sequences of both halves of this structure as different as possible. The modification of the nucleotide sequences of the two halves of the proto-ribosome presumably has tuned the structure of the peptidyl transferase center toward higher efficiency of the transpeptidation reaction. As a result, in the modern ribosome, a similarity between the nucleotide sequences of the two halves of the symmetric structure in domain V is not detectable, despite the fact that the symmetry is clearly seen in both secondary and tertiary structure.

To conclude, we can say that the secondary and tertiary structure of ribosomal RNA has been substantially more conserved than the nucleotide sequence, which makes the molecular palaeontology approach described here a valuable tool for analysis of the early ribosome evolution.

Acknowledgements

This work was supported by an operating grant from the Canadian Institutes of Health Research to Sergey V. Steinberg.

References

Agmon I, Bashan A, Zarivach R, Yonath A (2005) Symmetry at the active site of the ribosome: structural and functional implications. Biol Chem 386: 833–844

Ban N, Nissen P, Hansen J, Moore PB, Steitz TA (2000) The complete atomic structure of the large ribosomal subunit at 2.4 A resolution. Science 289: 905–920

Bokov K, Steinberg SV (2009) A hierarchical model for evolution of 23S ribosomal RNA. Nature 457: 977–980

Crick FH (1968) The origin of the genetic code. J Mol Biol 38: 367–369

Doherty EA, Batey RT, Masquida B, Doudna JA (2001) A universal mode of helix packing in RNA. Nature Struct Biol 8: 339–343

Doudna JA, Rath VL (2002) Structure and function of the eukaryotic ribosome: the next frontier. Cell 109: 153–156

Gilbert W (1986) Origin of life: The RNA world. Nature 319: 618

Gutell RR, Larsen N, Woese CR (1994) Lessons from an evolving rRNA: 16S and 23S rRNA structures from a comparative perspective. Microbiol Rev 58: 10–26

Hansen JL, Schmeing TM, Moore PB, Steitz TA (2002) Structural insights into peptide bond formation. Proc Natl Acad Sci USA 99: 11 670–11 675

Hury J, Nagaswami U, Larios-Sanz M, Fox GE (2006) Ribosome origins: The relative age of 23S rRNA Domains. Orig Life Evol Biosph 36: 421–429

Kim DF, Green R (1999) Base-pairing between 23S rRNA and tRNA in the ribosomal A site. Mol Cell 4: 859–864

Mears JA, Cannone JJ, Stagg SM, Gutell RR, Agrawal RK, Harvey SC (2002) Modeling a minimal ribosome based on comparative sequence analysis. J Mol Biol. 321: 215–234

Nissen P, Hansen J, Ban N, Moore PB, Steitz TA (2000) The structural basis of ribosome activity in peptide bond synthesis. Science 289: 920–930

Nissen P, Ippolito JA, Ban N, Moore PB, Steitz TA (2001) RNA tertiary interactions in the large ribosomal subunit: the A-minor motif. Proc Natl Acad Sci USA 98: 4899–4903

Noller HF (2004) The driving force for molecular evolution of translation. RNA 10: 1833–1837

O'Brien TW (2002) Evolution of a protein-rich mitochondrial ribosome: implications for human genetic disease. Gene 286: 73–79

Polacek N, Mankin AS (2005) The ribosomal peptidyl transferase center: structure, function, evolution, inhibition. Crit Rev Biochem Mol Biol 40: 285–311

Samaha RR, Green R, Noller HF (1995) A base pair between tRNA and 23S rRNA in the peptidyl transferase centre of the ribosome. Nature 377: 309–314

Schuwirth BS, Borovinskaya MA, Hau CW, Zhang W, Vila-Sanjurjo A, Holton JM, Cate JH (2005) Structures of the bacterial ribosome at 3.5 A° resolution. Science 310: 827–834

Selmer M, Dunham CM, Murphy FV, Weixlbaumer A, Petry S, Kelley AC, Weir JR, Ramakrishnan V (2006) Structure of the 70S ribosome complexed with mRNA and tRNA. Science 313: 1935–1942

Smit S, Widnan J, Knight R (2007) Evolutionary rates vary among rRNA structural elements. Nucleic Acids Res 35: 3339–3354

Smith TF, Lee JC, Gutell, RR, Hartman H (2008) The origin and evolution of the ribosome. Biol Direct 3: 16

Wimberly BT, Brodersen DE, Clemons WM Jr, Morgan-Warren RJ, Carter AP, Vonrhein C, Hartsch T, Ramakrishnan V (2000) Structure of the 30S ribosomal subunit. Nature 407: 327–339

Wuyts J, Van de Peer I, De Wachter R (2001) Distribution of substitution rates and location of insertion sites in the tertiary structure of ribosomal RNA. Nucleic Acid Res 29: 5017–5028

Zhang B, Cech TR (1997) Peptide bond formation by in vitro selected ribozymes. Nature 390: 96–100

Zhang B, Cech TR (1998) Peptidyl-transferase ribozymes: trans-reactions, structural characterization and ribosomal RNA-like features. Chem Biol 5: 539–553

List of contributors

Editors

Marina Rodnina
Department of Physical Biochemistry
Max Planck Institute for Biophysical Chemistry
37077 Göttingen, Germany
E-mail: rodnina@mpibpc.mpg.de

Wolfgang Wintermeyer
Department of Physical Biochemistry
Max Planck Institute for Biophysical Chemistry
37077 Göttingen, Germany
E-mail: wolfgang.wintermeyer@mpibpc.mpg.de

Rachel Green
Howard Hughes Medical Institute
Department of Molecular Biology and Genetics
Johns Hopkins University School of Medicine
Baltimore, MD 21205, USA
E-mail: ragreen@jhmi.edu

Authors

Rajendra K. Agrawal
Wadsworth Center
New York State Department of Health
Empire State Plaza
Albany, NY 12201-0509, USA
Department of Biomedical Sciences
State University of New York at Albany
Albany, NY 12222, USA
E-mail: agrawal@wadsworth.org

Colin Echeverría Aitken
Department of Biophysics and Biophysical Chemistry
Johns Hopkins University School of Medicine
725 North Wolfe Street
Baltimore, MD 21205, USA

Rashid Akbergenov
Institut für Medizinische Mikrobiologie
Universität Zürich
Gloriastraße 30/32
8006 Zürich, Switzerland

Roger B. Altman
Department of Physiology and Biophysics
Weill Cornell Medical College
New York, NY 10021, USA

Johan Åqvist
Department of Cell and Molecular Biology
Uppsala University, BMC
Box 596
751 24 Uppsala, Sweden

Haruichi Asahara
Center for Molecular Biology of RNA
and Department of MCD Biology
University of California, Santa Cruz
Santa Cruz, CA 95064, USA

Nenad Ban
Institute of Molecular Biology
and Biophysics
ETH Zürich, Schafmattstr. 20
8093 Zürich, Switzerland
E-mail: ban@mol.biol.ethz.ch

Anat Bashan
The Weizmann Institute of Science
Department of Structural Biology
76100 Rehovat, Israel

Thomas Becker
Gene Center and Department for Chemistry
and Biochemistry, and Center for integrated
Protein Science Munich (CiPSM)
University of Munich
Feodor-Lynen-Str. 25
81377 München, Germany

Roland Beckmann
Gene Center and Department for Chemistry
and Biochemistry, and Center for integrated
Protein Science Munich (CiPSM)
University of Munich
Feodor-Lynen-Str. 25
81377 München, Germany
E-mail: beckmann@lmb.uni-muenchen.de

Riccardo Belardinelli
Laboratory of Genetics
Department of Biosciences and Biotechnology
University of Camerino
62032 Camerino (MC), Italy

Adam Ben-Shem
Institute of Genetics and of Molecular
and Cellular Biology
1 rue Laurent Fries
BP10142, Illkirch, 67400 France
Université de Strasbourg
Strasbourg, 67000, France

Shashi Bhushan
Gene Center and Department for Chemistry
and Biochemistry, and Center for integrated
Protein Science Munich (CiPSM)
University of Munich
Feodor-Lynen-Str. 25
81377 München, Germany

Gregor Blaha
Howard Hughes Medical Institute,
Department of Molecular Biophysics
and Biochemistry
Yale University
New Haven, CT 06520-8114 USA

Scott C. Blanchard
Department of Physiology
and Biophysics
Weill Cornell Medical College
New York, NY 10021, USA
E-mail: scb2005@med.cornell.edu

Daniel Boehringer
Institute of Molecular Biology
and Biophysics
ETH Zürich
Schafmattstr. 20
8093 Zürich, Switzerland

Alexey Bogdanov
Department of Chemistry and A.N.
Belozersky Institute of Physico-Chemical Biology
Moscow State University
Moscow, 119899, Russia

Konstantin Bokov
Université de Montréal C.P. 6128
Succursale Centre-ville
H3C 3J7 Montréal, QC, Canada

Erik C. Böttger
Institut für Medizinische Mikrobiologie
Universität Zürich
Gloriastraße 30/32
8006 Zürich, Switzerland
E-mail: boettger@imm.uzh.ch

Anna Brandi
Laboratory of Genetics,
Department of Biosciences and Biotechnology
University of Camerino
62032 Camerino (MC), Italy

David Bulkley
Howard Hughes Medical Institute
Department of Molecular Biophysics
and Biochemistry
Yale University
New Haven, CT 06520-8114, USA

Dmitry Burakovsky
Department of Chemistry and A.N.
Belozersky Institute of Physico-Chemical Biology
Moscow State University
Moscow, 119899, Russia

Allen Buskirk
Department of Chemistry
and Biochemistry
Brigham Young University
C203 BNSN
Provo, UT 84602, USA
E-mail: buskirk@chem.byu.edu

Dale Cameron
Department of Molecular Biology
Cell Biology and Biochemistry
Brown University
Providence, RI 02912, USA

Jennifer F. Carr
Department of Molecular Biology,
Cell Biology and Biochemistry
Brown University
Providence, RI 02912, USA

Jamie H.D. Cate
Departments of Molecular and
Cell Biology and Chemistry
University of California at Berkeley
Physical Biosciences Division
Lawrence Berkeley National Laboratory
Berkeley, CA 94720, USA
E-mail: jcate@lbl.gov

Biswajoy Roy-Chaudhuri
Department of Biology
University of Rochester
Rochester, NY 14627, USA

Chunlai Chen
Pennsylvania Muscle Institute
University of Pennsylvania, School of Medicine
Philadelphia, PA 19104-6083, USA

Barry S. Cooperman
Department of Chemistry
University of Pennsylvania
Philadelphia, PA 19104-6323, USA
E-mail: cooprman@pobox.upenn.edu

Peter Cornish
Department of Physics
University of Illinois
Champaign-Urbana, IL 61801 USA

Gloria M. Culver
Department of Biology
University of Rochester
Rochester, NY 14627, USA
E-mail: gculver@mail.rochester.edu

Albert E. Dahlberg
Department of Molecular Biology
Cell Biology and Biochemistry
Brown University
Providence, RI 02912, USA
E-mail: ae_dahlberg@brown.edu

Maria del Carmen Ruiz Ruiz
Laboratorium für Organische Chemie
ETH Zürich
Wolfgang-Pauli-Straße 10
8093 Zürich, Switzerland

Natalia Demeshkina
Institut de Génétique et de Biologie Moléculaire
et Cellulaire (IGBMC)
Département de Biologie et de Génomique
Structurales
CNRS, UMR7104, INSERM, U964
University of Strasbourg
67400 Illkirch, France

Hasan Demirci
Department of Molecular Biology
Cell Biology and Biochemistry
Brown University
Providence, RI 02912, USA

Aishwarya Devaraj
Ohio State Biochemistry Program
Center for RNA Biology
The Ohio State University
484 W. 12th Ave
Columbus, OH 43210, USA

Fabio Di Pietro
Laboratory of Genetics
Department of Biosciences and Biotechnology
University of Camerino
62032 Camerino (MC), Italy

John Paul Donohue
Center for Molecular Biology of RNA
and Department of MCD Biology
University of California, Santa Cruz
Santa Cruz, CA 95064, USA

Olga Dontsova
Department of Chemistry and A.N.
Belozersky Institute of Physico-Chemical Biology
Moscow State University
Moscow, 119899, Russia
E-mail: dontsova@genebee.msu.ru

Srinivas R. Dubbaka
Laboratorium für Organische Chemie
ETH Zürich
Wolfgang-Pauli-Straße 10
8093 Zürich, Switzerland

Jack A. Dunkle
Departments of Molecular and
Cell Biology and Chemistry
University of California at Berkeley
Physical Biosciences Division
Lawrence Berkeley National Laboratory
Berkeley, CA 94720, USA

Stefan Duscha
Institut für Medizinische Mikrobiologie
Universität Zürich
Gloriastraße 30/32
8006 Zürich, Switzerland

Måns Ehrenberg
Department of Cell and Molecular Biology
Uppsala University, BMC
Box 596
751 24 Uppsala, Sweden
E-mail: ehrenberg@xray.bmc.uu.se

Dmitri N. Ermolenko
Center for Molecular Biology of RNA
and Department of MCD Biology
University of California
Santa Cruz, CA 95064, USA

Attilio Fabbretti
Laboratory of Genetics
Department of Biosciences and Biotechnology
University of Camerino
62032 Camerino (MC), Italy

Ian Farrell
Department of Chemistry
University of Pennsylvania
Philadelphia, PA 19104-6323, USA

Niels Fischer
3D Electron Cryomicroscopy Group
Max Planck Institute for Biophysical Chemistry
37077 Göttingen, Germany

Joachim Frank
Howard Hughes Medical Institute
Department of Biochemistry and Molecular Biophysics
and Department of Biological Sciences
Columbia University
650 W. 168th Street
New York, NY 10032, USA
E-mail: jf2192@columbia.edu

Kurt L. Fredrick
Ohio State Biochemistry Program
Department of Microbiology
Center for RNA Biology
The Ohio State University
484 W. 12th Ave.
Columbus, OH 43210, USA
E-mail: fredrick.5@osu.edu

Peter Geggier
Department of Physiology and Biophysics
Weill Cornell Medical College
New York, NY 10021, USA

Anna Maria Giuliodori
Laboratory of Genetics
Department of Biosciences and Biotechnology
University of Camerino
62032 Camerino (MC), Italy

Yale E. Goldman
Pennsylvania Muscle Institute
University of Pennsylvania, School of Medicine
Philadelphia, PA 19104-6083, USA

Anna Golovina
Department of Chemistry and A.N.
Belozersky Institute of Physico-Chemical Biology
Moscow State University
Moscow, 119899, Russia

Ruben L. Gonzalez, Jr.
Department of Chemistry
Columbia University
3000 Broadway
New York, NY 10027, USA
E-mail: rlg2118@columbia.edu

Basil Greber
Institute of Molecular Biology and Biophysics
ETH Zürich
Schafmattstr. 20
8093 Zürich, Switzerland

Steven T. Gregory
Department of Molecular Biology,
Cell Biology and Biochemistry
Brown University
Providence, RI 02912, USA

Claudio O. Gualerzi
Laboratory of Genetics
Department of Biosciences and Biotechnology
University of Camerino
62032 Camerino (MC), Italy
E-mail: claudio.gualerzi@unicam.it

Taekjip Ha
Department of Physics
University of Illinois
Champaign-Urbana, IL 61801 USA

Shinde Harish
Laboratorium für Organische Chemie
ETH Zürich
Wolfgang-Pauli-Straße 10
8093 Zürich, Switzerland

Shan L. He
Howard Hughes Medical Institute
Department of Molecular Biology and Genetics
Johns Hopkins University School of Medicine
Baltimore, MD 21205, USA

David Healey
Department of Chemistry and Biochemistry
Brigham Young University
C203 BNSN
Provo, UT 84602, USA

Christopher U. T. Hellen
Department of Cell Biology
State University of New York
Downstate Medical Center
450 Clarkson Ave.
Brooklyn, NY 11203, USA

Robyn Hickerson
Center for Molecular Biology of RNA
and Department of MCD Biology
University of California, Santa Cruz
Santa Cruz, CA 95064, USA

Alexander Hirschi
Center for Molecular Biology of RNA
and Department of MCD Biology
University of California
Santa Cruz, CA 95064, USA

Lucas Horan
Department of Molecular and Cell Biology
University of California, Berkeley
Berkeley, CA 94720, USA

C. Axel Innis
Howard Hughes Medical Institute,
Department of Molecular Biophysics
and Biochemistry
Yale University
New Haven, CT 06520-8114 USA

Lasse B. Jenner
Institut de Génétique et de Biologie Moléculaire
et Cellulaire (IGBMC)
Département de Biologie et de Génomique
Structurales
CNRS, UMR7104, INSERM, U964
University of Strasbourg
67400 Illkirch, France

Gerwald Jogl
Department of Molecular Biology
Cell Biology and Biochemistry
Brown University
Providence, RI 02912, USA

Magnus Johansson
Department of Cell and Molecular Biology
Uppsala University, BMC
Box 596
751 24 Uppsala, Sweden

Ka Weng Ieong
Department of Cell and Molecular Biology
Uppsala University, BMC
Box 596
751 24 Uppsala, Sweden

Vladimir I. Katunin
Petersburg Nuclear Physics Institute
Russian Academy of Sciences
188300 Gatchina, Russia

Jaskarin Kaur
Department of Chemistry
University of Pennsylvania
Philadelphia, PA 19104-6323, USA

Teresa Kelley
Department of Biochemistry, Biophysics
and Molecular Biology
Iowa State University
Ames, IA 50011, USA

Narayanaswamy Kirthi
Department of Biochemistry, Biophysics
and Molecular Biology
Iowa State University
Ames, IA 50011, USA

Bruno Klaholz
Institute of Genetics and of Molecular and
Cellular Biology (IGBMC)
Department of Integrative Structural Biology
Centre National de la Recherche Scientifique (CNRS)
UMR 7104
Institut National de la Santé de la Recherche
Médicale (INSERM) U964
University of Strasbourg
1 rue Laurent Fries
67404 Illkirch, France

Andrey L. Konevega
Department of Physical Biochemistry
Max Planck Institute for Biophysical Chemistry
37077 Göttingen, Germany

Andrei Korostelev
Center for Molecular Biology of RNA
and Department of MCD Biology
University of California
Santa Cruz, CA 95064, USA

Ivona Kudyba
Laboratorium für Organische Chemie
ETH Zürich
Wolfgang-Pauli-Straße 10
8093 Zürich, Switzerland

Indrajit Lahiri
Wadsworth Center
New York State Department of Health
Empire State Plaza
Albany, NY 12201-0509, USA
Department of Biomedical Sciences
State University of New York at Albany
Albany, NY 12222, USA

Laura Lancaster
Center for Molecular Biology of RNA
and Department of MCD Biology
University of California, Santa Cruz
Santa Cruz, CA 95064, USA

Martin Laurberg
Center for Molecular Biology of RNA
and Department of MCD Biology
University of California
Santa Cruz, CA 95064, USA

Joshua M. Leisring
Department of Microbiology
The Ohio State University
484 W. 12th Ave.
Columbus, OH 43210, USA

Hanqing Liu
Department of Chemistry
University of Pennsylvania
Philadelphia, PA 19104-6323, USA

Wei Liu
Department of Chemistry
University of Pennsylvania
Philadelphia, PA 19104-6323, USA

Jon R. Lorsch
Department of Biophysics and Biophysical Chemistry
Johns Hopkins University School of Medicine
725 North Wolfe Street
Baltimore, MD 21205, USA
E-mail: jlorsch@bs.jhmi.edu

Daniel D. MacDougall
Columbia University
Department of Chemistry
3000 Broadway
New York, NY 10027, USA

Alexander S. Mankin
Center for Pharmaceutical Biotechnology
University of Illinois at Chicago
900 S. Ashland Ave.
Chicago, IL 60607, USA
E-mail: shura@uic.edu

Stefano Marzi
Institute of Genetics and of Molecular
and Cellular Biology (IGBMC)
Department of Integrative Structural Biology
Centre National de la Recherche Scientifique (CNRS)
UMR 7104
Institut National de la Santé de la Recherche
Médicale (INSERM) U964
University of Strasbourg
1 rue Laurent Fries
67404 Illkirch, France

Tanja Matt
Institut für Medizinische Mikrobiologie
Universität Zürich
Gloriastraße 30/32
8006 Zürich, Switzerland

Sean P. McClory
Ohio State Biochemistry Program
Center for RNA Biology
The Ohio State University
484 W. 12th Ave.
Columbus, OH 43210, USA

Jean-François Menetret
Institute of Genetics and of Molecular and
Cellular Biology (IGBMC)
Department of Integrative Structural Biology
Centre National de la Recherche Scientifique (CNRS)
UMR 7104
Institut National de la Santé de la Recherche
Médicale (INSERM) U964
University of Strasbourg
1 rue Laurent Fries
67404 Illkirch, France

Martin Meyer
Institut für Medizinische Mikrobiologie
Universität Zürich
Gloriastraße 30/32
8006 Zürich, Switzerland

Mickey Miller
Department of Chemistry and Biochemistry
Brigham Young University
C203 BNSN, Provo, UT 84602, USA

Sarah F. Mitchell
Department of Biophysics and Biophysical Chemistry
Johns Hopkins University School of Medicine
725 North Wolfe Street
Baltimore, MD 21205, USA

James B. Munro
Department of Physiology and Biophysics
Weill Cornell Medical College
New York, NY 10021, USA

Frank Murphy
Northeastern Collaborative Access Team
Cornell University
Argonne, IL 60439, USA

Alexander G. Myasnikov
Institute of Genetics and of Molecular
and Cellular Biology (IGBMC)
Department of Integrative Structural Biology
Centre National de la Recherche Scientifique (CNRS)
UMR 7104
Institut National de la Santé de la Recherche
Médicale (INSERM) U964
University of Strasbourg
1 rue Laurent Fries
67404 Illkirch, France

Mikhail Nesterchuk
Department of Chemistry and A.N.
Belozersky Institute of Physico-Chemical Biology
Moscow State University
Moscow, 119899, Russia

Harry F. Noller
Center for Molecular Biology of RNA
and Department of MCD Biology
University of California, Santa Cruz
Santa Cruz, CA 95064, USA
E-mail: harry@nuvolari.ucsc.edu

José N. Onuchic
Center for Theoretical Biological Physics
and Department of Physics
University of California, San Diego
9500 Gilman Dr.
La Jolla, CA 92093, USA

Rodrigo F. Ortiz-Meoz
Howard Hughes Medical Institute
Department of Molecular Biology and Genetics
Johns Hopkins University School of Medicine
Baltimore, MD 21205, USA

Ilya Osterman
Department of Chemistry and A.N.
Belozersky Institute of Physico-Chemical Biology
Moscow State University
Moscow, 119899, Russia

Rashmi Pathak
Laboratorium für Organische Chemie
ETH Zürich
Wolfgang-Pauli-Straße 10
8093 Zürich, Switzerland

Michael Pavlov
Department of Cell and Molecular Biology
Uppsala University, BMC
Box 596
751 24 Uppsala, Sweden

Déborah Perez Fernandez
Laboratorium für Organische Chemie
ETH Zürich
Wolfgang-Pauli-Straße 10
8093 Zürich, Switzerland

Frank Peske
Department of Physical Biochemistry
Max Planck Institute for Biophysical Chemistry
37077 Göttingen, Germany

Tatyana Pestova
Department of Cell Biology
State University of New York
Downstate Medical Center
450 Clarkson Ave.
Brooklyn, NY 11203, USA
E-mail: tatyana.pestova@downstate.edu

Lolita Piersimoni
Laboratory of Genetics
Department of Biosciences and Biotechnology
University of Camerino
62032 Camerino (MC), Italy

Andrey V. Pisarev
Department of Cell Biology
State University of New York
Downstate Medical Center
450 Clarkson Ave.
Brooklyn, NY 11203, USA

Vera P. Pisareva
Department of Cell Biology
State University of New York
Downstate Medical Center
450 Clarkson Ave.
Brooklyn, NY 11203, USA

Cynthia L. Pon
Laboratory of Genetics
Department of Biosciences and Biotechnology
University of Camerino
62032 Camerino (MC), Italy

Irina Prokhorova
Department of Chemistry and A.N.
Belozersky Institute of Physico-Chemical Biology
Moscow State University
Moscow, 119899, Russia

Joseph D. Puglisi
Department of Structural Biology
Stanford University School of Medicine
Stanford CA 94305-5126, USA
E-mail: puglisi@stanford.edu

Daoming Qin
Ohio State Biochemistry Program
Center for RNA Biology
The Ohio State University
484 W. 12th Ave.
Columbus, OH 43210, USA

Vaishnavi Rajagopal
Department of Biophysics and
Biophysical Chemistry
Johns Hopkins University School of Medicine
725 North Wolfe Street
Baltimore, MD 21205, USA

Venki Ramakrishnan
MRC Laboratory of Molecular Biology
Hills Road
Cambridge CB2 0QH, UK
E-mail: ramak@mrc-lmb.cam.ac.uk

Haripriya Ramu
Center for Pharmaceutical Biotechnology
University of Illinois at Chicago
900 S. Ashland Ave.
Chicago, IL 60607, USA

Daniel Rodriguez-Correa
Department of Molecular Biology,
Cell Biology and Biochemistry
Brown University
Providence, RI 02912, USA

Gabriel Rosenblum
Department of Chemistry
University of Pennsylvania
Philadelphia, PA 19104-6323, USA

Margaret E. Saks
Department of Biochemistry, Molecular Biology,
and Cell Biology
Northwestern University
2205 Tech Drive Hogan 2-100
Evanston, IL 60208, USA

Sumantha Salian
Laboratorium für Organische Chemie
ETH Zürich
Wolfgang-Pauli-Straße 10
8093 Zürich, Switzerland

Karissa Y. Sanbonmatsu
Theoretical Biology and Biophysics
Theoretical Division
Los Alamos National Laboratory
MS K710
Los Alamos, NM 87545, USA

Suparna Sanyal
Department of Cell and Molecular Biology
Uppsala University, BMC
Box 596
751 24 Uppsala, Sweden

Andreas Savelsbergh
Institute for Medical Biochemistry
Department for Medicine, Faculty of Health
University of Witten/Herdecke
58448 Witten, Germany

Jared M. Schrader
Department of Biochemistry, Molecular Biology,
and Cell Biology
Northwestern University
2205 Tech Drive Hogan 2-100
Evanston, IL 60208, USA

Yuri P. Semenkov
Petersburg Nuclear Physics Institute
Russian Academy of Sciences
188300 Gatchina, Russia

Olga Sergeeva
Department of Chemistry and A.N.
Belozersky Institute of
Physico-Chemical Biology
Moscow State University
Moscow, 119899, Russia

Petr Sergiev
Department of Chemistry and A.N.
Belozersky Institute of Physico-Chemical Biology
Moscow State University
Moscow, 119899, Russia

Manjuli R. Sharma
Wadsworth Center
New York State Department of Health
Empire State Plaza
Albany, NY 12201-0509, USA
Department of Biomedical Sciences
State University of New York at Albany
Albany, NY 12222, USA

Dmitry Shcherbakov
Institut für Medizinische Mikrobiologie
Universität Zürich
Gloriastraße 30/32
8006 Zürich, Switzerland

Sandrina Silva
Laboratorium für Organische Chemie
ETH Zürich
Wolfgang-Pauli-Straße 10
8093 Zürich, Switzerland

Angelita Simonetti
Institute of Genetics and of Molecular and
Cellular Biology (IGBMC)
Department of Integrative Structural Biology
Centre National de la Recherche Scientifique (CNRS)
UMR 7104
Institut National de la Santé de la Recherche
Médicale (INSERM) U964
University of Strasbourg
1 rue Laurent Fries
67404 Illkirch, France

Maxim A. Skabkin
Department of Cell Biology
State University of New York
Downstate Medical Center
450 Clarkson Ave.
Brooklyn, NY 11203, USA

Olga V. Skabkina
Department of Cell Biology
State University of New York
Downstate Medical Center
450 Clarkson Ave.
Brooklyn, NY 11203, USA

Zeev Smilansky
Anima Cell Metrology Inc.
Bernardsville, NJ 07924-2270, USA

Christian M. T. Spahn
Institute for Medical Physics and Biophysics
Charité – Universitätsmedizin Berlin
10117 Berlin, Germany

Clint Spiegel
Department of Chemistry
Western Washington University
Bellingham, WA 98225, USA

Linda L. Spremulli
Department of Chemistry
Campus Box 3290
University of North Carolina
Chapel Hill, NC 27599-3290, USA

Holger Stark
3D Electron Cryomicroscopy Group
Max Planck Institute for Biophysical Chemistry
37077 Göttingen, Germany

Sergey V. Steinberg
Université de Montréal C.P. 6128
Succursale Centre-ville
H3C 3J7 Montréal, QC, Canada
E-mail: serguei.chteinberg@umontreal.ca

Thomas A. Steitz
Howard Hughes Medical Institute
Department of Molecular Biophysics and Biochemistry
Yale University
New Haven, CT 06520, USA
E-mail: thomas.steitz@yale.edu

Benjamin Stevens
Anima Cell Metrology Inc.
Bernardsville, NJ 07924-2270 USA

Daniel S. Terry
Department of Physiology and Biophysics
Weill Cornell Medical College
New York, NY 10021, USA

Jill R. Thompson
Department of Molecular Biology
Cell Biology and Biochemistry
Brown University
Providence, RI 02912, USA

Sergei Trakhanov
Institute for Organic Chemistry
and Chemical Biology
Goethe University
60325 Frankfurt, Germany

Sotaro Uemura
Department of Structural Biology
Stanford University School of Medicine
Stanford, CA 94305-5126, USA

Olke C. Uhlenbeck
Department of Biochemistry, Molecular Biology,
and Cell Biology
Northwestern University
2205 Tech Drive Hogan 2-100
Evanston, IL 60208, USA
E-mail: o-uhlenbeck@northwestern.edu

Andrea Vasella
Laboratorium für Organische Chemie
ETH Zürich
Wolfgang-Pauli-Straße 10
8093 Zürich, Switzerland

Nora Vazquez-Laslop
Center for Pharmaceutical Biotechnology
University of Illinois at Chicago
900 S. Ashland Ave.
Chicago, IL 60607, USA

Sarah E. Walker
Department of Biophysics and
Biophysical Chemistry
Johns Hopkins University School of Medicine
725 North Wolfe Street
Baltimore, MD 21205, USA

Paul C. Whitford
Theoretical Biology and Biophysics,
Theoretical Division
Los Alamos National Laboratory
MS K710
Los Alamos, NM 87545, USA

Daniel N. Wilson
Gene Center and Department for Chemistry
and Biochemistry, and Center for integrated
Protein Science Munich (CiPSM)
University of Munich
Feodor-Lynen-Str. 25
81377 München, Germany

Christopher Woolstenhulme
Department of Chemistry and Biochemistry
Brigham Young University
C203 BNSN
Provo, UT 84602, USA

Aymen Yassin
Wadsworth Center
New York State Department of Health
Empire State Plaza
Albany, NY 12201-0509, USA
Department of Biomedical Sciences
State University of New York at Albany
Albany NY 12222, USA

Ada Yonath
The Weizmann Institute of Science
Department of Structural Biology
76100 Rehovot, Israel
E-mail: ada.yonath@weizmann.ac.il

Marat Yusupov
Institut de Génétique et de Biologie Moléculaire
et Cellulaire (IGBMC)
Département de Biologie et de Génomique
Structurales
CNRS, UMR7104, INSERM, U964
University of Strasbourg
67400 Illkirch, France

Gulnara Yusupova
Institut de Génétique et de Biologie Moléculaire
et Cellulaire (IGBMC)
Département de Biologie et de Génomique
Structurales
CNRS, UMR7104, INSERM, U964
University of Strasbourg
67400 Illkirch, France

Hani S. Zaher
Howard Hughes Medical Institute
Department of Molecular Biology and Genetics
Johns Hopkins University School of Medicine
Baltimore, MD 21205, USA

Haibo Zhang
Department of Chemistry
University of Pennsylvania
Philadelphia, PA 19104-6323, USA

Wen Zhang
Departments of Molecular and Cell Biology
and Chemistry
University of California at Berkeley
Physical Biosciences Division
Lawrence Berkeley National Laboratory
Berkeley, CA 94720, USA

Jianyu Zhu
Center for Molecular Biology of RNA
and Department of MCD Biology
University of California, Santa Cruz
Santa Cruz, CA 95064, USA